ELECTRIC CIRCUITS

SECOND EDITION

Theodore F. Bogart, Jr.
The University of Southern Mississippi

GLENCOE

Macmillan/McGraw–Hill

Lake Forest, Illinois   Columbus, Ohio   Mission Hills, California   Peoria, Illinois

Electric Circuits, 2/E

International Editions 1992

Exclusive rights by McGraw Hill Book Co-Singapore for manufacture and export. This book cannot be re-exported from the country to which it is consigned by McGraw Hill.

PSpice and Probe are registered trademarks of the MicroSim Corporation.

Bogart, Theodore, F.
   Electric circuits / Theodore F. Bogart, Jr. —2nd ed.
      p.   cm.
   Includes index.
   ISBN 0-02-800662-3
   1. Electric circuit analysis. I. Title.
TK454.B589 1992   92-2571
621.319′2—dc20   CIP

   2 3 4 5 6 7 8 9 0 KHL SW 9 6 5 4 3 2

When ordering this title, use ISBN 0-07-112920-0.

Printed in Singapore.

Preface

Scope

Electric Circuits is a comprehensive treatment of traditional topics in dc and ac circuit analysis, suitable for 2- and 4-year programs in engineering technology. While the book emphasizes the development of analysis skills, most of the chapters contain optional examples and exercises that can be used to teach design and troubleshooting principles. Since most practical circuit analysis is performed in support of one of those two activities, it is certainly appropriate for technology students to gain experience developing the mental processes needed for that kind of problem solving. However, recognizing the considerable body of material that must be covered in a typical circuits course, the design and troubleshooting material has been incorporated in such a way that it can be covered as time permits, or not at all, without loss of continuity.

The organization of the material and the sequence of topics is somewhat different from current textbooks in one respect: the rigorous treatment of concepts that many beginning students find too abstract to assimilate is deferred to Chapter 8. Since technology students often begin their study of circuit theory without a thorough background in physics, it is beneficial to delay concepts such as electric potential (in terms of work per unit charge) until they have developed confidence in their problem-solving skills and their handling of units, terminology, notation, and circuit visualization. Furthermore, any concurrent coursework in physics will not progress rapidly enough to provide the immediate foundation that students would need to understand concepts introduced in terms of force, work, energy, power, etc. Thus, the student's initial exposure to voltage, current, and resistance in this book is conceptual rather than rigorous, allowing an early discussion of Ohm's law and a speedy transition to circuit concepts. The fundamental quantities are later refined, and endowed with more sophistication, when field theory is introduced.

Exercises and Examples

The book contains a very large number of end-of-chapter exercises, with answers provided for odd-numbered exercises. One reason for the large quantity of exercises is that every effort was made to include at least one odd-numbered exercise and one even-numbered exercise for every important topic or concept discussed in the book. Thus, instructors who prefer to assign exercises with answers and those who prefer exercises without answers will both find an ample selection.

To promote immediate reinforcement of problem-solving skills, each worked-out example is followed by a drill exercise. The drill exercises are typically slight variations of the examples that precede them, so students are forced to work through a new set of computations. This kind of activity is a valuable way for students to learn of any gaps that may exist in their understanding, or in their computational skills—learning that does not usually follow from a cursory reading of an example.

SPICE

The appendix contains a thorough treatment of how SPICE (Simulation Program with Integrated Circuit Emphasis) can be used to analyze passive dc and ac circuits. Although SPICE was originally developed to model integrated circuits, this powerful and versatile software can be used to analyze transients and to plot frequency response, as well as to print any dc or ac voltage or current, in circuits containing no semiconductor devices at all. The appendix contains numerous examples of these applications, beginning with simple resistive circuits. The sequence of topics is such that students can build computer modeling skills as they learn more circuit theory. In subsequent courses covering electronic device theory, they can readily incorporate device models into SPICE programs whose formats have already become familiar.

Features of the Second Edition

To widen the scope of the book, and in response to suggestions from many users of the first edition, *Electric Circuits* now contains a chapter on polyphase circuits and electric power distribution. This material was omitted from the first edition not from any lack of appreciation for its importance (on the contrary, the author's first rewarding experience as a professional engineer was in the electrical power industry). Rather, the first edition was designed to contain just slightly more material than could be covered in a two-semester course, in order to give instructors some discretion in selecting topics appropriate to individual programs. The second edition offers even greater scope to instructors, and the book can also support those programs that devote three quarters or semesters to the study of circuit theory. The new chapter has two features that are not found in most current texts: a treatment of single-phase, three-wire power distribution—a topic whose importance stems from the fact that it is found in most American households—and a coherent explanation of the two-wattmeter method for power measurement.

Another feature of the second edition is expanded coverage of material related to SPICE. New topics include transformer simulation and programming temperature changes. Also, SPICE examples and exercises have been added to most chapters, so programming skills can be developed at a pace consistent with the development of circuit analysis skills.

Finally, new material related to PSpice has been added, including a section on Probe and Control Shell. (PSpice and Probe are registered trademarks of MicroSim Corporation.)

Acknowledgments

I wish to express my thanks to my colleagues at the University of Southern Mississippi for comments and suggestions based upon their classroom experience. I appreciate my students, who continue to provide me with their thoughtful feedback concerning questions that arise from their studies and from their preparation of the homework assignments. I am especially grateful to the reviewers—Mysore Narayanan of Miami University, Joseph Baker and Mark Oliver of Monroe Community College, Teresa Bowen of Waycross-Ware Technical Institute, and Joseph Farren of the University of Dayton—for their professional critiques and helpful suggestions for improving the work. Special thanks are due the staff of Glencoe for their considerable expertise in publishing this new edition.

T.F.B.

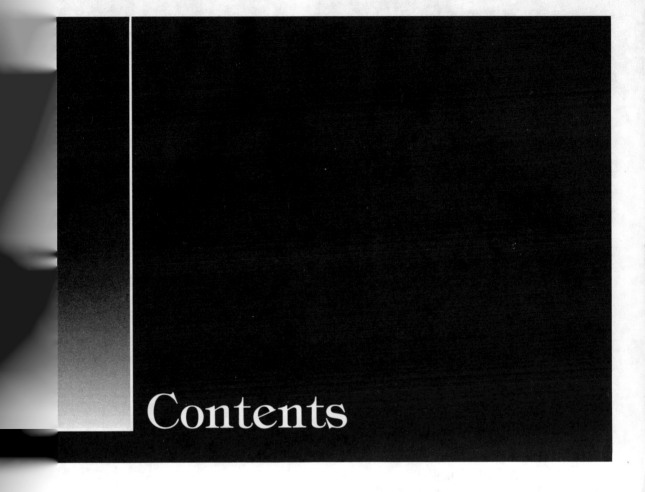

Contents

5

Series–Parallel Circuits 116

6

Network Transformations and Multisource Circuits 154

10

Magnetic Fields and Circuits   343

11

Inductance and Inductors   374

12

Alternating-Current Fundamentals   414

13

Complex Algebra and Phasors   472

14

Series and Parallel AC Circuits   496

15

Series–Parallel AC Circuits   554

16

AC Network Transformations
and Multisource Circuits   594

17

AC Network Theorems 634

18

Filters and Resonant Circuits 679

19

Transformers   745

20

Polyphase Circuits and
Electric Power Distribution   785

ELECTRIC CIRCUITS

SECOND EDITION

1 Units and Mathematical Notation

1.1 Systems of Units

Understanding electricity and its practical applications depends on our ability to describe, measure, and compare various physical quantities. The numerical values we use to describe these quantities must be expressed with certain *units* of measurement. For example, length can be expressed in meters or feet, and time can be expressed in seconds or minutes. Some quantities and their units are fundamental, in the sense that all other units can be derived from them. For example, if mass, length, time, and electrical charge are defined to be fundamental, then all other quantities, such as force, volume, velocity, and so on, have units that can be expressed in terms of the fundamental units. To illustrate, velocity is distance (length) divided by time, as in the units "miles per hour." The fundamental units are often called *base* units. A *system of units* is all the base units plus all the units derived from them.

Certain systems of units have been in use for many years. These include the *English* system, widely used in the United States and Great Britain, and the CGS (centimeter-gram-second) and MKS (meter-kilogram-second) systems. The latter two are *metric* systems, and their names are derived from the base units of the fundamental quantities: length, mass, and time.

The SI System

For the purpose of establishing a worldwide standard system of units, an international meeting, called the General Conference of Weights and Measures, was held in Sèvres, France, in 1960. This conference adopted the *Système International* (French: International System), or SI, system of units. The United States officially endorsed the use of SI units by passing the Metric Conversion Act of 1975. Reports in American press and television concerning conversion to "the metric" system actually refer to the SI system (not the CGS or MKS metric systems).

There are seven base units in the SI system, five of which are important to the material we will study in this book. Those five are: meters (the units of length), kilograms (the units of mass), seconds (the units of time), amperes (the units of electric current), and kelvin (the units of temperature). The SI base units and the units that are derived from them are shown in Figure 1.1. The SI system is quite similar to the MKS system, particularly for electrical quantities, so most SI electrical units are the same as the MKS units that have been in use for many years (amperes, volts, watts, etc.).

In Figure 1.1, note that *force* has the derived units called *newtons* (N). The arrows show that force is derived from mass and acceleration, and acceleration is derived from length and time. Thus, in terms of base units, force is derived from mass, length, and time. The notation beside the circle labeled "force" means that

$$1 \text{ newton (N)} = 1 \, \frac{\text{kilogram} \cdot \text{meter}}{\text{second}^2} \quad \left(\frac{\text{kg} \cdot \text{m}}{\text{s}^2}\right)$$

Similarly, we see that pressure has the derived units called *pascal* (Pa), where

$$1 \text{ Pa} = 1 \text{ N/m}^2$$

Therefore, in terms of base units alone,

$$1 \text{ Pa} = 1 \, \frac{\dfrac{\text{kg} \cdot \text{m}}{\text{s}^2}}{\text{m}^2} = 1 \, \frac{\text{kg}}{\text{m} \cdot \text{s}^2}$$

Also note in Figure 1.1 that temperature in kelvin (K) and temperature in degrees Celsius (°C) are related by

$$t(°C) = T_K - 273.15 \text{ K}$$

It is *not* correct to speak of "degrees kelvin"; the correct units are simply "kelvin."

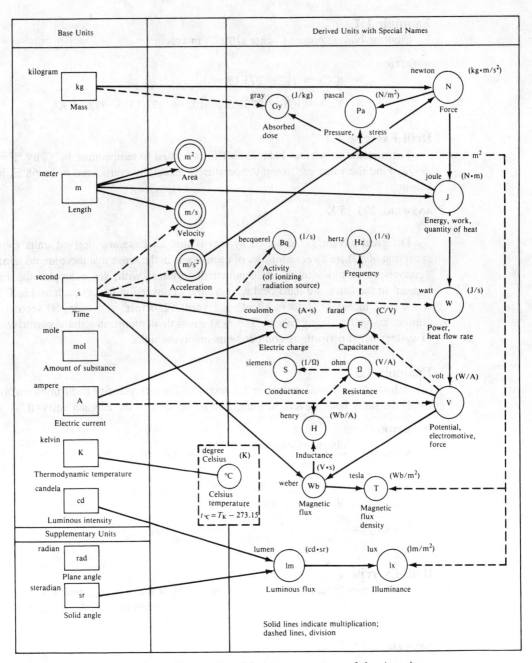

FIGURE 1.1 Reprinted from *Engineering Education*, courtesy of the American Society for Engineering Education.

Example 1.1

Find the boiling point of water (100°C) in kelvin.

SOLUTION

$$t(°C) = T_K - 273.15$$

$$T_K = t(°C) + 273.15 = 100 + 273.15 = 373.15 \text{ K}$$

Drill Exercise 1.1

Temperature in degrees Fahrenheit (°F) is related to temperature in °C by °F = $\frac{9}{5}$°C + 32. Find the value of "room temperature," generally considered to be 68°F, in kelvin.

ANSWER: 293.15 K. ☐

The great advantage of the SI system is that all base and derived units (with the exception of °C) are *direct multiples* of each other, in the sense that there are no numerical "conversion factors" involved. Contrast this system with, for example, the English system. In the latter, we must deal with arbitrary conversion factors such as 12, 3, 5280, and 60: 12 inches = 1 foot; 3 feet = 1 yard; 5280 feet = 1 mile; 60 seconds = 1 minute; as well as many others. The next example demonstrates the superiority of the SI system when performing computations involving units.

Example 1.2

(a) A body travels 14.2 meters in 7.1 seconds. Find its velocity in SI units (m/s).
(b) If the same body travels 50 mph, find its velocity in the English units (ft/s).

SOLUTION

(a) Velocity = $\dfrac{\text{distance (length)}}{\text{time}} = \dfrac{14.2 \text{ m}}{7.1 \text{ s}} = 2.0 \text{ m/s}$

(b) 50 miles = 50(5280) ft = 264,000 ft
   1 hour = 1(60)(60) s = 3600 s

   Velocity = $\dfrac{\text{distance}}{\text{time}} = \dfrac{264,000 \text{ ft}}{3600 \text{ s}} = 73.333 \text{ ft/s}$

Drill Exercise 1.2

As shown in Figure 1.1, the SI unit of energy is the *joule* (J), where 1 J = 1 N · m (force × distance). How much energy in joules is expended when a force of 85 kg · m/s² acts over a distance of $5 × 10^{-3}$ m?

ANSWER: 0.425 J. ☐

Example 1.2 shows that the computation of velocity in English system units requires unit conversions, a potential source of computational error as well as a time-consuming inconvenience. The same computation in the SI system is a direct application of the definition of velocity. In Section 1.3 we will learn that the only conversion factors that are required in computations involving SI units are different powers of 10.

1.2 Review of Exponents

Many of the computations in electrical circuit theory involve the algebraic manipulation of expressions containing exponents. For that reason, we present here a short review of some important relationships involving exponents. Equations (1.1) through (1.9) summarize these relationships:

$$X^n X^m = X^{n+m} \tag{1.1}$$

$$\frac{X^n}{X^m} = X^{n-m} \tag{1.2}$$

$$X^n = \frac{1}{X^{-n}} \tag{1.3}$$

$$X^{-n} = \frac{1}{X^n} \tag{1.4}$$

$$X^{1/n} = \sqrt[n]{X} \tag{1.5}$$

$$(X^n)^m = X^{nm} \tag{1.6}$$

$$X^{n/m} = (X^n)^{1/m} = (X^{1/m})^n \tag{1.7}$$

$$(XY)^n = X^n Y^n \tag{1.8}$$

$$\left(\frac{X}{Y}\right)^n = \frac{X^n}{Y^n} \tag{1.9}$$

$$X^0 = 1 \tag{1.10}$$

Several of the relationships listed above are special cases of others, and they are listed for reference only. The student who has the mathematical preparation necessary to study the electrical theory in this book will not find it necessary to memorize the equations, but will be able to apply them instinctively. Repeated practice is the key to acquiring that ability.

Example 1.3

Determine which of the following are valid equalities.

(a) $\left(\dfrac{X}{6}\right)^2 = \dfrac{X^2}{36}$

(b) $(3V)^3 = 3V^3$

(c) $\left(\dfrac{1}{4}\right)^{1/2} = \dfrac{1}{4^2}$

(d) $(A + 2)^3 = A^3 + 8$

(e) $(10^3)^{-3} = 1$

(f) $(27I^3)^{1/3} = 3I$

(g) $\left(\dfrac{8}{w}\right)^{2/3} = 4W^{-2/3}$

(h) $\sqrt{\dfrac{a^3}{16}} = \dfrac{a^{2/3}}{4}$

SOLUTION

(a) Valid: $\left(\dfrac{X}{6}\right)^2 = \dfrac{X^2}{6^2} = \dfrac{X^2}{36}$

(b) Invalid: $(3V)^3 = 3^3V^3 = 27V^3$

(c) Invalid: $\left(\dfrac{1}{4}\right)^{1/2} = \dfrac{1^{1/2}}{4^{1/2}} = \dfrac{\sqrt{1}}{\sqrt{4}} = \dfrac{1}{2}$

(d) Invalid: $(X + Y)^n \neq X^n + Y^n$ (See remarks below.)

(e) Invalid: $(10^3)^{-3} = 10^{(3)(-3)} = 10^{-9}$

(f) Valid: $(27I^3)^{1/3} = 27^{1/3}(I^3)^{1/3} = \sqrt[3]{27}\,I^{3/3} = 3I$

(g) Valid: $\left(\dfrac{8}{W}\right)^{2/3} = \dfrac{8^{2/3}}{W^{2/3}} = \dfrac{(8^{1/3})^2}{W^{2/3}} = \dfrac{(\sqrt[3]{8})^2}{W^{2/3}} = \dfrac{2^2}{W^{2/3}} = 4W^{-2/3}$

(h) Invalid: $\sqrt{\dfrac{a^3}{16}} = \dfrac{\sqrt{a^3}}{\sqrt{16}} = \dfrac{(a^3)^{1/2}}{4} = \dfrac{a^{3/2}}{4}$

The solution to part (d) is based on the fact that

$$(X + Y)^n \underset{\text{(does } not \text{ equal)}}{\neq} X^n + Y^n \tag{1.11}$$

The belief that equality *does* hold in (1.11) is an *error* frequently committed by students. Most students readily accept inequality (1.11) for the special case $n = 2$:

$$(X + Y)^2 = X^2 + 2XY + Y^2 \neq X^2 + Y^2$$

When in doubt, use the knowledge of this special case ($n = 2$) for the general case as well (any value of n).

Drill Exercise 1.3

Evaluate $[4\sqrt{(\tfrac{1}{2})^4}]^{1/2}$ without using a calculator.

ANSWER: 1.   □

1.3 Powers of 10

Many electrical computations are performed using quantities that are multiplied by powers of 10. It is common practice to omit writing powers of 10 and, instead, to attach certain prefixes to the units associated with the quantities. Each prefix signifies a different power of 10. The most commonly used prefixes, their corresponding powers of 10, and their standard (SI) abbreviations are shown in Table 1.1. The table should be memorized as soon as possible, since the prefixes will be used in nearly all of the examples and exercises that follow in this book.

The prefixes can be thought of as *multipliers*: Each multiplies a standard unit by a certain power of 10. For example,

TABLE 1.1

Prefix	Power of 10	Abbreviation
tera	10^{12}	T
giga	10^9	G
mega	10^6	M
kilo	10^3	k
deci	10^{-1}	d
centi	10^{-2}	c
milli	10^{-3}	m
micro	10^{-6}	μ
nano	10^{-9}	n
pico	10^{-12}	p

$$2 \text{ mm} = 2 \text{ millimeters} = (2)(10^{-3})(\text{meter}) = 2 \times 10^{-3} \text{ meter}$$
$$= 0.002 \text{ meter}$$

$$0.5 \text{ MW} = 0.5 \text{ megawatt} = (0.5)(10^6)(\text{watts}) = 0.5 \times 10^6 \text{ watts}$$
$$= 500,000 \text{ watts}$$

Also,

$$0.000016 \text{ ampere} = 16 \times 10^{-6} \text{ A} = 16 \text{ microamperes} = 16 \text{ }\mu\text{A}$$

$$27,500 \text{ volts} = 27.5 \times 10^3 \text{ V} = 27.5 \text{ kilovolts} = 27.5 \text{ kV}$$

It is often desirable or necessary to convert a quantity whose units have one prefix to another equal quantity having a different prefix. The conversion can be accomplished using the technique called *dimensional analysis*, wherein the quantity to be converted is multiplied by factor(s) that contain the prefixed units, and that are themselves equal to 1. Units in a numerator cancel the same units in a denominator, so if the factors are selected properly, the only units remaining are those having the desired prefix. The next example illustrates several such conversions.

Example 1.4
Convert

(a) 0.0065 km to meters
(b) 1,750,000 μV to kilovolts
(c) 5.02×10^3 ps to nanoseconds
(d) 3800 cm^2 to square meters

SOLUTION

(a) We use the fact that 1 km = 10^3 m, or

$$\frac{10^3 \text{ m}}{1 \text{ km}} = 1$$

$$(0.0065 \text{ k\cancel{m}})\left(\frac{10^3 \text{ m}}{1 \text{ k\cancel{m}}}\right) = 0.0065 \times 10^3 \text{ m} = 6.5 \text{ m}$$

(b) Again, each multiplying factor equals 1:

$$(1{,}750{,}000 \ \cancel{\mu V})\left(\frac{10^{-6} \ \cancel{V}}{1 \ \cancel{\mu V}}\right)\left(\frac{1 \ kV}{10^3 \ \cancel{V}}\right) = (1{,}750{,}000) \times 10^{-9} \ kV = 0.00175 \ kV$$

(c) $(5.02 \times 10^3 \ \cancel{ps})\left(\dfrac{10^{-12} \ \cancel{s}}{1 \ \cancel{ps}}\right)\left(\dfrac{1 \ ns}{10^{-9} \ \cancel{s}}\right) = (5.02 \times 10^3) \times 10^{-3} \ ns = 5.02 \ ns$

(d) Note that two factors containing cm units must be used to cancel the cm^2 units:

$$(3800 \ \cancel{cm^2})\left(\frac{10^{-2} \ m}{1 \ \cancel{cm}}\right)\left(\frac{10^{-2} \ m}{1 \ \cancel{cm}}\right) = 3800 \times 10^{-4} \ m^2 = 0.38 \ m^2$$

Drill Exercise 1.4

Convert 0.04×10^{-3} MV to millivolts.

ANSWER: 4×10^4 mV. □

Scientific calculators have an exponent key (usually labeled EE, EXP, or E) that allows the user to enter a number followed by the exponent of a power of 10 that multiplies that number. For example, 25.7×10^3 can be entered in the sequence 25.7, EE, 3. When using a calculator this way, beware of a common pitfall: 10 raised to a certain power, say n, is *not* entered as 10, EE, n. Remember that $10^n = 1 \times 10^n$. For example, the number 10^4 should be entered as 1, EE, 4. If 10, EE, 4 is entered, the result is $10 \times 10^4 = 10^5$.

Example 1.5

(a) Find the area of a square whose sides are 1 cm long. Express the result in square meters (m^2).
(b) Find the velocity of a body that travels 50 mm in 200 ns. Express the result in meters/second (m/s).
(c) Find the area of a circle whose diameter is 1 km. Express the result in square meters (m^2).
(d) Find the volume of a sphere whose radius is 10 mm. Express the result in cubic meters (m^3).

SOLUTION

(a) $A = l^2 = (1 \ cm)^2 = (10^{-2} \ m)^2 = 10^{-4} \ m^2$

(b) $V = \dfrac{l}{t} = \dfrac{50 \ mm}{200 \ ns} = \dfrac{50 \times 10^{-3} \ m}{200 \times 10^{-9} \ s} = 2.5 \times 10^5 \ m/s$

(c) $A = \dfrac{\pi d^2}{4} = \dfrac{\pi(1 \ km)^2}{4} = \dfrac{\pi(10^3 \ m)^2}{4} = \dfrac{\pi}{4} \times 10^6 \ m^2 = 7.85 \times 10^5 \ m^2$

(d) $V = \frac{4}{3}\pi r^3 = \frac{4}{3}\pi(10 \ mm)^3 = \frac{4}{3}\pi(10 \times 10^{-3} \ m)^3$
    $= \frac{4}{3}\pi(10^{-2} \ m)^3 = \frac{4}{3}\pi \times 10^{-6} \ m^3 = 4.189 \times 10^{-6} \ m^3$

Drill Exercise 1.5

Find the area of a triangle whose height is 0.18 m and whose base is 52 mm long. Express the answer in square centimeters (cm^2).

ANSWER: $46.8 \ cm^2$. □

Scientific and Engineering Notation

When a number is expressed with one digit to the left of the decimal point and is multiplied by an appropriate power of 10, it is said to be in *scientific notation*. In each of the following examples, the number on the left is converted to the scientific notation shown on the right:

$$432,000 = 4.32 \times 10^5$$
$$0.0071 = 7.1 \times 10^{-3}$$
$$86.3 \times 10^2 = 8.63 \times 10^3$$

Scientific notation is useful when it is necessary or desirable to express all numbers in a consistent manner. However, when the numbers in question have units associated with them, it is better practice to use the standard prefixes to express powers of 10, regardless of the location of the decimal point. For example, 40.2 kW is preferable to 4.02×10^4 W. This is an example of *engineering notation*: numbers are expressed as a value between 0.1 and 999 multiplied by a power of 10 equal to one of the standard multipliers in Table 1.1. In each of the following examples, the quantity on the left is converted to the engineering notation shown on the right:

$$0.00033 \text{ A} = 0.33 \text{ mA or } 330 \text{ }\mu\text{A}$$
$$40 \text{ V} = 40 \text{ V}$$
$$125,100 \text{ W} = 125.1 \text{ kW}$$

When two or more quantities that are multiplied by different powers of 10 are to be added or subtracted, the decimal points and powers of 10 must be adjusted so that all quantities are multiplied by the same power of 10. For example, to add 6.14 mm and 210 μm, we write

$$6.14 \text{ mm} = \quad 6.14 \times 10^{-3} \text{ m} = \quad 6.14 \times 10^{-3} \text{ m}$$
$$+ \underline{210 \text{ }\mu\text{m}} = + \underline{210 \times 10^{-6} \text{ m}} = + \underline{0.21 \times 10^{-3} \text{ m}}$$
$$6.35 \times 10^{-3} \text{ m}$$

Significant Digits, Precision, and Accuracy

The number of significant digits in a number is the number of digits used to express it, not counting any zeros that appear before the first nonzero digit (leading zeros). Following are some examples:

378	3 significant digits
0.56	2 significant digits
0.560	3 significant digits
0.003	1 significant digit
1.050	4 significant digits
90.000	5 significant digits

The more significant digits that are used to express a measured quantity, the greater the *precision* of that measurement. However, *accuracy* is not necessarily related to precision. The more accurate a measurement, the closer it is to the true value of the quantity being measured. For example, suppose that we are measuring a voltage whose true value is 12.5 V. Using one instrument, we measure 12.4 V, and using another, we measure 12.9765 V. The first instrument is more accurate, but the second has greater precision, because its measurements have six significant digits.

In practice, it is conventional to assume that the rightmost digit (the *least significant digit*) in a number used to express the value of a physical quantity may be in error by ± one-half unit. For example, a measurement of 12.3 inches may actually be between 12.25 inches and 12.35 inches. When performing arithmetic operations using measured values with different degrees of precision, we must round off so that the results of the operations are expressed with the same precision (number of significant digits) as the value having the *least* precision. Following is an example:

$$
\begin{array}{ll}
4.071 \text{ V} & \text{(4 significant digits)} \\
+\ 2.0\ \ \text{ V} & \text{(2 significant digits)} \\
\hline
6.071 \text{ V} & \text{round off to 6.1 V (2 significant digits)}
\end{array}
$$

In the computational examples in this book, we are dealing with *theoretical* rather than measured values, so we will not always round off results in accordance with the rule just described. In the interest of clarity and conservation of space, we may write, for example, 1 V/8 Ω = 0.125 A, rather than 1.000 V/8.000 Ω = 0.1250 A.

1.4 The Algebra of Inequalities

In many practical problems, we are required to find the minimum or maximum values that some quantity is permitted to have, based on certain physical constraints or on the range of values it may acquire in a situation where variations in some other quantity affect it. In these types of problems, it is necessary to solve *inequalities*: expressions involving the symbols > (greater than), ≥ (greater than or equal to), < (less than), and ≤ (less than or equal to). The algebraic manipulation of an inequality often creates the "opposite" type of inequality, in the following sense:

Inequality	Opposite
<	>
≤	≥
>	<
≥	≤

Following are three fundamental rules concerning the manipulation of inequalities Each is accompanied by an example.

1. The same quantity may be added to or subtracted from each side of an inequality without changing the inequality. For example, if $X + 7 > 9$, then subtracting 7 from both sides gives

$$X > 2$$

2. The same *positive* quantity may multiply or divide both sides of an inequality without changing the inequality. For example, if $X/3 \leq 4Y$, then multiplying both sides by 3 gives

$$X \leq 12Y$$

3. If the same *negative* quantity multiplies or divides both sides of an inequality, the inequality symbol is replaced by its opposite type. For example, if $-10Y > 20$, then dividing both sides by -10 gives

$$Y < -2$$

Example 1.6

The speed limit on an 800-m section of a certain highway is 50 km/h. What is the minimum time required to travel that section of highway without exceeding the speed limit?

SOLUTION Letting d = distance traveled and t = time of travel, we have

$$\frac{d}{t} \leq 50 \text{ km/h} = 50 \times 10^3 \text{ m/h}$$

Since t is positive, we may multiply both sides of the inequality by t to obtain

$$d \leq (50 \times 10^3 \text{ m/h})t$$

Dividing both sides by 50×10^3 gives

$$\frac{d}{50 \times 10^3 \text{ m/h}} \leq t \quad \text{or, equivalently,} \quad t \geq \frac{d}{50 \times 10^3 \text{ m/h}}$$

Since $d = 800$ m, we have

$$t \geq \frac{800 \text{ m}}{50 \times 10^3 \text{ m/h}} = 0.016 \text{ h} = 57.6 \text{ s}$$

Note that this inequality, stated as $t \geq 0.016$ h, is the same as stating that the *minimum* t is 0.016 h.

Drill Exercise 1.6

What is the minimum value of y that satisfies the inequality $14 \geq -6y - 1$?

ANSWER: $y = -2.5$. ☐

In many situations, we encounter relationships where one variable is *inversely* proportional to another variable. When the variable y is inversely proportional to the variable x, we write

$$y = \frac{K}{x}$$

where K is a constant. Notice that the value of y is maximum when the value of x is minimum, and vice versa.

Example 1.7

Given that $-R \geq -4/A$ and that A may vary from 0.5 to 10, what is the maximum value of R?

SOLUTION Multiplying both sides of the inequality by -1 gives (by rule 3)

$$R \leq \frac{4}{A}$$

Since we are seeking the maximum value of R, we consider the value of A that makes $4/A$ a maximum, namely, the minimum value of A, or $A = 0.5$:

$$R \leq \frac{4}{0.5} = 8$$

Stating that $R \leq 8$ is the same as stating that the maximum value of R is 8.

Drill Exercise 1.7

Assuming that y can be any positive value, in what range of values is

$$x = \frac{1}{4 + 1/y} \ ?$$

ANSWER: $0 < x < \frac{1}{4}$. □

Approximations

The symbols $>>$ and $<<$ mean "much greater than" and "much less than," respectively. They often appear in connection with the derivation of approximations. For example, consider the expression

$$Z = X + Y$$

where X, Y, and Z are variables. Suppose that $X = 100$ and $Y = 0.001$. Then $Z = 100 + 0.001 = 100.001$. Notice that X is much greater than Y and that Z is therefore approximately equal to X. Expressing this observation mathematically, we write

$$\text{if} \ \ X >> Y, \ \ \text{then} \ \ Z \approx X$$

where $\approx$ means "approximately equal to."

In many electrical computations it is a time-saving convenience to use approximations. A question often raised by students is "When can I use the approximation?" or "How much greater should 'much greater than' be before I can use the approximation?" These questions cannot be answered in a way that fits every situation. Whether or not an approximate calculation is acceptable in a given application depends on the accuracy desired in that application and on the accuracy that it is *possible* to achieve. Returning to the example $Z = X + Y$, suppose that X and Y represent the measured values of two quantities and that these measurements are only accurate to $\pm 10\%$. Thus, if

$X = 100 \pm 10\%$, the true value of X may be anywhere from 90 to 110. If $Y = 0.001$, it is clear that we can ignore Y and accept $Z \approx X$. In this case, the effect of ignoring Y creates less error in the computed value of Z than the error that may *already* be there due to the measurement error in X. In fact, we could accept the approximation in any situation where "$X \gg Y$" means that X is only about 10 times greater than Y.

In some applications, the use of an approximation is justified even when the known values are perfectly accurate. It may well be that there is no practical *need* for a highly accurate result. Using our example once again, suppose that X and Y represent perfectly accurate measurements of two lengths, each in centimeters. Suppose further that Z is the length of a section of special cable that can only be purchased in lengths of 50 cm, 100 cm, and 150 cm. We must purchase the length that is closest to the computed value of Z. Then if $X = 100$ cm and $Y = 0.001$ cm, there is no point in considering Y, because it will have no effect on the actual length Z of the cable that is purchased. Again, the approximation is acceptable under the condition that "$X \gg Y$" means that X is at least 10 times greater than Y.

Example 1.8

When A is very large, $1/A$ is very small; that is, if $A \gg 1$, $1/A \approx 0$. Therefore, V in the expression

$$V = \frac{55}{1 + 1/A}$$

is approximately equal to 55 when $A \gg 1$. That is,

$$\text{if} \quad A \gg 1, \quad \text{then} \quad V \approx 55$$

It is necessary to compute V with error no greater than 5%. Assuming no error in the value of A, can the approximation be used when $A = 10$? When $A = 100$?

SOLUTION When $A = 10$, the exact value of V is

$$V = \frac{55}{1 + \frac{1}{10}} = 50$$

Since the approximate value of V is 55, the percent error resulting from use of the approximation is

$$\% \text{ error} = \frac{55 - 50}{50} \times 100\% = 10\%$$

Thus, the approximation is not acceptable for $A = 10$, and we conclude that 10 is not sufficiently greater than 1 to satisfy $A \gg 1$.

Repeating the computations when $A = 100$, we find that the exact value of V is 54.455 and that

$$\% \text{ error} = \frac{55 - 54.455}{54.455} \times 100\% = 1\%$$

Thus, $A = 100$ is adequate to satisfy $A \gg 1$.

Drill Exercise 1.8

In Example 1.8, what is the smallest value of A that will satisfy the 5% error criterion in computing V?

ANSWER: $A = 20$.   □

Exercises

Section 1.1   Systems of Units

1.1 Using the chart of SI units shown in Figure 1.1, determine all base units that are involved in the SI unit of energy. Express that unit in SI base units.

1.2 Using the chart of SI units shown in Figure 1.1, determine all base units that are involved in the SI unit of power. Express that unit in SI base units.

1.3 A force of 240 N acts on a surface area that measures 1.2 m by 0.5 m. What is the pressure on that area, in SI units?

1.4 What is the freezing temperature of water (0°C), in kelvin?

Section 1.2   Review of Exponents

1.5 Evaluate each of the following expressions without using a calculator.

(a) $(4^2 + 2^4)^{1/5}$

(f) $\left(\dfrac{80}{\sqrt{2 \times 10^2 + 2 \times 10^2}} \right)^{1/2}$

(b) $(10^{-2})^3$

(g) $125^{2/3}$

(c) $(a^x)^{1/x}$

(h) $(10^6 + 10^5)^{-1}$

(d) $(5^2 8^{1/3})^2$

(i) $(47.5)^{0.5n - n/2}$

(e) $\left(\dfrac{1}{10 \times 10^4} \right)^{-2}$

(j) $(1,000,000)^{-3/2}$

1.6 Determine which of the following are valid equalities.

(a) $\sqrt{50^2 + 50^2} = 50$

(f) $(10^6)^{1/3} = 100$

(b) $\dfrac{\sqrt{2}}{2} = \dfrac{1}{\sqrt{2}}$

(g) $10^{\Gamma/2} 10^{-1/2} = 1$

(c) $(10^x)(100^y) = 10^{x + 2y}$

(h) $(AX)^m (BY)^n = (ABXY)^{m+n}$

(d) $X^A Y^B = (X + Y)^{A+B}$

(i) $(\frac{1}{8})^{-1/3} = 0.5$

(e) $(7X^2 + 9Y^2)^{1/2} = 4(X^2 + Y^2)^{1/2}$

(j) $\dfrac{M^x}{N^{1/x}} = \dfrac{M}{N}$

Section 1.3   Powers of 10

1.7 Convert

(a) 600 mm to meters

(b) 0.04 m to centimeters.

(c) 150 ns to microseconds

(d) 500 kW to megawatts

(e) 5×10^{-4} V to millivolts

(f) 2.4×10^{-2} mV to microvolts

(g) 30×10^3 pA to nanoamperes

(h) 0.007 MW to kilowatts

1.8 Convert

(a) 32 m^2 to square centimeters

(b) 10,042 cm to meters

(c) 1.55×10^{-7} kV to millivolts

(d) 4300 ns to microseconds

(e) 52,589 mW to kilowatts

(f) 10^{-6} mA to microamperes

(g) 1.75×10^3 dB to bels

(h) 0.4×10^{-4} MV to kilovolts

1.9 (a) Find the velocity of a body that travels 1.2×10^{-4} km in 600 ms. Express the result in meters per second.

(b) Find the area of a rectangle whose length is 420 mm and whose width is 50 cm. Express the result in square meters.

1.10 (a) Find the area of a circle whose radius is 1.6×10^3 cm. Express the result in square meters.

(b) A body travels with velocity 0.04×10^3 km/s. How far does it travel in 500 ms? Express the result in meters.

Section 1.4   The Algebra of Inequalities

1.11 Solve each of the following inequalities by finding a range of values of x that satisfies the inequality. Express your answer in the form of an inequality. (For example, if $x - 2 > 5$, the solution is $x > 7$.)

(a) $4 - x \leq 3$

(b) $\dfrac{1}{x - 5} > 2$

(c) $1.5 \times 10^4 - 25 \times 10^3 x < -200$

(d) $4R - 3x \leq 5x - 2R$

(e) $200a \leq \dfrac{10^5 a^2}{5x - x}$

1.12 Repeat Exercise 1.11 for each of the following.

(a) $24 > 3 - x$

(b) $30 \leq \dfrac{120}{2 - x}$

(c) $30 - 50x < -10(x + 1)$

(d) $4 \times 10^3 x - 4000 R_1 \leq 0.2 \times 10^5 R_1 - 0.4 \times 10^4 x$

(e) $\dfrac{-12W^2}{4x - 2W} < -10W$

1.13 **(a)** Derive an approximate relationship for R when $R_1 \gg R_2$ in the equation

$$R = \frac{R_1 R_2}{R_1 + R_2}$$

(b) Repeat part (a) when $R_1 \ll R_2$.

1.14 Find an approximate value for I when $x \ll a$ in the equation

$$I = \frac{(10a)^3}{a^2 + ax}$$

(*Hint*: Divide numerator and denominator by an appropriate quantity.)

1.15 **(a)** Find the percent error that results when the approximation in Exercise 1.13(b) is used for the case $R_1 = 100$ and $R_2 = 1000$.

(b) Repeat part (a) for $R_1 = 100$ and $R_2 = 1 \times 10^4$.

1.16 Find the percent error that results when the approximation in Exercise 1.14 is used for the case $a = 5$ and $x = 0.1$.

2 The Nature of Electricity

2.1 The Structure of Matter

The field of study we call *electricity* is the investigation of the forces created by *charged* particles, especially electrons, and the motion and interactions of those particles. The electron is a fundamental component of matter and is considered to have the smallest possible unit of *negative* charge. In comparison to ordinary visible objects in our environment, the electron is an extremely small particle, having a mass of only 9.109×10^{-31} kg.

All matter is composed of *atoms*, each of which has a central *nucleus* and one or more electrons that travel in orbits around the nucleus, like satellites around the earth. The nucleus contains one or more *positively* charged particles called *protons*. The positive charge of a proton is "opposite" to the negative charge of an electron, in the sense that the total, or net, charge of the combination is zero. Thus, an atom that has the same number of electrons in orbit as it has protons in its nucleus is *electrically neutral*. The nucleus of every atom except that of hydrogen also contains one or more *neutrons*, which carry no electrical charge. The number of protons and neutrons in the nucleus of an atom uniquely determines the element it represents—iron, copper, oxygen, and so on—and all the atoms of a given element have identical neuclei.

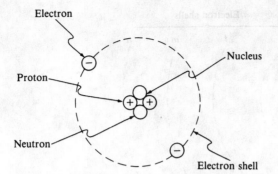

FIGURE 2.1 Structure of the helium atom.

Figure 2.1 is a diagram of the structure of the helium atom. Notice that the nucleus is a cluster of two protons and two neutrons and that there are two electrons in an orbit, called an *electron shell*, around the nucleus. The atom is electrically neutral because the two positively charged protons neutralize the two negatively charged electrons.

A very important fact that accounts for many of the electrical phenomena we will study in this book is that *there is a force of attraction between oppositely charged particles* and a *force of repulsion between similarly charged particles*. For example, two electrons in the vicinity of one another will each experience a force that drives them apart. An electron and a proton will each experience a force that draws them together. This behavior is summarized by the statement:

Like charges repel and opposite charges attract.

In Chapter 8 we will study the nature of the forces on charged particles in more detail. We will learn that the magnitude of the force is inversely proportional to the square of the distance separating the charged particles. In short, the force increases dramatically when the particles are brought closer together. We can see in Figure 2.1 that there is a strong force of attraction between the orbiting electrons and the positive nucleus of the helium atom. The *dynamic* forces resulting from the orbital motion of the electrons counteract the attractive force of the nucleus and prevent the electrons from falling into the nucleus. This phenomenon is similar to that which keeps an earth satellite from falling out of orbit due to gravitational attraction.

The atoms of other elements contain additional electron shells that are farther removed from the nucleus than the electron shell shown in Figure 2.1. Figure 2.2 shows the electron shells that surround the copper atom. Each shell is designated by a letter (k, l, m, and n) and each can contain only a certain number of electrons. If we regard the innermost (k) shell as shell number 1, the next (l) as shell number 2, and so on, the maximum number, M, of electrons that the nth shell can contain is given by

$$M = 2n^2 \text{ electrons} \tag{2.1}$$

Thus, the innermost shell can contain 2 electrons, the next can contain 8 electrons, the next 18 electrons, and the fourth, 32 electrons. Equation (2.1) is valid for determining the number of electrons in any of the *first four* shells of an atom, but is not generally applicable for complex atoms containing additional shells.

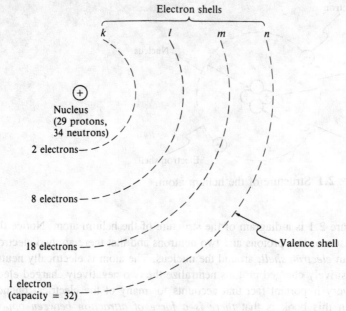

FIGURE 2.2 Diagram of the copper atom. Notice that the outermost (valence) shell has a capacity of 32 electrons but contains only one.

2.2 Conductors, Insulators, and Semiconductors

As shown in Figure 2.2, the nucleus of the copper atom contains 29 protons. A neutral copper atom must therefore have 29 electrons distributed among its various shells. Shells k, l, and m are filled to capacity with a total of 28 electrons, so there is only one electron in the n shell. The outermost shell of an atom, the n shell in this case, is called the *valence* shell, and the number of electrons it contains strongly influences the electrical properties of the element that the atom represents. Since electrons in the valence shell are the farthest removed from their nucleus, they experience the least force of attraction to the nucleus. Coupled with the fact that electrons in a nearly empty shell are easily dislodged from that shell, we find that the lone electron in the valence shell of copper can readily vacate that shell. When an electron breaks away from its "parent" atom, it is called a *free* electron, since it is then free to wander randomly through the material. An atom producing such a free electron acquires a net positive charge, because its total number of protons is then one greater than its total number of electrons. Such an atom is called a *positive ion*. To become free, an electron must acquire enough energy to overcome the force that binds it to its nucleus. In copper, there is enough heat energy at ordinary room temperatures to liberate a vast number of the weakly held valence electrons.

The presence of a large number of free electrons in copper, as in other metals, is what makes it a good *conductor* of electricity. As we shall presently learn, conduction is the transfer of electrical charge through a material, a process that is enhanced by the availability of large numbers of free electrons. In an isolated conductor having no electrical circuit connections (no battery connections, for example), the free electrons are continually subjected to random forces and move in all different directions as they come under the influence of other electrons and nuclei. Since there are, on average, as many electrons moving in any one direction as in any other, there is no long-term net transfer of charge from any one region of the conductor to any other.

The best conductors are silver, copper, and gold, in that order. Since copper is the least expensive, it is widely used in the electrical and electronic industries. A cubic centimeter of copper (about the size of a thimble) contains approximately 8.4×10^{22} free electrons at room temperature.

Materials whose valence electrons are tightly held to their parent atoms produce relatively few free electrons. Such materials are poor conductors of electricity and are called *insulators*. Most materials, including plastics, ceramics, rubber, paper, and most liquids and gases, fall into that category. Of course, there are many practical uses for insulators in the electrical and electronic industries, including wire coatings, safety enclosures, and power-line insulators.

Semiconductors are a special class of materials which, as their name implies, are neither good conductors nor good insulators. These materials have a crystalline structure in which the electrons in the valence shell of every atom are simultaneously in the valence shells of all neighboring atoms. As a consequence, free electrons are produced much less readily than in conductors. However, by controlling the impurity concentrations in these materials during their commercial manufacture, it is possible to control very closely the total number of free electrons that result. Thus, a semiconductor can be manufactured to behave more like a conductor or more like an insulator. *Silicon* is the principal semiconductor material, and it is widely used in the manufacture of electronic devices such as transistors and integrated circuits.

2.3 Electric Current

It is conventional to speak of the motion, or transfer, of electrical charge from one location to another with the understanding that it is actually charged *particles* that do the moving. In conductors, the transfer of charge occurs only as a result of the motion of electrons, and for that reason, electrons are often called *charge carriers*. The greater the number of electrons that move from one location to another, the greater the transfer of charge. In this book we will conform to convention, and refer to charge as if it were, by itself, a movable object.

In the SI system, the unit of charge is the *coulomb* (C). One coulomb of negative charge is the total charge carried by 6.242×10^{18} electrons. We can then find the total charge on one electron (the number of coulombs per electron) as follows:

$$\frac{1 \text{ coulomb (C)}}{6.242 \times 10^{18} \text{ electrons}} = 1.6 \times 10^{-19} \text{ C/electron}$$

We should note that charge can be either positive or negative. One coulomb of positive charge is the total charge of 6.242×10^{18} positively charged atoms, each of which has lost one electron. The symbol Q is used to represent a quantity of charge. For example, it would be correct to write $Q = 100 \ \mu\text{C}$. (Do not confuse the symbol Q for the quantity of charge with the symbol C for the units of charge.)

Electrical current exists in material when there is a net transfer of charge through the material, from one region to another. For example, if electrons are somehow injected into one end of a copper wire, travel through the wire, and emerge at the other end, we say that there is current in the wire. Current is measured in terms of the *rate* at which charge is transferred, that is, the amount of charge that moves past a point per unit time. The SI unit of current is the *ampere* (A), named after the eighteenth-century French physicist André Ampère. One ampere equals a rate of flow of charge equal to 1 coulomb per second.* Thus, 2 amperes of current exist when 2 coulombs of charge pass a point in 1 second (or when 4 coulombs pass in 2 seconds, etc.) The symbol for current is I, from the French word *intensité* (intensity). From the definition it follows that the general relationship between current, charge, and time is

$$I = \frac{Q}{t} \quad \text{amperes} \tag{2.2}$$

where Q is the number of coulombs of charge that pass a point in t seconds.

Example 2.1

The current in a certain conductor is 40 mA.

(a) Find the total charge in coulombs that passes through the conductor in 1.5 s.
(b) Find the total number of electrons that pass through the conductor in that time.

SOLUTION
(a) From equation (2.2),

$$Q = It = (40 \times 10^{-3} \text{ A})(1.5 \text{ s}) = 60 \times 10^{-3} \text{ C}$$

(b) $(60 \times 10^{-3} \text{ C})\left(\dfrac{6.242 \times 10^{18} \text{ electrons}}{1 \text{ C}}\right) = 3.745 \times 10^{17} \text{ electrons}$

Drill Exercise 2.1

Find the current in a conductor when 1×10^{17} electrons pass through it in 50 ms.

ANSWER: 0.3204 A. □

---

*In the SI system, the ampere is one of the base units, and charge in coulombs is a *derived* unit: $1 \text{ C} = (1\text{A}) \times (1\text{s}) = 1 \text{ A-s}$. However, it is easier to understand the concept of current by first defining coulombs and then using those units to define amperes. From a practical standpoint, it makes no difference which is defined first.

Strictly speaking, it is not correct to say that current "flows," or "goes" from one point to another, since current is a *rate* of flow. We would not say, for example, that a rate (speed) of 45 mph "goes" down the highway. However, it is a widely used practice to speak of current as flowing from one point to another (meaning that charge flows from one point to another), and we will follow that practice in this book.

2.4 Electromotive Force (Voltage)

To establish a flow of charge through a conductor, it is necessary to exert some sort of force on the electrons that carry the charge, thereby causing them to move through the conductor. The force, called *electromotive force* (emf), is produced by a body or object having a negative charge, that is, by a *reservoir* of negative charge, since that type of charge will repel electrons in the conductor and cause them to move. Imagine the reservoir to be a rod or bar in which there is a large, concentrated accumulation of *surplus* electrons, that is, many more electrons than would be necessary to neutralize the positive nuclei in the atoms of the rod. The rod thus has the required negative charge, and when it is connected to the conductor, it will force electrons to move.

To sustain a flow of charge through a conductor, it is necessary to provide a destination where the electrons moving through the conductor will be accepted. Without such a destination, the conductor is like a blocked water pipe: No "water" (electrons) will flow through it, regardless of the "pressure" (emf) exerted at one end. The appropriate destination for electrons is another rod which has a large *deficiency* of electrons, that is, one that has far fewer electrons than are necessary to neutralize the positive nuclei. This rod will have a large positive charge, and when it is connected to the other end of the conductor, electrons will be attracted to it.

Figure 2.3 shows the arrangement that we have described: two rods, called *electrodes*, and a conductor connected between them. The rods can be regarded as the positive and negative electrodes of a battery. As such, they are both embedded in a chemical mixture whose purpose is to gather the electrons arriving at the positive electrode and force them back to the negative electrode. In this way, a continuous flow of charge is maintained through the conductor. *Chemical energy* is expended in a battery to create the force necessary to restore electrons to the negative electrode. In a subsequent discussion, we will describe other means used to produce emf and to maintain the flow of electrons.

Whether or not current is established or maintained in a conductor, the basic ingredients for a source of electromotive force capable of producing that current are (1) a concentration of negative charge in some region, and (2) a concentration of positive charge in some other region. The greater the amounts of charge in each region, the greater the electromotive force the combination provides. The unit of measure of electromotive force is the *volt*, named after the eighteenth-century Italian physicist Alessandra Volta. In Chapter 8 we will define the volt more precisely and will relate it to the amount of energy required to move charge from one point to another. For our purposes now, it is sufficient to understand that the greater the voltage of a source of electromotive force, the greater the current it is capable of producing in a conductor.

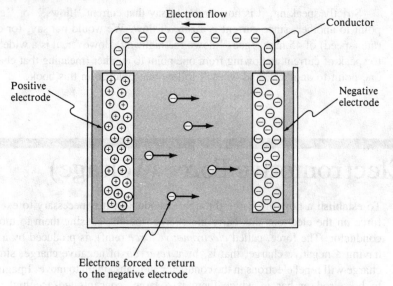

Electron flow

Conductor

Positive
electrode

Negative
electrode

Electrons forced to return
to the negative electrode

FIGURE 2.3 Current is sustained in a conductor by providing a
source of electrons (negative electrode) that repels free electons through
the conductor, a destination (positive electrode) that attracts the electrons,
and a process that restores electrons to the negative electrode.

A source of electromotive force such as we have described is called a *voltage source*.
The voltage is much like the pressure developed by a pump in a water system. We may
think of the rate of water flow as being like the rate of flow of charge (i.e., current).
Then, just as a greater pressure at the pump will produce a greater rate of water flow, a
greater voltage in the voltage source will produce a greater current. Notice that the
pressure, or voltage, *exists* at its source. It does not "flow" or "go" anywhere. Therefore, it is entirely incorrect to speak of voltage as "going" somewhere, and such terminology is used only by individuals who do not clearly understand electrical concepts.
Since electromotive force, in volts, is a measure of the capability, or *potential*, that a
source has to produce current, it is also called electric potential.

Figure 2.4 shows the symbol used to represent a voltage source. The shorter line in
the symbol is the negative electrode and the longer line is the positive electrode. The
point at which a conductor would be connected to the positive electrode is called the
positive terminal of the source, and the point of connection to the negative electrode is
the *negative terminal*. The terminals are shown by small circles, but these are often
omitted in electrical diagrams.

The symbols E (for electromotive force) and V are both used to represent voltage. A
voltmeter is an instrument used to measure voltage, and when a voltmeter is connected
across (between) the terminals of a voltage source, the reading is called the *terminal
voltage*. Figure 2.4 shows how the terminal voltage (E volts) *across* the terminals is
designated. Note that *two* terminals must always be specified (or understood) when
referring to the voltage of a voltage source. In electrical diagrams, the voltage is often
written beside the symbol for the source (e.g., 12 V), with the understanding that the
voltage across the terminals is meant.

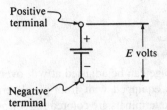

Positive terminal

E volts

Negative terminal

FIGURE 2.4 Symbol for a voltage source.

Types of Voltage Sources

We have described a battery as an example of a voltage source, in particular, one that utilizes chemical energy to furnish electric current. A battery has a fixed lifetime that depends on the amount of current it supplies; the greater the current, the shorter the life. The lifetime of a battery can be determined from its *ampere-hour* (A-h) *rating*, as follows:

$$\text{life (hours)} = \frac{\text{ampere-hour rating}}{\text{amperes of current produced}} \qquad (2.3)$$

For example, a battery with a 60-A-h rating can supply 1 A for 60 h, 0.5 A for 120 h, 2 A for 30 h, and so on. Rechargeable batteries, such as automobile batteries, can have their lifetime restored through a recharging procedure that forces electrons into the negative terminal.

A *generator* is another example of a voltage source, in this case, one that utilizes mechanical energy to produce the electromotive force. The mechanical energy is supplied to the generator by an external force that causes a cylindrical *armature* to rotate. As a result of the interaction that takes place between the rotating armature and a magnetic field, which we will study in Chapter 10, an emf is produced across the armature terminals.

In laboratory and experimental work, an *adjustable power supply* is often used as a voltage source. Figure 2.5 shows a typical power supply of that type. The advantage of

FIGURE 2.5 A typical laboratory type power supply. (Photo courtesy of Hewlett-Packard Company.)

FIGURE 2.6 Symbol for resistance.

a power supply is that the terminal voltage can be adjusted at will over a wide range of values. Note that the power supply is equipped with front-panel terminals to which external connections can be made. These terminals are colored red for positive and black for negative. Most adjustable power supplies also have a built-in voltmeter that displays the terminal voltage and aids in adjusting that voltage to a particular value.

2.5 Resistance

Resistance is a measure of the extent to which a material interferes with, or resists, the flow of current through it. From that definition, we know that a conductor has a small resistance and that an insulator has a large resistance. Returning to our water system analogy, resistance can be likened to the roughness of the interior walls of the water pipes and the presence of any other obstructions in them. Any impediment to the flow increases resistance and reduces the flow rate.

The unit of resistance is the *ohm* (Ω), named after Georg Ohm, a nineteenth-century German physicist. The symbol R is used to represent resistance, so to express the fact that a certain component has a resistance of 100,000 ohms, it would be correct to write $R = 100 \text{ k}\Omega$. A short piece of copper wire may have a resistance equal to a fraction of an ohm, while a similar piece of plastic may have a resistance of several hundred megohms (MΩ). A perfect conductor would have zero resistance and a perfect insulator would have infinite resistance.

In Chapter 8 we will learn how the resistance of a body is affected by temperature and by the physical dimensions of the body. For the time being, it is sufficient to understand that all materials and all electrical and electronic components have *some* resistance. Figure 2.6 shows the standard symbol used to represent resistance in an electrical diagram.

2.6 Conventional Current

Recall that current will not flow in a conductor unless there is a complete path from the negative terminal of a voltage source to the positive terminal. In other words, there must be a destination that will accept electrons as well as a source of electrons. Figure 2.7(a) shows a diagram of a voltage source with a conductor, such as a copper wire, and resistance R, connected in a path between the positive and negative terminals. The resistance could represent the resistance of the wire or could include the resistance of some other component connected in the path. Since there is a complete path, electrons will flow from the negative terminal of the source, through the conductor and resistance,

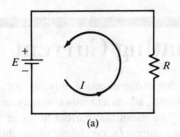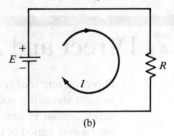

(a)   (b)

FIGURE 2.7 Electron current and conventional current. (a) Electron current flows
from − to +. (b) Conventional current flows from + to −

and into the positive terminal of the source. This current, flowing from − to +, is often
called *electron current*. Early investigators of electrical phenomena, notably Benjamin
Franklin, believed that current flow was caused by the motion of positive charge, rather
than negative charge. As a consequence, much of the early theory and technical literature
was based on the assumption that current flowed from + to −. That assumption, or
rather that usage, has persisted to this day and is called *conventional current*, as shown
in Figure 2.7(b).

It is important to realize that none of the practical consequences nor any of the results
of computations performed in the study of electricity and electronics are in any way
affected by the direction of current flow that one assumes. In this book, as in most books,
the direction of conventional current will always be assumed and will be used in all
diagrams where current directions are shown. By definition, *conventional* current is that
which has been adopted by the majority of practicing engineers, technologists, scientists,
and technical writers.

Figure 2.8 shows the water system analogy to a voltage source with resistance connected between its terminals. Note once again that the voltage produced by the source
is like the pressure produced at the water pump. The greater the pressure, the greater the
flow rate. The flow rate itself, say, in gallons per minute, is like the electrical current
in coulombs per second. The constriction in the pipe, which slows the flow rate, is like
the resistance in the conductor.

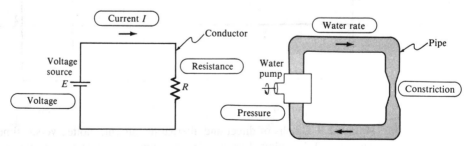

FIGURE 2.8 Water system analogy to a voltage source with resistance connected
between its terminals.

2.7 Direct and Alternating Current

Direct current (dc) is current that flows in one direction only, as, for example, from left to right through a conductor. Of course, all current flows in only one direction *at any single instant of time*, but by "dc" we mean that current flows in only one direction over a long period of time. Alternating current (ac) is current whose direction periodically reverses, as, for example, current that flows from left to right for a period of time, then from right to left, then again from left to right, and so on. We will study ac in more detail in Chapter 12.

We can arbitrarily assume that current flowing in a certain direction through a conductor, say from left to right, is *positive* current. Under that assumption, if current were ever to flow in the reverse direction, from right to left, we would regard it to be *negative* current. As shown in Figure 2.9(a), a graph of direct current plotted versus time would simply be a horizontal line, indicating that the current never reverses direction (i.e., is never negative). Figure 2.9(b) shows an example of a graph of alternating current. In this case the current is + 2 A for 1s, then reverses direction and is plotted as − 2 A during the next second. The continuous reversal of direction is shown graphically by the fact that the current periodically alternates between +2 A and −2 A.

All batteries produce direct current, that is, current whose (conventional) direction is always from the positive terminal, through an external conductor, and then to the negative terminal. A battery is therefore called a *dc voltage source*. In spite of the apparent contradiction of terms ("direct-current voltage"), it is conventional to use the terminology dc voltage, meaning a voltage that creates or results from the flow of direct current. The power supply shown in Figure 2.5 is called a dc power supply because it, too, produces direct current.

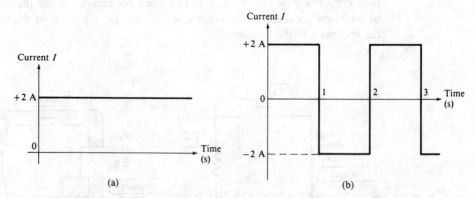

(a)   (b)

FIGURE 2.9 Graphs of direct and alternating current, plotted versus time. (a) A direct current of 2 A is plotted versus time as a horizontal line because the current never reverses direction. (b) Example of a plot of alternating current. Every second, the current reverses direction, so it is alternately positive and negative.

Exercises

Section 2.1   The Structure of Matter

2.1 How many electrons would together comprise a total mass of 1 kg?

2.2 The mass of a proton is approximately 1837 times greater than that of an electron. If the mass of a neutron is the same as that of a proton, approximately how many atoms are in a 1-kg mass of copper?

2.3 The element silicon has 14 protons and 14 neutrons in the nucleus of each atom. Find the number of electrons in each of the electron shells of a neutral atom.

2.4 The element germanium has 32 protons and 40 neutrons in the nucleus of each atom. Find the number of electrons in each of the electron shells of a neutral atom.

Section 2.3   Electric Current

2.5 What is the total charge, in coulombs, of 3.2×10^{15} copper atoms, each of which has lost one valence electron?

2.6 What is the total charge, in coulombs, of a 1-kg mass of electrons?

2.7 What is the current in a conductor if 450 μC of charge passes through it in 15 ms?

2.8 What is the current in a conductor if 3.141×10^{15} electrons pass through it in 0.25 s?

2.9 How many coulombs of charge pass through a conductor in 50 ms if the current in the conductor is 2.3 A?

2.10 How many electrons pass through a conductor in 1 min if the current in the conductor is 200 μA?

2.11 For how long must the current in a conductor equal 125 mA in order that 50 C of charge passes through it?

2.12 For how long must the current in a conductor equal 1 A in order that 6.242×10^{18} electrons pass through it?

Section 2.4   Electromotive Force (Voltage)

2.13 A certain battery has a rating of 150 A-h. For how long can it supply a current of 200 mA?

2.14 How much current can a battery with a rating of 80 A-h supply continuously for 150 min?

Section 2.5   Resistance

2.15 The resistance of a certain conductor is directly proportional to its length. If a 50-m length of the conductor has resistance 25 Ω, what is the resistance of a 75-m length?

2.16 The resistance of a cylinder made of a certain material is inversely proportional to the square of its diameter. If a cylinder having diameter 60 mm has resistance 1.5 kΩ, what is its resistance if the diameter is reduced to 30 mm?

2.17 Convert
(a) 87,532 Ω to kilohms
(b) 87,532 Ω to megohms
(c) 0.04 kΩ to ohms
(d) 1250 kΩ to megohms
(e) 8.4×10^{-4} MΩ to kilohms
(f) 1.885×10^5 Ω to kilohms

2.18 Convert
(a) 1.075 kΩ to ohms
(b) 94,617 Ω to kilohms
(c) 94,617 Ω to megohms
(d) 0.0032 MΩ to kilohms
(e) 24.83×10^3 kΩ to megohms
(f) 99.01×10^{-4} MΩ to kilohms

3 Fundamental Relations in Electric Circuits

3.1 Ohm's Law

In Chapter 2 we learned that voltage is a measure of the ability of a source of emf to produce current; the greater the voltage, the greater the current that the source can produce in a fixed resistance. In other words, the current produced in a resistance is *directly proportional* to the voltage of the source. We also observed that resistance *reduces* the flow of current: The greater the resistance, the smaller the current produced through it by a fixed voltage. In short, current is *inversely proportional* to resistance. The way in which voltage and resistance affect current can be combined in a single mathematical expression called *Ohm's law*, which is one of the most important laws in the theory of electricity:

$$I = \frac{E}{R} \tag{3.1}$$

In the SI system, I is in amperes when E is in volts and R is in ohms. Notice that Ohm's law correctly expresses the relationships we have described: For a fixed R, I increases when E increases, and for a fixed E, I decreases when R increases.

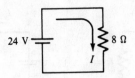

FIGURE 3.1 (Example 3.1)

Example 3.1

With reference to Figure 3.1:

(a) Find the current I in the resistance.
(b) Find the current when the voltage of the source is doubled.
(c) Find the current when the voltage is restored to its original value and the resistance is doubled.

SOLUTION

(a) $I = \dfrac{E}{R} = \dfrac{24 \text{ V}}{8 \text{ }\Omega} = 3\text{A}$

(b) $I = \dfrac{48 \text{ V}}{8 \text{ }\Omega} = 6 \text{ A}$

Note that doubling the voltage doubled the current.

(c) $I = \dfrac{24 \text{ V}}{16 \text{ }\Omega} = 1.5 \text{ A}$

Note that doubling the resistance reduced the current by one-half.

Drill Exercise 3.1

Find the current in a 0.04-kΩ resistance when it is connected across a 720-mV voltage source.

ANSWER: 18 mA.
☐

Figure 3.1, showing a resistance connected to a voltage source, is an example of an electric *circuit*. As we shall study in Chapter 4, a circuit is any configuration of electrical components connected in such a way that current can flow in the components.

In many practical circuits, we know values for two of the quantities I, E, or R, and we are required to find the third. Toward that end, Ohm's law can be expressed in the equivalent forms

$$E = IR \quad \text{and} \quad R = \frac{E}{I} \tag{3.2}$$

Example 3.2

Find the unknown quantity in each circuit shown in Figure 3.2.

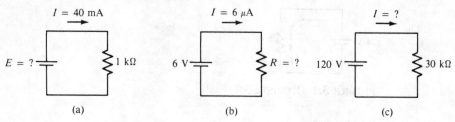

FIGURE 3.2 (Example 3.2)

SOLUTION
(a) $E = IR = (40 \text{ mA})(1 \text{ k}\Omega) = (40 \times 10^{-3} \text{ A})(1 \times 10^{3} \text{ }\Omega) = 40 \text{ V}$

(b) $R = \dfrac{E}{I} = \dfrac{6 \text{ V}}{6 \text{ }\mu\text{A}} = \dfrac{6 \text{ V}}{6 \times 10^{-6} \text{ A}} = 1 \text{ M}\Omega$

(c) $I = \dfrac{E}{R} = \dfrac{120 \text{ V}}{30 \text{ k}\Omega} = \dfrac{120 \text{ V}}{30 \times 10^{3} \text{ }\Omega} = 4 \text{ mA}$

Drill Exercise 3.2
How much voltage is necessary to create a flow of 0.24 C in 0.8 s through a resistance of 500 Ω?

ANSWER: 150 V. ☐

3.2 Measuring Voltage, Current, and Resistance

As mentioned in Chapter 2, a *voltmeter* is an instrument designed to measure voltage. In our discussion of that instrument, we stated that it can be connected *across* the terminals of a voltage source to measure the voltage produced by the source. Figure 3.3(b) shows a voltmeter, indicated by a circle with a V inside it, that is connected across a voltage source. Figure 3.3(c) and (d) show the voltmeter connections when resistance is also connected across the source. Note that the value of the voltage measured is the *same* (6 V) in every case. The voltmeter is connected across the resistance as well as the voltage source, and we say that the *voltage across the resistance* is 6 V. It is important to realize that voltage measurements are always made *across* components; that is, *it is not necessary to disconnect any components for the purpose of connecting a voltmeter* and making a voltage measurement. The + and − symbols on the voltmeter symbol show the *polarity* of the connections (red terminal to + and black terminal to −) and the polarity of the measured voltage. We say that the voltage on the + side is 6 V *with respect to* the − side (an interpretation that will be discussed in more detail

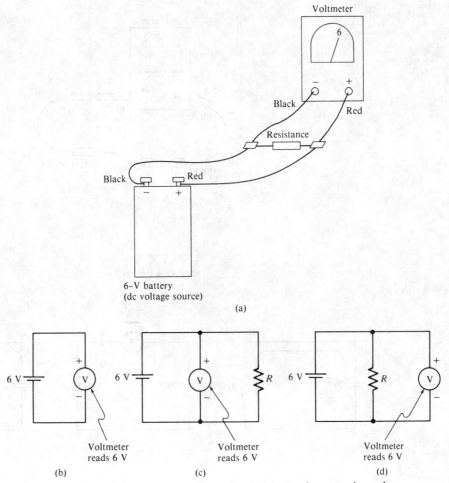

FIGURE 3.3 Voltmeter connections. The voltage measured across the voltage source is the same as that measured across the resistance. (a) Pictorial representation of a voltmeter connected *across* a battery and a resistance. (b) Voltmeter connected across voltage source alone. (c) Voltmeter connected across voltage source with resistance present. (d) Voltmeter connected across resistance [same as part (c)].

in Chapter 4). If the voltmeter connections were reversed (red terminal to − and black terminal to +), the reading would be − 6 V. Some voltmeters are not capable of indicating a negative voltage.

   Current is measured by an instrument called an *ammeter*. To measure the current flowing in a resistance, it is necessary to disconnect the resistance and insert an ammeter in such a way that all the current flowing in the resistance also flows through the ammeter. This connection is shown in Figure 3.4. As illustrated, the side of the resistor on which the ammeter is connected does not affect the reading, since the same current flows into the resistor as flows out of it. The polarity of an ammeter requires that it be connected

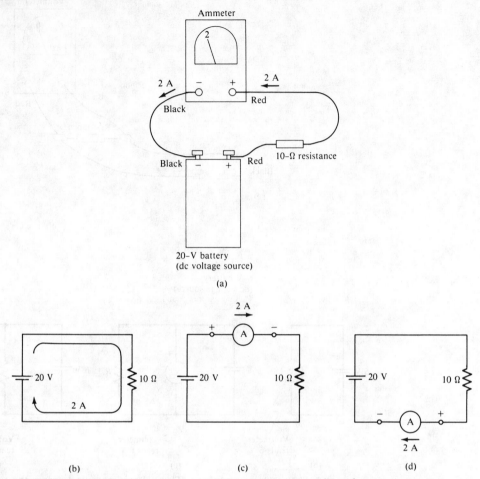

FIGURE 3.4 Ammeter connections. (a) Pictorial representation of an ammeter connected to measure the current in a 10-Ω resistance. (b) $I = E/R = 20$ V/10 Ω = 2 A. (c) Ammeter inserted. (d) The ammeter reading is the same as in part (c).

so that (conventional) current flows into the + terminal of the meter and out of the − terminal. If the connections are reversed, a negative value of current will be indicated. Some ammeters are not capable of showing negative current.

   The importance of proper voltmeter and ammeter connections, as shown in Figures 3.3 and 3.4, cannot be overstressed. If a voltmeter is connected as if it were an ammeter, or vice versa, the instrument may be severely damaged. Figure 3.5 shows the correct way to connect a voltmeter and an ammeter in a circuit for simultaneous measurement of the voltage across and current through a resistance.

   An *ohmmeter* is an instrument designed to measure resistance. It is basically a voltage source and an ammeter. The built-in voltage source is effectively connected across the resistance to be measured and the ammeter measures the current that flows. Since the

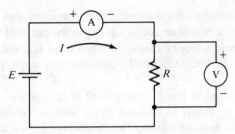

FIGURE 3.5 Simultaneous measurement
of the voltage across and current
through resistance R.

Portable, analog type VOM

Digital multimeter

FIGURE 3.6 Typical laboratory instruments for measuring voltage, current, and
resistance. (Courtesy of Simpson Electric
Company.)

current is inversely proportional to the resistance, the instrument can be calibrated to indicate the number of ohms of resistance corresponding to the different values of current it measures. Resistance should never be measured when there is a voltage source connected across it (other than the one built into the ohmmeter) or when there is any other component connected to it.

There is one type of instrument that is widely used in laboratories and for general-purpose testing because of its ability to measure either voltage, current, or resistance, depending on the setting of a selector switch. This instrument is available with either a *digital* display, like that of a calculator, or with a "needle" (pointer) that moves across a continuous scale. The latter is called an *analog*-type display. Figure 3.6 shows examples of each type of instrument. These instruments are known variously as multimeters, volt-ohm-ammeters (VOMs), and digital multimeters (DMMs). The same precautions discussed previously in regards to proper meter connections also apply to multimeters. In particular, the instrument must not be set to measure current when it is connected as a voltmeter, nor set to measure voltage when it is connected as an ammeter.

3.3 Work, Energy, and Power

Work is the expenditure of energy to overcome some restraint or to achieve some change in the physical state of a body. Examples include the work performed by a man when expending energy to lift a weight above his head, and the action of a steam-driven piston when water is made to boil by consumption of heat energy. Energy is often defined as the ability to do work. *Energy and work have the same units*. In the SI system, those units are *joules* (J). Heat is a particularly important form of energy in the study of electricity, not only because it affects the electrical properties of materials, but also because it is *liberated* whenever electrical current flows. This liberation of heat is in fact the conversion of electrical energy to heat energy. To gain some appreciation for the magnitude of a joule of heat energy, consider that it would require about 90,000 J to heat a cup of water from room temperature to boiling.

Power is the *rate* at which energy is expended, or the rate at which work is performed. Since energy and work both have the units of joules, it follows that power, being a rate, has the units joules/second. Returning to the boiling-water example, suppose that all the heat required (all 90,000 J) could be supplied to the cup of water in 1 s. Then the power would be the very large value 90,000 J/s. On the other hand, if the water were allowed to come to a very slow boil by supplying the same amount of heat over a period of 30 min, the power would be only 50 J/s. Note that the total amount of heat energy is the same in each case, and that in each case the water comes to a boil. However, the power is quite different, because the rates at which heat is supplied are different. In the SI system, the units joules/second are called *watts*. Thus, the watt is the SI unit of power. In general,

$$P = \frac{W}{t} \quad \text{watts} \tag{3.3}$$

where W is the total number of joules of work performed, or the total joules of energy expended, in t seconds. We will use the symbol W to stand for energy as well as work, to avoid confusion with the symbol E for voltage.

Example 3.3

How much heat energy is produced by a 1.5-kW electric heater when it is operated for 30 min?

SOLUTION From equation (3.3),

$$W = Pt = (1.5 \times 10^3 \text{ W})(30 \text{ min})\left(\frac{60 \text{ s}}{1 \text{ min}}\right) = 2.7 \times 10^6 \text{ J}$$

Drill Exercise 3.3

How long would it take the heater in Example 3.3 to produce 0.6 MJ of heat energy?

ANSWER: 400 s.

When electric current flows through resistance, electric energy is converted to heat energy at a rate that depends on the voltage across the resistance and on the value of current through it:

$$P = VI \qquad \text{watts} \tag{3.4}$$

where V is the voltage across the resistance and I is the current through it. Note that the product of volts and amperes is watts, or joules per second. Although not technically correct usage, it is conventional to say that resistance "dissipates power," meaning that it dissipates (liberates) heat at a certain rate.

By substituting $V = IR$ from Ohm's law into equation (3.4), we find another useful expression for electrical power:

$$P = VI = (IR)I = I^2R \qquad \text{watts} \tag{3.5}$$

We can also substitute $I = V/R$ into equation (3.4) and obtain still another expression for power:

$$P = VI = V\left(\frac{V}{R}\right) = \frac{V^2}{R} \qquad \text{watts} \tag{3.6}$$

Example 3.4

Find the power in the resistance shown in Figure 3.7 using each of equations (3.4), (3.5), and (3.6).

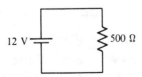

FIGURE 3.7 (Example 3.4)

SOLUTION By Ohm's law,

$$I = \frac{V}{R} = \frac{12 \text{ V}}{500 \text{ }\Omega} = 24 \text{ mA}$$

By equation (3.4),

$$P = VI = (12 \text{ V})(24 \text{ mA}) = 288 \text{ mW}$$

By equation (3.5),

$$P = I^2R = (24 \times 10^{-3}\text{A})^2(500 \text{ }\Omega) = 288 \text{ mW}$$

By equation (3.6),

$$P = \frac{V^2}{R} = \frac{(12 \text{ V})^2}{500 \text{ }\Omega} = 288 \text{ mW}$$

Drill Exercise 3.4

What voltage connected across the 500-Ω resistance in Figure 3.7 would cause the power dissipation to be 5 W?

ANSWER: 50 V.   □

The rate at which resistance liberates heat energy equals the rate at which energy is supplied to the resistance by a voltage source. Loosely speaking, the power "delivered" to the resistance equals the power dissipated in the resistance. There is no resistance associated with an (ideal) voltage source, so the power delivered by the source can only be calculated using equation (3.4): $P = EI$ watts, where E is the terminal voltage and I is the current furnished by the source. In Example 3.4, the power delivered by the 12-V source is $P = EI = (12 \text{ V})(24 \text{ mA}) = 288 \text{ mW}$, which equals the power dissipated in the 500-Ω resistance.

Kilowatt-hours

In the electrical power industry, the unit of power most often used is the kilowatt (1 kW = 1000 W). The unit of energy used is the total energy delivered or consumed in 1 hour when the rate of delivery or consumption is 1 kW. That amount of energy is 1 *kilowatt-hour* (kWh).

$$\text{kWh} = (P \text{ in kW}) \times (t \text{ in hours}) \tag{3.7}$$

The kilowatt-hour is *not* an SI unit.

Note that customers of a power company are billed for the total amount of *energy* consumed, in kWh, not for power. As illustrated in the next example, the total energy consumed by various electrical "loads" can be calculated using equation (3.7).

Example 3.5

If a power company charges $0.10 for each kWh of energy delivered to a customer, find the total cost of operating a 500-W television set for 2 h, six 75-W light bulbs for 4 h, a 1500-W clothes drier for 30 min, and a 2-kW electric heater for 45 min.

SOLUTION
The total energy consumed by each load is computed using equation (3.7):

Television: (0.5 kW)(2 h)	1 kWh
Light bulbs: (6)(0.075 kW)(4 h)	1.8 kWh
Clothes drier: (1.5 kW)(0.5 h)	0.75 kWh
Heater: (2 kW)(0.75 h)	1.5 kWh
Total	5.05 kWh

Total cost = (5.05 kWh)($0.10/kWh) = $0.505

Drill Exercise 3.5
If electrical energy costs $0.12/kWh, for how long could a 900-W oven be operated without costing more than 36 cents?

ANSWER: $3\frac{1}{3}$ h.

3.4 Resistors

Some earlier discussions may have left the impression that resistance is a consistently undesirable property for a material to have. On the contrary, in many practical applications, it is necessary to insert resistance in a circuit for the express purpose of limiting or controlling the flow of current. For example, if a 24-V voltage source is available and it is necessary to establish a current of 2 A, we would wish to connect $R = 24 \text{ V}/2 \text{ A} = 12 \ \Omega$ of resistance across the source to obtain that current. A *resistor* is a device that is manufactured so that it has a specific amount of resistance, and it is used in circuits for precisely the purposes we have described.

Resistors are available with a wide range of resistance values and in many different physical sizes. Figure 3.8 shows some examples. The physical size of a resistor is not necessarily related to its resistance value, but rather to its *power rating*. Recall that heat energy is produced when current flows through resistance, and that a resistor must be capable of dissipating that heat. If the power rating of a resistor is too small for a particular application, the resistor will not be able to dissipate heat at a rate rapid enough to prevent destructive temperature buildup. Resistors that have large power ratings (i.e., those that are capable of dissipating heat at a rapid rate) are physically large because a large surface area is required to promote the transfer of heat into the surrounding air.

Example 3.6
A 200-Ω resistor has a 2-W power rating. What is the maximum current that can flow in the resistor without exceeding the power rating?

SOLUTION From equation (3.5),

$$P = I^2R \le 2 \text{ W}$$
$$I^2(200) \le 2$$
$$I^2 \le 0.01$$
$$I \le 0.1 \text{ A} = 100 \text{ mA}$$

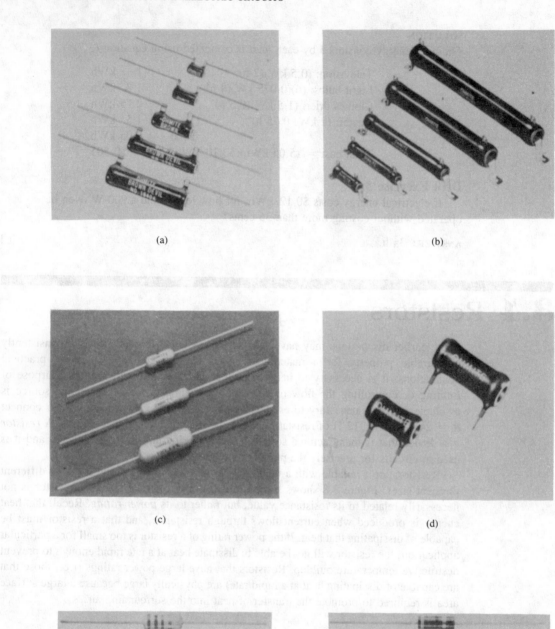

(a)

(b)

(c)

(d)

(e)

(f)

FIGURE 3.8 Examples of commercially available resistors. (Courtesy of OHMITE Manufacturing Company.) (a), (b) Power resistors. (c) Wirewound resistors. (d) Printed circuit board resistors. (e) Carbon film resistor. (f) Carbon composition resistor (see also Figure 3.9).

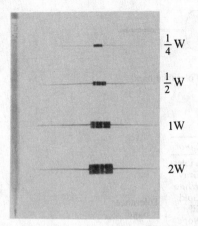

$\frac{1}{4}$ W

$\frac{1}{2}$ W

1W

2W

FIGURE 3.9 Carbon composition resistors and their power dissipation ratings.

Drill Exercise 3.6

A 1.5-kΩ resistor has a 1-W power rating. What maximum voltage can be connected across the resistor without exceeding the power rating?

ANSWER: 38.73 V.   □

In integrated circuits, resistors with specific resistance values are fabricated by controlling the dimensions and the number of free electrons in a tiny strip of semiconductor material. We will discuss those techniques in more detail in Chapter 8. Unlike integrated-circuit resistors, *discrete* resistors are individual components that can be connected at will to other components. The discrete resistors that are most commonly used in electronic circuits are of the *carbon-composition* type, as shown in Figure 3.9. These cylinder-shaped devices are available with resistance values ranging from 2.7 Ω to 22 MΩ and with power ratings of $\frac{1}{8}$, $\frac{1}{4}$, $\frac{1}{2}$, 1, and 2 W. Notice in Figure 3.9 that resistors with higher power ratings are physically larger than those with smaller ratings. The size of these resistors has no relation to their resistance values.

The Color Code

The resistance value of many resistors, including carbon-composition types, can be determined by "reading" a series of colored bands imprinted on the resistor body. In this scheme, called the resistor *color code*, each color represents a different decimal digit. Figure 3.10 shows the sequence of four (sometimes three) color bands on the body of a resistor and the digits corresponding to each color. Note that the fourth band is called a *tolerance* band and identifies a percentage rather than a decimal digit.

The first three bands of the color code are used to specify the *nominal* value of the resistance, and the fourth, or tolerance, band gives the percent variation from the nominal value that the *actual* resistance may have. In other words, due to manufacturing variations, the actual resistance may be anywhere in a range equal to the nominal value plus

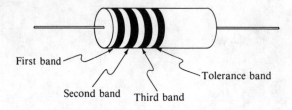

First band

Second band   Third band

Tolerance band

Digit	Color	Digit	Color
0	Black	7	Violet
1	Brown	8	Gray
2	Red	9	White
3	Orange	5%	Gold
4	Yellow	10%	Silver
5	Green	20%	None
6	Blue		

Tolerance band

FIGURE 3.10 Resistor color code.

or minus a certain percentage of that value. The first two color bands specify the first two digits of the nominal value, and the third band represents the power of 10 by which the first two digits are multiplied. The next example demonstrates these computations.

Example 3.7

Find the nominal resistance and the possible range of actual resistance values corresponding to each of the following color codes.

(a) yellow, violet, orange, silver
(b) brown, black, red
(c) blue, gray, black, gold

SOLUTION

(a) Yellow, violet, orange, silver

$$47 \times 10^3 \pm 10\%$$

Thus, the nominal resistance is 47 kΩ, and the possible range of actual values is 47 kΩ $\pm$ 0.1(47 kΩ) = 47 kΩ $\pm$ 4.7 kΩ, or

$$\text{from } 47 \text{ k}\Omega - 4.7 \text{ k}\Omega \text{ to } 47 \text{ k}\Omega + 4.7 \text{ k}\Omega$$
$$\text{that is, from } 42.3 \text{ k}\Omega \text{ to } 51.7 \text{ k}\Omega.$$

(b) Brown, black, red, (none)

$$10 \times 10^2 \pm 20\%$$
$$= 10^3 \ \Omega \pm 0.2 \times 10^3$$
$$= 1 \text{ k}\Omega \pm 200 \ \Omega$$
$$\text{range} = 800 \ \Omega \text{ to } 1.2 \text{ k}\Omega$$

(c) Blue, gray, black, gold

$$68 \times 10^0 \pm 5\%$$
$$= 68\ \Omega \pm 0.05(68\ \Omega)$$
$$= 68\ \Omega \pm 3.4\ \Omega$$
$$\text{range} = 64.6\ \Omega \text{ to } 71.4\ \Omega$$

Drill Exercise 3.7

What is the color code of a 2.7 MΩ 5% resistor?

ANSWER: red-violet-green-gold. □

Standard Values

Resistors having 5 and 10% tolerances are not manufactured in every possible nominal value. (20% resistors are rarely found in modern circuits.) Obviously, there would be no point in manufacturing a 1 kΩ ± 10% resistor and a 950 Ω ± 10% resistor, since the possible range of values of the 1-kΩ resistor includes 950 Ω. In many practical

TABLE 3.1 Standard values of 5% and 10% resistors[a]

Ohms (Ω)					Kilohms (kΩ)		Megohms (MΩ)	
0.10	**1.0**	**10**	**100**	**1000**	**10**	**100**	**1.0**	**10.0**
0.11	1.1	11	110	1100	11	110	1.1	11.0
0.12	**1.2**	**12**	**120**	**1200**	**12**	**120**	**1.2**	**12.0**
0.13	1.3	13	130	1300	13	130	1.3	13.0
0.15	**1.5**	**15**	**150**	**1500**	**15**	**150**	**1.5**	**15.0**
0.16	1.6	16	160	1600	16	160	1.6	16.0
0.18	**1.8**	**18**	**180**	**1800**	**18**	**180**	**1.8**	**18.0**
0.20	2.0	20	200	2000	20	200	2.0	20.0
0.22	**2.2**	**22**	**220**	**2200**	**22**	**220**	**2.2**	**22.0**
0.24	2.4	24	240	2400	24	240	2.4	
0.27	**2.7**	**27**	**270**	**2700**	**27**	**270**	**2.7**	
0.30	3.0	30	300	3000	30	300	3.0	
0.33	**3.3**	**33**	**330**	**3300**	**33**	**330**	**3.3**	
0.36	3.6	36	360	3600	36	360	3.6	
0.39	**3.9**	**39**	**390**	**3900**	**39**	**390**	**3.9**	
0.43	4.3	43	430	4300	43	430	4.3	
0.47	**4.7**	**47**	**470**	**4700**	**47**	**470**	**4.7**	
0.51	5.1	51	510	5100	51	510	5.1	
0.56	**5.6**	**56**	**560**	**5600**	**56**	**560**	**5.6**	
0.62	6.2	62	620	6200	62	620	6.2	
0.68	**6.8**	**68**	**680**	**6800**	**68**	**680**	**6.8**	
0.75	7.5	75	750	7500	75	750	7.5	
0.82	**8.2**	**82**	**820**	**8200**	**82**	**820**	**8.2**	
0.91	9.1	91	910	9100	91	910	9.1	

[a]5% resistors are available in all the values listed; 10% resistors are available only in the boldface values.

designs, the variation in circuit performance that might occur due to inexact resistor values is not a critical consideration. However, in some applications the resistors must have closer tolerances, and in those cases *precision* resistors must be selected. Precision resistors are available in a very wide range of values but are considerably more expensive than the standard 5 and 10% types. Table 3.1 shows standard 5 and 10% resistor values.

3.5 Conductance

Conductance is the *reciprocal* of resistance, that is, 1 divided by resistance. Thus, a large resistance corresponds to a small conductance, and vice versa. The symbol for conductance is G and its units are *siemens* (S):

$$G = \frac{1}{R} \quad \text{siemens} \tag{3.8}$$

Of course, it follows that $R = 1/G$ ohms.

Example 3.8
(a) Find the conductance of a resistor whose resistance is 40 Ω.
(b) Find the resistance of a resistor whose conductance is 0.5 mS.

SOLUTION
(a) $G = \dfrac{1}{R} = \dfrac{1}{40\ \Omega} = 0.025$ S

(b) $R = \dfrac{1}{G} = \dfrac{1}{0.5 \times 10^{-3}\ \text{S}} = 2$ kΩ

Drill Exercise 3.8
What is the maximum possible conductance of a 75-kΩ 10% resistor?

ANSWER: 14.81 μS.

Substituting $R = 1/G$ into Ohm's law, we find

$$I = \frac{E}{R} = \frac{1}{R}(E) = GE$$

Equivalently, $E = I/G$ and $G = I/E$.

3.6 Efficiency

Any device, circuit, system, or machine that utilizes electricity to perform some useful function does so by converting work or energy to some other form. The energy or work that must be supplied to achieve a desired outcome is called the *input*, and the modified form of work or energy that represents that outcome is the *output*. Following are some familiar examples:

1. An electric motor
   Input: Electrical energy.
   Output: Work, in the form of a force rotating a mechanical load.
2. An electric generator
   Input: Work, in the form of a force rotating an armature.
   Output: Electrical energy.
3. An electronic amplifier
   Input: Electrical energy, obtained from a dc power supply.
   Output: Electrical energy in the form of ac voltages and currents.
4. A dc power supply
   Input: Electrical energy in the form of ac voltage and current.
   Output: Electrical energy in the form of dc voltage and current.
5. An electric heater
   Input: Electrical energy.
   Output: Heat energy.
6. A public address (PA) system
   Input: Electrical energy.
   Output: Sound energy.
7. An incandescent lamp
   Input: Electrical energy.
   Output: Light energy.

   In the process of converting work or energy from one form to a different form, it is inevitably the case that some energy is *lost*, in the sense that it is converted to a form different from that which constitutes useful output. For example, in an electric motor, some of the electrical energy supplied as input is converted to heat, because the motor has resistance. That energy is not converted to useful ouput (work), and it is therefore called a *loss*. On the other hand, in an electric heater the desired output is heat energy, and the loss is light energy, manifesting itself in the glow of the heater element. The desired output of an incandescent lamp is light energy, and the heat it produces is a loss. We can see that the form of energy we call a "loss" depends entirely on how output is defined.

   When output and input are both expressed in the same units of work or energy, *efficiency* is defined to be the ratio of the two:

$$\eta = \frac{W_o}{W_i} \qquad (3.9)$$

where η is efficiency (often expressed as a percent by multiplying 3.9 by 100%), W_o is output work or energy, and W_i is input work or energy. Since losses are inevitable, W_o is always less than W_i, and we conclude that η *is always less than 1*. In other words, every system is less than 100% efficient. In spite of a long history of claims by amateur inventors to have developed machines capable of producing more energy than they consume ($W_o > W_i$), it is *never* possible for η to be greater than 1. The total energy supplied to a system equals the sum of the output energy and the losses:

$$W_i = W_o + W_L \tag{3.10}$$

where W_L is sum of all energy losses. Using (3.10), we can rewrite (3.9) in the equivalent forms:

$$\eta = \frac{W_o}{W_o + W_L} = \frac{W_i - W_L}{W_i} = 1 - \frac{W_L}{W_i} \tag{3.11}$$

If we divide input work or energy by the time t during which it is furnished to a system, we obtain the rate at which that work or energy is furnished, that is, the *input power P_i*:

$$P_i = \frac{W_i}{t} \quad \text{watts} \tag{3.12}$$

Similarly, output power, P_o, is

$$P_o = \frac{W_o}{t} \quad \text{watts} \tag{3.13}$$

Dividing numerator and denominator of equation (3.9) by t, we obtain

$$\eta = \frac{W_o/t}{W_i/t} = \frac{P_o}{P_i} \tag{3.14}$$

Thus, efficiency can be computed as the ratio of output power to input power. Also, regarding W_L/t as power loss P_L (the rate at which energy is lost), we can write equation (3.11) as

$$\eta = \frac{P_o}{P_o + P_L} = \frac{P_i - P_L}{P_i} = 1 - \frac{P_L}{P_i} \tag{3.15}$$

Example 3.9

Mechanical energy is supplied to a dc generator at the rate of 4200 J/s. The generator delivers 32.2 A at 120 V.

(a) What is the percent efficiency of the generator?
(b) How much energy is lost per minute of operation?

SOLUTION

(a) $P_i = 4200 \text{ J/s} = 4200 \text{ W}$

$P_o = EI = (120 \text{ V})(32.2 \text{ A}) = 3864 \text{ W}$

$\eta = \dfrac{P_o}{P_i} = \dfrac{3864 \text{ W}}{4200 \text{ W}} = 0.92$, or 92%

(b) $P_L = P_i - P_o = 4200 \text{ W} - 3864 \text{ W} = 336 \text{ W}$

$W_L = P_L t = (336 \text{ J/s})(60 \text{ s}) = 20{,}160 \text{ J}$

Drill Exercise 3.9

If the generator in Example 3.9 were 98% efficient, how much current would it supply at 120 V?

ANSWER: 34.3 A. ▫

Horsepower

Although not in the SI system, the units of horsepower (hp) are widely used to specify the output power of electric motors. The relationship between horsepower and watts is

$$1 \text{ hp} = 746 \text{ W} \tag{3.16}$$

When one speaks of a motor as having a certain horsepower, it is understood that the horsepower specification refers to *output* power. When using equation (3.14) to compute the efficiency of a motor, both P_i and P_o must be expressed in watts, or both must be expressed in hp.

Example 3.10

What is the efficiency of a $\frac{1}{2}$-hp motor that draws 3.5 A from a 120-V source?

SOLUTION

$$P_o = (0.5 \text{ hp})\left(\frac{746 \text{ W}}{1 \text{ hp}}\right) = 373 \text{ W}$$

$$P_i = EI = (120 \text{ V})(3.5 \text{ A}) = 420 \text{ W}$$

$$\eta = \frac{P_o}{P_i} = \frac{373 \text{ W}}{420 \text{ W}} = 0.888 \quad \text{or} \quad 88.8\%$$

Drill Exercise 3.10

How much energy is lost in the motor of Example 3.10 when it is operated for 1h?

ANSWER: 16.92×10^4 J. ▫

When the output of one device or system is the input to another, the two are said to be in *cascade*. The overall efficiency of several such cascaded components is the *product* of their individual efficiencies.

Example 3.11

Figure 3.11 shows an electric motor driving an electric generator. The 2-hp motor draws 14.6 A from a 120-V source and the generator supplies 56 A at 24 V.

(a) Find the motor efficiency and the generator efficiency.
(b) Find the overall efficiency.

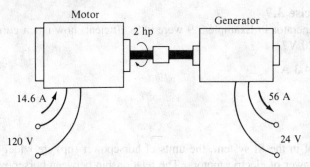

FIGURE 3.11 (Example 3.11)

SOLUTION

(a) $P_i(\text{motor}) = (120 \text{ V})(14.6 \text{ A}) = 1752 \text{ W}$

$P_o(\text{motor}) = (2 \text{ hp})(746 \text{ W/hp}) = 1492 \text{ W}$

$\eta(\text{motor}) = \dfrac{1492 \text{ W}}{1752 \text{ W}} = 0.8516$

$P_i(\text{generator}) = 2 \text{ hp} = 1492 \text{ W}$

$P_o(\text{generator}) = (24 \text{ V})(56 \text{ A}) = 1344 \text{ W}$

$\eta(\text{generator}) = \dfrac{1344 \text{ W}}{1492 \text{ W}} = 0.90$

(b) $\eta(\text{overall}) = \dfrac{P_o(\text{generator})}{P_i(\text{motor})} = \dfrac{1344 \text{ W}}{1752 \text{ W}} = 0.767$

Note that $\eta(\text{overall})$ is the product of the efficiencies of the individual machines:

$$\eta(\text{overall}) = (0.8516)(0.90) = 0.767$$

Drill Exercise 3.11

If the overall efficiency of the motor and generator in Example 3.11 were 48.02% and if the generator were twice as efficient as the motor, what would be the efficiency of each?

ANSWER: $\eta(\text{motor}) = 49\%$; $\eta(\text{generator}) = 98\%$. ☐

3.7 Real and Ideal Sources

An *ideal* voltage source is one that maintains a constant terminal voltage no matter how much current is drawn from it. For example, an ideal 12-V source would theoretically maintain 12 V across its terminals when a 1-MΩ resistor is connected (so $I = 12 \text{ V}/1 \text{ M}\Omega = 12 \ \mu\text{A}$), as well as when a 1-kΩ resistor is connected

($I = 12$ mA), or when a 1-Ω resistor is connected ($I = 12$ A), and even when a 0.01-Ω resistor is connected ($I = 1200$ A). It is not possible to construct an ideal voltage source, because all *real* voltage sources have *internal resistance* that causes the terminal voltage to drop if the current is made sufficiently large (i.e., if a small enough resistance is connected across the terminals). Nevertheless, it is convenient when analyzing electric circuits to assume that a real voltage source behaves like an ideal one. That assumption is justified by the fact that in circuit analysis we are not usually concerned with changing current over a wide range of values (if, in fact, we are concerned with any current change at all).

Voltage Regulation

In applications where the current drawn from a voltage source may vary over a wide enough range to cause the terminal voltage to change, we must have some knowledge of how great that change is liable to be. One measure of how well a voltage source maintains a constant output voltage is the percent change in voltage that occurs as a result of some specified variation in current. This measure is called percent *voltage regulation*. The percent change in voltage is that which occurs due to a change in current from zero amperes, called the *no-load* condition, to the maximum rated current of the source, called the *full-load* condition (see Figure 3.12). Percent voltage regulation (% VR) is thus computed as

$$\% \text{ VR} = \frac{V_{NL} - V_{FL}}{V_{FL}} \times 100\% \tag{3.17}$$

where V_{NL} is the no-load voltage and V_{FL} is the full-load voltage. It is clear that an ideal source has *zero* percent voltage regulation.

Example 3.12
A certain power supply is specified to have a voltage regulation of 1.2%. If its no-load voltage is 15 V, what is its full-load voltage?

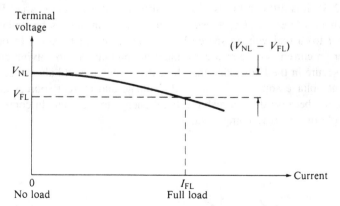

FIGURE 3.12 Typical plot of the variation in output voltage of a real voltage source as the current varies from no load to full load.

SOLUTION Using the decimal form of equation (3.17), we have

$$0.012 = \frac{15 - V_{FL}}{V_{FL}}$$

Solving for V_{FL} gives

$$V_{FL} = \frac{15}{1.012} = 14.822 \text{ V}$$

Drill Exercise 3.12

What is the no-load voltage of a power supply if its voltage regulation is 1% and its full-load voltage is 25 V?

ANSWER: 25.25 V.   □

Current Sources

An *ideal current source*, also called a constant-current source, will supply the same current to any resistance connected across its terminals. Figure 3.13 shows the symbol used to represent an ideal current source. The arrow shows the direction of the (conventional) current produced by the source.

Since an ideal current source supplies the same current to any resistance, it is clear that the voltage across the terminals of the source must change if the resistance is changed. For example, if a 2-A current source has 10 Ω across its terminals, the voltage is $E = (2 \text{ A})(10 \text{ Ω}) = 20$ V. If the resistance is changed to 100 Ω, the voltage becomes $E = (2 \text{ A})(100 \text{ Ω}) = 200$ V. An ideal current source cannot be constructed because a real current source always has some internal resistance that causes the current to drop if the voltage developed across the terminals becomes sufficiently large (i.e., if the resistance connected across the terminals is made too large). However, in practical circuit analysis it is convenient to assume that current sources are ideal, unless a wide range of resistance values must be considered.

Unlike a voltage source, which we can imagine as two oppositely charged electrodes (Figure 2.3), it is difficult to visualize the structure of a current source. However, as we will learn in a later chapter, a real current source can always be replaced by (i.e., is *equivalent* to) a real voltage source having certain characteristics. In other words, we can regard a current source as a convenient fiction that aids in solving circuit problems, yet feel secure in the knowledge that the current source could always be replaced by an equivalent voltage source, if so desired. Voltage and current sources are called *active* components, because they furnish electrical energy to a circuit. In contrast, a resistor is an example of a *passive* component.

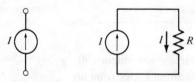

FIGURE 3.13 Ideal current source.

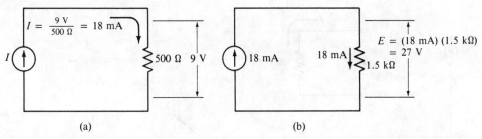

FIGURE **3.14** (Example 3.13)

Example 3.13

A constant-current source develops a terminal voltage of 9 V when a 500-Ω resistor is connected across its terminals. What is its terminal voltage when the 500-Ω resistor is replaced by a 1.5-kΩ resistor?

SOLUTION As shown in Figure 3.14(a), we use Ohm's law to find the value of the current when the 500-Ω resistor is connected:

$$I = \frac{E}{R} = \frac{9 \text{ V}}{500 \text{ }\Omega} = 18 \text{ mA}$$

Since the current remains constant at 18 mA when the 1.5-kΩ resistor is connected, as shown in Figure 3.14(b), the terminal voltage becomes

$$E = IR = (18 \text{ mA})(1.5 \text{ k}\Omega) = 27 \text{ V}$$

Drill Exercise 3.13

What is the maximum voltage that could be expected across a 120-Ω 5% resistor when it is connected across a 400-mA constant-current source?

ANSWER: 50.4 V.   □

3.8 Linearity

An electrical device is said to be *linear* if a graph of the voltage across it versus the current through it is a straight line. The concept of linearity is important because many circuit analysis techniques that we shall study can only be applied to circuits composed of linear devices. A resistor is an example of a linear device, because the graph of $V = IR$ is a straight line. This fact is illustrated in Figure 3.15 for the case of a 100-Ω resistor. The linearity property is equivalent to stating that the voltage across a device is *directly proportional* to the current through it. From Ohm's law, we know this to be the case for a resistor. For example, if the current through a 100-Ω resistor is 0.1 A, which makes

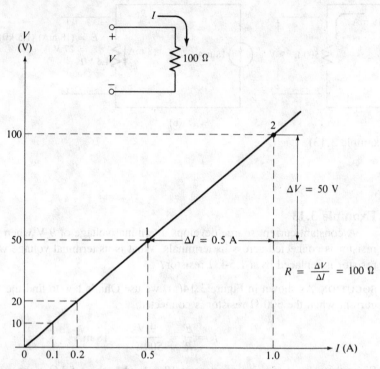

FIGURE 3.15 Plot of the voltage across a 100-Ω resistor versus the current through it.

the voltage 10 V, then doubling the current to 0.2 A will also double the voltage to 20 V.

Notice in Figure 3.15 that the *slope* of the graph is equal to the resistance of the resistor:

$$\text{slope} = \frac{\Delta V}{\Delta I} = R \tag{3.18}$$

where ΔV is the *change* in voltage corresponding to a change, ΔI, in current. The slope of a straight line is the same no matter where along the line it is computed. The figure illustrates computation of the slope over the portion of the line between points 1 and 2. In this case,

$$\Delta V = V_2 - V_1 = 100 \text{ V} - 50 \text{ V} = 50 \text{ V}$$

$$\Delta I = I_2 - I_1 = 1 \text{ A} - 0.5 \text{ A} = 0.5 \text{ A}$$

and

$$R = \frac{\Delta V}{\Delta I} = \frac{50 \text{ V}}{0.5 \text{ A}} = 100 \text{ }\Omega$$

3.9 SPICE Examples

Example 3.14   (SPICE)
Use SPICE to find the current I in the 2.2-kΩ resistor in Figure 3.16(a).

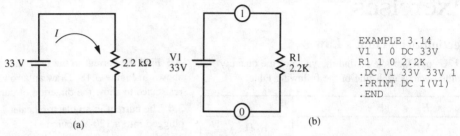

```
EXAMPLE 3.14
V1 1 0 DC 33V
R1 1 0 2.2K
.DC V1 33V 33V 1
.PRINT DC I(V1)
.END
```

(a)   (b)

FIGURE 3.16 (Example 3.14)

SOLUTION The circuit is redrawn in its SPICE format as shown in Figure 3.16(b). Note that there are only two *nodes* (connection points, or common terminals) in the circuit. (See the first paragraph of Section 6.6 for a detailed explanation of nodes.) The input data file is also shown in the figure. When the program is executed, the statement .PRINT DC I(V1) produces the result -15 mA. The negative sign appears because current is flowing out of the positive terminal of the 33-V source, the direction that SPICE (unconventionally) assumes to be negative. To obtain a positive result, we could have inserted a dummy (0-valued) voltage source in the circuit with its negative terminal joined to the positive terminal of the 33-V source and then requested the current in the dummy source. See Section A–4 of the Appendix.

Example 3.15   (SPICE)
Use SPICE to find the voltage across the 470-Ω resistor in Figure 3.17(a).

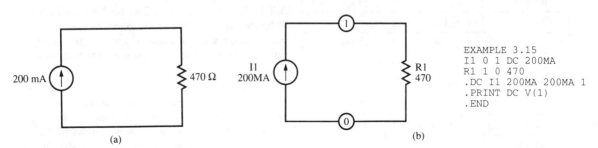

```
EXAMPLE 3.15
I1 0 1 DC 200MA
R1 1 0 470
.DC I1 200MA 200MA 1
.PRINT DC V(1)
.END
```

(a)   (b)

FIGURE 3.17 (Example 3.15)

SOLUTION The circuit is redrawn in its SPICE format as shown in Figure 3.17(b). Also shown is the input data file. Note that current source I1 has node 0 as its *positive* terminal. See Section A.4 and Figure A.4 of the Appendix. When the program is executed, the statement .PRINT DC V(1) produces the result 94 V, which we know to be correct (200 mA × 470 Ω = 94 V).

Exercises

Section 3.1   Ohm's Law

3.1 Using Ohm's law, find the value of the quantity omitted across each line of the following table.

E	I	R
64 V		16 Ω
	1.5 A	23 Ω
9 V	0.36 A	
36 V		25 kΩ
120 V		1.5 MΩ
	8 mA	4.7 kΩ
3.8 V	76 μA	
450 mV		0.03 kΩ
0.2 kV	800 mA	
	5200 pA	1.85 MΩ

3.2 Using Ohm's law, find the value of the quantity omitted across each line of the following table.

V	I	R
	0.42 A	200 Ω
48 V		25 Ω
160 V	6.4 A	
	3 mA	15 kΩ
24 V		1.2 MΩ
110 V	550 μA	
1.41 kV		470 kΩ
	33.3 mA	0.5 kΩ
90 mV	3.6 mA	
	1000 pA	0.44 MΩ

3.3 Find the current in the resistance in each circuit shown in Figure 3.18. Draw an arrow beside the resistance to show the direction of current through it.

3.4 The current in an electric heater is 0.6 A when it is plugged into a 120-V source.

(a) What is the resistance of the heater?

(b) What current would flow in the heater if the voltage dropped 10%?

3.5 The resistance of an electronic component changes from 860 Ω to 1.5 kΩ when its temperature changes over a certain range. If it is desired to maintain 30 mA of current in the component at all times, what range of voltages must a voltage source connected to it be capable of providing?

3.6 How many coulombs of charge pass through a 4.2-kΩ resistor in 2.5 s when the voltage across the resistor is 16.8 V?

3.7 What is the voltage across a 1.2-MΩ resistor if 4.5 μC of charge pass through it in 30 ms?

Section 3.2   Measuring Voltage, Current, and Resistance

3.8 Redraw each circuit of Exercise 3.3 showing how a voltmeter should be connected to measure the voltage across each resistance. Draw + and − signs on the voltmeter terminals to show the correct polarity of the connections.

3.9 Redraw each circuit of Exercise 3.3 showing how a voltmeter should be connected to measure the voltage across each resistance and how an ammeter should be connected for simultaneous measurement of the current in each resistance. Draw + and − signs on the terminals of each instrument to show the correct polarity of the connections.

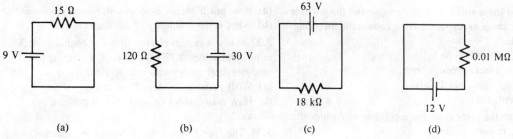

FIGURE 3.18 (Exercise 3.3)

Section 3.3   Work, Energy, and Power

3.10 How much power, in watts, is required by an engine that consumes 42 J of energy in 3 ms?

3.11 How long will it take a 1.5-kW clothes dryer to produce 2.025×10^4 J of heat energy?

3.12 Find the power
(a) dissipated by 50 Ω of resistance when the current through it is 0.1 A
(b) delivered by an 18-V voltage source when it supplies 3 A of current to a circuit
(c) dissipated by 1.5 kΩ of resistance connected across a 30-V source

3.13 Find the power
(a) delivered to an electric heater having resistance 250 Ω when a current of 2.2 A is supplied to it
(b) dissipated by 47 Ω of resistance when the voltage across it is 20 V
(c) delivered by a 100-V voltage source when it supplies 10 mA of current

3.14 How much current flows in 80 Ω of resistance when it dissipates 32 mW?

3.15 What is the voltage across a 1-kΩ resistor when it dissipates 0.625 W?

3.16 What is the total heat energy dissipated by a resistor in 1 min when it is connected across a 12-V source that supplies 75 mA of current to it?

3.17 How long would it take 0.5 MΩ of resistance to dissipate 1000 J of heat if the current in the resistance is 20 μA?

3.18 (a) How many joules does 1 kWh equal?
(b) How many kWh does 1 J equal?

3.19 A certain electrical appliance consumes energy at the rate of 540 J/s. How many kWh of energy does it consume in $3\frac{1}{2}$ h?

3.20 If electrical energy costs $0.12/kWh, find the total cost of operating a 5-W night light for 24 h, a 100-W radio for 6 h, a 250-W home computer for 2 h, and a 1.1-kW oven for 30 min.

3.21 If electrical energy costs $0.15/kWh, for how many 24-h days can a 5-W night light be operated without the cost exceeding $1.00?

Section 3.4   Resistors

3.22 What is the maximum resistance that a 2-W resistor can have without exceeding its power rating if the current through it is 10 mA?

3.23 What maximum voltage can be connected across a 0.5-W 270-Ω resistor without exceeding the rated power dissipation?

3.24 Carbon-composition resistors are available with power ratings of $\frac{1}{8}$, $\frac{1}{4}$, $\frac{1}{2}$, 1, and 2 W. Given the circuit values of voltage, current, and/or resistance in each of the following, determine the *minimum* power rating that the carbon-composition resistor can have.
(a) $R = 910$ Ω, $I = 0.02$ A
(b) $R = 10$ Ω, $V = 4$ V
(c) $V = 250$ mV, $I = 400$ mA
(d) $R = 1.5$ kΩ, $I = 20$ mA

3.25 Repeat Exercise 3.24 for each of the following.
(a) $R = 510$ Ω, $V = 10$ V
(b) $R = 470$ kΩ, $I = 1.95$ mA
(c) $R = 3.3$ kΩ, $V = 50$ V
(d) $V = 50$ mV, $I = 0.2$ A

3.26 Find the nominal resistance value and the possible resistance range of resistors having each of the following color codes.
(a) green, blue, red, silver
(b) brown, black, yellow, gold
(c) brown, red, brown, gold
(d) orange, orange, green, silver

3.27 Find the nominal resistance value and the possible resistance range of resistors having each of the following color codes.
(a) gray, red, red, gold
(b) brown, black, green, silver
(c) orange, white, black, silver
(d) red, red, red

3.28 Write the color code corresponding to each of the following resistor values.
(a) 68 kΩ ± 10%
(b) 1.2 kΩ ± 5%
(c) 390 Ω ± 10%
(d) 2.7 MΩ ± 5%

3.29 Write the color code corresponding to each of the following resistor values.
(a) 470 Ω ± 5%
(b) 1.8 kΩ ± 10%
(c) 820 kΩ ± 20%
(d) 1.5 MΩ ± 5%

3.30 A particular design requires that the current in a circuit be 3.45 mA. A 12-V source provides the current.
(a) Find the closest standard-value resistor with 10% tolerance whose *nominal* value would limit the current to the required value ± 10%.
(b) Find the maximum percent deviation from 3.45 mA that could result from using the standard resistor value found in part (a).

3.31 Repeat Exercise 3.30 if standard-value resistors with 5% tolerances can be used.

Section 3.5   Conductance

3.32 Find the maximum possible conductance of a 100-kΩ resistor having 10% tolerance.

3.33 Find the minimum possible conductance of a 4.7-kΩ resistor having 5% tolerance.

3.34 Derive an expression for the power dissipated in a resistor in terms of its conductance G and the current I through it.

3.35 Derive an expression for the power dissipated in a resistor in terms of its conductance G and the voltage V across it.

Section 3.6   Efficiency

3.36 When an electric heater is operated for 30 min, it consumes 0.4 kWh of energy and delivers 1.25×10^6 J of heat.

(a) How much energy is lost in 30 min of operation?
(b) What is the efficiency of the heater?

3.37 When a certain electric motor is operated for 30 min, it consumes 0.75 kWh of energy. During that time, its total energy loss is 3×10^5 J.
(a) What is the efficiency of the motor?
(b) How many joules of work does it perform in 30 min?

3.38 The specifications for a stereo (dual-channel) audio system state that it can deliver 60 W per channel, maximum. If the system draws 1.8 A from a 120-V source when delivering maximum power, what is its efficiency?

3.39 The total power supplied to an engine that drives an electric generator is 40.25 kW. If the generator delivers 15 A to a 100-Ω load, what is the efficiency of the system?

3.40 A $1\frac{1}{2}$-hp motor is 94.2% efficient. How much current does it draw from a 120-V source?

3.41 A certain motor draws 17 A from a 240-V source. If the motor is 91.5% efficient, what is its horsepower rating?

3.42 A certain amplifier develops 80 W of output power. A dc power supply furnishes the amplifier 3 A at 40 V. The power supply draws 1.6 A from a 120-V source.
(a) What are the efficiencies of the power supply and the amplifier?
(b) What is the overall efficiency?

3.43 A certain system consists of three identical devices in cascade, each having efficiency 0.85. The first device draws 3 A from a 20-V source. How much current does the third device deliver to a 50-Ω load?

Section 3.7   Real and Ideal Sources

3.44 An ideal voltage source supplies 30 mA to a 1.5-kΩ resistor connected across its terminals. How much current would it supply to a 5-kΩ resistor?

3.45 What range of currents would an ideal 36-V voltage source supply to resistors ranging in value from 144 Ω to 144 kΩ?

3.46 When there is no resistance connected across the terminals of a certain power supply, its terminal voltage is 24 V. It delivers a full-load current of 1 A to a 23.75-Ω resistor. Find the percent voltage regulation of the supply.

3.47 A certain power supply has a full-load voltage of 18 V and has 2% voltage regulation. What is its no-load voltage?

3.48 An ideal current source has a terminal voltage of 50 V when a resistor having conductance 0.5 mS is connected across its terminals. What is the voltage across its terminals when the resistor is replaced by one having conductance 2 mS?

3.49 What range of terminal voltages would be developed by an ideal 12-mA current source when resistors ranging from 50 Ω to 5 kΩ are connected across its terminals?

Section 3.8 Linearity

3.50 From the voltage versus current graph of a certain resistor, it is found that changing the current from 0.04 A to 0.09 A causes the voltage to change from 32.8 V to 73.8 V. What is the resistance of the resistor?

3.51 On the voltage versus current graph of a 10-kΩ resistor, what is the total change in voltage when the current changes from 1 mA to 5.5 mA?

3.52 If the current through a resistor is plotted on the vertical axis and the voltage across it is plotted on the horizontal axis, what does the slope of the resulting graph represent?

SPICE Exercises

3.1S Use SPICE to find the current in the circuits of Figures 3.18(a) and (c) (Exercise 3.3), p. 53.

3.2S Use SPICE to find the current in the circuits of Figures 3.18(b) and (d) (Exercise 3.3). Write each input data file in such a way that the .PRINT statement produces a positive result.

3.3S Use SPICE to find the voltage across a 1.5-MΩ resistance connected across the terminals of an ideal 40-μA current source.

3.4S Use SPICE to solve Exercise 3.49.

4 Series and Parallel Circuits

4.1 Electric Circuits

An *electric circuit* is any configuration of interconnected resistors, active sources, or other electrical components through which current flows, or can flow. To emphasize the fact that current *can* flow (i.e., that there is a path through which current originating at a source can return to that source), a circuit is sometimes referred to as a *complete* circuit. The term *network* is generally used in reference to an arbitrary configuration of passive components, while *circuit* usually implies the presence of active sources and current flow. However, there is no hard and fast rule for making these distinctions, and the terms ''network'' and ''circuit'' are often used interchangeably. Electric circuits exist in vastly diverse forms, ranging from electrical power lines that span the continent to microscopic electronic devices in modern integrated circuits.

In this book we will be using the standard method for representing a circuit on a piece of paper: a diagram, called a *schematic*, where resistance and other electrical properties are shown by symbols, and on which interconnections are shown by lines drawn between the symbols. Bear in mind that a resistance symbol could represent all the resistance in a mile-long wire, or the resistance of a single carbon resistor, or the resistance of a millimeter of silicon embedded in an integrated circuit. No distinction is made between resistance that exists because it is an inherent, unavoidable property of a device and resistance which is intentionally inserted in a circuit (a resistor) to alter the circuit's characteristics.

Analysis, Design, and Troubleshooting

Circuit *analysis* is the process of determining the current that flows in one or more components of an electric circuit and/or the voltage that exists at various points in the circuit. Returning to the water system analogy, circuit analysis is like finding the flow rate and the pressure in various sections of the system, given information on the size of the individual pipes, how they are interconnected, and the pressure generated at the pumping station. Circuit *synthesis*, or design, is the process of selecting components having certain properties, and interconnecting them in a way necessary to achieve a desired goal, such as delivering a specific amount of power to a loudspeaker. It is similar to selecting pumps and water pipes and connecting them in such a way that a certain amount of flow or a certain pressure is available at a location where it is needed. Circuit *troubleshooting* is identifying a defective component, or other circuit fault, based on a knowledge of the voltages and currents that are supposed to exist in a circuit and on the actual values measured in the defective circuit. It is like determining the location of a ruptured pipe, given pressure and flow measurements in a water system. Successful circuit design and troubleshooting depend heavily on the designer/troubleshooter's ability to perform accurate circuit analysis, so the emphasis in this book will be on that aspect of circuit theory. From time to time, we will provide design and troubleshooting examples to show how analysis skills can be applied to those important practical activities.

There are several well-known ways to interconnect electrical components that result in simple circuits whose currents and voltage levels can be readily predicted. Whether our goal is circuit analysis, design, or troubleshooting, it is essential to have a thorough understanding of the rules used to analyze these fundamental configurations, for they are the building blocks of all electric circuits. In this chapter we will consider the two most basic types of circuit connections: *series* and *parallel*.

4.2 Series Circuits

Two components are said to have a *common terminal* when there is a path of zero resistance joining a terminal of one to a terminal of the other. From a practical standpoint, components have a common terminal when there is a very low-resistance wire joining one terminal to another, or if they are soldered together, or if they are connected through some other path having negligibly small resistance. In these cases it is common practice to say that the terminals are *electrically the same*, or *electrically common*, because they have the same electrical properties, even though they may differ physically. On a schematic diagram, the zero-resistance path joining two terminals is shown by a solid line.

Two components are connected in series if they have exactly one common terminal and if no other component has a terminal that shares that common connection. Figure 4.1 shows two examples of series connections. It is easy to form a mental image of two series-connected components, since the word "series" conveys the idea of sequential,

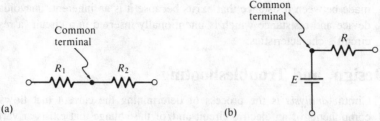

FIGURE 4.1 Examples of series-connected components. (The common terminal is darkened for emphasis, although it is not usually shown in a schematic diagram unless a third component is joined to the terminal.) (a) Resistors R_1 and R_2 are connected in series. (b) Voltage source E and resistor R are connected in series.

or one-after-the-other, and that is certainly the case here. However, beware of the simplicity of this notion; one of the most common errors in circuit analysis is to believe that two components are in series when they are not. When in doubt, make the tests implied by the formal definition: Do the components have one and only one terminal in common? Is any other component joined to that common point?

A *series path* is one in which every component in the path is in series with another component. It is conventional to say, simply, that components are ''in series'' if they are in the same series path, even though they may not have a common terminal. A *series circuit* is a (complete) circuit that consists *exclusively* of one series path. Lest the reader feel that we are belaboring this basically simple concept of a series connection, we emphasize again that failure to recognize it is the most frequent source of error in elementary circuit analysis.

Figure 4.2 shows some examples of series and nonseries connections, as well as some series and nonseries paths and circuits. Each example should be studied carefully and tested for compliance with the definitions we have given.

Analysis of Series Circuits

The most important property of a series path or circuit is that *the current is the same in every series-connected component*. We know from the definition that no third component can be connected to the common terminal of a series connection, so there is no way that current can be injected into or drawn away from the point where two series components are connected. Therefore, the current in one component must be exactly the same as the current in the component in series with it.

Another important fact about a series circuit is that its total resistance is the *sum* of all the series-connected resistances:

$$R_T = R_1 + R_2 + R_3 + \cdots \tag{4.1}$$

where R_T is the total resistance, and R_1, R_2, R_3, . . . , are the resistances of the series-connected components. When a voltage source is connected in a series circuit, the total current produced by that source is, from Ohm's law,

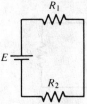

(a) E is in series with R_1. R_1 is in series with R_2. R_2 is in series with E. E, R_1, and R_2 form a series path and a series circuit. Every component is in series with every other component.

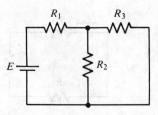

(b) E is in series with R_1. E and R_1 form a series path. R_2 and R_3 are not in series. R_1 is not in series with either R_2 or R_3. E is not in series with either R_2 or R_3. There is no series circuit.

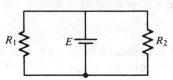

(c) E is in series with R_1. R_1 is in series with R_2. R_2 is in series with R_3. R_3 is in series with E. R_1, R_2, R_3 and E form a series path and a series circuit. Every component is in series with every other component.

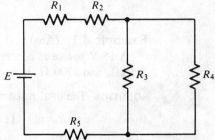

(d) E is in series with R_1. R_1 is in series with R_2. R_3 and R_4 are not in series with each other or with any other component. R_5 is in series with E. R_5, E, R_1, and R_2 form a series path, so each is in series with the others. There is no series circuit.

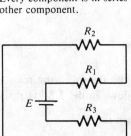

(e) There are no series–connected components. There are no series paths or series circuits.

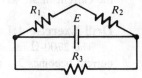

(f) R_1 is in series with R_2. R_1 and R_2 form a series path. No other components are in series. There is no series circuit.

FIGURE 4.2 Examples of series and nonseries connections, paths, and circuits.

$$I_T = \frac{E}{R_T} \qquad\qquad (4.2)$$

In other words, the current that flows in a series circuit is the same current that flows when a *single* resistor having resistance R_T is connected in series with the voltage source. We say that R_T is *equivalent* to all the series-connected resistors. As far as the voltage source is concerned, there is no difference between three series-connected 100-Ω resistors, two series-connected 150-Ω resistors, or a single 300-Ω resistor. All are equivalent to 300 Ω, and this idea is often conveyed by saying that the voltage source "sees" 300 Ω in all cases. Figure 4.3 illustrates these concepts.

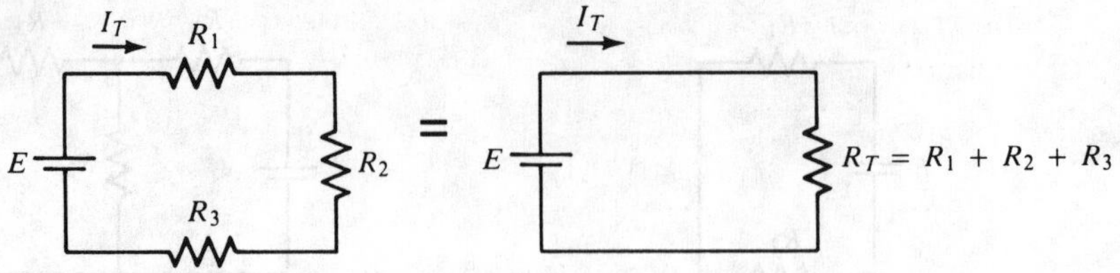

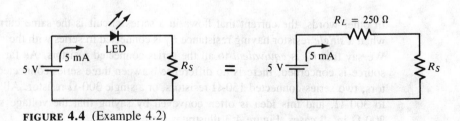

$$I_T = \frac{E}{R_1 + R_2 + R_3} = \frac{E}{R_T}$$

FIGURE 4.3 The total equivalent resistance of series-connected resistors is the sum of their resistance values.

Example 4.1 (Analysis)

A 15-V source is connected in series with the following resistances: 1 kΩ, 500 Ω, 3.3 kΩ, and 2700 Ω. How much current flows in the 3.3-kΩ resistance?

SOLUTION The total resistance of the circuit is

$$R_T = 1 \text{ k}\Omega + 0.5 \text{ k}\Omega + 3.3 \text{ k}\Omega + 2.7 \text{ k}\Omega = 7.5 \text{ k}\Omega$$

Therefore, the total current is

$$I_T = \frac{E}{R_T} = \frac{15 \text{ V}}{7.5 \times 10^3 \ \Omega} = 2 \times 10^{-3} \text{ A} = 2 \text{ mA}$$

Since the circuit is a series circuit, the current is the same in every component. Thus, the current in 3.3-kΩ resistor (and every other resistor) is 2 mA.

Drill Exercise 4.1

The 2700-Ω resistor in Example 4.1 is replaced by another resistor and the total current becomes 2.5 mA. What value of resistance replaced the 2700-Ω resistor?

ANSWER: 1.2 kΩ. □

Example 4.2 (Design)

It is necessary to limit the current in a certain light-emitting diode (LED) to 5 mA. The resistance of the LED is 250 Ω and it is connected in series with a 5-V source. How much resistance should be inserted in series with the LED? (See Figure 4.4.)

FIGURE 4.4 (Example 4.2)

SOLUTION Let R_S be the required series resistance and R_L be the LED resistance. The total resistance of the series combination will then be

$$R_T = R_S + R_L = R_S + 250$$

Therefore, $R_S = R_T - 250$. The total resistance, R_T, must limit the current to 5 mA:

$$I_T = \frac{E}{R_T}$$

$$5 \text{ mA} = \frac{5 \text{ V}}{R_T} \Rightarrow R_T = \frac{5 \text{ V}}{5 \text{ mA}} = 1 \text{ k}\Omega$$

Therefore,

$$R_S = R_T - 250 = 1000 - 250 = 750 \text{ }\Omega$$

Drill Exercise 4.2

What *additional* resistance should be inserted in series with the value of R_S found in Example 4.2 if it is desired to limit the current in the LED to 2.5 mA?

ANSWER: 1 kΩ.   □

Once we have found the current in a series circuit, we can find the voltage across each resistance in the circuit using Ohm's law. The next example illustrates this fact.

Example 4.3   (Analysis)

Find the current in and voltage across each resistor in the circuit of Figure 4.5.

SOLUTION
$$R_T = R_1 + R_2 + R_3 + R_4 = 12 \text{ }\Omega + 6 \text{ }\Omega + 10 \text{ }\Omega + 20 \text{ }\Omega = 48 \text{ }\Omega$$

$$I_T = \frac{E}{R_T} = \frac{24 \text{ V}}{48 \text{ }\Omega} = 0.5 \text{ A}$$

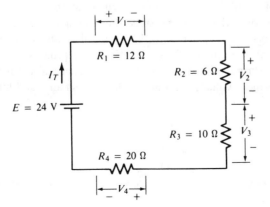

FIGURE 4.5 (Example 4.3)

Since the circuit is a series circuit, the current in each resistor is 0.5 A. Applying Ohm's law to each, we find

$$V_1 = (I_T)(R_1) = (0.5 \text{ A})(12 \text{ }\Omega) = 6 \text{ V}$$
$$V_2 = (I_T)(R_2) = (0.5 \text{ A})(6 \text{ }\Omega) = 3 \text{ V}$$
$$V_3 = (I_T)(R_3) = (0.5 \text{ A})(10 \text{ }\Omega) = 5 \text{ V}$$
$$V_4 = (I_T)(R_4) = (0.5 \text{ A})(20 \text{ }\Omega) = 10 \text{ V}$$

Notice that Figure 4.5 shows + and − polarity symbols attached to each of the voltage designations V_1 through V_4. These are the polarities that should be observed when connecting a voltmeter to measure the voltage across each resistor. In each case we say that the voltage is *referenced* to the side where the minus sign appears.

Drill Exercise 4.3

If the resistance of R_1 in Figure 4.5 were doubled, what would be the new voltage across it?

ANSWER: 9.6 V.   □

4.3 Kirchhoff's Voltage Law

Figure 4.6 shows a series circuit containing a voltage source and three resistors. Following the method demonstrated in Example 4.3, we can find the voltage across each resistor in the circuit:

$$I_T = \frac{E}{R_T} = \frac{20 \text{ V}}{(3 + 5 + 2) \text{ }\Omega} = 2 \text{ A}$$

$$V_1 = (3 \text{ }\Omega)(2 \text{ A}) = 6 \text{ V} \qquad V_2 = (5 \text{ }\Omega)(2 \text{ A}) = 10 \text{ V} \qquad V_3 = (2 \text{ }\Omega)(2 \text{ A}) = 4 \text{ V}$$

These voltages are shown on the figure. Also shown are the voltages measured between each resistor terminal and the minus side of the voltage source. Notice that each of these voltages equals the voltage of the source *minus* the sum of all the voltages across resistors lying to the left of the point where it is measured. For example, the voltage between the first and second resistors is 20 V − 6 V = 14 V, and the voltage between the second and third resistors is 20 V − (6 + 10) V = 4 V. As we progress through the circuit from left to right, measuring voltages between each terminal and the minus side of the source, we notice that these voltages become smaller and smaller and finally reach zero

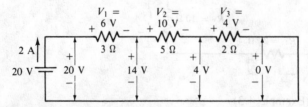

FIGURE 4.6

at the very end. We see that the voltage across each resistor causes a *voltage drop* (or potential drop) as we progress through the circuit.

In a sense, each voltage drop across a resistor in Figure 4.6 represents the amount of voltage "used up" in progressing from one resistor to the next. When we reach the end (rightmost side), we have used up the entire 20 V that was originally available. This concept is, again, very much like a water distribution system: At the pumping station, the maximum pressure in the system is available, but the pressure continually drops as we progress through the system, the lowest pressure being that at the most remote location.

If we imagine ourselves traveling around the circuit in Figure 4.6, beginning at the + terminal of the 20-V source, and measuring voltages with respect to the minus side of that source as we progress, we observe that the voltage drops *accumulate*, that is, the total drop at any point is the sum of all the drops up to that point. When we reach the end, the sum of all the drops equals the source voltage, 20 V. We can regard the 20-V source as a voltage *rise*, because if we continue around the circuit and pass through the source, we once again have 20 V to "work with."

The path we follow in traveling completely around the circuit of Figure 4.6, beginning at the + terminal of the 20-V source and ending up at the same point, is called a *closed loop*. A closed loop is any path that begins and ends at the same point. The example we have just discussed illustrates a very important fact about the voltage drops around any closed loop in an electric circuit, called *Kirchhoff's voltage law*:

> **The sum of the voltage drops around any closed loop equals the sum of the voltage rises around that loop.**

In our example, which contains only one voltage rise, we see that the sum of the drops across the resistors, $V_1 + V_2 + V_3 = 6\ V + 10\ V + 4\ V$, equals the 20-V rise in the source.

In our example, it is clear that the voltage drops are the voltages across the resistors and that the voltage rise occurs in the source. However, in more complex circuits, it is not always obvious which voltages represent drops and which represent rises. It is therefore convenient to establish a systematic procedure that allows us to apply Kirchhoff's voltage law in a way that does not depend on a physical interpretation of drops and rises. Figure 4.7 shows the circuit we have been discussing and shows a closed loop represented

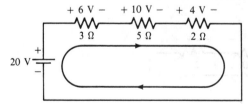

FIGURE 4.7 Closed loop. When the loop passes through a component from + to −, the voltage across that component is a *drop*. When the loop passes through a component from − to +, the voltage is a *rise*.

by a path drawn with arrows superimposed on it. The arrows show that we will travel around the loop in a *clockwise* direction. As we go around this loop, notice that we pass *through* each resistor *from the positive side* of its voltage drop *to the negative side*. On the other hand, when we pass *through* the voltage source, we travel *from its negative* terminal *to its positive* terminal. These observations will be the basis for our *definitions* of voltage drops and voltage rises in *any* component:

$$\text{From } + \text{ to } - \implies \text{voltage drop}$$

$$\text{From } - \text{ to } + \implies \text{voltage rise}$$

Now that we have established definitions for drops and rises based on a purely mechanical procedure, we can apply Kirchhoff's voltage law without regard to physical interpretations. Furthermore, the law is valid *regardless of the direction in which we travel around the loop*. Figure 4.8 shows our example circuit once again, this time with a counterclockwise loop. In accordance with the definitions, notice that the voltages across the resistors are now treated as rises and the source voltage becomes a drop. It is still true that the sum of the voltage rises equals the sum of the voltage drops:

$$\underbrace{6 \text{ V} + 10 \text{ V} + 4 \text{ V}}_{\text{rises}} = \underbrace{20 \text{ V}}_{\text{drop}}$$

It should be clear from this result that *the direction of a loop does not have to be the same as the direction of the current in a circuit*.

Example 4.4   (Analysis)
Use Kirchhoff's voltage law to find the source voltage E in Figure 4.9.

SOLUTION We arbitrarily choose to draw a clockwise loop, as shown in Figure 4.9. With this choice, notice that the voltage across each resistor is a rise and the unknown voltage E is a drop. Applying Kirchhoff's voltage law, we find

$$E = 23.4 + 10.8 + 6.6 + 13.2 = 54 \text{ V}$$

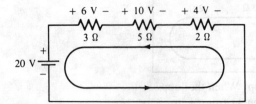

FIGURE 4.8 Same circuit as Figure 4.7, with the loop drawn in a counterclockwise direction. Notice that the drops and rises are the opposite of those in Figure 4.7.

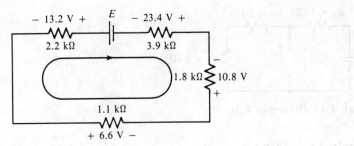

FIGURE 4.9 (Example 4.4)

Drill Exercise 4.4

Find the current in the circuit shown in Figure 4.9.

ANSWER: 6 mA.   □

We can apply Kirchhoff's voltage law to determine the voltage across *any* pair of terminals in a circuit, regardless of what is or what is not connected between those terminals. Remember that voltage at any point in a circuit is always measured *with respect to* some other point in the circuit, so when we speak of the voltage across a pair of terminals, we are assuming that one of the terminals is the reference point for the voltage at the other. It is convenient to assign labels to circuit terminals as an aid in remembering which terminal is the reference for the voltage at the other. For example, we might assign the labels a and b to a pair of terminals, as shown in Figure 4.10. The notation V_{ab} then means that we regard terminal a as "positive" with respect to terminal b, but only in the sense that we measure the voltage between a and b with b as the reference terminal. In other words, we would connect the high side of a voltmeter to terminal a and the low side to terminal b. It may well be that V_{ab} is a negative voltage, in which case our voltmeter reading would be negative. In that case, reversing the voltmeter terminals (connecting the low side to terminal a) would give us the voltage V_{ba}, a positive value. Whether V_{ab} is a positive or negative value, it is *always* true that

$$V_{ab} = -V_{ba} \tag{4.3}$$

For example, if $V_{ab} = 6$ V, then $V_{ba} = -6$ V. If $V_{ab} = -20$ V, then $V_{ba} = +20$ V.

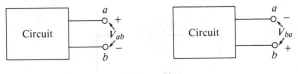

$$V_{ab} = -V_{ba}$$

FIGURE 4.10 The convention used to show which terminal is the *reference* for a voltage across a pair of terminals. The first subscript is the "high" side, where the positive lead of a voltmeter would be connected if a voltage measurement were made. The second subscript is the "low" side, or reference terminal.

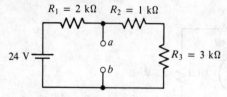

FIGURE 4.11 (Example 4.5)

Example 4.5 (Analysis)

Use Kirchhoff's voltage law to find the voltage V_{ab} in Figure 4.11.

SOLUTION We will first find the voltage drops across the resistors.

$$R_T = 2 \text{ k}\Omega + 1 \text{ k}\Omega + 3 \text{ k}\Omega = 6 \text{ k}\Omega$$

$$I_T = \frac{E}{R_T} = \frac{24 \text{ V}}{6 \text{ k}\Omega} = 4 \text{ mA}$$

$$V_1 = (4 \text{ mA})(2 \text{ k}\Omega) = 8 \text{ V} \qquad V_2 = (4 \text{ mA})(1 \text{ k}\Omega) = 4 \text{ V}$$
$$V_3 = (4 \text{ mA})(3 \text{ k}\Omega) = 12 \text{ V}$$

Figure 4.12 shows the circuit with the voltage drops across each resistor. Notice that there are *two* closed loops that pass through V_{ab}, both shown clockwise in the figure, so we can write Kirchhoff's voltage law around either one to solve for V_{ab}. Notice also that + and − polarity symbols are assigned to V_{ab} in accordance with our convention that *a* is "positive" with respect to *b* (*b* is the reference).

Writing Kirchhoff's voltage law around loop 1, we find

$$24 \text{V} = 8 \text{ V} + V_{ab}$$

or

$$V_{ab} = 16 \text{ V}$$

We obtain precisely the same result in loop 2. Notice that V_{ab} is a rise in loop 2, because we pass through it from − to +:

$$V_{ab} = 4 \text{ V} + 12 \text{ V} = 16 \text{ V}$$

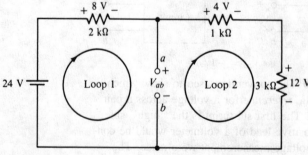

FIGURE 4.12

Drill Exercise 4.5

Find V_{ab} in Figure 4.11 when the voltage source is changed to 36 V.

ANSWER: 24 V.   ☐

4.4 Open Circuits

As the name implies, an *open* is a gap, break, or interruption in a circuit path. No current can flow through an open, so no current can flow in a series circuit containing an open. Since no current can flow through it, an open has infinite resistance ($R = \infty$), which is consistent with Ohm's law:

$$I = \frac{E}{\infty} = 0$$

A circuit containing an open is said to be an *open circuit*, or to be *open-circuited*. Now, our definition implies that an open is a fault condition, or the result of a circuit failure of some type, and it is certainly true that an open circuit is a common result of component failure, or of disintegration of a conducting path, such as the breaking of a wire. However, in many situations, an open circuit is a normal and useful concept, particularly in circuit analysis. All practical circuits are designed to produce a certain amount of current, voltage, or power in some type of *load*, such as a lamp, electric motor, measuring instrument, loudspeaker, or electronic device. For analysis purposes, the load is often represented by its resistance, designated R_L, and in many cases it is useful to study circuit conditions when the load is removed, that is, when the load is open: $R_L = \infty$. We hear, for example, phrases such as "the open-circuit load voltage," or "the output voltage when the load is open."

It is a common error to believe that since the current in an open is zero, the voltage across the open must also be zero. That is certainly not the case, and the reader should take time to reflect on why it is not. Many examples come to mind, one of the most convincing being the voltage across the terminals of a battery having no circuitry connected to it. Another example is the voltage across terminals *a–b* in Example 4.5, which we found to be 16 V. Certainly there is no current flowing from *a* to *b* in that example.

Example 4.6 (Analysis)

What is the voltage *V* across the switch terminals in Figure 4.13 when the switch is opened?

SOLUTION When the switch is open, there is no current in the circuit. Consequently, *there is zero voltage drop across the resistors*. Therefore, by Kirchhoff's voltage law, the voltage across the switch terminals must be the same as the source voltage:

$$V = 60 \text{ V}$$

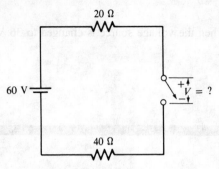

FIGURE 4.13 (Example 4.6)

Example 4.7 (Troubleshooting)

None of the bulbs in a string of series-connected Christmas tree lights illuminate when the string is plugged in. The voltage measured at the wall outlet is 120 V. What is the problem? What measurement(s) would further confirm the diagnosis? Are there any dangerous voltages in the circuit?

SOLUTION Since none of the bulbs illuminate, there is no current in the circuit. Therefore, one of the bulbs, or the wire connecting them, is open. (The fact that there is voltage at the wall outlet tells us that a circuit breaker or fuse has not opened the circuit.) The voltage measured across each bulb is 0 V because there is no voltage drop across any bulb.

Although there is zero voltage across each bulb, there is still 120 V between the low side of the outlet and any point in the string lying on the outlet side of the break (see Figure 4.14).

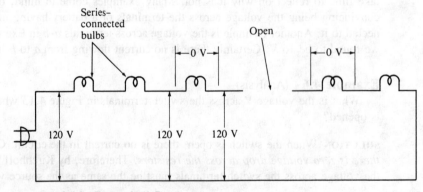

FIGURE 4.14 (Example 4.7)

4.5 Series-Connected Sources

Two or more voltage sources can be connected in series, or as components in a series path, or in a series circuit. If two series-connected sources tend to produce current in the same direction, they are said to be *series-aiding*, and their net effect on the circuit is the same as that of a single source whose voltage equals the *sum* of the two source voltages. As shown in Figure 4.15(a), the series-aiding connection is formed by connecting the positive terminal of one source to the negative terminal of the other.

When the two series sources are connected so that they tend to produce current in opposite directions, they are said to be *series-opposing*. The net effect on the circuit is the same as that of a single source equal in magnitude to the *difference* between the source voltages and having the same polarity as the larger of the two. This connection is formed by connecting terminals of like polarity ($+$ to $+$ or $-$ to $-$), as shown in Figure 4.15(b).

The results we have described can be extended in an obvious way to three or more sources. Note that Kirchhoff's voltage law can be used to determine the net voltage (called the *resultant*) of series-connected sources between any pair of terminals in a circuit. If voltage sources are in the same series path, but not necessarily connected directly to each other, the net effect on the circuit can still be determined by replacing them with a single equivalent source having the resultant voltage.

Example 4.8   (Analysis)
(a) Find the current in the circuit shown in Figure 4.16.
(b) Find V_{ab}.

SOLUTION
(a) Voltage sources E_1 and E_2 both tend to produce current in a clockwise direction in the circuit, while E_3 tends to produce current in a counterclockwise direction. The total resultant voltage, E_T, is therefore

$$E_T = 12 \text{ V} + 25 \text{ V} - 15 \text{ V} = 22 \text{ V}$$

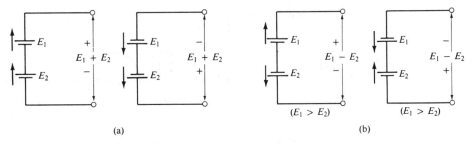

(a)   (b)

FIGURE 4.15 Series-connected voltage sources. (a) Series-aiding connections. Both sources tend to produce current in the same direction and the net voltage equals the sum of the two. (b) Series-opposing connections. The sources tend to produce current in opposite directions and the net voltage equals the difference between the two.

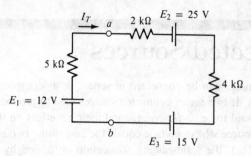

FIGURE 4.16 (Example 4.8)

E_T has the same polarity as E_1 and E_2, as shown in Figure 4.17(a). The total current in the series circuit is then

$$I_T = \frac{E_T}{R_T} = \frac{22\ \text{V}}{11\ \text{k}\Omega} = 2\ \text{mA}$$

(b) To determine V_{ab}, we must restore the original sources to the circuit, as shown in Figure 4.17(b). Since we have already found the current in the circuit, we can find the drop across the 5-kΩ resistor:

$$V_{5\ \text{k}\Omega} = (2\ \text{mA})(5\ \text{k}\Omega) = 10\ \text{V}$$

Figure 4.17(b) shows two closed loops, either one of which can be used to write Kirchhoff's voltage law and find V_{ab}. Using loop 1, we find

$$12 = 10 + V_{ab}$$

or

$$V_{ab} = 2\ \text{V}$$

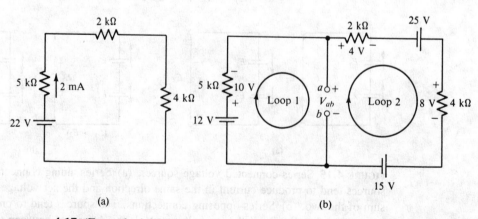

(a)   (b)

FIGURE 4.17 (Example 4.8)

Using loop 2, we find

$$V_{ab} + 25 = 4 + 8 + 15$$

or

$$V_{ab} = 2 \text{ V}$$

Drill Exercise 4.8

Repeat Example 4.8 when the polarity of E_2 in Figure 4.16 is reversed.

ANSWER: $I_T = 2.545$ mA, $V_{ab} = 24.727$ V. □

Current sources are not connected in series. Recall that an ideal current source produces the same, constant current in any circuit connected to it. It is an obvious contradiction to expect a current source that produces one value of current to maintain it in another current source that maintains a different value.

4.6 The Voltage-Divider Rule

Consider the series circuit shown in Figure 4.18. We derive expressions for the voltages V_1 and V_2 across resistors R_1 and R_2 as follows:

$$I = \frac{E}{R_T} = \frac{E}{R_1 + R_2}$$

$$V_1 = IR_1 = \frac{E}{R_1 + R_2} R_1 \qquad (4.4)$$

$$V_2 = IR_2 = \frac{E}{R_1 + R_2} R_2 \qquad (4.5)$$

Notice that equations (4.4) and (4.5) allow us to compute the voltage drop across either resistor without first computing the current in the circuit. The equations show that the voltage across a resistance is the source voltage E times the ratio of that resistance to the total series resistance. We can generalize this result to a series circuit containing any number of resistors, as illustrated in Figure 4.19(a). The result is called the *voltage-divider rule: The voltage across any resistance in a series circuit is the source voltage times the ratio of that resistance to the total resistance of the circuit*. A series circuit can be called a *voltage divider* because the total voltage is divided among the various resistors in direct proportion to the resistance of each. In equation form, the voltage-divider rule is expressed as

$$V_x = E \frac{R_x}{R_T} \qquad (4.6)$$

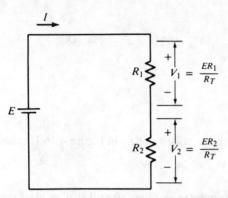

FIGURE 4.18 The voltage across each resistor in a series circuit can be found without knowing the current in the circuit. The voltage E divides between R_1 and R_2 in direct proportion to the resistance of each.

where V_x is the voltage drop across resistance R_x in the series circuit. Note that R_x may also represent the total resistance of any two or more successive resistors, in which case V_x is the voltage drop across the combination of those resistors. This interpretation is illustrated in Figure 4.19(b).

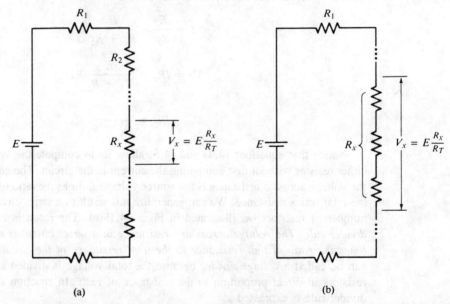

(a)

(b)

FIGURE 4.19 Voltage-divider rule. (a) V_x is the voltage across any single resistance in a series circuit. (b) V_x is the voltage across any number of successive resistances in a series circuit.

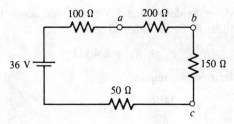

FIGURE 4.20 (Example 4.9)

Example 4.9   (Analysis)
Use the voltage-divider rule to find voltages V_{ab} and V_{ac} in Figure 4.20.

SOLUTION

$$R_T = 100 \ \Omega + 200 \ \Omega + 150 \ \Omega + 50 \ \Omega = 500 \ \Omega$$

$$V_{ab} = (36 \ \text{V})\left(\frac{200}{500}\right) = 14.4 \ \text{V}$$

$$V_{ac} = (36 \ \text{V})\left(\frac{200 + 150}{500}\right) = 25.2 \ \text{V}$$

Drill Exercise 4.9
Find V_{ca} and V_{bc} in Figure 4.20.

ANSWER: $V_{ca} = -25.2 \ \text{V}, \ V_{bc} = 10.8 \ \text{V}.$ □

Example 4.10   (Design)
A certain electronic device is activated when a voltage level of 5 V ± 10% is applied to it. In one application where it is used, the only dc power available is a 24-V source.

(a) Design a voltage divider that will provide the required activation voltage across a resistor. The voltage divider must not draw more than 10 mA from the 24-V source.
(b) Assuming that only standard-value resistors having 5% tolerance can be used in the circuit, draw the schematic diagram of the final design. Verify that it satisfies the design criteria.

SOLUTION
(a) Since the total current in the voltage divider cannot exceed 10 mA, we require

$$\frac{24 \ \text{V}}{R_T} \leq 10 \ \text{mA}$$

or

$$R_T \geq \frac{24 \ \text{V}}{10 \times 10^{-3} \ \Omega} = 2.4 \ \text{k}\Omega$$

We will design a two-resistor divider so that 5 V is dropped across one of them. Let R_2 be that resistor and let R_1 be the other. Then

$$R_T = R_1 + R_2 \geq 2.4 \text{ k}\Omega$$

and, by the voltage-divider rule, we require

$$\frac{24\,R_2}{R_T} = 5 \text{ V}$$

Let us choose $R_2 = 2.4$ kΩ, since we know that this choice will satisfy $R_1 + R_2 \geq 2.4$ kΩ, whatever the value of R_1. Then

$$\frac{(24)(2.4 \times 10^3)}{R_T} = 5 \text{ V}$$

or

$$R_T = \frac{(24)(2.4 \times 10^3)}{5} = 11.52 \text{ k}\Omega$$

Finally,

$$R_1 = R_T - R_2 = 11.52 \text{ k}\Omega - 2.4 \text{ k}\Omega = 9.12 \text{ k}\Omega$$

(b) From the table of standard resistor values given in Chapter 3, we see that 2.4 kΩ is a standard 5% value and that the standard value closest to 9.12 kΩ is 9.1 kΩ. Using these values, our completed design is shown in Figure 4.21. Checking the design, we find

$$I_T = \frac{24 \text{ V}}{2.4 \times 10^3 + 9.1 \times 10^3} = 2.09 \text{ mA} \leq 10 \text{ mA}$$

$$V_1 = \frac{(2.4 \times 10^3)24 \text{ V}}{2.4 \times 10^3 + 9.1 \times 10^3} = 5.009 \text{ V} = 5 \text{ V} \pm 10\%$$

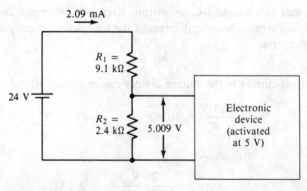

2.09 mA

$R_1 = 9.1$ kΩ

24 V

$R_2 = 2.4$ kΩ 5.009 V

Electronic device (activated at 5 V)

FIGURE 4.21 (Example 4.10)

(Note that our check computations assume that the 5% resistors have zero variation from their nominal values. It is left as an exercise at the end of the chapter to show that the design specifications are still satisfied if the resistors deviate by 5% from their nominal values.)

Drill Exercise 4.10

Repeat Example 4.10 if the electronic device requires 6 V ± 10% to be activated.

ANSWER: $R_1 = 7.5$ kΩ; $R_2 = 2.4$ kΩ. □

Potentiometers and Rheostats

A *potentiometer* is an adjustable voltage divider that is widely used in a variety of electronic circuit applications. Figure 4.22(a) shows the schematic symbol for a potentiometer and gives some idea of its principle of operation. Notice that it is a *three-terminal* device. A voltage source E is connected across the fixed resistance R_T between terminals a and c. A terminal labeled b is attached to a movable contact (called the *wiper arm*, or *slider*) which can be adjusted so that it contacts the resistance between a and c at any point along its entire length. Visualize the contact as sliding up or down between the extreme end positions at terminals a and c. As the contact is moved down, the resistance between b and c, R_{bc}, decreases and the resistance between a and b, R_{ab}, increases. No matter where the contact is positioned,

$$R_T = R_{ab} + R_{bc} \tag{4.7}$$

As illustrated in Figure 4.22(b), we can apply the voltage-divider rule to the potentiometer to determine the voltage V_{bc} between adjustable contact b and fixed contact c:

$$V_{bc} = \frac{R_{bc}}{R_T} E \tag{4.8}$$

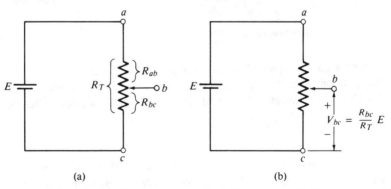

(a)   (b)

FIGURE 4.22 A potentiometer is an adjustable voltage divider. As the adjustable contact b is moved down, R_{bc} decreases and the voltage V_{bc} decreases. (a) $R_T = R_{ab} + R_{bc}$. (b) V_{bc} can be found using the voltage-divider rule.

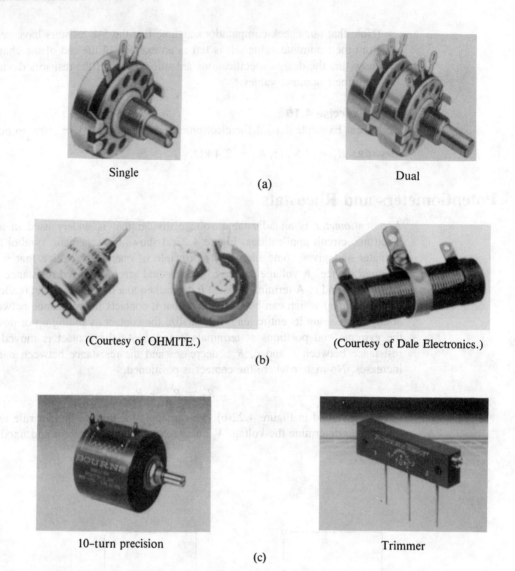

Single   (a)   Dual

(Courtesy of OHMITE.)   (Courtesy of Dale Electronics.)

(b)

10–turn precision   (c)   Trimmer

FIGURE 4.23 Examples of commercially available potentiometers. (a) Carbon composition. (Courtesy of Allen-Bradley, a Rockwell International Subsidiary.) (b) High-power, wirewound. (c) Multi-turn. (Courtesy of Bourns Precisions/Controls.)

Since R_T remains constant, the voltage V_{bc} is directly proportional to resistance R_{bc} and can be adjusted by moving the wiper arm to a desired position. When the arm is positioned at the "top," $R_{bc} = R_T$ and $V_{bc} = E$. When it is positioned at the bottom, $R_{bc} = 0$ and $V_{bc} = 0$. Thus, V_{bc} can be adjusted to any voltage between 0 and E.

Potentiometers are available in a wide variety of sizes, resistance and power ratings, and *tapers*. The latter term refers to the way the resistance between the wiper arm and

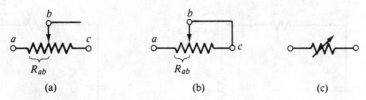

FIGURE 4.24 A rheostat is an adjustable resistance.

one end varies as the arm is moved. In some devices the variation is *linear*, meaning that R_{bc} is directly proportional to the position of the arm, while in others, the resistance may be logarithmically related to position. Figure 4.23 shows some typical, commercially available potentiometers. In practical devices, it is not possible to obtain exactly 0 V or exactly E volts at the extreme end positions, because there is always a small end resistance that the wiper arm cannot contact.

A *rheostat* is an adjustable resistance. Like a potentiometer, it has an adjustable contact that can be moved to different positions to change the resistance between a pair of terminals. However, since its sole purpose is to provide adjustable resistance, it has only two terminals: the two ends of the resistance. Rheostats are commercially available components, usually manufactured in large sizes for heavy power applications. In electronic circuits, adjustable resistance is obtained far more often by connecting a potentiometer as a rheostat. Figure 4.24 shows two ways this connection can be made. In Figure 4.24(a), terminal c is simply left open, and the adjustable resistance is R_{ab}. In Figure 4.24(b), terminal b is connected to terminal c. This connection *shorts out* R_{bc} and, as we shall see in a forthcoming discussion, the result is that the resistance between terminals b and c remains zero, regardless of wiper arm position. Thus, the adjustable resistance is again R_{ab}. Figure 4.24(c) shows the schematic symbol for a reheostat. An arrow drawn through a device symbol is the standard convention for representing adjustable components of all types.

Example 4.11   (Analysis)

Figure 4.25(a) shows a 10-kΩ potentiometer connected in a series circuit as an adjustable voltage divider, and Figure 4.25(b) shows the same device connected in a series circuit as a rheostat.

(a) What total range of voltage V_1 can be obtained in Figure 4.25(a) by adjusting the potentiometer through its entire range?
(b) What total range of voltage V_2 can be obtained in Figure 4.25(b) by adjusting the rheostat through its entire range?

SOLUTION

(a) We first find the total voltage E that appears across the end terminals of the potentiometer. By the voltage-divider rule,

$$E = \left(\frac{10\ k\Omega}{5\ k\Omega + 10\ k\Omega + 10\ k\Omega} \right) 24\ V = 9.6\ V$$

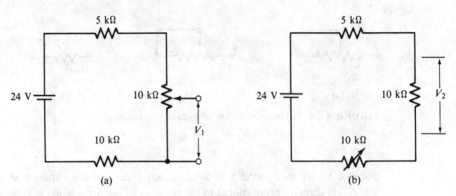

FIGURE 4.25 (Example 4.11)

When the wiper arm is at the top of the potentiometer,

$$V_1 = \left(\frac{10 \text{ k}\Omega}{10 \text{ k}\Omega}\right) 9.6 \text{ V} = 9.6 \text{ V}$$

When the wiper arm is at the bottom of the potentiometer,

$$V_1 = \left(\frac{0}{10 \text{ k}\Omega}\right) 9.6 \text{ V} = 0 \text{ V}$$

Thus, V_1 can be adjusted between 0 and 9.6 V. We express this range by writing $0 \text{ V} \leq V_1 \leq 9.6 \text{ V}$.

(b) Let R_a be the adjustable resistance of the rheostat. By the voltage-divider rule,

$$V_2 = \left(\frac{10 \text{ k}\Omega}{5 \text{ k}\Omega + 10 \text{ k}\Omega + R_a}\right) 24 \text{ V}$$

When $R_a = 0$,

$$V_2 = \left(\frac{10 \text{ k}\Omega}{15 \text{ k}\Omega}\right) 24 \text{ V} = 16 \text{ V}$$

When $R_a = 10 \text{ k}\Omega$,

$$V_2 = \left(\frac{10 \text{ k}\Omega}{25 \text{ k}\Omega}\right) 24 \text{ V} = 9.6 \text{ V}$$

Thus, V_2 can be adjusted between 9.6 and 16 V: $9.6 \text{ V} \leq V_2 \leq 16 \text{ V}$.

Drill Exercise 4.11

Repeat Example 4.11 if V_1 is measured across the 5-kΩ resistor in Figure 4.25(a) and V_2 is measured across the 5-kΩ resistor in Figure 4.25(b).

ANSWER: $V_1 = 4.8 \text{ V}$ at all settings of the potentiometer; (b) $4.8 \text{ V} \leq V_2 \leq 8 \text{ V}$.   ☐

4.7 Parallel Circuits

Two components are connected in *parallel* when they have two common terminals. A parallel circuit is one in which all components are connected in parallel. Figure 4.26 shows some examples of parallel connections and some parallel circuits. Also shown are some circuits which contain parallel circuits but are not themselves parallel circuits. Study these examples carefully. Some have been intentionally drawn in contorted ways to make it difficult (as often happens in practice) to discern the type of circuits they represent. *Always feel free to redraw a circuit in as many equivalent ways as necessary to convince yourself that it represents a circuit of a specific type* or that a series or parallel connection is truly present. Remember that solid lines represent zero-resistance paths, and they can be drawn as long, as short, or as twisted as desired, without affecting the basic nature of a circuit. It is good practice, particularly for a beginner, to redraw a circuit in a way that feels "comfortable," because only then does it become apparent

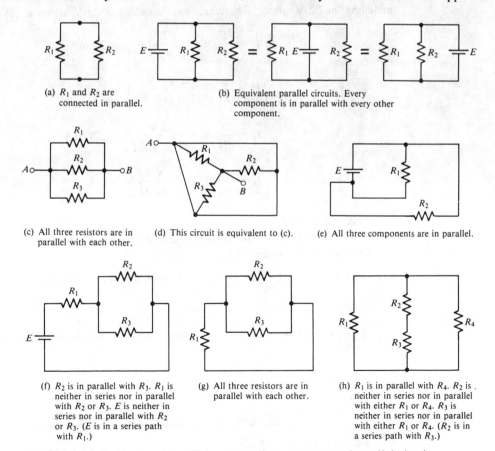

(a) R_1 and R_2 are connected in parallel.

(b) Equivalent parallel circuits. Every component is in parallel with every other component.

(c) All three resistors are in parallel with each other.

(d) This circuit is equivalent to (c).

(e) All three components are in parallel.

(f) R_2 is in parallel with R_3. R_1 is neither in series nor in parallel with R_2 or R_3. E is neither in series nor in parallel with R_2 or R_3. (E is in a series path with R_1.)

(g) All three resistors are in parallel with each other.

(h) R_1 is in parallel with R_4. R_2 is neither in series nor in parallel with either R_1 or R_4. R_3 is neither in series nor in parallel with either R_1 or R_4. (R_2 is in a series path with R_3.)

FIGURE 4.26 Examples of parallel-connected components and parallel circuits.

which circuit analysis tools should be applied to it. Note that it is not necessary for two components to be either in series or in parallel with each other. It is often the case that they are neither in series nor in parallel. Figure 4.26(f) and (h) show examples.

Analysis of Parallel Circuits

The most important property of parallel circuits is that *every parallel-connected component has the same voltage across it*. This property is obvious when we imagine how we would measure the voltage across parallel components, and then realize that measuring across one is the same as measuring across all of them. If we know the voltage across any one parallel component, we know the voltage across each of the others and can therefore find the current in each, using Ohm's law. This computation is illustrated in the next example.

Example 4.12   (Analysis)
Find the current in each resistor in Figure 4.27.

SOLUTION It is clear that all the components in Figure 4.27 are connected in parallel, including the 32-V source. Therefore, each resistor has 32 V across it. By Ohm's law,

$$I_1 = \frac{E}{R_1} = \frac{32 \text{ V}}{1.6 \times 10^3 \text{ }\Omega} = 20 \text{ mA}$$

$$I_2 = \frac{E}{R_2} = \frac{32 \text{ V}}{320 \text{ }\Omega} = 0.1 \text{ A}$$

$$I_3 = \frac{E}{R_3} = \frac{32 \text{ V}}{100 \times 10^3 \text{ }\Omega} = 320 \text{ }\mu A$$

$$I_4 = \frac{E}{R_4} = \frac{32 \text{ V}}{4 \times 10^3 \text{ }\Omega} = 8 \text{ mA}$$

Drill Exercise 4.12
Repeat Example 4.12 when E is changed to 48 V and each resistor has half the resistance shown in Figure 4.27.

ANSWER: $I_1 = 60 \text{ mA}$, $I_2 = 0.3 \text{ A}$, $I_3 = 0.96 \text{ mA}$, $I_4 = 24 \text{ mA}$.   ☐

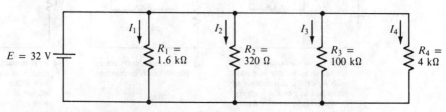

FIGURE 4.27 (Example 4.12)

Recall that the conductance G of a resistor is defined to be the reciprocal of its resistance: $G = 1/R$ siemens (S). *The total conductance of n parallel-connected resistors is the sum of the n individual conductances*:

$$G_T = G_1 + G_2 + \cdots + G_n \qquad (4.9)$$

For example, if a 10-Ω resistor is in parallel with a 20-Ω resistor, the total conductance of the combination is $G_T = \frac{1}{10} + \frac{1}{20} = 0.1 + 0.05 = 0.15$ S. Using equation (4.9), we can derive an expression for the total equivalent resistance of n parallel-connected resistors:

$$G_T = \frac{1}{R_T} = \frac{1}{R_1} + \frac{1}{R_2} + \cdots + \frac{1}{R_n} \qquad (4.10)$$

Inverting both sides, we find

$$\frac{1}{G_T} = R_T = \frac{1}{1/R_1 + 1/R_2 + \cdots + 1/R_n} \qquad (4.11)$$

Recall that Ohm's law expressed in terms of conductance is $I = EG$. Therefore, for a parallel circuit, we have

$$I_T = EG_T \qquad (4.12)$$

A special case of a parallel circuit that is often encountered in circuit analysis is the combination of just two resistors, R_1 and R_2. For that case, equation (4.11) is

$$R_T = \frac{1}{1/R_1 + 1/R_2} = \frac{1}{(R_2 + R_1)/R_1 R_2}$$

or

$$R_T = \frac{R_1 R_2}{R_1 + R_2} \qquad (4.13)$$

Equation (4.13) shows that the total equivalent resistance of two parallel resistances is their product divided by their sum. For example, if $R_1 = 3$ kΩ and $R_2 = 6$ kΩ, then

$$R_T = \frac{(3 \times 10^3)(6 \times 10^3)}{3 \times 10^3 + 6 \times 10^3} = \frac{18 \times 10^6}{9 \times 10^3} = 2 \text{ k}\Omega$$

An important fact about parallel-connected resistances is that the total equivalent resistance is *always* less than any one, including the smallest, of the resistances in parallel. This result is evident in the example just discussed: $R_1 = 3$ kΩ, $R_2 = 6$ kΩ, and $R_T = 2$ kΩ.

Another special case of a parallel circuit is the case where all the resistors have the same value: $R_1 = R_2 = \cdots = R_n = R$. In that case, the total equivalent resistance equals the resistance of any one of them, divided by n:

$$R_T = \frac{R}{n} \qquad (n \text{ parallel resistors, each having resistance } R) \qquad (4.14)$$

For example, three 15-kΩ resistors connected in parallel have a total equivalent resistance of $R_T = 15 \text{ k}\Omega/3 = 5 \text{ k}\Omega$.

Example 4.13   (Analysis)

Find the total conductance and the total equivalent resistance of each of the networks in Figure 4.28, with respect to terminals A–B (i.e., the resistance and conductance that a source connected across A–B would see).

SOLUTION

(a) From equation (4.11),

$$R_T = \frac{1}{\frac{1}{50} + \frac{1}{100} + \frac{1}{200}} = \frac{1}{0.02 + 0.01 + 0.005}$$

$$= \frac{1}{0.035} = 28.57 \ \Omega$$

$$G_T = \frac{1}{R_T} = \frac{1}{28.57 \ \Omega} = 0.035 \text{ S}$$

(b) From equation (4.13),

$$R_T = \frac{(15 \text{ k}\Omega)(30 \text{ k}\Omega)}{15 \text{ k}\Omega + 30 \text{ k}\Omega} = \frac{450}{45} \text{ k}\Omega = 10 \text{ k}\Omega$$

$$G_T = \frac{1}{R_T} = \frac{1}{10 \times 10^3 \ \Omega} = 0.1 \text{ mS}$$

(c) From equation (4.13),

$$R_T = \frac{27 \ \Omega}{3} = 9 \ \Omega$$

$$G_T = \tfrac{1}{9} = 0.111 \text{ S}$$

(d) From equation (4.14), the two 1.2-MΩ resistors are equivalent to 1.2 M$\Omega/2 = $ 0.6 MΩ = 600 kΩ. This 600-kΩ resistance is in parallel with the remaining two 600-kΩ resistors, so

$$R_T = \frac{600 \text{ k}\Omega}{3} = 200 \text{ k}\Omega$$

$$G_T = \frac{1}{200 \text{ k}\Omega} = 5 \text{ } \mu\text{S}$$

Drill Exercise 4.13

A parallel circuit consists of four parallel-connected 480-Ω resistors in parallel with six 360-Ω resistors. What is the total resistance and total conductance of the circuit?

ANSWER: $R_T = 40 \ \Omega$, $G_T = 0.025$ S. $\square$

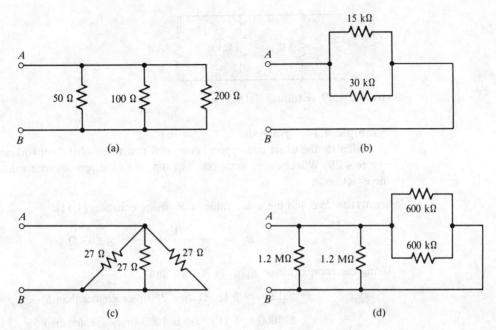

FIGURE 4.28 (Example 4.13)

When one resistance is much greater than another one connected in parallel with it, the equivalent resistance of the combination is very nearly equal to the *smaller* of the two:

$$\text{If} \quad R_2 >> R_1, \quad \text{then} \quad R_1 \| R_2 \approx R_1 \qquad (4.15)$$

where $\|$ means "in parallel with"
     $\approx$ means "approximately equal to"
     $>>$ means "much greater than"

To illustrate, suppose that $R_1 = 10 \ \Omega$ and $R_2 = 10 \ \text{k}\Omega$. Then

$$R_1 \| R_2 = \frac{10 \times 10^4}{10 + 10^4} = \frac{10^5}{10,010} = 9.99 \ \Omega \approx R_1$$

In this example, R_2 is 1000 times as great as R_1 and the approximation is very good. The approximation is widely used in practical design and analysis work. Even if R_2 is only 10 times as great as R_1, the assumption that $R_1 \| R_2 = R_1$ gives an error of less than 10%, which may be less than the error resulting from resistor tolerances. The standard student question is: "When can I use this approximation?" As discussed in Chapter 1, the only valid answer is that it depends on the situation. If a particular application requires calculation of a voltage or current to a high accuracy, and if all the component values are known to a high accuracy, the approximation probably should not be used. In situations such as preliminary design work, where only rough estimates are needed, the approximation can be a valuable time-saver.

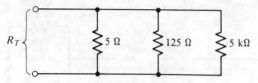

FIGURE 4.29 (Example 4.14)

Example 4.14 (Analysis)

Calculate the exact and approximate total resistance of the network shown in Figure 4.29. What percent error results from using the approximate value instead of the exact?

SOLUTION We find the exact value of R_T using equation (4.11):

$$R_T = \frac{1}{\frac{1}{5} + \frac{1}{125} + \frac{1}{5000}} = 4.803 \ \Omega$$

Using the approximation $R_1 \| R_2 \approx R_1$, we find

$$5 \| 125 \ \approx 5 \ \Omega \quad (125 \text{ is } 25 \text{ times greater than } 5)$$

$$5 \| 5000 \approx 5 \ \Omega \ (5000 \text{ is } 1000 \text{ times greater than } 5)$$

Thus, the approximate value is 5 Ω. The percent error is

$$\% \text{ error} = \frac{5 - 4.803}{4.803} \times 100\% = 4.1\%$$

Drill Exercise 4.14

Repeat Example 4.14 if the 5-Ω resistor is replaced by a 50-Ω resistor.

ANSWER: $R_T = 35.461 \ \Omega$; $R_T \approx 50 \ \Omega$; 41% error.   □

4.8 Kirchhoff's Current Law

Electrical charge is neither created nor destroyed in ordinary circuit operation. Consequently, when there is current in a component, all the charge that enters it over a period of time must equal all the charge that leaves it in the same period of time. It follows that the *rate* at which charge enters a component equals the rate at which it leaves. Recalling that electrical current is the rate of flow of charge, we conclude that the current entering a component must equal the current leaving it. This idea can be extended to include a network containing several components, where current enters and leaves via several different paths. In fact, if we draw a closed line around *any* portion of an electrical circuit, we can apply the same logic and conclude that the total current crossing that line as it enters the enclosed portion must equal the total current crossing the line in the opposite direction, regardless of the number of paths by which current can enter or leave. The portion of the circuit around which the boundary line is drawn can be as small as a

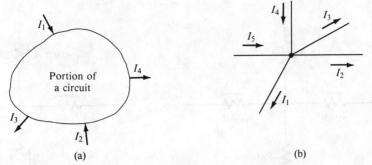

FIGURE 4.30 Examples of Kirchhoff's current law. In each case, the total current entering equals the total current leaving. (a) $I_1 + I_2 = I_3 + I_4$. (b) $I_4 + I_5 = I_1 + I_2 + I_3$.

single junction where two or more components are joined. Figure 4.30 illustrates these ideas, which are summarized by *Kirchhoff's current law*:

The sum of all currents entering a junction, or any portion of a circuit, equals the sum of all currents leaving the same.

Kirchhoff's current law is often used to find an unknown current, given the magnitudes and directions of several other currents in a circuit. It is important to realize that current, like voltage, can have a negative value. The implication of a negative current is simply that its "actual" (positive) direction is the opposite of that which was assumed in a computation. For example, to say that 1 ampere of current is flowing from left to right through a resistor is *entirely equivalent* to saying that *minus* 1 ampere is flowing from right to left. If we assume left-to-right as the *reference* direction for positive current in a particular circuit, then right-to-left current is negative, and we cannot say that a positive value is "correct" while a negative value is "incorrect." The next example illustrates this point.

Example 4.15 (Analysis)

Find the current in the 150-Ω resistor, in Figure 4.31(a).

SOLUTION As shown in Figure 4.31(b), we arbitrarily assume that the current in the 150-Ω resistor is entering the junction. In other words, we take that direction as our reference. Assigning the designations I_1, I_2, I_3, and I_4 to the various currents as shown, we write Kirchhoff's current law:

$$\underbrace{I_1 + I_4}_{\text{current entering}} = \underbrace{I_2 + I_3}_{\text{current leaving}}$$

$$0.8 + I_4 = 0.2 + 0.1$$

or

$$I_4 = 0.3 - 0.8 = -0.5 \text{ A}$$

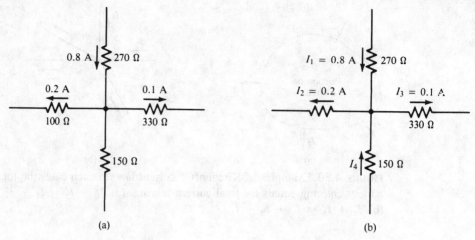

(a)   (b)

FIGURE 4.31 (Example 4.15)

Thus, the current entering the junction through the 150-Ω resistor is -0.5 A. That statement, as it stands, is perfectly correct. It is equally correct to say that the current leaving the junction through the 150-Ω resistor is $+0.5$ A.

Drill Exercise 4.15

Repeat Example 4.15 if the directions of the 0.8-A and 0.2-A currents in Figure 4.31(a) are reversed.

ANSWER: 0.7 A entering the junction.   □

Example 4.16   (Analysis)

Find the currents I_1, I_2, I_3 and I_4 in Figure 4.32.

SOLUTION Since all components are in parallel, we know that the voltage across each resistor is 48 V. We find the current in each resistor using Ohm's law:

$$I_{12\ k\Omega} = \frac{48\ V}{12\ k\Omega} = 4\ mA$$

$$I_{8\ k\Omega} = \frac{48\ V}{8\ k\Omega} = 6\ mA$$

$$I_{9.6\ k\Omega} = \frac{48\ V}{9.6\ k\Omega} = 5\ mA$$

These currents are shown in Figure 4.33(a). The current leaving terminal A equals the current entering it, so

$$I_1 = 5\ mA$$

The current entering junction B equals the sum of the currents leaving it, so

$$I_2 = I_1 + 6\ mA = 5\ mA + 6\ mA = 11\ mA$$

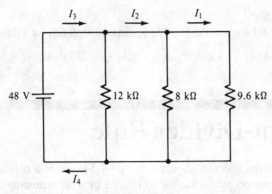

FIGURE 4.32 (Example 4.16)

Similarly, the current entering junction C is

$$I_3 = I_2 + 4 \text{ mA} = 11 \text{ mA} + 4 \text{ mA} = 15 \text{ mA}$$

Finally, the current entering the 48-V source must equal the current leaving it, so

$$I_4 = I_3 = 15 \text{ mA}$$

We can verify the last result by finding the total equivalent resistance of the parallel-connected resistors and applying Ohm's law [see Figure 4.33(b)]:

$$R_T = \frac{1}{1/12 \text{ k}\Omega + 1/8 \text{ k}\Omega + 1/9.6 \text{ k}\Omega} = 3.2 \text{ k}\Omega$$

Thus,

$$I_T = \frac{E_T}{R_T} = \frac{48 \text{ V}}{3.2 \times 10^3 \text{ }\Omega} = 15 \text{ mA}$$

Also,

$$I_T = EG_T = (48 \text{ V})\left(\frac{1}{12 \text{ k}\Omega} + \frac{1}{8 \text{ k}\Omega} + \frac{1}{9.6 \text{ k}\Omega}\right) = 15 \text{ mA}$$

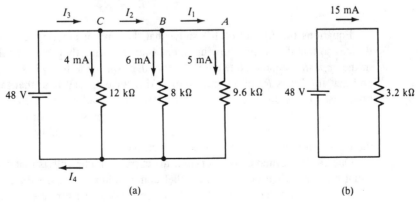

(a) (b)

FIGURE 4.33 (Example 4.16)

Drill Exercise 4.16

Repeat Example 4.16 when the 9.6-kΩ resistor is replaced by a 16-kΩ resistor.

ANSWER: $I_1 = 3$ mA; $I_2 = 9$ mA; $I_3 = 13$ mA; $I_4 = 13$ mA.   □

4.9 The Current-Divider Rule

Consider the two parallel resistors shown in Figure 4.34, where a certain known current I enters one of the common junctions. We will derive general expressions for the currents I_1 and I_2 in each resistor. The total equivalent resistance of the parallel combination is

$$R_T = \frac{R_1 R_2}{R_1 + R_2}$$

Therefore, the voltage across the combination is

$$V = IR_T = \frac{IR_1 R_2}{R_1 + R_2}$$

The current in R_1 is, by Ohm's law,

$$I_1 = \frac{V}{R_1} = \frac{IR_1 R_2}{R_1(R_1 + R_2)}$$

or

$$I_1 = I \frac{R_2}{R_1 + R_2} \tag{4.16}$$

Similarly,

$$I_2 = I \frac{R_1}{R_1 + R_2} \tag{4.17}$$

Equations (4.16) and (4.17) show that the current entering a parallel combination divides so that the portion flowing in one branch is directly proportional to the resistance in the *opposite* branch. Note carefully that the expression for I_1 has R_2 in the numerator, and that for I_2 has R_1 in the numerator. Also, take special note that the denominator in both cases is the *sum* of the resistances (*not* the parallel equivalent, a common error). The reader has probably heard the oversimplified expression "current always follows the path of least resistance." It should be clear now that, in reality, the greater *portion* of the current follows the path of lesser resistance.

One easily verified consequence of the current-divider rule is that current divides into equal portions when entering a parallel combination of equal-valued resistors. For example, if 6 mA enters a junction where three 1-kΩ resistors are connected in parallel, 6 mA/3 = 2 mA will flow in each 1-kΩ resistor.

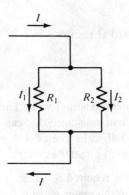

FIGURE 4.34 Current I entering a junction of two parallel resistors divides into two paths. The smaller the resistance of a path, the greater its share of the total current.

Example 4.17   (Analysis)
(a) Find the current in the 470-Ω resistor in Figure 4.35, using the current-divider rule.
(b) Find the current in the 330-Ω resistor using the current-divider rule, and verify the result using Kirchhoff's current law.

SOLUTION

(a) $I_1 = I \dfrac{R_2}{R_1 + R_2} = 160 \text{ mA} \dfrac{330}{330 + 470} = 66 \text{ mA}$

(b) $I_2 = I \dfrac{R_1}{R_1 + R_2} = 160 \text{ mA} \dfrac{470}{330 + 470} = 94 \text{ mA}$

Notice that the larger current (94 mA) flows in the smaller resistor (330 Ω). To verify part (b) using Kirchhoff's current law, note that

$$I = I_1 + I_2$$

so

$$I_2 = I - I_1 = 160 \text{ mA} - 66 \text{ mA} = 94 \text{ mA}$$

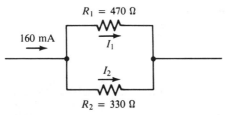

FIGURE 4.35 (Example 4.17)

Drill Exercise 4.17

Repeat Example 4.17 if R_2 is changed to a 470-Ω resistor.

ANSWER: $I_1 = I_2 = 80$ mA. □

Example 4.18 (Design)

A certain ammeter registers full scale (maximum reading) when 1 mA flows in it. The meter has a resistance of 50 Ω. Find a value of resistance that can be connected in parallel with the meter so that it will register full scale when 0.1 A enters the parallel combination. (The required resistor is called a meter *shunt*.)

SOLUTION Figure 4.36 shows the circuit, with the required resistance designated R_x. As shown in the figure, we want 1 mA to flow in the meter when 0.1 A enters the parallel combination. By the current-divider rule,

$$1 \text{ mA} = 0.1 \text{ A} \left(\frac{R_x}{R_x + 50} \right)$$

Solving for R_x, we find

$$10^{-3}(R_x + 50) = 0.1R_x$$

$$R_x = \frac{50 \times 10^{-3}}{0.099} = 0.505 \ \Omega$$

As this result indicates, a meter shunt typically has a very small resistance and must be a high-precision resistor.

Drill Exercise 4.18

What value of current entering the junction in Figure 4.36 will cause full-scale deflection of the meter if R_x is made equal to 5.556 Ω?

ANSWER: 10 mA. □

The current-divider rule can be generalized to include the case where current I enters a junction of an arbitrary number of parallel resistors:

$$I_x = \frac{IR_T}{R_x} \tag{4.18}$$

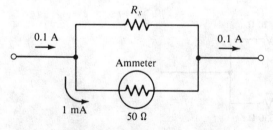

FIGURE 4.36 (Example 4.18)

where I_x is the current in resistor R_x and R_T is the total equivalent resistance of the parallel combination of all the resistors.

Example 4.19 (Design)

What should be the value of R in Figure 4.37 if the current in it must be 0.1 A?

SOLUTION The current from the 0.4-A current source divides among the three parallel resistors. By equation (4.18), the current in R is

$$I_x = 0.1 = \frac{0.4R_T}{R}$$

or

$$R_T = 0.25R$$

Now

$$R_T = 30\ \Omega \| 60\ \Omega \| R$$

But

$$30\ \Omega \| 60\ \Omega = \frac{(30)(60)}{30 + 60}\ \Omega = 20\ \Omega$$

so

$$R_T = 20 \| R = 0.25R$$

Thus

$$\frac{20R}{20 + R} = 0.25R$$

$$20R = 5R + 0.25R^2$$

$$15R = 0.25R^2$$

$$R = \frac{15}{0.25} = 60\ \Omega$$

Drill Exercise 4.19

What current will flow in resistor R in Figure 4.37 when R is made equal to 30 Ω?

ANSWER: 0.16 A.   □

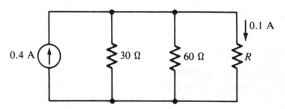

FIGURE 4.37 (Example 4.19)

4.10 Short Circuits

A short circuit, or "short," is a path of zero resistance. A component is said to be *short-circuited*, or "shorted out," when there is a short circuit connected in parallel with it. Figure 4.38 shows current I entering the junction where a short circuit is connected in parallel with a resistor. By the current-divider rule, the current in the resistor is

$$I_R = I \frac{0}{R + 0} = 0$$

Similarly, the current in the short circuit is

$$I_{SS} = I \frac{R}{R + 0} = I$$

We conclude that *no* current flows in a short-circuited component, and that *all* current is diverted through the path that shorts it.

The total equivalent resistance of a short-circuited resistor is

$$R_T = \frac{R(0)}{R + 0} = \frac{0}{R} = 0 \ \Omega$$

It follows that the total equivalent resistance of any number of parallel resistors is zero if any one of them is short-circuited. Since the resistance of a short circuit or of a short-circuited network is zero, the voltage across it must always be zero:

$$V = IR = I(0) = 0$$

An (ideal) switch is a device that produces a short circuit between its terminals when it is closed and an open circuit when it is opened.

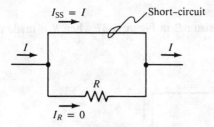

FIGURE 4.38 A short circuit has zero resistance. When a component is short-circuited, all current is diverted through the path that shorts it.

Example 4.20 (Analysis)

Find the total current I_T and the voltage across the 120-Ω resistor in Figure 4.39 when

(a) the switch is open
(b) the switch is closed

SOLUTION

(a) When the switch is open, the circuit consists of three series-connected resistors.

$$R_T = 80 + 120 + 300 = 500 \ \Omega$$

$$I_T = \frac{E}{R_T} = \frac{80 \ V}{500 \ \Omega} = 0.16 \ A$$

The voltage across the 120-Ω resistor is then

$$V = I_T R = (0.16 \ A)(120 \ \Omega) = 19.2 \ V$$

(b) When the switch is closed, the 120-Ω resistor is shorted out. The equivalent resistance of the shorted combination is therefore 0 Ω, and the total resistance of the series circuit becomes

$$R_T = 80 + 300 = 380 \ \Omega$$

Then

$$I_T = \frac{80 \ V}{380 \ \Omega} = 0.2105 \ A$$

The voltage across the 120-Ω resistor is 0 V because it is shorted.

Drill Exercise 4.20

Find the voltage across the 80-Ω resistor in Figure 4.39 when the switch is
(a) open; (b) closed.

ANSWER: (a) 12.8 V; (b) 16.84 V.   □

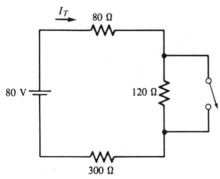

FIGURE 4.39 (Example 4.20)

When circuit components fail, they often become short circuits. Short circuits usually cause greater than normal current flows, which can result in damage to other components. For example, if a voltage source is connected in parallel with several resistors and one of them becomes shorted, the equivalent resistance seen by the source is 0 Ω. Since $I = E/R$, the current drawn from the source becomes extremely large, and, unless the source is protected by a fuse, circuit breaker, or some other current-limiting device, it can be damaged permanently. On the other hand, a short circuit, like an open circuit, is not necessarily an abnormal or destructive phenomenon. In many applications, short circuits can be intentionally inserted by mechanical or electronic switches to prevent current from flowing in a specific portion of a circuit. Also, in some circuit analysis techniques, short-circuit conditions are assumed at specific locations in order to derive important properties of the circuit under investigation.

Example 4.21   (Troubleshooting)
One of the resistors in the circuit of Figure 4.40 has failed and is shorted out. It is not known which resistor is defective. Describe how *one* voltage measurement across *any* one resistor could be used to determine which of the three resistors has failed.

SOLUTION If the one voltage measurement happens to be made across the shorted resistor, the reading will be 0 V, and it can be concluded that is the defective one.

If the voltage measurement is nonzero, the current in the series circuit can be determined from Ohm's law: $I = V/R$, where V is the measured voltage and R is the resistance across which the measurement is made. Knowing this current, the shorted resistor can be determined:

$$\text{If } I = 12 \text{ mA} = \frac{60 \text{ V}}{2 \text{ k}\Omega + 3 \text{ k}\Omega}, \text{ the 1-k}\Omega \text{ resistor is shorted.}$$

$$\text{If } I = 15 \text{ mA} = \frac{60 \text{ V}}{1 \text{ k}\Omega + 3 \text{ k}\Omega}, \text{ the 2-k}\Omega \text{ resistor is shorted.}$$

$$\text{If } I = 20 \text{ mA} = \frac{60 \text{ V}}{1 \text{ k}\Omega + 2 \text{ k}\Omega}, \text{ the 3-k}\Omega \text{ resistor is shorted.}$$

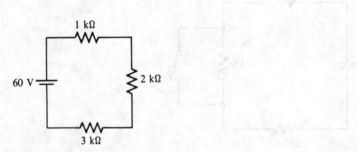

FIGURE 4.40 (Example 4.21)

Drill Exercise 4.21

Two resistors in the circuit shown in Figure 4.40 are shorted. The total current in the circuit is 30 mA. Which resistors are shorted?

ANSWER: 1 kΩ and 3 kΩ.   □

Example 4.22   (Design)

The circuit shown in Figure 4.41 is to be used to test the 15-V source by drawing three different values of current from it as the switch is put in the three positions shown. The test currents corresponding to the three switch positions are to be as follows:

$$\text{position 1:} \quad I = 10 \text{ mA}$$
$$\text{position 2:} \quad I = 30 \text{ mA}$$
$$\text{position 3:} \quad I = 150 \text{ mA}$$

Assuming that the ammeter has zero resistance, design the circuit. (Find the resistor values.)

SOLUTION When the switch is in position 1, the total resistance in series with the source is $R_1 + R_2 + R_3$. Therefore, we must have

$$\frac{15 \text{ V}}{R_1 + R_2 + R_3} = 10 \text{ mA}$$

or

$$R_1 + R_2 + R_3 = \frac{15 \text{ V}}{10 \text{ mA}} = 1.5 \text{ k}\Omega$$

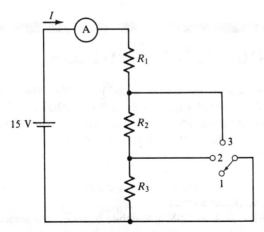

FIGURE 4.41 (Example 4.22)

When the switch is in position 2, R_3 is shorted out, so the total series resistance becomes $R_1 + R_2$. Thus,

$$\frac{15 \text{ V}}{R_1 + R_2} = 30 \text{ mA}$$

or

$$R_1 + R_2 = \frac{15 \text{ V}}{30 \text{ mA}} = 500 \text{ }\Omega$$

When the switch is in position 3, both R_2 and R_3 are shorted out, and the resistance in series with the source is R_1. Therefore,

$$\frac{15 \text{ V}}{R_1} = 150 \text{ mA}$$

or

$$R_1 = \frac{15 \text{ V}}{150 \text{ mA}} = 100 \text{ }\Omega$$

Then

$$R_1 + R_2 = 500 \text{ }\Omega \Rightarrow R_2 = 500 \text{ }\Omega - 100 \text{ }\Omega = 400 \text{ }\Omega$$

and

$$R_1 + R_2 + R_3 = 1500 \text{ }\Omega \Rightarrow R_3 = 1500 \text{ }\Omega - 500 \text{ }\Omega = 1 \text{ k}\Omega$$

Drill Exercise 4.22

Repeat Example 4.22 assuming that the ammeter has resistance 50 Ω.

ANSWER: $R_1 = 50 \text{ }\Omega$; $R_2 = 400 \text{ }\Omega$; $R_3 = 1 \text{ k}\Omega$. □

4.11 Parallel-Connected Sources

Different-valued voltage sources are not connected in parallel. Since components connected in parallel have the same voltage across them, it would be an obvious contradiction to require that two different-valued, ideal sources connected in parallel maintain the same voltage across their shared terminals. In practice, if two real voltage sources are connected in parallel, current will attempt to flow from the positive terminal of the larger-valued source into the positive terminal of the smaller-valued one, which may or may not damage one or both sources. The only resistance limiting current flow in this case is the very small *internal resistances* of the sources.

Rechargeable batteries, such as 12-V automobile batteries, are sometimes connected in parallel, even when one has a lower voltage than the other due to having been discharged. In these cases, the current that flows from the higher-voltage battery into the

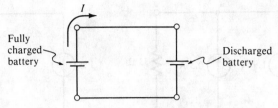

FIGURE 4.42 Rechargeable batteries connected in parallel. Current flows into the discharged battery until both batteries have equal terminal voltages. (Not a recommended procedure for recharging a battery.)

lower-voltage one recharges the latter (and discharges the former). Of course, the parallel connection must be made by connecting the positive terminals of the two batteries together and the negative terminals together (see Figure 4.42).

Two *identical* sources, having the same voltage and the same internal resistance, can be connected in parallel. This connection is sometimes made to obtain a source capable of supplying twice as much current as either single source alone. Of course, the voltage of the combination is the same as that of either single source.

Equal- or different-valued *current* sources can be connected in parallel. The parallel-connected sources can be replaced by a single equivalent current source that produces a constant current equal to the algebraic sum (resultant) of the several sources. The resultant is the value of current that would flow in a resistor connected across the parallel combination, where sources producing current in the same direction are added and those in opposite directions are subtracted. The direction of the equivalent source is the direction of a current source that would produce the same resultant current in the resistor. The next example illustrates these ideas.

Example 4.23   (Analysis)
Find the voltage V_{ab} across the 33-Ω resistor in Figure 4.43(a).

SOLUTION The 2.5-A and 0.5-A current sources both produce current from a to b in the resistor, while the 1.9-A current source produces current from b to a. Therefore, the resultant current is

$$2.5 + 0.5 - 1.9 = 1.1 \text{ A}$$

Since the sum of the currents flowing from a to b is greater than that flowing from b to a, the resultant source produces current from a to b, as shown in Figure 4.23(b). Thus, V_{ab} is positive, and

$$V_{ab} = (1.1 \text{ A})(33 \ \Omega) = 36.3 \text{ V}$$

Drill Exercise 4.23
Repeat Example 4.23 if the direction of the 2.5-A current source is reversed.

ANSWER: -128.7 V. $\square$

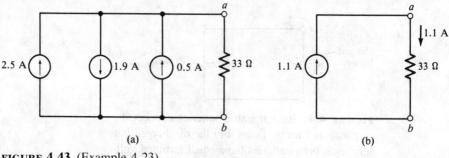

FIGURE 4.43 (Example 4.23)

4.12 SPICE Examples

Example 4.24 (SPICE)

For the circuit shown in Figure 4.11 (Example 4.5), p. 66, use SPICE to find
(a) The total equivalent resistance;
(b) The total current;
(c) The voltage drop across each resistor;
(d) The voltage V_{ab}.

SOLUTION

(a) If a circuit is connected to an ideal 1-A current source, then the voltage across the terminals of the current source will be numerically equal to the total equivalent resistance of the circuit, since $E = IR_T = (1)R_T$. Figure 4.44(a) shows the circuit of Figure 4.11 redrawn in a SPICE format with a 1-A current source replacing the original 24-V voltage source. Execution of the program produces the result V(1) = V(1,0) = 6000 V, so we conclude $R_T = 6$ kΩ.
(b) Figure 4.44(b) shows the circuit and input data file with the 24-V source restored. The total current is the current I(V1), which is printed as -4 mA when the program is executed, due once again to the polarity convention of SPICE.
(c) The voltage drops across the resistors are found by SPICE to be V(1,2) = 8 V (across R_1), V(2,3) = 4 V (across R_2), and V(3) = V(3,0) = 12 V (across R_3).
(c) Voltage V_{ab} is found by SPICE to be V(2) = V(2,0) = 16 V.

Example 4.25 (SPICE)

For the circuit shown in Figure 4.27 (Example 4.12), p. 80, use SPICE to find the total current and the current in each resistance.

SOLUTION The circuit is redrawn in a SPICE format as shown in Figure 4.45. Note that a dummy voltage source (V1, V2, V3, V4) is inserted in series with each resistance to determine the current in each. The polarity of each source is such that the current in each will be printed as a positive value. Note the following *defaults* in the input data file: (1) DC is omitted in the lines identifying the voltage sources; SPICE assumes a source is dc ("by default") if neither DC nor AC is specified. (2) The dummy voltage sources have no voltage values specified; in those cases, SPICE assumes the values are zero.

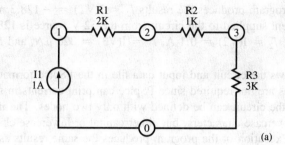

```
EXAMPLE 4.24(A)
I1 0 1 DC 1A
R1 1 2 2K
R2 2 3 1K
R3 3 0 3K
.DC I1 1A 1A 1
.PRINT DC V(1)
.END
```

(a)

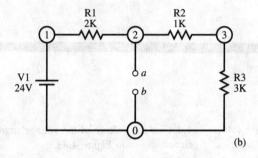

```
EXAMPLE 4.24(B,C,D)
V1 1 0 DC 24V
R1 1 2 2K
R2 2 3 1K
R3 3 0 3K
.DC V1 24V 24V 1
.PRINT DC I(V1)
.PRINT DC V(1,2) V(2,3) V(3)
.PRINT DC V(2)
.END
```

(b)

FIGURE 4.44 (Example 4.24)

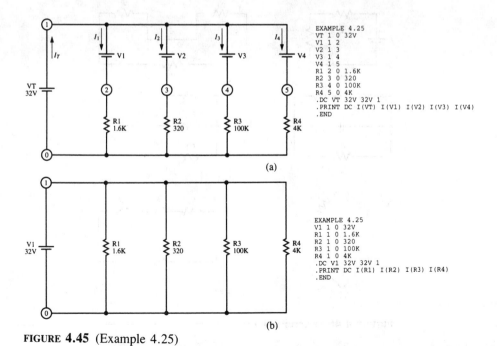

```
EXAMPLE 4.25
VT 1 0 32V
V1 1 2
V2 1 3
V3 1 4
V4 1 5
R1 2 0 1.6K
R2 3 0 320
R3 4 0 100K
R4 5 0 4K
.DC VT 32V 32V 1
.PRINT DC I(VT) I(V1) I(V2) I(V3) I(V4)
.END
```

(a)

```
EXAMPLE 4.25
V1 1 0 32V
R1 1 0 1.6K
R2 1 0 320
R3 1 0 100K
R4 1 0 4K
.DC V1 32V 32V 1
.PRINT DC I(R1) I(R2) I(R3) I(R4)
.END
```

(b)

FIGURE 4.45 (Example 4.25)

Execution of the program produces the results $I_T = \text{I(VT)} = -128.3$ mA, meaning the total current supplied to the circuit from the 32-V source is 128.3 mA, $I_1 = \text{I(V1)} = 20$ mA, $I_2 = \text{I(V2)} = 0.1$ A, $I_3 = \text{I(V3)} = 320$ μA, and $I_4 = \text{I(V4)} = 8$ mA.

Figure 4.45(b) shows the circuit and input data file in the PSpice format. Note that dummy voltage sources are not required since PSpice can print currents in specified resistors. As a result, the circuit can be defined with only two nodes. The input data file is written using lowercase characters, but either capital or lowercase characters (or both) could be used. Execution of the program produces the same results as the one written in Berkeley SPICE.

Exercises

Section 4.2 Series Circuits

4.1 For each network or circuit shown in Figure 4.46, list all components, if any, that are connected in series or in a series path with resistor R.

4.2 Repeat Exercise 4.1 for resistor R_1.

4.3 Find the total equivalent resistance and total current I_T in each circuit shown in Figure 4.47.

4.4 Find the voltage of the voltage source E in each circuit shown in Figure 4.48.

4.5 Find the voltage across each resistor in Figure 4.48.

4.6 Find the power dissipated in each resistor in Figure 4.47.

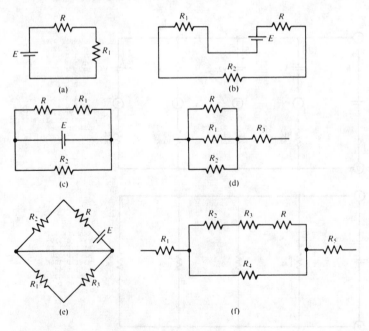

FIGURE 4.46 (Exercise 4.1)

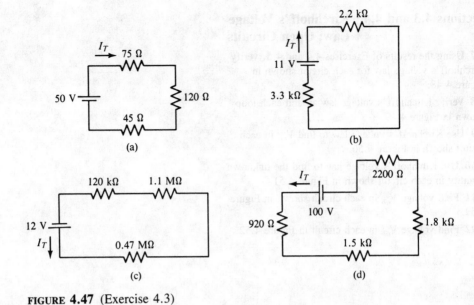

FIGURE **4.47** (Exercise 4.3)

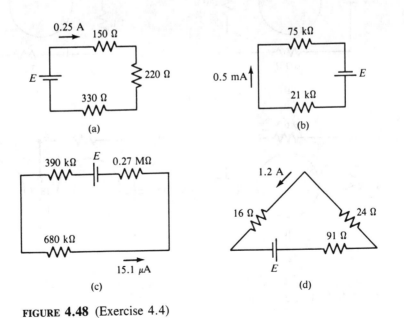

FIGURE **4.48** (Exercise 4.4)

Sections 4.3 and 4.4   Kirchhoff's Voltage Law; Open Circuits

4.7 Using the results of Exercises 4.4 and 4.5, verify Kirchhoff's voltage law for each circuit shown in Figure 4.48.

4.8 Verify Kirchhoff's voltage law around each loop shown in Figure 4.49.

4.9 Use Kirchhoff's voltage law to find V_{ab} in each circuit shown in Figure 4.50.

4.10 Use Kirchhoff's voltage law to find the unknown quantity in each circuit shown in Figure 4.51.

4.11 Find voltage V_{ab} in each circuit shown in Figure 4.52.

4.12 Find voltage V_{ab} in each circuit in Figure 4.52.

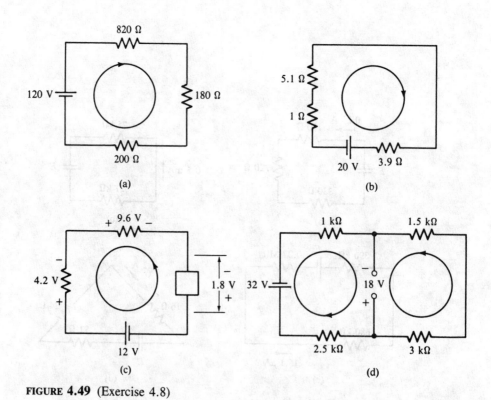

FIGURE 4.49 (Exercise 4.8)

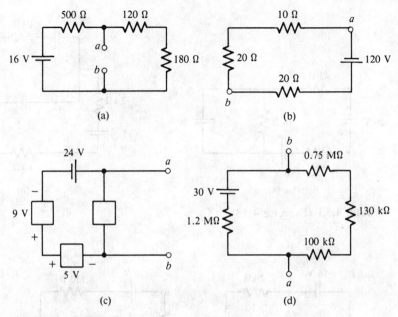

FIGURE **4.50** (Exercise 4.9)

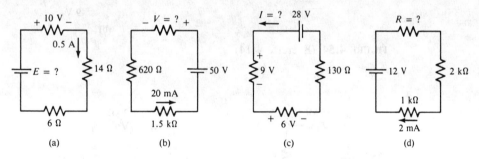

FIGURE **4.51** (Exercise 4.10)

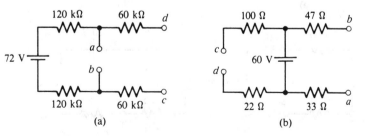

FIGURE **4.52** (Exercise 4.11)

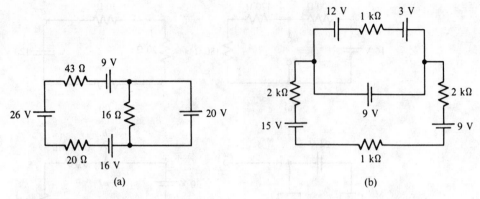

FIGURE **4.53** (Exercise 4.13)

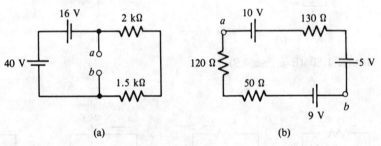

FIGURE **4.54** (Exercise 4.14)

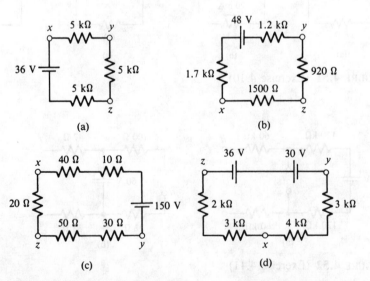

FIGURE **4.55** (Exercise 4.15)

Section 4.5   Series-Connected Sources

4.13 Reduce each circuit in Figure 4.53 to an equivalent circuit that has the minimum possible number of voltage sources.

4.14 Find voltage V_{ab} in each circuit shown in Figure 4.54.

Section 4.6   The Voltage-Divider Rule

4.15 Use the voltage-divider rule to find voltage V_{xy} in each circuit shown in Figure 4.55.

4.16 Use the voltage-divider rule to find voltage V_{xz} in each circuit shown in Figure 4.55.

4.17 In Figure 4.18, show that $V_1 + V_2 = E$.

4.18 In Example 4.10, show that $I_T \le 10$ mA and $V_1 = 5$ V $\pm$ 10% under worst-case conditions (when the 5% resistors deviate from their nominal values by the maximum amount).

4.19 Find the minimum and maximum value of voltage V in each circuit of Figure 4.56, as each potentiometer is adjusted through its full range. (The full potentiometer resistance is shown for each.)

4.20 Repeat Exercise 4.19 for each circuit in Figure 4.57.

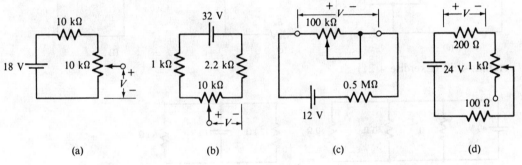

FIGURE **4.56** (Exercise 4.19)

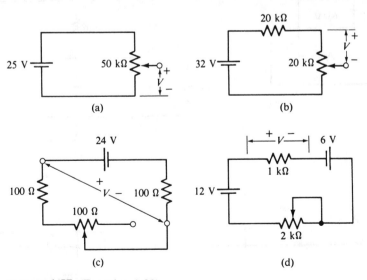

FIGURE **4.57** (Exercise 4.20)

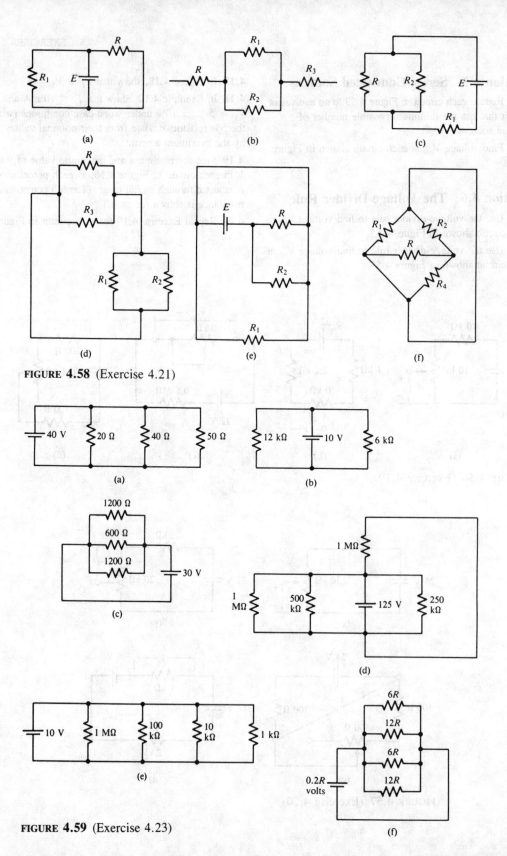

FIGURE 4.58 (Exercise 4.21)

FIGURE 4.59 (Exercise 4.23)

Section 4.7   Parallel Circuits

4.21 List all components, if any, that are connected in parallel with R in each of the circuits or networks shown in Figure 4.58.

4.22 Repeat Exercise 4.21 for resistor R_1.

4.23 Find the total conductance and total resistance of each circuit in Figure 4.59.

4.24 Find the total conductance and total resistance of each circuit in Figure 4.60.

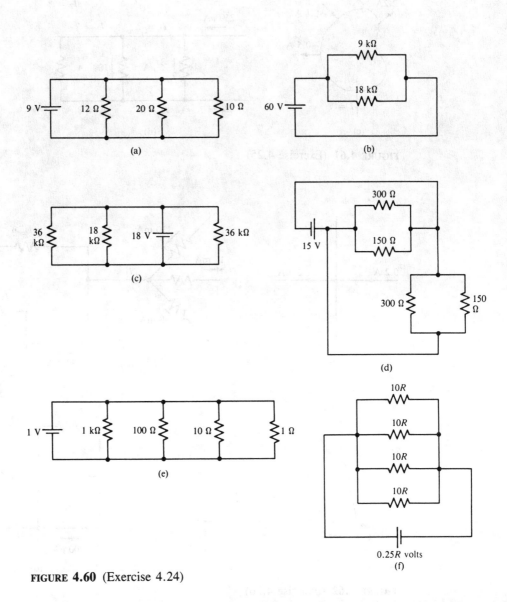

(a)

(b)

(c)

(d)

(e)

(f)

FIGURE 4.60 (Exercise 4.24)

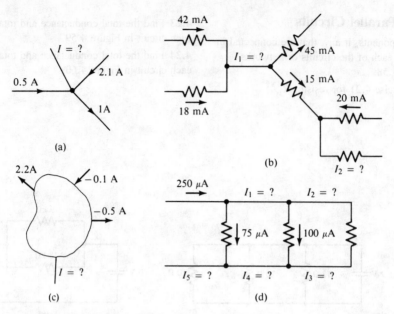

(a)

(b)

(c)

(d)

FIGURE 4.61 (Exercise 4.25)

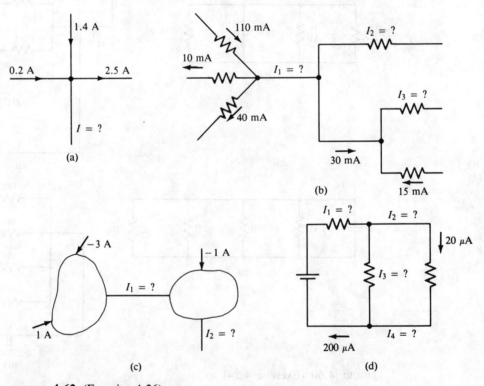

(a)

(b)

(c)

(d)

FIGURE 4.62 (Exercise 4.26)

Section 4.8   Kirchhoff's Current Law

4.25 Use Kirchhoff's current law to find the unknown current(s) in each part of Figure 4.61. Draw an arrow on each diagram to show the positive direction of each current.

4.26 Use Kirchhoff's current law to find the unknown current(s) in each part of Figure 4.62. Draw an arrow on each diagram to show the positive direction of each current.

Section 4.9   The Current-Divider Rule

4.27 For each circuit in Figure 4.60:
(i) Find the current in each resistor.
(ii) Using Kirchhoff's current law, find the total current supplied by the voltage source.
(iii) Verify part (b) using $I_T = E/R_T$ and $I_T = EG_T$.

4.28 Repeat Exercise 4.27 for Figure 4.57.

4.29 Use the current-divider rule to find the current I_1 in each part of Figure 4.61.

4.30 Use the current-divider rule to find the current I_2 in each part of Figure 4.61.

4.31 Use equations (4.16) and (4.17) to show that $I_1 + I_2 = I$ in Figure 4.34, p. 89.

4.32 Show that $I_1R_1 = IR_T = I_2R_2$ in Figure 4.34.

Section 4.10   Short Circuits

4.33 Find current I_1 and voltage V_1 in each part of Figure 4.64.

4.34 Find current I_2 and voltage V_2 in each part of Figure 4.64.

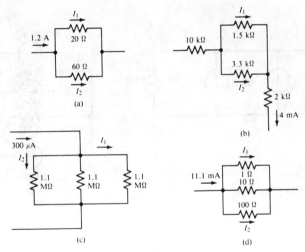

FIGURE 4.63 (Exercise 4.29)

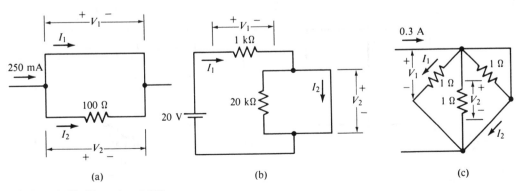

FIGURE 4.64 (Exercise 4.33)

Section 4.11   Parallel-Connected Sources

4.35 Two rechargeable batteries are connected as shown in Figure 4.65. The resistances represent the internal resistance of each. If $E_1 = 12.7$ V and $E_2 = 10.9$ V, how much current will flow between the batteries and which one will be charged?

4.36 What is the terminal voltage V_T across the parallel batteries in Exercise 4.35?

4.37 Find the current in the resistor and voltage V_{ab} in each circuit shown in Figure 4.66.

4.38 Find the current in the 150-Ω resistor in each circuit shown in Figure 4.67. Draw an arrow to show the direction of the current.

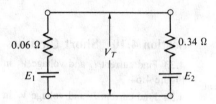

FIGURE 4.65 (Exercise 4.35)

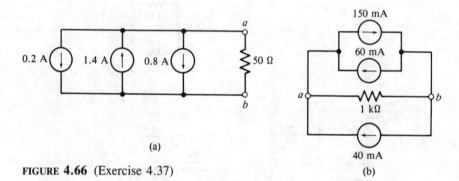

(a)

FIGURE 4.66 (Exercise 4.37)

(b)

(a)

(b)

FIGURE 4.67 (Exercise 4.38)

Design Exercises

4.1D In a certain application, two equal-valued resistors are to be inserted in a series circuit consisting of a 12-V voltage source, a 200-Ω resistor, and the two equal-valued resistors. If the current in the circuit must be 15 mA, what should be the values of the resistors?

4.2D If the 200-Ω resistor in Exercise 4.1D is exactly 200 Ω and the other two resistors have 10% tolerances, what is the total range of current (minimum to maximum) that might flow in the circuit?

4.3D Figure 4.68 shows a circuit that must be modified by the insertion of an additional series resistance so that the voltage drop across the 2-kΩ resistor will become 4 V. What value of series resistance should be inserted?

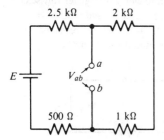

FIGURE 4.68 (Exercise 4.3D)

4.4D The voltage V_{ab} in Figure 4.69 must be 12 V. What source voltage E should be used?

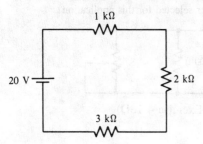

FIGURE 4.69 (Exercise 4.4D)

4.5D The voltage V_{ab} in Figure 4.70 must be 10 V.
(a) What value of resistance R should be used?
(b) If the standard 10% resistor value closest to the value calculated in part (a) is used, what will be the actual voltage V_{ab}? Assume that all resistors have exactly their nominal values.

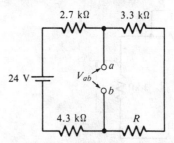

FIGURE 4.70 (Exercise 4.5D)

4.6D It is required to obtain a voltage drop of 16 V across the 4-kΩ resistor in Figure 4.71 by inserting a voltage source E in series with the circuit, as shown. What should be the value of E? Be sure to show the polarity required.

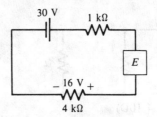

FIGURE 4.71 (Exercise 4.6D)

4.7D Repeat Exercise 4.6D if the voltage required across the 4-kΩ resistor is 28 V (with the same polarity shown in the figure for the 16-V drop).

4.8D The voltage V in the circuit shown in Figure 4.72 must be adjustable from 0 to 25 V. What should be the total resistance of the potentiometer selected for this application?

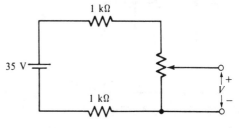

FIGURE 4.72 (Exercise 4.8D)

4.9D The voltage V in Figure 4.73 must be variable from 8 to 12 V as the potentiometer connected as a rheostat is adjusted through its entire range. What should be the total resistance of the potentiometer selected for this application?

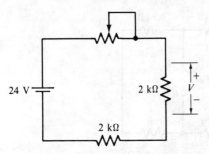

FIGURE 4.73 (Exercise 4.9D)

4.10D The total current drawn from the 12-V source shown in Figure 4.74 cannot exceed 24 mA. What is the minimum load resistance R_L that can be used in this design?

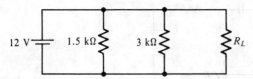

FIGURE 4.74 (Exercise 4.10D)

4.11D The total conductance of the network shown in Figure 4.75 must be at least 20 mS. What is the smallest resistance R that can be used in the network?

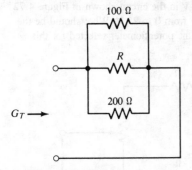

FIGURE 4.75 (Exercise 4.11D)

4.12D A resistance R is to be connected in parallel with the circuit shown in Figure 4.76 so that the current in the 12-kΩ resistor is 2 mA. What should be the value of R?

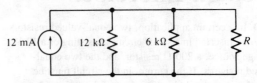

FIGURE 4.76 (Exercise 4.12D)

4.13D In the circuit shown in Figure 4.77, the potentiometer connected as a rheostat is to be used to adjust the current that flows in the 200-Ω resistor from 0 A to 0.1 A. What should be the total resistance of the potentiometer selected for this application?

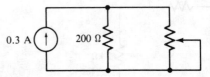

FIGURE 4.77 (Exercise 4.13D)

4.14D An ammeter registers full scale when 0.5 mA flows through it. The circuit shown in Figure 4.78 is to be designed so that the ammeter will register full scale when $I = 5$ mA (switch position 1) and when $I = 50$ mA (switch position 2). The ammeter resistance is 100 Ω. Find the values required for R_1 and R_2.

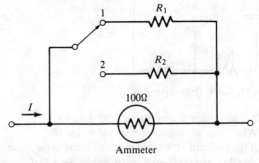

FIGURE 4.78 (Exercise 4.14D)

4.15D A current source I is to be inserted in parallel with the 10-mA source shown in Figure 4.79 so that voltage V is 36 V. What should be the value of the current source? Show the direction of the required source.

4.16D Repeat Exercise 4.15D if the required voltage is $V = 72$ V.

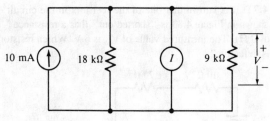

FIGURE 4.79 (Exercise 4.15D)

Troubleshooting Exercises

4.1T The circuit shown in Figure 4.80 is used to test 200-Ω resistors to determine whether they are within their 20% tolerance limits. The ammeter is used to measure the total current in the circuit when a 200-Ω test resistor is inserted for R. Assuming that the 100-Ω resistor is exactly 100-Ω, what are the minimum and maximum values of current I that correspond to the resistor being within tolerance?

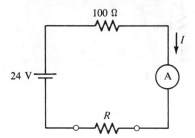

FIGURE 4.80 (Exercise 4.1T)

4.2T Repeat Exercise 4.1T if the test resistors are 100-Ω and are required to have a 10% tolerance.

4.3T It is known that an incorrect resistor has been inserted for one of those shown in Figure 4.81. (The resistor values shown are those that the correct circuit is supposed to have.) The current measured in the circuit is 4 mA and the voltages measured across the resistors are $V_1 = 8$ V, $V_2 = 12$ V, $V_3 = 16$ V. Which resistor is incorrect?

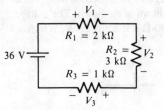

FIGURE 4.81 (Exercise 4.3T)

4.4T Repeat Exercise 4.3T if the voltages measured across the resistors are $V_1 = 14.4$ V, $V_2 = 14.4$ V, and $V_3 = 7.2$ V. The measured current is 7.2 mA.

4.5T The power supply voltage E in Figure 4.82 is required to be 12 V $\pm$ 10%. The voltage measured across the 2.2-kΩ resistor is 4.6 V. Assuming that all resistors have exactly the values shown, is the power supply within tolerance?

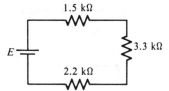

FIGURE 4.82 (Exercise 4.5T)

4.6T Repeat Exercise 4.5T if the voltage measured across the 2.2-kΩ resistor is 2.5 V.

4.7T It is known that one of the resistors in the circuit shown in Figure 4.83 is "shorted out" (has a resistance of 0 Ω). The measured value of V_{ab} is 6 V. Which resistor is defective?

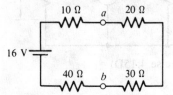

FIGURE 4.83 (Exercise 4.7T)

4.8T Repeat Exercise 4.7T if the measured value of V_{ab} is 13.33 V.

4.9T One of the batteries in the circuit shown in Figure 4.84 was connected backwards (polarity reversed). The battery polarities as they are supposed to appear in the correct circuit are shown in the figure. The voltage measured across the 200-Ω resistor is 2.4 V. Which battery is reversed?

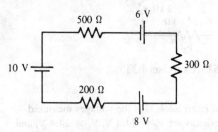

FIGURE 4.84 (Exercise 4.9T)

4.10T Repeat Exercise 4.9T if the voltage measured across the 200-Ω resistor is 1.6 V.

4.11T One of the resistors in Figure 4.85 has burned out and left an open circuit in its place. The total current, I_T, measured in the circuit is 17 mA. Which resistor is defective?

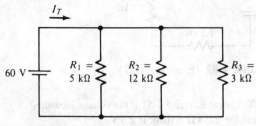

FIGURE 4.85 (Exercise 4.11T)

4.12T One of the resistors in Figure 4.86 has burned out and left an open circuit in its place. The voltage measured across the terminals of the current source is 1.6 V. Which resistor is defective?

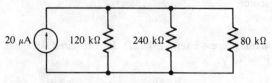

FIGURE 4.86 (Exercise 4.12T)

4.13T The potentiometer shown in Figure 4.87 is connected as a rheostat. Regardless of how it is adjusted, the voltage across the 1-kΩ resistor remains at 8 V. What is the problem with the rheostat (or the way it is connected)?

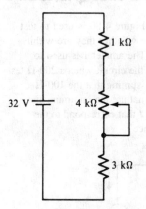

FIGURE 4.87 (Exercise 4.13T)

4.14T Repeat Exercise 4.13T if the voltage across the 1-kΩ resistor remains at 4 V regardless of how the rheostat is adjusted.

SPICE Exercises

4.1S Use SPICE to find the total equivalent resistance, the total current, and the voltage drop across each resistance in the circuit shown in Figure 4.47(d), p. 101.

4.2S Use SPICE to find the total equivalent resistance, the total current, the voltage drop across each resistance, and the voltage V_{ab} in Figure 4.50(d), p. 103.

4.3S Use SPICE to find the total equivalent resistance, the total current, and the current in each resistance in the circuit in Figure 4.60(c), p. 107.

4.4S Repeat Exercise 4.3S for the circuit shown in Figure 4.59(d), p. 106. Also use SPICE to print a numerical value equal to the total *conductance* of the circuit. (Hint: connect a voltage source that produces a current numerically equal to the total conductance.)

4.5S Use SPICE to solve Exercise 4.37 (Figure 4.66 (b), p. 110).

4.6S Solve Design Exercise 4.4D by trial-and-error runs of SPICE programs.

5 Series–Parallel Circuits

5.1 Analyzing Series–Parallel Circuits

A series–parallel circuit is one that contains combinations of series- and parallel-connected components, which are in turn connected in series and/or parallel with other such combinations. Figure 5.1 is an example of a simple series–parallel circuit. Note that the voltage source E and resistor R_1 are in a series path and that R_2 and R_3 are in parallel. As shown in the figure, we may regard the parallel combination of R_2 and R_3 as being in series with E and R_1.

There are no hard and fast rules for analyzing series–parallel circuits. In general, we apply any one or more of the rules and laws developed in preceding chapters (Ohm's law, Kirchhoff's laws, and the current- and voltage-divider rules), in any sequence that leads to the solution for a particular voltage or current. There are usually many different ways to find such a solution (i.e., many different sequences in which the rules can be applied), all of which produce the same solution.

One good approach to analyzing a series–parallel circuit is to replace any combination of series- and/or parallel-connected components by a single *equivalent* resistance, thereby simplifying the circuit. Complex circuits can be progressively simplified by repeated

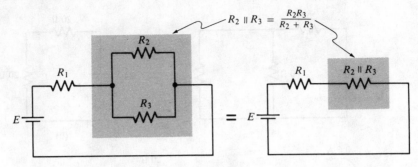

FIGURE 5.1 Example of a series–parallel circuit. The parallel combination of R_2 and R_3 is in series with E and R_1.

applications of this approach, that is, by reducing simplified equivalent circuits to even more simplified equivalent circuits, a procedure that will be demonstrated in many forthcoming examples. The next example illustrates the method and shows how two different analysis rules can be used to obtain the same solution.

Example 5.1

Find the voltage across the 60-Ω resistor in Figure 5.2 by

(a) using the current-divider rule
(b) using the voltage-divider rule

SOLUTION

(a) Figure 5.3 shows the sequence-of steps that leads to a solution using the current-divider rule. We first reduce the original circuit to a simplified equivalent circuit, shown in Figure 5.3(b), where the parallel combination of the 60- and 30-Ω resistors has been replaced by a single equivalent 20-Ω resistance:

$$\frac{30 \times 60}{30 + 60} = 20 \ \Omega$$

The resulting circuit is now clearly a series circuit, and can be reduced to a single equivalent resistance, as shown in Figure 5.3(c):

$$R_T = 20 \ \Omega + 20 \ \Omega = 40 \ \Omega$$

The total current drawn from the 24-V source is then

$$I_T = \frac{24 \ V}{40 \ \Omega} = 0.6 \ A$$

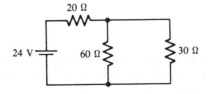

FIGURE 5.2 (Example 5.1)

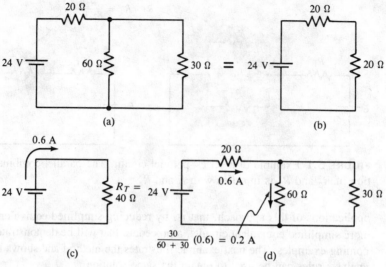

FIGURE 5.3 (Example 5.1)

Since each of the simplified circuits is equivalent to the original, the total current supplied by the 24-V source is 0.6 A in each case. Figure 5.3(d) shows the 0.6 A in the original circuit, and we see that we can apply the current-divider rule to determine the current in the 60-Ω resistor:

$$I_{60\Omega} = \left(\frac{30}{60 + 30}\right)(0.6 \text{ A}) = 0.2 \text{ A}$$

Thus, the voltage across the 60-Ω resistor is, by Ohm's law,

$$V_{60\Omega} = (0.2 \text{ A})(60 \text{ }\Omega) = 12 \text{ V}$$

(b) Figure 5.4 shows how the voltage-divider rule can be used to find the voltage across the 60-Ω resistor. Once again, Figure 5.4(b) is the simplified circuit that results when the parallel combination of the 30- and 60-Ω resistors are replaced by their 20-Ω equivalent. Figure 5.4(c) shows that we can use the voltage-divider rule to find the voltage across the 20-Ω equivalent resistance:

$$V_{20\Omega} = \left(\frac{20}{20 + 20}\right) 24 \text{ V} = 12 \text{ V}$$

Figure 5.4(d) shows that the 12-V drop we found across the 20-Ω resistance also appears across both the 60- and 30-Ω resistors in the original circuit, since the 20-Ω resistance is equivalent to that parallel combination. Thus, we find again that $V_{60\Omega} = 12$ V.

We should note that there are still other ways for finding the voltage across the 60-Ω resistor. For example, in Figure 5.3(d), we could have calculated the drop across the 20-Ω resistor [$V = (0.6 \text{ A})(20 \text{ }\Omega) = 12$ V] and then found the voltage

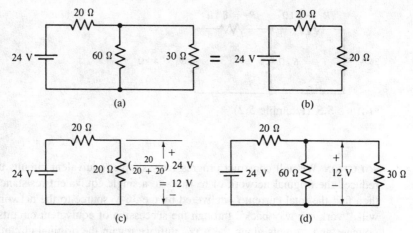

FIGURE 5.4 (Example 5.1)

across the 60-Ω resistor by applying Kirchhoff's voltage law around the loop consisting of the 24-V source, the 20-Ω resistor, and the 60-Ω resistor:

$$V_{60\Omega} = 24 \text{ V} - 12 \text{ V} = 12 \text{ V}$$

Drill Exercise 5.1

If the 20-Ω resistor in Figure 5.3(a) is replaced by a 10-Ω resistor, find the voltage across that resistor.

ANSWER: 8 V.   □

5.2 Series–Parallel Circuit Examples

In this section we will present numerous examples of methods used to analyze series–parallel circuits, as well as some design and troubleshooting examples. For most students, a great deal of repetition is required to become proficient at solving the types of problems these examples represent. The key is practice, practice, and more practice. A good way to gain practice is first to study the solutions to the examples and then, at a later time, to solve them again without referring to the solutions provided. Remember that there is no single "right" way to analyze many of these circuits.

Hereafter, we will say that we *solve* a network when we find the voltage across and current through every component in the network. The next example illustrates that type of analysis problem.

Example 5.2 (Analysis)

Find the voltage across and current through each resistor in the circuit shown in Figure 5.5.

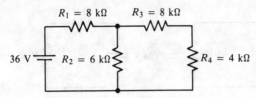

$R_1 = 8\ k\Omega$ $R_3 = 8\ k\Omega$

36 V $R_2 = 6\ k\Omega$ $R_4 = 4\ k\Omega$

FIGURE 5.5 (Example 5.2)

SOLUTION We will construct progressively simpler equivalent circuits until we have reduced the original network of resistors to a single equivalent resistance (R_T). We can then find the total current I_T delivered by the 36-V source to the network. Finally, we will "work our way back" through the succession of equivalent circuits, computing voltages and currents along the way, until we regain the original circuit with all currents and voltages determined. The succession of equivalent circuits is shown in Figure 5.6.

In Figure 5.6(a) we see that R_3 and R_4 are in series and can therefore be combined into the single equivalent resistance $R_3 + R_4 = 8\ k\Omega + 4\ k\Omega = 12\ k\Omega$, as shown in Figure 5.6(b). It is clear in Figure 5.6(b) that the 12-kΩ resistance is in parallel with R_2, so the combination can be replaced by

$$R_2 \| 12\ k\Omega = \frac{(6\ k\Omega)(12\ k\Omega)}{6\ k\Omega + 12\ k\Omega} = 4\ k\Omega$$

The new equivalent circuit that results is shown in Figure 5.6(c). It is now clear that the 4-kΩ resistance is in series with R_1, so the total resistance of the circuit is $R_T = R_1 + 4\ k\Omega = 8\ k\Omega + 4\ k\Omega = 12\ k\Omega$. Thus, the entire resistor network has been reduced to a single equivalent resistance of 12 kΩ, as shown in Figure 5.6(d).

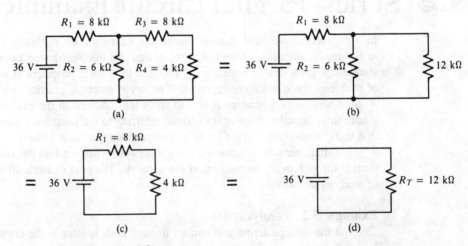

FIGURE 5.6 (Example 5.2)

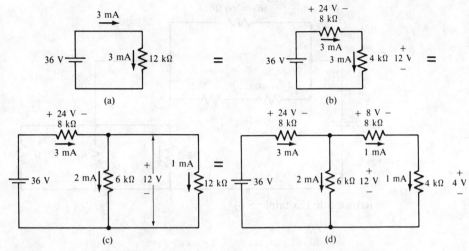

FIGURE 5.7 (Example 5.2)

Figure 5.7 shows how we work our way back through the equivalent circuits, solving for currents and voltages along the way, until the original circuit is completely solved. Note that the sequence of circuits is just the reverse of that shown in Figure 5.6. Figure 5.7(a) shows that the total current delivered by the source is

$$I_T = \frac{36 \text{ V}}{12 \text{ k}\Omega} = 3 \text{ mA}$$

As shown in Figure 5.7(b), the same 3 mA flows through the series combination of the 8-kΩ and 4-kΩ resistance, which together produced the equivalent 12 kΩ. The voltage drops across each resistance are therefore

$$V_{8k\Omega} = (3 \text{ mA})(8 \text{ k}\Omega) = 24 \text{ V}$$

$$V_{4k\Omega} = (3 \text{ mA})(4 \text{ k}\Omega) = 12 \text{ V}$$

As a check, note that Kirchhoff's voltage law is satisfied around this circuit (as it will prove to be around every loop in every equivalent circuit). Since there is 12 V across the 4-kΩ resistance, there is also 12 V across the parallel combination of the 6-kΩ and 12-kΩ resistance, because that combination produced the 4-kΩ equivalent. As shown in Figure 5.7(c), the current through each resistance can then be found:

$$I_{6k\Omega} = \frac{12 \text{ V}}{6 \text{ k}\Omega} = 2 \text{ mA}$$

$$I_{12k\Omega} = \frac{12 \text{ V}}{12 \text{ k}\Omega} = 1 \text{ mA}$$

Note that Kirchhoff's current law is satisfied in this circuit:

$$I_{8k\Omega} = 3 \text{ mA} = I_{6k\Omega} + I_{12k\Omega}$$

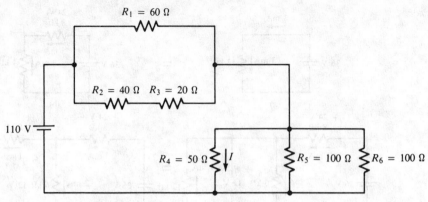

FIGURE 5.8 (Example 5.3)

Finally, as shown in Figure 5.7(d), the 1-mA current in the 12-kΩ resistor of Figure 5.7(c) also flows in the 8-kΩ and 4-kΩ resistance, since those two produced the 12-kΩ equivalent. Thus,

$$V_{8k\Omega} = (1 \text{ mA})(8 \text{ k}\Omega) = 8 \text{ V}$$

$$V_{4k\Omega} = (1 \text{ mA})(4 \text{ k}\Omega) = 4 \text{ V}$$

The circuit is now completely solved. As an exercise, verify Kirchhoff's voltage law around every closed loop and Kirchhoff's current law at every junction in the solved network.

Drill Exercise 5.2

Find the total current delivered by the source and the voltage across R_2 in Figure 5.5 when R_1 is changed to 4 kΩ.

ANSWER: $I_T = 4.5 \text{ mA}$; $V_2 = 18 \text{ V}$.   □

Example 5.3   (Analysis)

Find the current I in the 50-Ω resistor in the circuit shown in Figure 5.8.

SOLUTION R_2 and R_3 are in series and are equivalent to $R_2 + R_3 = 60 \text{ }\Omega$. R_5 and R_6 are in parallel and are equivalent to $100 \text{ }\Omega/2 = 50 \text{ }\Omega$. When these equivalents are substituted for the original components, the equivalent circuit shown in Figure 5.9(a) is obtained. We see that we now have two 60-Ω resistors in parallel and two 50-Ω resistors in parallel. Figure 5.9(b) shows the circuit that results when the two 60-Ω resistors are replaced by their 30-Ω equivalent and the two 50-Ω resistors are replaced by their 25-Ω equivalent. By the voltage-divider rule, the voltage across the 25-Ω is

$$V_{25\Omega} = \left(\frac{25 \text{ }\Omega}{30 \text{ }\Omega + 25 \text{ }\Omega}\right) 110 \text{ V} = 50 \text{ V}$$

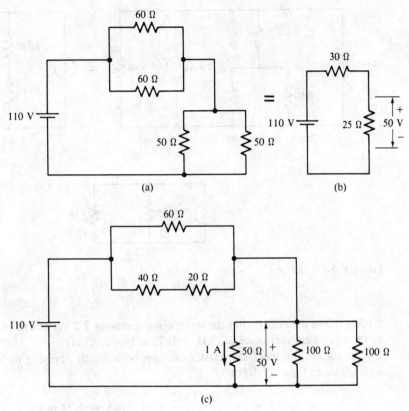

FIGURE 5.9 (Example 5.3)

This 50 V appears across the parallel combination of the 50-Ω and the two 100-Ω resistors in the original circuit, as shown in Figure 5.9(c), so

$$I_{50\Omega} = \frac{50 \text{ V}}{50 \text{ }\Omega} = 1 \text{ A}$$

Drill Exercise 5.3

Find the currents in R_3 and R_6 in Figure 5.8.

ANSWER: $I_{R3} = 1 \text{ A}$, $I_{R6} = 0.5 \text{ A}$.   □

Example 5.4   (Analysis)

Find the current in the 1-kΩ resistor in Figure 5.10(a).

SOLUTION Notice that the two current sources are in parallel. Figure 5.10(b) shows the circuit when these sources are replaced by a single equivalent current source having value 12 mA − 3 mA = 9 mA. Also, the two parallel-connected 1.5-kΩ resistors are replaced by the equivalent resistance 1.5 kΩ/2 = 0.75 kΩ. Figure

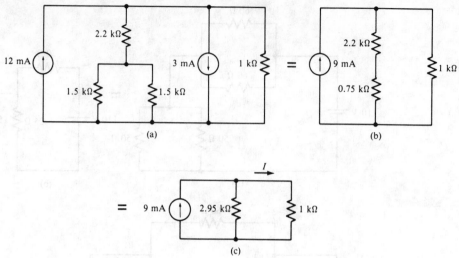

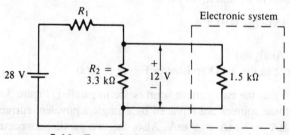

FIGURE **5.10** (Example 5.4)

5.10(c) shows the circuit after the series combination of 2.2 kΩ and 0.75 kΩ is replaced by the equivalent resistance 2.2 kΩ + 0.75 kΩ = 2.95 kΩ.

The current I in the 1-kΩ resistor can now be found by applying the current divider rule to Figure 5.10(c):

$$I = \left(\frac{2.95 \text{ k}\Omega}{2.95 \text{ k}\Omega + 1 \text{ k}\Omega} \right) 9 \text{ mA} = 6.72 \text{ mA}$$

Drill Exercise 5.4

Find the current in each of the 1.5-kΩ resistors in Figure 5.10(a).

ANSWER: 1.14 mA.

Example 5.5 (Design)

It is necessary to supply an electronic system with 12 V derived from a 28-V supply, as shown in Figure 5.11. The electronic system has an equivalent resistance of

FIGURE **5.11** (Example 5.5)

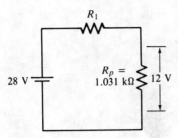

FIGURE 5.12 (Example 5.5)

1.5 kΩ. As shown in the figure, a voltage divider having $R_2 = 3.3$ kΩ will be connected across the 28-V supply so that the voltage across R_2, in parallel with the electronic system, is 12 V. What value of R_1 should be used?

SOLUTION Figure 5.12 shows the voltage divider when the 3.3-kΩ and 1.5-kΩ resistance are replaced by their parallel equivalent, R_p:

$$R_p = \frac{(3.3 \times 10^3)(1.5 \times 10^3)}{3.3 \times 10^3 + 1.5 \times 10^3} = 1.031 \text{ k}\Omega$$

By the voltage-divider rule,

$$\left(\frac{1.031 \times 10^3}{1.031 \times 10^3 + R_1} \right) 28 \text{ V} = 12 \text{ V}$$

Solving for R_1 gives

$$R_1 = 1.37 \text{ k}\Omega$$

Drill Exercise 5.5

If the standard-valued, 10% resistor closest in value to that calculated for R_1 in Example 5.5 is used in the circuit, what is the possible range of the voltage supplied to the electronic system?

ANSWER: 10.77 V − 12.12 V. □

Example 5.6 (Analysis)

Find the unknown quantities in the circuit shown in Figure 5.13.

SOLUTION Figure 5.14 shows three closed loops, around each of which we can write Kirchhoff's voltage law to determine the unknown voltages:

$$\text{loop 1:} \quad 9 + V_1 = 24 \Rightarrow V_1 = 15 \text{ V}$$
$$\text{loop 2:} \quad 24 = V_2 + 12 \Rightarrow V_2 = 12 \text{ V}$$
$$\text{loop 3:} \quad 12 = V_5 + 9 \Rightarrow V_5 = 3 \text{ V}$$

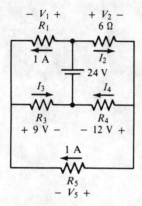

FIGURE 5.13 (Example 5.6)

Applying Ohm's law to $R_2 = 6\ \Omega$, we find

$$I_2 = \frac{V_2}{R_2} = \frac{12\ \text{V}}{6\ \Omega} = 2\ \text{A}$$

Writing Kirchhoff's current law at junction B in Figure 5.14, we have

$$I_2 = 2\ \text{A} = I_4 + 1\ \text{A}$$

or

$$I_4 = 1\ \text{A}$$

Applying Kirchhoff's current law at junction A gives us

$$I_3 = 1\ \text{A} + 1\ \text{A} = 2\ \text{A}$$

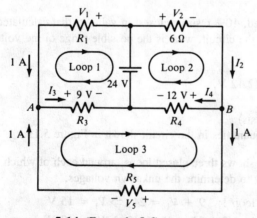

FIGURE 5.14 (Example 5.6)

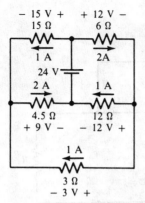

FIGURE 5.15 (Example 5.6)

Using Ohm's law, we can find the remaining unknown quantities:

$$R_1 = \frac{V_1}{I_1} = \frac{15\ V}{1\ A} = 15\ \Omega$$

$$R_3 = \frac{V_3}{I_3} = \frac{9\ V}{2\ A} = 4.5\ \Omega$$

$$R_4 = \frac{V_4}{I_4} = \frac{12\ V}{1\ A} = 12\ \Omega$$

$$R_5 = \frac{V_5}{I_5} = \frac{3\ V}{1\ A} = 3\ \Omega$$

The complete solution is shown in Figure 5.15. As an exercise, verify Kirchhoff's voltage law around every closed loop and Kirchhoff's current law at every junction.

Drill Exercise 5.6

Repeat Example 5.6 when the 24-V source in Figure 5.13 is changed to 36 V. Assume that the values of all other quantities shown in the figure remain the same.

ANSWER: $V_1 = 27\ V$; $V_2 = 24\ V$; $V_5 = 3\ V$; $I_2 = 4\ A$; $I_4 = 3\ A$; $I_3 = 2\ A$; $R_1 = 27\ \Omega$; $R_3 = 4.5\ \Omega$; $R_4 = 4\ \Omega$; $R_5 = 3\ \Omega$.   □

Example 5.7   (Troubleshooting)

One or more of the resistors in Figure 5.16 are suspected of being defective. The voltage measured across both the supply and R_4 is 48 V. Find all possible defects that would account for those measurements.

SOLUTION Figure 5.17(a) shows the equivalent circuit if there are no defective resistors. We see that the voltage across R_4 should be

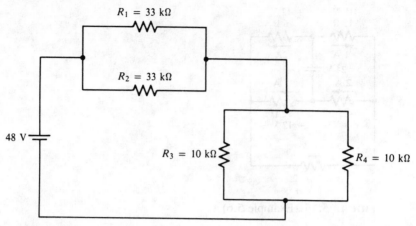

FIGURE 5.16 (Example 5.7)

$$V_4 = \left(\frac{5\ k\Omega}{5\ k\Omega\ +\ 16.5\ k\Omega}\right) 48\ V\ =\ 11.16\ V$$

If *either* R_1 or R_2 (or both) are shorted out, the parallel combination of R_1 and R_2 will equal $0\ \Omega$, as shown in Figure 5.17(b). In that case, the full 48 V would be dropped across the parallel combination of R_3 and R_4, which would account for the measurements obtained.

If *both* R_3 and R_4 are open-circuited, the equivalent circuit shown in Figure 5.17(c) results. In that case, no current flows in the circuit, so there is no drop across the parallel combination of R_1 and R_2, and the full 48 V is measured across the open-circuited combination of R_3 and R_4.

Drill Exercise 5.7

What are the possible defects in the circuit shown in Figure 5.16 if the voltage measured across R_4 is 6.316 V?

ANSWER: Either R_1 or R_2 is open. ☐

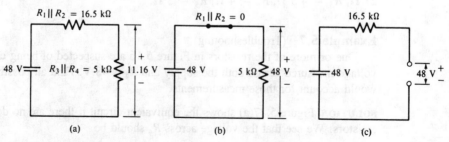

FIGURE 5.17 (Example 5.7)

5.3 Circuits with Ground, or Common, Connections

Instead of drawing lines to show component connections in a schematic diagram, it is a widespread practice to use a special symbol called the *ground*, or *common*, symbol, to designate terminals that are connected together in a circuit (see Fig. 5.18). It is understood that any terminal of any component having that symbol attached to it is connected to every other terminal having the symbol. Figure 5.18 is a simple example. Note that the negative terminal of E and one terminal of each of R_2 and R_4 are all connected together.

The principal reason for using the ground or common symbol is that it eliminates lines that might otherwise clutter a schematic diagram and make it difficult to visualize current flow between components. Beware of attaching more significance to the symbol than it warrants: Although the word ''ground'' implies that there is a connection to *earth* ground, such is not necessarily the case in many schematics where the symbol is used. The word ''common'' has the more accurate implication that terminals having the symbol are *electrically common* to each other (i.e., are connected together). Earth ground is the same as common only if there is an electrical connection that ultimately terminates in a buried pipe or *ground rod*, such as the connection that is made through the ground terminal of most modern electrical power outlets. We also hear of ''chassis ground,'' meaning that common terminals are all connected to a metal panel or enclosure in which the circuit components are mounted. In many electronic systems, chassis ground is connected to earth ground through the ground terminal of a power outlet. In some schematic diagrams, different symbols are used to distinguish between common, earth ground, and/or chassis ground, if such a distinction is necessary. In this book, we will use only the common symbol shown in Figure 5.18, and the symbol should be interpreted only as a convenient, shorthand way of showing connections between terminals.

In schematic diagrams where the common symbol is used, it is conventional to omit that symbol from the side of any voltage source that is connected to common. The terminal that is *not* connected to common is shown by a small circle and is labeled +

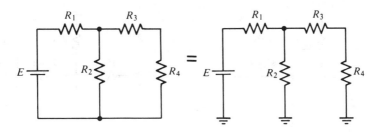

$\perp$ Ground or common symbol

FIGURE 5.18 Example of the use of the common symbol to show connections between component terminals.

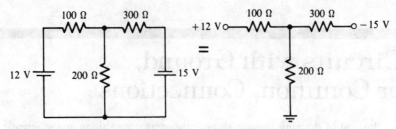

FIGURE 5.19 The common symbol is omitted from the terminal of any voltage source that is connected to common.

or − the value of the source voltage. Figure 5.19 shows an example. Notice that the negative terminal of the 12-V source is connected to common, so the source is labeled +12 V. The positive terminal of the 15-V source is connected to common, so that source is labeled −15 V. Any voltage source that does not have one side connected to common is said to be *floating*.

Another advantage of designating a common in a circuit is that it provides a "low side" to which all other voltages in a circuit can be referenced. For example, if we encounter a statement that the voltage at a certain point in a circuit is +5.5 V, it is understood that the voltage between that point and common is +5.5 V (i.e., that the voltage is +5.5 V with respect to common).

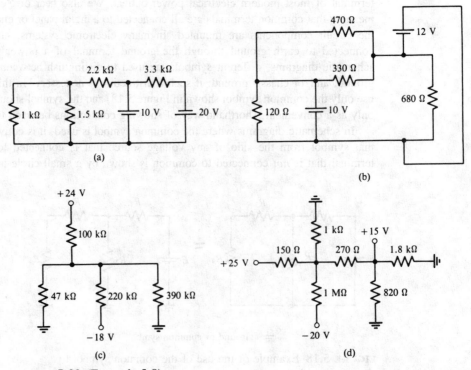

FIGURE 5.20 (Example 5.8)

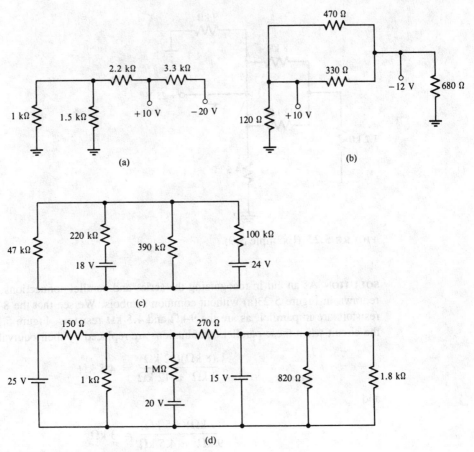

FIGURE 5.21 (Example 5.8)

Example 5.8
(a) Redraw each of the circuits in Figure 5.20(a) and (b) using the common symbol. Assume that the negative side of the 10-V source is common in each case.
(b) Redraw each of the circuits in Figure 5.20(c) and (d) with the common symbols eliminated (i.e., with all component connections shown explicitly).

SOLUTION The redrawn circuits are shown in Figure 5.21. Study each solution carefully and convince yourself that each is electrically equivalent to its original. Note that there are always numerous equivalent ways to draw a schematic diagram, so the solutions shown are not the only correct ones. In Section 5.6 we will have more to say about the benefits of redrawing schematic diagrams.

Example 5.9   (Analysis)
Find the total current I_T drawn from the 25-V source in Figure 5.22.

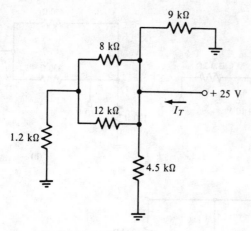

FIGURE 5.22 (Example 5.9)

SOLUTION As an aid in recognizing the series and parallel connections, the circuit is redrawn in Figure 5.23(a) without common symbols. We see that the 8-kΩ and 12-kΩ resistors are in parallel, as are the 9-kΩ and 4.5-kΩ resistors. Figure 5.23(b) shows the circuit when these parallel combinations are replaced by their equivalent resistances:

$$\frac{(8 \text{ k}\Omega)(12 \text{ k}\Omega)}{8 \text{ k}\Omega + 12 \text{ k}\Omega} = 4.8 \text{ k}\Omega$$

and

$$\frac{(9 \text{ k}\Omega)(4.5 \text{ k}\Omega)}{9 \text{ k}\Omega + 4.5 \text{ k}\Omega} = 3 \text{ k}\Omega$$

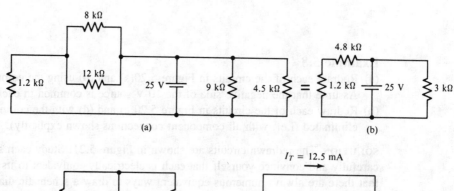

FIGURE 5.23 (Example 5.9)

It is now apparent that the 4.8-kΩ resistance is in series with the 1.2-kΩ resistance, giving a series equivalent of 4.8 kΩ + 1.2 kΩ = 6 kΩ. As shown in Figure 5.23(c), the 6 kΩ is in parallel with the 3 kΩ, so the total equivalent resistance of the circuit is

$$R_T = \frac{(6 \text{ k}\Omega)(3 \text{ k}\Omega)}{6 \text{ k}\Omega + 3 \text{ k}\Omega} = 2 \text{ k}\Omega$$

The total current is then

$$I_T = \frac{25 \text{ V}}{2 \text{ k}\Omega} = 12.5 \text{ mA}$$

Drill Exercise 5.9

Assume the 8-kΩ and 4.5-kΩ resistors in Figure 5.22 are interchanged. Find the current in the 1.2-kΩ resistor.

ANSWER: 5.59 mA. $\square$

Example 5.10   (Analysis)

Find the unknown quantities in the circuit shown in Figure 5.24.

SOLUTION Notice that $R_3 = 50 \ \Omega$ is in parallel with the 50-V source. Therefore,

$$V_3 = 50 \text{ V} \quad \text{and} \quad I_3 = \frac{50 \text{ V}}{50 \ \Omega} = 1 \text{ A}$$

It is clear that $I_4 = 1$ A. Writing Kirchhoff's current law at the junction labeled X in Figure 5.24, we find

$$3 \text{ A} = I_2 + I_3 + I_4$$
$$3 \text{ A} = I_2 + 1 \text{ A} + 1\text{A}$$
$$I_2 = 1 \text{ A}$$

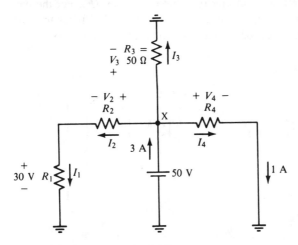

FIGURE 5.24 (Example 5.10)

Clearly, $I_2 = I_1 = 1$ A. Therefore,

$$R_1 = \frac{30 \text{ V}}{1 \text{ A}} = 30 \text{ }\Omega$$

Writing Kirchhoff's voltage law around the loop consisting of R_1, R_2, and the 50-V source, we find

$$50 = 30 + V_2$$
$$V_2 = 20 \text{ V}$$

Therefore,

$$R_2 = \frac{20 \text{ V}}{1 \text{ A}} = 20 \text{ }\Omega$$

Finally, R_4 is in parallel with the 50-V source, so

$$V_4 = 50 \text{ V}$$

$$R_4 = \frac{50 \text{ V}}{1 \text{ A}} = 50 \text{ }\Omega$$

Drill Exercise 5.10

Repeat Example 5.10 when the 50-V source in Figure 5.24 is replaced by a 75-V source. Assume that the values of all other quantities shown in the figure remain the same.

ANSWER: $I_1 = 0.5$ A; $I_2 = 0.5$ A; $I_3 = 1.5$ A; $I_4 = 1$ A; $V_2 = 45$ V; $V_3 = 75$ V; $V_4 = 75$ V; $R_1 = 60$ Ω; $R_2 = 90$ Ω; $R_4 = 75$ Ω.   □

Example 5.11 (Design)

Figure 5.25(a) shows a *transistor*, an electronic device having three terminals, called the base (B), the collector (C), and the emitter (E). For the particular device shown, the current flowing into the collector, I_C, is 100 times the current flowing into the base, I_B. The transistor is said to be *saturated* when the voltage between collector

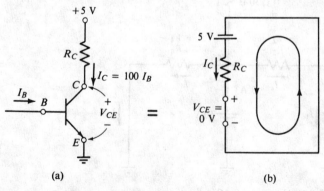

(a)   (b)

FIGURE 5.25 (Example 5.11)

and emitter, V_{CE}, is zero. What value of resistor R_C should be used if it is desired to saturate the transistor when $I_B = 20 \ \mu A$?

SOLUTION As shown in Figure 5.25(b), a closed loop can be drawn from the $+5$-V source, through R_C, across the collector–emitter terminals, and back to the negative side of the source. Writing Kirchhoff's voltage law around that loop, we find

$$5 = I_C R_C + V_{CE}$$

When $I_B = 20 \ \mu A$, $I_C = 100 I_B = 2$ mA. To achieve saturation, we want $V_{CE} = 0$. Thus,

$$5 \text{ V} = (2 \text{ mA}) R_C + 0$$

$$R_C = \frac{5 \text{ V}}{2 \text{ mA}} = 2.5 \text{ k}\Omega$$

Drill Exercise 5.11

What value of R_C should be used in Figure 5.25 if it is desired to make $V_{CE} = 2$ V when $I_B = 20 \ \mu A$?

ANSWER: 1.5 kΩ. □

5.4 Circuit Geometry

There is a strong link between the way the schematic diagram of a circuit is drawn and a particular person's ability to identify the series and parallel combinations it contains. The *geometry* of the circuit (i.e., the way component connections are shown in a schematic diagram, also called the *topology* of the circuit) can often be an impediment to visualizing voltage and current relations in the circuit. For that reason, complex or unusually contorted schematic diagrams should always be redrawn in as many ways as necessary to improve readability. Different individuals will generally require different geometries to feel "comfortable" with a schematic diagram and to be able to perceive the way that component interconnections affect current flow and voltage levels.

In practice, readability is seldom a primary objective when schematic diagrams of practical electronic and electrical systems are first drawn. It is therefore a useful skill to be able to redraw schematics in ways that clearly reveal the nature of the circuits they represent. Figure 5.26 shows three examples of distorted schematic diagrams that would be difficult for most students to analyze. Also shown in the figure are equivalent schematics that have been redrawn to improve circuit interpretation. Note how letters are used to identify corresponding points in each circuit diagram.

Example 5.12 (Analysis)
Find the total current I_T delivered by the 10-V source in Figure 5.27(a).

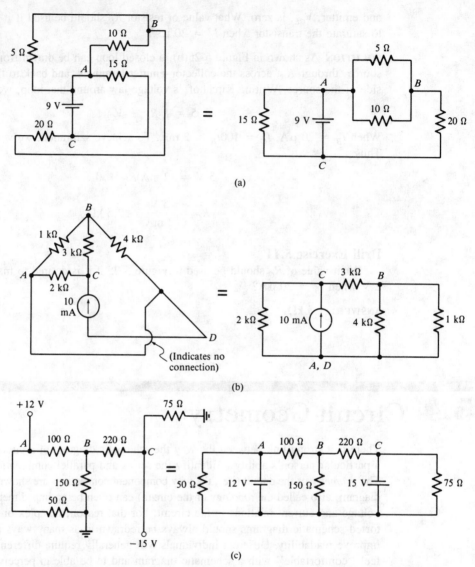

FIGURE 5.26 Examples of schematic diagrams redrawn to improve readability.

SOLUTION When the circuit is redrawn as shown in Figure 5.27(b), it is clear that the three resistors are in parallel. The parallel combination of the 6-kΩ and 3-kΩ resistors equals (6 kΩ)(3 kΩ)/(6 kΩ + 3 kΩ) = 2 kΩ. That 2-kΩ equivalent is in parallel with the other 2-kΩ resistor, so

$$R_T = \frac{2\ k\Omega}{2} = 1\ k\Omega \quad \text{and} \quad I_T = \frac{10\ V}{1\ k\Omega} = 10\ mA$$

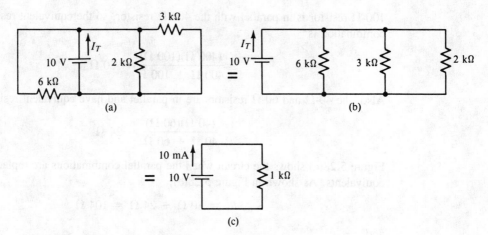

FIGURE 5.27 (Example 5.12)

Drill Exercise 5.12

What would be the current in the 2-kΩ resistor in Figure 5.27 if (a) the 3-kΩ resistor were open; (b) both the 3-kΩ and 6-kΩ resistors were open?

ANSWER: (a) 5 mA; (b) 5 mA.

Example 5.13   (Analysis)

Find the power dissipated in the 100-Ω resistor in Figure 5.28(a).

SOLUTION The schematic diagram is redrawn in Figure 5.28(b). Verify that it is equivalent to Figure 5.28(a). From the redrawn schematic, it is apparent that the

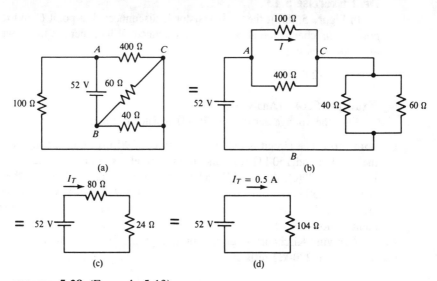

FIGURE 5.28 (Example 5.13)

100-Ω resistor is in parallel with the 400-Ω resistor, so the equivalent resistance of the combination is

$$\frac{(400 \ \Omega)(100 \ \Omega)}{400 \ \Omega + 100 \ \Omega} = 80 \ \Omega$$

Also, the 40-Ω and 60-Ω resistors are in parallel and have equivalent resistance

$$\frac{(40 \ \Omega)(60 \ \Omega)}{40 \ \Omega + 60 \ \Omega} = 24 \ \Omega$$

Figure 5.28(c) shows the circuit when the parallel combinations are replaced by their equivalents. As shown in Figure 5.28(d),

$$R_T = 80 \ \Omega + 24 \ \Omega = 104 \ \Omega$$

and

$$I_T = \frac{52 \ V}{104 \ \Omega} = 0.5 \ A$$

Refer now to Figure 5.28(b). Since the 0.5 A flows from the 52-V source into the parallel combination of 100 Ω and 400 Ω, we may find the current I in the 100-Ω resistor using the current-divider rule:

$$I = \left(\frac{400}{400 + 100}\right)(0.5 \ A) = 0.4 \ A$$

The power dissipated in the 100-Ω resistor is therefore

$$P = I^2R = (0.4)^2(100) = 16 \ W$$

Drill Exercise 5.13

In Figure 5.28(a), the 60-Ω resistor is disconnected at point C and reconnected at point A. Its connection at point B is unchanged. What, then, is the power dissipated in the 100-Ω resistor?

ANSWER: 12.02 W.   □

Example 5.14   (Analysis)

Find the voltage across the 270-kΩ resistor in Figure 5.29(a).

SOLUTION The circuit is redrawn in Figure 5.29(b). In this circuit it is apparent that the 100-kΩ and 150-kΩ resistors are in parallel, as are the two 220-kΩ resistors. In Figure 5.29(c), the 100-kΩ and 150-kΩ resistors are replaced by their equivalent, (100 kΩ)(150 kΩ)/(100 kΩ + 150 kΩ) = 60 kΩ, and the two 220-kΩ resistors are replaced by their equivalent, 220 kΩ/2 = 110 kΩ. Figure 5.29(d) shows the series combination of 60 kΩ and 110 kΩ replaced by 170 kΩ.

Applying Kirchhoff's current law in Figure 5.29(d), we find the current I that flows in the 270-kΩ resistor:

$$I = \left(\frac{170 \ k\Omega}{170 \ k\Omega + 270 \ k\Omega}\right) 40 \ \mu A = 15.45 \ \mu A$$

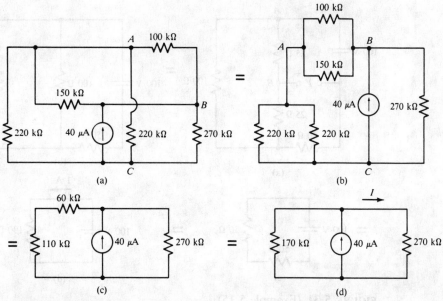

FIGURE **5.29** (Example 5.14)

Therefore, by Ohm's law, the voltage across the 270-kΩ resistor is

$$V = (15.45 \ \mu A)(270 \ k\Omega) = 4.17 \ V$$

Drill Exercise 5.14

In Figure 5.29(a), the 270-kΩ resistor is disconnected from point C and reconnected to point A. Its connection at point B is unchanged. What, then, is the voltage across the 270-kΩ resistor?

ANSWER: 1.96 V.

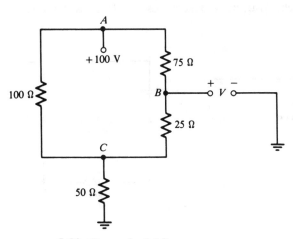

FIGURE **5.30** (Example 5.15)

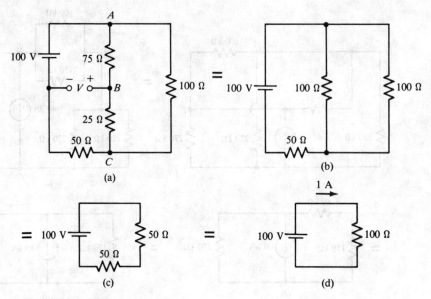

FIGURE 5.31 (Example 5.15)

Example 5.15   (Analysis)

Find the voltage V in Figure 5.30.

SOLUTION The circuit is redrawn as shown in Figure 5.31(a). It is clear that the 75-Ω and 25-Ω resistors are in series and are therefore equivalent to 100 Ω. The equivalent circuit containing that 100-Ω resistance is shown in Figure 5.31(b). We now have two 100-Ω resistances in parallel, equivalent to 50 Ω, as shown in Figure 5.31(c). Thus, the total resistance of the circuit is $R_T = 50 + 50 = 100\ \Omega$ and $I_T = 100\ \text{V}/100\ \Omega = 1$ A.

Figure 5.32(a) is the same equivalent circuit that is shown in Figure 5.31(b). We see that the 1-A total current divides equally between the two 100-Ω resistances, so 0.5 A flows in each. Also, the voltage across the 50-Ω resistance is (1 A)(50 Ω) =

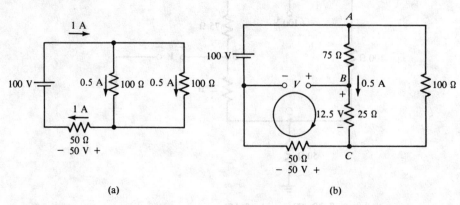

FIGURE 5.32 (Example 5.15)

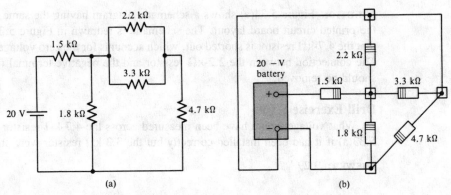

FIGURE 5.33 (Example 5.16)

50 V, with the polarity shown. Figure 5.32(b) is the original circuit. The 0.5 A flowing from A to C causes a voltage drop of $(0.5 \text{ A})(25 \ \Omega) = 12.5$ V across the 25-Ω resistor. Writing Kirchhoff's voltage law around the loop shown, we find

$$V = 12.5 + 50 = 62.5 \text{ V}$$

Drill Exercise 5.15
Find voltages V_{AB} and V_{AC} in Figure 5.30.

ANSWER: $V_{AB} = 37.5$ V; $V_{AC} = 50$ V.   □

Example 5.16   (Troubleshooting)
The circuit shown in Figure 5.33(a) was constructed on a circuit board using the component arrangement shown in Figure 5.33(b). When the completed circuit was tested, it was found that the voltage across the 4.7-kΩ resistor measured 0 V. Since it is clear from Figure 5.33(a) that this voltage should not be zero, it was concluded that there is an error in the board layout. Find the error.

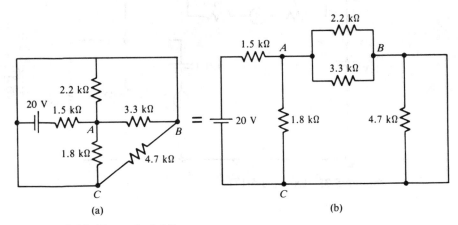

FIGURE 5.34 (Example 5.16)

SOLUTION Figure 5.34(a) shows a schematic diagram having the same geometry as the printed circuit board layout. The schematic is redrawn in Figure 5.34(b). It is clear that the 4.7-kΩ resistor is shorted out, which accounts for the zero voltage measurement. The connection between the 2.2-kΩ resistor and the negative terminal of the battery should be removed.

Drill Exercise 5.16

What voltage would have been measured across the 4.7-kΩ resistor in Figure 5.33(a) if it had been installed correctly but the 3.3-kΩ resistor were shorted?

ANSWER: 9.29 V.   □

5.5 SPICE Examples

Example 5.17   (SPICE)

Use SPICE to find the current I in the 50-Ω resistor in Figure 5.8 (Example 5.3), p. 122.

SOLUTION The circuit is redrawn in its SPICE format in Figure 5.35. Also shown is the input data file. Execution of the program gives I(VDUM) = 1 A, in agreement with Example 5.3.

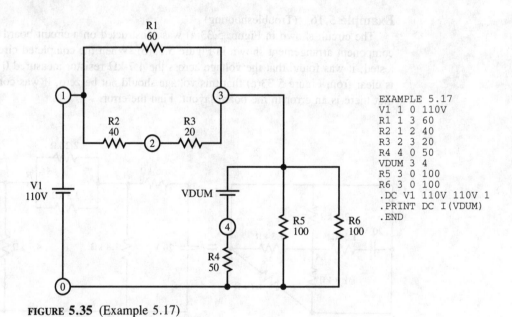

```
EXAMPLE 5.17
V1 1 0 110V
R1 1 3 60
R2 1 2 40
R3 2 3 20
R4 4 0 50
VDUM 3 4
R5 3 0 100
R6 3 0 100
.DC V1 110V 110V 1
.PRINT DC I(VDUM)
.END
```

FIGURE 5.35 (Example 5.17)

Exercises

Section 5.1 and 5.2   Analyzing Series–Parallel Circuits

5.1 Find the total equivalent resistance R_T and the total current I_T drawn from the voltage source in each circuit shown in Figure 5.36.

5.2 Find the total equivalent resistance R_T and the total current I_T drawn from the voltage source in each circuit shown in Figure 5.37.

5.3 Find the voltage across and current through each resistor in each circuit shown in Figure 5.37. On each schematic diagram, draw arrows to show the direction of the current in each resistor, and draw + and − signs to show the polarity of the voltage across each resistor.

5.4 Find the voltage across and current through each resistor in each circuit shown in Figure 5.36. On each

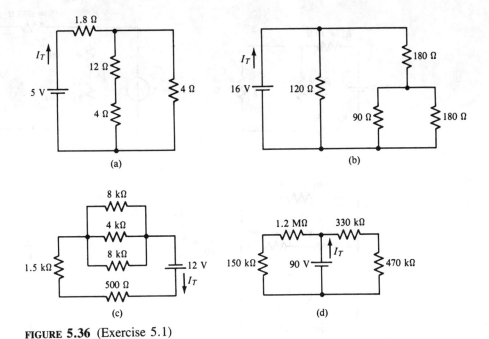

FIGURE 5.36 (Exercise 5.1)

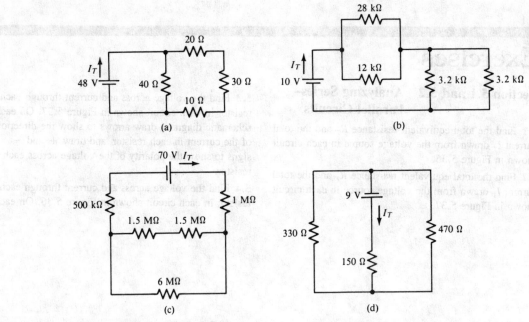

(a)

(b)

(c)

(d)

FIGURE 5.37 (Exercise 5.2)

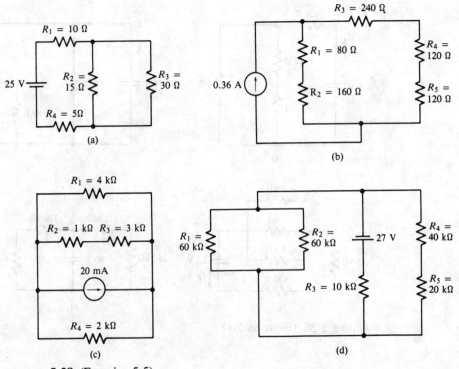

(a)

(b)

(c)

(d)

FIGURE 5.38 (Exercise 5.5)

schematic diagram, draw arrows to show the direction of the current in each resistor, and draw + and − signs to show the polarity of the voltage across each resistor.

5.5 Find the power dissipated in resistor R_1 and the voltage across resistor R_2 in each circuit in Figure 5.38.

5.6 Find the power dissipated in resistor R_3 and the current through resistor R_4 in each circuit shown in Figure 5.38.

5.7 Find V_{ab} in the circuit shown in Figure 5.39.

5.8 Find V_{ab} in the circuit shown in Figure 5.40.

5.9 Find the unknown quantities in each circuit shown in Figure 5.41.

5.10 Find the unknown quantities in each circuit shown in Figure 5.42.

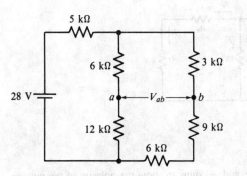

FIGURE 5.39 (Exercise 5.7)

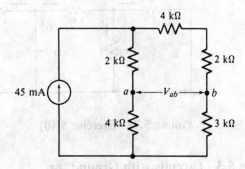

FIGURE 5.40 (Exercise 5.8)

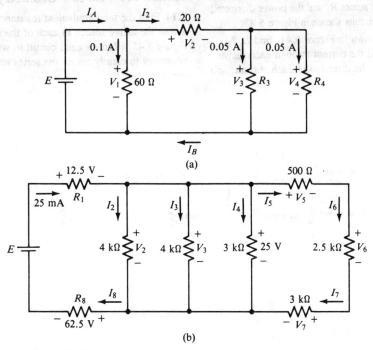

(a)

(b)

FIGURE 5.41 (Exercise 5.9)

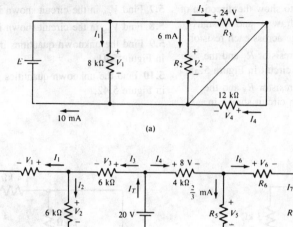

(a)

(b)

FIGURE 5.42 (Exercise 5.10)

Section 5.3   Circuits with Ground, or Common, Connections

5.11 Find the voltage across R_1 and the power dissipated by R_2 in each circuit shown in Figure 5.43.

5.12 Find the voltage across R_3 and the power dissipated by R_4 in each of the circuits shown in Figure 5.43.

5.13 In the circuit shown in Figure 5.44, find E, I_T, the voltage across and the current through each resistor. Draw arrows to show the direction of each current, and

draw + and − signs to show the polarity of the voltage drop across each resistor.

Section 5.4   Circuit Geometry

5.14 Find the total equivalent resistance connected across the active source in each of the circuits shown in Figure 5.45. (Redraw each circuit to whatever extent is necessary to clearly reveal the series and parallel connections.)

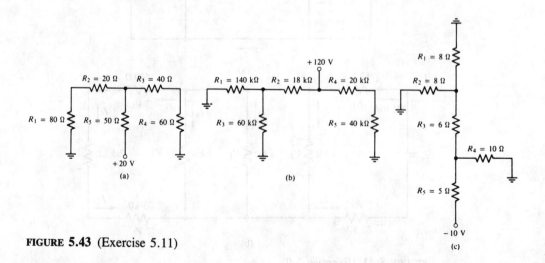

FIGURE 5.43 (Exercise 5.11)

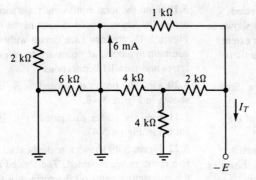

FIGURE 5.44 (Exercise 5.13)

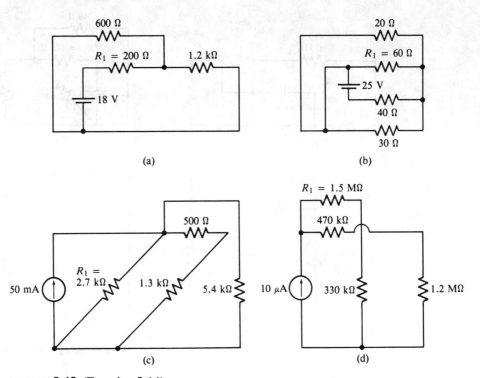

(a)

(b)

(c)

(d)

FIGURE 5.45 (Exercise 5.14)

5.15 Find the total equivalent resistance connected across the active source in each of the circuits shown in Figure 5.46. (Redraw each circuit to whatever extent is necessary to clearly reveal the series and parallel connections.)

5.16 Find the current through R_1 in each of the circuits shown in Figure 5.46.

5.17 Find the voltage drop across R_1 in each of the circuits shown in Figure 5.45.

5.18 Find the total equivalent resistance connected across the active source in each of the circuits shown in Figure 5.47. Redraw each circuit without using common symbols, as necessary to reveal clearly the series and parallel connections.

5.19 Find the total equivalent resistance connected across the active source in each of the circuits shown in Figure 5.48. Redraw each circuit without using common symbols, as necessary to reveal clearly the series and parallel connections.

5.20 Find the power dissipated by R_1 in each circuit shown in Figure 5.48.

5.21 Find the power dissipated by R_1 in each circuit shown in Figure 5.47.

5.22 Figure 5.49 shows a printed-circuit board on which five resistors are mounted. The shaded areas represent the conducting paths on the surface of the board. Find the total current drawn from the 20-V supply.

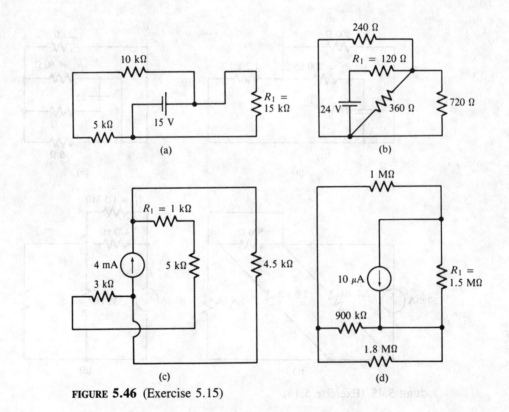

FIGURE 5.46 (Exercise 5.15)

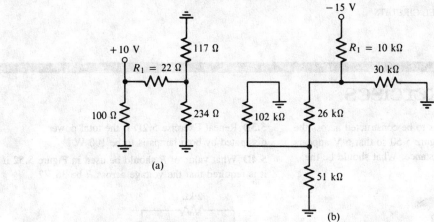

(a)

(b)

FIGURE 5.47 (Exercise 5.18)

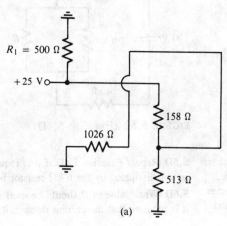

(a)

(b)

FIGURE 5.48 (Exercise 5.19)

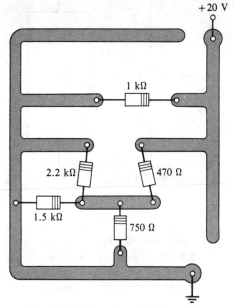

FIGURE 5.49 (Exercise 5.22)

Design Exercises

5.1D A voltage divider is to be constructed across the 12-V source shown in Figure 5.50 so that 6 V appears across the 2-kΩ load resistance. What should be the value of *R*?

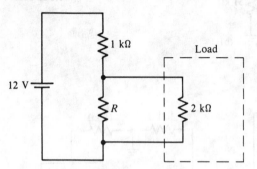

FIGURE 5.50 (Exercise 5.1D)

5.2D The two lamps shown in Figure 5.51(a) together draw a total current of 1.2 A from the 120-V source. What value of resistance *R* should be inserted in series, as shown in Figure 5.51(b), in order to limit the total current drawn from the source to 1 A?

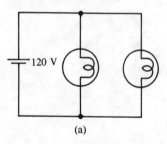

(a)

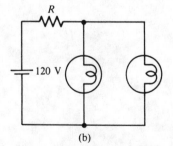

(b)

FIGURE 5.51 (Exercise 5.2D)

5.3D Repeat Exercise 5.2D if the total power dissipated by both lamps is to be 100 W.

5.4D What value of *R* should be used in Figure 5.52 if it is required that the voltage across *R* be 36 V?

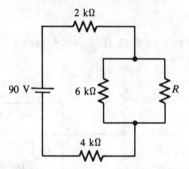

FIGURE 5.52 (Exercise 5.4D)

5.5D Repeat Exercise 5.4D if it is required that the power dissipated by the 6-kΩ resistor be 150 mW.

5.6D What value of *R* should be used in Figure 5.53 if it is required that the current through it be 0.3 A?

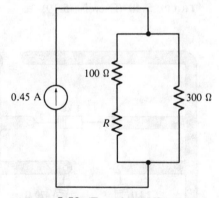

FIGURE 5.53 (Exercise 5.6D)

5.7D Repeat Exercise 5.6D if it is required that the power dissipated by the 100-Ω resistor be 2.25 W.

5.8D The current in adjustable resistance *R* in Figure 5.54 must not be allowed to exceed 6 mA. What is the minimum permissible value of *R*?

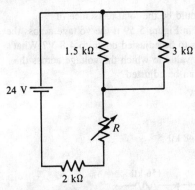

FIGURE 5.54 (Exercise 5.8D)

5.9D The voltage across adjustable resistance R in Figure 5.55 must not be allowed to exceed 250 V. What is the maximum permissible value of R?

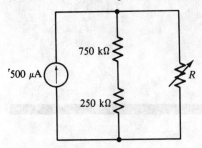

FIGURE 5.55 (Exercise 5.9D)

5.10D Find the value of R that should be used in Figure 5.56 in order that the voltage divider supply 12 V to the two parallel-connected loads.

5.11D Find the value of E in order that the current through the 1.18-kΩ resistor in Figure 5.57 be 4 mA.

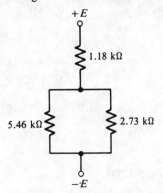

FIGURE 5.57 (Exercise 5.11D)

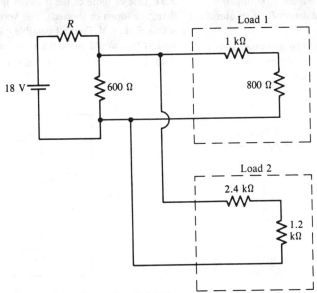

FIGURE 5.56 (Exercise 5.10D)

5.12D In the transistor shown in Figure 5.58, $I_C = 120 I_B$. What should be the value of V_{CC} to make $V_{CE} = 0$ when $I_B = 50$ μA?

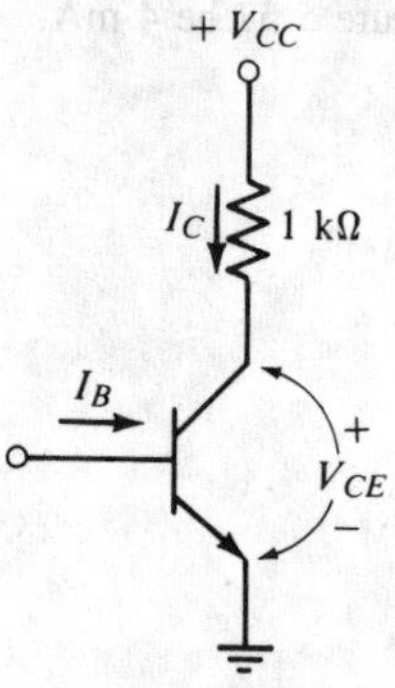

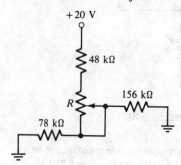

FIGURE 5.58 (Exercise 5.12D)

5.13D What should be the total resistance of potentiometer R in Figure 5.59 if the voltage across the 48-kΩ resistor is to be adjusted down to 4.8 V? What is the maximum value to which the voltage across the 48-kΩ resistor can be adjusted?

FIGURE 5.59 (Exercise 5.13D)

Troubleshooting Exercises

5.1T In each of the circuits of Figure 5.60, find the voltage that would be measured across R_3 if resistor R_1 failed by becoming a short.

5.2T Find the voltage that would be measured across R_3 in each of the circuits shown in Figure 5.60 if resistor R_2 failed by becoming open.

5.3T One or more of the resistors in Figure 5.61 is defective (open or shorted). The voltage measured across R_3 is 9 V. Find all possible combinations of defects that would account for that measurement.

5.4T Repeat Exercise 5.3T if the voltage measured across R_3 is 12 V.

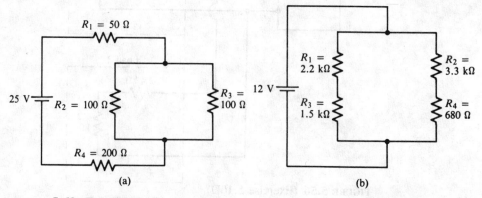

(a)

(b)

FIGURE 5.60 (Exercise 5.1T)

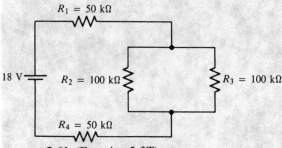

FIGURE 5.61 (Exercise 5.3T)

5.5T One or more of the resistors in Figure 5.62 is defective (open or shorted). The voltage measured across R_4 is 4 V. Find all possible combinations of defects that would account for that measurement.

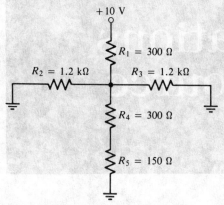

FIGURE 5.62 (Exercise 5.5T)

5.6T Repeat Exercise 5.5T if the voltage measured across R_4 is 10 V.

5.7T The 120-V source shown in Figure 5.63 is protected by a 6-A fuse. If the total current reaches or exceeds 6 A, the fuse will "blow" (open) and thus disconnect the source from the rest of the circuit. Assuming that only one resistor becomes shorted at any one time, which resistors will cause the fuse to blow if they become shorted?

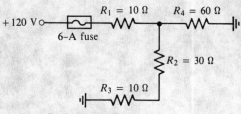

FIGURE 5.63 (Exercise 5.7T)

5.8T Repeat Exercise 5.7T if the 6-A fuse is replaced by a 4-A fuse.

5.9T Figure 5.64 shows a printed-circuit board on which the conducting paths are shown shaded. When the board was tested, it was found that the voltage across R_4 measured 12 V. On the figure, draw a line between two sections of the conducting paths showing where a short must have occurred.

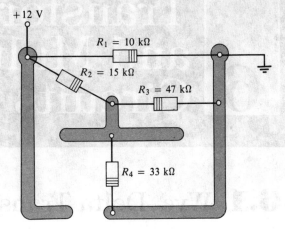

FIGURE 5.64 (Exercise 5.9T)

5.10T Repeat Exercise 5.9T if it was discovered that the board was defective by measuring a total current of 2 mA from the voltage source.

SPICE Exercises

5.1S Use SPICE to find the total current drawn from the voltage source and the current through and voltage across each resistance in the circuit of Figure 5.37(b), p. 144.

5.2S Repeat Exercise 5.1S for the circuit in Figure 5.38(d), p. 144.

5.3S Repeat Exercise 5.1S for the circuit in Figure 5.46(c) (p. 148), except instead of finding the total current, find the voltage across the terminals of the current source.

5.4S Repeat Exercise 5.1S for the circuit in Figure 5.48(a), p. 149.

Network Transformations and Multisource Circuits

6

6.1 Wye–Delta Transformations

Some electrical circuits have no components in series and no components in parallel. Figure 6.1 shows an example. These circuits cannot be analyzed directly using the methods of Chapter 5, because they cannot be reduced to simpler circuits containing the equivalent resistance of series or parallel combinations. However, in many cases it is possible to *transform* a portion of the circuit in such a way that the resulting configuration does contain series- and parallel-connected components. The transformation produces an equivalent circuit, in the sense that voltages and currents in the other (untransformed) components remain the same. Therefore, once the circuit has been transformed, voltages and currents in the unaffected components can be determined using conventional series–parallel analysis methods.

Circuits that must be analyzed using the transformation method often contain components connected in either a *wye* (Y) or a *delta* (Δ) configuration. Figure 6.2(a)–(c) show three ways that a wye configuration might appear in a circuit. Because the wye-connected components may appear in the equivalent form shown in Figure 6.2(b), the arrangement is also called a *tee* (T) configuration. Figure 6.2(d)–(f) show equivalent

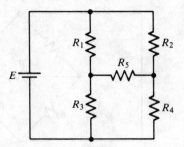

FIGURE 6.1 Example of a circuit that cannot be analyzed by combining series- or parallel-connected components. Notice that no two components are in series and that no two components are in parallel.

delta forms. Because the delta arrangement may appear in the equivalent form shown in Figure 6.2(f), it is also called a *pi* (Π) configuration. The figure shows only a few of the ways the wye and delta networks might be drawn in a schematic diagram. Many equivalent forms can be drawn by rotating these basic configurations through various angles. Notice that each network has three terminals.

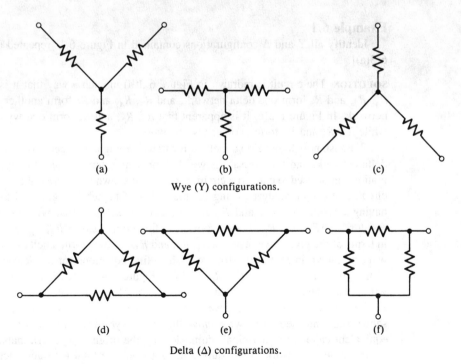

(a)   (b)   (c)

Wye (Y) configurations.

(d)   (e)   (f)

Delta (Δ) configurations.

FIGURE 6.2 Wye and delta networks.

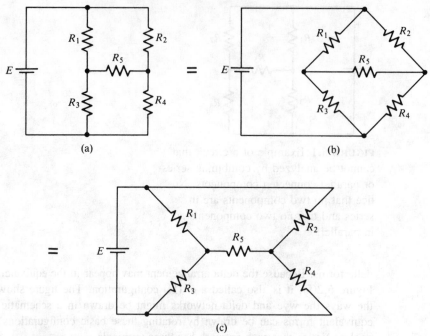

FIGURE 6.3 (Example 6.1)

Example 6.1

Identify all Y and Δ configurations contained in Figure 6.1 [repeated as Figure 6.3(a)].

SOLUTION The circuit is redrawn in Figure 6.3(b) in such a way that it is obvious that R_1, R_2, and R_5 form one delta network, and R_3, R_4, and R_5 form another delta network. In Figure 6.3(c) it is apparent that R_1, R_3, and R_5 form one wye network, while R_2, R_4 and R_5 form another wye network.

We wish now to develop a method for transforming a wye network into an equivalent delta network, and vice versa. As we shall presently demonstrate, this type of transformation can be used to convert a circuit such as that shown in Figure 6.3 to a series–parallel circuit that can be analyzed using the methods of Chapter 5. Figure 6.4(a) shows a wye having resistors R_1, R_2, and R_3 and the equivalent delta that will replace it, having resistors R_A, R_B, and R_C. Thus, we wish to find equations for R_A, R_B, and R_C expressed in terms of the given wye resistors R_1, R_2, and R_3. To transform a delta into an equivalent wye, as shown in Figure 6.4(b), we wish to find equations for R_1, R_2, and R_3 in terms of the given delta resistors R_A, R_B, and R_C. Figure 6.4(c) shows how one configuration will replace the other. Notice that the three terminals of each network, labeled X, Y, and Z, all coincide. For example, if we wish to replace the wye network by the delta network shown superimposed on it, we *remove* the given wye (discard it) and replace it with an equivalent delta whose terminals coincide with the original wye terminals.

For a delta network to be equivalent to a wye, the total resistance across each pair of terminals belonging to one network must be the same as the resistance across the

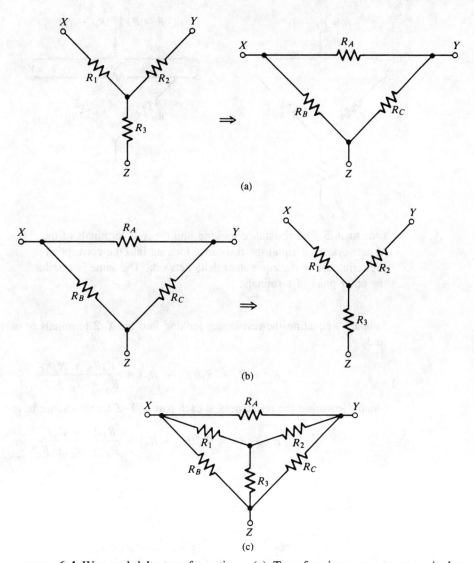

FIGURE 6.4 Wye and delta transformations. (a) Transforming a wye to an equivalent delta. (b) Transforming a delta to an equivalent wye. (c) Superimposing wye and delta networks to determine how one replaces the other.

corresponding pair of terminals in the other network. Referring to Figure 6.5, the resistance "looking into" the X–Y terminals of the wye network is $R_1 + R_2$, because one end of R_3 is open. The corresponding resistance acorss the X–Y terminals of the delta network is seen to be $R_A \| (R_B + R_C)$. Thus, for equivalency we require

$$R_1 + R_2 = R_A \| (R_B + R_C) = \frac{R_A R_B + R_A R_C}{R_A + R_B + R_C} \quad (6.1)$$

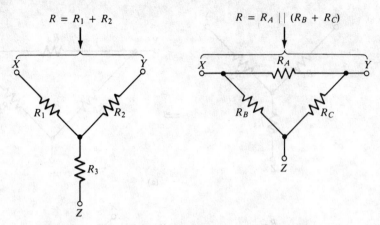

FIGURE 6.5 The resistance looking into the *X–Y* terminals of the Y network must equal the resistance looking into the corresponding terminals of the equivalent delta network. The same is true for the other pairs of terminals.

Similarly, equating the resistances looking into the *X–Z* terminals of each network, we find

$$R_1 + R_3 = R_B\|(R_A + R_C) = \frac{R_A R_B + R_B R_C}{R_A + R_B + R_C} \tag{6.2}$$

Finally, equating the resistances at each pair of *Y–Z* terminals, we have

$$R_2 + R_3 = R_C\|(R_A + R_B) = \frac{R_A R_C + R_B R_C}{R_A + R_B + R_C} \tag{6.3}$$

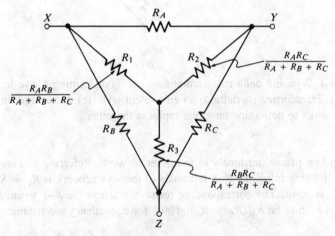

FIGURE 6.6 Transformation of a given delta network to an equivalent wye. Notice the pattern in the equations.

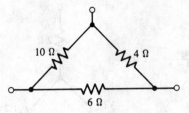

FIGURE 6.7 (Example 6.2)

Solving Equations (6.1), (6.2), and (6.3) simultaneously for R_1, R_2, and R_3 (Exercise 6.3) gives

$$R_1 = \frac{R_A R_B}{R_A + R_B + R_C} \qquad (6.4)$$

$$R_2 = \frac{R_A R_C}{R_A + R_B + R_C} \qquad (6.5)$$

$$R_3 = \frac{R_B R_C}{R_A + R_B + R_C} \qquad (6.6)$$

Equations (6.4), (6.5), and (6.6) are used to convert a given delta network having resistors R_A, R_B, and R_C into an equivalent wye network. There is no point in memorizing these equations. Instead, notice the pattern, as illustrated in Figure 6.6. Observe that each equation has the same denominator, namely, the sum of the delta resistors. To find the wye resistor that connects to terminal X, divide *the product of the two delta resistors connected to X* by the sum of the delta resistors. Similarly, the wye resistor that connects to terminal Y is the product of the two delta resistors connected to Y, divided by the sum. The same scheme is used to find the wye resistor that connects to terminal Z.

If the pattern illustrated in Figure 6.6 is understood, there is no need to assign designations R_A, R_B, . . . , R_1, R_2, and so on, to any of the resistors involved in a transformation. The next example illustrates this point.

Example 6.2

Convert the delta network shown in Figure 6.7 to an equivalent wye network.

SOLUTION Refer to Figure 6.8. We first draw the desired wye inside the given delta, as shown in Figure 6.8(a). The computations are performed following the pattern illustrated in Figure 6.6, and the original delta is discarded. The result, shown in Figure 6.8(b), is the equivalent wye network.

Drill Exercise 6.2

Find the resistor values in a wye network that is equivalent to a delta containing three 12-kΩ resistors.

ANSWER: 4 kΩ each.   □

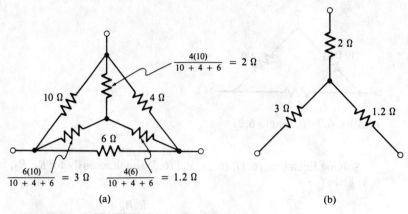

$$\frac{4(10)}{10 + 4 + 6} = 2 \ \Omega$$

10 Ω   4 Ω

6 Ω

$$\frac{6(10)}{10 + 4 + 6} = 3 \ \Omega \qquad \frac{4(6)}{10 + 4 + 6} = 1.2 \ \Omega$$

(a)

2 Ω

3 Ω   1.2 Ω

(b)

FIGURE 6.8 (Example 6.2)

Equations (6.4), (6.5), and (6.6) can be solved simultaneously for R_A, R_B, and R_C to give

$$R_A = \frac{R_1 R_2 + R_2 R_3 + R_1 R_3}{R_3} \tag{6.7}$$

$$R_B = \frac{R_1 R_2 + R_2 R_3 + R_1 R_3}{R_2} \tag{6.8}$$

$$R_C = \frac{R_1 R_2 + R_2 R_3 + R_1 R_3}{R_1} \tag{6.9}$$

Equations (6.7), (6.8), and (6.9) are used to convert a given wye into an equivalent delta. Once again, notice the pattern. Every numerator is the sum of the products of all

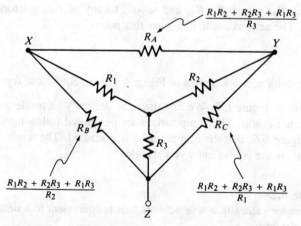

$$\frac{R_1 R_2 + R_2 R_3 + R_1 R_3}{R_3}$$

X R_A Y

R_1 R_2

R_B R_C

R_3

$$\frac{R_1 R_2 + R_2 R_3 + R_1 R_3}{R_2} \qquad \qquad \frac{R_1 R_2 + R_2 R_3 + R_1 R_3}{R_1}$$

Z

FIGURE 6.9 Transformation of a given wye network to an equivalent delta. Notice the pattern in the equations.

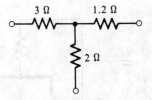

FIGURE 6.10 (Example 6.3)

distinct pairs of wye resistors. The denominator in each equation is the wye resistor lying *opposite* the desired delta resistor. The pattern can be seen in Figure 6.9.

Example 6.3

Convert the T network shown in Figure 6.10 to an equivalent delta network.

SOLUTION Notice that the given T network has the same resistors obtained from the transformation illustrated in Figure 6.8 (Example 6.2). Therefore, when we transform this T back to a Δ, we should obtain the same resistors that we started with in the delta of Figure 6.7. Figure 6.11 shows the computations and confirms that result. Note that the first step is to draw the required delta around the given T. We then perform the computations observing the pattern described earlier.

Drill Exercise 6.3

Find the resistor values in a delta network that is equivalent to a wye containing three 120-Ω resistors.

ANSWER: 360 Ω each.   □

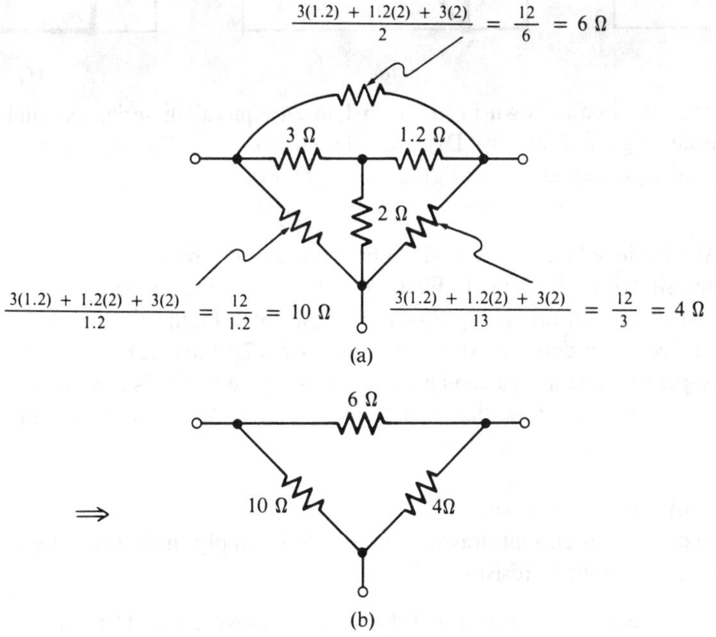

$$\frac{3(1.2) + 1.2(2) + 3(2)}{2} = \frac{12}{6} = 6 \ \Omega$$

3 Ω   1.2 Ω

2 Ω

$$\frac{3(1.2) + 1.2(2) + 3(2)}{1.2} = \frac{12}{1.2} = 10 \ \Omega \qquad \frac{3(1.2) + 1.2(2) + 3(2)}{13} = \frac{12}{3} = 4 \ \Omega$$

(a)

6 Ω

10 Ω   4Ω

⟹

(b)

FIGURE 6.11 (Example 6.3)

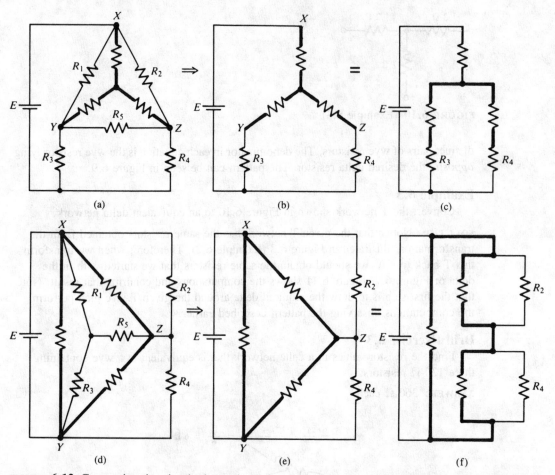

FIGURE 6.12 Converting the circuit shown in Figure 6.1 to an equivalent series–parallel circuit. (a) Draw new wye inside original delta. (b) Discard original delta. (c) Equivalent to (b). (d) Draw new delta around original wye. (e) Discard original wye. (f) Equivalent to (e).

As mentioned earlier, wye–delta transformations are useful for analyzing circuits like the one shown in Figure 6.1. Figure 6.12(a) shows that same circuit with an equivalent wye superimposed on the R_1–R_2–R_5 delta [see also Figure 6.3(b)]. When the equivalent wye replaces the delta, as shown in Figures 6.12(b) and (c), we see that the result is a series–parallel circuit that can be analyzed using the methods of Chapter 5. Alternatively, Figure 6.12(d)–(f) show that transforming the R_1–R_3–R_5 wye into an equivalent delta also produces a series–parallel circuit.

Example 6.4   (Analysis)

Find the total current drawn from the 48-V supply in Figure 6.13 and the voltage drop across the 40-Ω resistor.

SOLUTION Since we wish to find the voltage across the 40-Ω resistor, any wye–delta transformation we perform must not involve that resistor. In other words, the 40-Ω

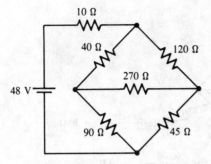

FIGURE 6.13 (Example 6.4)

resistor must remain intact. Figure 6.14(a) and (b) show the conversion of the circuit to series–parallel form by transformation of the lower delta to an equivalent wye. This choice does not affect the 40-Ω resistor. Figure 6.14(c)–(e) show how conventional series–parallel circuit analysis is used to determine the total resistance and total current:

$$I_T = \frac{E}{R_T} = \frac{48 \text{ V}}{80 \text{ }\Omega} = 0.6 \text{ A}$$

In Figure 6.14(f) we use the current-divider rule to determine the portion of the 0.6-A total current that flows in the branch containing the 40-Ω resistor:

$$I = \left(\frac{150}{100 + 150}\right)(0.6 \text{ A}) = 0.36 \text{ A}$$

Finally, Figure 6.14(g) shows that the 0.36-A current in the 40-Ω resistor produces the voltage drop

$$V = (0.36 \text{ A})(40 \text{ }\Omega) = 14.4 \text{ V}$$

Drill Exercise 6.4

Find the voltage across the 90-Ω resistor in Figure 6.13.

ANSWER: 27.45 V.   □

When all three resistors of a wye or delta network are equal, it is particularly easy to transform the network to one of the opposite type. The transformed network will also contain three equal resistors. Let

$$R_\Delta = \text{the value of the three equal delta resistors}$$
$$R_Y = \text{the value of the three equal wye resistors}$$

Then, using equations (6.4) through (6.6), it is easy to show (Exercise 6.4) that

$$R_Y = \frac{R_\Delta}{3} \tag{6.10}$$

or

$$R_\Delta = 3R_Y \tag{6.11}$$

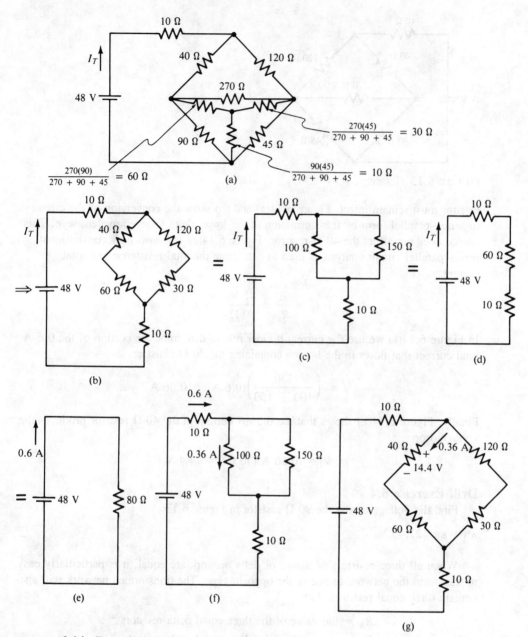

$$\frac{270(90)}{270 + 90 + 45} = 60 \ \Omega$$

$$\frac{270(45)}{270 + 90 + 45} = 30 \ \Omega$$

$$\frac{90(45)}{270 + 90 + 45} = 10 \ \Omega$$

(a)

(b)

(c)

(d)

(e)

(f)

(g)

FIGURE 6.14 (Example 6.4)

Example 6.5 (Design)

A Π network is to be constructed as shown in Figure 6.15(a) so that the resistance R_{XZ} looking into the X-Z terminals (with Y-Z open) equals the resistance R_{YZ} looking into the Y-Z terminals (with X-Z open). If that resistance must equal 1 kΩ, find the value of R_Δ that should be used in the Π network.

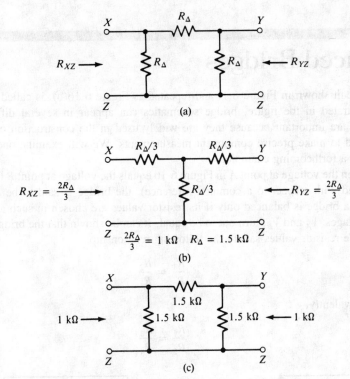

FIGURE 6.15 (Example 6.5)

SOLUTION As shown in Figure 6.15(b), a wye–delta transformation produces a T network that, by equation (6.10), has the equal-valued resistors $R_\Delta/3$. It is clear from this figure that

$$R_{XZ} = R_{YZ} = \frac{R_\Delta}{3} + \frac{R_\Delta}{3} = \frac{2R_\Delta}{3}$$

Therefore,

$$1 \text{ k}\Omega = \frac{2R_\Delta}{3}$$

or

$$R_\Delta = 1.5 \text{ k}\Omega$$

Thus, the Π network must have three 1.5-kΩ resistors, as shown in Figure 6.15(c).

Drill Exercise 6.5

What value of resistance connected across terminals *Y-Z* (in parallel with the 1.5-kΩ resistor) in Figure 6.15(c) will make R_{XZ} equal to 875 Ω?

ANSWER: 1 kΩ.

6.2 Balanced Bridges

The circuit shown in Figure 6.1, and repeated as Figure 6.16(b), is called a *bridge*. As demonstrated in the figure, bridge schematics can appear in several different forms. Bridges are important because they are widely used in the construction of instruments designed to make precise component measurements. We will examine one such instrument in a forthcoming example.

When the voltage at point A in Figure 6.16 equals the voltage at point B (both voltages measured with respect to a common reference), the bridge is said to be *balanced*. Of course, a bridge is balanced only if its resistor values are chosen in such a way that the two voltages, V_A and V_B, turn out to be equal. It can be shown that the bridge is balanced when the resistor values satisfy the following relationship:

$$\frac{R_1}{R_3} = \frac{R_2}{R_4} \tag{6.12}$$

or, equivalently,

$$\frac{R_1}{R_2} = \frac{R_3}{R_4} \tag{6.13}$$

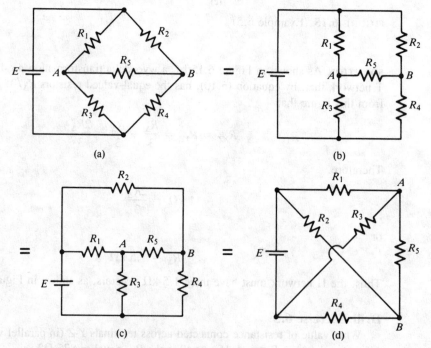

FIGURE 6.16 Equivalent forms of bridge circuits.

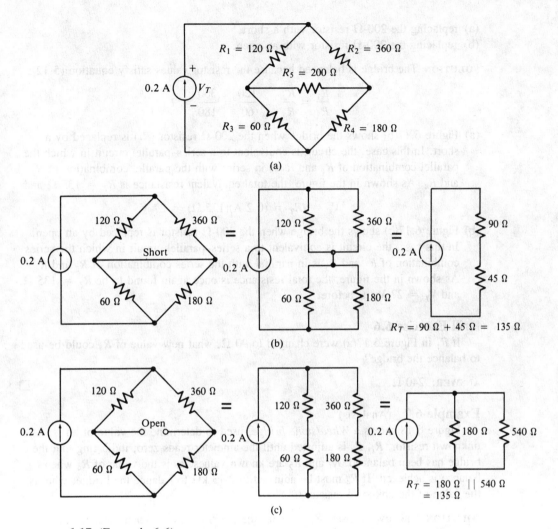

FIGURE 6.17 (Example 6.6)

When a bridge is balanced, $V_A = V_B$, so the voltage *across* R_5 in Figure 6.16 is zero: $V_{AB} = V_A - V_B = 0$. When there is zero voltage across R_5, there is also zero current through it. Consequently, in a balanced bridge R_5 *can be replaced by either a short circuit or an open circuit* without affecting the voltages or currents anywhere else in the circuit. This fact is very helpful in analyzing a balanced bridge, because it eliminates the necessity for performing a wye–delta transformation, as illustrated in the next example.

Example 6.6 (Analysis)

Verify that the bridge shown in Figure 6.17(a) is balanced. Then find the voltage V_T across the 0.2-A current source by

(a) replacing the 200-Ω resistor with a short,

(b) replacing the 200-Ω resistor with an open.

SOLUTION The bridge is balanced because the resistor values satisfy equation (5.12):

$$\frac{R_1}{R_3} = \frac{R_2}{R_4} \qquad \frac{120}{60} = \frac{360}{180} = 2$$

(a) Figure 6.17(b) shows the bridge when the 200-Ω resistor (R_5) is replaced by a short. In this case, the circuit is equivalent to a series–parallel circuit in which the parallel combination of R_1 and R_2 is in series with the parallel combination of R_3 and R_4. As shown in the figure, the total equivalent resistance is $R_T = 135\ \Omega$ and

$$V_T = IR_T = (0.2\ \text{A})(135\ \Omega) = 27\ \text{V}$$

(b) Figure 6.17(c) shows the bridge when the 200-Ω resistor is replaced by an open. In this case, the circuit is equivalent to a series–parallel circuit in which the series combination of R_1 and R_3 is in parallel with the series combination of R_2 and R_4. As shown in the figure, the total resistance is once again found to be $R_T = 135\ \Omega$, and $V_T = 27\ \text{V}$, as before.

Drill Exercise 6.6

If R_1 in Figure 6.17(a) were changed to 90 Ω, what new value of R_4 could be used to balance the bridge?

ANSWER: 240 Ω.   ▫

Example 6.7   (Analysis)

Figure 6.18 shows a *Wheatstone bridge*, used to determine the value of an unknown resistor, R_x. R_4 is adjusted until the ammeter reads zero, indicating that the bridge has been balanced. R_1 and R_2 are known values, as is the value of R_4 when balance is achieved. If R_4 must be adjusted to 2.73 kΩ to balance the bridge, what is the value of the unknown resistor?

SOLUTION Unknown resistor R_x occupies the position of R_3. When balance is achieved, we have, from equation (6.13),

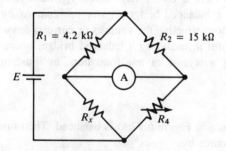

FIGURE 6.18 (Example 6.7)

$$R_x = R_3 = \frac{R_1}{R_2} R_4 = \left(\frac{4.2 \text{ k}\Omega}{15 \text{ k}\Omega}\right) 2.73 \text{ k}\Omega = 764.4 \ \Omega$$

Drill Exercise 6.7

If the unknown resistor in Figure 6.18 had been 910 Ω, what value of R_4 would have been necessary to balance the bridge?

ANSWER: 3250 Ω.

6.3 Voltage and Current Source Conversions

Real Sources

In Chapter 3 we discussed the fact that a real (nonideal) voltage source has *internal resistance* that causes its terminal voltage to decrease when current is drawn from it. A real voltage source can be represented as an ideal voltage source in series with a resistance equal to its internal resistance, R_{int}, as shown in Figure 6.19. This representation can be used to calculate the true terminal voltage of a source when current is drawn from it. Understand, however, that internal resistance is an inherent property of a source; it is not a discrete component that can be measured with an ohmmeter. Only the external terminals are accessible to a user, so any and all measurements must be made at those terminals.

When a *load resistance* R_L is connected across the terminals, the terminal voltage can be computed using conventional analysis methods. The next example illustrates this computation.

Example 6.8 (Analysis)

A 12-V voltage source has an internal resistance of 2 Ω.

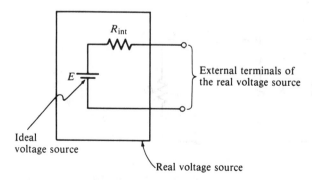

FIGURE 6.19 Representation of a real voltage source as an ideal voltage source E in series with internal resistance.

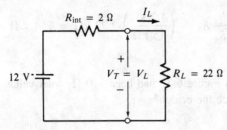

FIGURE 6.20 (Example 6.8)

(a) Find the terminal voltage when a 22-Ω load is connected across its terminals.
(b) Find the current that flows in the load.

SOLUTION

(a) Figure 6.20 shows the source represented as an ideal 12-V source in series with
the 2 Ω of internal resistance. When the 22-Ω load is connected, the result is
a simple series circuit, and the terminal voltage can be found using the voltage-
divider rule:

$$V_T = \left(\frac{22}{22 + 2}\right) 12 \text{ V} = 11 \text{ V}$$

Note that the terminal voltage is the same as the voltage across the 22-Ω load, so
that voltage is also called the *load voltage*, V_L.

(b) $I_L = \dfrac{V_L}{R_L} = \dfrac{11 \text{ V}}{22 \ \Omega} = 0.5 \text{ A}$

Drill Exercise 6.8

If the load voltage across the 22-Ω resistor in Figure 6.20 were measured to be
10.56 V, what would be the internal resistance of the voltage source?

ANSWER: 3 Ω.

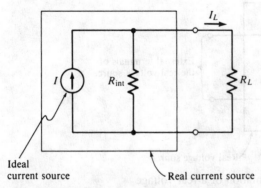

Ideal
current source

Real current source

FIGURE 6.21 Representation of a real current
source.

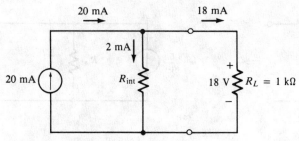

FIGURE 6.22 (Example 6.9)

A real (nonideal) current source can be represented as an ideal current source in *parallel* with internal resistance, as shown in Figure 6.21. When load resistance R_L is connected across the terminals, the current produced by the source divides between R_{int} and R_L. Consequently, the load current is less than it would be if the source were ideal. Note that an ideal current source would have *infinite* internal resistance (i.e., R_{int} in Figure 6.21 would be replaced by an open circuit). In that case, all of the source current would be delivered to the load.

Example 6.9   (Analysis)

When a 1-kΩ load is connected across a 20-mA current source, it is found that only 18 mA flows in the load. What is the internal resistance of the source?

SOLUTION As shown in Figure 6.22, the current source with the 1-kΩ load connected can be analyzed as a simple parallel circuit. The voltage across the load is

$$V_L = I_L R_L = (18 \text{ mA})(1 \text{ k}\Omega) = 18 \text{ V}$$

By Kirchhoff's current law, the current in R_{int} is

$$I_{int} = 20 \text{ mA} - 18 \text{ mA} = 2 \text{ mA}$$

Since R_{int} is in parallel with R_L, the voltage across R_{int} is the same as V_L (i.e., 18 V). Therefore,

$$R_{int} = \frac{V_L}{I_{int}} = \frac{18 \text{ V}}{2 \text{ mA}} = 9 \text{ k}\Omega$$

Drill Exercise 6.9

What would be the terminal voltage of the current source in Figure 6.22 (with R_L connected) if the internal resistance of the source were 1.5 kΩ?

ANSWER: 12 V.   □

Source Conversions

A real voltage source can be converted to an *equivalent* real current source, and vice versa. When the conversion is made, the sources are equivalent in every sense of the word: It is impossible to make any measurement or perform any test at the external

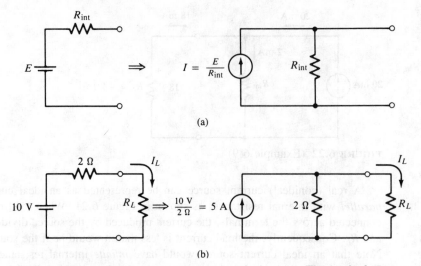

(a)

(b)

FIGURE 6.23 Voltage-to-current source conversion. (a) The equivalent sources have the same value of internal resistance. E and I are related by Ohm's law. (b) Notice that the polarities of the equivalent sources are such that current flows through external load R_L in the same direction.

terminals that would reveal whether either source is a voltage source or its equivalent current source.

Figure 6.23 shows how a voltage source is converted to an equivalent current source. The first step is to draw the current source, as shown in Figure 6.23(a). Notice that the internal resistance of the equivalent current source has the same value as the internal resistance of the original voltage source. Thus, the next step is simply to label the (parallel) internal resistance with that value. Finally, compute the equivalent current using Ohm's law: $I = E/R_{int}$. This last step is easy to remember: Compute current the only way it is possible to compute current using the given quantities (i.e., the given voltage divided by the given resistance). Figure 6.23(b) shows an example. This example demonstrates that the polarity of the equivalent current source (current flowing up) produces current in load R_L that flows in the same direction as the current produced in R_L by the voltage source (downward). The polarity of the equivalent current source must always be such that it produces current in an external load in the same direction as the original voltage source produces current.

Figure 6.24(a) shows how a current source is converted to an equivalent voltage source. In this case, the first step is to draw the voltage source. Once again, the series resistance of the voltage source has the same value as the parallel resistance of the original current source. The value of the equivalent voltage is computed using Ohm's law: $E = IR_{int}$. Figure 6.24(b) shows an example. Notice again that current is produced in the same direction through an external load, in this case, upward.

The reason one type of source is converted to the other type is that it often simplifies the analysis of a circuit containing both types. Also, some of the analysis techniques we will study later require that all sources be of the same type. Any resistance that is in

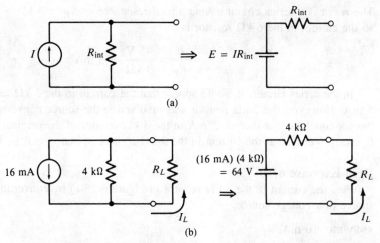

(a)

(b)

FIGURE 6.24 Current-to-voltage source conversion. (a) The equivalent sources have the same value of internal resistance. E and I are related by Ohm's law. (b) The polarities of the equivalent sources are such that current flows through external load R_L in the same direction.

series with a voltage source, whether it be internal or external resistance, can be included in its conversion to an equivalent current source. Similarly, any resistance in parallel with a current source can be included when it is converted to an equivalent voltage source. However, the voltage across or current through any such resistance cannot be computed if that resistance is included in the source conversion. The next example illustrates these points.

Example 6.10   (Analysis)

Find the current in the 6-kΩ resistor in Figure 6.25(a) by converting the current source to a voltage source.

SOLUTION Since we want to find the current in the 6-kΩ resistor, we use the 3-kΩ resistor to convert the current source to an equivalent voltage source. Thus, as shown in Figure 6.25(c),

$$E = IR = (15 \text{ mA})(3 \text{ k}\Omega) = 45 \text{ V}$$

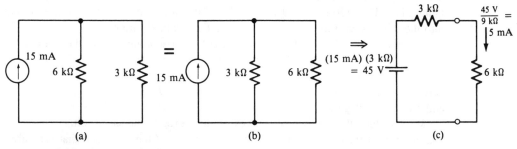

(a)   (b)   (c)

FIGURE 6.25 (Example 6.10)

The result is a series circuit having a total resistance of $R_T = 3 \text{ k}\Omega + 6 \text{ k}\Omega = 9 \text{ k}\Omega$, so the current in the 6-kΩ resistor is

$$I = \frac{E}{R_T} = \frac{45 \text{ V}}{9 \text{ k}\Omega} = 5 \text{ mA}$$

In the series circuit, it would appear that the current in the 3-kΩ resistor is also 5 mA. However, the 3-kΩ resistor was involved in the source conversion, so we *cannot* conclude that there is 5 mA in the 3-kΩ resistor of the original circuit [Figure 6.25(a)]. Verify that the current in the 3-kΩ resistor in that circuit is, in fact, 10 mA.

Drill Exercise 6.10

Find the current in the 3-kΩ resistor in Figure 6.25(a) by converting the current source to a voltage source.

ANSWER: 10 mA. ☐

The next example demonstrates that a voltage source is indistinguishable from its equivalent current source, insofar as currents and voltages at its external terminals are concerned.

Example 6.11

Show that the equivalent sources in Figure 6.26(a) have exactly the same terminal voltage and produce exactly the same external current when the terminals

(a) are shorted
(b) are open
(c) have a 500-Ω load connected

SOLUTION
(a) See Figure 6.26(b). The terminal voltage is 0 V in both circuits because the terminals are shorted. The current produced by the voltage source is determined by Ohm's law: $I = 15 \text{ V}/500 \ \Omega = 30 \text{ mA}$. The same current flows in the shorted terminals of the current source because the short diverts all of the source current around the 500-Ω resistor.
(b) See Figure 6.26(c). The voltage across the open terminals of the voltage source is 15 V because no current flows and there is no drop across the 500-Ω resistor. The voltage across the open terminals of the current source is also 15 V, by Ohm's law: $V = (30 \text{ mA})(500 \ \Omega) = 15 \text{ V}$. The current flowing from one terminal into the other is zero in both cases, because the terminals are open.
(c) See Figure 6.26(d). The voltage across the 500-Ω load connected to the voltage source is found by the voltage-divider rule: $V_L = (500/1000)15 \text{ V} = 7.5 \text{ V}$. The load current is found by Ohm's law: $I_L = 7.5 \text{ V}/500 \ \Omega = 15 \text{ mA}$. In the current source, the source current divides equally, so $I_L = (\frac{1}{2})(30 \text{ mA}) = 15 \text{ mA}$ and the load voltage is found by Ohm's law: $V_L = (15 \text{ mA})(500 \ \Omega) = 7.5 \text{ V}$.

We conclude that equivalent sources produce exactly the same voltages and currents at their external terminals, no matter what the load, and that they are therefore indistinguishable.

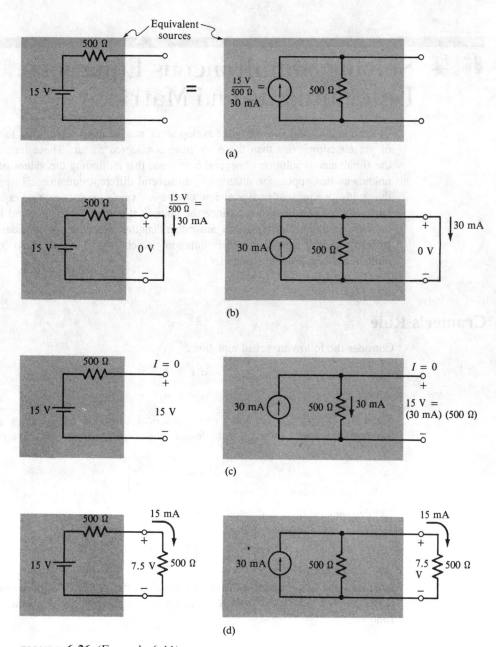

FIGURE 6.26 (Example 6.11)

Drill Exercise 6.11

Show that the same value of resistance must be connected across the terminals of the equivalent sources in Figure 6.26(a) to produce a terminal voltage of 10 V.

ANSWER: 1 kΩ in each case.

6.4 Solving Simultaneous Equations: Determinants and Matrices

In Sections 6.5 and 6.6 we will develop some new analysis techniques to solve circuits of greater complexity than those we have considered so far. These techniques require the simultaneous solution of several equations, that is, finding the values of one or more unknowns that appear simultaneously in several different equations. In preparation for that study, we present here a review of *Cramer's rule* for solving two equations simultaneously for two unknowns. Since the numerical computations required to solve three or more simultaneous equations are rather lengthy, we will not consider those cases. Instead, we will introduce *matrix* notation, which is now widely used for solving simultaneous equations by computer.

Cramer's Rule

Consider the following set of equations:

$$a_1 x + b_1 y = c_1 \qquad (6.14)$$

$$a_2 x + b_2 y = c_2 \qquad (6.15)$$

where a_1, a_2, b_1, b_2, c_1, and c_2 are numerical constants and x and y are variables (unknowns) whose values are to be found. The *coefficient matrix* is the array

$$\mathbf{A} = \begin{bmatrix} a_1 & b_1 \\ a_2 & b_2 \end{bmatrix}$$

The *determinant* of $\mathbf{A}$ is defined to be

$$\det \mathbf{A} = \det \begin{bmatrix} a_1 & b_1 \\ a_2 & b_2 \end{bmatrix} = a_1 b_2 - a_2 b_1 \qquad (6.16)$$

Note that the determinant is a number equal to the product of the constants on one diagonal of $\mathbf{A}$ subtracted from the product of the constants on the other diagonal, as follows:

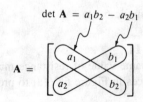

$$\det \mathbf{A} = a_1 b_2 - a_2 b_1$$

$$\mathbf{A} = \begin{bmatrix} a_1 & b_1 \\ a_2 & b_2 \end{bmatrix}$$

Example 6.12

Find the determinant of the coefficient matrix for the equations

$$3x - 4 = 2y$$
$$y + 2x = -5$$

SOLUTION We must first rearrange the equations so they appear in the format of equations (6.14) and (6.15), with the x-coefficients in the first column and the y-coefficients in the second column:

$$3x - 2y = 4$$
$$2x + y = -5$$

Then

$$\mathbf{A} = \begin{bmatrix} 3 & -2 \\ 2 & 1 \end{bmatrix}$$

and

$$\det \mathbf{A} = (3)(1) - (2)(-2) = 3 + 4 = 7$$

Drill Exercise 6.12

Find the determinant of the coefficient matrix corresponding to the equations $x = 2y$ and $y = 4 - x$.

ANSWER: 3. □

The determinant of the coefficient matrix is given the special symbol Δ (delta):

$$\Delta = \det \mathbf{A} = \det \begin{bmatrix} a_1 & b_1 \\ a_2 & b_2 \end{bmatrix}$$

According to Cramer's rule, the solutions for x and y are found by performing the following computations:

$$x = \frac{\det \begin{bmatrix} c_1 & b_1 \\ c_2 & b_2 \end{bmatrix}}{\Delta} \tag{6.17}$$

$$y = \frac{\det \begin{bmatrix} a_1 & c_1 \\ a_2 & c_2 \end{bmatrix}}{\Delta} \tag{6.18}$$

Note that the numerator of (6.17) is the determinant of the matrix that results when the constants c_1 and c_2 [see equations (6.14) and (6.15)] are substituted into the *first* column of coefficient matrix $\mathbf{A}$. Similarly, the numerator of (6.18) is the determinant of the matrix that results when the constants c_1 and c_2 are substituted into the *second* column of $\mathbf{A}$.

Of course, the unknowns can be *any* two variables, not just x and y. The way to remember which column of $\mathbf{A}$ is replaced by c_1 and c_2 is to observe that the coefficients of the first (leftmost) variable appearing in the simultaneous equations are substituted into the first column of $\mathbf{A}$ in order to solve for the first unknown. Similarly, the coefficients of the second variable in the equations are substituted into the second column of $\mathbf{A}$ to solve for the second unknown. The next example illustrates this procedure.

Example 6.13

Using Cramer's rule, solve the following equations simultaneously for I_1 and I_2:

$$3I_2 - 5 = I_2 - 3I_1$$
$$2(I_1 + I_2) = 6 + 6I_2$$

SOLUTION We must first rearrange the equations, simplify, and collect terms, so that the I_1 coefficients are in one column and the I_2 coefficients are in the other column. Performing the necessary algebra leads to

$$3I_1 + 2I_2 = 5$$
$$2I_1 - 4I_2 = 6$$

The determinant of the coefficient matrix is

$$\Delta = \det \mathbf{A} = \det \begin{bmatrix} 3 & 2 \\ 2 & -4 \end{bmatrix} = (3)(-4) - (2)(2) = -16$$

Note that $c_1 = 5$ and $c_2 = 6$. By Cramer's rule [equations (6.17) and (6.18)],

first
unknown   — substitute c_1 and c_2 in first column
$\downarrow$

$$I_1 = \dfrac{\det \begin{bmatrix} 5 & 2 \\ 6 & -4 \end{bmatrix}}{\Delta} = \dfrac{(5)(-4) - (6)(2)}{-16} = \dfrac{-32}{-16} = 2$$

and

second
unknown   — substitute c_1 and c_2 in second column
$\downarrow$

$$I_2 = \dfrac{\det \begin{bmatrix} 3 & 5 \\ 2 & 6 \end{bmatrix}}{\Delta} = \dfrac{(6)(3) - (2)(5)}{-16} = \dfrac{8}{-16} = -0.5$$

Drill Exercise 6.13

Using Cramer's rule, solve the following equations simultaneously for V_1 and V_2:
$3V_1 - 2V_2 = 2V_2 - 29$; $6V_2 + 2V_1 = 27 + V_1$.

ANSWER: $V_1 = -3$; $V_2 = 5$.   □

Matrix Formulation

As noted earlier, simultaneous equations having more than two unknowns are usually solved using a computer, because of the complexity of the computations involved. For computer formulation, a set of simultaneous equations must be expressed as a single matrix equation. To understand how a set of equations can be expressed using matrices, we must introduce some fundamental definitions from the theory of *linear algebra*.

An m by n ($m \times n$) matrix is an array of numbers having m rows and n columns. For example, the coefficient matrix $\mathbf{A}$ that we defined in connection with two simultaneous equations is a 2×2 matrix because it has two rows of numbers and two columns of numbers. A *square* matrix has the same number of rows as columns. The 2×2 coefficient matrix is an example. An *identity* matrix is a square matrix having the number 1 in every diagonal position, and 0 everywhere else. Following are examples of 2×2 and 3×3 identity matrices:

$$\mathbf{I}_{(2 \times 2)} = \begin{bmatrix} 1 & 0 \\ 0 & 1 \end{bmatrix} \qquad \mathbf{I}_{(3 \times 3)} = \begin{bmatrix} 1 & 0 & 0 \\ 0 & 1 & 0 \\ 0 & 0 & 1 \end{bmatrix}$$

Some matrices, called *column matrices*, have only a single column. For example, we can regard the two constants c_1, c_2 in equations (6.14) and (6.15) as the 2×1 column matrix

$$\mathbf{C} = \begin{bmatrix} c_1 \\ c_2 \end{bmatrix}$$

Notice that $\mathbf{C}$ has two rows and one column.

Two matrices $\mathbf{A}$ and $\mathbf{B}$ can be multiplied to produce a third (product) matrix $\mathbf{AB}$, *provided* that the number of columns of the first (left-hand) matrix equals the number of rows of the second (right-hand) matrix. For example, if $\mathbf{A}$ is a 2×2 matrix and $\mathbf{B}$ is a 2×1 matrix, it is possible to form the product matrix $\mathbf{AB}$. The product matrix has the same number of rows as $\mathbf{A}$ and the same number of columns as $\mathbf{B}$. It may be permissible to form the product $\mathbf{AB}$, but not the product $\mathbf{BA}$. Following is an example:

$$\begin{array}{ccc} \mathbf{A} & \mathbf{B} & = & \mathbf{C} \\ (3 \times 3) & (3 \times 1) & & (3 \times 1) \end{array}$$

(equal)

$$\begin{array}{cc} \mathbf{B} & \mathbf{A} \\ (3 \times 1) & (3 \times 3) \end{array}$$

Not equal → product not defined

In our study of circuit analysis, it will not be necessary to know how the values in the product matrix are obtained.

Note that it is permissible to multiply any $m \times m$ square matrix by an $m \times m$ identity matrix. We can form both the products $\mathbf{AI}$ and $\mathbf{IA}$. An important property of $\mathbf{I}$

is that

$$AI = IA = A \tag{6.19}$$

The *inverse* of any square matrix A is a square matrix A^{-1} that satisfies

$$AA^{-1} = A^{-1}A = I \tag{6.20}$$

An inverse can only be defined for a square matrix, and then only if the determinant of the matrix is nonzero.

Let us now express a set of two simultaneous equations in matrix form. We will let the unknowns be x_1 and x_2:

$$a_1x_1 + b_1x_2 = c_1 \tag{6.21}$$

$$a_2x_1 + b_2x_2 = c_2 \tag{6.22}$$

The coefficient matrix is

$$A = \begin{bmatrix} a_1 & b_1 \\ a_2 & b_2 \end{bmatrix}$$

Let the matrix X of unknowns be the column matrix

$$X = \begin{bmatrix} x_1 \\ x_2 \end{bmatrix}$$

Let the matrix C of constants c_1 and c_2 be the column matrix

$$C = \begin{bmatrix} c_1 \\ c_2 \end{bmatrix}$$

Then equations (6.1) and (6.2) are written in matrix form as

$$\underset{(2 \times 2)}{A} \underset{(2 \times 1)}{X} = \underset{(2 \times 1)}{C} \tag{6.23}$$

We will now show that the solution matrix X containing the values of the unknowns x_1 and x_2 can be obtained by multiplying both sides of 6.23 by A^{-1}:

$$\underbrace{A^{-1}A}X = A^{-1}C$$

$$IX = A^{-1}C \tag{6.24}$$

$$X = A^{-1}C$$

This very important result expresses the fact that the matrix X containing the solutions to a set of two simultaneous equations is found by multiplying the inverse of the coefficient matrix by the matrix C of constants. Note that the multiplication is permissible and consistent:

$$\underset{(2 \times 1)}{X} = \underset{(2 \times 2)}{A^{-1}} \underset{(2 \times 1)}{C} \tag{6.25}$$

The same procedure can be applied to any set of m simultaneous equations containing m unknowns:

$$\underset{(m \times 1)}{\mathbf{X}} = \underset{(m \times m)}{\mathbf{A}^{-1}} \underset{(m \times 1)}{\mathbf{C}} \tag{6.26}$$

In each case, we assume that the equation *can* be solved, which means that it is *possible* to find the inverse matrix $\mathbf{A}^{-1}$. The principal computational task of a computer programmed to solve simultaneous equations is computing the inverse of the coefficient matrix.

Example 6.14

Express the following set of three simultaneous equations in matrix form. The unknowns are V_1, V_2, and V_3. Define each matrix, and write the matrix equation for the solution matrix.

$$V_2 + 2V_3 = 13 - 3V_1$$

$$2(V_1 + V_2 - V_3) + (V_3 - 2V_2 + V_1) - 2 = 0$$

$$2V_1 - V_3 = -4(1 + V_2)$$

SOLUTION We must first rearrange the equations, simplify, and collect terms, so that the V_1, V_2, and V_3 coefficients are aligned in their respective columns. Performing the necessary algebra leads to

$$3V_1 + V_2 + 2V_3 = 13$$

$$3V_1 \quad\quad - V_3 = 2$$

$$2V_1 + 4V_2 - V_3 = -4$$

The 3×3 coefficient matrix is then

$$\mathbf{A} = \begin{bmatrix} 3 & 1 & 2 \\ 3 & 0 & -1 \\ 2 & 4 & -1 \end{bmatrix}$$

Note that there is no V_2 term in the second equation, so zero is entered in that position of the matrix. The $\mathbf{V}$ matrix and the matrix of constants are

$$\mathbf{V} = \begin{bmatrix} V_1 \\ V_2 \\ V_3 \end{bmatrix} \quad\quad \mathbf{C} = \begin{bmatrix} 13 \\ 2 \\ -4 \end{bmatrix}$$

In matrix form, we write

$$\begin{bmatrix} 3 & 1 & 2 \\ 3 & 0 & -1 \\ 2 & 4 & -1 \end{bmatrix} \begin{bmatrix} V_1 \\ V_2 \\ V_3 \end{bmatrix} = \begin{bmatrix} 13 \\ 2 \\ -4 \end{bmatrix}$$

or

$$AV = C$$

The solution matrix is then computed by

$$V = A^{-1}C$$

As a point of interest, solving this problem by computer reveals that

$$A^{-1} = \begin{bmatrix} 0.1081 & 0.2432 & -0.0270 \\ 0.0270 & -0.1892 & 0.2432 \\ 0.3243 & -0.2702 & -0.0811 \end{bmatrix}$$

and

$$V = \begin{bmatrix} 2 \\ -1 \\ 4 \end{bmatrix}$$

Thus, the solution is $V_1 = 2$, $V_2 = -1$, and $V_3 = 4$.

Drill Exercise 6.14

Write the coefficient matrix and the constant matrix for the following set of equations: $3(I_1 - 2I_2) + 10 = 14 - I_3$; $I_2 = 16I_3$; $I_3 - 5(I_1 - I_3) = -21 + 2I_2$.

ANSWER:

$$A = \begin{bmatrix} 3 & -6 & 1 \\ 0 & 1 & -16 \\ -5 & -2 & 6 \end{bmatrix} \qquad C = \begin{bmatrix} 4 \\ 0 \\ -21 \end{bmatrix} \qquad \square$$

To illustrate how a computer is programmed to perform matrix computations, following is a typical BASIC program for solving the equations given in Example 6.14.

```
10 DIM A(3,3), B(3,3), V(3), C(3)
20 MAT READ A
30 MAT READ C
40 MAT B = INV(A)
50 MAT V = B*C
60 MAT PRINT V
70 DATA 3, 1, 2, 3, 0, -1, 2, 4, -1
80 DATA 13, 2, -4
90 END
```

As can be seen in the program, BASIC statements that involve matrix operations contain the prefix MAT. Statement 10 dimensions all the matrices that are used in the program. Statements 20 and 30 assign the values given in the DATA statements to matrices **A** and **C**. Note that the data values are read from left to right across the first row of **A**, then across the second row, and so on. Statement 40 computes the inverse of **A** and sets it equal to $\mathbf{B}$ (i.e., $\mathbf{B} = \mathbf{A}^{-1}$). Statement 50 computes the solution matrix: $\mathbf{V} = \mathbf{B} * \mathbf{C} = \mathbf{A}^{-1} * \mathbf{C}$. Finally, statement 60 prints the solution matrix.

Most microcomputers are not capable of performing the matrix operations illustrated above. However, most mini and mainframe computers can be programmed to perform matrix operations.

6.5 Mesh Analysis

If all the active sources in a circuit are either series-connected voltage sources or parallel-connected current sources, we can reduce the circuit to one that has a single equivalent source, as discussed in Chapter 4. However, it is often the case that the sources in a multisource circuit are neither in series nor in parallel. Figure 6.27 shows an example.

There are special techniques for analyzing multisource circuits of the type shown in Figure 6.27, one of which is called *mesh* analysis. This technique requires that we write Kirchhoff's voltage law around two or more loops and solve the resulting equations simultaneously. The unknowns are the so-called *loop currents*. Once the loop currents are known, we can find the actual current in any component and the voltage across it.

Mesh analysis requires that all the sources in a circuit be voltage sources. Therefore, if a circuit contains any current sources, the first step is to convert them to equivalent voltage sources.

The next step is to draw closed loops in the circuit, each loop representing a path around which Kirchhoff's voltage law will be written. The direction of each loop is arbitrary: It may be either clockwise or counterclockwise. It is convenient to think of each loop as representing a current that flows around the loop, and we designate each by an appropriate symbol: I_1, I_2 and so on. These loop currents are the unknowns in the set of simultaneous equations that result when Kirchhoff's voltage law is written around each loop. Thus, the number of unknowns (loop currents) is the same as the number of equations.

The method is best understood by way of an example. Purely for convenience and for the sake of consistency, *we will hereafter draw all loops in a clockwise direction*. In the example that follows, take special note of how we represent the *net* current in a component that appears in *two* loops.

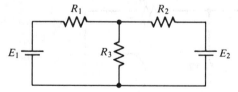

FIGURE 6.27 Voltage sources E_1 and E_2 are neither in series nor in parallel, so they cannot be combined into a single equivalent source.

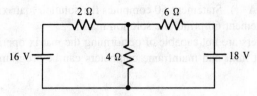

FIGURE 6.28 (Example 6.15)

Example 6.15 (Analysis)

Use mesh analysis to find the current in each resistor in Figure 6.28.

SOLUTION In Figure 6.29(a), we show two clockwise loops and the corresponding loop currents, I_1 and I_2. By Ohm's law, the voltage drop across the 2-Ω resistor is $(2\ \Omega)(I_1) = 2I_1$, with the polarity shown. Similarly, the voltage drop across the 6-Ω resistor is $6I_2$.

Note that the 4-Ω resistor is in both loops. Also note that I_1 flows downward through that resistor, while I_2 flows through it in the *opposite* direction: upward. When writing Kirchhoff's voltage law around loop 1, we regard the *net* current in the 4-Ω resistor to be $I_1 - I_2$, as shown in Figure 6.29(b). Consequently, the voltage drop in loop 1 is $4(I_1 - I_2)$. However, in loop 2, we regard the net current in the 4-Ω resistor to be $I_2 - I_1$. Thus, in loop 2, the voltage drop is $4(I_2 - I_1)$. This distinction is

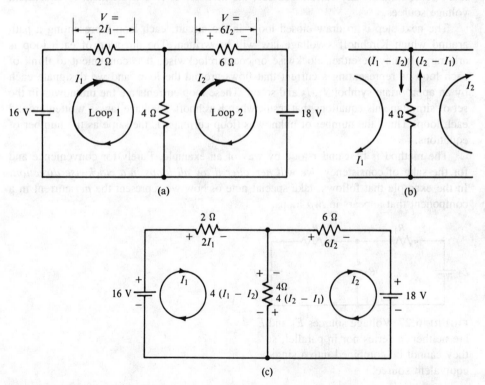

(a)

(b)

(c)

FIGURE 6.29 (Example 6.15)

very important. If we had drawn I_1 clockwise and I_2 counterclockwise, I_1 and I_2 would have been in the *same* direction through the 4-Ω resistor, and the net current would have been $I_1 + I_2$ in both loops.

Figure 6.29(c) shows the voltage across each component in each loop. When writing Kirchhoff's voltage law around a loop, recall that we treat the voltage across a component as a *rise* if we pass through it from $-$ to $+$, and we treat it as a *drop* if we pass through it from $+$ to $-$. Applying this principle, we write the two equations corresponding to Kirchhoff's voltage law around the two loops:

$$\overbrace{\hspace{2.5cm}}^{\text{drops}} \qquad = \overbrace{\hspace{1cm}}^{\text{rises}}$$

$$\text{loop 1:} \quad 2I_1 + 4(I_1 - I_2) \qquad = \quad 16 \tag{6.27}$$

$$\text{loop 2:} \quad 6I_2 + 4(I_2 - I_1) + 18 = \quad 0 \tag{6.28}$$

Simplifying and rearranging equations (6.27) and (6.28) leads to

$$6I_1 - 4I_2 = 16$$

$$-4I_1 + 10I_2 = -18$$

The determinant of the coefficient matrix is

$$\Delta = \det \begin{bmatrix} 6 & -4 \\ -4 & 10 \end{bmatrix} = (6)(10) - (-4)(-4) = 44$$

By Cramer's rule,

$$I_1 = \frac{\det \begin{bmatrix} 16 & -4 \\ -18 & 10 \end{bmatrix}}{\Delta} = \frac{(16)(10) - (-18)(-4)}{44} = 2 \text{ A}$$

$$I_2 = \frac{\det \begin{bmatrix} 6 & 16 \\ -4 & -18 \end{bmatrix}}{\Delta} = \frac{6(-18) - (-4)(16)}{44} = -1 \text{ A}$$

Since I_1 is the current in the 2-Ω resistor, we see that the current in that resistor equals 2 A. Notice that I_2 equals -1 A. As discussed in Chapter 4, we conclude that *minus* 1 A flows from left to right through the 6-Ω resistor (the assumed direction of I_2), or, equivalently, that *plus* 1 A flows from right to left (the opposite direction of I_2):

$$\begin{array}{ccc} -1 \text{ A} & & +1 \text{ A} \\ \longrightarrow & & \longleftarrow \\ -\!\!\wedge\!\!\wedge\!\!\wedge\!\!- & = & -\!\!\wedge\!\!\wedge\!\!\wedge\!\!- \\ 6\,\Omega & & 6\,\Omega \end{array}$$

We can find the current in the 4-Ω resistor in any one of three equivalent ways (see Figure 6.30). First, since we assumed in *loop 1* that the net current in the 4-Ω resistor was $I_1 - I_2$, we have, as shown in Figure 6.30(a),

$$I_{4\Omega} = I_1 - I_2 = 2 \text{ A} - (-1 \text{ A}) = 3 \text{ A}$$

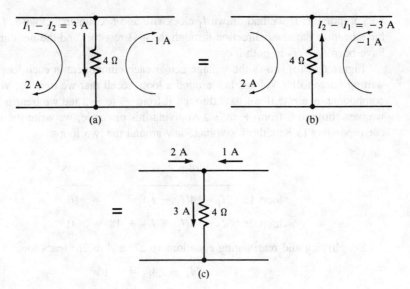

FIGURE 6.30 (Example 6.15)

We conclude that 3 A flows *downward* through the 4-Ω resistor (i.e., in the same direction as loop 1 current). We also assumed in *loop 2* that the net current in the 4-Ω resistor was $I_2 - I_1$. Therefore, as shown in Figure 6.30(b),

$$I_{4\Omega} = I_2 - I_1 = -1 \text{ A} - (2 \text{ A}) = -3 \text{ A}$$

This result tells us that -3 A flows through the 4-Ω resistor in the same direction as I_2, that is, upward. But -3 A upward is the same as $+3$ A downward, so we have the same result as before. The third and final way we can find $I_{4\Omega}$ is simply to apply Kirchhoff's current law at the junction shown in Figure 6.30(c). We know that I_1 is a positive 2 A entering the junction, and I_2 is a positive 1 A entering the junction, so $I_{4\Omega}$ is the current leaving the junction:

$$I_{4\Omega} = 2 \text{ A} + 1 \text{ A} = 3 \text{ A}$$

Drill Exercise 6.15

Repeat Example 6.15 when the polarity of the 16-V source in Figure 6.28 is reversed and the positions of the 4-Ω and 6-Ω resistors are interchanged. Draw arrows showing the direction of each positive current.

ANSWER: $I_2 = 6.09 \overleftarrow{\text{A}}$; $I_4 = 5.45 \overleftarrow{\text{A}}$; $I_6 = 0.64 \text{ A} \uparrow$. ☐

Example 6.16 (Analysis)

Find the voltage drop across each resistor in Figure 6.31.

SOLUTION Figure 6.32(a) shows the clockwise loop currents I_1 and I_2 and the voltages across each component. As in Example 6.15, the net current in the 2.2-kΩ resistor is $I_1 - I_2$ in loop 1 and $I_2 - I_1$ in loop 2. Note carefully that the 6-V source is a voltage drop in loop 1, but is a voltage rise in loop 2.

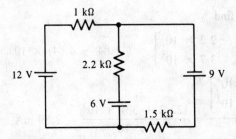

FIGURE 6.31 (Example 6.16)

Writing Kirchhoff's current law around each loop, we obtain

$$\text{loop 1:} \quad 10^3 I_1 + 2.2 \times 10^3 (I_1 - I_2) + 6 \quad = 12$$
$$\text{loop 2:} \quad 1.5 \times 10^3 I_2 + 2.2 \times 10^3 (I_2 - I_1) = 6 + 9$$

Simplifying and collecting terms gives

$$3.2 \times 10^3 I_1 - 2.2 \times 10^3 I_2 = 6$$
$$-2.2 \times 10^3 I_1 + 3.7 \times 10^3 I_2 = 15$$

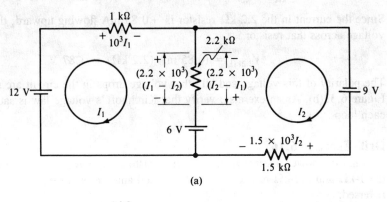

(a)

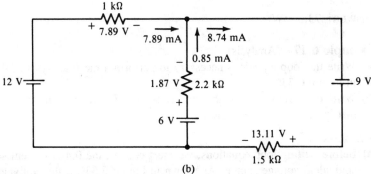

(b)

FIGURE 6.32 (Example 6.16)

Applying Cramer's rule, we find

$$\Delta = \det \begin{bmatrix} 3.2 \times 10^3 & -2.2 \times 10^3 \\ -2.2 \times 10^3 & 3.7 \times 10^3 \end{bmatrix} = (11.84 - 4.84) \times 10^6 = 7 \times 10^6$$

$$I_1 = \frac{\det \begin{bmatrix} 6 & -2.2 \times 10^3 \\ 15 & 3.7 \times 10^3 \end{bmatrix}}{\Delta} = \frac{(22.2 + 33) \times 10^3}{7 \times 10^6} = 7.89 \text{ mA}$$

$$I_2 = \frac{\det \begin{bmatrix} 3.2 \times 10^3 & 6 \\ -2.2 \times 10^3 & 15 \end{bmatrix}}{\Delta} = \frac{(48 + 13.2) \times 10^3}{7 \times 10^6} = 8.74 \text{ mA}$$

The current in the 2.2-kΩ resistor is

$$I_{2.2\text{k}\Omega} = I_1 - I_2 = 7.89 \text{ mA} - 8.74 \text{ mA} = -0.85 \text{ mA}$$

We conclude that 0.85 mA flows *upward* through the 2.2-kΩ resistor.

The voltage drop across each resistor is found using Ohm's law:

$$V_{1\text{k}\Omega} = (I_1)(1 \text{ k}\Omega) = (7.89 \text{ mA})(1 \text{ k}\Omega) = 7.89 \text{ V}$$

$$V_{1.5\text{k}\Omega} = (I_2)(1.5 \text{ k}\Omega) = (8.74 \text{ mA})(1.5 \text{ k}\Omega) = 13.11 \text{ V}$$

Since the current in the 2.2-kΩ resistor is +0.85 mA flowing upward, the (positive) voltage across that resistor is

$$V_{2.2\text{k}\Omega} = (0.85 \text{ mA})(2.2 \text{ k}\Omega) = 1.87 \text{ V}$$

The polarity of this voltage and all other voltage drops in the circuit are shown in Figure 6.32(b). As an exercise, verify that Kirchhoff's voltage law is satisfied around each loop.

Drill Exercise 6.16

Find the power dissipated in the 1-kΩ resistor in Figure 6.31 when the positions of the 1-kΩ and 2.2-kΩ resistors are interchanged and the polarity of the 12-V source is reversed.

ANSWER: 73.6 mW. ◻

Example 6.17   (Analysis)

(a) Write the loop equations necessary to perform a mesh analysis of the circuit shown in Figure 6.33(a).
(b) Write the matrix form of the loop equations. Define each matrix and write the matrix equation for the solution.

SOLUTION

(a) Before writing loop equations, we must convert the 0.4-A current source to an equivalent voltage source. As shown in Figure 6.33(b), the equivalent voltage is

$$E = (0.4 \text{ A})(60 \text{ }\Omega) = 24 \text{ V}$$

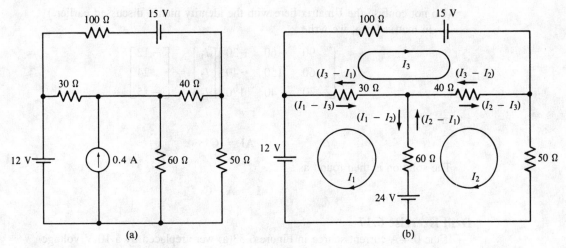

FIGURE 6.33 (Example 6.17)

Note that the polarity of the equivalent voltage source is such that it produces current in the same direction as the original current source.

The figure shows the three loop currents and the net currents in the resistors that appear in more than one loop. For example, the net current in the 40-Ω resistor is $I_2 - I_3$ when writing Kirchhoff's voltage law around loop 2, and is $I_3 - I_2$ when writing the law around loop 3. The three equations are

$$\text{loop 1:} \quad (I_1 - I_3)30 + (I_1 - I_2)60 + 24 = 12$$

$$\text{loop 2:} \quad (I_2 - I_1)60 + (I_2 - I_3)40 + 50I_2 = 24$$

$$\text{loop 3:} \quad 100I_3 + (I_3 - I_2)40 + (I_3 - I_1)30 + 15 = 0$$

Simplifying and collecting terms gives

$$90I_1 - 60I_2 - 30I_3 = -12$$

$$-60I_1 + 150I_2 - 40I_3 = 24$$

$$-30I_1 - 40I_2 + 170I_3 = -15$$

(b) The coefficient matrix is

$$\mathbf{A} = \begin{bmatrix} 90 & -60 & -30 \\ -60 & 150 & -40 \\ -30 & -40 & 170 \end{bmatrix}$$

The solution matrix $\mathbf{I}$ and the constant matrix $\mathbf{C}$ are the 3×1 column matrices

$$\mathbf{I} = \begin{bmatrix} I_1 \\ I_2 \\ I_3 \end{bmatrix} \qquad \mathbf{C} = \begin{bmatrix} -12 \\ 24 \\ -15 \end{bmatrix}$$

(Do not confuse the **I** matrix here with the identity matrix discussed earlier.)

In matrix form, we write

$$
\begin{bmatrix}
90 & -60 & -30 \\
-60 & 150 & -40 \\
-30 & -40 & 170
\end{bmatrix}
\begin{bmatrix}
I_1 \\
I_2 \\
I_3
\end{bmatrix}
=
\begin{bmatrix}
-12 \\
24 \\
-15
\end{bmatrix}
$$

or

$$
\mathbf{AI = C}
$$

The solution is then found as

$$
\mathbf{I = A^{-1}C}
$$

Drill Exercise 6.17

If the 0.4-A current source in Figure 6.33(a) were replaced by a 10-V voltage source in series with a 20-Ω resistor, how many rows and how many columns would be in the coefficient matrix resulting from a mesh analysis of the circuit?

ANSWER: 4×4.   □

6.6 Nodal Analysis

Another method used to analyze multisource circuits is called *nodal analysis*. As we shall presently demonstrate, this method requires that we solve the simultaneous equations that result when Kirchhoff's *current* law is applied at various *nodes* in a circuit. A node is any portion of a circuit where component terminals are electrically common. Practically speaking, a node is all of that region where two or more component terminals are joined, soldered, wired, or otherwise connected through negligibly small resistance. On a schematic diagram, a node is all of the solid connection line that can be drawn between component terminals without passing through any component. In Figure 6.34, several methods are used to identify the nodes in each of three different circuits. We can locate all the nodes in a circuit by repeatedly entering the schematic diagram at different junctions and tracing out all the connection lines that are common to that junction. (Think of a node as all of the path over which molten solder would flow without passing through or around any component.) As shown in Figure 6.34(c), a node is usually represented simply by a number, with the understanding that it refers to all the terminals that are electrically common at a given point.

In nodal analysis, we select one node as a *reference* node, with respect to which the voltages at all other nodes are measured. Thus, the reference node serves as a *ground*, or common, for the circuit. Notice that the voltage measured at any one node is independent of the point on the node where it is measured; that is, all terminals joined to a node are at the same voltage level with respect to the reference node.

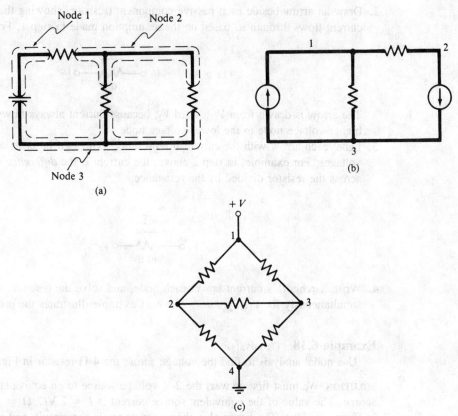

FIGURE 6.34 Identifying the nodes on a schematic diagram. (a) Nodes encircled and drawn heavy. (b) Nodes drawn heavy. (c) Nodes numbered at junctions.

The first step in performing a nodal analysis is to *convert all voltage sources to equivalent current sources*. The next step is to identify all the nodes in the circuit and to choose a reference node. This choice is arbitrary, but it is usually convenient to select the reference node as the one having the most components connected to it. All the nodes except the reference node are then numbered, and their corresponding voltages are designated V_1, V_2, ... Note that the reference node is at zero volts. Our task is to find the values of the other node voltages.

As mentioned earlier, we obtain simultaneous equations involving the unknowns V_1, V_2, . . . , by applying Kirchhoff's current law at each node. Thus, we obtain the same number of equations as unknowns. To apply the current law, we must know (or assume) which currents are *entering* each node and which currents are *leaving*. Following is *one* systematic procedure for making that determination (there are other, equivalent, procedures):

1. Make an (arbitrary) assumption about the relative magnitudes of the node voltages. In this book, we will always assume that $V_1 > V_2 > V_3 > \cdots > 0$. Note that all node voltages are assumed to be greater than the reference voltage (zero).

2. Draw an arrow beside each passive component (resistor) showing the direction that current flows through it, based on the assumption made in step 1. For example:

$$(V_1 > V_2) \qquad V_1 \; \mathrm{o} \!\!-\!\!\!\bigwedge\!\!\!-\!\! \mathrm{o} \, V_2$$
$$10 \; \Omega$$

The arrow is drawn from V_1 toward V_2, because current always flows from the higher-voltage node to the lower-voltage node.

3. Label each arrow with the current it represents, expressed in terms of the node voltages. For example, in step 2 above, the current is the *difference* in voltage across the resistor divided by the resistance:

$$\frac{V_1 - V_2}{10 \; \Omega}$$
$$V_1 \; \mathrm{o} \!\!-\!\!\!\bigwedge\!\!\!-\!\! \mathrm{o} \, V_2$$
$$10 \; \Omega$$

4. Write Kirchhoff's current law at each node, and solve the resulting equations simultaneously for $V_1, V_2, \ldots$. The next example illustrates the procedure.

Example 6.18   (Analysis)

Use nodal analysis to find the voltage across the 4-Ω resistor in Figure 6.35(a).

SOLUTION We must first convert the 2-V voltage source to an equivalent current source. The value of the equivalent source current is $I = 2 \; \mathrm{V}/2 \; \Omega = 1$ A, as shown in Figure 6.35(b). The figure also shows the nodes in the circuit, and node voltages V_1 and V_2. The reference node has a ground symbol attached to it.

We assume that $V_1 > V_2 > 0$. Under that assumption, a current will flow through the 2-Ω resistor in the direction from node 1 to the reference node, as shown by the arrow in Figure 6.35(c). Also, current flows through the 4-Ω resistor from node 1 to node 2. Finally, current flows through the 8-Ω resistor from node 2 to the reference node.

By Ohm's law, the current in each resistor is

$$I_{2\Omega} = \frac{V_1}{2} \qquad \text{amperes}$$

$$I_{4\Omega} = \frac{V_1 - V_2}{4} \qquad \text{amperes}$$

$$I_{8\Omega} = \frac{V_2}{8} \qquad \text{amperes}$$

The arrows in Figure 6.35(c) are labeled with these values. Note that the current produced by the 1-A current source is entering node 1, and the current produced by the 2-A current source is entering node 2.

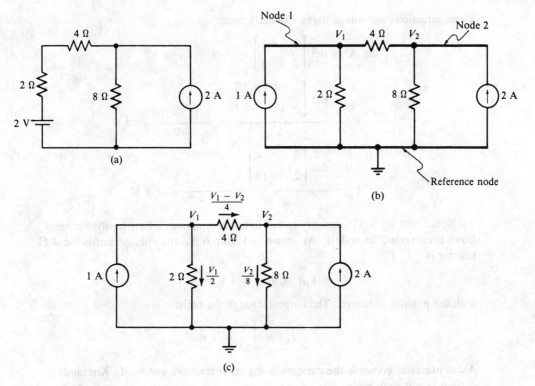

FIGURE 6.35 (Example 6.18)

Writing Kirchhoff's current law at each node, we find

$$\text{current entering node} \quad = \text{current leaving node}$$

node 1:
$$1 = \frac{V_1}{2} + \frac{V_1 - V_2}{4} \qquad (6.29)$$

node 2:
$$\frac{V_1 - V_2}{4} + 2 = \frac{V_2}{8} \qquad (6.30)$$

It is convenient to simplify these equations by first multiplying each by its least common denominator. Multiplying equation (6.29) by 4 and (6.30) by 8 gives

$$4 = 2V_1 + V_1 - V_2$$

$$2(V_1 - V_2) + 16 = V_2$$

Rearranging and collecting terms, we obtain

$$3V_1 - V_2 = 4$$

$$2V_1 - 3V_2 = -16$$

These equations are solved using Cramer's rule:

$$\det \mathbf{A} = \det \begin{bmatrix} 3 & -1 \\ 2 & -3 \end{bmatrix} = -9 + 2 = -7$$

$$V_1 = \frac{\det \begin{bmatrix} 4 & -1 \\ -16 & -3 \end{bmatrix}}{\Delta} = \frac{-12 - 16}{-7} = 4 \text{ V}$$

$$V_2 = \frac{\det \begin{bmatrix} 3 & 4 \\ 2 & -16 \end{bmatrix}}{\Delta} = \frac{-48 - 8}{-7} = 8 \text{ V}$$

Notice that $V_2 > V_1$, contrary to our initial assumption. Thus, positive current flows from node 2 to node 1. As shown in Figure 6.36, the voltage across the 4-Ω resistor is

$$V_{4\Omega} = 8 \text{ V} - 4 \text{ V} = 4 \text{ V}$$

with the polarity indicated. The current through the resistor is

$$I_{4\Omega} = \frac{4 \text{ V}}{4 \text{ }\Omega} = 1 \text{ A}$$

As an exercise, compute the currents in the other resistors and verify Kirchhoff's current law at each node.

Drill Exercise 6.18

Use nodal analysis to find the current through the 2-Ω resistor in Figure 6.35(b) when the positions of the 4-Ω and 8-Ω resistors are interchanged. Draw an arrow showing the direction of positive current through the resistor.

ANSWER: 1.43 A $\downarrow$.

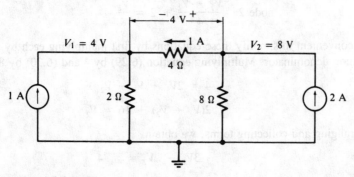

FIGURE 6.36 (Example 6.18)

Expressing node currents in terms of *conductance* is sometimes more convenient than expressing them in terms of resistance. Following is an example:

$$I = G(V_1 - V_2) = 0.1 (V_1 - V_2)$$

$$G = \frac{1}{10\ \Omega} = 0.1\ S$$

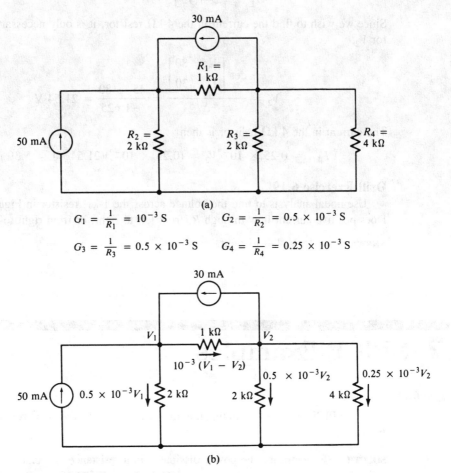

$$V_1 \circ\!\!-\!\!\!\bigwedge\!\!\!-\!\!\circ V_2$$
$$10\ \Omega$$

The advantage of this approach is that it eliminates fractional coefficients in the current equations.

Example 6.19   (Analysis)

Use nodal analysis to find the current in the 4-kΩ resistor shown in Figure 6.37(a).

SOLUTION Figure 6.37(b) shows the circuit with node voltages V_1 and V_2. The conductance of each resistor is computed as shown in the figure, and the currents are

$$G_1 = \frac{1}{R_1} = 10^{-3}\ S \qquad G_2 = \frac{1}{R_2} = 0.5 \times 10^{-3}\ S$$

$$G_3 = \frac{1}{R_3} = 0.5 \times 10^{-3}\ S \qquad G_4 = \frac{1}{R_4} = 0.25 \times 10^{-3}\ S$$

FIGURE 6.37 (Example 6.19)

expressed in terms of those conductances. Writing Kirchhoff's current law at each node, we obtain

$$\text{current entering} = \text{current leaving}$$

node 1:
$$50 \times 10^{-3} + 30 \times 10^{-3} = 0.5 \times 10^{-3}V_1 + 10^{-3}(V_1 - V_2)$$

node 2:
$$10^{-3}(V_1 - V_2) = 30 \times 10^{-3} + 0.5 \times 10^{-3}V_2$$
$$+ 0.25 \times 10^{-3}V_2$$

Dividing through by 10^{-3}, simplifying, collecting terms, and rearranging gives

$$1.5V_1 - V_2 = 80$$

$$V_1 - 1.75V_2 = 30$$

$$\det \mathbf{A} = \det \begin{bmatrix} 1.5 & -1 \\ 1 & -1.75 \end{bmatrix} = -2.625 + 1 = -1.625 = \Delta$$

Since we wish to find the current in the 4-kΩ resistor, it is only necessary to solve for V_2:

$$V_2 = \frac{\det \begin{bmatrix} 1.5 & 80 \\ 1 & 30 \end{bmatrix}}{\Delta} = \frac{45 - 80}{-1.625} = 21.54 \text{ V}$$

The current in the 4-kΩ resistor is then

$$I_{4k\Omega} = 0.25 \times 10^{-3}V_2 = (0.25 \times 10^{-3})(21.54 \text{ V}) = 5.39 \text{ mA}$$

Drill Exercise 6.19

Use nodal analysis to find the voltage across the 1-kΩ resistor in Figure 6.37(a). Does positive current flow through R_1 from left to right, or from right to left?

ANSWER: 46.15 V; left to right.   □

6.7 SPICE Examples

Example 6.20   (SPICE)

Use SPICE as an aid in finding the power dissipated by the 40-Ω resistance in Figure 6.33(a).

SOLUTION To determine the power dissipated in a resistance, we can find the current in it, the voltage across it, or both, and then use one of the three power equations (3.4), (3.5), or (3.6). We choose to find the voltage and use the equation $P = V^2/R$, since that eliminates the need for inserting a dummy voltage source in the SPICE simulation.

It is not necessary to convert the current source in Figure 6.33(a) to an equivalent voltage source, as was done in Example 6.17, since we are not performing a mesh analysis. Figure 6.38 shows the SPICE circuit and input data file. Note that we can cite any one of the three dc sources in the .DC control statement; we arbitrarily choose the 12-V voltage source, V1. Also note that we have no way of knowing in advance whether the voltage V(3,2) across the 40-Ω resistor is positive or negative, but that is totally irrelevant, and we arbitrarily choose to print V(3,2) rather than V(2,3). Execution of the program reveals that V(3,2) = −7.286 V, so we compute $P = (7.286 \text{ V})^2 / 40 \text{ }\Omega = 1.327$ W.

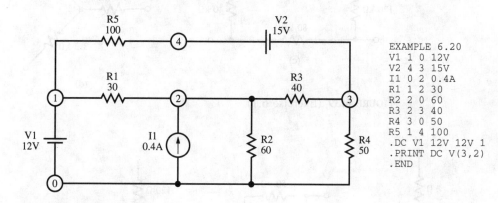

```
EXAMPLE 6.20
V1 1 0 12V
V2 4 3 15V
I1 0 2 0.4A
R1 1 2 30
R2 2 0 60
R3 2 3 40
R4 3 0 50
R5 1 4 100
.DC V1 12V 12V 1
.PRINT DC V(3,2)
.END
```

FIGURE 6.38 (Example 6.20)

Exercises

Section 6.1   Wye–Delta Transformations

6.1 Convert each wye network in Figure 6.39 to an equivalent delta network, and each delta to an equivalent wye. Be certain to identify the terminals in the transformed network that correspond to those in the original network.

6.2 Convert each wye network in Figure 6.40 to an equivalent delta network, and each delta to an equivalent wye. Be certain to identify the terminals in the transformed network that correspond to those in the original network.

6.3 Solve equations (6.1), (6.2), and (6.3) simultaneously for R_1, R_2, and R_3 to obtain equations (6.4), (6.5), and (6.6). [*Hint*: To find R_1, add equations (6.1) and (6.2), and then subtract equation (6.3).]

6.4 Derive equation (6.10) using equations (6.4), (6.5), and (6.6).

6.5 Find the total current drawn from the voltage source and the current through R_1 in each circuit shown in Figure 6.41.

6.6 Find the voltage across the current source and the voltage drop across R_1 and R_2 in each circuit shown in Figure 6.42.

Section 6.2   Balanced Bridges

6.7 Determine which of the bridges shown in Figure 6.43 are balanced. In those that are balanced, find the current in resistor R.

6.8 In each of the bridges that are *not* balanced in Exercise 6.7, find a value for R that will balance the

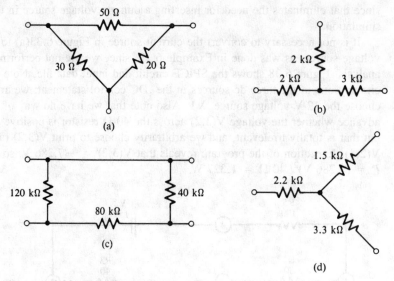

FIGURE 6.39 (Exercise 6.1)

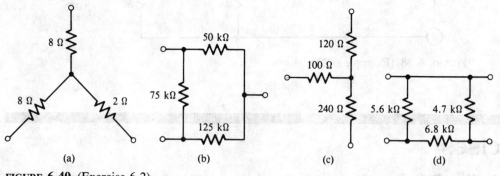

FIGURE 6.40 (Exercise 6.2)

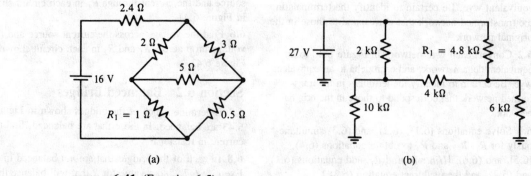

FIGURE 6.41 (Exercise 6.5)

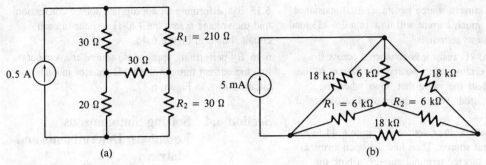

FIGURE 6.42 (Exercise 6.6)

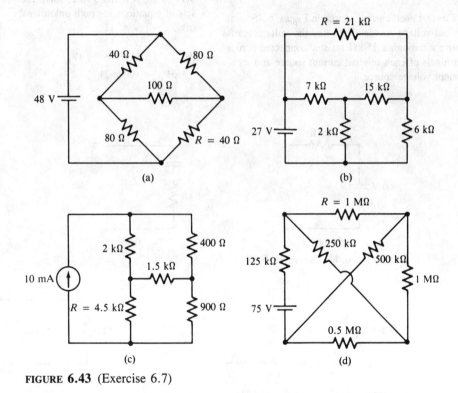

FIGURE 6.43 (Exercise 6.7)

bridge. Then find the current in R when that value of R replaces the original one.

Section 6.3 Voltage and Current Source Conversions

6.9 A voltage source has a terminal voltage of 28 V when its terminals are open-circuited. When a 12-Ω

load is connected across the terminals, the terminal voltage drops to 24 V. What is the internal resistance of the source?

6.10 A voltage source has an internal resistance of 6.2 Ω. What load connected across its terminals will cause the terminal voltage to drop to one-half of the open-circuit terminal voltage?

6.11 A 16-mA current source has an internal resistance of 10 kΩ. How much current will flow in a 2.5-kΩ load connected across its terminals?

6.12 When a 40-kΩ resistor is connected across the terminals of a certain current source, the current in the resistor is one-half the value that flows when the terminals are shorted. What is the internal resistance of the current source?

6.13 Convert each voltage source in Figure 6.44 to an equivalent current source. Then find the open terminal voltage and the shorted terminal current in both the original voltage source and its equivalent current source.

6.14 Convert each current source in Figure 6.45 to an equivalent voltage source. Then find the voltage across and current through a 15-kΩ resistor connected across the terminals of each original current source and its equivalent voltage source.

6.15 By performing an appropriate source conversion, find the voltage across the 120-Ω resistor in each circuit shown in Figure 6.46.

6.16 By performing appropriate source conversion(s), find the current through the 1-kΩ resistor in each circuit shown in Figure 6.47.

Section 6.4 Solving Simultaneous Equations: Determinants and Matrices

6.17 Using Cramer's rule, solve each of the following sets of equations for both unknowns.

(a) $3x - y = 11$
$2x + 3y = 11$

(b) $V_2 = 6 - 4V_1$
$3V_1 - 10 = 2V_2$

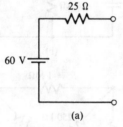
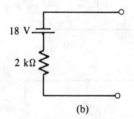

FIGURE **6.44** (Exercise 6.13)

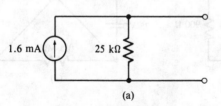
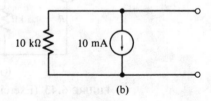

FIGURE **6.45** (Exercise 6.14)

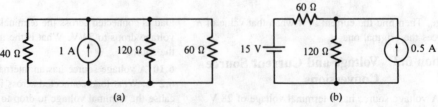

FIGURE **6.46** (Exercise 6.15)

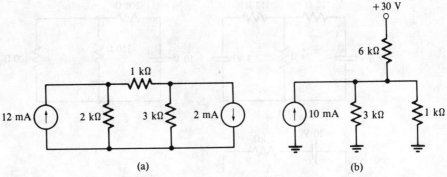

FIGURE **6.47** (Exercise 6.16)

(c) $10(10I_1 - I_2) + 16.6 = 0$

$\quad 10(4I_2 - I_1) - 19.6 = 0$

6.18 Using Cramer's rule, solve each of the following sets of equations for both unknowns.

(a) $3x_1 + 6x_2 = \quad 1.8$

$\quad 2x_1 - \quad x_2 = -10.8$

(b) $4(630I_1 - 901I_2) + 732 = 0$

$\quad 8005I_2 - 3604I_1 = 9599$

(c) $\quad V_2 + 4.2V_1 + 20 = -20.8 + 2.1V_1 - 3.4V_2$

$\quad 3.8V_1 + 16 + 8.5V_2 = -3V_1 + 52.6 + 5.5V_2$

6.19 Write each of the following sets of equations in matrix form. Define each matrix used in the equation, and write the matrix equation for the solution.

(a) $\quad 7I_2 = 3(I_1 - I_3) + 40$

$\quad I_2 + I_3 = 6$

$\quad 2(I_1 - 7) = 5(I_2 + 2I_3 - 4)$

(b) $\quad 0.02V_4 - 0.4V_3 =$

$\quad\quad\quad\quad\quad 6(0.01V_1 - 0.02V_2 - 0.04)$

$\quad 0.01V_2 + 0.9V_1 + 0.55V_3 = V_4$

$\quad\quad\quad\quad 0.04V_4 =$

$\quad\quad -0.8V_1 + 0.5(0.5V_2 + 1.5) - V_3$

$\quad 0.022V_3 + 0.5 = V_1 + 0.01V_4$

6.20 Write each of the following sets of equations in matrix form. Define each matrix used in the equation, and write the matrix equation for the solution.

(a) $\quad 4.2 \times 10^3a - 2 \times 10^3 =$

$\quad\quad 4(0.2 \times 10^4a + 1.8 \times 10^3b) - 10^4c$

$\quad 2.5 \times 10^4b + 1.5 \times 10^3c =$

$\quad\quad\quad\quad\quad 0.3 \times 10^3(10b + 60)$

$\quad 2200(a - b) + 8.2 \times 10^3c = 0$

(b) $\quad 4V_1 - .3V_3 + V_2 = -6$

$\quad\quad V_3 + 2V_2 - V_4 = 1$

$\quad 8V_5 - 7V_4 + 3V_3 = 12$

$\quad -6V_2 + 5V_1 - V_5 = 0$

$\quad 4V_5 - 8V_1 + 3V_4 = -8$

Section 6.5 Mesh Analysis

6.21 Use mesh analysis to find the current in each resistor of the circuits shown in Figure 6.48. After finding the currents, draw arrows on the schematic diagram showing the direction and value of the positive current through each resistor.

6.22 Use mesh analysis to find the voltage across each resistor in the circuits shown in Figure 6.49. Label each resistor on the schematic diagram with the value of the voltage across it, and use + and − signs to show the polarity.

6.23 Write the equations necessary to analyze each of the circuits in Figure 6.50 by mesh analysis. Express the equations in matrix form, and define each matrix used. Write the matrix form of the solution equation. (It is not necessary to find numerical values for the solutions.)

6.24 Repeat Exercise 6.23 for the circuits shown in Figure 6.51.

Section 6.6 Nodal Analysis

6.25 Using nodal analysis, find the voltage at the nodes of each of the circuits in Figure 6.52, with

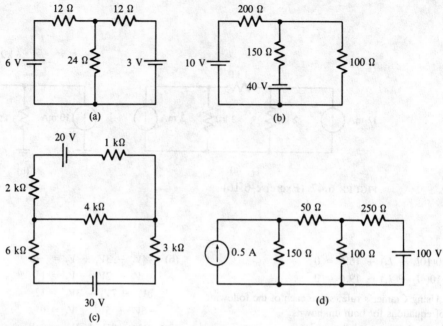

(a)

(b)

(c)

(d)

FIGURE **6.48** (Exercise 6.21) [For (d), it is not necessary to find the current in the 150 Ω resistor.]

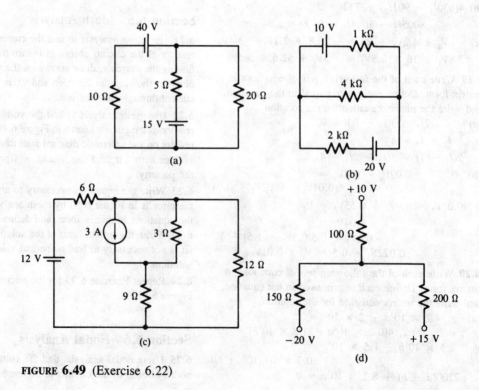

(a)

(b)

(c)

(d)

FIGURE **6.49** (Exercise 6.22)

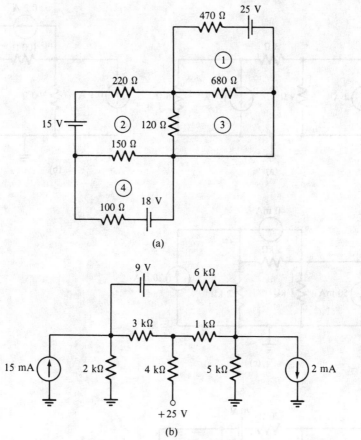

(a)

(b)

FIGURE **6.50** (Exercise 6.23) [For (a), loop numbers to be used are shown in circles.]

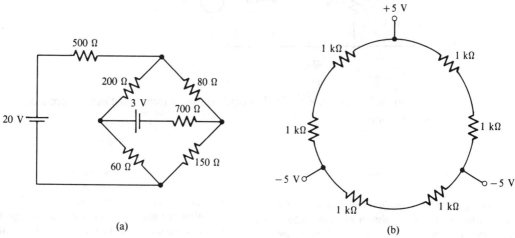

(a)

(b)

FIGURE **6.51** (Exercise 6.24)

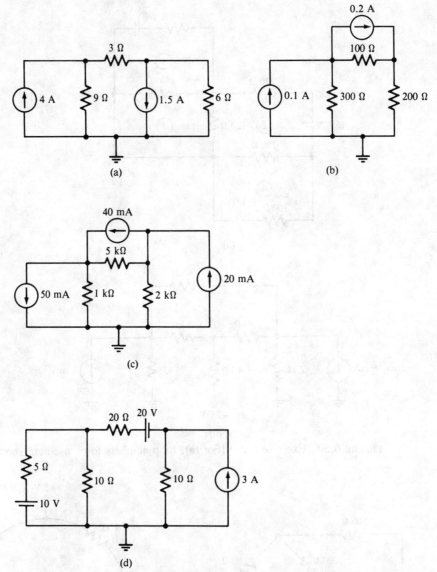

FIGURE 6.52 (Exercise 6.25) [For (c), express current equations in terms of conductance.]

respect to the reference node shown by the ground symbol.

6.26 Using nodal analysis, find the current through each resistor in the circuits shown in Figure 6.53. Then draw arrows on the schematic diagram showing the direction and value of each positive current.

6.27 Write the equations necessary to analyze each of the circuits in Figure 6.54 by nodal analysis. Express the equations in matrix form, and define each matrix used. Write the matrix form of the solution equation. (It is not necessary to find numerical values for the solutions.)

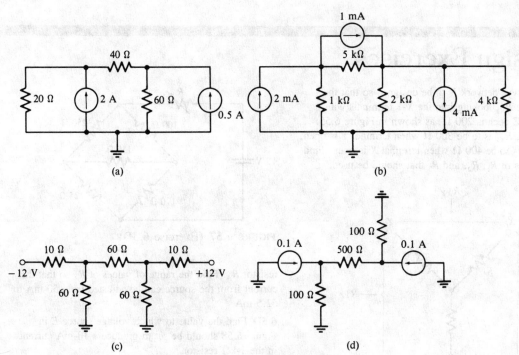

FIGURE 6.53 (Exercise 6.26) [For (b), express current equations in terms of conductance. For (c), find the currents in the 60-Ω resistors only.]

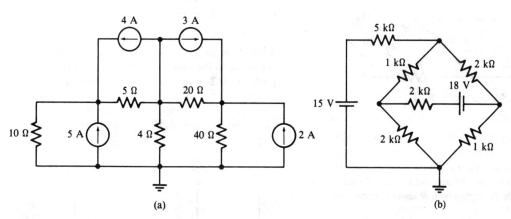

FIGURE 6.54 (Exercise 6.27) [For (b), express current equations in terms of conductance.]

Design Exercises

6.1D A wye network is to be designed so that the resistance R_{XY} looking into the X-Y terminals with terminal Z open is 200 Ω, as shown in Figure 6.55. Similarly, R_{XY} is to be 300 Ω when terminal Y is open, and R_{YZ} is to be 400 Ω when terminal X is open. Find the values of R_1, R_2, and R_3 that should be used.

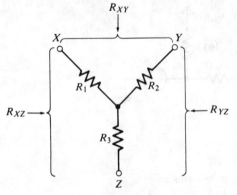

FIGURE 6.55 (Exercise 6.1D)

6.2D A resistor R is to be connected in parallel with the 40-Ω resistor in the delta network shown in Figure 6.56 so that the total current drawn from the voltage source is 250 mA. What value of R should be used?

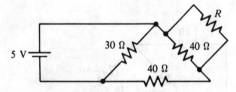

FIGURE 6.56 (Exercise 6.2D)

6.3D Adjustable resistance R is to be inserted in the circuit shown in Figure 6.57 so that the current drawn from the voltage source can be adjusted from 50 mA to 100 mA. What are the minimum and maximum values of resistance through which R should be capable of adjustment?

6.4D Repeat Exercise 6.3D if R is set (fixed) at 90 Ω and the 170-Ω resistor is replaced by an adjustable

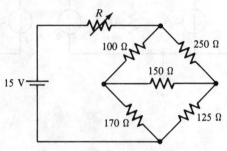

FIGURE 6.57 (Exercise 6.3D)

resistor R_A. Find the range of values of R_A so that the current from the source can be adjusted from 50 mA to 62.5 mA.

6.5D Find the value to which voltage source E in Figure 6.58 should be set to produce a 10-mA current in the 1-kΩ resistor.

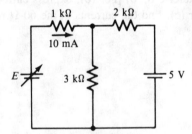

FIGURE 6.58 (Exercise 6.5D)

6.6D What value of constant current I must be connected to the circuit in Figure 6.59 in order to produce a 0.2-A current in the 30-Ω resistor?

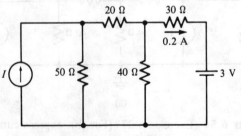

FIGURE 6.59 (Exercise 6.6D)

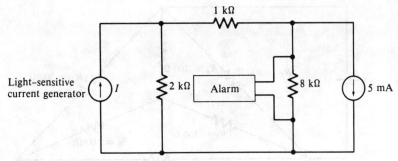

FIGURE **6.60** (Exercise 6.7D)

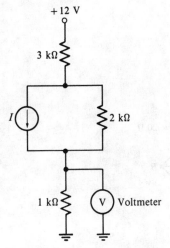

FIGURE **6.61** (Exercise 6.8D)

6.7D Figure 6.60 shows an alarm circuit that is activated by light falling on a photoelectric cell. The photoelectric cell behaves like a current generator; the greater the light, the greater the current I. When the voltage across the 8-kΩ resistor reaches 2.5 V, the alarm is activated. Assuming that the alarm itself has a very large resistance (so its parallel combination with the 8 kΩ is approximately 8 kΩ), find the value of current I that must be generated to activate the alarm.

6.8D Use nodal analysis to determine what value of constant current I should be used in Figure 6.61 to cause the voltmeter to read 3 V.

Troubleshooting Exercises

6.1T The current drawn from the source in Figure 6.62 is 40 mA. Which resistor(s) may be shorted?

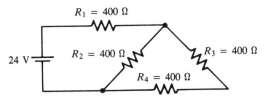

FIGURE **6.62** (Exercise 6.1T)

6.2T When one of the resistors in the circuit shown in Figure 6.63 shorted out, it was observed that there was no change in the total current drawn from the voltage source. Which resistor shorted? (*Hint*: Convert the delta to an equivalent wye.)

6.3T The bridge in Figure 6.64 is supposed to be balanced. In fact, one of the resistors has opened. If the voltage at A is positive with respect to B, which resistor(s) or combinations of resistors may be open?

6.4T The resistors in a certain bridge were installed on a printed-circuit board in the pattern shown in Figure 6.65. The bridge is supposed to be balanced. Assuming

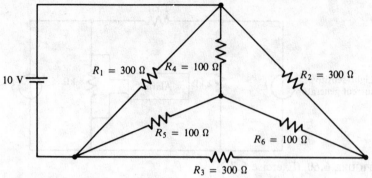

FIGURE **6.63** (Exercise 6.2T)

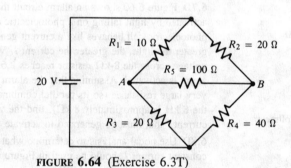

FIGURE **6.64** (Exercise 6.3T)

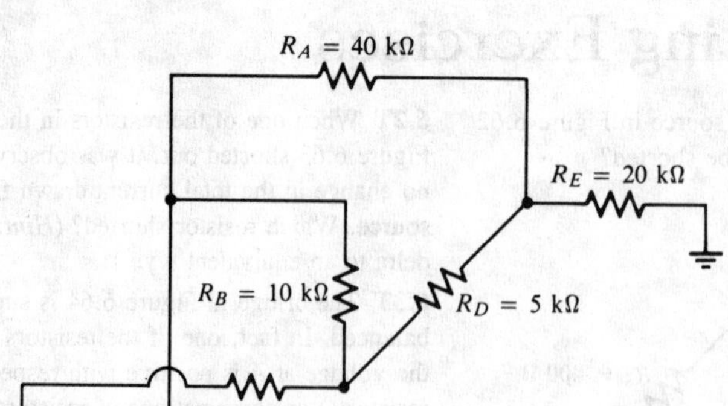

FIGURE **6.65** (Exercise 6.4T)

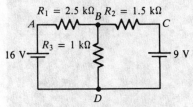

$R_1 = 2.5\ \text{k}\Omega$ $R_2 = 1.5\ \text{k}\Omega$

$R_3 = 1\ \text{k}\Omega$

16 V   9 V

FIGURE 6.66 (Exercise 6.5T)

that R_A is installed correctly, which two resistors were mistakenly interchanged?

6.5T One of the resistors in Figure 6.66 is shorted. Test measurements revealed that the current from A to B is 6.4 mA and the current from C to B is 6 mA. Which resistor is shorted?

6.6T A troubleshooting guide is to be prepared for the circuit in Figure 6.66. The following table, showing the voltage at node B with respect to node D under various fault conditions, is to be included in the guide. Complete the table.

Fault	V_{BD}
None	
R_1 shorted	
R_1 open	
R_2 shorted	
R_2 open	
R_3 shorted	
R_3 open	

SPICE Exercises

6.1S Use SPICE to find the voltage across each resistor of the circuit shown in Figure 6.49(c), p. 202.

6.2S Use SPICE to find the current in each resistor of the circuit shown in Figure 6.48(c), p. 202.

6.3S Use SPICE as an aid in finding the power dissipated in the 150-Ω resistor in Figure 6.51(a), p. 203.

6.4S Solve Design Exercise 6.6D by performing trial-and-error runs of SPICE programs.

7 Network Theorems

7.1 The Superposition Principle

Superposition is a general principle that forms the basis for a very powerful technique used to analyze multisource electric circuits. The principle is widely used in other technical fields to study physical phenomena of all types, including heat transfer, mechanical forces, electromagnetism, light, sound, and many others. In essence, the superposition principle allows us to determine the effect of several energy sources acting simultaneously on a system by analyzing the effect of each source acting *alone*, and then combining (superimposing) those effects.

Application of the superposition principle in circuit analysis is restricted to *linear circuits*: those that contain only linear components. Recall from Section 3.8 that resistors are linear components, so all of the multisource circuits we have studied (and will study) in this book can be analyzed by superposition.

To determine the current through or voltage across any component of a multisource circuit using superposition, we first compute that current or voltage due to each source acting alone. Each such computation requires that we *remove* all sources except one. Sources are removed by making their contributions equal to zero; that is, we

1. *Replace ideal voltage sources by short circuits ($E = 0$).*
2. *Replace ideal current sources by open circuits ($I = 0$).*

Notice that these steps are equivalent to replacing *real* sources by their internal resistances.

The procedure for analyzing a circuit by superposition is as follows:

1. Select any one source in the circuit and remove all others (by replacing them with shorts or opens, as described above).
2. Compute the desired voltage or current in a component when the only source present is the one selected in step 1.
3. Repeat steps 1 and 2, after selecting a *new* source to be the only one present. Continue until the desired voltage or current has been computed due to each and every source acting alone.
4. Add all the computed values obtaned from analyzing the circuit with each source acting alone. This sum is the actual voltage or current when all sources are acting simultaneously (i.e., when all sources are present).

In step 4, note that the addition must take into account the *polarities* of the quantities being added. For example, if one source produces a current of 5 A from left to right through a resistor, and another source produces a current of 1 A from right to left, the total (net) current is 5 A + (-1 A) = 4 A, or, 4 A from left to right.

Example 7.1   (Analysis)

Using the superposition principle, find the current in the 15-Ω resistor in Figure 7.1(a).

SOLUTION It makes no difference which source we choose to remove first. Figure 7.1(b) shows the circuit when the 54-V source is replaced first by a short circuit. With the short circuit in place, the 30-Ω and 15-Ω resistors are in parallel and have equivalent resistance $(30)(15)/45 = 10\ \Omega$. The total resistance of the circuit is then $R_T = 10\ \Omega + 10\ \Omega = 20\ \Omega$, and the total current is $I_T = 30\ \text{V}/20\ \Omega = 1.5$ A. As shown in the figure, we can then find the current I_1 in the 15-Ω resistor due only to the 30-V source by using the current-divider rule:

$$I_1 = \left(\frac{\cdot 30}{15 + 30}\right) 1.5\ \text{A} = 1\text{A} \downarrow$$

The arrow on this result reminds us that the current due to the 30-V source flows downward through the 15-Ω resistor.

Figure 7.1(c) shows the circuit when the 54-V source is restored and the 30-V source is replaced by a short circuit. It can be seen that the 10-Ω and 15-Ω resistors are now in parallel and have equivalent resistance $(10)(15)/25 = 6\ \Omega$. The total resistance of the circuit is $30\ \Omega + 6\ \Omega = 36\ \Omega$ and the total current is $I_T = 54\ \text{V}/36\ \Omega = 1.5$ A. Applying the current-divider rule, we find the current I_2 in the 15-Ω resistor due only to the 30-V source:

$$I_2 = \left(\frac{10}{10 + 15}\right) 1.5\ \text{A} = 0.6\ \text{A} \downarrow$$

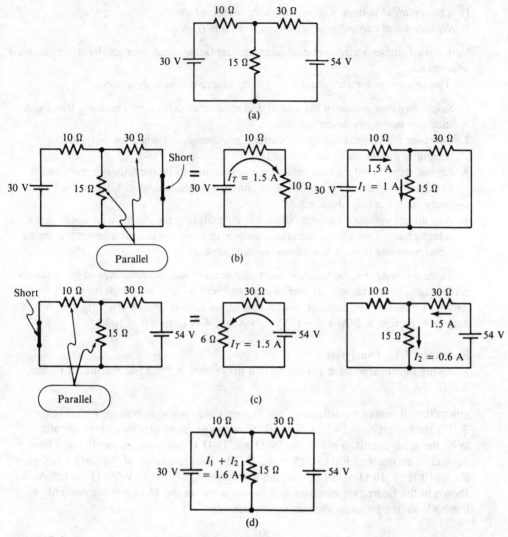

FIGURE 7.1 (Example 7.1) (b) Computing the current (I_1) due to the 30-V source. (c) Computing the current (I_2) due to the 54-V source. (d) Total current due to both sources.

As shown in Figure 7.1(d), the actual current in the 15-Ω resistor due to both sources acting together is the sum of I_1 and I_2:

$$I_{15} = I_1 + I_2 = 1\,A\downarrow\ + 0.6\,A\downarrow\ = 1.6\,A\downarrow$$

The currents are added directly because each flows downward, as does their sum.

Drill Exercise 7.1

Using the superposition principle, find the current in the 10-Ω resistor in Figure 7.1(a).

ANSWER: $\overrightarrow{0.6}$ A.

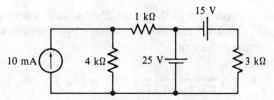

FIGURE 7.2 (Example 7.2)

Example 7.2   (Analysis)

Using the superposition principle, find the voltage across the 1-kΩ resistor in Figure 7.2.

SOLUTION The voltage across the 1-kΩ resistor due to the current source acting alone is found by replacing the 25-V and 15-V sources by short circuits, as shown in Figure 7.3(a). Since the 3-kΩ resistor is shorted out, the current in the 1-kΩ resistor is, by the current-divider rule,

$$I_{1k\Omega} = \left(\frac{4 \text{ k}\Omega}{1 \text{ k}\Omega + 4 \text{ k}\Omega}\right) 10 \text{ mA} = 8 \text{ mA}$$

Thus, the voltage V_1 across the 1-kΩ resistor is

$$V_1 = (8 \text{ mA})(1 \text{ k}\Omega) = {}^+8 \text{ V}^-$$

The + and − symbols indicate the polarity of the voltage due to the current source acting alone, as shown in Figure 7.3(a).

The voltage across the 1-kΩ resistor due to the 25-V source acting alone is found by replacing the 10-mA current source by an open circuit and the 15-V source by a

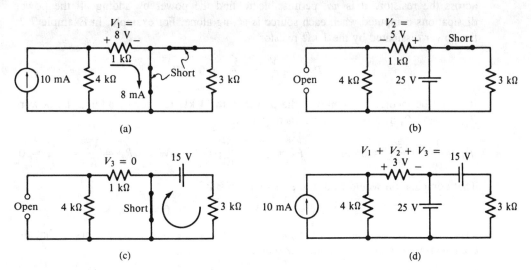

FIGURE 7.3 (Example 7.2) (a) Voltage across the 1-kΩ resistor due to the 10-mA current source acting alone. (b) Voltage across the 1-kΩ resistor due to the 25-V voltage source acting alone. (c) Voltage across the 1-kΩ resistor due to the 15-V voltage source acting alone. (d) Voltage across the 1-kΩ resistor due to all sources acting simultaneously.

short circuit, as shown in Figure 7.3(b). Since the 25-V source is across the series combination of the 1-kΩ and 4-kΩ resistors, the voltage V_2 across the 1-kΩ resistor can be found by the voltage-divider rule:

$$V_2 = \left(\frac{1\ k\Omega}{4\ k\Omega + 1\ k\Omega}\right) 25\ V = {}^-5\ V^+$$

Note that the 3-kΩ resistor has no effect on this computation.

Finally, the voltage V_3 across the 1-kΩ resistor due to the 15-V source acting alone is found by replacing the 25-V source by a short circuit and the 10-mA current source by an open circuit, as shown in Figure 7.3(c). The short circuit prevents any current from flowing in the 1-kΩ resistor, so $V_3 = 0$.

Applying the superposition principle, the voltage across the 1-kΩ resistor due to all three sources acting simultaneously is

$$V_{1k\Omega} = V_1 + V_2 + V_3 = {}^+8\ V^- + {}^-5\ V^+ + 0\ V = {}^+3\ V^-$$

Note that V_1 and V_2 have opposite polarities, so the sum (net) voltage is actually 8 V − 5 V = 3 V, with the same polarity as V_1.

Drill Exercise 7.2

Using the superposition principle, find the voltage across the 4-kΩ resistor in Figure 7.2.

ANSWER: 28 V_-^+.   □

The superposition principle can be used to find the power dissipated in a resistor, provided that computation is performed *after* finding the actual current through or voltage across the resistor. It is *not* permissible to find the power by adding all the power dissipations computed when each source is acting alone. For example, in Example 7.2, the power dissipated by the 1-kΩ resistor is

$$P_{1k\Omega} = \frac{V^2}{R} = \frac{3^2}{10^3} = 9\ mW$$

It would be incorrect to compute the power in the 1-kΩ resistor by adding the powers P_1, P_2, and P_3 due to each source acting alone:

$$P_1 = \frac{V_1^2}{R} = \frac{8^2}{10^3} = 64\ mW \qquad P_2 = \frac{V_2^2}{R} = \frac{5^2}{10^3} = 25\ mW \qquad P_3 = \frac{V_3^2}{R} = \frac{0}{10^3} = 0$$

That computation would lead to the *incorrect* result

$$P_1 + P_2 + P_3 = 64\ mW + 25\ mW + 0 = 89\ mW$$

The reason that superposition cannot be used to add the individual powers is that the computation of power is a *nonlinear* mathematical operation.

Example 7.3   (Design)

To what voltage should adjustable source E be set in order to produce a current of 0.3 A in the 400-Ω resistor shown in Figure 7.4?

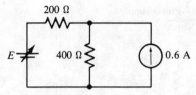

FIGURE 7.4 (Example 7.3)

SOLUTION We first find the current I_1 in the 400-Ω resistor due to the 0.6-A current source. As shown in Figure 7.5(a), that current is found by replacing E by a short circuit and applying the current-divider rule:

$$I_1 = \left(\frac{200}{200 + 400}\right) 0.6 \text{ A} = 0.2 \text{ A}$$

In order for the actual current in the 400-Ω resistor to equal 0.3 A, the current produced in the resistor by the voltage source acting alone must therefore equal 0.3 A $-$ 0.2 A $=$ 0.1 A. The current in the 400-Ω resistor due to the voltage source acting alone is found by open-circuiting the current source, as shown in Figure 7.5(b). In this situation it can be seen that

$$I = \frac{E}{200 \ \Omega + 400 \ \Omega} = \frac{E}{600 \ \Omega}$$

$$0.1 \text{ A} = \frac{E}{600 \ \Omega}$$

$$E = (0.1 \text{ A})(600 \ \Omega) = 60 \text{ V}$$

Drill Exercise 7.3

If E in Figure 7.4 is set to 30 V, to what value should the current source be changed in order that the total current in the 400-Ω resistor be 0.3 A?

ANSWER: 0.75 A. □

Example 7.4 (Troubleshooting)

Figure 7.6(a) shows a dual power supply and voltage-divider arrangement used to furnish an electronic system with supply voltage V_S. In an effort to economize,

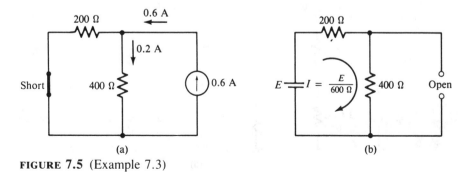

(a) (b)

FIGURE 7.5 (Example 7.3)

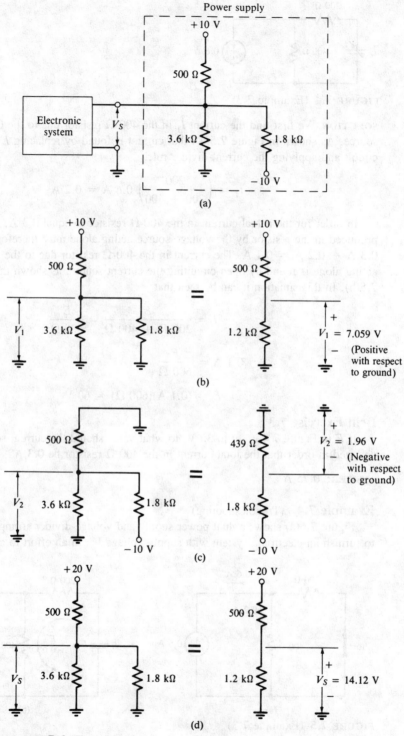

FIGURE 7.6 (Example 7.4)

the manufacturer decided to eliminate the -10-V power supply. He reasoned that the circuit would be equivalent if the -10-V connection were grounded and the $+10$-V supply were increased to $+20$ V. When the modification was completed, it was discovered that the electronic system malfunctioned. Why?

SOLUTION We will use the superposition principle to determine the value of V_S in the original design. Figure 7.6(b) shows the circuit when the -10-V supply is grounded. The 3.6-kΩ and 1.8-kΩ resistors are then in parallel and have an equivalent resistance of 3.6 k$\Omega\|1.8$ k$\Omega = 1.2$ kΩ. The value of V_S due to the $+10$-V supply acting alone is the voltage V_1 across this 1.2-kΩ resistance:

$$V_1 = \left(\frac{1.2 \text{ k}\Omega}{500 \text{ }\Omega + 1.2 \text{ k}\Omega}\right) 10 \text{ V} = 7.059 \text{ V}$$

To find the contribution of the -10-V supply, we ground the $+10$-V supply, as shown in Figure 7.6(c). In this case, the 3.6-kΩ and 500-Ω resistors are in parallel and have equivalent resistance 500 $\Omega\|3.6$ k$\Omega = 439$ Ω. Thus, the value of V_S due to the -10-V supply acting alone is

$$V_2 = \left(\frac{439 \text{ }\Omega}{439 \text{ }\Omega + 1.8 \text{ k}\Omega}\right)(-10 \text{ V}) = -1.96 \text{ V}$$

Notice that V_1 and V_2 have *opposite* polarities when both are referenced to ground. Applying superposition, we find

$$V_S = V_1 + V_2 = 7.059 \text{ V} - 1.96 \text{ V} = 5.1 \text{ V}$$

Figure 7.6(d) shows the computation of V_S when the -10-V supply is removed and the $+10$-V supply is increased to $+20$ V. Note that this computation is similar to that for finding V_1. As shown in the figure,

$$V_S = \left(\frac{1.2 \text{ k}\Omega}{500 \text{ }\Omega + 1.2 \text{ k}\Omega}\right) 20 \text{ V} = 14.12 \text{ V}$$

The modification produced nearly a threefold increase in the supply voltage for the electronic system, causing it to fail.

Drill Exercise 7.4

Find the power dissipated in the 1.8-kΩ resistor in Figure 7.6(a).

ANSWER: 0.127 W. □

7.2 Thévenin's Theorem

Thévenin's theorem is not by itself an analysis tool but rather the basis for a very useful method of simplifying complex circuits. The theorem simply states that any linear circuit is equivalent to a single voltage source in series with a single resistance. (We will study a more general version of the theorem in a later chapter.) Consider how truly remarkable the implications of this theorem are: No matter how complex the circuit and no matter how many voltage and/or current sources it contains, it is equivalent to a single real voltage source!

Notice that the statement of Thévenin's theorem does not tell us how to *find* the single equivalent voltage source and resistance. There is, however, a standard procedure by which those values can be computed. First, we must realize that *the equivalency is with respect to a selected pair of terminals.* In other words, we replace all of the original circuit lying on one side or the other of a pair of terminals by its Thévenin equivalent circuit. The Thévenin equivalent circuit consists of the Thévenin equivalent voltage, E_{TH}, in series with the Thévenin equivalent resistance, R_{TH}, and the terminals of this combination coincide with the terminals of the portion of the original circuit that was replaced. Figure 7.7 illustrates this point. In Figure 7.7(a), the circuitry lying to the left of terminals $a-b$ is replaced by a Thévenin equivalent circuit. In Figure 7.7(b), the circuitry lying to the left of terminals $x-y$ is replaced by its Thévenin equivalent. Although the original circuits are identical in each example, the Thévenin equivalent circuits are different because they replace different portions of the original circuits.

The procedure for finding a Thévenin equivalent circuit is as follows:

1. Open-circuit the terminals with respect to which the Thévenin equivalent circuit is desired. In other words, remove all of the circuitry that will not be replaced by a Thévenin equivalent, leaving the terminals where it was connected open-circuited.

2. The Thévenin equivalent resistance, R_{TH}, is the total resistance at the open-circuited terminals when all voltage sources are replaced by short circuits and all current sources are replaced by open circuits.

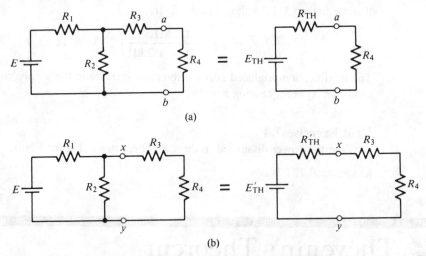

(a)

(b)

FIGURE 7.7 A Thévenin equivalent circuit replaces a portion of an original circuit at a selected pair of terminals. (a) The Thévenin equivalent circuit replaces the circuitry lying to the left of terminals $a-b$. (b) A different Thévenin equivalent circuit replaces the circuitry lying to the left of terminals $x-y$.

3. The Thévenin equivalent voltage, E_{TH}, is the voltage across the open-circuited terminals. Depending on the complexity of the circuit, this voltage may be computed using any of the methods we have studied earlier, including series–parallel circuit analysis, mesh and nodal analysis, or superposition.

4. Replace the original circuitry by its Thévenin equivalent circuit, with the Thévenin terminals occupying the same position as the original terminals. The external circuitry that was removed in step 1 may now be reconnected. Be certain that the polarity of E_{TH} is such that it produces current in the external circuitry in the same direction as the original circuit produced it.

Example 7.5

Find the Thévenin equivalent circuit of the circuitry lying to the left of terminals $x–y$ in Figure 7.8(a).

SOLUTION Figure 7.8(b) shows the circuit after removal of the 50-Ω resistor, leaving the $x–y$ terminals open.

In Figure 7.8(c) we replace the 24-V source with a short circuit to calculate the Thévenin equivalent resistance. The short circuit places the 120 Ω and 60-Ω resistors in parallel, making their equivalent resistance equal to 120 Ω ‖60 Ω = 40 Ω. R_{TH} is the resistance looking into the open terminals $x–y$, which is 100 Ω in series with 40 Ω, or 140 Ω.

In Figure 7.8(d) we calculate E_{TH} by finding the voltage across the open $x–y$ terminals. *Note that there is no current in the 100-Ω resistor, since it is open on one side, and therefore no voltage drop across it.* Consequently, E_{TH} is the same as the voltage across the 120-Ω resistor. (Write Kirchhoff's voltage law around the loop containing the $x-y$ terminals to convince yourself of this fact.) By the voltage-divider rule,

$$E_{TH} = \left(\frac{120}{120 + 60} \right) 24 \text{ V} = 16 \text{ V}$$

Figure 7.8(e) shows the final Thévenin equivalent circuit. Note that the polarity of E_{TH} is such that terminal x is positive with respect to terminal y as in the original circuit [Figure 7.8(d)]. The 50-Ω resistor may now be reconnected across the $x–y$ terminals, as shown.

Drill Exercise 7.5

Find the Thévenin equivalent circuit lying to the left of terminals $x–y$ in Figure 7.8(a) if the 24-V source is changed to 36 V and the 60-Ω resistor is changed to 120 Ω.

ANSWER: R_{TH} = 160 Ω; E_{TH} = 18 V. $\square$

Example 7.6

Find the Thévenin equivalent circuit lying to the right of terminals $x–y$ in Figure 7.9(a).

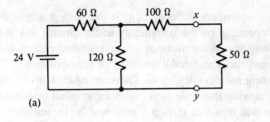

(a)

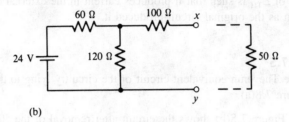

(b)

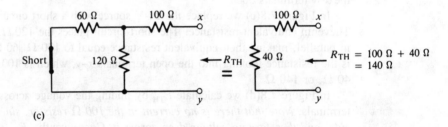

(c)

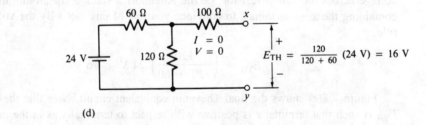

(d)

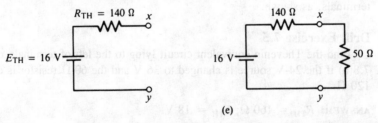

(e)

FIGURE 7.8 (Example 7.5) (b) Open-circuit the terminals. (c) Calculate the Thévenin equivalent resistance. (d) Calculate the Thévenin equivalent voltage.

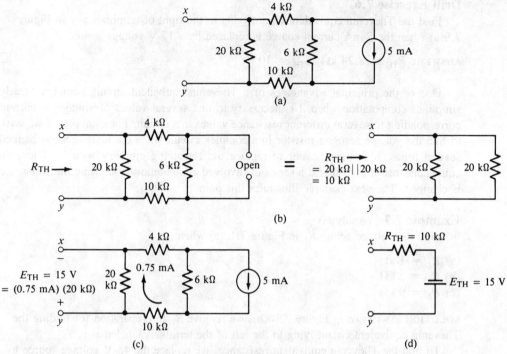

FIGURE 7.9 (Example 7.6)

SOLUTION In this example, there is no external circuitry connected to the x–y terminals. To find the Thévenin equivalent resistance, we open-circuit the current source, as shown in Figure 7.9(b). Notice that the 4-kΩ, 6-kΩ, and 10-kΩ resistors are then in series, and have a total resistance of 20 kΩ. Thus, R_{TH} is the parallel combination of that 20-kΩ resistance and the other 20-kΩ resistor, that is,

$$R_{TH} = 20\ k\Omega \| 20\ k\Omega = \frac{20\ k\Omega}{2} = 10\ k\Omega$$

Figure 7.9(c) shows the computation of the Thévenin equivalent voltage. Notice that E_{TH} is the voltage drop across the 20-kΩ resistor. The current from the 5-mA source divides between the 6-kΩ resistor and the series string 10 kΩ + 20 kΩ + 4 kΩ = 34 kΩ. Thus, by the current-divider rule, the current in the 20-kΩ resistor is

$$I_{20k\Omega} = \left(\frac{6\ k\Omega}{34\ k\Omega + 6\ k\Omega}\right) 5\ mA = 0.75\ mA$$

The voltage across the 20-kΩ resistor is then

$$E_{TH} = (0.75\ mA)(20\ k\Omega) = 15\ V$$

Notice that terminal y is positive with respect to terminal x.

Figure 7.9(d) shows the Thévenin equivalent circuit. The polarity of E_{TH} is such that terminal y is positive with respect to terminal x, as required.

Drill Exercise 7.6

Find the Thévenin equivalent circuit lying to the right of terminals x–y in Figure 7.9(a) when the 5-mA current source is replaced by a 17-V voltage source.

ANSWER: $R_{TH} = 8.24 \text{ k}\Omega$; $E_{TH} = 10 \text{ V}$. □

One of the principal advantages of a Thévenin equivalent circuit is that it greatly simplifies computation when it is necessary to find several values of voltage or current corresponding to several different resistance values in a circuit. For example, if we wish to find the voltage across a resistor in a complex circuit when the resistance is changed several times, it is much easier to replace the remaining circuitry with its Thévenin equivalent than it is to repeat a series of involved computations each time the resistance is changed. The next example illustrates this point.

Example 7.7 (Analysis)

Find the voltage across R_L in Figure 7.10(a) when

(a) $R_L = 1 \text{ k}\Omega$
(b) $R_L = 2 \text{ k}\Omega$
(c) $R_L = 9 \text{ k}\Omega$

SOLUTION As shown in Figure 7.10(b), we remove R_L in preparation for finding the Thévenin equivalent circuit lying to the left of the terminals labeled x–y.

To find the Thévenin equivalent resistance, we replace the 45-V voltage source by a short circuit and the 12-mA current source by an open circuit, as shown in Figure 7.10(c). As can be seen in the figure, R_{TH} is then the parallel equivalent resistance of the 1.5-kΩ and 3-kΩ resistors:

$$R_{TH} = 1.5 \text{ k}\Omega \| 3 \text{ k}\Omega = 1 \text{ k}\Omega$$

We will use the superposition principle to find the Thévenin voltage appearing across the open x–y terminals. With the current source removed (opened), we find the voltage E_1 due to the 45-V voltage source acting alone, as shown in Figure 7.10(d). Since E_1 is the voltage across the 3-kΩ resistor, we have, by the voltage-divider rule,

$$E_1 = \left(\frac{3 \text{ k}\Omega}{1.5 \text{ k}\Omega + 3 \text{ k}\Omega} \right) 45 \text{ V} = 30 \text{ V}^+_-$$

The voltage E_2 due to the current source acting alone is found by shorting the 45-V voltage source, as shown in Figure 7.10(e). By the current-divider rule, the current in the 3-kΩ resistor is

$$I_{3k\Omega} = \left(\frac{1.5 \text{ k}\Omega}{1.5 \text{ k}\Omega + 3 \text{ k}\Omega} \right) 12 \text{ mA} = 4 \text{ mA}$$

Therefore, by Ohm's law, $E_2 = (4 \text{ mA})(3 \text{ k}\Omega) = 12 \text{ V}$.

Notice that E_1 and E_2 have opposite polarities. Applying superposition, we find

$$E_{TH} = E_1 + E_2 = 30 \text{ V}^+_- + 12 \text{ V}^-_+ = 18 \text{ V}^+_-$$

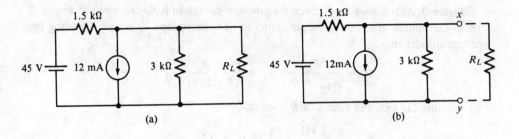

(a)   (b)

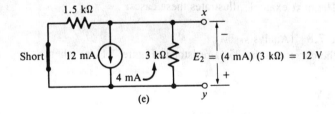

(c)

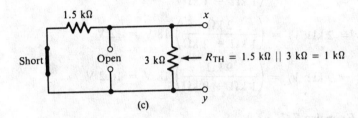

(d)

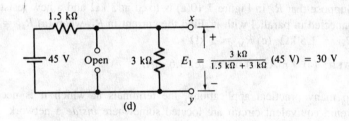

(e)

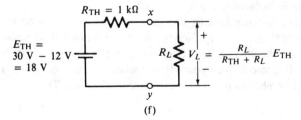

(f)

FIGURE 7.10 (Example 7.7)

Figure 7.10(f) shows the Thévenin equivalent circuit with R_L reconnected across the x–y terminals. It is now a simple matter to find the voltage V_L across R_L, using the voltage-divider rule:

$$V_L = \frac{R_L}{R_{TH} + R_L} E_{TH} = \left(\frac{R_L}{1\ k\Omega + R_L}\right) 18\ V$$

Substituting the required values of R_L, we find

(a) $R_L = 1\ k\Omega;\ V_L = \left(\dfrac{1\ k\Omega}{1\ k\Omega + 1\ k\Omega}\right) 18\ V = 9\ V$

(b) $R_L = 2\ k\Omega;\ V_L = \left(\dfrac{2\ k\Omega}{1\ k\Omega + 2\ k\Omega}\right) 18\ V = 12\ V$

(c) $R_L = 9\ k\Omega;\ V_L = \left(\dfrac{9\ k\Omega}{1\ k\Omega + 9\ k\Omega}\right) 18\ V = 16.2\ V$

Drill Exercise 7.7

Suppose that R_L in Figure 7.10(a) is fixed at 3 kΩ and a new resistor, R_M, is connected in parallel with it. Find the current in R_M when (a) $R_M = 750\ \Omega$; (b) $R_M = 1.5\ k\Omega$; (c) $R_M = 3\ k\Omega$.

ANSWER: (a) 9 mA; (b) 6 mA; (c) 3.6 mA.   □

In many practical applications, the terminals at which it is necessary to find a Thévenin equivalent circuit are located somewhere *inside* a network. Also, it is often the case that the circuitry external to those terminals consists of more than just a single resistor. The next example illustrates these cases.

Example 7.8   (Analysis)

Find the current in the 25-Ω resistor in Figure 7.11(a) when

(a) $E = 3\ V$
(b) $E = 6\ V$

SOLUTION As shown in Figure 7.11(b), we remove voltage source E and the 25-Ω resistor, leaving the terminals labeled x–y open-circuited.

To find the Thévenin resistance looking into the x–y terminals, we short-circuit the 18-V source, as shown in Figure 7.11(c). Notice that the 30-Ω and 60-Ω resistors are then in parallel, as are the 90-Ω and 45-Ω resistors. Study the schematic diagrams in Figure 7.11(c) carefully, to convince yourself that when looking into the x–y terminals, those parallel combinations are in series with each other. Thus,

$$R_{TH} = (30\ \Omega \| 60\ \Omega) + (90\ \Omega \| 45\ \Omega) = 20\ \Omega + 30\ \Omega = 50\ \Omega$$

Figure 7.11(d) shows the computation of the Thévenin equivalent voltage. We use the voltage-divider rule to find the voltage drops across the 60-Ω and 45-Ω resistors:

$$V_{60\Omega} = \left(\frac{60}{60 + 30}\right) 18\ V = 12\ V \qquad V_{45\Omega} = \left(\frac{45}{90 + 45}\right) 18\ V = 6\ V$$

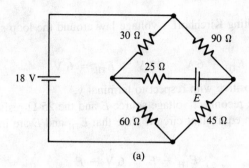

(a)

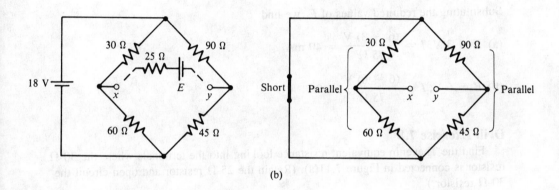

(b)

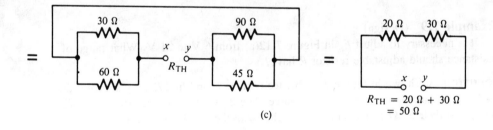

$R_{TH} = 20\ \Omega + 30\ \Omega$
$= 50\ \Omega$

(c)

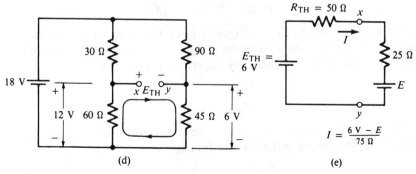

(d)

$I = \dfrac{6\ V - E}{75\ \Omega}$

(e)

FIGURE 7.11 (Example 7.8)

E_{TH} is then found by writing Kirchhoff's voltage law around the loop shown in the figure:

$$E_{TH} + 6 \text{ V} = 12 \text{ V} \Rightarrow E_{TH} = 6 \text{ V}$$

Note that terminal x is positive with respect to terminal y.

In Figure 7.11(e), we reconnect voltage source E and the 25-Ω resistor to the x–y terminals of the Thévenin equivalent circuit. Note that E_{TH} and E are in series opposition, so

$$I = \frac{E_{TH} - E}{R_{TH} + 25} = \frac{6 \text{ V} - E}{75 \text{ } \Omega}$$

Substituting the required values of E, we find

(a) $E = 3$ V; $I = \dfrac{(6 - 3) \text{ V}}{75 \text{ } \Omega} = 40$ mA

(b) $E = 6$ V; $I = \dfrac{(6 - 6) \text{ V}}{75 \text{ } \Omega} = 0$

Drill Exercise 7.8

Find the Thévenin equivalent resistance looking into the terminals where the 30-Ω resistor is connected in Figure 7.11(a). (Retain the 25-Ω resistor and open-circuit the 30-Ω resistor.)

ANSWER: 28.696 Ω.   □

Example 7.9   (Design)

It is necessary to adjust E_o in Figure 7.12(a) from 6 V to 21 V. What range of resistance should adjustable resistor R have?

SOLUTION As shown in Figure 7.12(b), we remove R and find R_{TH} by shorting the 24-V source and opening the 27-mA source. The 2-kΩ and 4-kΩ resistors are then in series and their 6-kΩ equivalent is in parallel with the 3-kΩ resistor. Thus,

$$R_{TH} = 1 \text{ k}\Omega + 6 \text{ k}\Omega \| 3 \text{ k}\Omega$$
$$= 1 \text{ k}\Omega + 2 \text{ k}\Omega = 3 \text{ k}\Omega$$

We use superposition to find E_{TH}. In Figure 7.12(c), the voltage E_1 due to the voltage source acting alone is the same as the voltage across the 2-kΩ resistor, since there is no drop across the 1-kΩ resistor. By the voltage-divider rule,

$$E_1 = \left(\frac{3 \text{ k}\Omega}{2 \text{ k}\Omega + 3 \text{ k}\Omega + 4 \text{ k}\Omega} \right) 24 \text{ V} = 8 \text{ V} \overset{+}{\_}$$

In Figure 7.12(d), we find the voltage E_2 due to the current source acting alone. By the current-divider rule, the current in the 3-kΩ resistor is

$$I_{3k\Omega} = \left(\frac{2 \text{ k}\Omega}{3 \text{ k}\Omega + 4 \text{ k}\Omega + 2 \text{ k}\Omega} \right) 27 \text{ mA} = 6 \text{ mA}$$

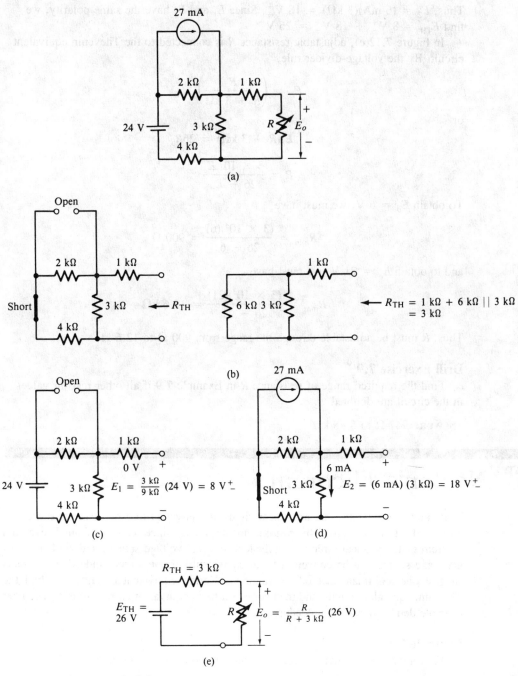

FIGURE 7.12 (Example 7.9)

Thus, $E_2 = (6 \text{ mA})(3 \text{ k}\Omega) = 18 \text{ V}^+_-$. Since E_1 and E_2 have the same polarity, we find $E_{TH} = 8 \text{ V}^+_- + 18 \text{ V}^+_- = 26 \text{ V}^+_-$.

In Figure 7.12(e), adjustable resistance R is connected to the Thévenin equivalent circuit. By the voltage-divider rule,

$$E_o = \left(\frac{R}{R + 3 \text{ k}\Omega} \right) 26 \text{ V}$$

or

$$E_o(R + 3 \text{ k}\Omega) = 26R$$

$$R = \frac{3 \times 10^3 \, E_o}{26 - E_o}$$

To obtain $E_o = 6 \text{ V}$, we must have

$$R_{min} = \frac{(3 \times 10^3)(6)}{26 - 6} = 900 \ \Omega$$

and to obtain $E_o = 21 \text{ V}$, we must have

$$R_{max} = \frac{(3 \times 10^3)(21)}{26 - 21} = 12.6 \text{ k}\Omega$$

Thus, R must be adjustable through the range from 900 Ω to 12.6 kΩ.

Drill Exercise 7.9

Find the required range of resistance R in Example 7.9 if all other resistor values in the circuit are doubled.

ANSWER: 947 Ω to 5.48 kΩ. ▫

7.3 Norton's Theorem

Norton's theorem states that any linear circuit is equivalent to a real *current* source, at a selected set of terminals. This result is not surprising, since we know from Thévenin's theorem that any linear circuit is equivalent to a real voltage source, and we know that a voltage source can be converted to an equivalent current source. Indeed, these facts suggest one legitimate method for finding a Norton equivalent circuit: First find the Thévenin equivalent circuit and then convert it to an equivalent current source. The next example demonstrates the procedure.

Example 7.10

Find the Norton equivalent current source at terminals x–y in Figure 7.13(a).

SOLUTION To find R_{TH}, we short both voltage sources, as shown in Figure 7.13(b). Notice that the 10-Ω and 20-Ω resistors are then in parallel, and

$$R_{TH} = 10 \ \Omega \| 20 \ \Omega = 6.67 \ \Omega$$

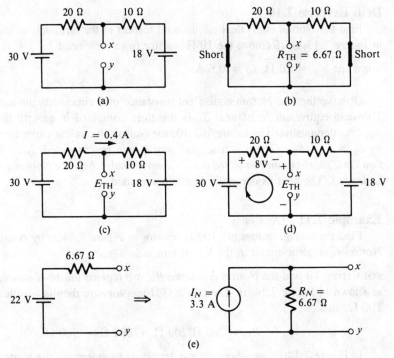

FIGURE 7.13 (Example 7.10)

Since the 30-V and 18-V sources are in series opposition, we can find the current I, as shown in Figure 7.13(c):

$$I = \frac{30 \text{ V} - 18 \text{ V}}{20 \text{ }\Omega + 10 \text{ }\Omega} = \frac{12 \text{ V}}{30 \text{ }\Omega} = 0.4 \text{ A}$$

The voltage drop across the 20-Ω resistor is therefore

$$V_{20\Omega} = (0.4 \text{ A})(20 \text{ }\Omega) = 8 \text{ V}$$

Writing Kirchhoff's voltage law around the loop shown in Figure 7.13(d), we find

$$E_{TH} + 8 \text{ V} = 30 \text{ V}$$

$$E_{TH} = 22 \text{ V}$$

The Thévenin equivalent circuit is shown in Figure 7.13(e). To find the Norton equivalent circuit, we convert the Thévenin voltage source to an equivalent current source:

$$I_N = \frac{E_{TH}}{R_{TH}} = \frac{22 \text{ V}}{6.67 \text{ }\Omega} = 3.3 \text{ A}$$

$$R_N = R_{TH} = 6.67 \text{ }\Omega$$

Drill Exercise 7.10

Find the Norton equivalent circuit with respect to the terminals of the 10-Ω resistor in Figure 7.13(a). (Remove the 10-Ω resistor from the circuit.)

ANSWER: $R_N = 20\ \Omega;\ I_N = 0.6$ A. ◻

Observe that the Norton equivalent resistance of a circuit has the same value as the Thévenin equivalent resistance. R_N is therefore computed in exactly the same way as R_{TH}. As demonstrated in Example 7.10, the Norton equivalent current can be computed by $I_N = E_{TH}/R_{TH}$. There is a second way I_N can be computed: I_N *is the current that flows in a short-circuit connected across the terminals where the Norton equivalent circuit is desired.* The next example illustrates this method.

Example 7.11   (Analysis)

Find the voltage across the 100-Ω resistor in Figure 7.14(a) by constructing a Norton equivalent circuit to the left of terminals x-y.

SOLUTION To find the Norton resistance R_N, we replace the 18-V source by a short, as shown in Figure 7.14(b). The two 200-Ω resistors are then in parallel, giving 100 Ω, so

$$R_N = 200\ \Omega \| 200\ \Omega + 800\ \Omega = 900\ \Omega$$

In Figure 7.14(c), we short the x-y terminals to determine the Norton current, I_N. The 800-Ω and 200-Ω resistors are then in parallel, giving 800 $\Omega \| 200\ \Omega = 160\ \Omega$. Thus, the total equivalent resistance seen by the 18-V source is $R_T = 200\ \Omega + 160\ \Omega = 360\ \Omega$, and $I_T = 18$ V$/360\ \Omega = 0.05$ A. Applying the current-divider rule to determine the portion of I_T that flows through the 800-Ω resistor and through the shorted x-y terminals, we find

$$I_N = \left(\frac{200}{800 + 200}\right)(0.05\ \text{A}) = 0.01\ \text{A}$$

Figure 7.14(d) shows the Norton equivalent circuit with the 100-Ω resistor restored to the x-y terminals. The parallel equivalent resistance of 900 Ω and 100 Ω is 90 Ω, so

$$V_{100\Omega} = I_N(90\ \Omega) = (0.01\ \text{A})(90\ \Omega) = 0.9\ \text{V}$$

Note carefully that the 0.1-A current flows *downward* through the shorted x-y terminals [Figure 7.14(c)], but the Norton current source produces $I_N = 0.1$ A *upward*. The direction of I_N is such that it produces current in an external short in the same direction that current flows when the original circuit terminals are shorted.

Drill Exercise 7.11

Find the Norton equivalent current in Example 7.11 by first finding the Thévenin equivalent voltage of the circuit lying to left of terminals x-y in Figure 7.11(a).

ANSWER: $E_{TH} = 9$ V; $I_N = 9$ V$/900\ \Omega = 0.01$ A. ◻

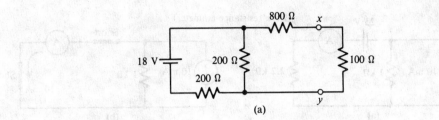

(a)

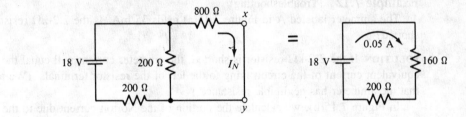

(b)

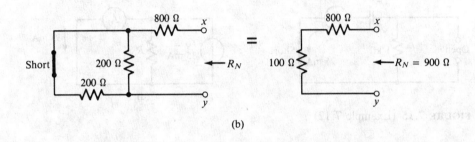

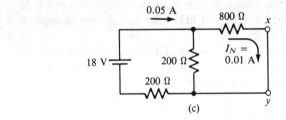

(c)

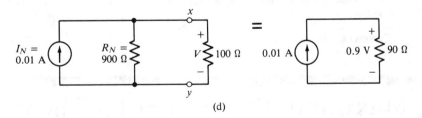

(d)

FIGURE 7.14 (Example 7.11)

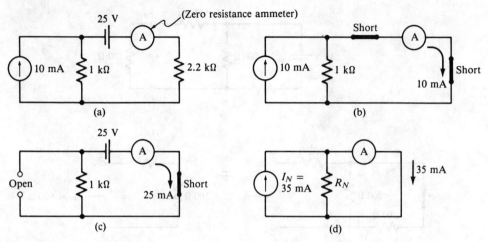

FIGURE 7.15 (Example 7.12)

Example 7.12 (Troubleshooting)

The ammeter labeled A in Figure 7.15(a) reads 35 mA. Is the 2.2-kΩ resistor shorted?

SOLUTION If the 2.2-kΩ resistor is shorted, the ammeter current will equal the Norton equivalent current of the circuit lying to the left of the resistor terminals. (We assume that the ammeter has negligible resistance.)

In Figure 7.15(b), we calculate the portion of the Norton current due to the 10-mA current source acting alone. This value is clearly 10 mA. In Figure 7.15(c), we open the current source to determine the contribution of the 25-V voltage source to I_N. This current is 25 V/1 kΩ = 25 mA. By the superposition principle I_N = 10 mA + 25 mA = 35 mA.

Figure 7.15(d) shows that a short connected across the terminals of the Norton equivalent circuit draws the entire Norton current of 35 mA through the ammeter. We conclude that the 2.2-kΩ resistor is shorted.

Drill Exercise 7.12

What should the ammeter in Figure 7.15(a) read when the 2.2-kΩ resistor is not shorted?

ANSWER: 10.94 mA. □

7.4 Maximum Power Transfer Theorem

In many electrical and electronic applications, we are interested in the amount of power that can be furnished to a particular *load*, such as a loudspeaker, electric motor, antenna, or other useful device. It is convenient to think of electrical systems as being composed

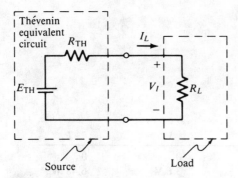

FIGURE 7.16 The load "receives" power from the source, which is represented by its Thévenin equivalent circuit.

of a *source* of power and of a load which is connected to that source. The source may be an amplifier, a generator, a power supply, or similar device. Often, the source can be reduced to a Thévenin equivalent circuit, as shown in Figure 7.16. In dc circuits, the load can be represented by its resistance, R_L.

Although not strictly correct terminology, we say that the source in Figure 7.16 "delivers," or "transfers," power to the load. In fact, the source develops a voltage V_L across the load and enables a current I_L to flow into it. As we know, the power in the load is then

$$P_L = V_L I_L = I_L^2 R_L = \frac{V_L^2}{R_L}$$

The power delivered to the load resistance depends on the value of R_L itself. As a demonstration of this fact, consider the example shown in Figure 7.17. Here, the source voltage is fixed at 36 V and the Thévenin equivalent resistance of the source is a fixed 18 Ω. The table accompanying the figure shows the load current, load voltage, and load power for different values of load resistance. For example, when $R_L = 2\ \Omega$, we find

$$I_L = \frac{36\ \text{V}}{18\ \Omega + 2\ \Omega} = 1.8\ \text{A}$$

$$V_L = I_L R_L = (1.8\ \text{A})(2\ \Omega) = 3.6\ \text{V}$$

and

$$P_L = I_L V_L = (1.8\ \text{A})(3.6\ \text{V}) = 6.48\ \text{W}$$

Notice that the load current is maximum (2 A) when $R_L = 0$ (terminals shorted) and that the load voltage is maximum (36 V) when $R_L = \infty$ (terminals open). However, the load power is *zero* in both of these extreme cases.

The graph in Figure 7.17 is a plot of load power versus load resistance. It is clear from the table and the graph that load power is a maximum (18 W) when $R_L = 18\ \Omega$.

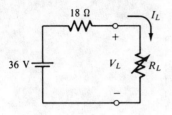

R_L	I_L	V_L	P_L
0 (short)	2 A	0	0
2 Ω	1.8 A	3.6 V	6.48 W
6 Ω	1.5 A	9 V	13.5 W
12 Ω	1.2 A	14.4 V	17.28 W
18 Ω	1 A	18 V	18 W
27 Ω	0.8 A	21.6 V	17.28 W
54 Ω	0.5 A	27 V	13.5 W
72 Ω	0.4 A	28.8 V	11.52 W
162 Ω	0.2 A	32.4 V	6.48 W
342 Ω	0.1 A	34.2 V	3.42 W
∞ (open)	0	36 V	0

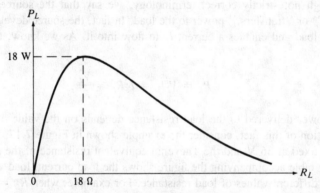

FIGURE 7.17 Load power P_L versus load resistance R_L.

Not coincidentally, load power is maximum when R_L equals the Thévenin resistance of the source. This fact is called the maximum power transfer theorem:

Maximum power is developed in a load when the load resistance equals the Thévenin resistance of the source to which it is connected.

The concept of maximum power transfer is exceptionally important. When source and load have the same resistance, they are said to be *matched*, and in many practical applications the major part of a circuit designer's effort is to ensure that components are matched. A familiar example is an audio amplifier and a loudspeaker. In many such systems, matching is achieved by connecting the speaker to an appropriate output terminal on the amplifier.

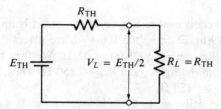

FIGURE 7.18 When $R_L = R_{TH}$, the voltage across R_L is $E_{TH}/2$.

When $R_L = R_{TH}$, as shown in Figure 7.18, the (maximum) power delivered to the load can be calculated by realizing that $V_L = E_{TH}/2$:

$$P_L(\text{max}) = \frac{V_L^2}{R_L} = \frac{(E_{TH}/2)^2}{R_{TH}} = \frac{E_{TH}^2}{4\,R_{TH}} \qquad (7.1)$$

Rather than memorizing this equation, it is recommended that any of the standard methods for calculating power be applied to this very simple dc circuit. For example, it is easy to calculate the total current I in the circuit and then use the relationship $P_L = I^2 R_L$.

Example 7.13   (Design)
(a) Find the value of R_L in Figure 7.19(a) necessary to obtain maximum power in R_L.
(b) Find the maximum power in R_L.

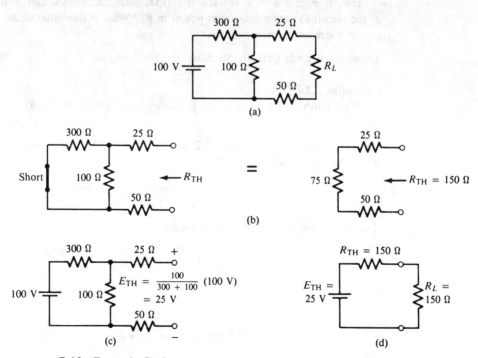

FIGURE 7.19 (Example 7.13)

SOLUTION

(a) We must find the Thévenin equivalent resistance of the circuit lying to the left of R_L, as shown in Figure 7.19(b). Shorting the 100-V source places the 300-Ω and 100-Ω resistors in parallel, giving $300\|100 = 75\ \Omega$. We then have three resistors in series as shown in the figure, and $R_{TH} = 150\ \Omega$. Thus, maximum power is delivered when $R_L = 150\ \Omega$.

(b) To find the maximum power delivered to R_L, it is necessary to find E_{TH}. As shown in Figure 7.19(c), there is no voltage drop across either the 25-Ω or 50-Ω resistors, so E_{TH} is the same as the voltage drop across the 100-Ω resistor. By the voltage-divider rule, $E_{TH} = (100/400)100\ \text{V} = 25\ \text{V}$.

Figure 7.19(d) shows the Thévenin equivalent circuit with the 150-Ω load connected for maximum power transfer. By equation (7.1),

$$P_L(\text{max}) = \frac{E_{TH}^2}{4\ R_{TH}} = \frac{25^2}{4(150)} = 1.042\ \text{W}$$

Alternatively,

$$I_L = \frac{25\ \text{V}}{300\ \Omega} = 0.0833\ \text{A}$$

$$P_L(\text{max}) = I_L^2 R_L = (0.0833)^2 150 = 1.042\ \text{W}$$

Drill Exercise 7.13

If R_L in Figure 7.19(a) is fixed at 100 Ω, what alteration(s) can be made in the rest of the circuit to obtain maximum power in R_L? What is the value of the maximum power under those circumstances?

ANSWER (ONE SOLUTION): Short out the 50-Ω resistor; $P_L(\text{max}) = 1.5625\ \text{W}$. ☐

Example 7.14 (Analysis)

What percent of the maximum possible power is delivered to R_L in Figure 7.20(a) when $R_L = 2R_{TH}$?

SOLUTION As shown in Figure 7.20(b), when $R_L = 2R_{TH}$, the load voltage is found by the voltage-divider rule to be

$$V_L = \frac{2\ R_{TH}}{R_{TH} + 2\ R_{TH}}\ E_{TH} = \frac{2}{3}\ E_{TH}$$

The power delivered to the load is then

$$P_L = \frac{V_L^2}{R_L} = \frac{(\frac{2}{3}E_{TH})^2}{R_L} = \frac{\frac{4}{9}E_{TH}^2}{2\ R_{TH}} = \frac{4\ E_{TH}^2}{18\ R_{TH}}$$

Since $P_L(\text{max}) = E_{TH}^2/4\ R_{TH}$, the ratio of P_L to $P_L(\text{max})$ is

$$\frac{P_L}{P_L(\text{max})} = \frac{4\ E_{TH}^2/18\ R_{TH}}{E_{TH}^2/4\ R_{TH}} = \frac{16}{18}$$

Thus,

$$P_L = \left[\tfrac{16}{18}P_L(\text{max})\right] \times 100\% = 88.89\%\ P_L(\text{max})$$

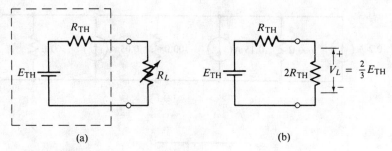

FIGURE 7.20 (Example 7.14)

Drill Exercise 7.14

What percent of the maximum possible power is delivered to R_L in Figure 7.20(a) when $R_L = R_{TH}/2$?

ANSWER: 88.89%.   □

7.5 Millman's Theorem

In its simplest form, Millman's theorem states that parallel-connected current sources can be replaced by a single equivalent current source. Recall that we have already discussed the fact that ideal current sources connected in parallel are equivalent to a single ideal current source that produces the *resultant* of the parallel sources (Section 4.11). Since a real current source has a parallel-connected resistance, it follows that several real sources in parallel have a total resistance equal to the parallel equivalent of the several resistances. Thus, a Millman equivalent current source consists of an ideal current source and a parallel equivalent resistance.

Example 7.15

Find the Millman equivalent current source with respect to terminals x–y in Figure 7.21(a).

SOLUTION The resultant current of the three sources is

$$0.2 \text{ A} \uparrow + 0.15 \text{ A} \downarrow + 0.05 \text{ A} \uparrow = 0.1 \text{ A} \uparrow$$

The equivalent resistance of the three parallel resistors is

$$600 \text{ }\Omega \| 100 \text{ }\Omega \| 600 \text{ }\Omega = 75 \text{ }\Omega$$

Thus, the single equivalent current source has value 0.1 A and resistance 75 Ω, as shown in Figure 7.21(b).

Drill Exercise 7.15

Find the Thévenin equivalent circuit of Figure 7.21(a).

ANSWER: $E_{TH} = 7.5 \text{ V}$; $R_{TH} = 75 \text{ }\Omega$.   □

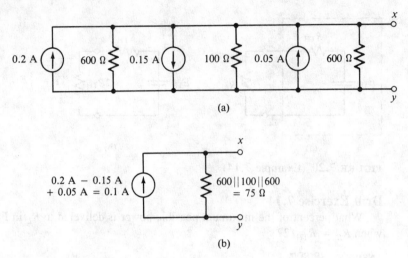

(a)

(b)

FIGURE 7.21 (Example 7.15)

A more general form of Millman's theorem states that parallel-connected *voltage* sources can be replaced by a single equivalent voltage source. Of course, we already know this statement to be true, by virtue of Thévenin's theorem. We may regard Millman's theorem as simply a special case of Thévenin's theorem, one that permits us to use a particularly simple procedure to compute the equivalent voltage source. The procedure is as follows:

1. Convert each parallel-connected voltage source to an equivalent current source. The result is a set of parallel-connected current sources.
2. Replace the parallel-connected current sources by a single equivalent current source.
3. Convert the single current source to an equivalent voltage source.

Example 7.16   (Analysis)

Find the current in the 1-kΩ resistor in Figure 7.22(a) by finding the Millman equivalent voltage source with respect to terminals x–y.

SOLUTION As shown in Figure 7.22(b), each of the three voltage sources is converted to an equivalent current source. For example, the 36-V source in series with the 18-kΩ resistor becomes a 36 V/18 kΩ = 2 mA current source in parallel with 18 kΩ. Note that the polarity of each current source is such that it produces current in the same direction as the voltage source it replaces.

The resultant of the three current sources is

$$2 \text{ mA} \uparrow + 3 \text{ mA} \uparrow + 2 \text{ mA} \downarrow = 3 \text{ mA} \uparrow$$

The parallel equivalent resistance of the three resistors is

$$18 \text{ k}\Omega \parallel 9 \text{ k}\Omega \parallel 3 \text{ k}\Omega = 6 \text{ k}\Omega \parallel 3 \text{ k}\Omega = 2 \text{ k}\Omega$$

Figure 7.22(c) shows the single equivalent current source.

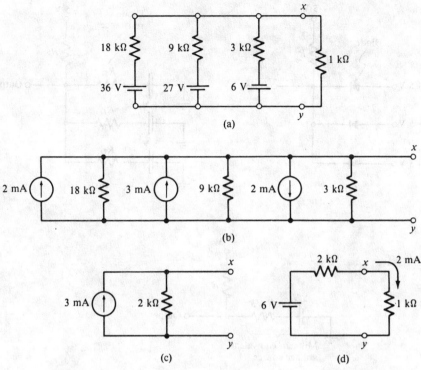

FIGURE 7.22 (Example 7.16)

Figure 7.22(d) shows the voltage source that is equivalent to the current source in Figure 7.22(c):

$$E = (3 \text{ mA})(2 \text{ k}\Omega) = 6 \text{ V}$$

When the 1-kΩ resistor is connected across the x–y terminals, the current is

$$I = \frac{6 \text{ V}}{2 \text{ k}\Omega + 1 \text{ k}\Omega} = 2 \text{ mA}$$

Drill Exercise 7.16

Repeat Example 7.16 when the polarity of the 6-V source is reversed.

ANSWER: 4.67 mA. ☐

Millman's theorem is particularly useful for analyzing electronic circuits that have two or more *inputs*, that is, two or more voltage sources whose values determine the output of the circuit. A common example is a *digital logic gate*, which may have as many as five (or more) inputs. Figure 7.23(a) shows an example of a logic gate having three inputs. The *diodes* are equivalent to resistors, as shown in Figure 7.23(b), so the three 5-V sources are effectively in parallel. Figure 7.23(c) shows that the gate can be

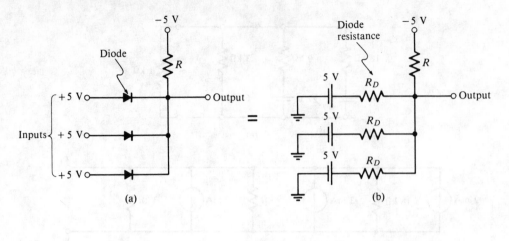

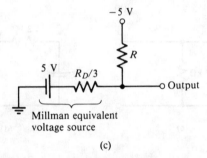

FIGURE 7.23 A Millman equivalent voltage source simplifies the analysis of a *digital logic gate*.

analyzed as if a single Millman equivalent voltage source were connected to it, a fact that greatly simplifies the analysis. Exercise 7.14D deals with such a case.

It is important to remember that Millman's theorem can be applied only in the special situation where real sources are in parallel, as in Figure 7.22(a). However, it is often the case that multisource circuits can be redrawn and resistors combined in such a way that a Millman equivalent circuit can be constructed. The next example illustrates such a case.

Example 7.17 (Analysis)

By constructing a Millman equivalent voltage source with respect to terminals x–y, find the voltage across the 40-Ω resistor in Figure 7.24(a).

SOLUTION Notice that the 120-Ω and 180-Ω resistors are in a series path and can therefore be combined into an equivalent resistance of 300 Ω. The circuit is redrawn as shown in Figure 7.24(b). It makes no difference on which side of each voltage source its series resistance is drawn, and it can be seen that the redrawn circuit consists of three parallel-connected voltage sources. The x–y terminals are relocated in the drawing to enhance identification of the parallel sources.

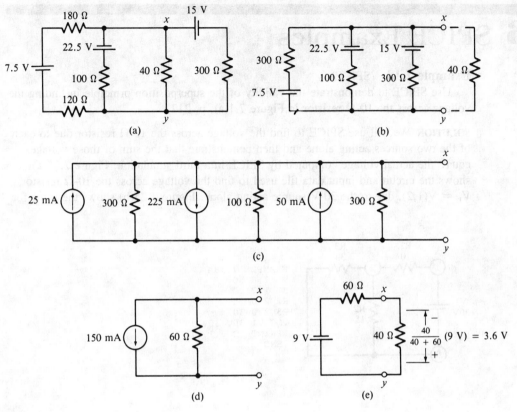

FIGURE 7.24 (Example 7.17)

Figure 7.24(c) shows the circuit after each voltage source is converted to an equivalent current source. Figure 7.24(d) is the single equivalent current source:

$$I = 25 \text{ mA} \uparrow + 225 \text{ mA} \downarrow + 50 \text{ mA} \uparrow = 150 \text{ mA} \downarrow$$

$$R = 300 \ \Omega\|300 \ \Omega\|100 \ \Omega = 150 \ \Omega\|100 \ \Omega = 60\Omega$$

Figure 7.24(e) shows the equivalent voltage source:

$$E = (150 \text{ mA})(60 \ \Omega) = 9 \text{ V}$$

With the 40-Ω resistor restored to the x–y terminals, we use the voltage-divider rule to find

$$V_{40\Omega} = \left(\frac{40}{40 + 60}\right) 9 \text{ V} = 3.6 \text{ V}$$

Drill Exercise 7.17

To what value should the 15-V source in Figure 7.24(a) be changed in order to make the current in the 40-Ω resistor equal zero?

ANSWER: 60 V.

7.6 SPICE Examples

Example 7.18 (SPICE)

Use SPICE to demonstrate the validity of the superposition principle in finding the voltage across the 10-Ω resistor in Figure 7.1(a), p. 212.

SOLUTION We will use SPICE to find the voltage across the 10-Ω resistor due to each of the two sources acting alone and then demonstrate that the sum of those voltages equals the actual voltage computed by SPICE in a third simulation. Figure 7.25(a) shows the circuit and input data file used to find the voltage across the 10-Ω resistor, $V_1 = V(1,2)$, due to the 30-V source acting alone. Figure 7.25(b) shows the same for

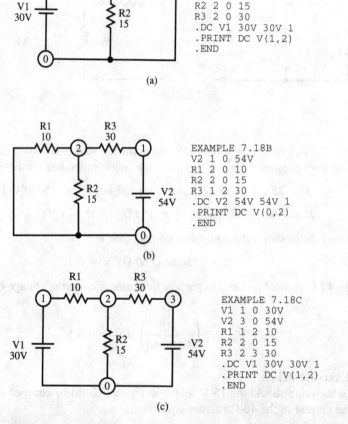

(a)

(b)

(c)

FIGURE 7.25 (Example 7.18) (a) (54-V source shorted). (b) (30-V source shorted). (c) (Both sources in circuit).

finding the voltage $V_2 = V(0,2)$, due to the 54-V source. Execution of the two programs reveals that $V_1 = 15$ V and $V_2 = -9$ V. Since V_1 and V_2 are referenced the same way (positive on left), the actual voltage due to both sources should be $V_1 + V_2 = 15$ V $- 9$ V $= 6$ V. A SPICE simulation of the circuit with both sources present [Figure 7.25(c)] shows that V(1,2) = 6 V, confirming the superposition principle.

Example 7.19   (SPICE)

Use SPICE to find the Thévenin equivalent resistance and the Thévenin equivalent voltage with respect to terminals x,y in Figure 7.11(b) (Example 7.8), p. 225.

SOLUTION To find the Thévenin equivalent resistance, we find the total resistance at terminals x,y with the 18-V source shorted. In the SPICE simulation, this resistance is determined by finding the voltage across a 1-A current source connected to terminals x,y, as shown in Figure 7.26(a). (See Example 4.24, p. 99, for a discussion of this technique.) Execution of the program reveals that the voltage across the terminals is $V_{xy} = V(1,2) = 50$ V, so we conclude that $R_{TH} = 50$ Ω. Figure 7.26(b) shows the SPICE circuit used to find the Thévenin equivalent voltage, the voltage $V_{xy} = V(1,2)$ across the open-circuited terminals x,y with the 18-V source restored. Execution of the program shows that $E_{TH} = V(1,2) = 6$ V.

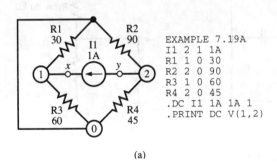

```
EXAMPLE 7.19A
I1 2 1 1A
R1 1 0 30
R2 2 0 90
R3 1 0 60
R4 2 0 45
.DC I1 1A 1A 1
.PRINT DC V(1,2)
```

(a)

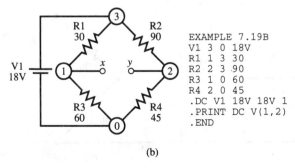

```
EXAMPLE 7.19B
V1 3 0 18V
R1 1 3 30
R2 2 3 90
R3 1 0 60
R4 2 0 45
.DC V1 18V 18V 1
.PRINT DC V(1,2)
.END
```

(b)

FIGURE 7.26 (Example 7.19) (a) [R_{TH} numerically equals V(1,2)]. (b) [$E_{TH} = V(1,2)$].

Exercises

Section 7.1 The Superposition Principle

7.1 Use the superposition principle to find the current through R_1 in each circuit of Figure 7.27. Be sure to indicate the direction of each current.

7.2 Use the superposition principle to find the voltage across R_2 in each circuit of Figure 7.27. Be sure to

indicate the polarity of each voltage.

7.3 Use the superposition principle to find the voltage across R_1 in each circuit shown in Figure 7.28. Be sure to indicate the polarity of the voltage.

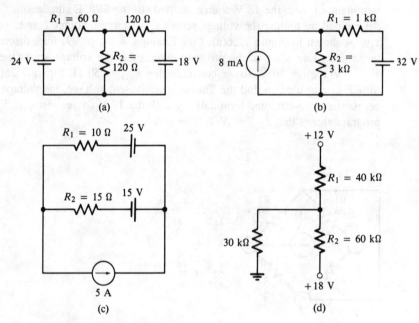

FIGURE 7.27 (Exercise 7.1)

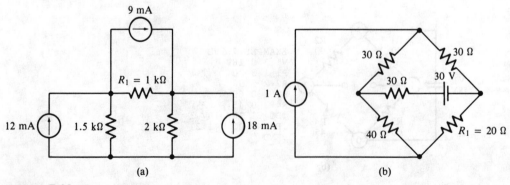

FIGURE 7.28 (Exercise 7.3)

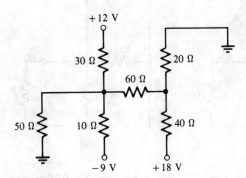

FIGURE 7.29 (Exercise 7.4)

7.4 Use the superposition principle to find the power dissipated in the 60-Ω resistor in Figure 7.29.

Section 7.2   Thévenin's Theorem

7.5 Find the Thévenin equivalent circuit of the circuitry, excluding R_1, connected to terminals $x-y$ in each part of Figure 7.30.

7.6 Find the Thévenin equivalent circuit of the circuitry, excluding R_1, connected to terminals $x-y$ in each part of Figure 7.31.

7.7 Find the voltage across R_1 in each of the circuits shown in Figure 7.32 by constructing a Thévenin equivalent circuit at the R_1 terminals. Be certain to indicate the polarity of each voltage.

7.8 In the circuits shown in Figure 7.33, find the current in resistor R_1 for each of the values of R_1 shown. Be sure to indicate the direction of each current.

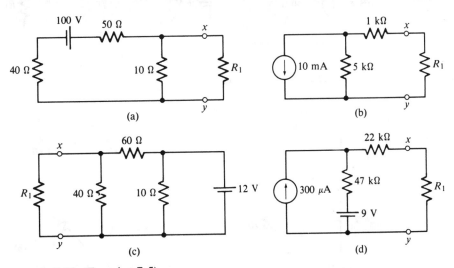

FIGURE 7.30 (Exercise 7.5)

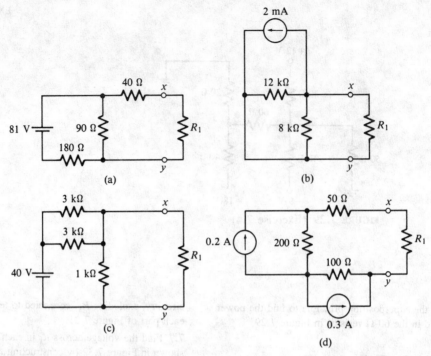

(a)

(b)

(c)

(d)

FIGURE 7.31 (Exercise 7.6)

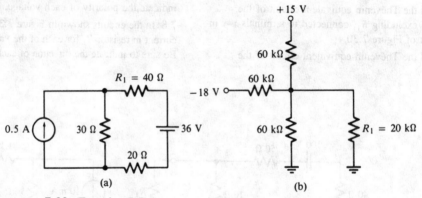

$R_1 = 40 \ \Omega$

(a)

(b)

FIGURE 7.32 (Exercise 7.7)

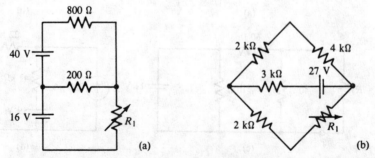

(a)

(b)

FIGURE 7.33 (Exercise 7.8) (a) $R_1 = 80 \ \Omega$, $160 \ \Omega$, $240 \ \Omega$. (b) $R_1 = 2 \ k\Omega$, $5 \ k\Omega$, $14 \ k\Omega$.

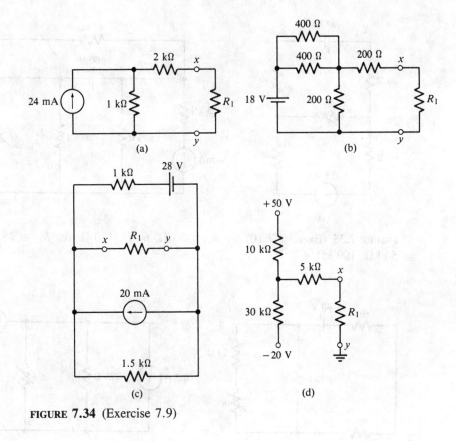

FIGURE 7.34 (Exercise 7.9)

Section 7.3   Norton's Theorem

7.9 Find the Norton equivalent circuit of the circuitry, excluding R_1, connected to terminals x–y in each part of Figure 7.34

(a) by first finding the Thévenin equivalent circuit
(b) by finding the current when the x–y terminals are shorted.

7.10 For all the values of R_1 in each circuit shown in Figure 7.35, find the power dissipated in R_1 by constructing Norton equivalent circuits.

Section 7.4   Maximum Power Transfer Theorem

7.11 For each circuit shown in Figure 7.36:
(a) Find the value of R_L necessary to obtain maximum power in R_L.
(b) Find the maximum power in R_L.

7.12 For each circuit shown in Figure 7.37:
(a) Find the value of R_L necessary to obtain maximum power in R_L.
(b) Find the maximum power in R_L.

7.13 What percent of the maximum possible power is delivered to a load if the load resistance is 10 times greater than the Thévenin resistance of the source to which it is connected?

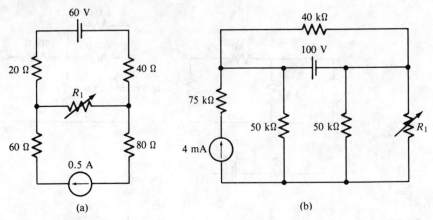

FIGURE 7.35 (Exercise 7.10) (a) $R_1 = 30\ \Omega,\ 60\ \Omega,\ 120\ \Omega$. (b) $R_1 = 25\ \mathrm{k}\Omega$, $50\ \mathrm{k}\Omega,\ 100\ \mathrm{k}\Omega$.

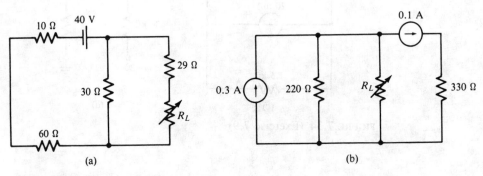

FIGURE 7.36 (Exercise 7.11)

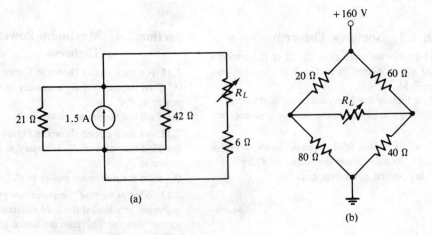

FIGURE 7.37 (Exercise 7.12)

7.14 Find the value of R_L in Figure 7.38 necessary to obtain maximum power in R_L. (The answer will be expressed in terms of R.)

Section 7.5 Millman's Theorem

7.15 Find single equivalent current sources for each of the circuits shown in Figure 7.39.

7.16 Find single equivalent current sources for each of the circuits in Figure 7.40.

7.17 By constructing a Millman equivalent voltage source at terminals x–y, find the voltage across R_1 in each circuit shown in Figure 7.41. Be sure to indicate the polarity of each voltage.

7.18 By constructing a Millman equivalent voltage source at terminals x–y, find the current through R_1 in each circuit shown in Figure 7.42. Be sure to indicate the direction of each current.

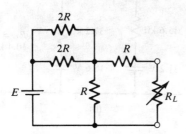

FIGURE 7.38 (Exercise 7.14)

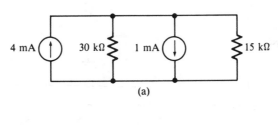

(a)

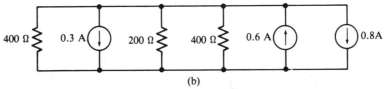

(b)

FIGURE 7.39 (Exercise 7.15)

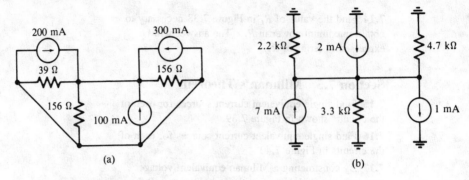

FIGURE 7.40 (Exercise 7.16)

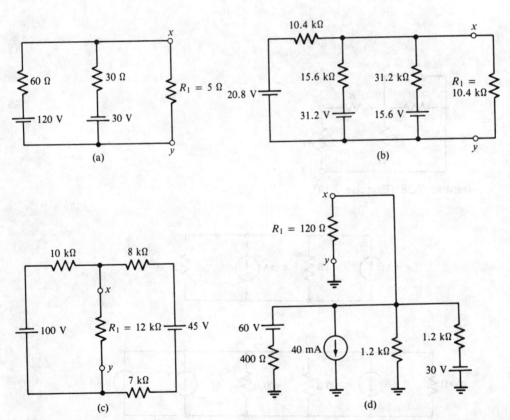

FIGURE 7.41 (Exercise 7.17)

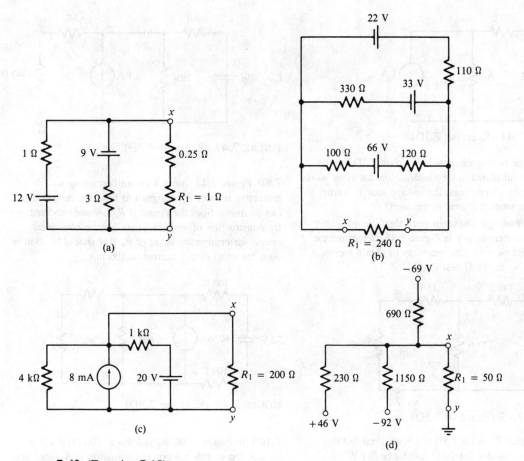

FIGURE **7.42** (Exercise 7.18)

Design Exercises

7.1D Figure 7.43 shows a *balancing* circuit that is to be designed so that the voltage V across the 120-Ω resistor can be zeroed by adjusting E when the value of R is changed. To what value must E be capable of being adjusted if it is necessary to achieve zeroing when $R = 60\ \Omega$?

7.2D What range of E is required in Exercise 7.1D if the values of R may range from 30 Ω to 240 Ω?

7.3D A small heating element in an oven used to test electronic components must dissipate 50 W. The

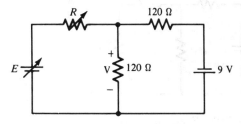

FIGURE **7.43** (Exercise 7.1D)

element is designated R_H in Figure 7.44. What should be the resistance of the element?

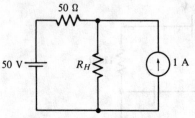

FIGURE **7.44** (Exercise 7.3D)

7.4D If the heating element in Exercise 7.3D has the resistance calculated in the exercise, to what value would it be necessary to increase the voltage source in order to double the amount of heat in the oven?

7.5D By finding a Thévenin equivalent circuit with respect to terminals x–y in Figure 7.45, determine the value of E that would be necessary to obtain a current of 75 mA in the 10-Ω resistor.

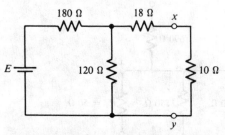

FIGURE **7.45** (Exercise 7.5D)

7.6D Repeat Exercise 7.5D if it is required that the power dissipated in the 10-Ω resistor be 0.1 W.

7.7D What value of R should be used in Figure 7.46 if it is desired to develop 6 V across the 100-Ω resistor? (Construct a Thévenin equivalent circuit with respect to terminals x–y.)

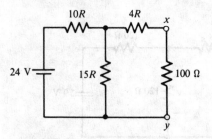

FIGURE **7.46** (Exercise 7.7D)

7.8D What should be a value of E in Figure 7.47 if it is desired to produce a current of 0.1 A in the 1340-Ω resistor? (Construct a Thévenin equivalent circuit with respect to terminals x–y.)

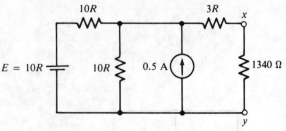

FIGURE **7.47** (Exercise 7.8D)

7.9D Figure 7.48 shows a circuit containing a protective resistor R_P designed to limit the current that can be drawn from the circuit if R_L becomes shorted. By construction of an appropriate Norton equivalent circuit, determine the value of R_P that should be used to limit the short-circuit current to 100 mA.

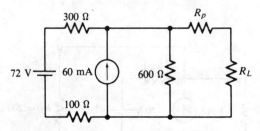

FIGURE **7.48** (Exercise 7.9D)

7.10D In Figure 7.48, R_L is fixed at 100 Ω, R_P is to be selected so that the power dissipation in R_L does not exceed 40 mW. What value of R_P should be used?

7.11D Figure 7.49 shows the equivalent circuit of an amplifier having an adjustable resistor R used to match the amplifier to load R_L. What should be the value of R when R_L is 12 Ω?

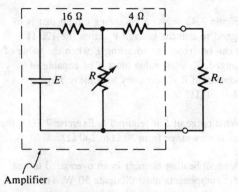

Amplifier

FIGURE **7.49** (Exercise 7.11D)

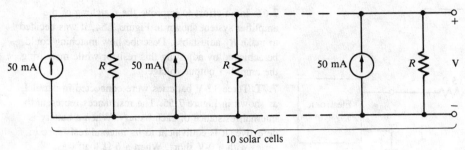

FIGURE 7.50 (Exercise 7.13D)

7.12D Over what range should R in Figure 7.49 be adjustable if the amplifier must be capable of delivering maximum power to loads ranging from 6 Ω to 18 Ω?

7.13D Figure 7.50 shows the equivalent of 10 identical solar cells connected in parallel. Each cell produces 50 mA of current. What is the maximum permissible value of each cell's internal resistance R, if the open-circuit terminal voltage V must be 1 V?

7.14D Figure 7.51 shows the equivalent circuit of a diode logic gate in which the diodes are represented by their resistances. What value of R should be used if the output must be +3 V?

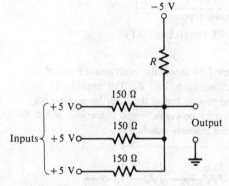

FIGURE 7.51 (Exercise 7.14D)

Troubleshooting Exercises

7.1T Figure 7.52 shows the correct schematic diagram of a certain circuit. When the circuit was constructed, the polarity of one of the voltage sources was mistakenly reversed. When troubleshooting the circuit, it was found that the voltage across the 150-Ω resistor measured 3.75 V, with the polarity shown in the figure. Which source is reversed?

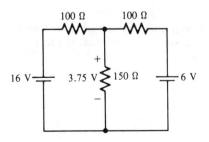

FIGURE 7.52 (Exercise 7.1T)

7.2T One of the voltage sources in Figure 7.52 is shorted. The voltage across the 150-Ω resistor is 2.25 V, with the same polarity as the 3.75 V shown in the figure. Which source is shorted?

7.3T Figure 7.53 shows an electronic system that receives power from the dual voltage source configuration shown. When the system failed to operate, it was not known whether the fault occurred in the power supply or in the electronic system itself. The electronic system was removed and the open-circuit voltage V_S produced by the supply was measured to be +8 V. By finding the Thévenin equivalent voltage at the supply terminals, determine if the supply is operating properly.

7.4T As another check on the power supply shown in Figure 7.53, it was decided to measure the Thévenin resistance at the supply terminals. Describe how this measurement should be made and the value that should be obtained if there are no resistor faults.

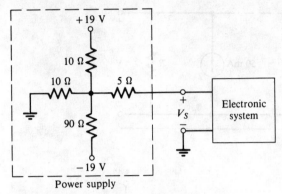

FIGURE 7.53 (Exercise 7.3T)

7.5T Figure 7.54 shows the equivalent circuit of an amplifier connected to load R_L. The amplifier is supposed to be matched to the load. Measurements revealed that the power delivered to the load is 8.5 W. Is the system properly matched?

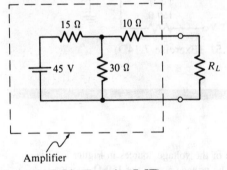

FIGURE 7.54 (Exercise 7.5T)

7.6T In an effort to improve the matching of the amplifier system shown in Figure 7.54, it was decided to make R_L adjustable. Describe how matching could be achieved by adjusting this resistor while measuring the amplifier output voltage.

7.7T Three 12-V batteries were connected in parallel, as shown in Figure 7.55. The resistances represent the internal resistance of each battery. When a battery is defective, it is equivalent to its internal resistance in series with a 0-V short. When a 6-Ω load was connected across the parallel combination, the terminal voltage measured 9 V. Is one of the batteries defective?

7.8T What terminal voltage would be measured in Exercise 7.7T if one of the batteries was defective? If two of the batteries were defective?

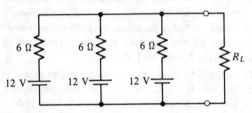

FIGURE 7.55 (Exercise 7.7T)

SPICE Exercises

7.1S Use SPICE to demonstrate the validity of the superposition principle for finding the current in R_1 in Figure 7.28(b), p. 244. (Follow the procedure of Example 7.18, p. 242).

7.2S Use SPICE to find the Norton equivalent current and the Norton equivalent resistance with respect to terminals x,y, excluding R_1, in Figure 7.34(d), p. 247.

Introduction to Fields and Electrical Physics

8.1 Coulomb's Law

In Chapter 2 we discussed the very important fact that oppositely charged particles experience a force of attraction, and that similarly charged particles experience a force of repulsion. Figure 8.1 illustrates these facts. The arrows labeled F represent the forces developed on the charged particles, and Q_1 and Q_2 represent the magnitudes of their charges. Note that each particle experiences the *same* force as its counterpart, regardless of which has the greater charge. If there are no restraints or other external forces acting on the charges, they will move in the directions indicated by the force arrows.

In Figure 8.1, r represents the distance separating the charged particles. The magnitude of the force F is directly proportional to the product of the charges, Q_1Q_2, and is inversely proportional to the *square* of the distance r separating them. These facts constitute *Coulomb's law*, which is expressed in equation form by

$$F = \frac{k\, Q_1Q_2}{r^2} \tag{8.1}$$

255

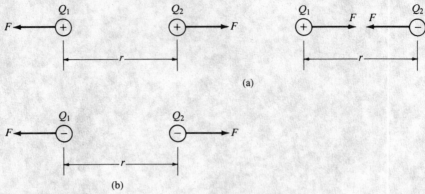

FIGURE 8.1 Forces on charged particles. (a) Opposite charges attract. (b) Like charges repel.

where F = force, newtons
k = constant, 9×10^9 N $\cdot$ m^2/C^2
Q_1, Q_2 = charges, coulombs
r = separating distance, meters

Example 8.1

In Figure 8.1(b), Q_1 is a 40-µC positive charge, Q_2 is a 100-µC negative charge, and r = 50 mm. Find the force of attraction between the charges.

SOLUTION From equation (8.1),

$$F = \frac{(9 \times 10^9)(40 \times 10^{-6} \text{ C})(100 \times 10^{-6} \text{ C})}{(50 \times 10^{-3} \text{ m})^2}$$
$$= 14.4 \times 10^3 \text{ N}$$

Drill Exercise 8.1

What charge on each particle in Example 8.1 would develop the same force at the same distance if the two charges have equal magnitudes?

ANSWER: 63.24 µC.   □

8.2 Electric Fields

Imagine a negatively charged particle fixed and isolated in space, as shown in Figure 8.2(a). If a positively charged particle is placed on the right side of the negative charge, the positive charge will be drawn to the left, as shown by the arrow, because of the force of attraction. Similarly, if the positive charge is placed above the negative charge, the positive charge will be attracted downward. The figure shows that a positive charge placed anywhere in the vicinity of the negative charge is drawn inward toward the

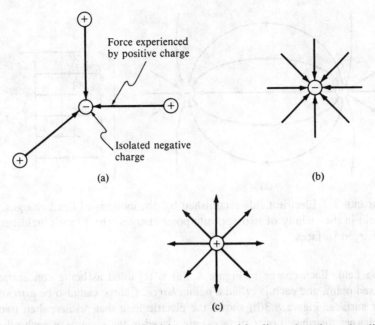

FIGURE 8.2 An electric field is represented by lines that show the direction of the force on a positive charge placed anywhere in the field. (a) Positive charges are drawn toward the isolated negative charge. (b) Electric field lines in the vicinity of an isolated negative charge. (c) Electric field in the vicinity of an isolated positive charge.

negative charge. In Figure 8.2(b), we eliminate the positive charges and simply show the lines along which positive charges would be drawn to the negative charge. The lines are a pictorial way of visualizing the reaction of any positive charge to the presence of the fixed negative charge. Of course, there are an infinite number of such lines, but we can only show several. We say that the negative charge is responsible for an *electric field*, represented by the lines in the diagram. We can use the electric field lines to predict the behavior of a *positive* charge placed anywhere in the field: the charge will experience a force in the direction shown by a line at the point where the charge is placed. Figure 8.2(c) shows the electric field established by an isolated positive charge. In this case, the lines radiate outward, because any positive charge placed in the vicinity of the isolated charge will be repelled away from it.

When two or more charged particles occupy fixed locations in space, the pattern of the electric field in the region surrounding them depends on the magnitudes of the charges and on their locations with respect to each other. For example, Figure 8.3(a) shows the electric field established by a fixed positive charge in the vicinity of a fixed negative charge. Note that lines always originate at a positive charge and terminate at a negative charge, in the sense that the direction of each line is from + to −. This direction is, again, the same as the direction that a positive charge would move if it were placed in

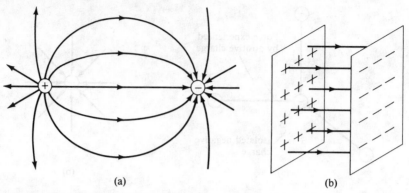

(a)   (b)

FIGURE 8.3 Electric fields established by distributions of fixed charges. (a) Electric field in the vicinity of two opposite point charges. (b) Electric field between two charged surfaces.

the field. Each charge in Figure 8.3(a) is regarded as being concentrated at a single, fixed point, and each is called a *point charge*. Charge can also be *distributed* over a line or surface. Figure 8.3(b) shows the electric field that results when two *sheet charges* (charges distributed over surfaces) are placed in the vicinity of each other.

Electric Flux Density

We have seen that an electric field pattern shows the *direction* of the force on a positive charge placed in the field. The pattern can also be interpreted to reveal the relative *magnitude* of the force experienced by a positive charge placed at any point in the field. In any region where the lines are close together (dense), the force on a positive charge is greater than it is in a region where the lines are less dense. Consider again the field pattern around an isolated negative charge, as shown in Figure 8.4. The closer we approach the charge, the more closely spaced are the lines. Thus, as confirmed by Coulomb's law, the closer we move a positive charge to the fixed negative charge, the greater the force of attraction on it. On the other hand, at distances well removed from the fixed negative charge, the field lines are less dense and the force is correspondingly smaller.

In one sense, the number of lines appearing on an electric field diagram is purely arbitrary: There are, after all, an infinite number of paths along which a positive charge could move. However, to develop a quantitative basis for comparing fields and for performing field computations, it is convenient to assume that the number of lines produced by a charge is the same as the charge in coulombs. Instead of "lines," the term electric *flux* is used, and is designated by the symbol ψ. Thus, ψ has the units of coulombs, and

$$\psi = Q \qquad \text{coulombs} \qquad (8.2)$$

Understand that the notion of flux, and its units, serves only as a convenient basis for mathematical computations. We could not draw a field diagram showing the one one-millionth of a line corresponding to a flux of 1 μC!

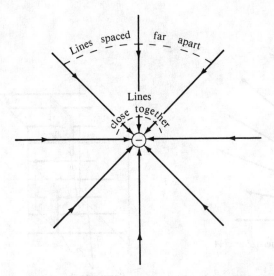

FIGURE 8.4 The more dense the field lines (the closer they are together), the greater the force on a positive charge. In the example shown, the force is greatest in the immediate vicinity of the negative charge.

We now have a basis for defining a numerical quantity that reflects how closely spaced the lines are in an electric field, that is, a measure of the *density* of the flux. We must visualize flux lines as occupying three-dimensional space around a charge, so we can picture a certain number of lines as penetrating a surface. This idea is illustrated in Figure 8.5. *Flux density D* is defined to be flux per unit area:

$$D = \frac{\psi}{A} \quad \text{coulombs/meters}^2 \tag{8.3}$$

In this definition, the surface area used in a computation must be *perpendicular* to the flux lines at every point where the flux penetrates the area. As a consequence, depending on the field pattern, the area A may in fact be that of a *curved* surface. Figure 8.5(a) shows such a case. In the figure it can be seen that the flux density is large in the region near the positive charge, because a large amount of flux penetrates the curved surface area shown. Farther away from the charge, there is less flux penetrating the same size area, so the flux density is smaller. Figure 8.5(b) shows that the flux density is the same everywhere in the region between two charged surfaces.

Example 8.2

One of the areas between the two charged surfaces in Figure 8.5(b) measures 6 mm by 8 mm, and the flux penetrating it is 96 μC. Each of the charged surfaces measures 2.5 cm by 4 cm.

(a) What is the flux density in the region between the charged surfaces?
(b) What is the total flux in the region between the charged surfaces?

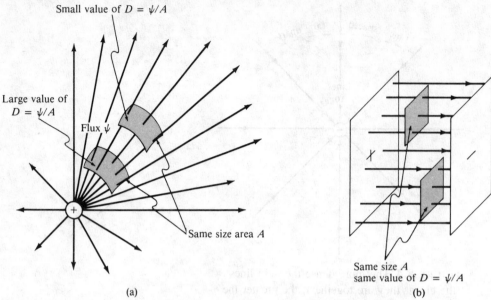

FIGURE 8.5 Flux density is flux divided by the area the flux penetrates. (a) The flux density is greatest in the region nearest the point charge. (b) The field between the charged plates is *uniform*, and the flux density is everywhere the same.

SOLUTION

(a) $A = (6 \text{ mm})(8 \text{ mm}) = (6 \times 10^{-3} \text{ m})(8 \times 10^{-3} \text{ m}) = 48 \times 10^{-6} \text{ m}^2$

$$D = \frac{\psi}{A} = \frac{96 \times 10^{-6} \text{ C}}{48 \times 10^{-6} \text{ m}^2} = 2 \text{ C/m}^2$$

(b) $A = (2.5 \text{ cm})(4 \text{ cm}) = (2.5 \times 10^{-2} \text{ m})(4 \times 10^{-2} \text{ m}) = 10^{-3} \text{ m}^2$

$\psi = DA = (2 \text{ C/m}^2)(10^{-3} \text{ m}^2) = 2 \times 10^{-3} \text{ C}$

(In the latter computation, we neglect the phenomenon called *fringing*, whereby some of the flux along the edges of the charged surfaces "escapes" from the region immediately between the surfaces, that is, bulges outward rather than remaining parallel to the other lines.)

Drill Exercise 8.2

Suppose that *each* dimension of the two charged surfaces in Figure 8.5(b) is doubled, but the total charge on each surface remains the same. What, then, is the flux density in the region between the surfaces?

ANSWER: 0.5 C/m². ☐

Electric Field Intensity

Electric field intensity, also called electric field *strength*, is the ratio of the force experienced by a charge placed in the field to the magnitude of the charge itself:

$$\mathscr{E} = \frac{F}{Q} \qquad \text{newtons/coulomb} \qquad (8.4)$$

For example, if a 40-μC charge placed in a certain field experiences a force of 20 N, the electric field intensity at that point is $\mathscr{E} = 20\text{ N}/40 \times 10^{-6}\text{ C} = 5 \times 10^5$ N/C. Notice that field intensity is a characteristic of the field itself: the "stronger" the field, the greater the force a given charge will experience when placed in the field. Furthermore, $\mathscr{E}$ can be expected, in general, to have different values at different points in the field. It is obvious that the field intensity in a region close to the positive charge in Figure 8.5(a) is greater than it is at a long distance from the charge, because a given charge will experience greater force in the region near the fixed charge, where the flux is denser.

The foregoing remarks suggest that there is a relationship between D and $\mathscr{E}$, because each is related to the magnitude of the force experienced by a charge placed in an electric field. The more dense the flux (the greater the value of D), the greater the force on such a charge. Similarly, the greater the field intensity $\mathscr{E}$, the greater the force on the charge. In fact, D and $\mathscr{E}$ are proportional to each other, as expressed by the equation

$$D = \epsilon \mathscr{E} \qquad (8.5)$$

where ϵ is a constant whose value depends on the material in which the field is established (air, glass, water, etc.). ϵ is called the *permittivity* of the material and may range in value from $10^{-9}/36\pi \approx 8.84 \times 10^{-12}$ (for a vacuum) to 6.6×10^{-8} (for certain ceramics).

Example 8.3

When a 1000-μC charge is placed at a certain point in a certain electric field, it experiences a force of 28.2 N. If the field exists in a vacuum:

(a) Find the field intensity at the point where the charge is placed.
(b) Find the flux density at the point where the charge is placed.

SOLUTION
(a) From equation (8.4),

$$\mathscr{E} = \frac{F}{Q} = \frac{28.2\text{ N}}{1000 \times 10^{-6}\text{ C}} = 28{,}200\text{ N/C}$$

(b) From equation (8.5),

$$D + = \epsilon \mathscr{E} = (8.84 \times 10^{-12})(28.2 \times 10^3) = 2.5 \times 10^{-7}\text{ C/m}^2$$

Drill Exercise 8.3

What force will be experienced by a 200-μC charge placed in the field at the same point as the charge in Example 8.3 was placed?

ANSWER: 5.64 N.   □

We noted that the flux density of the electric field in the region between two charged surfaces [Figure 8.3(b)] is everywhere the same. The field is said to be *uniform*. From the equation $D = \epsilon \mathscr{E}$, it follows that the electric field intensity $\mathscr{E}$ is everywhere the

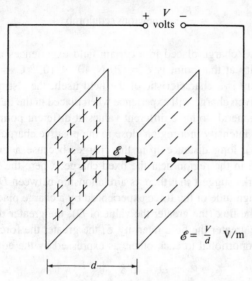

FIGURE 8.6 The electric field intensity between two charged, parallel surfaces can be computed from the voltage difference V across the surfaces.

same in the region between the charged surfaces. (These conclusions are based on ideal conditions and on two perfectly parallel surfaces, which we will assume in future discussions.) Recall from Section 2.4 that a voltage difference always exists between two oppositely charged regions. Figure 8.6 shows two charged surfaces and the voltage V that exists between them. It can be shown that the units of $\mathscr{E}$, N/C, are equivalent to volts/meter (V/m), and in the special case of Figure 8.6, the electric field intensity in the region between the two surfaces can be computed from

$$\mathscr{E} = \frac{V}{d} \quad \text{V/m} \tag{8.6}$$

where V is the voltage between the surfaces and d is their separation in meters.

Example 8.4

Two parallel surfaces are separated by 12 mm. A 600-μC charge placed between them experiences a force of 7.2 N. What is the voltage difference between the surfaces?

SOLUTION From equation (8.4),

$$\mathscr{E} = \frac{F}{Q} = \frac{7.2 \text{ N}}{600 \text{ }\mu\text{C}} = 12 \times 10^3 \text{ N/C} = 12 \times 10^3 \text{ V/m}$$

From equation (8.6),

$$V = \mathscr{E}d = (12 \times 10^3 \text{ V/m})(12 \times 10^{-3} \text{ m}) = 144 \text{ V}$$

Drill Exercise 8.4

If the total amount of charge on each surface in Example 8.4 remains the same and the distance between them is doubled, what is the force on a 200-μC charge placed between the surfaces?

ANSWER: 1.2 N.   □

8.3 Breakdown Strength

In our discussion of the electric field between two charged surfaces, we assumed that the material between the surfaces is an insulator. If that were not the case, the electrons would travel to the positive surface under the influence of the force of attraction, and the charge there would be neutralized.

If the field intensity in the insulator between the surfaces becomes too great, the insulator will *break down*, in the sense that it will act as a conductor rather than as an insulator. The reason for breakdown is that electrons in the insulator atoms break loose from their parent atoms under the influence of the strong electric force created by the field. The insulator then becomes *ionized* (as discussed in Chapter 2) and serves as a good conductor of electric current. Different materials break down at different values of electric field intensity. The value of $\mathscr{E}$ at which breakdown occurs in a material is called its *breakdown strength*. Table 8.1 lists approximate (typical) breakdown strengths for several different materials.

A familiar example of insulator breakdown occurs when lightning strikes the earth. In this situation, a very intense electric field is developed between a charged cloud and the surface of the earth, due to the buildup of charge in the cloud. Air, which is normally an insulator, breaks down and conducts charge to the earth. To gain an appreciation for the magnitude of the electrical quantities involved, the altitude of a cloud on a stormy day may be about 100 m, so from equation (8.6) we find the voltage between cloud and earth to be

$$V = d\mathscr{E}_{air}(\text{breakdown}) = (100 \text{ m})(3 \times 10^6 \text{ V/m}) = 3 \times 10^8 \text{ V}$$

$$\text{or} \quad 300 \text{ million volts!}$$

The current during such a discharge may be as great as 200,000 A.

TABLE 8.1 Breakdown Strengths of Materials

Material	Typical Breakdown Strength (V/m)
Air	3×10^6
Porcelain	7×10^6
Bakelite	16×10^6
Rubber	27×10^6
Teflon	60×10^6
Glass	12×10^7
Mica	20×10^7

Example 8.5

The voltage across two oppositely charged surfaces separated by a 0.12 mm thickness of porcelain is being increased by increasing the amount of charge on them. At what voltage will the electrons on the negative surface penetrate the porcelain and neutralize the positive surface?

SOLUTION The negative surface *discharges* when the voltage is sufficient to break down the porcelain insulator. By equation (8.6),

$$\mathscr{E} = \frac{V}{d}$$

From Table 8.1, porcelain breaks down at $\mathscr{E} = 7 \times 10^6$ V/m, so

$$7 \times 10^6 \text{ V/m} = \frac{V}{0.12 \times 10^{-3} \text{ m}}$$

$$V = (7 \times 10^6 \text{ V/m})(0.12 \times 10^{-3} \text{ m}) = 840 \text{ V}$$

Drill Exercise 8.5

What thickness of porcelain would be required in Example 8.5 if breakdown were not permitted to occur for voltages up to 35 kV?

ANSWER: 5 mm.   □

8.4 Voltage

In the SI system, voltage is derived from power and current. Specifically, 1 volt is defined to be the potential difference that exists between two points when the current between the points is 1 ampere and the rate of energy consumption (power) is 1 watt. Of course, this definition follows directly from the now familiar power equation, $P = VI$:

$$V = \frac{P}{I} \qquad 1 \text{ V} = \frac{1 \text{ W}}{1 \text{ A}}$$

However, a definition based on power reveals little about the true nature of voltage, or potential difference. We wish now to develop the notion of voltage in a more insightful way, using some of the concepts that have been introduced in connection with electric fields.

Recall from Chapter 2 that we described voltage as a measure of the ability of a source to produce current. In order to produce current, a two-terminal voltage source creates a force that repels electrons from its negative terminal and attracts them to its positive terminal. As we now know, there is an electric field between the two oppositely charged terminals. The greater the field intensity, the greater the force on a charge placed in the field. One way we could measure how effective that field is in forcing charge to move through it would be to determine how much work would have to be performed

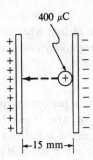

400 µC

|←15 mm→|

FIGURE 8.7 (Example 8.6)

(i.e., how much energy would have to be expended) to move a charge in *opposition* to the field. For example, the greater the work required to move a positive charge from the negative to the positive terminal of a voltage source, the greater the voltage of the source. We say that the potential difference between two points is 1 volt when 1 joule of work must be performed to move 1 coulomb of charge from one point to the other. In other words, voltage is the ratio of work to charge:

$$\text{(volts) } V = \frac{W}{Q} = \frac{\text{joules of work}}{\text{coulombs of charge}} \tag{8.7}$$

We see that 1 volt is the same as 1 joule per coulomb.

Regarding voltage as the ratio of work to charge helps build an intuitive understanding of the concept, because it conveys the idea of potential as an ability to accomplish something useful: Voltage can put charge in motion. Similarly, a voltage *drop* is the loss of ability, or potential, to move charge. The greater the drop, the less energy is available to move charge from one point to another. Understand, however, that this notion of voltage is *equivalent* to 1 volt = 1 watt/ampere; it is not an alternative definition. The units volts, watts/ampere, and joules/coulomb are all equivalent and can all be reduced to the same SI *base* units.

Example 8.6

24×10^{-3} J of energy is required to move the 400-µC positive charge shown in Figure 8.7 from the negatively charged surface to the positively charged surface.

(a) What is the potential difference between the two charged surfaces?
(b) What is the electric field intensity in the region between the surfaces?

SOLUTION

(a) The number of joules of energy required to move the charge equals the amount of work that must be performed on it. By equation (8.7),

$$V = \frac{W}{Q} = \frac{24 \times 10^{-3} \text{ J}}{400 \times 10^{-6} \text{ C}} = 60 \text{ V}$$

(b) By equation (8.6),

$$\mathscr{E} = \frac{V}{d} = \frac{60 \text{ V}}{15 \times 10^{-3} \text{ m}} = 4000 \text{ V/m}$$

Drill Exercise 8.6

How much work is performed on a 50-μC negative charge when it is moved from the positively charged surface in Example 8.6 to the negatively charged surface?

ANSWER: 3 mJ. ☐

We must be careful to observe sign conventions when performing potential calculations. For example, if $V_{ab} = + 10$ V, then 10 J of work must be performed to move a negative charge of 1 C from b to a, or to move a positive charge of 1 C from a to b (against the field, in both cases). On the other hand, if $V_{ab} = - 10$ V, then $V_{ba} = + 10$ V, and 10 J of work must be performed to move a positive charge of 1 C from a to b, or a negative charge of 1 C from b to a. These ideas are illustrated in Figure 8.8.

Inside a voltage source, electrical charge is indeed forced to move in opposition to the field, that is, electrons are forced from the positive terminal to the negative terminal, as discussed in Chapter 2. The energy necessary to accomplish this movement is obtained from the source itself, as, for example, from chemical energy in a battery or from mechanical energy in a generator. On the other hand, when charge moves through the passive components of a circuit (resistors), energy is *lost* through conversion to heat, and there is a corresponding drop in potential.

8.5 Resistance of Conductors

Recall that resistance is a measure of the extent to which a material interferes with, or resists, the flow of charge through it. When an electric field is established in a conductor by connecting a voltage source across opposing ends of it, free electrons in the material experience a force and move in a direction opposite to that of the field. (Remember that the direction of a field is the same as the direction that *positive* charge moves; it is also the same as the direction of *conventional* current.) As electrons move through the conductor, they collide with each other and with other atoms. These collisions hinder the smooth flow of charge and reduce the flow rate. Thus, resistance in a conductor is primarily a consequence of random collisions of free electrons. As we shall see, the

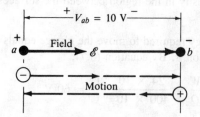

FIGURE 8.8 Work is required to move a negative charge from a to b or to move a positive charge from b to a.

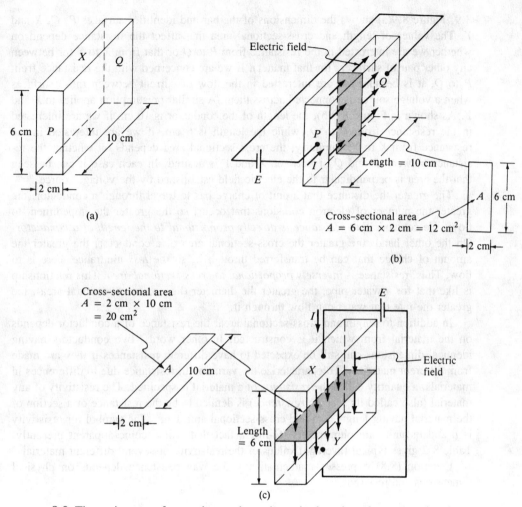

FIGURE 8.9 The resistance of a conductor depends on its length and cross-sectional area. The length and cross-sectional area depend on the direction through the conductor that the resistance is computed. (a) Dimensions' of a bar conductor. (b) Current flow perpendicular to surfaces P and Q. (c) Current flow perpendicular to surfaces X and Y.

physical dimensions of the conductor and its temperature influence the actual value of that resistance.

The resistance of a body is affected by both its length and *cross-sectional area*. Cross-sectional area is the area of a surface lying *perpendicular* to the direction of the electric field through the body, and length is the dimension *parallel* to the electric field. In other words, cross-sectional area is the area through which charge flows, and length is the total distance that charge travels through a conductor. When performing resistance computations, we must be very careful to use the correct physical dimensions, in reference to the electric field. Consider, for example, the rectangular bar shown in Figure

8.9. Figure 8.9(a) shows the dimensions of the bar and identifies surfaces P, Q, X and Y. The values of length and cross-sectional area that affect the resistance depend on whether we are interested in the resistance from P to Q or that from X to Y (or between any other pair of surfaces, for that matter). If we are concerned with the resistance from P to Q, it is because we are interested in the flow of current between those surfaces when a voltage source is connected across them. A similar presumption applies to X and Y. As shown in Figure 8.9(b), the length of the conductor is 10 cm if we are interested in the resistance from P to Q, while the length is 6 cm, if we are interested in the resistance from X to Y. Similarly, the cross-sectional area depends on whether the resistance between P and Q or between X and Y is desired. In each case it can be seen that the area is perpendicular to the electric field established by the voltage source.

The greater the distance that a unit of charge has to travel through a conductor, the greater the number of electron collisions that occur, so the greater the impediment to flow. We conclude that *resistance is directly proportional to the length of a conductor*. On the other hand, the greater the cross-sectional area of a conductor, the greater the amount of charge that can be transferred through it, so the *less* hindrance there is to flow. Thus, resistance is *inversely proportional to cross-sectional area*. This relationship is like that for a water pipe: the greater the diameter (hence, cross-sectional area), the greater the rate that water can flow through it.

In addition to length and cross-sectional area, the resistance of a conductor depends on the material from which it is constructed. In other words, two conductors having identical dimensions can still be expected to have different resistances if they are made from different materials. To characterize the variation in resistance due to differences in materials, a quantity called the *resistivity* of a material is specified. The resistivity of any material (also called its *specific resistance*) is defined to be the resistance of a section of the material having length 1 m and cross-sectional area 1 m^2. The symbol for resistivity is ρ and its units are ohm-meters (Ω-m), a fact that will become apparent presently. Table 8.2 gives typical (average) values of the resistivity of several different materials.

Equation (8.8) expresses mathematically the way resistance depends on physical dimensions and resistivity:

$$R = \frac{\rho l}{A} \quad \text{ohms} \tag{8.8}$$

TABLE 8.2 Resistivity of Materials

Material	Typical Resistivity, ρ (Ω-m)
Silver	1.64×10^{-8}
Copper	1.72×10^{-8}
Gold	2.45×10^{-8}
Aluminum	2.83×10^{-8}
Tungsten	5.5×10^{-8}
Nickel	7.8×10^{-8}
Constantan	49×10^{-8}
Nichrome	115×10^{-8}
Carbon	3500×10^{-8}

where R = resistance, Ω

ρ = resistivity, Ω-m

l = length, m

A = cross-sectional area, m^2

Solving (8.8) for ρ shows that resistivity has the units of ohm-meters:

$$\rho = \frac{RA}{l}$$

(8.9)

$$\frac{\Omega\text{-}m^2}{m} = \Omega\text{-}m$$

Note that equation (8.8) correctly expresses the fact that resistance is directly proportional to length and inversely proportional to cross-sectional area, as discussed previously. Furthermore, when $l = 1$ m and $A = 1$ m^2, $R = \rho$, confirming the definition of ρ.

Example 8.7

Find the resistance between surfaces P and Q of the bar conductor shown in Figure 8.9, assuming that it is made of aluminum.

SOLUTION Example 8.6 shows that $l = 10$ cm $= 0.1$ m and $A = 12$ $cm^2 = 12 \times 10^{-4}$ m^2. From Table 8.2, we find the resistivity of aluminum to be $\rho = 2.83 \times 10^{-8}$ Ω-m. Thus, using equation (8.8), we find

$$R = \frac{\rho l}{A} = \frac{(2.83 \times 10^{-8})(0.1)}{12 \times 10^{-4}} = 2.36 \ \mu\Omega$$

Drill Exercise 8.7

Find the resistance between surfaces X and Y of the bar conductor shown in Figure 8.9, assuming that it is made of copper.

ANSWER: $0.516 \ \mu\Omega$. □

Example 8.8

What length should a 22-Ω cylindrical carbon resistor have if its diameter is 0.15 mm?

SOLUTION The cross-sectional area is the area of a circle:

$$A = \frac{\pi d^2}{4} = \frac{\pi(0.15 \times 10^{-3} \ m)^2}{4} = 1.767 \times 10^{-8} \ m^2$$

From Table 8.2, the resistivity of carbon is 3500×10^{-8} Ω-m. From equation (8.8),

$$l = \frac{AR}{\rho} = \frac{(1.767 \times 10^{-8} \ m^2)(22 \ \Omega)}{3500 \times 10^{-8} \ \Omega\text{-}m} = 0.011 \ m = 11 \ mm$$

Drill Exercise 8.8

What would be the resistance of a copper wire having the same dimensions as the carbon resistor in Example 8.8?

ANSWER: $0.0107 \ \Omega$. □

Wire Resistance

Resistance computations involving conductors with circular cross sections, such as wire and cable, are often performed using the unit of cross-sectional area called a *circular mil* (CM). The circular mil is not an SI unit, but it is still widely used in the electrical power industry in the United States. One *mil* is 0.001 in. and 1 CM is the area of a circle having a diameter of 1 mil. The cross-sectional area, in circular mils, of a circle having diameter d mils is simply d^2 CM:

$$A = d^2 \text{(mils)} \qquad \text{CM} \tag{8.10}$$

Note carefully that d in equation (8.10) must be expressed in mils (thousandths of inches).

Example 8.9

Find the cross-sectional area, in CM, of a wire whose diameter is 0.02 in.

SOLUTION

$$(0.02 \text{ in})\left(\frac{1 \text{ mil}}{0.001 \text{ in.}}\right) = 20 \text{ mils}$$

$$A = d^2 \text{(mils)} = 20^2 = 400 \text{ CM}$$

Drill Exercise 8.9

What is the diameter, in inches, of a wire whose cross-sectional area is 1225 CM?

ANSWER: 0.035 in.   ☐

Equation (8.8), $R = \rho l/A$, can be used to compute wire resistance when A is expressed in CM, provided that the other variables in the equation have consistent units. If l is in feet, then resistivity ρ must be in CM · Ω/ft:

$$\rho = \frac{AR}{l} \frac{\text{CM} \cdot \Omega}{\text{ft}} \tag{8.11}$$

Table 8.3 shows typical values of ρ in CM · Ω/ft for several different materials.

TABLE 8.3 Resistivity of Materials

Material	Resistivity (CM · Ω/ft)
Silver	9.9
Copper	10.37
Gold	14.7
Aluminum	17.0
Tungsten	33.0
Nickel	47.0
Iron	74.0
Carbon	2100

Example 8.10

Find the resistance of 1 mile of copper wire having a diameter of 0.30 in.

SOLUTION

$$(0.30 \text{ in.})\left(\frac{1 \text{ mil}}{0.001 \text{ in.}}\right) = 300 \text{ mils}$$

$$A = 300^2 \text{ CM} = 9 \times 10^4 \text{ CM}$$

From Table 8.3, $\rho = 10.37$ CM $\cdot$ Ω/ft, and since 1 mile $=$ 5280 ft, we find

$$R = \frac{\rho l}{A} = \frac{(10.37)(5280)}{9 \times 10^4} = 0.608 \ \Omega$$

Drill Exercise 8.10

How long is a copper wire whose total resistance is 1 Ω and whose diameter is 0.4 in.?

ANSWER: 15,429 ft, or 2.92 miles.

The American Wire Gage

Wire sizes have been standardized in the United States through use of American Wire Gage (AWG) numbers. In the numbering system used for this standard, wires having larger cross-sectional areas are assigned smaller numbers. For example, an AWG No. 20 wire has a cross-sectional area approximately 10 times *greater* than that of an AWG No. 30 wire. To help develop a mental image of the sizes corresponding to various wire gages, Figure 8.10 shows the actual sizes of several different gages. Very large sizes are designated by 0, 00, 000, and so on, where additional zeros correspond to larger sizes. Table 8.4 lists AWG wire sizes ranging from No. 0000 to No. 40, their cross-sectional areas in circular mils, and their approximate diameters in mils. Also shown is the resistance per thousand feet of *copper* wire having each of the AWG sizes in the table.

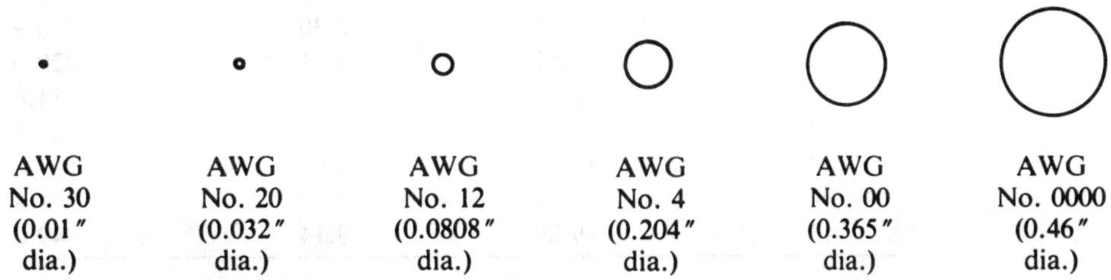

| AWG No. 30 (0.01″ dia.) | AWG No. 20 (0.032″ dia.) | AWG No. 12 (0.0808″ dia.) | AWG No. 4 (0.204″ dia.) | AWG No. 00 (0.365″ dia.) | AWG No. 0000 (0.46″ dia.) |

FIGURE 8.10 Actual sizes of some American wire gages (AWG numbers).

TABLE 8.4 American Wire Gage (AWG) Sizes

AWG No.	Area (CM)	Approximate Diameter (mils)	Ω/1000 ft at 20°C (copper wire)
0000	211,600	460	0.0490
000	167,810	410	0.0618
00	133,080	365	0.0780
0	105,530	325	0.0983
1	83,694	289	0.1240
2	66,373	258	0.1563
3	52,634	229	0.1970
4	41,742	204	0.2485
5	33,102	182	0.3133
6	26,250	162	0.3951
7	20,816	144	0.4982
8	16,509	128	0.6282
9	13,094	114	0.7921
10	10,381	102	0.9989
11	8,234.0	90.7	1.260
12	6,529.0	80.8	1.588
13	5,178.4	72.0	2.003
14	4,106.8	64.1	2.525
15	3,256.7	57.1	3.184
16	2,582.9	50.8	4.016
17	2,048.2	45.3	5.064
18	1,624.3	40.3	6.385
19	1,288.1	35.9	8.051
20	1,021.5	32.0	10.15
21	810.10	28.5	12.80
22	642.40	25.3	16.14
23	509.45	22.6	20.36
24	404.01	20.1	25.67
25	320.40	17.9	32.37
26	254.10	15.9	40.81
27	201.50	14.2	51.47
28	159.79	12.6	64.90
29	126.72	11.3	81.83
30	100.50	10.0	103.2
31	79.70	8.92	130.1
32	63.21	7.95	164.1
33	50.13	7.08	206.9
34	39.75	6.30	260.9
35	31.52	5.61	329.0
36	25.00	5.00	414.8
37	19.83	4.45	523.1
38	15.72	3.96	659.6
39	12.47	3.53	831.8
40	9.89	3.14	1049.0

Example 8.11

A 60-ft copper cable connecting two electrical installations will carry a current of 40 A.

(a) If the voltage drop in the cable is not allowed to exceed 2 V, what is the smallest AWG size that can be used?
(b) What is the resistance of 250 ft of copper cable having the AWG size found in part (a)?

SOLUTION

(a) $R \le \dfrac{2 \text{ V}}{40 \text{ A}} = 0.05 \ \Omega$

From equation (8.8),

$$0.05 = \frac{\rho l}{A} = \frac{(10.37)(60)}{A}$$

$$A = 12,444 \text{ CM}$$

From Table 8.4, the smallest AWG wire with cross-sectional area greater than 12,444 CM is AWG No. 9 ($A = 13,094$ CM).

(b) From Table 8.4, the resistance of AWG No. 9 wire is 0.7921 Ω per 1000 ft. Thus, the resistance of 250 ft is

$$R = (250 \text{ ft})\left(\frac{0.7921 \ \Omega}{1000 \text{ ft}}\right) = 0.198 \ \Omega$$

Drill Exercise 8.11

If AWG No. 12 wire must be used for the application in Example 8.11, what is the greatest length it can have?

ANSWER: 31.48 ft.   □

8.6 Resistance of Semiconductors

As discussed in Chapter 2, a semiconductor is a material that is neither a good conductor nor a good insulator. It is nevertheless an extremely useful material and is the fundamental ingredient of most modern electronic devices, such as transistors and integrated circuits. *Silicon* is the most widely used semiconductor. In the design and construction of electronic devices, it is important to be able to calculate the resistance of various semiconductor shapes, and equation (8.8), $R = \rho l/A$, can be used for that purpose. However, in these applications, resistivity ρ is considered a *variable*, because its value can be controlled during the manufacturing process. By altering the number of impurity atoms added to the semiconductor material (a process called *doping*), the resistivity can be adjusted over a wide range. A typical problem in the design of a semiconductor resistor is to specify the resistivity and the dimensions of a silicon bar, or strip, necessary to achieve a desired resistance.

In the semiconductor industry, *conductivity* is sometimes specified instead of resistivity. Conductivity, σ, is simply the reciprocal of resistivity:

$$\sigma = \frac{1}{\rho} \qquad (8.12)$$

Since $1/\Omega$ = siemens, the units of σ are siemens/meter:

$$\sigma = \frac{1}{\Omega \cdot m} = S/m \qquad (8.13)$$

Do not confuse conductivity with conductance (nor resistivity with resistance). We should note that the resistivity and conductivity of semiconductor materials depends not only on the number of free electrons in the material, as it does in conductors, but also on the number of *holes* in the material. A hole is essentially the absence of an electron in the crystalline structure of the material, and it has many interesting properties that we will leave for other studies of semiconductor theory.*

Figure 8.11 shows a strip of semiconductor material with dimensions l (length), W (width), and t (thickness). We are interested in the resistance of the strip in the direction parallel to the field, so the cross-sectional area is

$$A = tW \qquad \text{square meters} \qquad (8.14)$$

Substituting for A in equation (8.8), we obtain

$$R = \frac{\rho l}{A} = \frac{\rho l}{tW} \qquad \text{ohms} \qquad (8.15)$$

---

*See, for example, T. F. Bogart, *Electronic Devices and Circuits*, 3rd ed., Merrill/Macmillan Publishing Company, Columbus, Ohio, 1993.

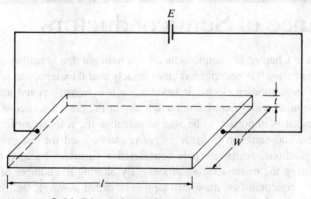

FIGURE 8.11 Dimensions of a semiconductor resistor.

The quantity l/W is called the *aspect ratio*, a:

$$a = \frac{l}{W} \quad \text{(dimensionless)} \tag{8.16}$$

Aspect ratio is a widely used variable in the semiconductor industry. Substituting (8.16) into (8.15), we obtain an expression for resistance in terms of aspect ratio:

$$R = \frac{\rho a}{t} \tag{8.17}$$

If $l = W$, then $a = 1$, and the cross section is square. In that circumstance, equation (8.17) gives $R = \rho/t$. The ratio ρ/t is called the *sheet resistivity R_s* of the material and has the units *ohms/square*:

$$R_s = \frac{\rho}{t} \quad \text{ohms/square} \tag{8.18}$$

Sheet resistivity is another widely used variable in the semiconductor industry. Substituting (8.18) into (8.17), we obtain resistance in terms of aspect ratio and sheet resistivity:

$$R = R_s a \tag{8.19}$$

Example 8.12

A 1.2-kΩ semiconductor resistor is to be designed using silicon whose conductivity is 2000 S/m. If the silicon has a thickness of 5 microns (1 micron = 10^{-6} m):

(a) Find the sheet resistivity of the material.
(b) Find the required aspect ratio.
(c) Find the length of the resistor, if its width is to be 25 μm.

SOLUTION

(a) $\rho = \dfrac{1}{\sigma} = \dfrac{1}{2000 \text{ S/m}} = 5 \times 10^{-4}$ Ω-m

$R_s = \dfrac{\rho}{t} = \dfrac{5 \times 10^{-4} \text{ Ω-m}}{5 \times 10^{-6} \text{ m}} = 100$ Ω/square

(b) From equation (8.19),

$$1.2 \times 10^3 \text{ Ω} = R_s a = 100a$$
$$a = 12$$

(c) $a = 12$, $12 = l/25$ microns
$l = (12)(25 \text{ microns}) = 300 \text{ microns} = 0.3 \text{ mm}$

Drill Exercise 8.12

What would be the resistance of the resistor whose dimensions were determined in Example 8.12 if the conductivity of the material was changed to 1500 S/m?

ANSWER: 1600 Ω.

8.7 Temperature Dependence of Resistance

Heat energy strongly influences the behavior of electrons in both conductors and semiconductors, so the resistance of both types of materials is affected by changes in temperature. In conductors, the dominant affect of increased temperature is an increase in the random motions of free electrons, which causes more frequent electron collisions, and which therefore makes it more difficult to transfer charge. Thus, *the resistance of conductors increases with temperature*. In semiconductors, the dominant effect of an increase in temperature is the production of greater numbers of free electrons. Increased heat gives valence electrons the additional energy necessary to escape from their parent atoms and to become charge carriers (i.e., free electrons). Since a semiconductor has vastly fewer free electrons than a conductor, any increase in their number causes the semiconductor to behave more like a conductor. Thus, *the resistance of semiconductors decreases with temperature*.

Figure 8.12 shows a typical plot of the resistance of a conductor versus temperature. The plot clearly shows that resistance increases with temperature. Over the range of temperatures encountered in most practical applications, the plot is *linear* (i.e., resistance is directly proportional to temperature). At *absolute zero* (0 kelvin, or $-273.15°C$) all electron activity theoretically ceases and resistance becomes zero. As can be seen in Figure 8.12, the plot is not linear in the region near absolute zero, but levels out as it approaches that temperature. Since we do not ordinarily deal with such low temperatures, we can make the simplifying assumption that the plot continues to descend linearly as shown by the dashed line, to a point called the *inferred absolute zero*, T_Z, where it

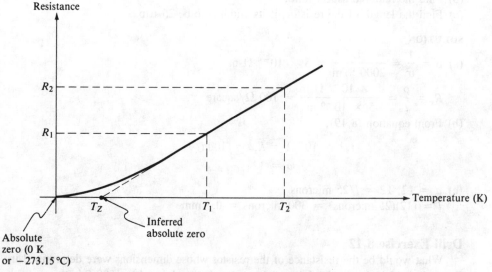

FIGURE 8.12 Typical plot of the resistance of a conductor versus temperature, in kelvin.

intersects the horizontal axis. Under that assumption, it is easy to show, using plane geometry, that the relationship between any two temperatures and the resistances at those temperatures is (see Figure 8.12)

$$\frac{T_1 - T_Z}{R_1} = \frac{T_2 - T_Z}{R_2} \tag{8.20}$$

where T_Z = inferred absolute zero temperature, kelvin (dependent on the material)
R_1 = resistance at temperature T_1, kelvin
R_2 = resistance at temperature T_2, kelvin

Table 8.5 lists approximate values of the inferred absolute zero temperatures for several different materials. By substituting the appropriate value for T_Z into equation (8.20), we can find the resistance of a material at any temperature, given its resistance at some other temperature. The next example demonstrates such a computation. Remember that temperature T_K in kelvin is related to temperature T_C in °C by

$$T_K = T_C + 273.15 \tag{8.21}$$

Example 8.13

At 20°C, the resistance of a copper conductor is 1.5 Ω. What is its resistance at 100°C?

SOLUTION

$$T_1 = 20 + 273.15 = 293.15 \text{ K}$$

$$T_2 = 100 + 273.15 = 373.15 \text{ K}$$

From Table 8.5, the inferred absolute zero of copper is $T_Z = 38.65$ K. By equation (8.20),

$$\frac{293.15 - 38.65}{1.5 \ \Omega} = \frac{373.15 - 38.65}{R_2}$$

TABLE 8.5 Inferred Absolute Zero Temperatures

Material	Inferred Absolute Zero, T_Z (K)
Silver	30.15
Copper	38.65
Gold	−0.85
Aluminum	37.15
Tungsten	69.15
Nickel	126.15
Iron	111.15
Constantan	−124,727
Nichrome	−1977

Solving for R_2 gives

$$R_2 = \left(\frac{373.15 - 38.65}{293.15 - 38.65} \right) 1.5\,\Omega = 1.97\,\Omega$$

Drill Exercise 8.13

What is the resistance of the conductor in Example 8.13 at 0°C?

ANSWER: 1.382 Ω.   □

Temperature Coefficient of Resistance

Solving equation (8.20) for R_2 gives

$$R_2 = \frac{1}{T_1 - T_Z} R_1 (T_2 - T_Z) \qquad (8.22)$$

The quantity $1/(T_1 - T_Z)$ is called the *temperature coefficient of resistance at temperature T_1* and is designated by α_1:

$$\alpha_1 = \frac{1}{T_1 - T_Z} \qquad (8.23)$$

The subscript 1 in α_1 reminds us that the value of this coefficient depends on the temperature T_1 at which it is calculated. By substituting equation (8.23) into (8.22) and performing some algebraic manipulations, we find that (8.22) can be expressed as

$$R_2 = R_1 [1 + \alpha_1 (T_2 - T_1)] \qquad (8.24)$$

Notice that $T_2 - T_1$ has the same value whether temperatures T_1 and T_2 are expressed in kelvin or in °C, so equation (8.24) can be used to find the resistance R_2 at temperature T_2 using either set of temperature units. Table 8.6 lists typical values for the temperature coefficients of resistance of several materials at $T_1 = 20°C = 293.15$ K.

TABLE 8.6 Temperature Coefficients of Resistance at 20°C (293.15 K)

Material	α_1
Silver	3.8×10^{-3}
Copper	3.93×10^{-3}
Gold	3.4×10^{-3}
Aluminum	3.91×10^{-3}
Tungsten	5×10^{-3}
Nickel	6×10^{-3}
Iron	5.5×10^{-3}
Constantan	8×10^{-6}
Nichrome	5×10^{-4}

Example 8.14

(a) A copper conductor has resistance 50 Ω at 20°C. Use the temperature coefficient of resistance to find the resistance at 5°C.
(b) Using the temperature coefficient of resistance at 5°C, find the resistance of the conductor at 100°C.

SOLUTION

(a) From Table 8.6, the temperature coefficient of copper at $T_1 = 20$°C is $\alpha_1 = 3.93 \times 10^{-3}$. Therefore, by equation (8.24), the resistance R_2 at $T_2 = 5$°C is

$$R_2 = 50[1 + 3.93 \times 10^{-3}(5° - 20°)]$$
$$= 50(1 - 0.05895) = 47.05 \ \Omega$$

Notice that T_2 is less than T_1 in this case, so the resistance at T_2 is less than the resistance at T_1.

(b) Since Table 8.6 only gives values for α_1 at $T_1 = 20$°C, we must use equation (8.23) to calculate α_1 at $T_1 = 5$°C = 278.15 K:

$$\alpha_1 = \frac{1}{T_1 - T_Z} = \frac{1}{278.15 - 38.65} = 4.175 \times 10^{-3}$$

Then, from equation (8.24), using $R_1 = 47.05 \ \Omega$ at $T_1 = 5$°C, we find

$$R_2 = 47.05[1 + 4.175 \times 10^{-3}(100 - 5)] = 65.71 \ \Omega$$

Drill Exercise 8.14

Using the temperature coefficient of resistance at 100°C, find the resistance of the conductor in Example 8.14 at 0°C.

ANSWER: 46.06 Ω. □

Equation (8.24) may be written in the form

$$R_2 = R_1(1 + \alpha_1 \ \Delta T) \tag{8.25}$$

where $\Delta T = T_2 - T_1 = $ *change* in temperature. If α_1 were equal to zero, we see that R_2 would equal R_1 regardless of the value of ΔT (i.e., regardless of how much the temperature changed). Similarly, if α_1 were very small, the quantity $\alpha_1 \ \Delta T$ would be correspondingly small and, from equation (8.25), we would have $R_2 \approx R_1$. We conclude that the smaller the value of α_1, the smaller the change in resistance due to a change in temperature. Table 8.6 shows that constantan, with $\alpha_1 = 8 \times 10^{-6}$, is quite insensitive to temperature changes.

Equation (8.25) also shows that R_2 is greater than R_1 when the quantity $\alpha_1 \ \Delta T$ is positive. Now, $\alpha_1 \ \Delta T$ is positive when α_1 is positive and when T_2 is greater than T_1:

$$\left.\begin{array}{r} \alpha_1 > 0 \\ T_2 - T_1 > 0 \end{array}\right\} \Rightarrow \alpha_1 \ \Delta T > 0 \Rightarrow R_2 > R_1 \tag{8.26}$$

This is simply a mathematical statement of the fact that the resistance of a conductor increases when temperature increases. On the other hand, the resistance of a semiconductor (and some other materials) decreases when temperature increases, as we have already discussed. Consequently, *the temperature coefficient of a semiconductor is negative* ($\alpha_1 < 0$):

$$\left.\begin{array}{c} \alpha_1 < 0 \\ T_2 - T_1 > 0 \end{array}\right\} \Rightarrow \alpha_1 \, \Delta T < 0 \Rightarrow R_2 < R_1 \qquad (8.27)$$

A graph of semiconductor resistance versus temperature is quite nonlinear, so the use of a temperature coefficient of resistance to calculate resistance changes should be restricted to situations where the change in temperature is small. The smaller the change in temperature from T_1, the more accurate a computation of resistance will be when α_1 is used in the computation.

Example 8.15

The temperature coefficient of a certain silicon resistor is $\alpha_1 = -0.015$ at $T_1 = 20°C$. If its resistance at 20°C is 4 kΩ, find its approximate resistance at 30°C.

SOLUTION Fron equation (8.24),

$$R_2 = 4 \text{ k}\Omega[1 + (-0.015)(30° - 20°)]$$
$$= 4 \text{ k}\Omega(1 - 0.15) = 4 \text{ k}\Omega(0.85) = 3.4 \text{ k}\Omega$$

We see that the resistance of this semiconductor decreases significantly with a small increase in temperature. Some semiconductor resistors, called *thermistors*, are specially designed to exhibit a large change in resistance in response to temperature changes. They are used in practical applications such as temperature controllers, where temperature changes are sensed by circuits containing the devices.

Drill Exercise 8.15

Find the resistance of the silicon resistor in Example 8.15 at $-5°C$.

ANSWER: 5.5 kΩ. □

8.8 SPICE Example

Example 8.16 (SPICE)

To investigate how temperature changes affect a circuit, use SPICE to simulate the circuit in Figure 7.29, p. 245, when the temperature is 27°C and again when it is 70°C. Find the voltage across the 40-Ω resistance and the current supplied by each voltage source at both temperatures. Assume the temperature coefficient of resistance at 27°C is 0.004 for every resistance in the circuit.

SOLUTION Figure 8.13(a) shows the original circuit, and 8.13(b) shows it in the SPICE format. Note how the temperature coefficient of resistance is given after each resistor specification in the input data file: TC = .004. (We are allowing the second-order temperature coefficient to default to zero, as discussed in the Appendix.)

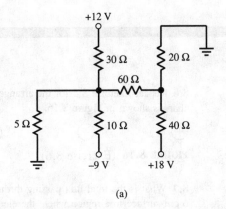

(a)

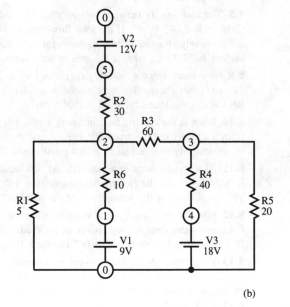

(b)

```
EXAMPLE 8.16
V1 0 1 9V
V2 5 0 12V
V3 4 0 18V
R1 2 0 5 TC=.004
R2 2 5 30 TC=.004
R3 2 3 60 TC=.004
R4 3 4 40 TC=.004
R5 3 0 20 TC=.004
R6 1 2 10 TC=.004
.TEMP 27 70
.DC V1 9V 9V 1
.PRINT DC V(3,4)
.PRINT DC I(V1) ·I(V2) I(V3)
.END
```

FIGURE 8.13 (Example 8.16)

The .TEMP statement causes SPICE to perform the simulation at 27°C and again, automatically, at 70°C. When the program is executed, SPICE prints a list of "temperature-adjusted values" for all the resistors at each temperature. We find that the voltage across the 40-Ω resistor, V(3,4) is −13.31 V at both temperatures. This result is to be expected, since all resistance values increase at the higher temperature by the same percentage. However, the magnitudes of the currents supplied by the sources decrease from I(V1) = 0.779 A, I(V2) = 0.440 A, and I(V3) = 0.333 A to I(V1) = 0.665 A, I(V2) = 0.376 A, and I(V3) = 0.284 A. The reduction in total current is a direct result of the increase in every resistance at the higher temperature.

Exercises

Section 8.1   Coulomb's Law

8.1 Find the force on each charge in each part of Figure 8.14. Draw arrows on the charges to show the directions of the forces on them.

8.2 The force on each of two charges is 2400 N. If the distance between the charges is halved, what is the force on each?

8.3 How far apart would it be necessary to move two 1-C charges in order to reduce the force on each to 1 N?

8.4 A valence electron of the copper atom is approximately 1.2×10^{-10} m from its nucleus. The nucleus contains 29 protons. What is the force of attraction between the electron and the nucleus?

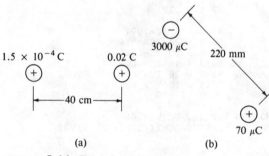

1.5×10^{-4} C   0.02 C

|——— 40 cm ———|

(a)

3000 μC   220 mm

(b)

70 μC

FIGURE 8.14 (Exercise 8.1)

Section 8.2   Electric Fields

8.5 Using your intuitive understanding of the nature of electric field lines, sketch the approximate appearance of the electric field around the fixed charges in each part of Figure 8.15. Include arrows to show the directions of the field lines.

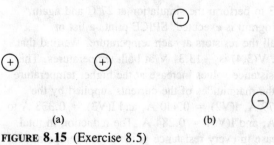

(a)   (b)

FIGURE 8.15 (Exercise 8.5)

8.6 Repeat Exercise 8.5 for the arrangement of fixed charges shown in Figure 8.16.

+   −   +

FIGURE 8.16 (Exercise 8.6)

8.7 What is the total flux passing through a 10 cm × 6 cm surface in a region where the electric flux density is 2700 μC/m^2?

8.8 The flux density between two parallel, charged surfaces is 0.042 C/m^2. If the total flux crossing from the positively charged surface to the negatively charged surface is 77.7 μC, what is the area of each surface?

8.9 How much force would be experienced by a 220-μC point charge if it were placed in an electric field where the intensity is 1.3×10^4 V/m?

8.10 What is the electric field intensity 1 mm away from a single fixed electron? (*Hint:* First find the force on an arbitrary charge placed at that point, say 1 C.)

8.11 At a certain point in a material, the flux density is 0.09 C/m^2 and the electric field intensity is 1.2×10^9 V/m. What is the permittivity of the material?

8.12 Find the flux density in a vacuum at a point 6.42 mm away from a fixed charge of 5000 μC. (*Hint:* Use an approach similar to that in Exercise 8.10.)

8.13 The voltage difference between two parallel, charged surfaces is 9 V and the electric field intensity in the region between them is 1600 V/m. How far apart are the surfaces?

8.14 Two parallel, charged surfaces are separated by 0.8 mm of a material that has a permittivity of 1.25×10^{-11}. The flux density in the material is 50 μC/m^2. What is the voltage difference between the surfaces?

Section 8.3   Breakdown Strength

8.15 Two parallel, charged surfaces are separated by air and have a potential difference of 250 V across them. At what separating distance will the air between the surfaces break down?

8.16 Mica has a permittivity that is five times greater than that of air. At what value of flux density will mica break down?

Section 8.4 Voltage

8.17 Find the voltage V_{xy} between points x and y in a circuit if
(a) 23 J of work must be performed on a 0.025-C negative charge to move it from point x to point y
(b) 147 mJ of energy is required to move a 300-μC positive charge from y to x
(c) 64 J of work must be performed on a 16-C negative charge to move it from y to x

8.18 Find the amount of work that must be performed
(a) to move a 400-μC negative charge from x to y if $V_{xy} = 250$ V
(b) to move a 0.33-C positive charge from x to y if $V_{xy} = -20$ V
(c) to move a 1000-μC negative charge from x to y if $V_{yx} = 52$ V

Section 8.5 Resistance of Conductors

8.19 Find the resistance between surfaces A and B of the bar shown in Figure 8.17, assuming that it is made of copper.

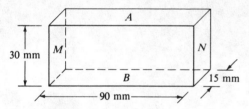

FIGURE 8.17 (Exercise 8.19)

8.20 Find the resistance between surfaces M and N of the bar shown in Figure 8.17, assuming that it is made of nickel.

8.21 The tungsten bar shown in Figure 8.18 has a square cross section ($d \times d$). If the current through the bar is to be 20 mA when the 3-V source is connected as shown, what should be dimension d?

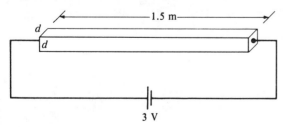

FIGURE 8.18 (Exercise 8.21)

8.22 A heater element is to be constructed from nichrome wire having a diameter of 3 mm. If the element is to dissipate 200 W when a 24-V source is connected between its ends, how long should the wire be?

8.23 Find the resistance of 250 ft of copper wire whose diameter is 0.08 in.

8.24 What is the radius of an aluminum cable if a 2000-ft length of it has a resistance of 212.5 mΩ?

8.25 An AWG standard wire size is to be selected for an application in which 20 A must be conducted over a 1000-ft distance. If the wire is copper and the power loss in it cannot exceed 70 W, what is the smallest size that can be used?

8.26 The resistance of a 6-in. length of AWG No. 35 wire is measured to be 1.174 Ω. Of what material is the wire made?

Section 8.6 Resistance of Semiconductors

8.27 A silicon bar is 2 mm long and has a cross-sectional area of 0.5×10^{-7} m^2. If the conductivity of the silicon is 800 S/m, what is the resistance of the bar?

8.28 If the thickness of the bar in Exercise 8.27 is 0.1 mm, what is its aspect ratio? What is its sheet resistivity?

8.29 A 500-Ω semiconductor resistor is to be designed using silicon whose sheet resistivity is 200 Ω/square. If the width of the resistor is 20 microns, how long should it be?

8.30 An 800-Ω semiconductor resistor is to be constructed using silicon whose conductivity is 2000 S/m. If the resistor is to be 6 mm long by 1 mm wide, how thick should the silicon layer be, in microns?

Section 8.7 Temperature Dependence of Resistance

8.31 An aluminum conductor has resistance 12 Ω at 50°C. Find its resistance at
(a) 85°C
(b) −5°C

8.32 A copper conductor has resistance 3.5 Ω at 30°C. At what temperature would its resistance be 3.7 Ω?

8.33 Using the temperature coefficient of resistance, find the resistance of a nichrome wire at 200°C if its resistance at 20°C is 300 Ω.

8.34 Using the temperature coefficient of resistance, find the resistance of an iron bar at 90°C if its resistance at 0°C is 2.8 Ω.

8.35 Using the temperature coefficient of resistance, find the temperature at which a copper conductor will have resistance 0.5 Ω if its resistance at 40°C is 0.75 Ω.

8.36 The resistance of a certain semiconductor resistor is 900 Ω at 45°C and 932 Ω at 40°C. What is its (approximate) temperature coefficient of resistance at 40°C?

SPICE Exercises

8.1S The resistors in Figure 7.32(b) each have a first-order temperature coefficient of resistance of 5×10^{-3} at 27°C. Use SPICE as an aid in finding the change in the total *power* dissipation in the circuit when the temperature changes from 27°C to 80°C. (The total power is the sum of the powers delivered by the voltage sources.)

8.2S A certain semiconductor resistor has resistance 1.2 kΩ and temperature coefficient of resistance equal to -0.02 at 27°C. Use SPICE to find its resistance at 40°C. (Hint: Use a 1-A current source.)

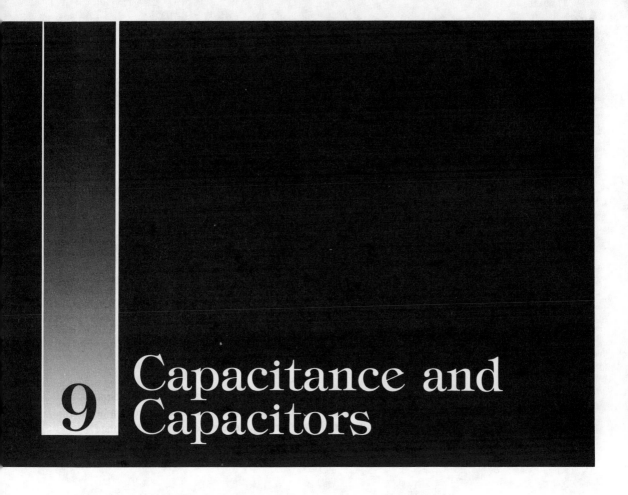

9 Capacitance and Capacitors

9.1 The Nature of Capacitance

All passive components have three, and only three, *electrical* properties. One of those properties, resistance, we have already studied in considerable detail. The second, *capacitance*, is the subject of this chapter, and the third, *inductance*, will be studied in Chapter 11.

Capacitance is a measure of the ability of a component to *store* charge. In fact, the word is derived from *capacity*, the conventional term meaning a measure of how much storage is available. Just as a resistor is a component that has been intentionally manufactured to have a certain amount of resistance, a *capacitor* is a device specially designed to have a certain amount of capacitance. The concept of capacitance is best introduced through the study of a specific kind of capacitor: the *parallel-plate capacitor*.

Figure 9.1(a) shows two parallel surfaces, or *plates*, made of conducting material and separated by an insulator. This is the basic structure of a parallel-plate capacitor. The insulator between the plates is called the *dielectric*, a term that was introduced in Chapter 8. Imagine that a voltage source, such as a battery, is connected across the two

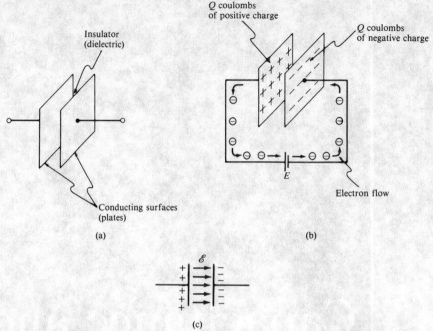

FIGURE 9.1 Parallel-plate capacitor. (a) Structure of a parallel-plate capacitor. (b) Electrons flow from the negative terminal of the battery and produce a negative charge on one plate. An equal positive charge is developed on the other plate. (c) End view of charged capacitor, showing electric field $\mathscr{E}$ between the plates.

plates, as shown in Figure 9.1(b). Electrons will flow from the negative terminal of the battery and accumulate on one of the plates, developing a negative charge on that plate. Electrons on that plate repel an equal number of electrons from the other plate, leaving it with a positive charge. In other words, an electric field is established in the dielectric between the plates, and the direction of the field is such that it drives electrons off the positively charged plate [see Figure 9.1(c)]. The electrons are attracted to and received by the positive terminal of the battery. The amount of negative charge Q stored on the negative plate exactly equals the positive charge on the positive plate.

Figure 9.2 shows the schematic diagram of the capacitor circuit shown in Figure 9.1(b). The curved line (arc) in the symbol for capacitor C usually represents the negative plate, but in many applications it is not necessary to make that distinction. (As we shall learn presently, some capacitors are designed so that one side *must* be the negative plate, or terminal, while in other types it makes no difference which side is negative.) Recall that a voltage exists between any two regions containing opposite types of charge. The figure shows that voltage V is developed across the capacitor as a result of the charge stored on its plates. Notice that the polarity of V *opposes* the polarity of voltage source E. As charge builds up on the plates, the voltage V increases. When the charge becomes so large that $V = E$, no more electrons can flow from the battery, because V is then equal and opposite to E. The capacitor is said to be *fully charged* at that time.

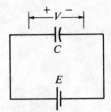

FIGURE 9.2 Schematic diagram of the charged capacitor, showing the voltage V developed across it.

Capacitance is defined to be *the ratio of the charge stored by a capacitor to the voltage V across it*:

$$C = \frac{Q}{V} \qquad (9.1)$$

The units of capacitance are *farads* (F), named after the English physicist Michael Faraday (1791–1867). Equation 9.1 shows that 1 farad is the same as 1 coulomb/volt. One farad is an exceptionally large amount of capacitance, and the units most often used for practical capacitors are microfarads (μF) and picofarads (pF). Note that Q in equation (9.1) is the (equal) charge on *either* capacitor plate, not the sum of the charges on both plates.

Let us use equation (9.1) to gain some insight into the nature of capacitance. We indicated earlier that capacitance is a measure of the ability to store charge. This measure is meaningful only if we divide the amount of charge stored by the voltage developed across the capacitor (i.e., by the voltage of the source that supplied the charge). After all, there is theoretically no limit to the amount of charge that can be stored on any capacitor if there is no limit to how large a voltage source can be used. On the other hand, if we limit the amount of voltage, we can say that one capacitor is larger than another (has greater capacity) if the *same* voltage produces more charge on the plates of one than it does on the other. In other words, the ratio Q/V (i.e., the capacitance) will be larger for one than for the other. As we shall see, the physical dimensions and the dielectric material determine just how large that capacity is.

Example 9.1

A 6-V source is required to store 24 μC of charge on a certain capacitor.

(a) What is the capacitance of the capacitor?
(b) How much charge is stored on the capacitor when a 9-V source is connected across it?
(c) What is the voltage across the capacitor when 16 μC is stored on it?

SOLUTION

(a) From equation (9.1),

$$C = \frac{Q}{V} = \frac{24 \times 10^{-6}\ C}{6\ V} = 4 \times 10^{-6}\ F = 4\ \mu F$$

(b) From equation (9.1),

$$Q = CV = (4 \ \mu F)(9 \ V) = 36 \ \mu C$$

(c) From equation (9.1),

$$V = \frac{Q}{C} = \frac{16 \ \mu C}{4 \ \mu F} = 4 \ V$$

Drill Exercise 9.1

How much charge is stored by a 220-pF capacitor when a 50-V source is connected across it?

ANSWER: $0.011 \ \mu C$.   ☐

9.2 Capacitor Dimensions and Dielectrics

The Effect of Dielectric Material on Capacitance

In Chapter 8 we discussed the permittivity ϵ of a material in connection with the relationship between field intensity and flux density: $D = \epsilon \ \mathscr{E}$. We will see that the permittivity of the dielectric material between the plates of a capacitor strongly influences the value of its capacitance.

Since the dielectric material in a capacitor is an insulator, electrons are strongly bound to their parent atoms and are not free to travel under the influence of an electric field (unless the breakdown strength is exceeded). When an electric field is established in the dielectric by charging the capacitor plates, the electrons in the dielectric can only *shift position*, leaving one side of each atom positive and the other side negative. Figure 9.3 illustrates this phenomenon. As can be seen in Figure 9.3(a), each atom in the interior of the dielectric has a positive side that is neutralized by the negative side of an adjacent atom. However, atoms along the extreme edges of the dielectric, at the boundaries of the capacitor plates, have no such adjacent atoms to neutralize their sides. As a consequence, they establish an electric field, $\mathscr{E}_d$, within the dielectric that *opposes* the field set up by the charged plates [see Figure 9.3(b)]. The *net* field intensity between the plates is therefore smaller than it would be if there were no dielectric present. A smaller value of $\mathscr{E}$ means that a smaller value of V is required to deposit a fixed amount of charge Q on the plates, so from $C = Q/V$, we conclude that the capacitance C is increased.

The prefix *di* in dielectric means *opposing* or *against*, and we have seen that a dielectric is effective in increasing capacitance because it sets up an electric field in opposition to that created by the charged plates. ϵ is called the permittivity of the dielectric because it is a measure of how well the material *permits* the establishment of an electric field. The greater the permittivity ϵ of a material, the greater the capacitance of a capacitor that uses the material as a dielectric.

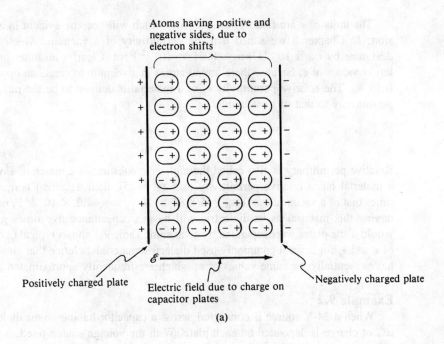

Atoms having positive and
negative sides, due to
electron shifts

Positively charged plate

Electric field due to charge on
capacitor plates

Negatively charged plate

(a)

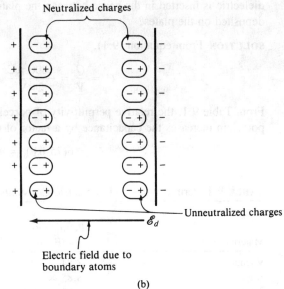

Neutralized charges

Unneutralized charges

Electric field due to
boundary atoms

(b)

FIGURE 9.3 Electric fields in the dielectric of a capacitor. (a) The electric field set up by the charged capacitor forces electron shifts in the atoms of the dielectric material. (b) The unneutralized atoms along the boundaries of the dielectric set up an electric field opposing that in part (a).

The units of ϵ are farads/meter (F/m), which will become evident in a later discussion. In Chapter 8 we stated that the permittivity of a vacuum, which we will now designate by ϵ_0, is $10^{-9}/36\pi \approx 8.84 \times 10^{-12}$ F/m. Clearly, all other insulators have larger values of ϵ, because there are no atoms in a vacuum to create an opposing electric field $\mathcal{E}_d$. The *relative* permittivity ϵ_r of a material is defined to be the ratio of its actual permittivity to that of a vacuum:

$$\epsilon_r = \frac{\epsilon}{\epsilon_0} \tag{9.2}$$

Relative permittivity is also called the *dielectric constant* of a material. For example, if a material has a relative permittivity of 5 ($\epsilon_r = 5$), then its actual permittivity is five times that of a vacuum: $\epsilon = \epsilon_r \epsilon_0 = 5(8.84 \times 10^{-12}) = 44.2 \times 10^{-12}$ F/m. A capacitor having this material as its dielectric will have a capacitance five times greater than it would if the plates were separated by a vacuum. Table 9.1 shows typical (average) values of ϵ and ϵ_r for several commonly used dielectric materials. Notice that air and a vacuum have essentially the same values of ϵ, which are frequently approximated as equal.

Example 9.2

When a 24-V source is connected across a capacitor having an air dielectric, 1800 μC of charge is deposited on each plate. With the voltage source fixed, a porcelain dielectric is inserted in the space between the plates. How much charge is then deposited on the plates?

SOLUTION From equation (9.1),

$$C = \frac{Q}{V} = \frac{1800 \ \mu C}{24 \ V} = 75 \ \mu F$$

From Table 9.1, the relative permittivity of porcelain is $\epsilon_r = 6$. Thus, inserting the porcelain increases the capacitance by a factor of 6:

$$C = 6(75 \ \mu F) = 450 \ \mu F$$

TABLE 9.1 Permittivities and Dielectric Constants of Dielectric Materials

Material	Permittivity, ϵ (F/m)	Relative Permittivity (Dielectric Constant), $\epsilon_r = \epsilon/\epsilon_0$
Vacuum	8.84×10^{-12}	1
Air	8.85×10^{-12}	1.0006
Teflon	17.68×10^{-12}	2
Paper, paraffined	22.10×10^{-12}	2.5
Mica	44.20×10^{-12}	5
Porcelain	53.04×10^{-12}	6
Bakelite	61.88×10^{-12}	7
Glass	66.30×10^{-12}	7.5
Water	707×10^{-12}	80
Barium–strontium–titanite	$66,300 \times 10^{-12}$	7500

Since V remains 24 V, the capacitor charges to

$$Q = CV = (450\ \mu F)24\ V = 10,800\ \mu C$$

Drill Exercise 9.2

Suppose that the 24-V source in Example 9.2 is removed after the air-dielectric capacitor is charged with 1800 μC. What does the voltage across the capacitor become when the porcelain dielectric is inserted?

ANSWER: 4 V.

Effect of Capacitor Dimensions on Capacitance

The physical dimensions that affect the capacitance of a parallel-plate capacitor are the area A of its plates and the distance d between them, as illustrated in Figure 9.4. The greater the area of the plates, the greater the surface on which charge can be stored, so the greater the storage capacity of the capacitor. We conclude that *capacitance is directly proportional to the area A of each parallel plate*. Recall from Chapter 8 that the electric field intensity $\mathscr{E}$ between two parallel, charged surfaces is V/d, where V is the voltage across the surfaces and d is the distance between them. Applying that relationship to the parallel-plate capacitor, we have $V = \mathscr{E}d$. Substituting $V = \mathscr{E}d$ in $C = Q/V$, we find

$$C = \frac{Q}{V} = \frac{Q}{\mathscr{E}d} \tag{9.3}$$

All other factors being equal, the ratio of $Q/\mathscr{E}$ is a constant, so we conclude from equation (9.3) that *capacitance is inversely proportional to the distance d between the parallel plates*.

The Capacitance Equation

We can express the dependency of capacitance on the permittivity ϵ of the dielectric, the area A of each parallel plate, and the distance d between the plates in a single equation, which we will call the *capacitance equation*:

$$C = \frac{\epsilon A}{d} \tag{9.4}$$

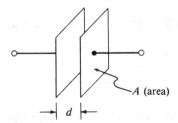

FIGURE 9.4 The area A of each parallel plate and the distance d between them affects the value of capacitance.

As an exercise, verify that the units of ϵ are F/m, when C is in farads, A is in square meters, and d is in meters. Notice that the capacitance equation correctly expresses the fact that C is directly proportional to ϵ and A, and is inversely proportional to d. Since $\epsilon = \epsilon_r\epsilon_0$, the equation can be written in the equivalent form

$$C = \frac{\epsilon_r\epsilon_0 A}{d} \tag{9.5}$$

We should note that equation (9.4) can be derived mathematically from earlier definitions, without having to rely on intuitive conclusions about dependency relations. From equation (8.5),

$$\mathcal{E} = \frac{D}{\epsilon} \tag{9.6}$$

But, by definition, $D = \psi/A = Q/A$, so equation (9.6) can be written

$$\mathcal{E} = \frac{Q}{A\epsilon} \tag{9.7}$$

Substituting equation (9.7) into equation (9.3) gives

$$C = \frac{Q}{(Q/A\epsilon)d} = \frac{\epsilon A}{d}$$

which is the same as equation (9.4).

Example 9.3

Each plate of a parallel-plate capacitor measures 1 cm $\times$ 2 cm. The plates are separated by a 0.2-mm thickness of mica.

(a) Find the capacitance of the capacitor.
(b) Find the capacitance if each dimension of each plate is doubled.
(c) Find the capacitance under the conditions of part (b) if the thickness of the mica separation is also doubled.

SOLUTION

(a) From Table 9.1, the permittivity of mica is $\epsilon = 44.2 \times 10^{-12}$. The area of each plate is

$$A = (1 \times 10^{-2} \text{ m})(2 \times 10^{-2} \text{ m}) = 2 \times 10^{-4} \text{ m}^2$$

From equation (9.4),

$$C = \frac{\epsilon A}{d} = \frac{(44.2 \times 10^{-12})(2 \times 10^{-4})}{0.2 \times 10^{-3}} = 44.2 \text{ pF}$$

(b) Doubling the dimensions of each plate increases the area of each by a factor of 4:

$$A = 2 \text{ cm} \times 4 \text{ cm} = 8 \times 10^{-4} \text{ m}^2$$

Since the capacitance is directly proportional to A, we know that increasing A by a factor of 4 will increase C by a factor of 4:

$$C = 4(44.2 \text{ pF}) = 176.8 \text{ pF}$$

(c) Since C is inversely proportional to d, we know that doubling the value of d will reduce C by one-half:

$$C = (\tfrac{1}{2})(176.8 \text{ pF}) = 88.4 \text{ pF}$$

Drill Exercise 9.3

What should be the area of each plate of a parallel-plate capacitor if it is to have a capacitance of 1 F when the plates are separated by 1 mm of air?

ANSWER: $1.13 \times 10^8 \text{ m}^2$ (or about 43 square miles!). □

9.3 Capacitor Types and Ratings

Capacitor Specifications

The rated capacitance of a capacitor is the first characteristic of interest in most practical applications where a commercially available capacitor must be selected to perform a specific circuit function. Capacitors are available in a wide range of *nominal* values, from 1 pF to several thousand microfarads, but, like resistors, the actual value of any specific capacitor is subject to the manufacturer's *tolerance* specification. Typical capacitor tolerances range from ±5% to ±20%, although precision capacitors with smaller tolerances are available at greater costs. In some applications, usually involving very large capacitors, we wish to insert a certain *minimum* amount of capacitance in a circuit, and any larger amount is acceptable. Capacitors designed for these applications may have tolerance ratings such as −10% to +150%.

The second most important consideration in selecting a capacitior for a specific application is the voltage it can tolerate across its terminals without breaking down. A capacitor breaks down when the voltage across it becomes so great that the breakdown strength of its dielectric is reached, as discussed in Section 8.3. Recall that the electric field intensity in a capacitor is

$$\mathcal{E} = \frac{V}{d}$$

where V is the voltage across the plates of the capacitor and d is the distance between them. Since d is fixed, the field intensity increases as V increases, so the breakdown value of $\mathcal{E}$ sets a limit on how large the voltage V can become. Notice this *trade-off* in capacitor design: We would like to have a large value of d to ensure a large breakdown voltage, but by the capacitor equation ($C = \epsilon A/d$), we need a small value of d to achieve a large capacitance. Thus, commercial capacitors that have very large capacitance values generally have small breakdown voltages, and vice versa. The maximum voltage at which a capacitor is designed to operate continuously without breaking down is called its *dc*

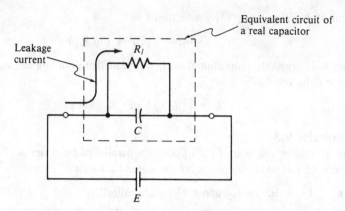

FIGURE 9.5 A real capacitor has leakage resistance R_1 that allows a small leakage current to flow between the plates. The capacitor can be analyzed by regarding it as an ideal capacitor in parallel with leakage resistance.

working voltage (DCWV). Manufacturers' specifications for this value may range from a few volts for very large capacitors to several thousand volts for small capacitors.

The dielectric in an *ideal* capacitor is a perfect insulator (i.e., it has infinite resistance), and zero current flows through it when a voltage is applied across its terminals. The dielectric in a *real* capacitor has a large but finite resistance, so a small current, called *leakage current*, does flow between the capacitor plates when a voltage is applied. The resistance of the dielectric is called *leakage resistance*, and its value is another important specification that must be considered in many practical applications. Figure 9.5 shows the equivalent circuit of a real capacitor, consisting of an ideal capacitor in parallel with leakage resistance R_l. Typical values of leakage resistance may range from about 1 MΩ (considered a very "leaky" capacitor) to greater than 100,000 MΩ.

Capacitor Types

Capacitors are most often classified by the materials used for their dielectrics. We hear, for example, of air capacitors, mica capacitors, paper capacitors, and so on, all of which refer to the type of dielectric used. Dielectric materials are selected by manufacturers to endow capacitors with specific properties, such as large capacitance, large leakage resistance, large breakdown voltage, or small physical size. Often the choice of dielectric and the method used to embed it between the plates will enhance one or more capacitor properties at the expense of others. In later paragraphs we will discuss the construction and properties of several popular types of capacitors using different dielectric materials.

Capacitors are also classified as being *electrolytic* or *nonelectrolytic*. An electrolytic capacitor is said to be *polarized*, because one of its terminals must always be positive with respect to the other. Electrolytic capacitors have the advantage that they provide a large capacitance in a physically small package, but they are generally leaky and are restricted in their applications by the polarization requirement.

Finally, capacitors are classified as being *fixed* or *variable* (adjustable). Variable capacitors are used in applications where it is necessary or desirable to adjust circuit characteristics for calibration purposes, or in experimental work, or simply to achieve versatility. (For example, the *tuning dial* on most radio sets is a variable capacitor.)

Air Capacitors

Most air-dielectric capacitors are of the variable type and are constructed as shown in Figure 9.6(a). Notice that each capacitor plate is actually a *set* of metal plates that are electrically common. Thus, one set of plates comprises the positive plate of the capacitor and the other set comprises the negative plate. One set is mounted on a shaft that rotates the entire set to make it mesh to a greater or lesser extent with the plates of the other set. When the plates are fully meshed, the capacitance is maximum. The effect is the same as a set of parallel-connected capacitors whose total capacitance (as we shall learn in Section 9.4) is the sum of the individual capacitances. When a portion of one set of plates is not meshed with the other set, that portion does not contribute to the total capacitance because it does not oppose any oppositely charged surface across the air dielectric. When the plates are completely unmeshed, there is no portion of a positive plate opposing any portion of a negative plate and the capacitance is essentially zero. In effect, adjusting a variable capacitor changes the value of area A in the capacitor equation, $C = \epsilon A/d$. Variable air capacitors typically have small values of maximum capacitance, on the order of 100 pF. Figure 9.6(b) shows the symbol for a variable capacitor, which, like the symbols for other variable devices, has an arrow drawn through it.

Paper Capacitors

The construction of a paper capacitor is illustrated in Figure 9.7. Here, two metal foils—the plates—are separated by a sheet of paper dielectric. The arrangement is then rolled into a cylinder and dipped in plastic or wax. Of course the "plates" in this type of construction remain physically parallel, but the structure is called *tubular*. The parallel-plate capacitance equation is still applicable to the tubular capacitor. In one variation of this construction, the paper is coated with metal film to form the opposing plates, and

(a)   (b)

FIGURE 9.6 Variable air capacitor and its symbol. (Courtesy of E. F. Johnson Co.) (a) Rotating the shaft causes one set of plates to mesh with the other. (b) Schematic symbol.

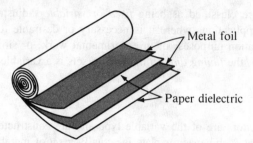

FIGURE 9.7 Construction of a tubular paper capacitor. Leads are attached to the two metal foils, which serve as parallel "plates."

the device is called a *metallized paper* capacitor. The terminal to which the outermost plate of a paper capacitor is connected is often identified by a black band around the body of the capacitor. In some applications it is necessary to connect this side to a lower voltage than the other side. Paper capacitors are inexpensive but are not noted for their reliability in circuits where stresses such as wide temperature variations or electrical surges are likely to occur. They are available with capacitance values ranging from about 500 pF to 50 µF and have dc working voltage ratings up to 600 V.

Plastic Film Capacitors

The structure of a plastic film capacitor is similar to that of a paper capacitor in which the paper dielectric is replaced by a thin film of plastic, usually Mylar® or polystyrene. The principal advantages of plastic film capacitors are that they can be manufactured with relatively small tolerances, such as ±2.5%, and they are less sensitive than some other types to temperature variations. These capacitors are generally smaller than paper types and are more expensive. They are available with capacitance values from about 5 pF to 0.5 µF and with dc working voltages up to 600 V.

Mica Capacitors

Mica capacitors are constructed by assembling alternating layers of mica and metal foil, as illustrated in Figure 9.8. Like the air-dielectric capacitor, one set of electrically common foils serves as one plate and the other set serves as the second plate. Terminals are attached to each set of foils and the entire structure is encapsulated in plastic. These capacitors can be constructed to have precise capacitance values and very large breakdown voltages, up to 35,000 V. They also have a very large leakage resistance, on the order of 1000 MΩ. Typical capacitances range from 1 pF to 0.1 µF.

Ceramic Capacitors

Ceramic capacitors are formed by depositing a metal film on each side of a ceramic base that serves as the dielectric. They may be constructed in the form of a single disk, as shown in Figure 9.9(a), or in the multilayered form shown in Figure 9.9(b). A ceramic material with a very large permittivity is used in some designs to create capacitors that are physically much smaller than capacitors of other designs having comparable capacitance values. Ceramic capacitors are also noted for their large leakage resistance (on the order of 1000 MΩ) and high working voltages, up to 5000 V. They are available in

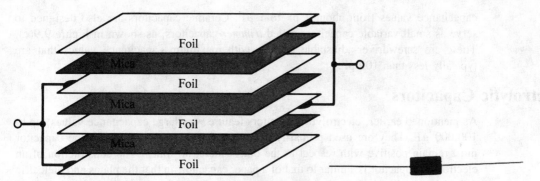

FIGURE 9.8 Typical mica capacitor.

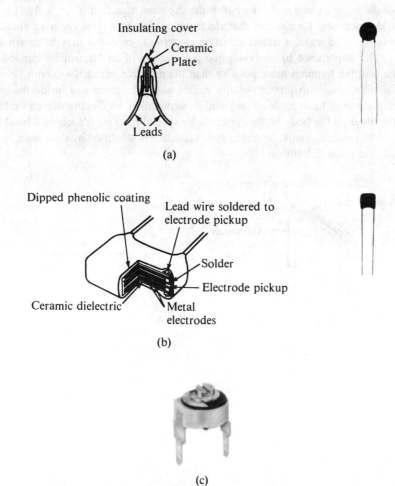

FIGURE 9.9 Ceramic capacitors. (a) Single disk. (b) Multilayered. (c) Screwdriver adjustable trimmer capacitor. (Courtesy of Johanson Manufacturing Corporation.)

capacitance values from about 1 pF to 1 μF. Ceramic capacitors are also designed to serve as small variable capacitors called *trimmer* capacitors, as shown in Figure 9.9(c). These are screwdriver-adjustable devices with maximum capacitance values that are typically less than 100 pF.

Electrolytic Capacitors

As mentioned earlier, electrolytic capacitors feature very large capacitance values, up to 100,000 μF. They are used primarily in dc circuits because one side of the capacitor must remain positive with respect to the other side at all times. The construction of an electrolytic capacitor is similar to that of a paper capacitor, in that the plates and dielectric material are rolled into a tubular shape. Figure 9.10(a) shows a cross-sectional view of a typical electrolytic capacitor. An aluminum foil serves as the positive plate and a very thin layer of aluminum dioxide serves as the dielectric. The thinness of this layer accounts for the large capacitance, since d in the capacitor equation ($C = \epsilon A/d$) is very small. It also accounts for the fact that electrolytic capacitors typically have small breakdown voltages. The leakage resistance of an electrolytic capacitor may be as small as 1 MΩ.

The importance of observing the polarization of an electrolytic capacitor—keeping the positive terminal more positive than the negative terminal—cannot be overstressed. Failure to maintain proper polarity causes gas to be generated inside the capacitor and may cause it to explode. The positive terminal of an electrolytic capacitor is always identified on the body of the capacitor by special markings, such as a band of + signs. In schematic diagrams, an electrolytic capacitor is identified by drawing a + sign alongside the positive terminal.

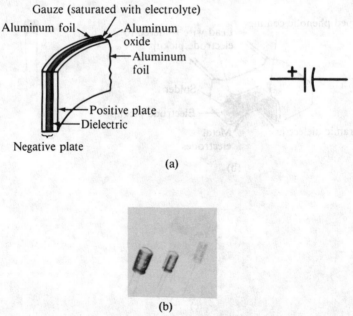

(a)

(b)

FIGURE 9.10 Electrolytic capacitors. (a) Construction and schematic symbol. (b) Typical capacitors. (Courtesy of United Chemi-Con, Inc.)

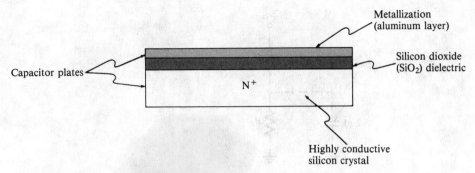

FIGURE **9.11** Cross-sectional view of an integrated-circuit capacitor.

Tantalum Capacitors

A tantalum capacitor is another type of electrolytic capacitor featuring large capacitance in a small package. Powdered tantalum is pressed into a cylindrical or rectangular shape and baked at a very high temperature. The result is a very porous material that is dipped in acid to produce the dielectric. A typical tantalum capacitor about the size of a 1-in. piece of chalk may have a capacitance of 100 μF and a working voltage of 20 V. A comparable electrolytic capacitor of the aluminum foil design is about the size of a doorknob.

Integrated Circuit Capacitors

An integrated circuit is a tiny piece (*chip*) cut from a semiconductor wafer, usually made of silicon, in which are embedded many electrical and electronic components. One way that a capacitor is formed in a chip is by depositing a thin layer of silicon dioxide over a region of highly conductive semiconductor material (one that has been heavily doped, as discussed in Section 8.6). The conductive semiconductor serves as one of the capacitor plates. A thin layer of metal, usually aluminum, is deposited on the other side of the silicon dioxide layer to form the other plate. The structure is illustrated in Figure 9.11. It is not practical to construct integrated-circuit capacitors with large capacitance values because the dimensions on a chip are so small. Designers create circuits that avoid the need for capacitors, and most have values less than 20 pF. Integrated-circuit capacitors are also constructed in the form of *reverse-biased diodes*, electronic devices that are studied in more advanced courses.*

Stray Capacitance

It is important to distinguish between *capacitance*—an inherent property of every electrical component—and *capacitors*—devices that are constructed for the express purpose of inserting capacitance in a circuit. *Stray capacitance* is capacitance that exists between terminals or conducting paths in a circuit simply by virtue of the physical structure of the circuit: the arrangement of components, the wiring between them, and the location of terminals, solder junctions, and so on. *Capacitance exists between any two conductors that are separated by an insulator*. The conductors serve as the "plates" and the insulator

*See, for example, T. F. Bogart, *Electronic Devices and Circuits*, 3rd ed., Merrill/Macmillan Publishing Company, Columbus, Ohio, 1993.

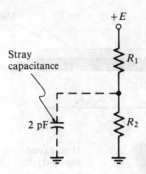

FIGURE 9.12 Stray capacitance is distinguished from a capacitor by using a dashed line to indicate its presence. In this example, there is 2 pF of stray capacitance across the terminals of R_2.

serves as the dielectric. For example, there is capacitance between two insulated wires placed alongside each other, and there is capacitance between the two terminals of a resistor. Stray capacitance is generally quite small, on the order of a few picofarads, but its presence is undesirable in many high-frequency (ac) circuit applications. It is called stray capacitance because it is distributed throughout a circuit and is difficult to predict or compute. In schematic diagrams where it is necessary to indicate the presence of stray capacitance, a dashed line is used to distinguish the stray capacitance from a capacitor. An example is shown in Figure 9.12.

9.4 Series and Parallel Capacitor Circuits

Capacitors in Series

Figure 9.13 shows two capacitors connected in series with voltage source E. As in the case of a single capacitor, electrons flow from the negative terminal of the source and accumulate on one side of capacitor C_2. An equal number of electrons are driven off the other side of C_2 and accumulate on one side of C_1. The same number of electrons are again driven off the other side of C_1 and return to the positive terminal of the source. Thus, both capacitors have the same negative charge on one side and the same positive charge on the other side. This common charge equals the total charge Q_T delivered by the source to the series circuit. We can extend this reasoning to any number of series-connected capacitors and conclude that *the total charge delivered to series-connected capacitors equals the charge on each capacitor.* Notice that every capacitor in a series circuit has the same charge, regardless of the individual capacitance values.

Consider a circuit containing any number (n) of series-connected capacitors, as shown in Figure 9.14. The capacitors may have arbitrary capacitance values, and we wish to determine the *total equivalent capacitance C_T* of the circuit. In other words, we wish to find a single capacitance that draws the same total charge Q_T from the source as does

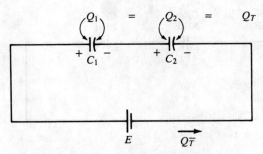

FIGURE 9.13 The charge on each series-connected capacitor is the same and equals the total charge, Q_T. (Q_T^- represents negative charge, or electrons, delivered to the circuit by the voltage source.)

the entire combination of series capacitors. The total equivalent capacitance will thus satisfy

$$C_T = \frac{Q_T}{E} \tag{9.8}$$

Writing Kirchhoff's voltage law around the series circuit in Figure 9.14, we obtain

$$E = V_1 + V_2 + \cdots + V_n \tag{9.9}$$

Dividing both sides of this equation by Q_T gives

$$\frac{E}{Q_T} = \frac{V_1}{Q_T} + \frac{V_2}{Q_T} + \cdots + \frac{V_n}{Q_T} \tag{9.10}$$

But, since $Q_1 = Q_2 = \cdots = Q_T = Q_n$, we have

$$\frac{E}{Q_T} = \frac{V_1}{Q_1} + \frac{V_2}{Q_2} + \cdots + \frac{V_n}{Q_n} \tag{9.11}$$

or

$$\frac{1}{C_T} = \frac{1}{C_1} + \frac{1}{C_2} + \cdots + \frac{1}{C_n} \tag{9.12}$$

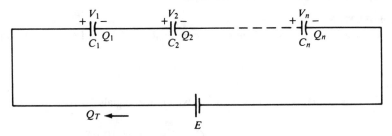

FIGURE 9.14 Series-connected capacitors.

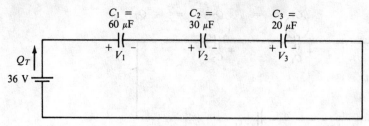

FIGURE 9.15 (Example 9.4)

Taking the reciprocal of both sides gives

$$C_T = \frac{1}{1/C_1 + 1/C_2 + \cdots + 1/C_n} \tag{9.13}$$

Equation (9.13) shows that we find the total equivalent capacitance of *series*-connected capacitors in exactly the same way as we find the total equivalent reistance of *parallel*-connected resistors. (The quantity $1/C$ is called *elastance* and is analogous to conductance in resistor circuits.)

The same relationships that apply to special cases of parallel-connected resistors also apply to special cases of series-connected capacitors. For example, it is an exercise at the end of this chapter to show that the total equivalent capacitance of two series-connected capacitors is

$$C_T = \frac{C_1 C_2}{C_1 + C_2} \tag{9.14}$$

Another exercise at the end of the chapter is to show that n equal capacitors, $C_1 = C_2 = \cdots = C_n = C$, connected in series have the total equivalent capacitance

$$C_T = \frac{C}{n} \tag{9.15}$$

Example 9.4 (Analysis)
(a) Find the total equivalent capacitance and the total charge delivered by the voltage source to the circuit in Figure 9.15.
(b) Find the voltage across each capacitor.

SOLUTION
(a) See Figure 9.16(a). The series combination of C_1 and C_2 is equivalent to

$$\frac{C_1 C_2}{C_1 + C_2} = \frac{(60 \ \mu\text{F})(30 \ \mu\text{F})}{60 \ \mu\text{F} + 30 \ \mu\text{F}} = 20 \ \mu\text{F}$$

This 20 μF is in series with C_3, which also equals 20 μF, so

$$C_T = \frac{20 \ \mu\text{F}}{2} = 10 \ \mu\text{F}$$

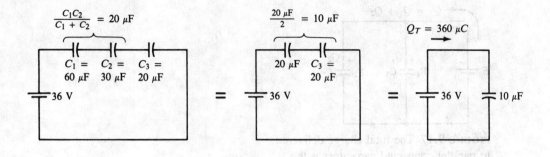

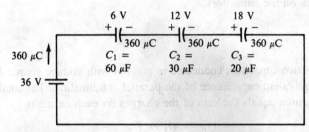

FIGURE 9.16 (Example 9.4)

The total charge delivered by the source is

$$Q_T = C_T E = (10 \ \mu\text{F})(36 \ \text{V}) = 360 \ \mu\text{C}$$

(b) See Figure 9.16(b). The total charge equals the charge on each capacitor. Thus, $Q_1 = Q_2 = Q_3 = 360 \ \mu\text{C}$. The voltage across each capacitor is therefore

$$V_1 = \frac{Q_1}{C_1} = \frac{360 \ \mu\text{C}}{60 \ \mu\text{F}} = 6 \ \text{V}$$

$$V_2 = \frac{Q_2}{C_2} = \frac{360 \ \mu\text{C}}{30 \ \mu\text{F}} = 12 \ \text{V}$$

$$V_3 = \frac{Q_3}{C_3} = \frac{360 \ \mu\text{C}}{20 \ \mu\text{F}} = 18 \ \text{V}$$

Notice that the sum of the capacitor voltage equals 36 V, confirming Kirchhoff's voltage law around the circuit.

Drill Exercise 9.4

Repeat Example 9.4 when $C_4 = 15 \ \mu\text{F}$ is added to the series circuit in Figure 9.15.

ANSWER: (a) $C_T = 6 \ \mu\text{F}$, $Q_T = 216 \ \mu\text{C}$; (b) $V_1 = 3.6 \ \text{V}$, $V_2 = 7.2 \ \text{V}$, $V_3 = 10.8 \ \text{V}$, $V_4 = 14.4 \ \text{V}$.   □

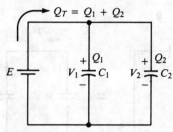

FIGURE 9.17 The total charge delivered to parallel-connected capacitors is the sum of the charges on the capacitors.

Capacitors in Parallel

Figure 9.17 shows two capacitors connected in parallel with voltage source E. We wish to find the total equivalent capacitance of the parallel combination. The total charge Q_T delivered by the source equals the sum of the charges on each capacitor:

$$Q_T = Q_1 + Q_2 \tag{9.16}$$

Dividing both sides by E gives

$$\frac{Q_T}{E} = \frac{Q_1}{E} + \frac{Q_2}{E} \tag{9.17}$$

Since all components are in parallel, the voltage is the same across each: $E = V_1 = V_2$. Therefore,

$$\frac{Q_T}{E} = \frac{Q_1}{V_1} + \frac{Q_2}{V_2} \tag{9.18}$$

or

$$C_T = C_1 + C_2 \tag{9.19}$$

We can extend this reasoning to any number of parallel-connected capacitors and conclude that *the total equivalent capacitance of parallel-connected capacitors is the sum of the capacitances*. In other words, we find the equivalent capacitance of *parallel*-connected capacitors in the same way we find the equivalent resistance of *series*-connected resistors.

Example 9.5   (Analysis)

(a) Find the total equivalent capacitance of the capacitors in Figure 9.18.
(b) Verify that the total charge delivered by the source equals the sum of the charges on the capacitors.

SOLUTION

(a) $C_T = C_1 + C_2 + C_3 = 100$ pF $+ 220$ pF $+ 50$ pF $= 370$ pF
(b) The total charge delivered by the source is

$$Q_T = C_T E = (370 \text{ pF})(10 \text{ V}) = 3700 \text{ pC}$$

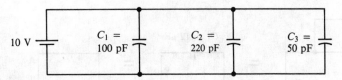

FIGURE 9.18 (Example 9.5)

The charge on each capacitor is

$$Q_1 = C_1V_1 = (100 \text{ pF})(10 \text{ V}) = 1000 \text{ pC}$$

$$Q_2 = C_2V_2 = (220 \text{ pF})(10 \text{ V}) = 2200 \text{ pC}$$

$$Q_3 = C_3V_3 = (50 \text{ pF})(10 \text{ V}) = 500 \text{ pC}$$

$$Q_T = Q_1 + Q_2 + Q_3 = 1000 \text{ pC} + 2200 \text{ pC} + 500 \text{ pC} = 3700 \text{ pC}$$

Drill Exercise 9.5

How much *additional* capacitance should be connected in parallel with the circuit in Figure 9.18 if the total charge stored by the circuit is to be 5000 pC?

ANSWER: 130 pF.   □

Series–Parallel Capacitor Circuits

Series–parallel capacitor circuits can be analyzed using the same general approach that we used for series–parallel resistor circuits: Combine series and parallel components to obtain progressively simpler equivalent circuits, and then work backward through the equivalent circuits until the original circuit is completely solved. The relationship $C = Q/V$ is used repeatedly to solve for unknown quantities, just as Ohm's law is used repeatedly in series–parallel resistor circuits. It is important to remember how charge is distributed among series- and parallel-connected capacitors, as discussed in the preceding paragraphs.

Example 9.6   (Analysis)

Find the voltage across and charge on each capacitor in Figure 9.19(a).

SOLUTION C_2 and C_3 are in parallel and therefore have an equivalent capacitance of

$$C_2 + C_3 = 10 \text{ }\mu\text{F} + 8 \text{ }\mu\text{F} = 18 \text{ }\mu\text{F}$$

As shown in Figure 9.19(b), this 18-μF equivalent is in series with C_1. Thus, the total equivalent capacitance of the circuit is

$$C_T = \frac{(12 \text{ }\mu\text{F})(18 \text{ }\mu\text{F})}{12 \text{ }\mu\text{F} + 18 \text{ }\mu\text{F}} = 7.2 \text{ }\mu\text{F}$$

The total charge delivered by the source is therefore

$$Q_T = EC_T = (5 \text{ V})(7.2 \text{ }\mu\text{F}) = 36 \text{ }\mu\text{C}$$

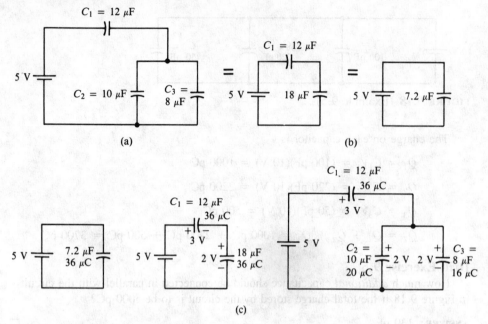

FIGURE 9.19 (Example 9.6)

As shown in Figure 9.19(c), the total charge of 36 μC is deposited both on the 18-μF equivalent capacitance and on $C_1 = 12$ μF, because the series combination of those two capacitances produced the 7.2-μF total capacitance. (This step is like concluding that the current in each of two series resistors is the same as the current in their series equivalent resistance.) The voltage across C_1 is then found to be

$$V_1 = \frac{Q_1}{C_1} = \frac{36 \ \mu C}{12 \ \mu F} = 3 \ V$$

Similarly, the voltage across the 18-μF equivalent capacitance is

$$\frac{36 \ \mu C}{18 \ \mu F} = 2 \ V$$

(Note that this result could also have been obtained from Kirchhoff's voltage law.)

Since the 18-μF equivalent capacitance was produced by the parallel combination of C_2 and C_3, the voltage across these two capacitors is the same as the 2 V across the 18 μF. Therefore, as shown in Figure 9.19(c),

$$Q_2 = C_2 V_2 = (10 \ \mu F)(2 \ V) = 20 \ \mu C$$

$$Q_3 = C_3 V_3 = (8 \ \mu F)(2 \ V) = 16 \ \mu C$$

Drill Exercise 9.6

Repeat Example 9.6 if a 40-μF capacitor, C_4, is connected in series with C_2 in Figure 9.19(a).

ANSWER: $Q_1 = 34.29\ \mu\text{C}$, $V_1 = 2.86\ \text{V}$; $Q_2 = 17.14\ \mu\text{C}$, $V_2 = 1.71\ \text{V}$; $Q_3 = 17.14\ \mu\text{C}$, $V_3 = 2.14\ \text{V}$; $Q_4 = 17.14\ \mu\text{C}$, $V_4 = 0.43\ \text{V}$. □

Example 9.7 (Design)

During the development of a certain electronic circuit, it was necessary to test the performance with 900 pF of capacitance connected to the circuit. The only capacitors available to the designers were two 1200-pF capacitors and a 2400-pF capacitor. How could these be connected to provide the necessary 900 pF of capacitance?

SOLUTION There are a limited number of series–parallel combinations that can be constructed from the three capacitors. Trial and error leads to the solution shown in Figure 9.20. The parallel combination of the 2400-pF and 1200-pF capacitors is equivalent to their sum, 3600 pF. This value in series with the other 1200-pF capacitor gives the required capacitance:

$$C_T = \frac{(3600\ \text{pF})(1200\ \text{pF})}{3600\ \text{pF} + 1200\ \text{pF}} = 900\ \text{pF}$$

Drill Exercise 9.7

What combination of the capacitors in Example 9.7 could be used to produce 480 pF of capacitance?

ANSWER: all three capacitors in series. □

9.5 Transients in Series RC Networks

In the capacitor circuits we have studied up to this point, we have shown a voltage source connected directly across a capacitor or across a combination of capacitors. In practice, this sort of connection would not be made, because there is no resistance to limit the flow of current into the capacitor(s). *An uncharged capacitor is equivalent to a short circuit*, as far as a dc voltage source is concerned. Therefore, unless a resistor is connected in series with a capacitor, the flow of charging current when a voltage source is first connected to it is limited only by the (usually) small internal resistance of the source and any wiring resistance present. The surge of current that flows when no resistor is present may be great enough to damage the capacitor, the source, or both.

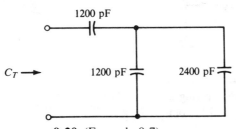

FIGURE 9.20 (Example 9.7)

A resistor and capacitor connected in series is called a series *RC network*. Besides the practical necessity we have mentioned for using a resistor to limit charging current, the series RC network is an exceptionally important configuration in its own right, one that is widely used and studied in all manner of practical circuits. For example, wiring resistance and stray capacitance form an RC network that strongly influences the speed at which a digital computer can operate.

The word *transient* means temporary, or short-lived. When a voltage source is first connected to an RC network, charging current flows only until the capacitor is fully charged, as discussed in Section 9.1. Thus, the charging current is called a transient current. In connection with dc circuits, a transient is a voltage or current that *changes* with time for a short duration of time. As a capacitor charges, its voltage builds up (i.e., changes) until the capacitor is fully charged and its voltage equals the source voltage that supplied the charge. After that time, there is no further change in capacitor voltage. Thus, the voltage across a capacitor during the time it is being charged is an example of a transient voltage.

Consider the circuit shown in Figure 9.21. We will assume that the capacitor is uncharged and we will close the switch to commence the charging process. Notice that charging current will flow through resistor R after the switch is closed. We will present equations for the transient voltages and currents in the circuit as functions of the time t that has elapsed after the switch is closed. Thus, we assume that the switch is closed at $t = 0$. In other words, for our purposes, all time begins with the closing of the switch, and all values of t are measured from that instant.

Transient Current

Using calculus, it can be shown that the equation for the current that flows in the circuit of Figure 9.21 after the switch is closed at $t = 0$ is

$$i(t) = \frac{E}{R} e^{-t/RC} \qquad \text{amperes} \qquad (9.20)$$

where e is the constant (approximately 2.7183) used as the base of the *natural logarithm*. It is important to know how to perform computations involving e on a scientific calculator. On many calculators, there is a key labeled e^x that can be used to raise e to any power: Enter the value of x (change its sign if necessary), then press the e^x key. On other

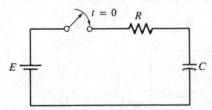

FIGURE 9.21 RC network showing a switch that is closed at $t = 0$ to commence charging the capacitor.

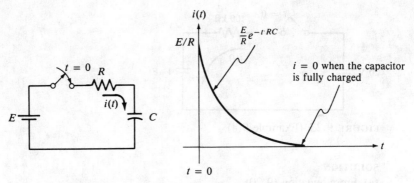

FIGURE 9.22 Plot of the transient charging current $i(t)$ that flows in the RC network after the switch is closed at $t = 0$.

calculators, the inverse of the natural logarithm (ln) must be computed to find a value for e^x: Enter the value of x, press the inverse function key, and then press the ln key. As an exercise, use a calculator to verify the following: $e^2 = 7.389056$, $e^{-1} = 0.3678795$, $e^{-8/11} = 0.4832252$, $e^0 = 1$.

We emphasize again that equation (9.20) gives current as a function of *time t*. It is conventional to use lowercase letters to designate currents or voltages whose values change with time, and the notation $i(t)$ in equation (9.22) further emphasizes that a value of transient current depends on the time t at which the equation is evaluated. For example, $i(1)$ means the value of i at $t = 1$, which is computed by substituting $t = 1$ in equation (9.20). Figure 9.22 shows a graph of equation (9.20) plotted versus time t. Notice that the current immediately "jumps" to a maximum value of E/R amperes at the instant the switch is closed (i.e., at $t = 0$). This fact can be verified by substituting $t = 0$ in equation (9.20):

$$\underbrace{i(0)}_{\displaystyle \substack{\big\lfloor \\ i(t) \text{ evaluated at } t = 0}} = \frac{E}{R} e^{-0/RC} = \frac{E}{R} e^0 = \frac{E}{R} (1) = \frac{E}{R}$$

As time goes by, the current gradually *decays* (diminishes) until it reaches zero when the capacitor is fully charged. Recall that the current must eventually reach zero, because the capacitor voltage reaches E volts when fully charged and has a polarity that opposes source voltage E.

Example 9.8   (Analysis)
(a) Find the value of the transient charging current in the circuit shown in Figure 9.23 at
   (i) $t = 0$ s
   (ii) $t = 0.05$ s
   (iii) $t = 0.1$ s
   (iv) $t = 0.2$ s
   (v) $t = 0.5$ s
(b) Sketch a graph of $i(t)$ versus t.

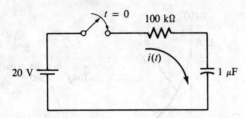

FIGURE 9.23 (Example 9.8)

SOLUTION

(a) From equation (9.20),

$$i(t) = \frac{E}{R} e^{-t/RC}$$

The quantity RC in this equation equals

$$(100 \times 10^3 \ \Omega)(1 \times 10^{-6} \ F) = 0.1$$

Thus,

$$i(t) = \frac{20 \ V}{100 \ k\Omega} e^{-t/0.1} \ A = 0.2e^{-t/0.1} \ mA$$

(i) Setting $t = 0$, we find

$$i(0) = 0.2e^{-0} \ mA = 0.2(1) \ mA = 0.2 \ mA$$

(ii) Setting $t = 0.05$, we have

$$i(0.05) = 0.2e^{-0.05/0.1} \ mA = 0.2e^{-0.5} \ mA = 0.121 \ mA$$

(iii) Setting $t = 0.1$ yields

$$i(0.1) = 0.2e^{-0.1/0.1} \ mA = 0.2e^{-1} \ mA = 0.0736 \ mA$$

(iv) Setting $t = 0.2$ gives us

$$i(0.2) = 0.2e^{-0.2/0.1} \ mA = 0.2e^{-2} \ mA = 0.0271 \ mA$$

(v) Setting $t = 0.5$, we find

$$i(0.5) = 0.2e^{-0.5/0.1} \ mA = 0.2e^{-5} \ mA = 0.0013 \ mA$$

(b) Figure 9.24 shows a plot of $i(t)$ versus t.

RC Time Constant

The quantity RC in equation 9.20 is called the *time constant* of the series RC network. It can be shown that the units of the product of resistance and capacitance (ohms $\times$ farads) are equivalent to *seconds*, so the RC time constant has the units of time, and the quantity t/RC in equation (9.20) is dimensionless. The conventional symbol for a time constant is the Greek lowercase letter τ (tau):

$$\tau = RC \text{ seconds} \tag{9.21}$$

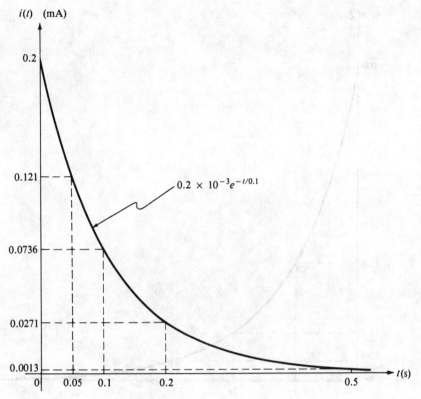

FIGURE 9.24 (Example 9.8)

Equation (9.20) can be written in terms of τ as

$$i(t) = \frac{E}{R} e^{-t/\tau} \tag{9.22}$$

The time constant is a measure of how long it takes the capacitor to charge. The larger the value of R or C, or both, the longer it takes for the capacitor to charge, because R limits the charging current and C determines how much charge must be stored.

After the switch in Figure 9.21 has been closed for a length of time equal to one time constant (i.e., at the instant of time $t = \tau$ seconds), we find from equation (9.22),

$$i(\tau) = \frac{E}{R} e^{-\tau/\tau} = \frac{E}{R} e^{-1} \approx 0.368 \frac{E}{R}$$

Thus, *after one time constant, the charging current has decayed to approximately 36.8% of its value at $t = 0$ (0.368E/R)*. This statement is true regardless of the particular value of the time constant. After the switch has been closed for a period of time equal to five time constants, at $t = 5\tau$, we find

$$i(5\tau) = \frac{E}{R} e^{-5\tau/\tau} = \frac{E}{R} e^{-5} \approx 0.0067 \frac{E}{R}$$

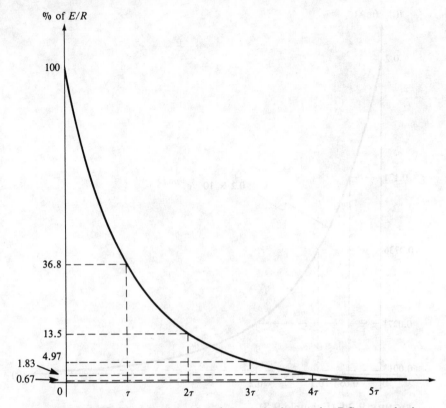

FIGURE 9.25 The transient charging current in a series RC network plotted as percent of its maximum value versus the number of time constants after charging commences.

This value is quite small compared to the maximum value, E/R at $t = 0$, so *for most practical purposes we may assume that the capacitor is fully charged after five time constants.* It is common practice to use this assumption. Figure 9.25 shows $i(t)$ plotted as a percent of its maximum value (percent of E/R) versus the number of time constants that have elapsed since the switch was closed. Notice that this plot applies to *any* series RC network.

Example 9.9 (Analysis)

With reference to the circuit shown in Figure 9.26:

(a) Find the time constant.
(b) Find the value of $i(t)$ after the switch has been closed for 1.5 time constants.
(c) Find the voltage $v_R(t)$ at $t = 1.5\tau$.
(d) After how many time constants will the current have decayed to 10% of its maximum value?

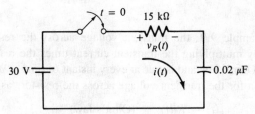

FIGURE 9.26 (Example 9.9)

SOLUTION

(a) $\tau = RC = (15 \text{ k}\Omega)(0.02 \text{ }\mu\text{F}) = (15 \times 10^3)(0.02 \times 10^{-6}) = 0.3 \text{ ms}$

(b) At $t = 1.5\tau$, we have, from equation (9.22),

$$i(1.5\tau) = \frac{30 \text{ V}}{15 \text{ k}\Omega} e^{-1.5\tau/\tau} = 2 \times 10^{-3}e^{-1.5} \text{ A} = 0.446 \text{ mA}$$

(c) Since we have a series circuit, the current $i(t)$ is the same in every component at every instant of time. Therefore, the current through the resistor at $t = 1.5\tau$ is the same as the current found in part (b): 0.446 mA. Thus, the voltage $v_R(t)$ at $t = 5\tau$ is, by Ohm's law,

$$v_R(1.5\tau) = (0.446 \text{ mA})(15 \text{ k}\Omega) = 6.69 \text{ V}$$

(d) The maximum current is $E/R = 30 \text{ V}/15 \text{ k}\Omega = 2 \text{ mA}$ and 10% of that value is $0.1 (2 \text{ mA}) = 0.2 \text{ mA}$. We substitute $i(t) = 0.2 \times 10^{-3}$ in equation (9.22) to obtain

$$0.2 \times 10^{-3} = 2 \times 10^{-3}e^{-t/0.3 \times 10^{-3}}$$

This equation must be solved for t. Dividing both sides by 2×10^{-3} gives $0.1 = e^{-t/0.3 \times 10^{-3}}$. We now take the natural log (ln) of both sides, recognizing that $\ln(e^x) = x$:

$$\ln(0.1) = \ln(e^{-t/0.3 \times 10^{-3}})$$

$$-2.3026 = \frac{-t}{0.3 \times 10^{-3}}$$

$$t = 0.6908 \text{ ms}$$

Thus, the current will decay to 10% of its maximum value 0.6908 ms after the switch is closed. Since $\tau = 0.3 \text{ ms}$, this time is equivalent to

$$\frac{0.6908 \text{ ms}}{0.3 \text{ ms}/\tau} = 2.3\tau \quad \text{or} \quad 2.3 \text{ time constants}$$

Drill Exercise 9.9

Find the value of $v_R(t)$ in Example 9.9 at (a) $t = 0$; (b) $t = 3.2\tau$; (c) $t = 5\tau$.

ANSWER: (a) 30 V; (b) 1.22 V; (c) 0.202 V. □

Transient Voltages

As demonstrated in Example 9.9, the transient voltage across the resistor in an RC network can be found by multiplying the transient current times the resistance. This is a direct consequence of Ohm's law and is true at every instant of time t. We can therefore derive a general equation for the transient voltage across the resistor, as follows:

$$v_R(t) = Ri(t) \qquad \text{(Ohm's law)} \tag{9.23}$$

$$= R\left(\frac{E}{R} e^{-t/RC}\right)$$

$$v_R(t) = Ee^{-t/RC} \qquad \text{volts} \tag{9.24}$$

This equation shows that the maximum value of $v_R(t)$ is E volts at $t = 0$, as would be expected, since the current is maximum at $t = 0$. In other words, the entire source voltage is dropped across the resistor the instant the switch is closed. This result also follows from the fact that the uncharged capacitor is a short circuit at $t = 0$ and therefore has zero voltage drop across it. Kirchhoff's voltage law is satisfied around the circuit at every instant of time, including at $t = 0$. Since $v_R(t)$ is simply a multiple (by R) of $i(t)$, the graph of $v_R(t)$ has the same appearance as that of $i(t)$. It is shown in Figure 9.27.

As mentioned earlier in this section, the voltage across the capacitor, $v_C(t)$, is also a transient, because it changes with time for only as long as it takes the capacitor to become fully charged. Writing Kirchhoff's voltage law around the loop shown in Figure 9.28(a), we find

$$E = v_R(t) + v_C(t) \tag{9.25}$$

or

$$v_C(t) = E - v_R(t) \tag{9.26}$$

Substituting equation (9.24) for $v_R(t)$ into equation (9.26) gives

$$v_C(t) = E - Ee^{-t/RC} = E(1 - e^{-t/RC}) \qquad \text{volts} \tag{9.27}$$

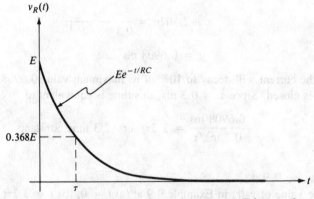

FIGURE 9.27 Transient voltage $v_R(t)$ across R in a series RC network when the capacitor is charging.

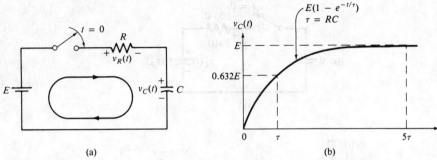

(a)   (b)

FIGURE 9.28 Derivation of the transient voltage across a charging capacitor in a series RC network. (a) Kirchhoff's voltage law around the loop is valid at every instant of time t. Thus, $v_C(t) = E - v_R(t)$. (b) The transient voltage $v_C(t)$ across the capacitor during the time it is charging toward its maximum value of E volts.

In terms of the time constant $\tau = RC$,

$$v_C(t) = E(1 - e^{-t/\tau}) \qquad \text{volts} \qquad (9.28)$$

Figure 9.28(b) shows a plot of $v_C(t)$ versus τ. Notice that $v_C(t) = 0$ at $t = 0$, which follows from the fact that C is initially uncharged. Also notice that $v_C(t)$ reaches its final, fully charged value of E volts after essentially five time constants. One time constant after the switch is closed, at $t = \tau$, we find

$$v_C(\tau) = E(1 - e^{-\tau/\tau}) = E(1 - e^{-1}) \approx 0.632E$$

Thus, *the capacitor voltage reaches approximately 63.2% of its fully charged value one time constant after the switch is closed.*

Example 9.10   (Analysis)

With reference to the circuit shown in Figure 9.29(a):

(a) Write the mathematical expression for the voltage $v_C(t)$ after the switch is closed.
(b) Find the value of $v_C(t)$ 0.25 s after the switch is closed.
(c) Find the value of $v_C(t)$ at $t = 5\tau$.
(d) Sketch $v_C(t)$ and $v_R(t)$.
(e) Verify that Kirchhoff's voltage law is satisfied at $t = 0.5$ s.

SOLUTION

(a) $\tau = RC = (500 \ \Omega)(500 \times 10^{-6} \ \text{F}) = 0.25$ s

From equation (9.27),

$$v_C(t) = E(1 - e^{-t/RC}) = 100(1 - e^{-t/0.25}) \ \text{V}$$

(b) At $t = 0.25$ s,

$$v_C(t) = 100(1 - e^{-0.25/0.25}) = 100(1 - e^{-1}) = 63.2 \ \text{V}$$

(c) At $t = 5\tau$,

$$v_C(t) = 100(1 - e^{-5\tau/\tau}) = 100(1 - e^{-5}) = 99.32 \ \text{V}$$

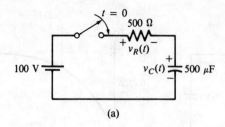

(a)

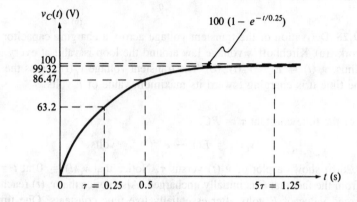

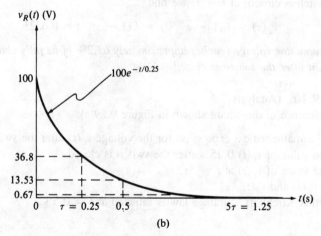

(b)

FIGURE 9.29 (Example 9.10)

(d) $v_C(t)$ and $v_R(t)$ are plotted in Figure 9.29(b). Note that the sum of these two voltages at any instant of time is $E = 100$ V, verifying Kirchhoff's voltage law.

(e) At $t = 0.5$,

$$v_C(0.5) = 100(1 - e^{-0.5/0.25}) = 100(1 - e^{-2}) = 86.47 \text{ V}$$

$$v_R(0.5) = 100e^{-0.5/0.25} = 100e^{-2} = 13.53 \text{ V}$$

$$v_C(0.5) + v_R(0.5) = 86.47 \text{ V} + 13.53 \text{ V} = 100 \text{ V} = E$$

Drill Exercise 9.10

How many seconds after the switch is closed in Example 9.10 does it take for $v_C(t)$ to reach 50 V?

ANSWER: 0.173 s. □

Example 9.11 (Design)

The electronic "trigger" circuit shown in Figure 9.30(a) sounds an alarm when its input voltage is 3.5 V. The alarm is sounded by closure of switch S. It is desired to provide a 1-s *time delay* between the time the switch is closed and the time the alarm is sounded. This delay is to be achieved by inserting the RC network shown in the figure. Given that $C = 500$ μF, what value of R should be used?

SOLUTION The input voltage to the trigger circuit after the switch is closed is the capacitor voltage shown in Figure 9.30(b). We want that voltage to equal 3.5 V at $t = 1$ s, so we must find the RC time constant τ that makes $v_C(t)$ equal to 3.5 V when $t = 1$:

$$v_C(t) = 5(1 - e^{-t/\tau})$$

$$3.5 = 5(1 - e^{-1/\tau})$$

$$0.3 = e^{-1/\tau}$$

Taking the natural logarithm of both sides gives

$$-1.20 = \frac{-1}{\tau}$$

or

$$\tau = 0.831 \text{ s}$$

To find the necessary value of resistance, we solve for R in the time-constant equation:

$$\tau = 0.831 = RC$$

$$0.831 = R(500 \times 10^{-6}\text{F})$$

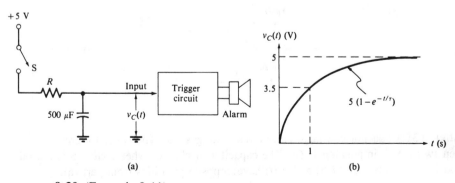

(a) (b)

FIGURE 9.30 (Example 9.11)

$$R = \frac{0.831}{500 \times 10^{-6}} = 1.66 \text{ k}\Omega$$

Drill Exercise 9.11

The time delay in Example 9.11 is to be made adjustable by making R an adjustable resistance. What range of values of R is necessary to adjust the time delay from 0.5 s to 10 s?

ANSWER: 831 Ω to 16.6 kΩ.   □

Discharge Transients

Consider the circuit shown in Figure 9.31(a). S is a two-position switch that allows the capacitor to charge when placed in position 1. Assume that the switch has been in position 1 for a long period of time, so the capacitor is fully charged to E volts and the current

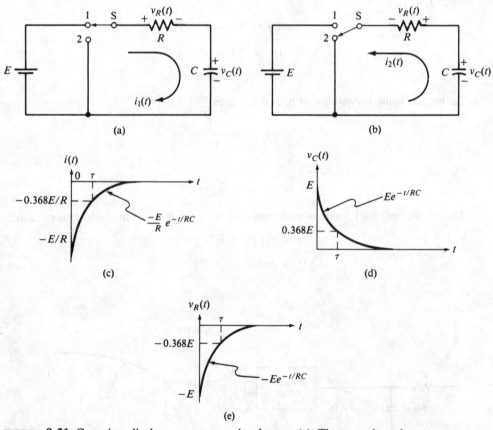

(a)   (b)

(c)   (d)

(e)

FIGURE 9.31 Capacitor discharge current and voltages. (a) The capacitor charges when switch S is in position 1. (b) The capacitor discharges when switch S is placed in position 2. Notice that $i_2(t)$ and $v_R(t)$ have opposite polarities from part (a). (c) Discharge current versus t. (d) Capacitor discharge voltage. (e) $v_R(t)$ during discharge.

$i_1(t)$ is zero. When the switch is placed in position 2, as shown in Figure 9.31(b), the voltage source is no longer in the circuit and there is a path for (conventional) current to flow from the positive side of the capacitor to the negative side. In other words, electrons on the negative side are drawn to the positive side, and the charge on each side is dissipated. This transfer of charge produces current $i_2(t)$, *which is in the opposite direction from the original charging current,* $i_1(t)$. Eventually, all of the charge on both sides of the capacitor will be neutralized, so both the current $i_2(t)$ and the voltage $v_C(t)$ will decay to zero. The process is called *discharging* the capacitor, and the decaying voltage and current are the discharge transients. The equations for the discharge transients are:

$$i(t) = -\frac{E}{R}e^{-t/RC} = -\frac{E}{R}e^{-t/\tau} \quad \text{amperes} \tag{9.29}$$

$$v_C(t) = Ee^{-t/RC} = Ee^{-t/\tau} \quad \text{volts} \tag{9.30}$$

$$v_R(t) = -Ee^{-t/RC} = -Ee^{-t/\tau} \quad \text{volts} \tag{9.31}$$

Notice that $i(t)$ and $v_R(t)$ are negative because the discharge current flows in the opposite direction from the charging current, which we assumed to be positive. Also notice that $v_R(t) + v_C(t) = 0$ at every instant of time, confirming Kirchhoff's voltage law, since there is no voltage source in the loop. The time reference for both equations ($t = 0$) is the instant at which the switch is put in position 2. The time constant $\tau = RC$ is defined the same way during discharge as it is during charge, because in both cases the resistor and capacitor are in series. Plots of the discharge current and voltages are shown in Figure 9.31(c)–(e).

Example 9.12   (Analysis)
Switch S in Figure 9.32 has been in position 1 long enough for the capacitor to be fully charged. When it is placed in position 2:

(a) Write the mathematical expressions for the discharge transients $i(t)$, $v_C(t)$, and $v_R(t)$.
(b) Find $v_C(t)$ 15 μs after the switch is placed in position 2.

SOLUTION

$$\tau = RC = (40 \times 10^3 \ \Omega)(200 \times 10^{-12} \ \text{F}) = 8 \ \mu s$$

(a) $i(t) = -\dfrac{E}{R}e^{-t/\tau} = -\dfrac{5 \ \text{V}}{40 \ \text{k}\Omega}e^{-t/8 \times 10^{-6}} = -125e^{-t/8 \times 10^{-6}} \ \mu A$

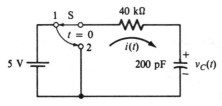

FIGURE 9.32 (Example 9.12)

$$v_C(t) = Ee^{-t/\tau} = 5e^{-t/8\times 10^{-6}} \text{ V}$$
$$v_R(t) = -Ee^{-t/\tau} = -5e^{-t/8\times 10^{-6}} \text{ V}$$

(b) $v_c(15 \ \mu s) = 5e^{-15\times 10^{-6}/8\times 10^{-6}} = 5e^{-1.875} = 0.767$ V

Drill Exercise 9.12

Repeat Example 9.12 if the switch is placed in position 2 3.125 μs after it is placed in position 1. (*Hint*: First find the voltage to which the capacitor charges, and then use that value of E in the discharge equations.)

ANSWER: (a) $i(t) = -40.42e^{-t/8\times 10^{-6}} \ \mu A$; $v_C(t) = 1.617e^{-t/8\times 10^{-6}}$ V; $v_R(t) = -1.617e^{-t/8\times 10^{-6}}$ V; (b) 0.248 V.   □

In some RC networks, the time constant during charge is different from that during discharge. The next example illustrates such a case. Remember that the resistance used in computing a transient is the total series equivalent resistance through which the capacitor charges or discharges. It is also important to note that E in equations (9.29) through (9.31) is the value to which the capacitor has charged at the time the discharge process begins. Unless the capacitor has been allowed to charge *fully*, the value used for E in the equations will be less than the value E of the voltage source.

Example 9.13 (Analyis)

With reference to the circuit shown in Figure 9.33:

(a) Write the mathematical expressions for $i(t)$ and $v_C(t)$ when the switch is placed in position 1.
(b) Write the mathematical expressions for $i(t)$ and $v_C(t)$ when the switch is placed in position 2, after having been in position 1 for 1 s.
(c) Sketch $i(t)$ in parts (a) and (b) using a single time axis. Repeat for $v_C(t)$.

SOLUTION
(a) When the switch is placed in position 1, the capacitor charges through R_1 only, so the charging time constant is

$$\tau \text{ (charge)} = R_1 C = (100 \ \Omega)(1000 \ \mu F) = 0.1 \text{ s}$$

Thus,

$$i(t) = \frac{E}{R_1} e^{-t/\tau} = \frac{20}{100} e^{-t/0.1} = 0.2e^{-t/0.1} \text{ A}$$

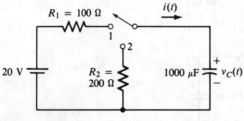

FIGURE 9.33 (Example 9.13)

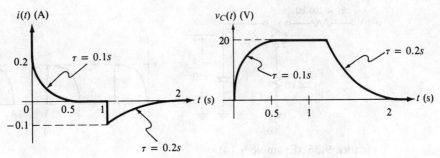

FIGURE 9.34 (Example 9.13) Notice that the capacitor charges fully in about 0.5 s, but takes about 1 s to discharge.

and

$$v_C(t) = E(1 - e^{-t/\tau}) = 20(1 - e^{-t/0.1}) \text{ V}$$

(b) Since the switch remains in position 1 for 1s, or 10 time constants, the capacitor charges fully to 20 V. When the switch is placed in position 2, the capacitor discharges through R_2 only, so the time constant during discharge is

$$\tau \text{ (discharge)} = R_2C = (200 \ \Omega)(1000 \ \mu\text{F}) = 0.2 \text{ s}$$

The discharge equations are therefore

$$i(t) = -\frac{E}{R_2} e^{-t/\tau} = -\frac{20}{200} e^{-t/0.2} = -0.1 e^{-t/0.2} \text{ A}$$

and

$$v_C(t) = E e^{-t/\tau} = 20 e^{-t/0.2} \text{ V}$$

(c) The plots of $i(t)$ and $v_C(t)$ are shown in Figure 9.34. It is evident that the charging time is less than the discharge time.

Drill Exercise 9.13
Repeat Example 9.13 if the switch remains in position 1 for 0.1 s only.

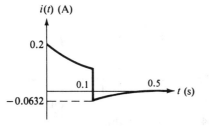

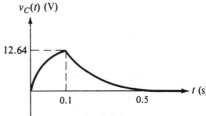

FIGURE 9.35 (Drill Exercise 9.13)

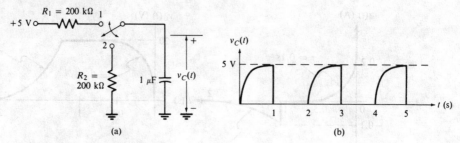

FIGURE 9.36 (Example 9.14)

ANSWER: (a) Same as Example 9.13; (b) $i(t) = -0.0632e^{-t/0.2}$ A; $v_C(t) = 12.64e^{-t/0.2}$ V; (c) See Figure 9.35.

Example 9.14 (Troubleshooting)

The switch in Figure 9.36(a) is alternately placed in position 1 and position 2 at 1-s intervals. The resulting capacitor voltage is shown in Figure 9.36(b). Which component in the circuit has failed, and what is the nature of the failure?

SOLUTION The charge and discharge time constants should both equal $(200 \text{ k}\Omega)(1 \text{ }\mu\text{F}) = 0.2$ s. Therefore, the capacitor should take a full five time constants, or $5(0.2) = 1$ s, to charge to 5 V, and another full second to discharge to 0 V. The graph of $v_C(t)$ shows that the capacitor is charging properly, but is discharging instantly. The discharge time constant is essentially zero, and we conclude that R_2 is shorted ($R_2 C = 0$).

Drill Exercise 9.14

Sketch $v_C(t)$ versus t in Example 9.14 when (a) R_1 alone is shorted; (b) both R_1 and R_2 are shorted; (c) the capacitor is open.

ANSWER: See Figure 9.37.

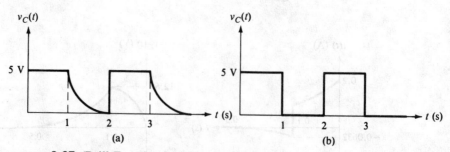

FIGURE 9.37 (Drill Exercise 9.14) (c) Same as (b).

9.6 Series–Parallel RC Networks

Transients in Series–Parallel RC Networks

When a capacitor charges or discharges through a series–parallel resistor network, the methods of Section 9.5 can still be used to determine transient voltages and currents in the capacitor, provided that the series–parallel circuitry is first converted to its Thévenin equivalent. Recall that the Thévenin equivalent circuit is a single voltage source E_{TH} in series with a single resistance R_{TH}, so this approach effectively converts the series–parallel RC network to a series RC network. As usual when finding a Thévenin equivalent circuit, the first step is to *remove* the capacitor (i.e., open-circuit the terminals where the capacitor is connected). Once we have found the Thévenin equivalent of the remaining circuitry, the capacitor is restored. Then the values E_{TH} and R_{TH} are used for E and R in the transient equations that were given for series RC networks. The next example illustrates this procedure.

Example 9.15   (Analysis)

With reference to the circuit shown in Figure 9.38(a):

(a) Write the mathematical expressions for $i(t)$ and $v_C(t)$ after the switch is closed.
(b) Sketch $i(t)$ and $v_C(t)$ versus t.

SOLUTION

(a) To find the Thévenin equivalent circuit at the terminals where the capacitor is connected, we first remove the capacitor, as shown in Figure 9.38(b). Replacing the 16-V source by a short circuit, we see that the 200-Ω and 600-Ω resistors are in parallel, so $R_{TH} = 250\ \Omega\ +\ 200\ \Omega \| 600\ \Omega\ =\ 400\ \Omega$. The Thévenin equivalent voltage is the same as the voltage across the 600-Ω resistor, because the 250-Ω resistor is open and there is no voltage drop across it. By the voltage-divider rule,

$$E_{TH} = \left(\frac{600\ \Omega}{600\ \Omega + 200\ \Omega}\right)16\ V = 12\ V$$

Restoring the capacitor to the Thévenin equivalent circuit, we see that we have a series RC network with time constant

$$\tau = R_{TH}C = (400\ \Omega)(300\ \mu F) = 0.12\ s$$

We can now use the equations for the transient current and voltage in a series RC network, after substituting $E = E_{TH} = 12\ V$ and $R = R_{TH} = 400\ \Omega$:

$$i(t) = \frac{E_{TH}}{R_{TH}}\,e^{-t/\tau} = \frac{12\ V}{400\ \Omega}\,e^{-t/0.12} = 0.03e^{-t/0.12}\ A$$

$$v_C(t) = E_{TH}(1 - e^{-t/\tau}) = 12(1 - e^{-t/0.12})\ V$$

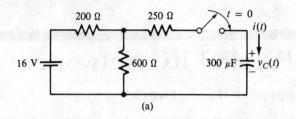

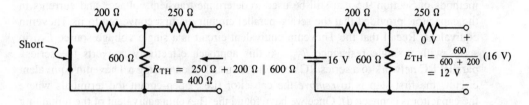

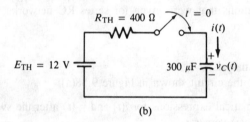

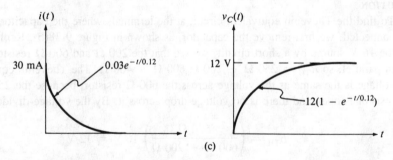

FIGURE 9.38 (Example 9.15)

(b) Figure 9.38(c) shows plots of $i(t)$ and $v_C(t)$ versus t. Notice that the maximum voltage to which the capacitor charges is $E_{TH} = 12$ V, not $E = 16$ V.

Drill Exercise 9.15

Suppose that a 250-Ω resistor is connected in parallel with the 250-Ω resistor in Figure 9.38(a). With that modification, find (a) the maximum voltage to which the capacitor charges; (b) the maximum value of the charging current.

ANSWER: (a) 12 V; (b) 0.0436 A.

□

Initial and Steady-State Currents and Voltages

The current that flows in a circuit the instant a voltage source is switched into it is called the *initial current*. We have already mentioned that an uncharged capacitor is equivalent to a short circuit at $t = 0$, so we can find the initial current(s) anywhere in a circuit by replacing all capacitors with short circuits. Any voltage produced by an initial current is called an initial voltage and can also be found using conventional analysis techniques, after all capacitors are replaced by short circuits.

When a source has been connected to a circuit for a long period of time (i.e., after all transients have expired), the currents and voltages in the circuit are said to have their *steady-state* values. *Under steady-state conditions, all capacitors are fully charged and are equivalent to open circuits.* Thus, we can find steady-state voltages and currents by replacing all capacitors with open circuits and using conventional analysis methods. In performing such an analysis, we find the voltage across a fully charged capacitor by finding the voltage across the terminals of the open circuit that replaces it.

Example 9.16   (Analysis)

(a) Find the initial current through and initial voltage across every component in the circuit shown in Figure 9.39(a).

(b) Find the steady-state current through and the steady-state voltage across every component in the circuit.

SOLUTION

(a) The *initial circuit*, with capacitors replaced by short circuits, is shown in Figure 9.39(b). Since R_1 is shorted out by C_1, the voltage is zero across each of those components. The total initial resistance and the total initial current in the circuit are then

$$R_T = R_2 \| R_3 = 30\ \Omega \| 30\ \Omega = 15\ \Omega$$

and

$$I_T = \frac{60\ \text{V}}{15\ \Omega} = 4\ \text{A}$$

The 4 A divides equally between the two 30-Ω resistors, so 2 A flows in each, and the voltage across each is

$$(2\ \text{A})(30\ \Omega) = 60\ \text{V}$$

As an exercise, verify that Kirchhoff's voltage law is satisfied around every closed loop in the initial circuit.

(b) The steady-state circuit, with capacitors replaced by open circuits, is shown in Figure 9.39(c). Since C_2 is open, there is zero current through R_2 and therefore zero voltage across it. The total steady-state resistance and the total steady-state current are then

$$R_T = R_1 + R_3 = 10\ \Omega + 30\ \Omega = 40\ \Omega$$

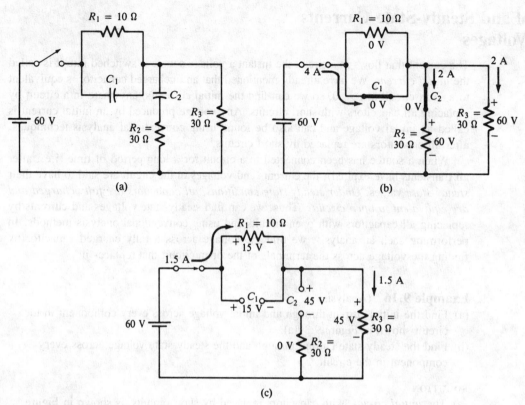

FIGURE 9.39 (Example 9.16)

$$I_T = \frac{60 \text{ V}}{40 \text{ Ω}} = 1.5 \text{ A}$$

The 1.5A flows through R_1, developing a voltage of (1.5 A)(10 Ω) = 15 V across it. Since C_1 is in parallel with R_1, it is fully charged to 15 V.

The 1.5 A also flows through R_3, developing a voltage of (1.5 A)(30 Ω) = 45 V across it. Writing Kirchhoff's voltage law around the loop containing R_2, C_2, and R_3, we find that capacitor C_2 is fully charged to 45 V. As an exercise, verify that Kirchhoff's voltage law is satisfied around every closed loop in the steady-state circuit.

Drill Exercise 9.16

Find (a) the initial, and (b) the steady-state currents and voltages in Figure 9.39(a) if a capacitor, C_3, is inserted in series with R_3.

ANSWER: (A) Same as part (a) of Example 9.16. $V_{C3} = 0$ V, $I_{C3} = 2$ A; (b) All currents equal zero and all resistor voltages equal zero. $V_{C1} = 0$ V, $V_{C2} = V_{C3} = 60$ V. □

9.7 Current Through and Voltage Across a Capacitor

By now, we know that current does not actually flow "through" a capacitor. Rather, electrons accumulate on one plate and drive an equal number of electrons off the other plate. When a voltage source is charging a capacitor, the source receives the same amount of charge that it delivers, so the effect is the same as if the charge did indeed flow through the capacitor. In subsequent discussions, we will speak of current as flowing *through* a capacitor, with the understanding that it does so only in the context we have described.

The current through a capacitor depends on the *rate of change* of the voltage across it. This very important concept is one we will encounter again and is the foundation of many useful applications of *differential calculus*. The rate of change of voltage can be determined from a graph of voltage versus time by calculating the *slope* of the graph (see Figure 9.40):

$$\text{slope} = \text{rate of change} = \frac{\Delta v}{\Delta t} = \frac{v_2 - v_1}{t_2 - t_1} \quad \text{volts/second} \qquad (9.32)$$

where Δ means "change in"

t_1, t_2 = two points in time, $t_2 > t_1$

v_1, v_2 = voltages at times t_1 and t_2

Equation (9.32) can be used only over a time interval, t_1 to t_2, where the graph is a straight line segment. In other words, a new rate of change must be calculated over each line segment that has a different slope. If the voltage is *constant* over a time interval, its rate of change is *zero* over that interval. (The latter statement should be obvious if

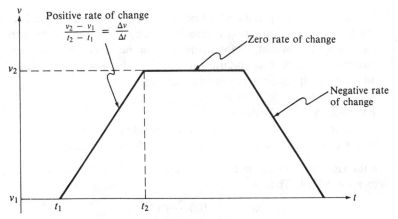

FIGURE 9.40 Calculating the rate of change of voltage, $\Delta v/\Delta t$.

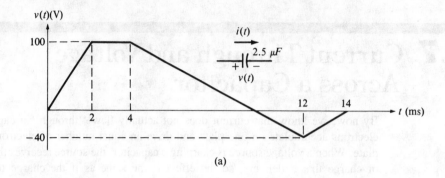

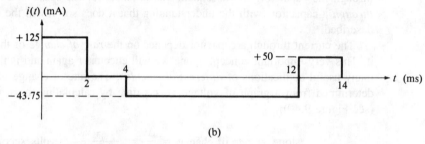

FIGURE 9.41 (Example 9.17)

the notion of rate of change is thoroughly understood. Think about it.) If voltage *decreases* with time, the rate of change (and slope of the graph) is negative. All these cases are illustrated in Figure 9.40.

The current through a capacitor in terms of the rate of change of the voltage across it is

$$i = C \frac{\Delta v}{\Delta t} \tag{9.33}$$

(In calculus, this equation would be written $i = C \, dv/dt$.) Equation (9.33) shows that the faster the voltage across a capacitor *changes*, the greater the current through it. If the voltage is constant, as for example when the capacitor is fully charged, the current through it is zero. This confirms our previous discussion of the fact that a capacitor is an open circuit under steady-state conditions.

Example 9.17   (Analysis)

Figure 9.41(a) shows a graph of the voltage across a 2.5-μF capacitor plotted versus time. Find and plot the current through the capacitor.

SOLUTION In the interval from $t_1 = 0$ to $t_2 = 2$ ms, the voltage changes from $v_1 = 0$ to $v_2 = 100$ V. Thus

$$\frac{\Delta v}{\Delta t} = \frac{(100 - 0) \text{ V}}{(2 - 0) \times 10^{-3} \text{ s}} = 50 \times 10^3 \text{ V/s}$$

From equation (9.33), the current over that interval is

$$i = C \frac{\Delta v}{\Delta t} = (2.5 \times 10^{-6})(50 \times 10^3) = 125 \text{ mA}$$

Since the rate of change is constant in the interval from 0 to 2 ms, the current is a constant 125 mA over the same interval, as shown in Figure 9.41(b).

In the interval from 2 to 4 ms, the voltage is a constant 100 V, the rate of change is zero, and the current is therefore zero:

$$\frac{\Delta v}{\Delta t} = \frac{100 \text{ V} - 100 \text{ V}}{4 \text{ ms} - 2 \text{ ms}} = \frac{0 \text{ V}}{2 \text{ ms}} = 0 \qquad i = C \frac{\Delta v}{\Delta t} = 0$$

In the interval from $t_1 = 4$ ms to $t_2 = 12$ ms, the voltage changes from $v_1 = 100$ V to $v_2 = -40$ V. Therefore,

$$\frac{\Delta v}{\Delta t} = \frac{v_2 - v_1}{t_2 - t_1} = \frac{-40 \text{ V} - 100 \text{ V}}{12 \text{ ms} - 4 \text{ ms}} = -\frac{140 \text{ V}}{8 \times 10^{-3} \text{s}} = -17.5 \times 10^3 \text{ V/s}$$

Notice that the voltage decreases with time over this interval, and the rate of change is negative. The current is the constant negative value

$$i = C \frac{\Delta v}{\Delta t} = (2.5 \times 10^{-6})(-17.5 \times 10^3) = -43.75 \text{ mA}$$

In the interval from $t_1 = 12$ ms to $t_2 = 14$ ms, the rate of change and the current are again positive:

$$\frac{\Delta v}{\Delta t} = \frac{0 \text{ V} - (-40 \text{ V})}{14 \text{ ms} - 12 \text{ ms}} = 20 \times 10^3 \text{ V/s}$$

$$i = C \frac{\Delta v}{\Delta t} = (2.5 \times 10^{-6})(20 \times 10^3) = 50 \text{ mA}$$

After $t = 14$ ms, the voltage remains constant at 0 V, so the rate of change is zero and the current is zero. The complete plot of i versus t is shown in Figure 9.41(b).

Drill Exercise 9.17

The voltage across a 0.015-μF capacitor starts at zero and increases at a constant rate. During this time, the current through the capacitor is a constant 3 mA. What is the voltage across the capacitor after 50 μs?

ANSWER: 10 V. $\square$

9.8 Energy Stored by a Capacitor

In Chapter 3 we discussed the fact that energy is dissipated when current flows through a resistance. Recall that electrical energy is converted to heat energy in the process. Resistance, *and only resistance*, dissipates energy in this manner. Thus, an ideal capacitor—"pure" capacitance—does not dissipate, or consume, any energy; instead it *stores* energy. As convincing evidence of this fact, consider an ideal capacitor that has been fully charged by a voltage source. When the voltage source is removed, leaving the capacitor terminals open, the capacitor remains charged indefinitely, because there is no path through which discharge current can flow. If a resistor is then connected across the terminals, discharge current flows and electrical energy is dissipated in the resistance. Clearly, this energy was stored in the capacitor for whatever length of time the terminals were left open.

Using calculus, it can be shown that the energy W stored by capacitance C when it is storing charge Q is

$$W = \frac{Q^2}{2C} \tag{9.34}$$

Multiplying the numerator and denominator of (9.34) by C gives

$$W = \frac{Q^2 C}{2C^2} = \frac{C}{2}\left(\frac{Q}{C}\right)^2 = \frac{1}{2}CV^2 \tag{9.35}$$

where V is the voltage across the capacitor.

9.9 SPICE Examples

Example 9.18 (SPICE)

Use SPICE to obtain a plot of the current in the RC circuit in Figure 9.26 (Example 9.9), p. 313, after the switch is closed at $t = 0$.

SOLUTION To obtain a plot of current values at different times, we must use the .TRAN and .PLOT control statements, as discussed in the Appendix (Section A.6). The closing of the switch that connects the 30-V source to the RC network is modeled in SPICE by a PULSE input that remains at 30 V for the duration of the analysis. Figure 9.42(a) shows the SPICE circuit and input data file. Since the time-constant of the circuit is 0.3 ms, we obtain a plot of $i(t)$ over a full five time-constants by making TSTOP in the .TRAN statement equal to 5×0.3 ms = 1.5 ms. The duration of the PULSE input will then also be 1.5 ms. By making TSTEP equal to 1.5 ms/20 = 75 μs, we obtain 21 values of $i(t)$ (counting the value at $t = 0$). Note that we must specify TRAN as the analysis type in the .PLOT statement. When the program is executed, the 21 values of the I(VDUM) = $i(t)$ are plotted as shown in Figure 9.42 (b). Theoretically, $i(0) = 30$ V/15 kΩ = 2 mA, but the resolution of this SPICE plot is such that the value closest to $i(0)$ that is obtained is 1.77 mA at $t = 75$ μs. See Appendix Example A.5 for further discussion and explanation of this point. See Appendix Example A.4 for another example of a SPICE simulation of an RC circuit, in this case the use of a PRINT statement to obtain transient values of a capacitor voltage.

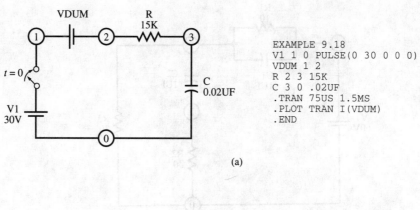

```
EXAMPLE 9.18
V1 1 0 PULSE(0 30 0 0 0)
VDUM 1 2
R 2 3 15K
C 3 0 .02UF
.TRAN 75US 1.5MS
.PLOT TRAN I(VDUM)
.END
```

(a)

```
   TIME       I(VDUM)

                   0.000D+00    5.000D-04    1.000D-03    1.500D-03  2.000D-03
                   - - - - - - - - - - - - - - - - - - - - - - - - - - - - - - -
0.000D+00    0.000D+00  *         .            .            .            .
7.500D-05    1.770D-03  .         .            .            .            .      *
1.500D-04    1.379D-03  .         .            .            .      *      .
2.250D-04    1.075D-03  .         .            .      *      .            .
3.000D-04    8.358D-04  .         .      *      .            .            .
3.750D-04    6.516D-04  .      *   .            .            .            .
4.500D-04    5.067D-04  .    *     .            .            .            .
5.250D-04    3.950D-04  .   *      .            .            .            .
6.000D-04    3.072D-04  .  *       .            .            .            .
6.750D-04    2.395D-04  . *        .            .            .            .
7.500D-04    1.863D-04  .*         .            .            .            .
8.250D-04    1.452D-04  .*         .            .            .            .
9.000D-04    1.129D-04  .*         .            .            .            .
9.750D-04    8.803D-05  . *        .            .            .            .
1.050D-03    6.846D-05  . *        .            .            .            .
1.125D-03    5.337D-05  .*         .            .            .            .
1.200D-03    4.151D-05  .*         .            .            .            .
1.275D-03    3.236D-05  .*         .            .            .            .
1.350D-03    2.516D-05  .*         .            .            .            .
1.425D-03    1.962D-05  .*         .            .            .            .
1.500D-03    1.526D-05  *          .            .            .            .
                   - - - - - - - - - - - - - - - - - - - - - - - - - - - - - - -
```

(b)

FIGURE 9.42 (Example 9.18)

Example 9.19 (SPICE)

Use SPICE to find the steady-state voltage across each component in Figure 9.39(a) (Example 9.16), p. 326.

SOLUTION When the analysis type in a SPICE simulation is .DC, the dc voltages and currents it computes are steady-state values. The circuit and input data file are shown in Figure 9.43. Note that the steady-state results will be the same regardless of the capacitance values specified. In the example, we arbitrarily set $C_1 = C_2 = 1$ μF. Execution of the program reveals that the voltage V(1,2) across R_1 and C_1 is 15 V, the voltage V(2,3) across C_2 is 45 V, the voltage V(2) across R_3 is 45 V, and the voltage V(3) across R_2 is 0 V, in agreement with the values calculated in Example 9.16.

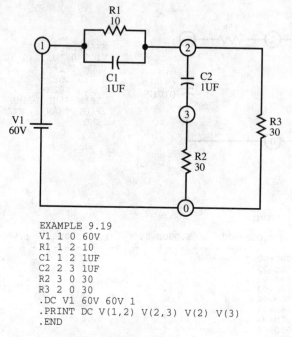

```
EXAMPLE 9.19
V1 1 0 60V
R1 1 2 10
C1 1 2 1UF
C2 2 3 1UF
R2 3 0 30
R3 2 0 30
.DC V1 60V 60V 1
.PRINT DC V(1,2) V(2,3) V(2) V(3)
.END
```

FIGURE 9.43 (Example 9.19)

Exercises

Section 9.1   The Nature of Capacitance

9.1 (a) How much charge must be stored on a 40-µF capacitor to make the voltage across it equal 20 V?
(b) What will be the voltage across the capacitor if the amount of charge stored on it is double that found in part (a)?

9.2 (a) How much capacitance is required for a 16-V source to store 64 µC of charge?
(b) What voltage would be required to store twice as much charge on that capacitance?

9.3 It is found that 2400 pC of charge are stored by a certain capacitor when a 6-V source is connected across it. What is the charge on the capacitor when its voltage is 1.5 V?

9.4 The electric field intensity between the plates of a certain parallel-plate capacitor is 12,000 V/m when it is storing 18 µC of charge. If the plates are 0.5 mm apart, what is the capacitance of the capacitor?

Section 9.2   Capacitor Dimensions and Dielectrics

9.5 A certain parallel-plate capacitor has a capacitance of 0.002 µF with an air dielectric. What is its capacitance when a dielectric material having permittivity 30.94×10^{-12} F/m is inserted between the plates?

9.6 A capacitor with a mica dielectric has a capacitance of 25 µF. When a new dielectric material is used, the capacitance increases to 37.5 µF. What is the relative permittivity of the new dielectric?

9.7 A certain capacitor with a paraffined paper dielectric stores 300 µC of charge when a 6-V source is connected across it. What should be the relative permittivity of a new dielectric if it is desired to store 450 µC when the same voltage is connected?

9.8 A capacitor with a mica dielectric is fully charged by a 20-V source that deposits 1 µC of charge on its plates. The 20-V source is removed and a Teflon dielectric replaces the mica. What then is the voltage across the capacitor?

9.9 Find the capacitance of a parallel-plate capacitor whose plates have dimensions 4 cm × 15 cm. The plates are separated by a 0.1-mm thickness of mica.

9.10 Each plate of a parallel-plate capacitor has an area of 0.025 m^2 and they are separated by a 0.5 × 10^{-4} m thickness of a dielectric whose relative permittivity is 50. Find the capacitance of the capacitor.

9.11 To what thickness should the dielectric in Exercise 9.10 be changed if it is desired to change the capacitance to 0.1 μF?

9.12 The capacitor in Exercise 9.10 is to be redesigned so that its plates are square and its capacitance is 0.15 μF. What should be the dimensions of each plate?

Section 9.3 Capacitor Types and Ratings

9.13 Two sheets of aluminum foil separated by a 0.05-mm thickness of paper are rolled into a cylinder to fabricate a paper capacitor. The paper has a relative permittivity of 2.8, and each sheet of aluminum foil has dimensions 4 cm × 120 cm. What is the capacitance of the capacitor?

9.14 The dielectric of a 0.01-μF mica capacitor is 0.12 mm thick. What is the breakdown voltage of the capacitor?

9.15 An electrolytic capacitor is to be connected between terminals A and B in Figure 9.44.
(a) To which terminal should the positive side of the capacitor be connected?
(b) For what minimum working voltage should the capacitor be rated?

9.16 An integrated-circuit capacitor is constructed by embedding a 0.05-μm-thick layer of silicon dioxide between a layer of aluminum and a bar of highly conductive semiconductor. The permittivity of the silicon dioxide is 15 × 10^{-12} F/m and the width of the layers in which it is embedded is 250 μm. What should be the length of the layers if the capacitance is to be 10 pF?

Section 9.4 Series and Parallel Capacitor Circuits

9.17 Find the total equivalent capacitance and the voltage across each capacitor in the circuits shown in Figure 9.45.

9.18 The 10-V source in Figure 9.46 delivers a total charge of 10 μC to the circuit. Find the value of capacitance C.

9.19 Using equation (9.13) for the total equivalent capacitance of series-connected capacitors, derive equation (9.14) for the equivalent capacitance of two series-connected capacitors.

9.20 Using equation (9.13) for the total equivalent capacitance of series-connected capacitors, derive equation (9.15) for the equivalent capacitance of n equal-valued capacitors.

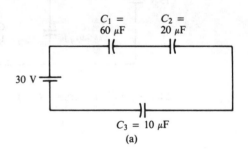

(a)

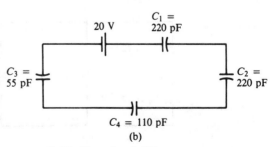

(b)

FIGURE 9.45 (Exercise 9.17)

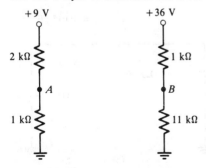

FIGURE 9.44 (Exercise 9.15)

9.21 Find the total charge delivered by the voltage source and the charge on each capacitor in Figure 9.47.

9.22 The total charge delivered by the 20-V source in Figure 9.48 is 1440 μC. Find the value of capacitance C.

9.23 Find the voltage across and charge on each capacitor in the circuits shown in Figure 9.49.

9.24 Find the voltage across and charge on each capacitor in the circuits shown in Figure 9.50.

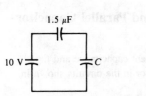

FIGURE **9.46** (Exercise 9.18)

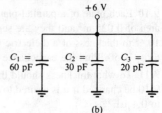

FIGURE **9.47** (Exercise 9.21)

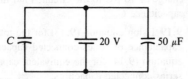

FIGURE **9.48** (Exercise 9.22)

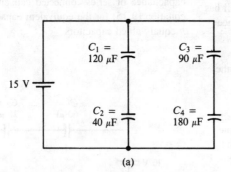

FIGURE **9.49** (Exercise 9.23)

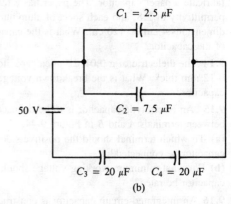

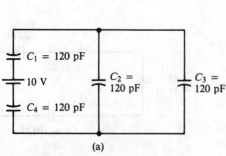

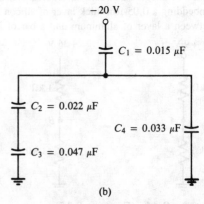

FIGURE **9.50** (Exercise 9.24)

Section 9.5 Transients in Series RC Networks

9.25 For each of the circuits shown in Figure 9.51:

(i) Write the mathematical expression for the charging current $i(t)$ after the switch is closed.

(ii) Find the value of the current at $t = 0$, $t = 0.2$, $t = 0.4$, $t = 0.8$, and $t = 1.2$ s.

(iii) Sketch $i(t)$ versus t.

9.26 For each of the circuits shown in Figure 9.52:

(i) Write the mathematical expression for the charging current $i(t)$ after the switch is closed.

(ii) Find the value of the current at $t = 0$, $t = 5$ ms, $t = 10$ ms, $t = 20$ ms, and $t = 40$ ms.

(iii) Sketch $i(t)$ versus t.

9.27 For each of the circuits shown in Figure 9.52:

(i) Find the time constant.

(ii) Find the value of the charging current one, two, three, and five time constants after the switch is closed.

(iii) After how many time constants will the current have decayed to one-half its maximum value?

9.28 For each of the circuits shown in Figure 9.51:

(i) Find the time constant.

(ii) Find the voltage across the resistor one, two, three, and five time constants after the switch is closed.

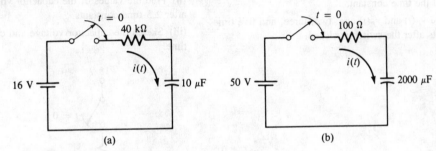

FIGURE **9.51** (Exercise 9.25)

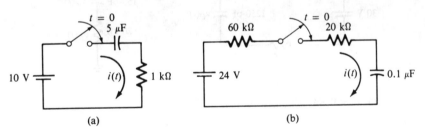

FIGURE **9.52** (Exercise 9.26)

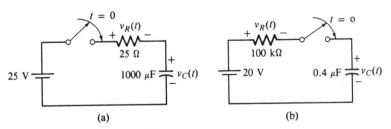

FIGURE **9.53** (Exercise 9.29)

9.29 For each of the circuits shown in Figure 9.53:

(i) Write the mathematical expressions for $v_C(t)$ and $v_R(t)$ after the switch is closed.

(ii) Find the values of $v_C(t)$ and $v_R(t)$ at $t = 0$, $t = 25$ ms, $t = 50$ ms, $t = 100$ ms, and $t = 150$ ms.

(iii) Sketch $v_C(t)$ and $v_R(t)$ versus t.

9.30 For each of the circuits shown in Figure 9.54:

(i) Write the mathematical expressions for $v_C(t)$ and $v_R(t)$ after the switch is closed. [Note carefully how $v_R(t)$ is defined in the figure.]

(ii) Find the values of $v_C(t)$ and $v_R(t)$ at $t = 0$, $t = 15$ μs, $t = 30$ μs, $t = 60$ μs, and $t = 100$ μs.

(iii) Sketch $v_C(t)$ and $v_R(t)$.

9.31 For each of the circuits in Figure 9.54:

(i) Find the time constant.

(ii) Find $v_C(t)$ and $v_R(t)$ one, two, three, and five time constants after the switch is closed.

(iii) After how many time constants will the capacitor voltage have risen to 70% of its maximum value?

9.32 For each of the circuits in Figure 9.53:

(i) Find the time constant.

(ii) Find $v_C(t)$ and $v_R(t)$ one, two, three, and five time constants after the switch is closed.

(iii) Verify that Kirchhoff's voltage law is satisfied around each circuit at the times specified in part (b).

9.33 In each of the circuits shown in Figure 9.55, switch S has been in position 1 long enough for the capacitor to charge fully. After the switch is placed in position 2:

(i) Write the mathematical expressions for the capacitor voltage and current.

(ii) Find the values of the capacitor voltage and current after 2.5 time constants.

(iii) Sketch the capacitor voltage and current versus time.

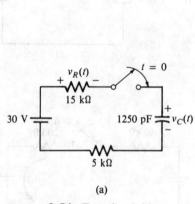

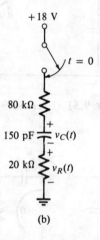

(a)　　　　　　　　　　(b)

FIGURE 9.54 (Exercise 9.30)

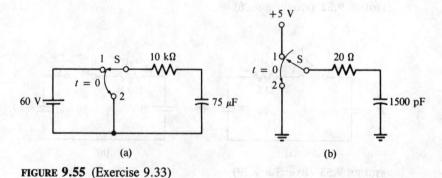

(a)　　　　　　　　　　(b)

FIGURE 9.55 (Exercise 9.33)

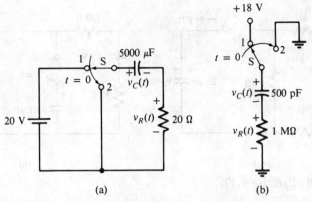

(a)   (b)

FIGURE 9.56 (Exercise 9.34)

9.34 In each of the circuits shown in Figure 9.56, switch S has been in position 1 long enough for the capacitor to charge fully. After the switch is placed in position 2:

(i) Write the mathematical expressions for $v_C(t)$ and $v_R(t)$.

(ii) Sketch $v_C(t)$ and $v_R(t)$ versus time.

(iii) Verify that Kirchhoff's voltage law is satisfied around the closed loop at $t = 0$, $t = \tau$, and $t = 5\tau$.

9.35 With reference to the circuit shown in Figure 9.57:

(a) Write the mathematical expressions for $v_C(t)$ and $i(t)$ after the switch is placed in position 1.

(b) Assuming the switch remains in position 1 long enough for the capacitor to become fully charged, write the mathematical expressions for $v_C(t)$ and $i(t)$ after the switch is placed in position 2. (For these expressions, $t = 0$ is the instant the switch is placed in position 2.)

(c) Sketch $v_C(t)$ found in parts (a) and (b) on the same time axis. Repeat for $i(t)$.

9.36 Repeat Exercise 9.35 for the circuit shown in Figure 9.58.

Section 9.6   Series–Parallel RC Networks

9.37 For each of the circuits shown in Figure 9.59:
(i) Write the mathematical expressions for $v_C(t)$ and $i(t)$ after the switch is closed.
(ii) Sketch $v_C(t)$ and $i(t)$ versus time.

9.38 For the circuit shown in Figure 9.60:
(a) Write the mathematical expressions for $v_C(t)$ and $i(t)$ after the switch is placed in position 1.
(b) Write the mathematical expressions for $v_C(t)$ and $i(t)$ after the switch is placed in position 2, assuming that it remained in position 1 long enough for the capacitor to charge fully.
(*Hint*: Find different Thévenin equivalent circuits for each switch setting.)

9.39 For each of the circuits shown in Figure 9.61:
(i) Find the initial current through the initial voltage across every resistor and capacitor at the instant the switch is closed:
(ii) Find the steady-state current through and the steady-voltage across every resistor and capacitor.

In each of parts (i) and (ii), draw arrows to show

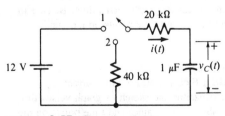

FIGURE 9.57 (Exercise 9.35)

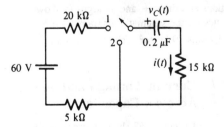

FIGURE 9.58 (Exercise 9.36)

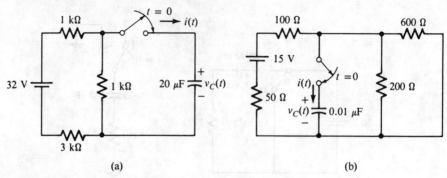

FIGURE 9.59 (Exercise 9.37)

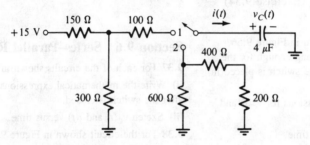

FIGURE 9.60 (Exercise 9.38)

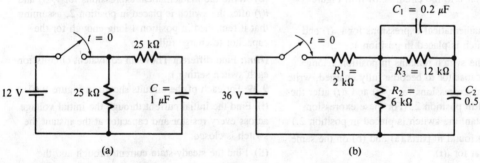

FIGURE 9.61 (Exercise 9.39)

current directions and draw + and − signs to show voltage polarities on the schematic diagram.

9.40 Repeat Exercise 9.39 for each of the circuits in Figure 9.62.

Section 9.7 Current Through and Voltage Across a Capacitor

9.41 Each part of Figure 9.63 shows a graph of the voltage $v(t)$ across a 5-µF capacitor as a function of

time. In each case, find the current $i(t)$ through the capacitor and sketch its graph versus time. Be sure to label current values and time points on the graphs.

9.42 Figure 9.64 shows a graph of the voltage $v(t)$ across a 0.01-µF capacitor. Find the current $i(t)$ through the capacitor and sketch its graph versus time. Be sure to label current values and time points on the graph.

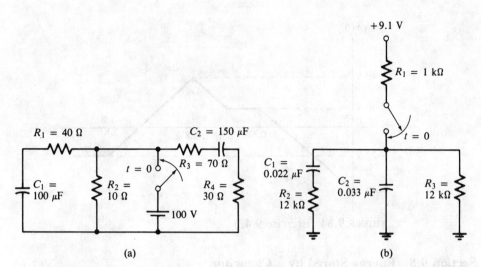

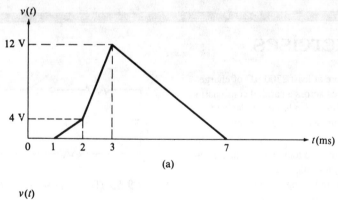

FIGURE **9.62** (Exercise 9.40)

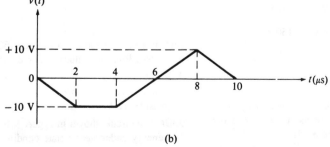

FIGURE **9.63** (Exercise 9.41)

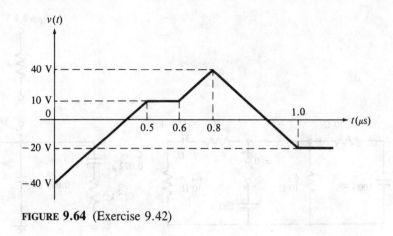

FIGURE 9.64 (Exercise 9.42)

Section 9.8   Energy Stored by a Capacitor

9.43 **(a)** A 100-μF capacitor is fully charged by a 12-V battery. How much energy is stored by the capacitor?

(b) How much energy is lost from the capacitor if 400 μC of charge are drained from it after it has been fully charged?

9.44 A capacitor is to be selected so it will store 0.16 J of energy when it is charged by a 40-V source.

(a) What should be the capacitance of the capacitor?

(b) How much charge must it store?

9.45 Find the energy stored by each capacitor in Figure 9.62 under steady-state conditions.

9.46 Find the energy stored by each capacitor in Figure 9.61 under steady-state conditions.

Design Exercises

9.1D It is necessary to store at least 2700 μC of charge by connecting a 20-V source across a parallel combination of 10-μF capacitors, each of which has a tolerance of $\pm 10\%$. What minimum number of capacitors must be connected in parallel?

9.2D It is necessary to connect a total equivalent capacitance of 15 μF across a 180-V source using a series–parallel combination of the following capacitors:

$$C_1 = 20 \ \mu\text{F, DCWV} = 50 \text{ V}$$

$$C_2 = 20 \ \mu\text{F, DCWV} = 150 \text{ V}$$

$$C_3 = 40 \ \mu\text{F, DCWV} = 100 \text{ V}$$

Design the circuit, draw the schematic diagram, and label capacitors C_1, C_2, and C_3 in the diagram.

9.3D When the switch in Figure 9.65 is closed, the voltage across the capacitor must rise to 6.59 V in 0.1 ms. What value of R should be used?

9.4D When the switch in Figure 9.66(a) is alternately placed in position 1 and 2 at 300-ms intervals, the

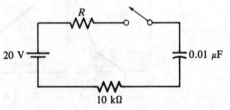

FIGURE 9.65 (Exercise 9.3D)

capacitor voltage shown in Figure 9.66(b) is to be generated. Find the values required for R_1 and R_2.

9.5D In the circuit shown in Figure 9.67 the initial current i_0 must be 18 mA and the steady-state voltage V_{ss} must be 12 V. Find the required values of R_1 and R_2.

9.6D The circuit shown in Figure 9.68 must store 0.34 J of energy under steady-state conditions. What should be the value of C? (The 100-μF capacitor stores part of the total required energy.)

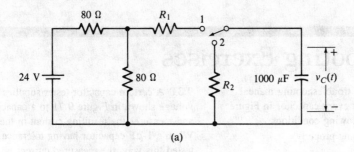

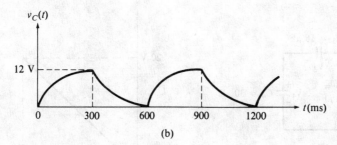

(a)

(b)

FIGURE 9.66 (Exercise 9.4D)

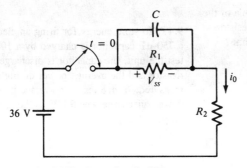

FIGURE 9.67 (Exercise 9.5D)

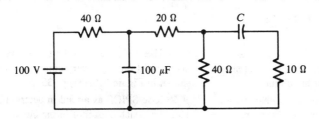

FIGURE 9.68 (Exercise 9.6D)

Troubleshooting Exercises

9.1T Prepare a table for a troubleshooting manual showing the voltage across every capacitor in Figure 9.69 under each of the following conditions:
(a) All capacitors functioning properly
(b) C_1 shorted
(c) C_2 shorted
(d) C_3 shorted
(e) C_4 shorted

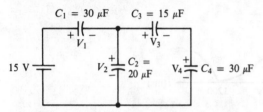

FIGURE 9.69 (Exercise 9.1T)

9.2T When the switch in Figure 9.70 is closed, it is found that it takes 0.5 s for voltage V to reach essentially 12 V. How long should it take if none of the components have failed? What failures in the resistors or capacitors would account for the observed time?

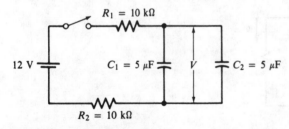

FIGURE 9.70 (Exercise 9.2T)

9.3T A certain capacitor tester supplies the ramp voltage shown in Figure 9.71 to a capacitor under test and measures the resulting current in the capacitor. When a 1-μF capacitor having tolerance $\pm$ 10% is tested this way, the measured current is 16 mA. Is the capacitor within tolerance?

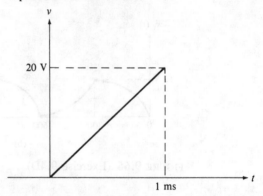

FIGURE 9.71 (Exercise 9.3T)

9.4T To store energy for firing an electronic flash unit, a 750-μF capacitor is charged by a 100-V source. To test the unit, the capacitor is discharged through a small resistor and the average power during discharge is measured. In one test, the average power during a 0.5-ms discharge was 5 kW. Is the unit operating properly? Explain.

SPICE Exercises

9.1S Use SPICE to obtain a plot of the voltage across the capacitor in Figure 9.54(a), p. 336. The plot should extend over five time constants of the circuit.

9.2S Use SPICE to obtain plots of the voltage across the capacitor and the current $i(t)$ in the circuit shown in Figure 9.59(a), p. 338. The plot should extend over five time constants of the (Thévenin equivalent) circuit.

9.3S Use SPICE to find the steady-state voltage across and current through every component in the circuit shown in Figure 9.61(b), p. 338.

9.4S Use SPICE as an aid in determining the energy stored by each capacitor in the circuit shown in Figure 9.62(a), p. 339, under steady-state conditions.

10 Magnetic Fields and Circuits

10.1 Bar Magnets

Most students are familiar with bar magnets, often bent into U-shapes and called *horseshoe* magnets, and most have observed the forces developed between the ends of such magnets. One end of a bar magnet is called its *north* (N) pole and the opposite end is called its *south* (S) pole. When two bar magnets are placed near each other, it is found that a force of attraction is developed between the north pole of one and the south pole of the other. It is also true that the two north poles of the bar magnets repel each other, as do the two south poles. In other words, *opposite poles attract and like poles repel*.

Bar magnets are examples of so-called *permanent* magnets, made of iron and certain metal alloys, that retain their magnetic properties for extended periods of time. In spite of their name, permanent magnets can lose their magnetism if subjected to severe mechanical shock or heating. As we shall learn in a later discussion, magnetic properties stem from the orientation of tiny regions called *domains* within a material, and that orientation can be altered or lost under certain conditions.

10.2 Magnetic Fields

Lines are drawn in the vicinity of a magnet to show the direction of the force on a *north* pole placed anywhere in the region around the magnet. These lines depict the nature of force variations in the same way that electric field lines show the force on a positive charge in an electric field. The pattern of lines around a magnet is a pictorial way of representing the *magnetic field* created by the magnet. Figure 10.1 shows the magnetic field around a bar magnet. Note that the lines enter the south pole, because a north pole is attracted to a south pole, and the lines leave the north pole, because a north pole repels another north pole. The concept of an isolated north pole that can be moved around a stationary magnet is fictitious, but it serves to illustrate the nature of a magnetic field. We can think of the south pole as being so far removed from the latter that it has no influence on the force developed. Note that magnetic field lines do not originate or terminate at poles. Instead, they are continuous loops that pass completely through the magnet.

Electromagnetism

One of the most important contributions to the theory of electricity was made in 1820 by the Danish physicist Hans Oersted, who discovered that electrical *current* creates a magnetic field. Recall that *stationary* electrical charge creates an *electric* field, as discussed in Chapter 8. Oersted showed that charge in *motion* (i.e., electrical current) creates a magnetic field. This phenomenon, called *electromagnetism,* and the relationship between electric and magnetic fields, was further studied and refined by theorists such as André Ampère, Michael Faraday, Karl Gauss, and James Maxwell. Their discoveries

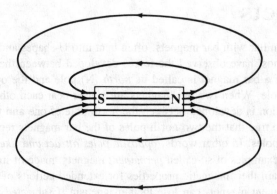

FIGURE 10.1 The magnetic field lines around a bar magnet show the direction of the force on an isolated north pole placed anywhere in the field.

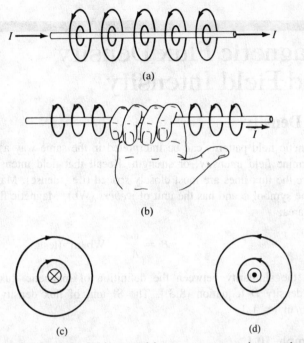

(a)

(b)

(c)   (d)

FIGURE 10.2 Magnetic field created by a current-carrying conductor. (a) Current flowing in a conductor creates a magnetic field around the conductor. (b) Right-hand rule for determining the direction of the magnetic field around a current-carrying conductor. (c) End view of conductor with current flowing into the page. (d) End view of conductor with current flowing out of the page.

were eventually responsible for the development of a host of practical and useful devices, such as electric motors and generators, transformers, loudspeakers, relays, antennas, and many others.

Figure 10.2(a) shows the magnetic field developed around a current-carrying conductor. Note that the lines encircle the conductor like concentric rings. These lines convey the same information as the magnetic field lines around a permanent magnet—the direction of the force on a north pole placed in the field. Figure 10.2(b) shows how the direction of the field can be determined using the so-called *right-hand rule:* Point the thumb in the direction of the (conventional) current in the conductor; the fingers then wrap around the conductor in the direction of the field. Figure 10.2(c) shows an end view of a conductor carrying current into the page. Applying the right-hand rule to this view shows that the field lines encircle the conductor in a clockwise direction. In Figure 10.2(d) current flows out of the page and the field has a counterclockwise direction. The symbols shown in these end views are the conventional means for showing current into and out of a page. Think of the cross as representing the tail feathers of an arrow and the dot as representing the point of an arrowhead.

10.3 Magnetic Flux Density and Field Intensity

Magnetic Flux Density

Magnetic field patterns can be interpreted in the same way as electric field patterns to determine field intensity, or strength. Recall that field intensity is greatest in regions where the flux lines are most closely spaced (i.e., dense). Magnetic flux is represented by the symbol ϕ and has the unit of *webers* (Wb). Magnetic flux density, B, is flux per unit area:

$$B = \frac{\phi}{A} \qquad \text{Wb/m}^2 \text{ (tesla)} \tag{10.1}$$

Note the similarity between the definition of magnetic flux density B and electric flux density D [equation (8.3)]. The SI unit of flux density is the *tesla* (T), where $1 \text{ Wb/m}^2 = 1 \text{ T}$.

Example 10.1

Figure 10.3 shows a bar magnet in which the flux density is 0.14 T. Find the total flux in the magnet.

SOLUTION The area perpendicular to the flux is the 1 cm × 2 cm cross section is

$$A = (10^{-2} \text{ m})(2 \times 10^{-2} \text{ m}) = 2 \times 10^{-4} \text{ m}^2$$

From equation (10.1)

$$\phi = BA = (0.14 \text{ T})(2 \times 10^{-4} \text{ m}^2) = 28 \times 10^{-6} \text{ Wb}$$

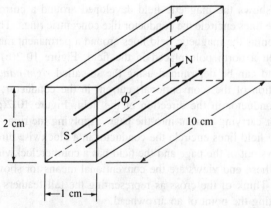

FIGURE 10.3 (Example 10.1)

Drill Exercise 10.1

What should be the flux density in the bar in Example 10.1 if the total flux in it is to be 0.42×10^{-3} Wb?

ANSWER: 2.1 T.

The flux density in air surrounding a long, straight wire carrying current I is given by

$$B = 2 \times 10^{-7} \frac{I}{r} \qquad (10.2)$$

where r is the distance from the wire to the point where B is to be determined. When I is in amperes and r is in meters, B is in webers/m², or tesla. Figure 10.4 illustrates the situation. Equation (10.2) shows that the flux density at a fixed point is directly proportional to the current in the wire. Although the equation is theoretically accurate only for a wire of infinite length and zero diameter, it is valid in practical situations where the distance r from the wire is small compared to the length of the wire and large compared to its diameter.

Example 10.2

What is the flux density 10 cm away from a wire 10 m long and 1 mm in diameter when the current in the wire is 1 mA?

SOLUTION Since $r = 10$ cm $= 0.1$ m is small compared to the 10-m length of the wire and large compared to its 1 mm diameter, we can use equation (10.2) to find B:

$$B = 2 \times 10^{-7} \left(\frac{10^{-3} \text{ A}}{10^{-1} \text{ m}} \right) = 2 \times 10^{-9} \text{ T}$$

Drill Exercise 10.2

At what distance from the wire in Example 10.2 is the flux density equal to 1×10^{-9} T?

ANSWER: 20 cm.

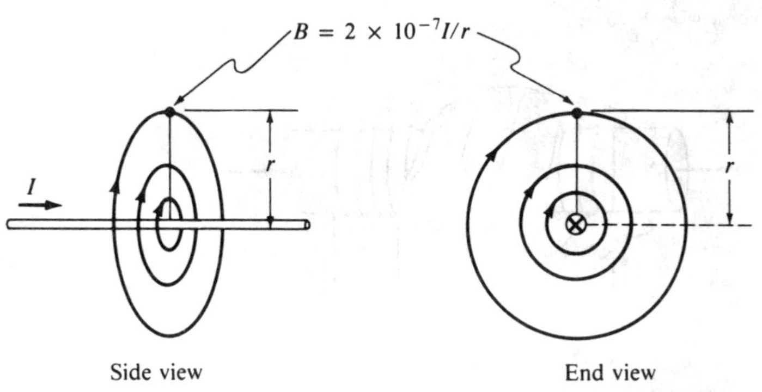

FIGURE 10.4 The flux density at distance r from a straight conductor carrying current I is directly proportional to I and inversely proportional to r.

Magnetomotive Force

Electromagnets are constructed by wrapping insulated wire around a *core* (i.e., a rod, bar, or some other shape), usually made of an iron alloy. Figure 10.5(a) shows wire wrapped around a cylindrical core. Each complete wrap around the core is called a *turn*, and the turns are collectively called a *winding*. When current flows through the winding, a magnetic field is produced in the core. Imagine the right-hand rule applied to each turn in Figure 10.5(a): We can see that the fingers will point to the left *inside* the core, when current flows in the direction shown. As illustrated in Figure 10.5(b), the right hand can be used in another way to determine the direction of the field in the core of an electromagnet. When the fingers are wrapped around the core in the direction that current flows around the winding, the thumb points in the direction of the field. An electromagnet formed by wrapping turns around a cylindrical core, such as that shown in Figure 10.5, is called a *solenoid*.

We have seen that the flux density in the vicinity of a single current-carrying conductor depends directly on the magnitude of the current in the conductor (equation 10.2). Similarly, flux density in the core of an electromagnet is directly proportional to the current flowing in the winding. Each turn in the winding is a conductor carrying the same current, so each adds the same flux to the core. Thus, a winding consisting of two turns will produce twice as much flux as a winding having only one turn, when both windings carry the same current. We conclude that for a fixed value of current, the flux produced in a core is directly proportional to the number of turns in the winding. Sim-

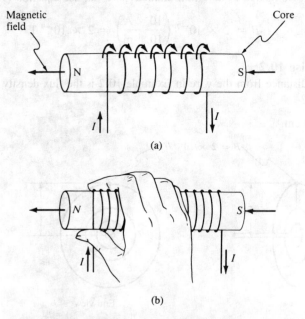

FIGURE 10.5 A magnetic field is produced in an electromagnet (solenoid) by wrapping current-carrying turns around a core. (a) Current flowing in the windings around a core produces a magnetic field in the core. (b) Using the right hand to determine the direction of the field in an electromagnet.

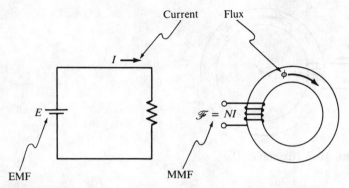

Current Flux

EMF MMF

FIGURE 10.6 In the analogy between electric circuits and magnetic circuits, magnetomotive force is like electromotive force, and flux is like current.

ilarly, for a fixed number of turns, the flux is directly proportional to the current in the winding. Thus, the product of the number of turns in a winding and the current flowing in it is a measure of how much magnetic "excitation" is applied to an electromagnet. This product is called *magnetomotive force* (mmf) and is given the symbol $\mathscr{F}$:

$$\mathscr{F} = NI \qquad \text{amperes or ampere-turns} \tag{10.3}$$

where N is the number of turns and I is the current in the winding. The units of $\mathscr{F}$ are often given as *ampere-turns* (At), in agreement with the definition of $\mathscr{F}$ as the product of current and turns. However, N is properly regarded to be a dimensionless multiplying factor, and the correct SI units of $\mathscr{F}$ are simply amperes.

Magnetomotive force is like electromotive force in an electric circuit, and flux is like electric current. This *analogy* is illustrated in Figure 10.6. In each case, the greater the force, the greater the "flow" that results. The circular core in which the flux is established is an example of a *magnetic circuit*.

Magnetic Field Intensity

Magnetic field intensity is magnetomotive force per unit length. It is designated by H and has the units of amperes per meter (A/m). Expressed in equation form, magnetic field intensity is

$$H = \frac{\mathscr{F}}{l} = \frac{NI}{l} \qquad \text{amperes/meter} \tag{10.4}$$

Magnetic field intensity is also called *magnetic field strength*, or *magnetizing force*. The latter name is derived from the fact that H represents magnetomotive force per unit length. Since $\mathscr{F}$ is often expressed in ampere-turns, the units of H are frequently expressed as ampere-turns per meter.

Example 10.3
How much current must flow in the 15-turn winding shown in Figure 10.7 to establish a magnetic field intensity of 60 A/m in the circular core?

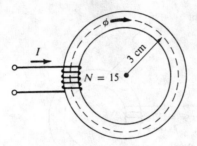

FIGURE 10.7 (Example 10.3)

SOLUTION The dashed line in Figure 10.7 shows the *mean* (average) length of the path in which the flux is established, and we use this length in our computation. The length of the path is the circumference of a circle with radius 3 cm, or 0.03 m:

$$l = 2\pi r = 2\pi(0.03 \text{ m}) \doteq 0.1885 \text{ m}$$

From equation (10.4),

$$I = \frac{Hl}{N} = \frac{(60 \text{ A/m})(0.1885 \text{ m})}{15} = 0.754 \text{ A}$$

Drill Exercise 10.3

How many turns would be required to establish the magnetic field intensity in Example 10.3 if the current supplied to the winding is 500 mA? Express the answer as the minimum integer (whole number) of turns.

ANSWER: 23.

10.4 Permeability

Recall that flux density in an electric field is directly proportional to field intensity ($D = \epsilon \mathscr{E}$), the constant of proportionality being the permittivity ϵ of the material. In the same way, magnetic flux density is directly proportional to magnetic field intensity, except that the constant of proportionality in this case is a magnetic property of the material called *permeability* (μ):

$$B = \mu H \tag{10.5}$$

The units of permeability are Wb/A-m:

$$\mu = \frac{B}{H} = \frac{\text{Wb/m}^2}{\text{A/m}} = \text{Wb/A-m} \tag{10.6}$$

The permeability of a body can be thought of as a measure of how well magnetic field lines concentrate inside its boundaries, under the influence of a fixed field intensity.

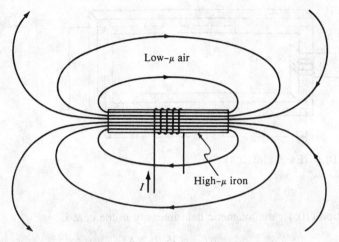

FIGURE 10.8 Magnetic field lines are dense in the high-μ
iron and sparse in the low-μ air.

A so-called *magnetic material* has a large permeability, and the flux density in such a
material is much greater than that in a nonmagnetic material when both experience the
same field intensity. Figure 10.8 shows that the flux in an iron bar, which has a high
permeability, is densely concentrated (large value of B), whereas the lines disperse in a
less dense pattern in the low-permeability air surrounding the bar.

The permeability of a *vacuum*, or *free space*, designated μ_0, is

$$\mu_0 = 4\pi \times 10^{-7} \text{ Wb/A·m}$$

The permeability of air is for all practical purposes the same as μ_0. Materials having
permeabilities slightly less than that of a vacuum are said to be *diamagnetic*. Copper and
silver are examples. Materials having permeabilities slightly greater than a vacuum are
called *paramagnetic*. Aluminum and platinum are examples. The important magnetic
materials, iron, steel, nickel, cobalt, and certain of their alloys, have permeabilities that
are several hundred times greater than that of a vacuum and are called *ferromagnetic*
materials.

The relative permeability, μ_r, of a material is the ratio of its permeability to the
permeability of a vacuum:

$$\mu_r = \frac{\mu}{\mu_0} \tag{10.7}$$

Example 10.4

The relative permeability of the rectangular core shown in Figure 10.9 is 750.
What is the flux density in the core?

SOLUTION The total length of the flux path in the core is twice its width plus twice its
height:

$$l = 2(0.05 \text{ m}) + 2(0.02 \text{ m}) = 0.14 \text{ m}$$

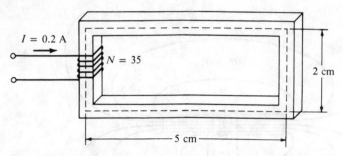

FIGURE 10.9 (Example 10.4)

By equation (10.4), the magnetic field intensity in the core is

$$H = \frac{NI}{l} = \frac{(35)(0.2 \text{ A})}{0.14 \text{ m}} = 50 \text{ A/m}$$

From equation (10.7), the permeability of the core is

$$\mu = \mu_r \mu_0 = 750(4\pi \times 10^{-7}) = 9.425 \times 10^{-4} \text{ Wb/A-m}$$

By equation (10.5),

$$B = \mu H = (9.425 \times 10^{-4})(50) = 0.0471 \text{ T}$$

Drill Exercise 10.4

What relative permeability should the core in Example 10.4 have if the same magnetomotive force is required to develop a flux density of 0.06 T in the core?

ANSWER: 955.4.   □

10.5 Reluctance

In the analogy between electric and magnetic circuits, we have seen that magnetomotive force is analogous to electromotive force and that flux is analogous to current. To complete the analogy, we introduce the concept of magnetic *reluctance* $\mathscr{R}$, which, like resistance, is a measure of the extent to which a material impedes flow, in this case, the "flow" of flux. Ohm's law for magnetic circuits expresses the fact that flux is directly proportional to magnetomotive force $\mathscr{F}$ and inversely proportional to reluctance $\mathscr{R}$:

$$\phi = \frac{\mathscr{F}}{\mathscr{R}} \tag{10.8}$$

The units of reluctance are amperes/weber:

$$\mathscr{R} = \frac{\mathscr{F}}{\phi} \qquad \text{A/Wb} \tag{10.9}$$

(The non-SI unit *rel* is also used.) The actual value of the reluctance of a body is seldom used in practical calculations involving the analysis or design of magnetic circuits, but it is instructive to realize that such a quantity can be defined to complete the analogy. The reluctance of a body in terms of its physical characteristics (length *l*, cross-sectional area *A*, and permeability μ) is given by

$$\mathscr{R} = \frac{l}{\mu A}$$ (10.10)

Compare equation (10.10) with the equation for the electrical resistance of a body: $R = \rho l/A$. We see that resistivity ρ is analogous to $1/\mu$.

10.6 Domain Theory of Magnetism

Certain properties shared by magnetic materials can be explained in terms of the modern *domain* theory of magnetism. For example, we shall see how the theory accounts for the fact that four elements, iron, nickel, cobalt, and gadolinium, have the ability to *retain* magnetism after they have been subjected to a magnetic field. It is this property that allows permanent magnets to be constructed from magnetic materials and their alloys: An unmagnetized bar is placed in a magnetic field, and when the field is removed, the bar is magnetized.

Every atom is a tiny electromagnet, because it contains charge (electrons) in motion. According to modern theory, the atoms of magnetic materials are clustered in tiny regions called domains, in each of which the atoms produce magnetic fields in the same direction. Thus, each domain is itself a magnet whose field is many times stronger than that produced by a single atom. Collectively, the domains in a bar of unmagnetized material produce fields whose directions are *randomly oriented,* as shown in Figure 10.10(a). Each arrow in this figure represents the direction of the field within a domain. (For simplicity, only two-dimensional directions are shown; in reality, there are also field components directed into and out of the page.) Notice that the random orientations of the fields prevent the bar from having one overall magnetic field extending throughout its length. For each domain producing a field in one direction, there is another domain producing a field in the opposite direction, so the domains effectively cancel each other out, resulting in zero overall field strength.

Figure 10.10(b) shows an externally produced magnetic field in which we now suppose that the unmagnetized bar shown in Figure 10.10(a) is placed. One method that could be used to produce the external field would be to wrap current-carrying wire around the bar, as shown in Figure 10.5. The external field and the fields in the magnetic domains interact in such a way that a rotational force is exerted on the atoms within each domain. (This is the same type of force that causes an electric motor to rotate when magnetic fields interact within its structure.) The atoms rotate so that the magnetic field within each domain tends to align itself with the external field, as shown in Figure 10.10(c).

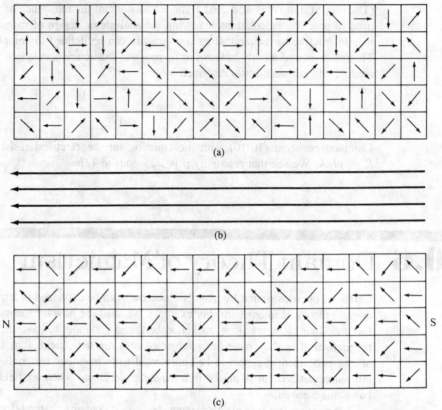

FIGURE 10.10 Effect of placing a bar of unmagnetized magnetic material in an external magnetic field. (a) The fields produced by the domains of an unmagnetized bar of magnetic material have random directions. There is no *net* field through the bar. (b) External magnetic field in which the bar shown in part (a) is placed. (c) After being subjected to the field in part (b), the fields produced by the domains all have a right-to-left component.

Not every domain is perfectly aligned in this case, but note that each has a *component* of magnetic field directed from right to left. In other words, a domain may still produce a field that has an upward or downward component, but at the same time it has a component directed from right to left. When the external field is removed, the domains retain their new alignment. As a consequence, the field in each domain now has a component that is reinforced by a similar component in every other domain, and there is a net field developed throughout the bar. In the example shown, the left end becomes the north pole of the bar magnet, and the right end becomes the south pole.

The more intense the external magnetic field, the greater the extent to which the magnetic domains align themselves with the field. When all domains are perfectly aligned, the magnetic material is said to be *saturated,* a condition that we will discuss further in the next section.

10.7 B–H Curves and Hysteresis

Imagine the several turns of wire are wrapped around an *air core,* that is, a hollow cylinder made of some nonmagnetic material such as paper or plastic. Since the number of turns and the length of the core is fixed, we can control the magnetic field intensity applied to the core by adjusting the current that flows in the windings: $H = NI/l$. Suppose further that we can measure the flux density B inside the core. Figure 10.11(a) shows a hypothetical experiment in which we gradually increase H by increasing the current in the winding of the core, and measure the flux density B that results as H is increased. As expected, B increases in direct proportion to H, since $B = \mu_0 H$. Figure 10.11(b) shows the graph that results when B is plotted versus H. The plot is called a *B–H curve,* and its slope equals the permeability μ_0 of the air in the core:

$$\mu_0 = \frac{\Delta B}{\Delta H} = 4\pi \times 10^{-7}$$

Note that μ_0 also equals the ratio of the coordinates (B_0 to H_0) at *any* point on the line, in accordance with the definition ($\mu = B/H$). If H is made to increase indefinitely, B also increases indefinitely along the line shown. As we shall presently see, such is not the case for magnetic materials.

Suppose now that we reverse the direction of the current I in the winding. Using the right-hand rule discussed earlier, we see that the direction of the field in the core will also reverse direction. We can represent this reversal on the *B–H* curve by plotting negative values of B versus negative values of H, as shown in Figure 10.11(c). Of course, the plot has the same slope (μ_0) as before, since the negative values represent nothing more than reversals in the directions of H and B.

Suppose now that the experiment we have just described is performed using a core made of a magnetic material instead of air. If the core is initially unmagnetized, then it has zero flux density when there is zero current in the winding ($B = 0$ and $H = 0$), that is, the *B–H* curve starts at the origin of the *B–H* axes, as in the case of the air core. If we gradually increase the value of H by increasing the current, we find once again that B rises in direct proportion to H, as shown in Figure 10.12. The rising portion of the graph is similar to that for the air core, except of course the slope is much steeper, because the permeability μ of the magnetic material is much greater. If we continue to increase H, we find that the *B–H* curve begins to level off and eventually becomes a horizontal line. A horizontal line indicates that there is no increase in B for further increases in H. This behavior is called magnetic *saturation* and is explained by the fact that all of the domains in the core material have become perfectly aligned. In that situation, increasing the magnetic field intensity can no longer improve the alignment and can, therefore, no longer increase the flux density in the core.

If the current in the winding is now gradually reduced, we find that the flux density does not decrease as rapidly as it increased when we first applied current to the winding. In other words, the flux density is greater when H is *reduced* to a certain value than it was when H was first increased to the same value. The *B–H* curve has the appearance

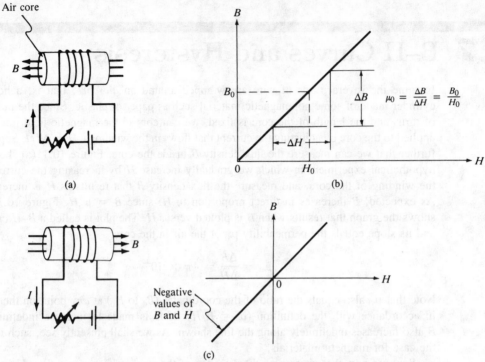

(a)

(b)

(c)

FIGURE 10.11 Plots of B versus H in an air core. (a) The field intensity H in the air core is adjusted by adjusting the current in the winding. (b) Plot of B versus H in the air core. (c) When the direction of I in the winding is reversed, the values of H and B are considered negative.

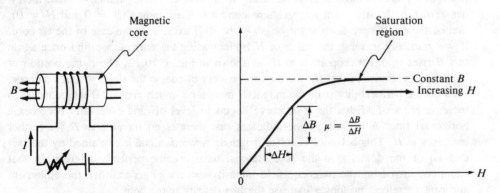

FIGURE 10.12 The B–H plot of a magnetic material shows that B does not increase indefinitely with H. Instead, *magnetic saturation* limits the maximum value of B.

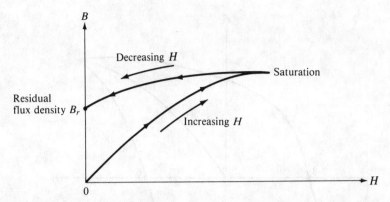

FIGURE 10.13 After magnetic saturation has been reached, decreasing H does not reduce B along the same curve that it followed when H was increased.

shown in Figure 10.13, where arrows are drawn to indicate the portion of the curve over which H is increasing and that over which H is decreasing. This phenomenon is again explained by the domain theory: The magnetic domains retain their alignment to a certain extent after the external field is reduced or removed. As can be seen in the figure, a certain flux density, called *residual* flux density, remains in the core after the current (and value of H) is reduced to zero. The degree to which a material retains its magnetism after the field intensity is reduced to zero is called its *retentivity*.

Hysteresis is the term used to describe the behavior of a variable whose value changes in one way when it is increasing and in a different way when it is decreasing. Hysteresis is associated with many different physical phenomena, and we see that the flux density in a magnetic material is an example. If we are given a particular value of H, it can be seen in Figure 10.13 that we cannot determine the corresponding value of B unless we also know whether B (and H) are increasing or decreasing.

Returning to the B–H experiment at the point where H was reduced to zero, suppose we now reverse the direction of H by reversing the current in the winding. As shown in Figure 10.14, negative values of H causes B to decrease (between points 2 and 3 in the figure). However, it can be seen that B remains positive, at least until H is made sufficiently negative to reduce B to zero at point 3 in the figure. (The negative value of H required to reduce B to zero is called the *coercive force*.) As H is increased further in the negative direction, B also reverses direction and increases negatively, in proportion to H. Eventually, H becomes so large in the reverse direction that the core again saturates, at point 4. If H is now reduced to zero, B decreases along the curve between points 4 and 5. Notice that the core has residual flux density $-B_r$ at point 5. If the direction of H is then reversed so that it is once again positive, we see that B remains negative until H is made sufficiently positive to bring B back to zero (point 6). Further increases in H cause B to increase toward positive saturation at point 1. If H is again alternately reduced and increased, following the same cycle we have described, then B again increases and decreases along the B–H curve, as shown by the arrows in the figure. The complete B–H curve is called a *hysteresis loop*.

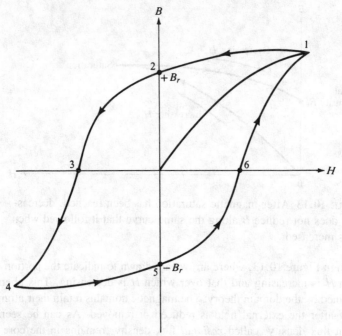

FIGURE 10.14 Hysteresis loop of a magnetic material.

The shape of a hysteresis loop is a characteristic of the magnetic material from which a core is constructed. For example, a material having a high retentivity will produce a nearly square hysteresis loop, as shown in Figure 10.15(a) for a *ferrite* core. Notice that the residual flux density in this case is very nearly equal to the saturation flux density. On the other hand, a material having low retentivity has a narrow hysteresis loop, such as that shown in Figure 10.15(b) for a *soft-iron* core.

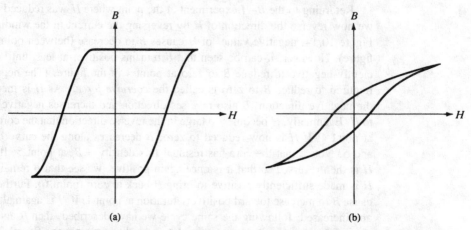

(a)   (b)

FIGURE 10.15 Magnetic materials are characterized by the shape of their hysteresis loops. (a) Ferrite has a nearly square hysteresis loop. (b) Hysteresis loop for soft iron.

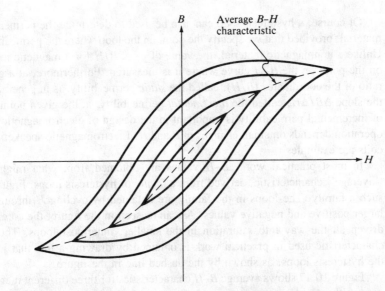

FIGURE 10.16 Family of hysteresis curves. The average *B–H* characteristic of a particular material is obtained by joining the tips of the curves, as shown by the dashed line.

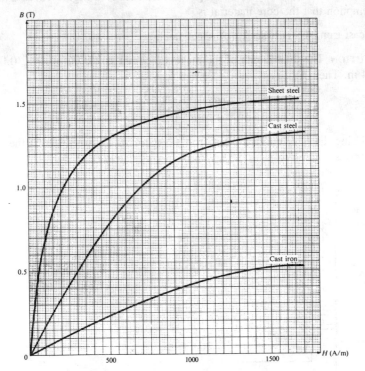

FIGURE 10.17 Typical *B–H* characteristics of three magnetic materials.

Of course, a hysteresis loop can also be used to determine the permeability of a core material, provided that we specify the point on the loop where the permeability is desired. Unlike a nonmagnetic material, the value of $\mu = B/H$ for a magnetic material depends on the point on the $B–H$ curve where it is measured. Furthermore, at a given point, the ratio of the coordinates (B/H), called the *static* permeability, is in general different from the slope $\Delta B/\Delta H$, called the *incremental* permeability, at the given point. A knowledge of incremental permeability is important in the design of electromagnetic devices whose operation depends on small changes in B and H. Electromagnetic speakers and deflection coils are examples.

In most practical work, $B–H$ values are obtained from what might be called an "average" characteristic, derived from a *family* of hysteresis loops. Figure 10.16 shows such a family. The loops in the family are obtained by cycling H through successively larger positive and negative values. As can be seen in the figure, the core material is not driven all the way into saturation in the smaller, innermost loops. The average $B–H$ characteristic used in practical work is obtained by drawing a line that joins the tips of the hysteresis loops, as shown by the dashed line in the figure.

Figure 10.17 shows average $B–H$ characteristics for three different magnetic materials used in the construction of practical electromagnetic devices.

Example 10.5

Find the flux density in the magnetic core shown in Figure 10.18 under the assumption that the core material is

(a) cast iron; (b) cast steel; (c) sheet steel.

SOLUTION The total length of the magnetic circuit is 2(0.06 m) + 2(0.06 m) = 0.24 m. Then

$$H = \frac{NI}{l} = \frac{(100)(1.2 \text{ A})}{0.24 \text{ m}} = 500 \text{ A/m}$$

(a) Using the $B–H$ curve for cast iron in Figure 10.17, we find that the value of B corresponding to $H = 500$ A/m is approximately 0.23 T.

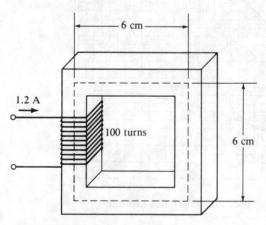

FIGURE 10.18 (Example 10.5)

(b) From the *B–H* curve for cast steel, the value of *B* corresponds to $H = 500$ A/m is approximately 0.78 T.

(c) From the *B–H* curve for sheet steel, the value of *B* corresponding to $H = 500$ A/m is approximately 1.28 T.

Drill Exercise 10.5

To what value should the current in the winding in Figure 10.18 be increased in order to produce a flux density of 1.0 T in a cast steel core?

ANSWER: 1.44 A.   □

10.8 Magnetic Circuits

We have seen that a magnetic circuit is like an electric circuit in the sense that flux (like current) is produced in a core (conductor) by a magnetomotive force (electromotive force). A typical problem in practical magnetic circuits is to determine the amount of current that must be supplied to a winding in order to produce a specified flux in a core. The next example illustrates such a problem.

Example 10.6

Find the current *I* that must be supplied to the 50-turn winding in Figure 10.19 in order to produce a flux of 3×10^{-5} Wb in the core. The cross-sectional area of the cast steel core is 4×10^{-5} m² throughout its length.

SOLUTION The flux density that must be established in the core is

$$B = \frac{\phi}{A} = \frac{3 \times 10^{-5} \text{ Wb}}{4 \times 10^{-5} \text{ m}^2} = 0.75 \text{ T}$$

From the *B–H* curve for cast steel in Figure 10.17, the magnetic field intensity corresponding to $B = 0.75$ T is approximately $H = 470$ A/m. Thus, we require

$$H = \frac{NI}{l}$$

that is,

$$470 \text{ A/m} = \frac{50I}{l}$$

Since the mean length of the core is $l = 2\pi r = 2\pi(0.03 \text{ m}) = 0.1885$ m, we have

$$470 = \frac{50I}{0.1885}$$

$$I = \frac{(0.1885)(470)}{50} = 1.772 \text{ A}$$

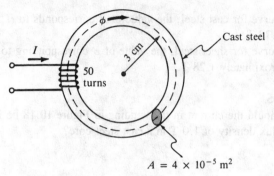

$$A = 4 \times 10^{-5} \, \text{m}^2$$

FIGURE 10.19 (Example 10.6)

Drill Exercise 10.6

Find the current required to produce a flux of 2×10^{-5} Wb in the core shown in Figure 10.19 if the core is made of cast iron.

ANSWER: 5.278 A.

Ampère's Circuital Law

In Example 10.6, the magnetomotive force applied to the core is $\mathcal{F} = NI = (50)(1.772 \, \text{A}) = 88.6 \, \text{A}$. Note that multiplying the magnetic field intensity in the core by the length of the core also gives 88.6 A: $Hl = (470 \, \text{A/m})(0.1885 \, \text{m}) = 88.6 \, \text{A}$. This result follows from the fact that

$$H = \frac{NI}{l} \Rightarrow Hl = NI$$

The quantity Hl is called the *magnetomotive drop* in the circuit. The example verifies that the magnetomotive force $\mathcal{F} = NI$ applied to the circuit equals the magnetomotive drop Hl around the circuit. In general, a magnetic circuit may contain a number of different materials, each having different lengths and different magnetomotive drops. Like Kirchhoff's voltage law in an electric circuit, Ampère's circuital law states that *the sum of the magnetomotive forces around a magnetic circuit (closed loop) equals the sum of the magnetomotive drops around that circuit*. When traversing a closed magnetic loop, more than one winding may be encountered. If the windings produce flux in the same direction, the magnetomotive forces due to each are simply added. Otherwise, the algebraic resultant of the forces must be used when applying Ampère's circuital law.

Example 10.7

How much current should be supplied to the 40-turn winding in Figure 10.20 to produce a flux of 4×10^{-4} Wb throughout the core? The cross section of the core is a 2 cm $\times$ 2 cm square throughout its length.

SOLUTION The cross-sectional area of the core is

$$A = (0.02 \, \text{m})(0.02 \, \text{m}) = 4 \times 10^{-4} \, \text{m}^2$$

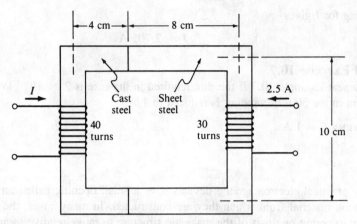

FIGURE 10.20 (Example 10.7)

Therefore, the flux density that must be established throughout the core is

$$B = \frac{\phi}{A} = \frac{4 \times 10^{-4} \text{ Wb}}{4 \times 10^{-4} \text{ m}^2} = 1 \text{ T}$$

From the B–H curves in Figure 10.17, the value of H that is required to produce $B = 1$ T in the cast steel section is approximately $H_{\text{cast steel}} = 700$ A/m. Using the sheet steel curve, we find the value of H necessary to produce $B = 1$ T in sheet steel is approximately $H_{\text{sheet steel}} = 220$ A/m.

We can now find the sum of the magnetomotive drops around the circuit by computing

$$(H_{\text{cast steel}})(l_{\text{cast steel}}) + (H_{\text{sheet steel}})(l_{\text{sheet steel}})$$

Note that $l_{\text{cast steel}} = 4$ cm $+ 10$ cm $+ 4$ cm $= 0.18$ m and $l_{\text{sheet steel}} = 8$ cm $+ 10$ cm $+ 8$ cm $= 0.26$ m. The computations are summarized in the following table:

Section	ϕ (Wb)	A (m²)	$B = \phi/A$ (T)	H (A/m)	l (m)	Hl (A)
Cast steel	4×10^{-4}	4×10^{-4}	1	700	0.18	126
Sheet steel	4×10^{-4}	4×10^{-4}	1	220	0.26	57.2
					Sum	183.2 A

We see that the sum of the magnetomotive drops is 183.2 A.

Using the right-hand rule, we see that each of the two windings produces flux in a clockwise direction around the core. Therefore, the sum of the applied magnetomotive forces is

$$40I + (30 \text{ turns})(2.5 \text{ A}) = 40I + 75$$

Using Ampère's circuital law, we equate the sum of the magnetomotive forces to the sum of the magnetomotive drops:

$$40I + 75 = 183.2$$

Solving for I gives

$$I = 2.705 \text{ A}$$

Drill Exercise 10.7

Repeat Example 10.7 if the flux required in the core is 2×10^{-4} Wb and the current in the 30-turn winding is reduced to 1 A.

ANSWER: $I = 1$ A.   □

Air Gaps

Many practical electromagnetic devices have a small opening called an *air gap* in the magnetic material from which they are constructed. In many cases, the air gap is necessary to permit one part of the magnetic structure to move relative to another part. An example is found in an electric motor, as illustrated in Figure 10.21. An electromagnetic field is produced by windings on a stationary core called the *stator*. For the motor to operate, flux produced by the field must be coupled into a rotating cylinder called the *rotor*. Since the rotor moves and the stator does not, there must be an air gap between the two. The magnetic circuit consists of two sections of magnetic material separated by the nonmagnetic air gap. As we shall see, it is desirable to make the air gap as small as possible, a fact that accounts for the very tight fit of the rotor inside a typical motor. Another example of air gaps in magnetic circuits is found in *relays,* which we will study in detail in a later discussion.

When an air gap occurs in a magnetic circuit, the flux at the edges of the gap bends outward, as shown in Figure 10.22. Thus, the flux density in the gap is not uniformly equal to the flux density in the surrounding magnetic material. However, in most practical cases, the gap is small enough that this *fringing* effect can be neglected, and for analysis purposes we usually assume that the flux density in the air gap is the same as that in the magnetic material.

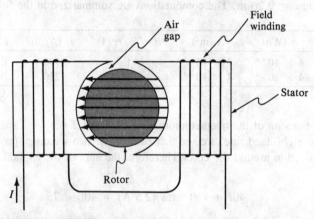

FIGURE 10.21 Cross-sectional view of an electric motor. An air gap must exist between the stationary stator and the rotating rotor.

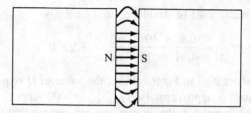

FIGURE 10.22 Flux at the edges of an air gap bends outward and reduces the flux density in those vicinities.

Since the permeability of air, μ_0, is much smaller than that of a ferromagnetic material, the value of H required to produce a certain flux density in an air gap is much greater than that required to produce the same flux density in the adjacent magnetic material ($H = B/\mu$). Therefore, the magnetomotive drop Hl in an air gap can be quite large, unless the width l of the gap is very small. It is uneconomical to provide a large winding with many turns and heavy current flow simply to make the applied magneto-motive force great enough to overcome a large magnetomotive drop in an air gap. This situation is like an electric circuit in which it is necessary to produce a certain current in a small resistance (low-reluctance magnetic material), but we must furnish a large voltage source because there is a large resistance (high-reluctance air gap) connected in series. The next example illustrates such a situation.

Example 10.8

Find the magnetomotive force that must be applied by the winding in Figure 10.23 in order to develop a flux of 0.6×10^{-5} Wb throughout the core. The cross section of the core is circular and has radius 1.25 mm. Neglect fringing in the air gap. The mean length of the sheet steel core is 30 cm.

SOLUTION The cross-sectional area of the core is the area of a circle with radius 1.25 mm = 1.25×10^{-3} m:

$$A = \pi r^2 = \pi(1.25 \times 10^{-3})^2 = 4.91 \times 10^{-6} \text{ m}^2$$

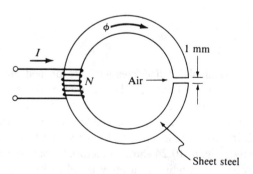

FIGURE 10.23 (Example 10.8)

Therefore, the flux density that must be developed in the core is

$$B = \frac{\phi}{A} = \frac{0.6 \times 10^{-5} \text{ Wb}}{4.91 \times 10^{-6} \text{ m}^2} = 1.22 \text{ T}$$

From the B–H curve for sheet steel in Figure 10.17, the value of H required to produce 1.22 T in sheet steel is approximately $H_{\text{sheet steel}} = 400$ A/m.

Since we are neglecting fringing in the air gap, we can assume that the flux density required in the gap is the same as that required in the core: $B_{\text{air}} = 1.22$ T. To find the corresponding value of H in the air gap, we must use equation (10.5), with $\mu = \mu_0 = 4\pi \times 10^{-7}$:

$$H_{\text{air}} = \frac{B_{\text{air}}}{\mu_0} = \frac{1.22 \text{ T}}{4\pi \times 10^{-7}} = 9.71 \times 10^5 \text{ A/m}$$

The foregoing computations, and the calculation of the magnetomotive drops, are summarized in the following table:

Section	ϕ (Wb)	A (m²)	$B = \phi/A$ (T)	H (A/m)	l (m)	Hl (A)
Sheet steel	0.6×10^{-5}	4.91×10^{-6}	1.22	400	0.3	120
Air gap	0.6×10^{-5}	4.91×10^{-6}	1.22	9.71×10^5	1×10^{-3}	971
					Sum	1091 A

We see that the sum of the magnetomotive drops is 1091 A. Therefore, the winding must furnish a magnetomotive force of 1091 A.

Notice that the magnetomotive drop in the air gap is much greater than that in the sheet steel (971 A versus 120 A). Fully 89% of the magnetomotive force supplied by the winding is necessary to overcome the drop in the air gap. This example demonstrates why it is so important to make air gaps very small in practical magnetic circuits.

Drill Exercise 10.8

What is the maximum permissible width of the air gap in Example 10.8 if the same flux must be produced in the core but the magnetomotive force supplied by the winding cannot exceed 500 A?

ANSWER: 0.39 mm.   ☐

Relays

A *relay* is a switch whose terminals, called *contacts*, are opened and closed by an electromagnet. When current is passed through the winding of the electromagnet, the resulting force actuates a metallic lever that causes one or more switch contacts to open and/or close. The lever, called an *armature*, is attached to a spring, so when the current in the winding is removed, the armature is retracted by the spring force and the contacts revert to their original state(s). Figure 10.24 shows a pictorial representation of a typical relay and schematic symbols. The winding in a relay is called its *coil*, and the coil is said to be *energized* when current flows through it. The switch contacts that are controlled

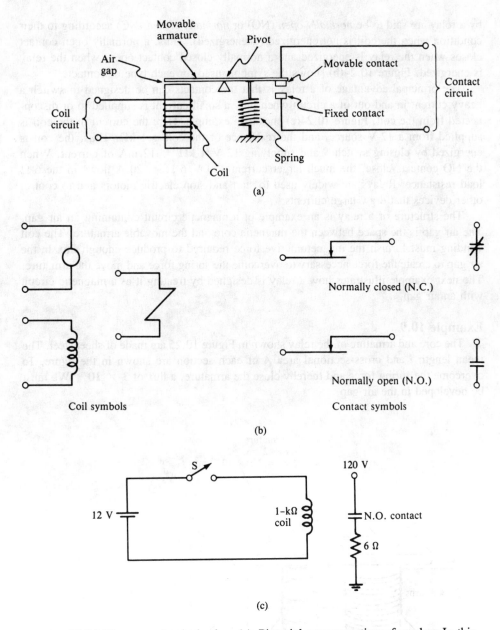

FIGURE 10.24 Electromechanical relay. (a) Pictorial representation of a relay. In this diagram, energizing the coil opens the contacts, and the spring force closes them when the coil is deenergized. The contacts are said to be *normally closed* (N.C.). The air gap is shown for clarity, but it disappears when the contacts open. (b) Symbols used for relay coils and contacts. (c) Typical relay circuit.

by a relay are said to be *normally open* (NO) or *normally closed* (NC) according to their condition when the coil is not energized (deenergized). Thus, a normally open contact closes when the relay is energized and a normally closed contact opens when the relay is energized. Figure 10.24(b) shows the symbols used for each type of contact.

The principal advantage of a relay is that the contacts can be designed to switch a heavy current in and out of a circuit when only a small current is supplied to or disconnected from the coil. Figure 10.24(c) shows an example. Here the current in the coil is supplied from a 12-V source, and the resistance of the coil is 1 kΩ. Thus, the coil is energized by closing switch S and supplying 12 V/1 kΩ = 12 mA of current. When the NO contacts close, the much larger current 120 V/6 Ω = 20 A flows in the 6-Ω load resistance. Relays are widely used to start and stop electric motors and to control other devices that draw large currents.

The structure of a relay is an example of a magnetic circuit containing an air gap. The air gap is the space between the magnetic core and the movable armature. The coil winding must furnish the magnetomotive force required to produce enough flux in the air gap to create the force necessary to overcome the spring force and move the armature. The next example illustrates how a relay is designed by treating it as a magnetic circuit with an air gap.

Example 10.9

The core and armature of the relay shown in Figure 10.25 are made of sheet steel. The mean length l and cross-sectional area A of each section are shown in the figure. To overcome the spring force and thereby close the armature, a flux of 3×10^{-5} Wb must be developed in the air gap.

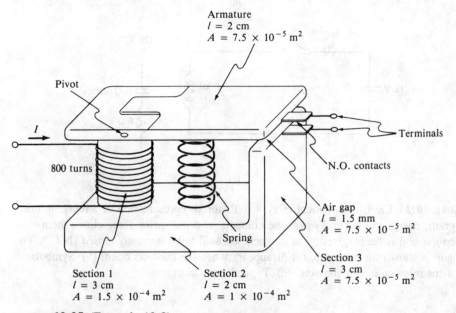

FIGURE 10.25 (Example 10.9)

(a) Find the current that must be supplied to the coil to close the armature. (This current is called the *pull-in* current.)

(b) After the armature has closed, the current in the coil is reduced until the armature springs back to its original position. The value of current at which this action occurs is called the *release* current, or the *drop-out* current. Find that value of current.

SOLUTION

(a) The flux density that must be developed in the air gap is

$$B_{air} = \frac{\phi}{A} = \frac{3 \times 10^{-5} \text{ Wb}}{0.75 \times 10^{-4} \text{ m}^2} = 0.4 \text{ T}$$

The magnetic field intensity in the air gap is therefore

$$H_{air} = \frac{B_{air}}{\mu_0} = \frac{0.4 \text{ T}}{4\pi \times 10^{-7}} = 318.3 \times 10^3 \text{ A/m}$$

The following table summarizes the computation of the total magnetomotive force that must be furnished by the coil:

Section	ϕ (Wb)	A (m²)	$B = \phi/A$ (T)	H (A/m)	l (m)	Hl (A)
1	3×10^{-5}	1.5×10^{-4}	0.2	20	3×10^{-2}	0.6
2	3×10^{-5}	1×10^{-4}	0.3	30	2×10^{-2}	0.6
3	3×10^{-5}	7.5×10^{-5}	0.4	40	3×10^{-2}	1.2
Armature	3×10^{-5}	7.5×10^{-5}	0.4	40	2×10^{-2}	0.8
Air gap	3×10^{-5}	7.5×10^{-5}	0.4	318.3×10^3	1.5×10^{-3}	477.5
					Sum	480.6 A

By Ampère's circuital law,

$$800I = 480.6$$

$$I = 0.6 \text{ A}$$

(b) When the armature is closed, the air gap is no longer in the magnetic circuit. Thus, to keep the armature closed, it is only necessary for the coil to supply sufficient magnetomotive force to overcome the magnetomotive drops in the steel. From the table in part (a), the sum of the magnetomotive drops in the steel is 3.2 A. The armature will remain closed so long as $NI > 3.2$ A. Therefore, the release current is the value of current that makes NI drop to 3.2 A, namely:

$$I = \frac{3.2 \text{ A}}{800} = 4 \text{ mA}$$

We see that the release is much smaller than the pull-in current, due to the absence of the air gap. Relay operation is another example of the general concept of hysteresis: The current must increase to a large value to energize the relay, but must fall to a much smaller value to deenergize it.

Exercises

Section 10.3   Magnetic Flux Density and Field Intensity

10.1 What flux density is necessary to produce 6×10^{-5} Wb of flux in a cylindrical rod having a diameter of 4 cm?

10.2 The flux density in a solid bar having a square cross section is 0.5 T. If the total flux in the bar is 1.125×10^{-4} Wb, what are the dimensions of the cross section?

10.3 A cable 100 m long and 1 cm in diameter is carrying a current of 50 A. At what distance from the cable is the flux density in air equal to 10^{-5} T?

10.4 A very thin conducting path on the surface of a printed-circuit board is carrying charge at the rate of 40 μC/s. What is the flux density (in air) on the board 1 cm away from the conducting path?

10.5 The current in the winding shown in Figure 10.26 is 3.2 A. If the winding has 60 turns, what is the magnetic field intensity in the core?

10.6 What magnetomotive force should be provided by the winding in Figure 10.26 if it is necessary to establish a magnetic field intensity of 2000 A/m in the core?

Section 10.4   Permeability

10.7 The permeability of the core in Figure 10.27 is 6×10^{-5} Wb/A-m. What is the flux density in the core?

10.8 What should be the relative permeability of the core in Figure 10.27 if it is desired to establish a flux density of 0.15 T in it?

10.9 What is the magnetic field intensity in a material whose relative permeability is 1 when the flux density is 0.005 T?

10.10 Derive the following equation for the flux density B in a core having relative permeability μ_r and length l

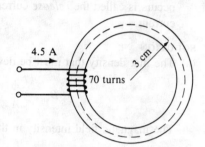

FIGURE 10.27 (Exercise 10.7)

meters when I amperes of current flow in its N-turn winding:

$$B = \frac{\mu_r \mu_0 NI}{l}$$

Section 10.5   Reluctance

10.11 When 0.8 A flows in the 55-turn winding of a certain solenoid, a flux of 1.1×10^{-5} Wb is established in the core. What is the reluctance of the core?

10.12 If the core in Exercise 10.11 is 8 cm long and has a cross-sectional area of 8×11^{-4} m², what is the permeability of the core material?

Section 10.7   B–H Curves and Hysteresis

10.13 The magnetic field intensity in the air core of the solenoid shown in Figure 10.28 is changed by 20 A/m. By what amount is the flux density in the core changed?

10.14 It is necessary to increase the flux density in the core in Figure 10.28 by 10^{-6} T. By what amount must the current I be increased?

10.15 At which of the numbered points on the hysteresis curve in Figure 10.14 is the incremental permeability greatest? At which points is it smallest?

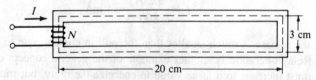

FIGURE 10.26 (Exercise 10.5)

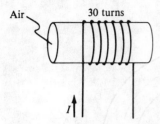

FIGURE 10.28 (Exercise 10.13)

10.16 Using the *B–H* characteristic given in Figure 10.17, find the approximate value of the incremental permeability of cast steel when $B = 0.5$ T.

10.17 What magnetic field intensity is necessary to produce a flux density of 1.25 T in sheet steel? If that value of magnetic field intensity is doubled, is the flux density also doubled? Explain why or why not.

10.18 What flux density will be established in cast iron by a magnetic field intensity of 200 A/m? If the magnetic field intensity is doubled, is the flux density also doubled? Explain why your answer is different from that in Exercise 10.17.

Section 10.8   Magnetic Circuits

10.19 Find the current *I* that must be supplied to the winding in Figure 10.29 in order to produce a flux of 0.5×10^{-4} Wb in the core. The core is made of cast iron and has a cross-sectional area of 1×10^{-4} m^2 throughout.

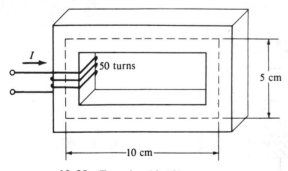

FIGURE 10.29 (Exercise 10.19)

10.20 Repeat Exercise 10.19 if the core is made of cast steel and it is required to produce a flux of 1×10^{-4} Wb in it.

10.21 Find the number of turns that should be used in the winding in Figure 10.30 if it is necessary to produce a flux of 2×10^{-4} Wb throughout the core. The cross section of the core is circular and has a diameter of 3 cm.

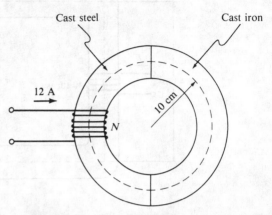

FIGURE 10.30 (Exercise 10.21)

10.22 Find the current that should be provided to the winding in Figure 10.31 if the flux in the core must be 6.82×10^{-4} Wb throughout. The cross-sectional area of the core is 6.2×10^{-4} m^2 throughout.

10.23 Find the current that must be supplied to the winding in Figure 10.32 in order to produce a flux of 6×10^{-3} Wb throughout the core. The mean length of the cast steel is 1.2 m and its cross section has dimensions 10 cm × 5 cm. Neglect fringing in the air gap.

10.24 Find the current that must be supplied to the winding in Figure 10.33 to produce a flux of 2×10^{-4} Wb throughout the core. The mean length *l* and cross-sectional area *A* of each section are shown in the figure. Neglect fringing in the air gap.

10.25 **(a)** If the magnetomotive drops in the ferromagnetic sections of the core in Exercise 10.24 are neglected, what current must be supplied to produce the required flux (based solely on the drop in the air gap)?
(b) What percentage of the actual current required is the current calculated under the assumption that the ferromagnetic sections can be neglected?

10.26 **(a)** If the magnetomotive drop in the cast steel of the core in Exercise 10.23 is neglected, what current must be supplied to produce the required flux (based solely on the drop in the air gap)?
(b) What percentage of the actual current required is the current calculated under the assumption that the drop in the cast steel can be neglected?

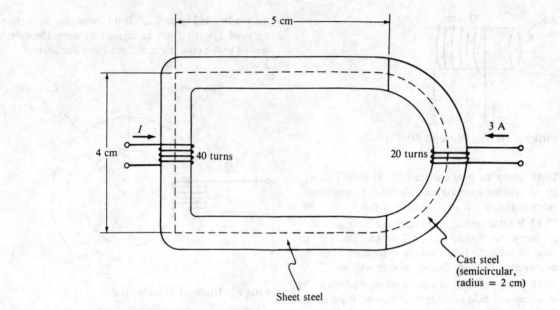

FIGURE 10.31 (Exercise 10.22)

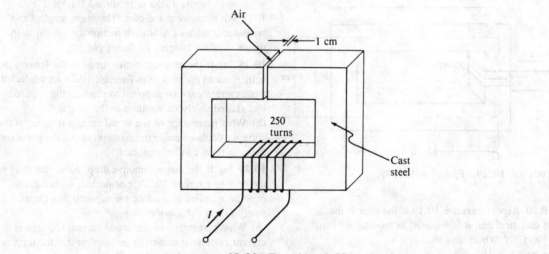

FIGURE 10.32 (Exercise 10.23)

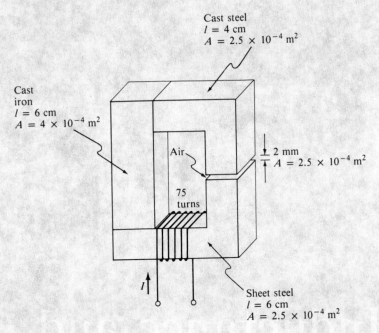

Cast steel
l = 4 cm
A = 2.5 × 10^{-4} m^2

Cast
iron
l = 6 cm
A = 4 × 10^{-4} m^2

Air

2 mm
A = 2.5 × 10^{-4} m^2

75
turns

I

Sheet steel
l = 6 cm
A = 2.5 × 10^{-4} m^2

FIGURE 10.33 (Exercise 10.24)

11 Inductance and Inductors

11.1 Electromagnetic Induction

In Chapter 10 we discussed the fact that a current-carrying conductor produces a magnetic field. We are ready now to explore another fundamental relationship between conductors and magnetic fields. Figure 11.1(a) shows a loop of wire, the lower portion of which is located in a magnetic field. As the loop is drawn upward, the lower portion moves vertically through the magnetic field. There is no voltage source connected to the loop, yet we notice that a current flows around the loop, in the direction shown, as long as the motion is sustained. Since a voltage must exist in order for current to flow, we conclude that *the motion of a conductor through a magnetic field creates a voltage across the ends of the conductor*. In other words, the portion of the loop moving through the magnetic field in Figure 11.1(a) is *itself* a voltage source, and the rest of the loop is an external circuit that carries the current produced by the source. Notice that the ends of the voltage-producing section are labeled + and − in accordance with the usual convention that current flows out of the + terminal and returns to the − terminal. *Inside* the voltage source (in the section of the loop moving through the field), current flows from − to +, just as it does in any voltage source. We say that a voltage is *induced* in

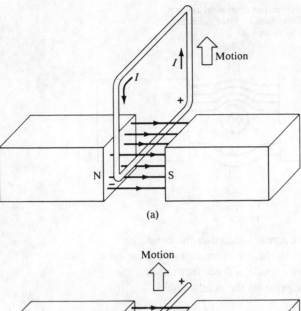

(a)

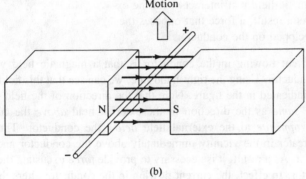

(b)

FIGURE 11.1 Electromagnetic induction. (a) Moving a loop of wire through a magnetic field causes a voltage to be generated (*induced*), which in turn causes a currrent to flow around the loop. (b) A voltage is induced across the ends of the moving conductor, even though there is no closed loop around which current can flow.

a conductor as a consequence of its motion through a magnetic field. Similarly, current that flows as a result of an induced voltage is called induced current. The creation of a voltage and/or current in this manner is a phenomenon called *electromagnetic induction*.

Figure 11.1(b) shows a straight section of conductor moving through a magentic field. In this case, there is no external loop through which current can flow, but a voltage is nonetheless induced across the ends of the conductor. The moving conductor is simply an open-circuited voltage source.

Lenz's Law

Figure 11.2 shows a cross-sectional view of the lower section of a conducting loop moving through a magnetic field. The cross on the conductor shows the direction of the induced current, into the page, in this case. As we know from Chapter 10, the conse-

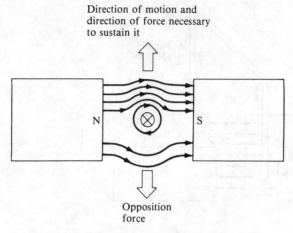

Direction of motion and
direction of force necessary
to sustain it

N   S

Opposition
force

FIGURE 11.2 Current induced in the conductor
creates a magnetic field that interacts with the external field. As a result, a force that opposes the
motion is developed on the conductor.

quence of current flowing in the conductor is that a magnetic field will be produced
around the conductor. Using the right-hand rule, we can see that this field has a clockwise
direction, as indicated in the figure. Note that the direction of the field produced by the
current is the *same* as the direction of the external field *above* the conductor (left to
right), but is *opposite* to the external field *below* the conductor. Thus, the net field
intensity is greater in the vicinity immediately above the conductor and smaller immediately below it. As a result, it is necessary to provide *force* to sustain the upward motion
of the conductor. In effect, the current flowing in the conductor alters the magnetic field
around the conductor in such a way that an *opposition* (downward) force is created on
the conductor. An upward force must therefore be applied to overcome the opposition
and to maintain the upward motion.

Lenz's law states that *the direction of the current induced by moving a conductor
through a magnetic field is always such that it creates opposition to the motion that
produced it*. In Figure 11.2, current flows into the page because that direction results in
a downward force, opposing the upward motion of the conductor. We deduce from Lenz's
law that moving the conductor *downward* through the field will produce a current that
flows *out* of the page. This case is illustrated in Figure 11.3.

Of course, the direction of the current induced in a moving conductor depends on
the direction of the external field as well as on the direction of motion. For example, if
the north and south poles in Figure 11.3 were interchanged, then downward motion of
the conductor would once again produce current into the page. Notice also that Lenz's
law can be used to determine the polarity of the *voltage* induced in a moving conductor.
It is not necessary that current actually flow through the conductor and thereby create an
opposition force on it. If the conductor consists simply of a straight wire instead of a
loop, no current will flow, but the polarity of the induced voltage is such that it *could*
produce current in the direction necessary to create an opposing force, if a complete loop
were present.

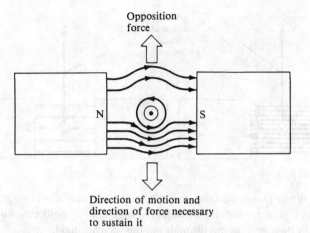

Opposition
force

N   S

Direction of motion and
direction of force necessary
to sustain it

FIGURE 11.3 The induced current flows out of the
page when the conductor moves downward, because that is the direction of current flow that will
cause an upward (opposition) force.

Faraday's Law

When a conductor moves through a magnetic field, we say that the conductor *cuts* the
flux lines. This terminology is helpful in visualizing the concept, but we of course realize
that magnetic field lines are themselves nothing more than a visual aid to help us understand and quantify certain field concepts. No lines are physically broken when a
conductor ''cuts'' through them. In order for a voltage to be induced in a moving
conductor, the motion must be such that successive flux lines are intersected (i.e., cut)
by the conductor. If the conductor moves *parallel* to the magnetic field, as shown in
Figure 11.4, no flux lines are cut and no voltage is induced.

Faraday's law states that *the magnitude of the voltage induced in a conductor moving
through a magnetic field is directly proportional to the rate at which flux lines are cut.*

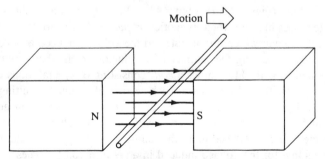

Motion

N   S

FIGURE 11.4 When a conductor is moved parallel to a
magnetic field, no flux lines are cut and no voltage is
induced.

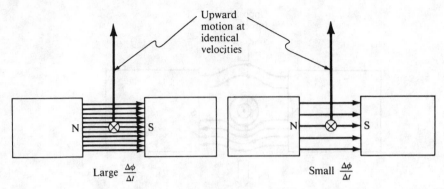

FIGURE 11.5 When two conductors move at identical velocities through magnetic fields, the conductor moving through the high-density field cuts flux at a greater rate than the one moving through the low-density field.

Letting $\Delta\phi$ represent the total flux that is cut by a conductor in the time interval Δt, the rate at which flux is cut is $\Delta\phi/\Delta t$. When $\Delta\phi$ is in webers and Δt is in seconds, Faraday's law states that the induced voltage is exactly equal to that ratio:

$$E = \frac{\Delta\phi}{\Delta t} \qquad \text{volts} \qquad (11.1)$$

[Some authors insert a minus sign in front of the quantity $\Delta\phi/\Delta t$ to convey the fact that the polarity of the induced voltage is such that it opposes the mechanism that created it (by Lenz's law). This is a needless complication that confuses practical computations.]

There are numerous ways to increase or decrease the rate at which flux is cut. The most obvious way is to increase or decrease the speed at which the conductor is moved through a field. Another way is to increase or decrease the flux density. As illustrated in Figure 11.5, a conductor moving at a fixed speed through a magnetic field will cut more lines in a high-density field than it will in a low-density field in the same interval of time.

The *angle* at which a conductor moves through a magnetic field also influences the value of $\Delta\phi/\Delta t$. We have already seen that $\Delta\phi/\Delta t = 0$ when a conductor moves parallel to a magnetic field (Figure 11.4). Figure 11.6 illustrates that a conductor cuts flux lines at the greatest possible rate when its motion is *perpendicular* to the field.

Finally, the rate at which flux is cut, and the voltage induced as a consequence, can be increased by increasing the number of conducting paths that simultaneously move through the field. Figure 11.7 shows a *coil*, consisting of several loops of wire, moving through a magnetic field. In this case, each section of conductor cutting flux lines adds to the total voltage induced in the coil. In other words, each loop acts as a voltage generator, and each generator is in series with all of the others, so the total voltage is the sum of the voltages induced in all the loops. An *N*-loop coil is said to have *N turns*, and Faraday's law for the voltage induced in an *N*-turn coil becomes

$$E = N\frac{\Delta\phi}{\Delta t} \qquad \text{volts} \qquad (11.2)$$

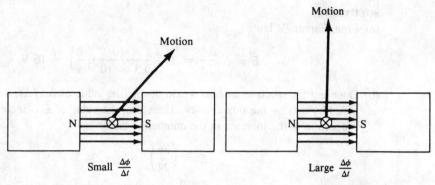

Small $\frac{\Delta\phi}{\Delta t}$ Large $\frac{\Delta\phi}{\Delta t}$

FIGURE 11.6 Speed of motion and flux density being equal, a conductor cuts flux lines at the greatest rate when it moves perpendicularly through a field.

Example 11.1

The coil shown in Figure 11.7 has 24 turns and cuts 2×10^{-4} Wb of flux in 0.3 ms.

(a) What voltage is induced in the coil?

(b) If the speed at which the coil moves through the magnetic field is doubled, the flux density is halved, and the number of turns is increased to 36, what voltage is then induced in the coil?

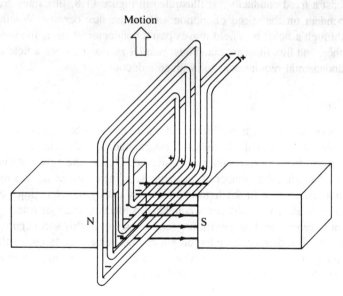

FIGURE 11.7 When an N-turn coil moves through a magnetic field, a voltage is induced in each of the turns. The total voltage is the sum of the voltages induced in the N turns.

SOLUTION

(a) From Faraday's law,

$$E = N \frac{\Delta \phi}{\Delta t} = 24 \left(\frac{2 \times 10^{-4} \text{ Wb}}{0.3 \times 10^{-3} \text{ s}} \right) = 16 \text{ V}$$

(b) Doubling the speed of motion would double the value of $\Delta\phi/\Delta t$, but this effect is offset by halving the flux density. Therefore, the value of E is increased in direct proportion to the increase in the number of turns:

$$E = \left(\frac{36}{24} \right) 16 \text{ V} = 24 \text{ V}$$

Drill Exercise 11.1

If the number of turns and the speed of the coil in Example 11.1 remain the same as in part (a), and if the flux density is 0.12 T when 16 V is generated, to what value should the flux density be changed if it is desired to generate 21 V?

ANSWER: 0.1575 T.   □

Relative Motion

Moving a conductor through a magnetic field is not the only way to cut flux lines and induce a voltage in the conductor. Electromagnetic induction also occurs when a magnetic field is moved past a stationary conductor. For example, when a permanent magnet is moved past a fixed conductor, as illustrated in Figure 11.8, flux lines are again cut at a rate dependent on the speed of motion and on the flux density. Whether a conductor moves through a field, or a field moves past a conductor, there is motion of one relative to the other, and flux lines are cut. Thus, *relative motion* between a field and a conductor is the fundamental requirement for voltage induction.

Transformer Action

Relative motion between a conductor and a field can also be created simply by *changing* the flux density of the field. In this case, neither the conductor nor the component producing the field need be physically moved. During the time interval that the flux density is changing, the number of magnetic field lines is either increasing or decreasing, so a conductor located in the field is effectively cut by the changing flux. We say that the field *expands* and *collapses* when its flux density increases and decreases. It is important to remember that electromagnetic induction of this sort occurs *only* during the time that the flux density is undergoing a change. One way to create that change is to change the current supplied to a winding on a core. The resulting change in H produces a corresponding change in the value of B in the core:

$$\Delta B = \mu \, \Delta H \tag{11.3}$$

where ΔB is the total change in flux density due to the change, ΔH, in field intensity. As illustrated in Figure 11.9, a second winding on the core will then be cut by the changing flux and a voltage will be induced in it. The arrangement shown in the figure

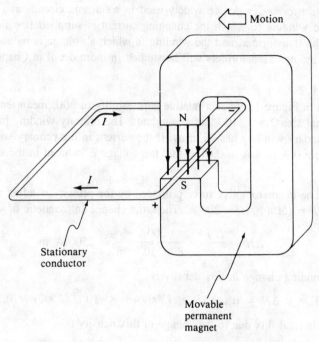

FIGURE 11.8 Flux lines are cut and a voltage is induced in a conductor when the magnetic field is moved past the conductor.

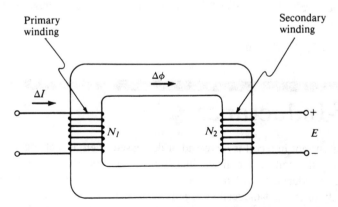

FIGURE 11.9 A transformer consists of two windings on a single core. Flux changes caused by changing the current in the primary winding induce a voltage across the secondary winding.

is called a *transformer*, a device widely used in electronic circuits and electric power systems. The winding to which the changing current is supplied is called the *primary* winding of the transformer, and the winding in which a voltage is induced is called the *secondary* winding. Transformers will be studied in more detail in Chapter 19.

Example 11.2

The core in Figure 11.9 has a relative permeability of 300, mean length 8 cm, and cross-sectional area 2.5×10^{-4} m^2 throughout. The primary winding has 50 turns and the secondary winding has 75 turns. If the current in the primary winding is increased from 0 A to 4 A in 0.01 s, find the voltage E induced in the secondary winding.

SOLUTION The magnetomotive force produced by the primary winding changes from 0 to $\mathcal{F} = NI = (50)(4 \text{ A}) = 200$ A. Thus, the change in magnetic field intensity is

$$\Delta H = \frac{\Delta \mathcal{F}}{l} = \frac{200 \text{A}}{8 \times 10^{-2} \text{ m}} = 2500 \text{ A/m}$$

The corresponding change in flux density is

$$\Delta B = \mu \, \Delta H = \mu_r \mu_0 \, \Delta H = (300)(4\pi \times 10^{-7})(2500) = 0.942 \text{ T}$$

The change in total flux due to the change in flux density is

$$\Delta \phi = (\Delta B)A = (0.942 \text{ T})(2.5 \times 10^{-4} \text{ m}^2) = 2.355 \times 10^{-4} \text{ Wb}$$

By Faraday's law, the voltage induced in the secondary winding is

$$E = N \frac{\Delta \phi}{\Delta t} = 75 \left(\frac{2.355 \times 10^{-4} \text{ Wb}}{0.01 \text{ s}} \right) = 1.77 \text{ V}$$

Drill Exercise 11.2

If all other quantities remain the same, to what value should the cross-sectional area of the core in Example 11.2 be changed if the induced voltage on the secondary winding is to be 3 V?

ANSWER: 4.25×10^{-5} m^2.

11.2 Self-Inductance

When the current in a coil is increased or decreased, there is a change in the flux density surrounding the coil. The coil winding *itself* is therefore cut by the changing flux, and a voltage is induced in it. This phenomenon is called *self-inductance*. By Lenz's law, the polarity of the self-induced voltage is such that it opposes the attempt to *change* the current flowing through it. This concept is illustrated in Figure 11.10. In Figure 11.10(a), closing the switch is an attempt to increase the current in the coil (i.e., to change it from zero to some positive value). As a consequence, a voltage V is induced in the coil with a polarity that opposes voltage source E, as shown in the figure. In Figure 11.10(b),

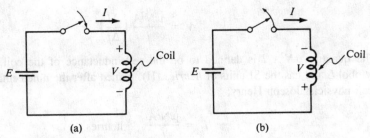

FIGURE 11.10 The voltage induced in a coil has a polarity that opposes any attempt to change the current through it.

opening the switch is an attempt to reduce the current in the coil (i.e., to change it from some positive value to zero). Consequently, the induced voltage V has a polarity that attempts to sustain the original direction of current flow. In this case, maintaining the original current constitutes opposition to the attempt to reduce it to zero. In both cases, the induced voltage eventually diminishes to zero becasue there is no continuing attempt to *change* the current beyond the instant that the switch is closed or opened.

A coil is said to have a large self-inductance if it produces a large opposition voltage when the current through it is changed. We wish now to quantify this important notion, so that we have a basis for determining how physical characteristics of a coil affect its self-inductance, and so that we have a means for calculating the magnitude of the opposition voltage produced by a given coil. Toward that end, we will apply Faraday's law to an N-turn coil whose core has permeability μ. We will assume that a certain change in current, ΔI, has occurred over the time interval Δt. The resulting change in magnetic field intensity is

$$\Delta H = \frac{N(\Delta I)}{l} \tag{11.4}$$

where l is the length of the core. By equation (11.3),

$$\Delta B = \mu \, \Delta H = \frac{\mu N(\Delta I)}{l} \tag{11.5}$$

The change in flux, $\Delta\phi$, due to the change in flux density, ΔB, is

$$\Delta\phi = (\Delta B)A = \frac{\mu N(\Delta I)}{l} A \tag{11.6}$$

where A is the cross-sectional area of the core. Dividing both sides of (11.6) by Δt gives

$$\frac{\Delta\phi}{\Delta t} = \frac{\mu N(\Delta I)A}{l(\Delta t)} \tag{11.7}$$

Multiplying both sides by N gives

$$N\frac{\Delta\phi}{\Delta t} = \frac{\mu N^2 A}{l}\left(\frac{\Delta I}{\Delta t}\right) \tag{11.8}$$

By Faraday's law, the left side of (11.8) is the induced voltage. Therefore,

$$V = \frac{\mu N^2 A}{l} \left(\frac{\Delta I}{\Delta t} \right) \tag{11.9}$$

The quantity $\mu N^2 A/l$ is defined to be the self-inductance of the coil. It is given the symbol L and has the SI units of *henries* (H), named after the nineteenth-century American physicist Joseph Henry:

$$L = \frac{\mu N^2 A}{l} \qquad \text{henries} \tag{11.10}$$

Substituting (11.10) into (11.9), we obtain the very important relationship

$$V = L \frac{\Delta I}{\Delta t} \tag{11.11}$$

Equation (11.11) shows that the magnitude of the voltage induced in a coil depends directly on the rate of change of the current through it, $\Delta I/\Delta t$.

It is conventional to use the term *inductance* with the understanding that *self-inductance* is meant.

Example 11.3

Find the inductance of the coil shown in Figure 11.11. The core has a relative permeability of 400.

SOLUTION The cross-sectional area of the core is

$$A = \frac{\pi d^2}{4} = \frac{\pi (0.01 \text{ m})^2}{4} = 7.854 \times 10^{-5} \text{ m}^2$$

The permeability of the core is

$$\mu = \mu_r \mu_0 = 400(4\pi \times 10^{-7}) = 5.026 \times 10^{-4}$$

By equation (11.10),

$$L = \frac{\mu N^2 A}{l} = \frac{5.026 \times 10^{-4}(100)^2 7.854 \times 10^{-5}}{4 \times 10^{-2}} = 9.87 \text{ mH}$$

Drill Exercise 11.3

How many turns should the coil in Example 11.3 have if an inductance of 15 mH is required?

ANSWER: 124. ☐

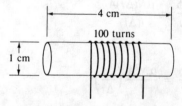

FIGURE 11.11 (Example 11.3)

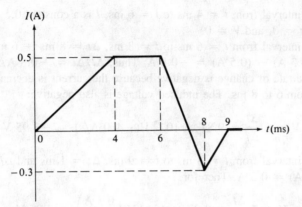

FIGURE 11.12 (Example 11.4)

Example 11.4

Sketch the voltage across the coil in Example 11.3 when the current through it changes with time as shown in Figure 11.12.

SOLUTION In the time interval from $t = 0$ to $t = 4$ ms, $\Delta t = 4$ ms and $\Delta I = 0.5$ A. Therefore, from equation (11.11),

$$V = L\frac{\Delta I}{\Delta t} = 9.87 \times 10^{-3} \text{ H} \left(\frac{0.5 \text{ A}}{4 \times 10^{-3} \text{ s}} \right) = 1.23 \text{ V}$$

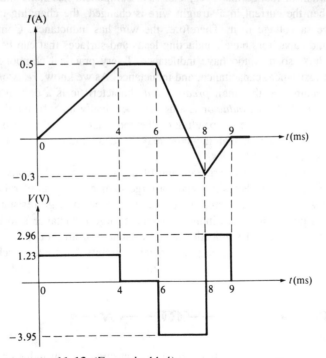

FIGURE 11.13 (Example 11.4)

In the time interval from $t = 4$ ms to $t = 6$ ms, I is a constant 0.5 A. Therefore, $\Delta I = 0$, $\Delta I/\Delta t = 0$, and $V = 0$.

In the time interval from $t = 6$ ms to $t = 8$ ms, $\Delta t = 8$ ms $- 6$ ms $= 2$ ms and $\Delta I = (-0.3$ A$) - (0.5$ A$) = -0.8$ A. Thus, $\Delta I/\Delta t = -0.8$ A/2 ms $= -400$ A/s. The rate of change is negative because the current is decreasing over the time interval from 6 to 8 ms. The induced voltage is also negative:

$$V = L\frac{\Delta I}{\Delta t} = 9.87 \times 10^{-3}\, H(-400\ A/s) = -3.95\ V$$

In the time interval from $t = 8$ ms to $t = 9$ ms, $\Delta t = 1$ ms and $\Delta I = 0$ A $- (-0.3$ A$) = 0.3$ A. Therefore,

$$V = L\frac{\Delta I}{\Delta t} = 9.87 \times 10^{-3}\, H\left(\frac{0.3\ A}{1\ ms}\right) = 2.96\ V$$

I remains constant at 0 A for all time after $t = 9$ ms, so $\Delta I = 0$, $\Delta I/\Delta t = 0$, and $V = 0$ thereafter. The voltage across the coil is plotted versus time in Figure 11.13.

11.3 Inductors

Although we have developed the notion of inductance in the context of an N-turn coil, it is important to realize that every electrical component has *some* inductance. For example, when the current in a straight wire is changed, the changing flux cuts the wire and induces a voltage in it. Therefore, the wire has inductance. Components such as resistors and capacitors have conducting leads and surfaces that can be similarly cut by changing flux, so they too have inductance. Every practical component has a certain amount of resistance, capacitance, and inductance. As we know, *resistors* and *capacitors* are manufactured so that their *predominant* characteristic is a certain resistance or capacitance. Similarly, an *inductor* is a device specifically manufactured to have a certain amount of self-inductance. Depending on the application for which it is intended and on its physical size, a practical inductor may have an inductance ranging from a few microhenries to several henries.

If an inductor having a large amount of inductance must be constructed in a small physical size, it must necessarily have a large number of turns of relatively fine wire. Since fine wire has a small cross-sectional area, it has a large resistance ($R = \rho l/A$). Therefore, a physically small inductor having a large inductance also has a large resistance. In many practical applications, the resistance of an inductor must be taken into account when selecting, designing, or analyzing circuits that contain such devices. Figure 11.14 shows the schematic symbol for an inductor and its equivalent circuit. Figure

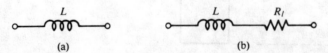

(a)   (b)

FIGURE 11.14 Schematic symbol and equivalent circuit of an inductor. (a) Pure inductance. (b) Equivalent circuit of a real inductor.

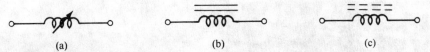

FIGURE **11.15** Special inductor symbols. (a) Adjustable inductor. (b) Iron-core inductor. (c) Ferrite-core inductor.

11.14(a) shows an ideally *pure* inductor, while Figure 11.14(b) shows that a real inductor can be represented by pure inductance in series with the resistance R_l of the inductor. In most practical dc circuits, the capacitance of an inductor can be neglected.

Inductors having a fixed or a variable (adjustable) inductance are available commercially. Figure 11.15(a) shows the schematic symbol for an adjustable inductor. Inductors of this type are usually constructed with a screwdriver-adjustable core. The farther the core is inserted into the winding by rotating the screw, the greater the inductance of the inductor. Special schematic symbols are also used to designate iron-core and ferrite-core inductors. These are shown in Figure 11.15 (b) and (c). Figure 11.16 shows examples of commercially available inductors.

Inductors used in high-frequency (alternating current) circuits are often called *chokes*, or simply *coils*. A familiar example of a practical application employing an inductor is an automobile *ignition coil*, illustrated in Figure 11.17. The objective of this coil is to generate the high voltage that is necessary to produce arcing across the air gaps of the spark plugs. The coil is actually a transformer, but its operation depends fundamentally on the fact that the magnitude of the induced voltage is directly porportional to the rate of change of current ($\Delta i/\Delta t$) through it. To achieve a large $\Delta i/\Delta t$, the dc voltage supplied to the coil from the automobile's battery is switched into and out of the circuit at a high rate. This switching is accomplished by the ''points'' (switch contacts) in the distributor. Rotating the distributor shaft causes the points to open and close, thereby switching voltage into and out of the coil. Thus, there are repeated attempts to change the current in the coil and a high voltage is induced in it.

11.4 Transients in RL Circuits

Figure 11.18(a) shows a series *RL circuit* connected through a switch to a dc voltage source. Closing the switch is an attempt to change the current through the inductor, so a voltage $v_L(t)$ is induced across its terminals. (We use lowercase *v* to represent this voltage because it is a time-varying quantity, and it is conventional to represent such variables with lowercase letters. The subscript L associates the voltage with the inductor.) The induced voltage at the instant the switch is closed equals the voltage E of the source, and has a polarity that opposes E. As time passes, the voltage v_L decays and eventually reaches zero. Thus, the voltage is a *transient*, in accordance with the definition presented in Chapter 9: a quantity that changes for a certain interval of time and remains constant thereafter. Letting $t = 0$ represent the time at which the switch is closed, the equation for $v_L(t)$ is

$$v_L(t) = Ee^{-t/(L/R)} \qquad \text{volts} \qquad (11.12)$$

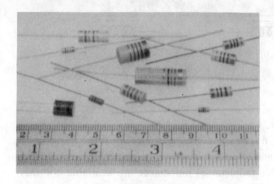

Molded, unshielded, radio–frequency coils

Shielded toroid inductors

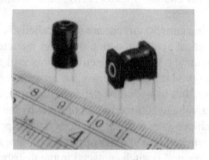

Variable radio–frequency coils

Micro inductors

High current filter chokes

High power inductor

FIGURE 11.16 Examples of commercially available coils and inductors. (Courtesy of the Delevan Division of American Precision Industries, Inc.)

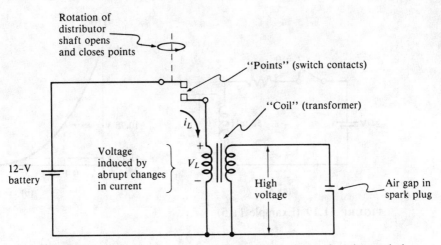

FIGURE 11.17 Principle of an automobile ignition system. Opening and closing the switch contacts (points) causes abrupt changes in the current through the coil, and a large voltage is induced as a consequence. (Not shown are the switch contacts in the distributor, used to route the high voltage to different spark plugs in succession.)

Figure 11.18(b) shows a plot of $v_L(t)$ versus t. The quantity L/R is the *time constant* of the RL circuit. It is designated by τ and has the units of seconds:

$$\tau = \frac{L}{R} \quad \text{seconds} \tag{11.13}$$

In terms of τ, equation (11.12) becomes

$$v_L(t) = Ee^{-t/\tau} \tag{11.14}$$

As indicated in Figure 11.18, the voltage decays to approximately $0.368E$ volts after the switch has been closed for a length of time equal to one time constant. This fact can be confirmed by substituting $t = \tau$ in equation (11.14):

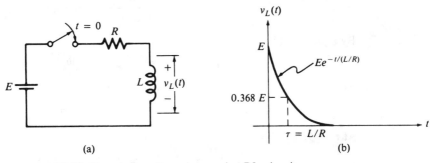

FIGURE 11.18 Transient voltage in a series RL circuit.

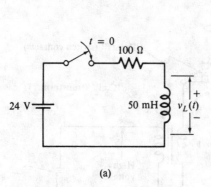

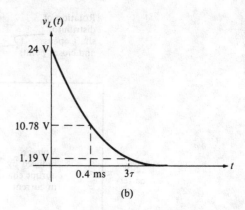

FIGURE 11.19 (Example 11.5)

$$v_L(\tau) = Ee^{-\tau/\tau} = Ee^{-1} = 0.36788E$$

The equation also confirms that $v_L(t) = E$ at $t = 0$:

$$v_L(0) = Ee^{-0} = E(1) = E$$

After five time constants have passed, $v_L(t)$ is quite small and can be assumed to be zero for most practical applications:

$$v_L(5\tau) = Ee^{-5\tau/\tau} = Ee^{-5} = 0.0067E$$

Example 11.5

The switch in Figure 11.19(a) is closed at $t = 0$.

(a) Write the equation for $v_L(t)$.
(b) Find the value of $v_L(t)$ at $t = 0$, $t = 0.4$ ms, and $t = 3\tau$.
(c) Sketch $v_L(t)$ versus t.

SOLUTION

(a) The time constant of the circuit is

$$\tau = \frac{L}{R} = \frac{50 \times 10^{-3} \text{H}}{100 \ \Omega} = 0.5 \text{ ms}$$

By equation (11.14),

$$v_L(t) = Ee^{-t/\tau} = 24e^{-t/0.5 \times 10^{-3}} \text{ V}$$

(b) At $t = 0$, $v_L(0) = 24e^{-0} = 24$ V. At $t = 0.4$ ms,

$$v_L(0.4 \times 10^{-3}) = 24e^{-(0.4 \times 10^{-3}/0.5 \times 10^{-3})} = 24e^{-0.8} = 10.78 \text{ V}$$

At $t = 3\tau$,

$$v_L(3\tau) = 24e^{-3\tau/\tau} = 24e^{-3} = 1.19 \text{ V}$$

(c) The transient voltage $v_L(t)$ is plotted versus t in Figure 11.19(b).

Drill Exercise 11.5

If the series resistance in the RL circuit of Example 11.5 is doubled, find the value of $v_L(t)$ at (a) $t = 0$; (b) $t = 0.4$ ms; (c) $t = 3\tau$.

ANSWER: (a) 24 V; (b) 4.85 V; (c) 1.19 V. □

Writing Kirchhoff's voltage law around the circuit in Figure 11.20(a), we have

$$E = v_R(t) + v_L(t) \tag{11.15}$$

or

$$v_R(t) = E - v_L(t) \tag{11.16}$$

Substituting $v_L(t) = Ee^{-t/\tau}$ into (11.16) gives

$$v_R(t) = E - Ee^{-t/\tau} = E(1 - e^{-t/\tau}) \quad \text{volts} \tag{11.17}$$

where $\tau = L/R$ seconds. Equation (11.17) is the transient voltage across the resistor in a series RL circuit, after the switch is closed at $t = 0$. It is plotted versus t in Figure 11.20(c). Notice that $v_R(t)$ starts at zero and reaches a *steady-state* value of E volts. For all practical purposes, it reaches this value after five time constants. Comparing $v_L(t)$, shown in Figure 11.20(b), with $v_R(t)$, we see that the sum of the two is equal to E *at every instant of time t*. For example, at $t = 0$, $v_L(t) = E$ and $v_R(t) = 0$. This is of course a consequence of Kirchhoff's voltage law.

The current in the series RL circuit is the same as the current in the resistor:

$$i(t) = \frac{v_r(t)}{R} = \frac{E}{R}(1 - e^{-t/\tau}) \quad \text{amperes} \tag{11.18}$$

The current is plotted in Figure 11.20(d). Note that both $v_R(t)$ and $i(t)$ reach approximately 63.2% of their steady-state values after one time constant has elapsed.

Example 11.6

The switch in Figure 11.21(a) is closed at $t = 0$.

(a) Write the expressions for $v_R(t)$ and $i(t)$.
(b) Find the value of $v_R(t)$ at $t = 0.1$ s.
(c) Find the value of $i(t)$ at $t = 2.5\tau$.
(d) Sketch $v_R(t)$ and $i(t)$ versus t.
(e) Verify Kirchhoff's voltage law around the circuit at the instant $t = 0.16$ s.

SOLUTION

(a) The time constant is $\tau = L/R = 0.8 \text{ H}/10 \text{ }\Omega = 0.08$ s.
    Therefore,

$$v_R(t) = E(1 - e^{-t/\tau}) = 120(1 - e^{-t/0.08}) \text{ V}$$

$$i(t) = \frac{E}{R}(1 - e^{-t/\tau}) = \frac{120}{10}(1 - e^{-t/0.08})$$

$$= 12(1 - e^{-t/0.08}) \text{ A}$$

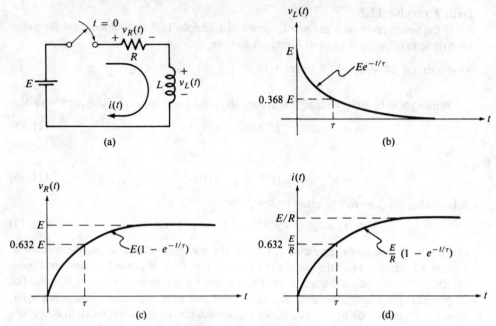

(a)

(b)

(c)

(d)

FIGURE 11.20 Transient voltages and current in a series RL circuit.

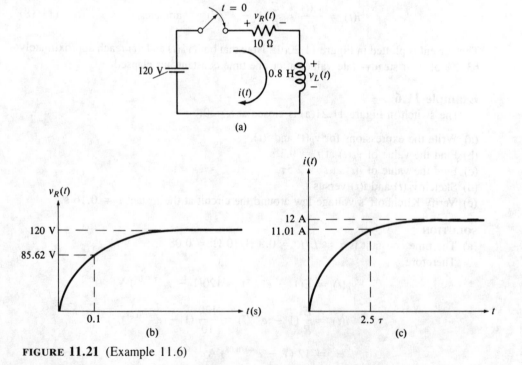

(a)

(b)

(c)

FIGURE 11.21 (Example 11.6)

(b) $v_R(0.1) = 120(1 - e^{-0.1/0.08}) = 120(1 - 0.286) = 85.62$ V

(c) $i(2.5\tau) = 12(1 - e^{-2.5\tau/\tau}) = 12(1 - 0.082) = 11.01$ A

(d) Plots of $v_R(t)$ and $i(t)$ are shown in Figure 11.21(b) and (c).

(e) $v_R(0.16) = 120(1 - e^{-0.16/0.08}) = 103.76$ V

$v_L(0.16) = 120e^{-0.16/0.08} = 16.24$ V

$v_R(0.16) + v_L(0.16) = 103.76$ V $+ 16.24$ V $= 120$ V $= E$

Drill Exercise 11.6

At what value of t is the current in Example 11.6 equal to 6 A?

ANSWER: 0.055 s.   □

Notice the similarity in the forms of the equations for transients in RL circuits and those for transients in RC circuits. The transient voltage in an RL circuit has the same form as the transient current in an RC circuit: $Ke^{-t/\tau}$, where K is a constant. Furthermore, the transient current in an RL circuit is like the transient voltage in an RC circuit: $K(1 - e^{-t/\tau})$. An important difference between RC and RL circuits is the effect of resistance on the duration of the transients in each. In an RC circuit, a large resistance prolongs the transient, because it makes the time constant, $\tau = RC$, large. In an RL circuit, a large resistance shortens the transient, because it makes $\tau = L/R$ small.

Figure 11.22(a) shows an RL circuit with a switch that has been in position 1 for a long period of time—long enough for the current in the circuit to have reached its steady-state value of E/R amperes. We assume now that the switch is placed in position 2 at $t = 0$. Since this action disconnects the voltage source, it is an attempt to change the current in the inductor from E/R amperes to zero, so a voltage is induced in the inductor. The polarity of the voltage is such that it attempts to sustain current flow in the direction

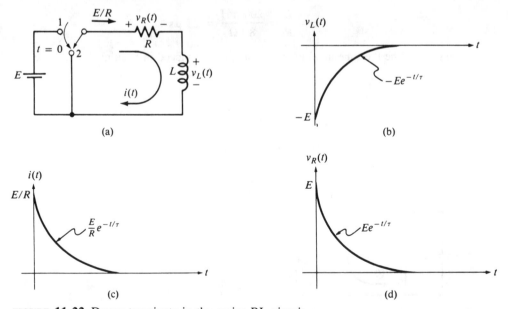

(a)   (b)

(c)   (d)

FIGURE 11.22 Decay transients in the series RL circuit.

of the original steady-state current. Thus, the polarity of the voltage is the *opposite* of that which was induced when the switch was first placed in position 1 (see Figure 11.18). Because it has the opposite polarity, we now regard $v_L(t)$ as a negative voltage and plot it below the time axis, as shown in Figure 11.22(b). Note that $v_L(t)$ eventually returns to zero because the current must eventually decay to zero, meaning that there is no further change in current. The equation for $v_L(t)$ after the switch is placed in position 2 is

$$v_L(t) = -Ee^{-t/\tau} \quad \text{volts} \tag{11.19}$$

where $\tau = L/R$ seconds. The equation for the transient current as it decays to zero is

$$i(t) = \frac{E}{R} e^{-t/\tau} \quad \text{amperes} \tag{11.20}$$

where $\tau = L/R$ seconds. The current is plotted versus t in Figure 11.22(c). As the current decays to zero, so does the voltage $v_R(t)$ across the resistance. By Ohm's law,

$$v_R(t) = Ri(t) = R\left(\frac{E}{R} e^{-t/\tau}\right) = Ee^{-t/\tau} \quad \text{volts} \tag{11.21}$$

where $\tau = L/R$ seconds. This voltage is plotted versus t in Figure 11.22(d). Notice that Kirchhoff's voltage law is satisfied at every instant of time because $v_L(t) + v_R(t) = 0$ at every instant of time.

Example 11.7

The switch in Figure 11.23 is moved between positions 1 and 2 at 250-μs intervals. Sketch $v_L(t)$ and $i(t)$ over a total time of 1 ms.

SOLUTION The time constant of the RL circuit is

$$\tau = \frac{L}{R} = \frac{200 \text{ mH}}{8 \text{ k}\Omega} = 25 \text{ μs}$$

Letting $t = 0$ be the time at which the switch is first placed in position 1, we have

$$v_L(t) = Ee^{-t/\tau} = 16e^{-t/25 \times 10^{-6}} \text{ V}$$

$$i(t) = \frac{E}{R} (1 - e^{-t/\tau}) = \frac{16 \text{ V}}{8 \text{ k}\Omega} (1 - e^{-t/25 \times 10^{-6}})$$

$$= 2(1 - e^{-t/25 \times 10^{-6}}) \text{ mA}$$

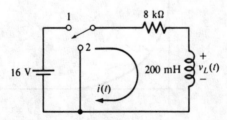

FIGURE 11.23 (Example 11.7)

The switch remains in position 1 for 250 μs, which equals 10 time constants, so $v_L(t)$ and $i(t)$ both reach their steady-state values of 0 V and 2 mA, respectively. The switch is then placed in position 2, and using that time as the new reference for $t = 0$, the decay transients are

$$v_L(t) = -16e^{-t/25 \times 10^{-6}} \text{ V}$$

$$i(t) = 2e^{-t/25 \times 10^{-6}} \text{ mA}$$

Since the switch remains in position 2 for another 10 time constants, the voltage and current both decay completely to zero. When the switch is returned to position 1, $v_L(t)$ and $i(t)$ behave exactly as when the switch was first placed in position 1. Alternating the switch between positions 1 and 2 at 250-μs intervals causes the current and voltage to repeat the same cycle of values, as shown in Figure 11.24.

Drill Exercise 11.7

What would be the maximum voltage reached by $v_L(t)$ and the maximum current reached by $i(t)$ in Example 11.7 if the 8-kΩ resistor were replaced by a 12-kΩ resistor?

ANSWER: 16 V and 1.33 mA.   □

Instantaneous Changes in Voltage and Current

An instantaneous change is one that occurs in zero time, that is, a voltage or current that "jumps" from one value to another without the passage of any time. An example is the

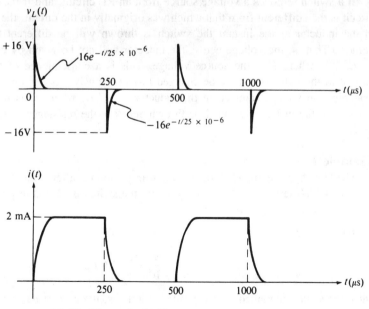

FIGURE 11.24 (Example 11.7)

voltage induced across an inductor at the instant a switch connects it to a voltage source, as shown in Figure 11.18(b). At $t = 0$, $v_L(t)$ changes instantaneously from 0 volts to E volts. Another example is the current in a capacitor at the instant a switch connects it to a voltage source (see Figure 9.22). At $t = 0$, $i(t)$ jumps instantaneously from 0 amperes to E/R amperes. In reality, it is not possible for any voltage or current to change instantaneously, but in practical circuits containing capacitors or inductors, the changes we have mentioned occur in such short times that they may be regarded as instantaneous.

On the other hand, even under ideal circumstances, it is impossible for some voltages and currents to change instantaneously. In particular, *it is not possible to change the current through an inductor instantaneously*. The reason it is impossible stems from the relationship between the voltage v_L across an inductor and the rate of change of current through it:

$$v_L = L \frac{\Delta i}{\Delta t}$$

By definition, an instantaneous change of current is one for which $\Delta t = 0$. Thus, an instantaneous change would mean that $\Delta i/\Delta t$ is infinitely large, which would require the voltage $v_L(t)$ to be infinitely large. Since infinite voltages are not possible, neither are instantaneous current changes in an inductor. Similarly, recall that the current through a capacitor is related to the voltage across it by

$$i = C \frac{\Delta v}{\Delta t}$$

An instantaneous change in voltage would make i infinitely large. Therefore, *it is not possible to change the voltage across a capacitor instantaneously*.

If a switch removes a voltage source from an RL circuit, and if the resistance in the circuit is then different from that which was originally in the circuit, the voltage induced in the inductor at the instant the switch is thrown will be different from the source voltage. That is, the voltage across the inductor will not be given by equation (11.14): $- Ee^{-t/\tau}$, where E is the source voltage. This is a consequence of the fact that the current in the inductor cannot be changed instantaneously. As demonstrated in the next example, the voltage induced in the inductor is $-I_L R$, where I_L is the current in the inductor at the instant the switch is thrown and R is the resistance in the circuit at that time.

Example 11.8

The switch in Figure 11.25(a) is placed in position 2 after it has been in position 1 for 1 ms. Write the equation for $v_L(t)$ after the switch is placed in position 2 (at $t = 0$), and sketch $v_L(t)$ versus t.

SOLUTION The time constant of the circuit when the switch is in position 1 is

$$\tau = \frac{L}{R} = \frac{9 \text{ mH}}{100 \text{ }\Omega} = 90 \text{ }\mu s$$

Since the switch remained in that position for 1 ms, the current has reached its steady-state value of

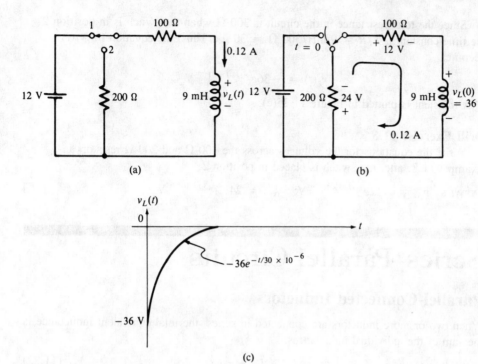

FIGURE **11.25** (Example 11.8)

$$\frac{E}{R} = \frac{12\ \text{V}}{100\ \Omega} = 0.12\ \text{A}$$

When the switch is placed in position 2, as shown in Figure 11.25(b), the 0.12-A current in the inductor cannot be changed instantaneously. Therefore, the same current must flow in the series loop consisting of the inductor, the 100-Ω resistor, and the 200-Ω resistor. By Ohm's law, the voltage drops across the resistors at the instant the switch is placed in position 2 are

$$(0.12\ \text{A})(100\ \Omega) = 12\ \text{V}$$

$$(0.12\ \text{A})(200\ \Omega) = 24\ \text{V}$$

Writing Kirchhoff's voltage law around the loop at $t = 0$ in Figure 11.25(b), we see that

$$v_L(0) = 12\ \text{V} + 24\ \text{V} = 36\ \text{V}$$

In other words,

$$v_L(0) = -(0.12\ \text{A})(300\ \Omega) = -36\ \text{V}$$

where the minus sign is inserted because $v_L(t)$ now has a polarity opposite to that of the voltage that was induced when the switch was first placed in position 1 [Figure 11.25(a)].

Since the total resistance in the circuit is 300 Ω when the switch is in position 2, the time constant is $L/R = 9 \text{ mH}/300\ \Omega = 30\ \mu s$. Thus, the equation for $v_L(t)$ becomes

$$v_L(t) = -36e^{-t/30 \times 10^{-6}}\ \text{V}$$

This transient is plotted in Figure 11.25(c).

Drill Exercise 11.8

Write the equations for the voltages across the 100-Ω and 200-Ω resistors in Example 11.8, after the switch is placed in position 2.

ANSWER: $v_{100\Omega} = 12e^{-t/30 \times 10^{-6}}\ \text{V}$; $v_{200\Omega} = 24e^{-t/30 \times 10^{-6}}\ \text{V}$.   □

11.5 Series–Parallel Circuits

Series- and Parallel-Connected Inductors

When two or more inductors are connected in series, the total equivalent inductance is the sum of the individual inductances:

$$L_T = L_1 + L_2 + \cdots + L_n \tag{11.22}$$

Thus, series-connected inductances are combined like series-connected resistances. Similarly, parallel-connected inductances are combined like parallel-connected resistances:

$$L_T = \frac{1}{1/L_1 + 1/L_2 + \cdots + 1/L_n} \tag{11.23}$$

The total equivalent inductance of two parallel-connected inductors is

$$L_T = \frac{L_1 L_2}{L_1 + L_2} \tag{11.24}$$

Equations (11.22) through (11.24) are based on the assumption that each inductor is *isolated* from every other inductor, in the sense that none of the flux produced by one cuts any of the windings of the others. We say that there are no flux *linkages* between the inductors. If flux linkages are present, as for example when two windings share the same core, then the total inductance must be calculated using the theory of *mutual* inductance, which we will discuss in a later chapter.

Transient Computations Using Thévenin Equivalent Circuits

To find the transient voltage across and the transient current through an inductor in a series–parallel circuit, it is necessary to construct a Thévenin equivalent circuit at the terminals of the inductor. The next example illustrates the procedure.

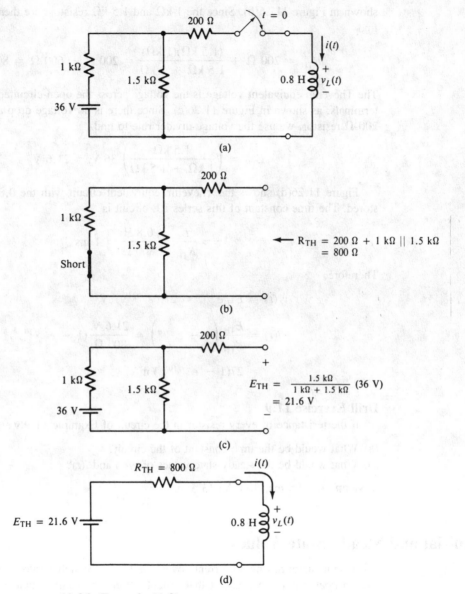

FIGURE 11.26 (Example 11.9)

Example 11.9

Write the equations for $v_L(t)$ and $i(t)$ in Figure 11.26(a), after the switch is closed at $t = 0$.

SOLUTION To find the Thévenin equivalent resistance of the circuit connected to the inductor terminals, we remove the inductor and short-circuit the 36-V source, as shown in Figure 11.26(b). Since the 1-kΩ and 1.5-kΩ resistors are then in parallel, we find

$$R_{TH} = 200 \ \Omega + \frac{(1.5 \ k\Omega)(1 \ k\Omega)}{1.5 \ k\Omega + 1 \ k\Omega} = 200 \ \Omega + 600 \ \Omega = 800 \ \Omega$$

The Thévenin equivalent voltage is the voltage across the open-circuited inductor terminals, as shown in Figure 11.26(c). Since there is no voltage drop across the 200-Ω resistor, we use the voltage-divider rule to find

$$E_{TH} = \left(\frac{1.5 \ k\Omega}{1 \ k\Omega + 1.5 \ k\Omega} \right) 36 \ V = 21.6 \ V$$

Figure 11.26(d) shows the Thévenin equivalent circuit with the 0.8-H inductor restored. The time constant of this series RL circuit is

$$\tau = \frac{L}{R_{TH}} = \frac{0.8 \ H}{800 \ \Omega} = 1 \ ms$$

Therefore,

$$v_L(t) = E_{TH}e^{-t/\tau} = 21.6e^{-t/10^{-3}} \ V$$

$$i(t) = \frac{E_{TH}}{R_{TH}} \left(1 - e^{-t/\tau} \right) = \frac{21.6 \ V}{800 \ \Omega} (1 - e^{-t/10^{-3}})$$

$$= 27(1 - e^{-t/10^{-3}}) \ mA$$

Drill Exercise 11.9

If the resistance of every resistor in the circuit of Example 11.9 were doubled:

(a) What would be the time constant of the circuit?
(b) What would be the steady-state values of $v_L(t)$ and $i(t)$?

ANSWER: (a) 0.5 ms; (b) 0 V; 13.5 mA.   □

Initial and Steady-State Values

When an inductor having no current flowing in it is first switched into a circuit, it behaves like an *open circuit*, because at that instant the current cannot change instantaneously from its initial value of zero. We can therefore determine the *initial* voltages and currents everywhere in a series–parallel RL circuit by replacing all inductors with open circuits. After steady-state conditions have been reached, an inductor behaves like a *short circuit*, because the steady-state voltage across it is zero. Here we are assuming that the inductor consists of pure inductance, or that its resistance is negligibly small. Thus, we can determine the steady-state voltages and currents everywhere in a series–parallel RL circuit by replacing all inductors with short circuits. The next example illustrates how conventional series–parallel circuit analysis is used to determine initial and steady-state currents and voltages after the inductors have been replaced by their initial or steady-state equivalents.

Example 11.10

Find the initial and steady-state voltage across and current through every component shown in Figure 11.27(a) after the switch is thrown at $t = 0$.

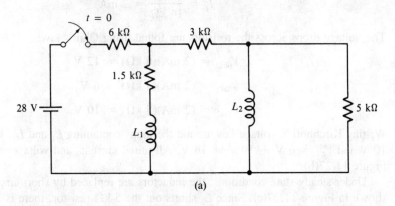

(a)

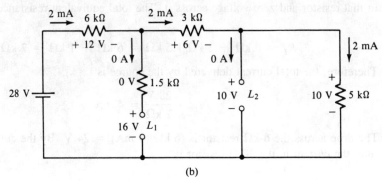

(b)

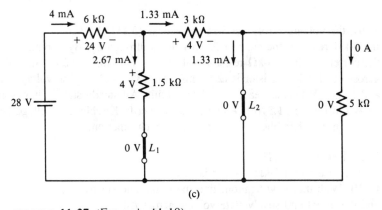

(c)

FIGURE 11.27 (Example 11.10)

SOLUTION The equivalent circuit at $t = 0$, with both inductors replaced by open circuits, is shown in Figure 11.27(b). Since L_1 and L_2 are both open, no current flows through either, and the voltage across the 1.5-kΩ resistor is 0 V. The total resistance of the circuit is the sum of the series-connected resistors: $R_T = 6 \text{ k}\Omega + 3 \text{ k}\Omega + 5\text{k}\Omega = 14 \text{ k}\Omega$. Thus,

$$I_T = \frac{28 \text{ V}}{14 \text{ k}\Omega} = 2 \text{ mA}$$

The voltage drops across the resistors are found using Ohm's law:

$$V_{6k\Omega} = (2 \text{ mA})(6 \text{ k}\Omega) = 12 \text{ V}$$

$$V_{3k\Omega} = (2 \text{ mA})(3 \text{ k}\Omega) = 6 \text{ V}$$

$$V_{5k\Omega} = (2 \text{ mA})(5 \text{ k}\Omega) = 10 \text{ V}$$

Writing Kirchhoff's voltage law around the loops containing L_1 and L_2, we find $V_{L2} = 10 \text{ V}$ and $V_{L1} = 6 \text{ V} + 10 \text{ V} = 16 \text{ V}$. All initial currents and voltages are shown in Figure 11.27(b).

Under steady-state conditions, the inductors are replaced by short circuits, as shown in Figure 11.27(c). Since L_2 shorts out the 5-kΩ resistor, there is zero current in that resistor and zero voltage across it. The total equivalent resistance of the circuit is

$$R_T = 6 \text{ k}\Omega + 3 \text{ k}\Omega \| 1.5 \text{ k}\Omega = 6 \text{ k}\Omega + 1 \text{ k}\Omega = 7 \text{ k}\Omega$$

Therefore, the total current delivered by the source is

$$I_T = \frac{28 \text{ V}}{7 \text{ k}\Omega} = 4 \text{ mA}$$

The drop across the 6-kΩ resistor is (6 kΩ)(4 mA) = 24 V. By the current-divider rule, the current in the 3-kΩ resistor is

$$I_{3k\Omega} = \left(\frac{1.5 \text{ k}\Omega}{1.5 \text{ k}\Omega + 3 \text{ k}\Omega} \right) 4 \text{ mA} = 1.33 \text{ mA}$$

Thus, the voltage across the 3-kΩ resistor is (3 kΩ)(1.33 mA) = 4 V. Since L_2 shorts the 5-kΩ resistor, the entire 1.33 mA flows through L_2. By Kirchhoff's current law, the current in the 1.5-kΩ resistor is 4 mA $-$ 1.33 mA = 2.67 mA. The drop across the 1.5-kΩ resistor is (2.67 mA)(1.5 kΩ) = 4 V. The voltage across both L_1 and L_2 is 0 V because each is a short circuit. All steady-state voltages and currents are shown in Figure 11.27(c). As an exercise, verify Kirchhoff's voltage law around each loop and Kirchhoff's current law at each junction.

Drill Exercise 11.10

Suppose that a 5-kΩ resistor is inserted in series with L_2 in the circuit of Example 11.10. With that modification, find the initial and steady-state current in L_2. Also find the initial and steady-state voltage across L_2.

ANSWER: initial: $i = 0$ A, $v = 10$ V; steady state: $i = 0.418$ mA, $v = 0$ V.   □

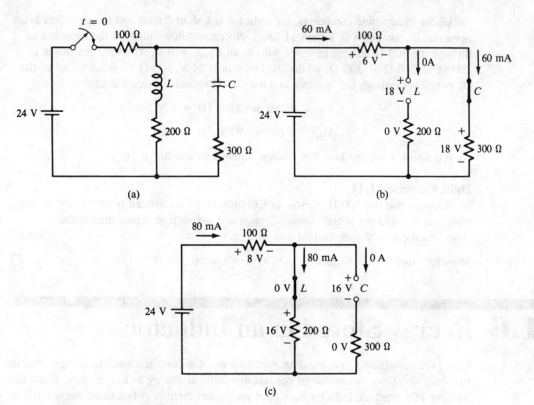

FIGURE 11.28 (Example 11.11)

Recall that an uncharged capacitor is equivalent to a short circuit at $t = 0$. Also recall that a capacitor is equivalent to an open circuit under steady-state conditions. Using these facts, we can find the initial and steady-state currents and voltages in a series–parallel RLC circuit, as demonstrated in the next example.

Example 11.11

Find the initial and steady-state voltage across and current through every component shown in Figure 11.28(a), after the switch is thrown at $t = 0$.

SOLUTION At $t = 0$, the inductor is an open circuit and the capacitor is a short circuit, as shown in Figure 11.28(b). No current flows through the inductor or through the 200-Ω resistor in series with it, so $V_{200\Omega} = 0$ V. The total resistance is 100 Ω + 300 Ω = 400 Ω, so $I_T = 24$ V/400 = 60 mA. All of the 60 mA flows through the capacitor and the 300-Ω resistor in series with it.

$$V_{300\Omega} = (60 \text{ mA})(300 \ \Omega) = 18 \text{ V}$$

$$V_{100\Omega} = (60 \text{ mA})(100 \ \Omega) = 6 \text{ V}$$

By Kirchhoff's voltage law, the voltage across the inductor is 18 V.

Under steady-state conditions, the inductor is a short circuit and the capacitor is an open circuit, as shown in Figure 11.28(c). No current flows through the capacitor or through the 300-Ω resistor in series with it, so $V_{300\Omega} = 0$ V. The total resistance is 100 Ω + 200 Ω = 300 Ω, so the total current is 24 V/300 Ω = 80 mA. All of the 80 mA flows through the inductor and the 200-Ω resistor in series with it.

$$V_{100\Omega} = (80 \text{ mA})(100 \text{ }\Omega) = 8 \text{ V}$$

$$V_{200\Omega} = (80 \text{ mA})(200 \text{ }\Omega) = 16 \text{ V}$$

By Kirchhoff's voltage law, the voltage across the capacitor is 16 V.

Drill Exercise 11.11

Suppose that the 300-Ω resistor in Example 11.11 is changed to 400 Ω. With that modification, find the initial current through and the voltage across the capacitor. Also find the steady-state current and voltage.

ANSWER: initial: $i = 48$ mA, $v = 0$ V; steady state: $i = 0$ A, $v = 16$ V.   □

11.6 Energy Stored in an Inductor

Like pure capacitance, pure inductance does not dissipate or consume energy. Recall that only resistance is capable of converting electrical energy to heat energy. Since the winding of a practical inductor does have resistance, there is in fact some energy loss in the device. However, the purely inductive component only stores energy when current flows through it. It is customary to say that the "energy is stored in the field." The energy is given by

$$W = \tfrac{1}{2}LI^2 \quad \text{joules} \tag{11.25}$$

where L is the inductance in henries and I is the current through it, in amperes.

Example 11.12

The inductor in Figure 11.29(a) has an inductance of 0.2 H. Its winding has a resistance of 400 Ω.

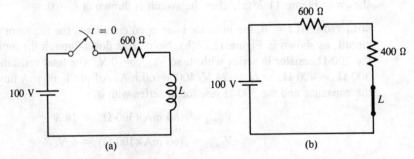

FIGURE 11.29 (Example 11.12)

(a) Find the energy stored in the inductor under steady-state conditions.

(b) Find the rate at which energy is dissipated by the winding under steady-state conditions.

SOLUTION

(a) Figure 11.29(b) shows the steady-state equivalent circuit when the inductor is replaced by a short-circuit in series with its winding resistance. The steady-state current is

$$I = \frac{100 \text{ V}}{600 \text{ } \Omega + 400 \text{ } \Omega} = 0.1 \text{ A}$$

By equation (11.25),

$$W = (\tfrac{1}{2})LI^2 = (\tfrac{1}{2})(0.2)(0.1)^2 = 10^{-3} \text{ J}$$

(b) The rate of energy dissipation in the winding is, by definition, the power

$$P = I^2R = (0.1)^2(400) = 4 \text{ W} = 4 \text{ J/s}$$

Drill Exercise 11.12

What additional resistance should be connected in series with the inductor in Figure 11.29 if it is desired to reduce the energy stored in the inductor to 0.5 mJ?

ANSWER: 414 Ω. ◻

11.7 SPICE Examples

Appendix Example A.5 illustrates how SPICE can be used to obtain plots of transient voltage and current in a series–parallel RL circuit. The next example demonstrates the use of SPICE to obtain steady-state values.

Example 11.13   (SPICE)

Use SPICE to obtain the steady-state voltage and current in inductor L in Figure 11.28 (Example 11.11), p. 403.

SOLUTION In Example 9.19, we showed that a dc analysis in SPICE produces steady-state values. In the present example, a dc analysis would likewise be the most direct way to obtain the required steady-state values. Another approach would be to perform a .TRAN analysis and examine voltages and currents a long time after the switch is closed. (We might wish to perform a transient analysis to gain insight into the way voltages and currents change with time, or to observe initial as well as steady-state values.) The difficulty in a circuit as complex as that in Figure 11.28 is in determining just what a "long time" means. Trial-and-error runs can be used to help resolve that question: We can examine the variation of voltages and currents over different periods of time and determine how long it takes for them to settle down to constant values. In the present case, an idea of the magnitude of the time constants

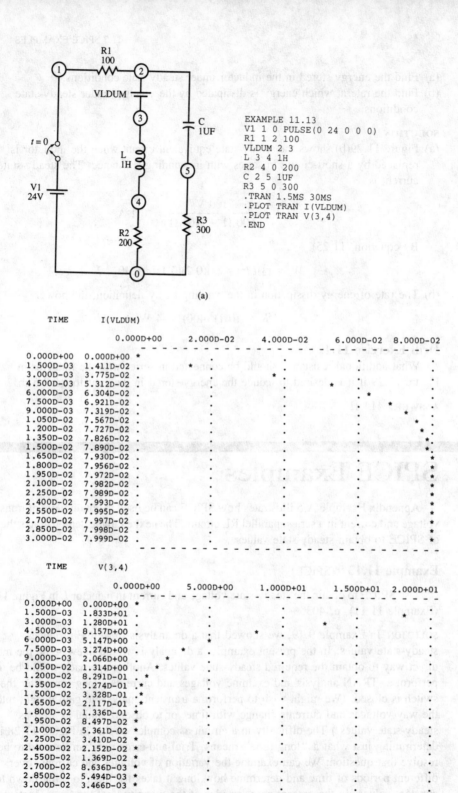

```
EXAMPLE 11.13
V1 1 0 PULSE(0 24 0 0 0)
R1 1 2 100
VLDUM 2 3
L 3 4 1H
R2 4 0 200
C 2 5 1UF
R3 5 0 300
.TRAN 1.5MS 30MS
.PLOT TRAN I(VLDUM)
.PLOT TRAN V(3,4)
.END
```

(a)

```
        TIME        I(VLDUM)

                    0.000D+00    2.000D-02   4.000D-02   6.000D-02  8.000D-02
                  - - - - - - - - - - - - - - - - - - - - - - - - - - - - - -
     0.000D+00    0.000D+00   *            .            .            .
     1.500D-03    1.411D-02      *         .            .            .
     3.000D-03    3.775D-02         *      .            .            .
     4.500D-03    5.312D-02            .   *            .            .
     6.000D-03    6.304D-02            .        *       .            .
     7.500D-03    6.921D-02            .            *   .            .
     9.000D-03    7.319D-02            .            .      *         .
     1.050D-02    7.567D-02            .            .          *     .
     1.200D-02    7.727D-02            .            .            *   .
     1.350D-02    7.826D-02            .            .             *  .
     1.500D-02    7.890D-02            .            .              * .
     1.650D-02    7.930D-02            .            .              * .
     1.800D-02    7.956D-02            .            .               *.
     1.950D-02    7.972D-02            .            .               *.
     2.100D-02    7.982D-02            .            .               *.
     2.250D-02    7.989D-02            .            .               *.
     2.400D-02    7.993D-02            .            .               *.
     2.550D-02    7.995D-02            .            .               *.
     2.700D-02    7.997D-02            .            .               *.
     2.850D-02    7.998D-02            .            .               *.
     3.000D-02    7.999D-02            .            .               *.
                  - - - - - - - - - - - - - - - - - - - - - - - - - - - - - -
```

```
        TIME        V(3,4)

                    0.000D+00    5.000D+00   1.000D+01   1.500D+01  2.000D+01
                  - - - - - - - - - - - - - - - - - - - - - - - - - - - - - -
     0.000D+00    0.000D+00   *            .            .            .
     1.500D-03    1.833D+01            .            .            .       *
     3.000D-03    1.280D+01            .            .    *       .
     4.500D-03    8.157D+00            .        *   .            .
     6.000D-03    5.147D+00            .   *        .            .
     7.500D-03    3.274D+00         *  .            .            .
     9.000D-03    2.066D+00       * .            .            .
     1.050D-02    1.314D+00      * .            .            .
     1.200D-02    8.291D-01     *  .            .            .
     1.350D-02    5.274D-01    *.            .            .
     1.500D-02    3.328D-01    .*            .            .
     1.650D-02    2.117D-01    .*            .            .
     1.800D-02    1.336D-01    *.            .            .
     1.950D-02    8.497D-02    *.            .            .
     2.100D-02    5.361D-02    *.            .            .
     2.250D-02    3.410D-02    *.            .            .
     2.400D-02    2.152D-02    *.            .            .
     2.550D-02    1.369D-02    *.            .            .
     2.700D-02    8.636D-03    *.            .            .
     2.850D-02    5.494D-03    *.            .            .
     3.000D-02    3.466D-03    *.            .            .
                  - - - - - - - - - - - - - - - - - - - - - - - - - - - - - -
```

(b)

FIGURE 11.30 (Example 11.13)

involved can be obtained by finding the time constants of the Thévenin equivalent circuits with respect to L and C under steady-state conditions. Of course, the value of $\tau = L/R_{TH}$ depends on L, which was not specified in Example 11.11. For this example, we arbitrarily choose $L = 1$ H. Then, referring to Figure 11.28(c), we see that $\tau = 1$ H/(100 $\Omega \parallel$ 200 Ω) = 4.44 ms.

The steady-state value of the capacitor time constant is $\tau = R_{TH}C = (300 \ \Omega +$ 100 $\Omega \parallel$ 200 Ω)C. If we choose $C = 1 \ \mu$F, the capacitive time constant is 367 μs. These calculations give us a rough idea of what might constitute a "long time" in the circuit, and we perform our first trial run over a period of $TSTOP = 30$ ms (about seven times the inductive time constant.) The circuit and input data file are shown in Figure 11.30(a). The plots produced by SPICE are shown in Figure 11.30(b). We see that the inductor current, I(VLDUM), reaches (for all practical purposes) its steady-state value of 80 mA in 30 ms. Additional simulations with larger values of $TSTOP$ are probably not necessary, but would further confirm the steady-state values inferred here.

Exercises

Section 11.1   Electromagnetic Induction

11.1 Draw a cross or a dot on the end of the conductor in each part of Figure 11.31 to show the direction of the current induced in the conductor (assuming that there is a complete path through which current can flow).

11.2 Draw + and − signs on the ends of the conductor in each part of Figure 11.32 to show the polarity of the voltage induced in it.

11.3 At what rate would it be necessary for a single conductor to cut flux in order that

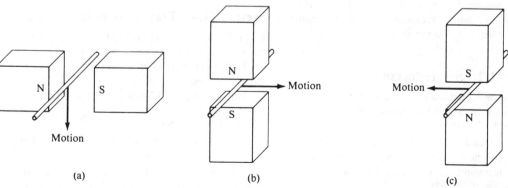

FIGURE 11.31 (Exercise 11.1)

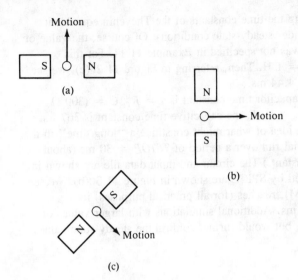

FIGURE 11.32 (Exercise 11.2)

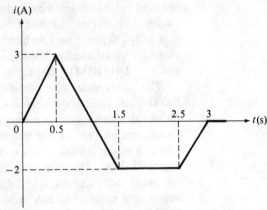

FIGURE 11.33 (Exercise 11.9)

(a) a voltage of 51.2 mV be induced in it?
(b) a current of 1.2 mA flow through it when a 10-Ω resistor is connected across its ends?

11.4 How many turns should a coil have if a voltage of 18 V is to be induced across it when it cuts flux at the rate of 0.15 Wb/s?

11.5 A cylindrical core is 4 cm long and has a diameter of 0.75 cm. The relative permeability of the core is 210 and it has a 60-turn winding. If the current in the winding changes from 0 A to 100 mA in 3 ms, what is the rate of change of flux in the core?

11.6 If the core in Exercise 11.5 has a secondary winding, how many turns should it have if 0.5 V is to be induced in it when the current in the 60-turn primary winding changes by 150 mA in 2 ms?

Section 11.2   Self-Inductance

11.7 What is the inductance of a 200-turn coil whose air core has a cross-sectional area of 4.4×10^{-4} m² and a length of 3 cm?

11.8 A cylindrical core has a diameter of 0.5 cm and is 2 cm long. If the core has a relative permeability of 590, how many turns should be wrapped on it to develop an inductance of 75 mH?

11.9 The current in a 2.5-H inductor changes with time as shown in Figure 11.33. Sketch the voltage across the inductor versus time.

11.10 The current through an 80-μH inductor changes with time as shown in Figure 11.34. Sketch the voltage across the inductor as a function of time.

Section 11.3   Inductors

11.11 The dc voltage V across the 200-mH inductor in Figure 11.35 is 0.25 V. What is the resistance of the winding?

11.12 In a certain application, the current through an inductor increases at the rate of 25 A/s. It is necessary to adjust the voltage across the inductor from 0.75 V to 1.25 V by adjusting the inductance. What range of inductance values should an adjustable inductor be able to provide for this application?

Section 11.4   Transients in RL Circuits

11.13 The switch in Figure 11.36 is closed at $t = 0$.
(a) What is the time constant of the circuit?
(b) Write the mathematical expression for $v_L(t)$ after the switch is closed.
(c) Find the value of $v_L(t)$ at $t = 0.1$ ms, 1.5τ, 0.5 ms, and 5τ.
(d) Sketch $v_L(t)$ versus t.

11.14 (a) At what time after the switch is closed in Figure 11.36 does $v_L(t)$ reach 15 V?

(b) How many time constants after the switch is closed in Figure 11.36 does $v_L(t)$ reach 17 V?

11.15 The switch in Figure 11.37 is closed at $t = 0$.
(a) Write the mathematical expressions for $v_L(t)$, $i(t)$, and $v_R(t)$ after the switch is closed.

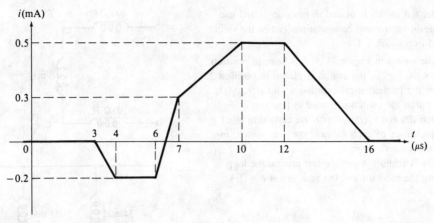

FIGURE **11.34** (Exercise 11.10)

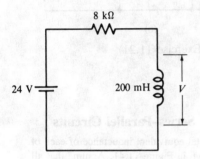

FIGURE **11.35** (Exercise 11.11)

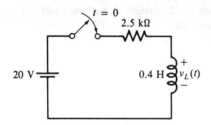

FIGURE **11.36** (Exercise 11.13)

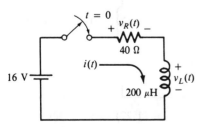

FIGURE **11.37** (Exercise 11.15)

(b) Sketch $v'_L(t)$, $i(t)$, and $v_R(t)$ versus t. Be certain to label significant values of each variable on your sketch, including the value of each at $t = \tau$.

(c) Verify that Kirchhoff's voltage law is satisfied around the loop at the instant $t = 7.5$ μs.

11.16 To what value would it be necessary to change the 200-μH inductance in Figure 11.37 if it is desired that $i(t) = 0.1$ A after the switch has been closed for 5 μs?

11.17 The switch in Figure 11.38 has been in position 1 for 20 ms. At $t = 0$, the switch is placed in position 2.
(a) Write the mathematical expressions for $i(t)$, $v_L(t)$, and $v_R(t)$ after the switch is placed in position 2.
(b) Sketch $i(t)$, $v_L(t)$, and $v_R(t)$ versus t. Be certain to label significant values of each variable on your sketch, including the value of each at $t = \tau$.
(c) Verify Kirchhoff's voltage law around the loop containing the inductor and the resistor at $t = 3$ ms.

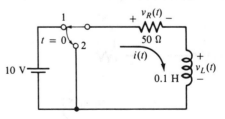

FIGURE **11.38** (Exercise 11.17)

11.18 Repeat Exercise 11.17 if the switch remains in position 1 for 1 ms (instead of 20 ms) before it is moved to position 2. [*Hint*: Compute the value of $i(t)$

1 ms after the switch is placed in position 1, and use that value in subsequent computations, when the switch is placed in position 2.]

11.19 The switch in Figure 11.39 has been in position 1 for 20 s. At $t = 0$, the switch is placed in position 2.
(a) Write the mathematical expressions for $i(t)$, $v_L(t)$, and $v_R(t)$ after the switch is placed in position 2.
(b) Sketch $i(t)$ and $v_L(t)$ versus t. Be certain to label significant values of each variable on your sketch, including the values at $t = \tau$.
(c) Verify Kirchhoff's voltage law around the loop containing the inductor and the resistors at $t = 1$ s.

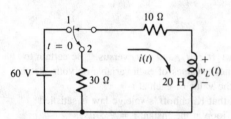

FIGURE 11.39 (Exercise 11.19)

11.20(a) The switch in Figure 11.40 is placed in position 1 at $t = 0$. Write the mathematical expressions for $v_L(t)$ and $i(t)$.
(b) After the switch has been in position 1 for 20 μs, it is placed in position 2. Assuming that the instant it is placed in position 2 is the new time reference ($t = 0$), write the mathematical expressions for $v_L(t)$ and $i(t)$.
(c) Sketch $v_L(t)$ obtained from parts (a) and (b) on the same time axis. Also sketch $i(t)$ from parts (a) and (b) on the same time axis. Be certain to label significant values on the sketches, including the values at $t = \tau$.

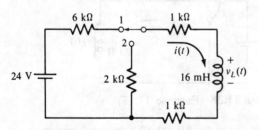

FIGURE 11.40 (Exercise 11.20)

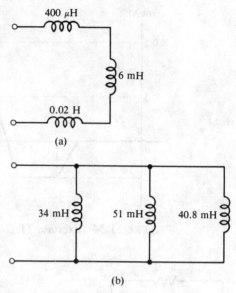

FIGURE 11.41 (Exercise 11.21)

Section 11.5 Series–Parallel Circuits

11.21 Find the total equivalent inductance of each of the networks shown in Figure 11.41. Assume that all inductors are isolated from each other.

11.22 Find the total equivalent inductance of the network in Figure 11.42. Assume that all inductors are isolated from each other.

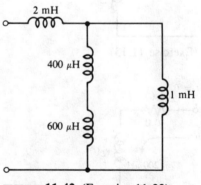

FIGURE 11.42 (Exercise 11.22)

11.23 The switch in Figure 11.43 is closed at $t = 0$. Write the mathematical expressions for $v_L(t)$ and $i(t)$.

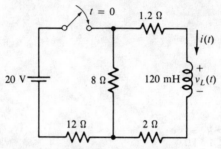

FIGURE 11.43 (Exercise 11.23)

11.24 The switch in Figure 11.44 is closed at $t = 0$. Write the mathematical expressions for $v_L(t)$ and $i(t)$.

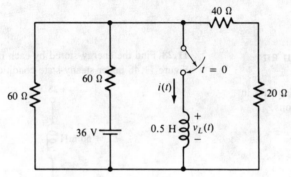

FIGURE 11.44 (Exercise 11.24)

11.25 Find the initial and steady-state values of the current through and voltage across every component in the circuit shown in Figure 11.45.

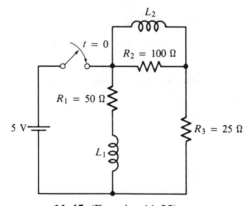

FIGURE 11.45 (Exercise 11.25)

11.26 Find the initial and steady-state values of the current through and voltage across every component in the circuit shown in Figure 11.46.

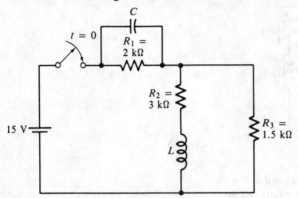

FIGURE 11.46 (Exercise 11.26)

Section 11.6   Energy Stored in an Inductor

11.27 Find the energy stored by each inductor in Figure 11.47 under steady-state conditions.

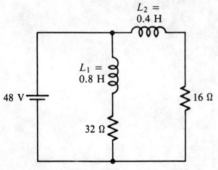

FIGURE 11.47 (Exercise 11.27)

11.28 Find the energy stored by each inductor in Figure 11.48 under steady-state conditions.

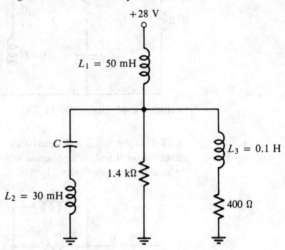

FIGURE 11.48 (Exercise 11.28)

SPICE Exercises

11.1S Use SPICE to obtain plots of $i(t)$ and $v_L(t)$ in Figure 11.37 after the switch is closed at $t = 0$. The plots should extend over five time constants.

11.2S Use SPICE to obtain a plot of $i(t)$ in Figure 11.38. The plot should cover and include the following time points: $t = 0$ when the switch is placed in position 1; $t = 10$ ms when the switch is placed in position 2; $t = 20$ ms, the end of the plot. (Hint: the input pulse must have pulse width 10 ms.)

11.3S In Figure 11.45, $L_1 = 0.5$ H and $L_2 = 1$ H. Using SPICE to perform a transient analysis, find the steady-state values of the voltage across and current through R_3.

11.4S Use SPICE as an aid in solving Exercise 11.28. Assume $C = 0.01 \ \mu F$.

12.1 AC Waveforms

Recall from Chapter 2 that alternating current (ac) is current that periodically *reverses direction*. As an illustration, consider the current that flows in the 10-Ω resistor in Figure 12.1(a) when the switch is moved between positions 1 and 2 at 1-s intervals. When the switch is in position 1, a current of 10 V/10 Ω = 1 A flows from left to right through the resistor. When the switch is moved to position 2, it can be seen that 1 A flows in the opposite direction through the resistor, (i.e., from right to left). Figure 12.1(b) shows a plot of the current versus time. Assuming that left-to-right current is considered positive, then right-to-left current is negative, and the current alternates between +1 A and −1 A at 1-s intervals.

It is important to remember that current must truly reverse direction (i.e., go negative) to be considered ac. The definition is not satisfied by current that merely changes its value at periodic intervals. Figure 12.2 shows some more examples of alternating current, as well as some examples that do not satisfy the definition. The plot, or graph, of a current versus time is called a *waveform*. Note that the waveforms in Figure 12.2(b) and

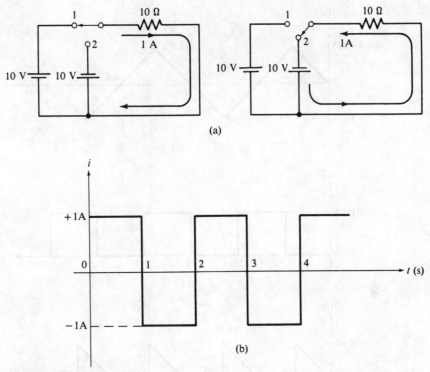

FIGURE 12.1 Example of alternating current. (a) Moving the switch between positions 1 and 2 reverses the direction of the current through the 10-Ω resistor. (b) Plot of the current through the 10-Ω resistor when the switch is moved between position 1 and 2 at 1-s intervals.

(d) undergo changes in *magnitude*, but not in direction, since all their values remain positive. These types of waveforms are often referred to as *pulsating dc*. They can also be regarded as the *superposition* (addition) of an ac waveform and a dc level, but are not properly considered ac.

An ac *voltage* is one that periodically reverses *polarity*. For example, the voltage across the 10-Ω resistor in Figure 12.1 reverses polarity at 1-s intervals, because the current through it reverses direction every second. An ac *voltage source* produces an electromotive force whose polarity reverses at periodic intervals. Thus, the positive and negative terminals of such a source are effectively interchanged at regular intervals: Current flows out of one terminal and returns to the second, then flows out of the second and returns to the first. Examples are ac *generators* (often called *alternators*), *oscillators*, and laboratory instruments called *signal generators*, or *function* generators.

The ac waveform studied most frequently in electrical circuit theory is shown in Figure 12.3. Called a *sinusoidal* waveform, or a *sine wave*, its value at any instant of time can be determined using the trigonometric *sine* function. Figure 12.3 shows a sinusoidal current; a sinusoidal voltage has the same general shape. Notice that the current

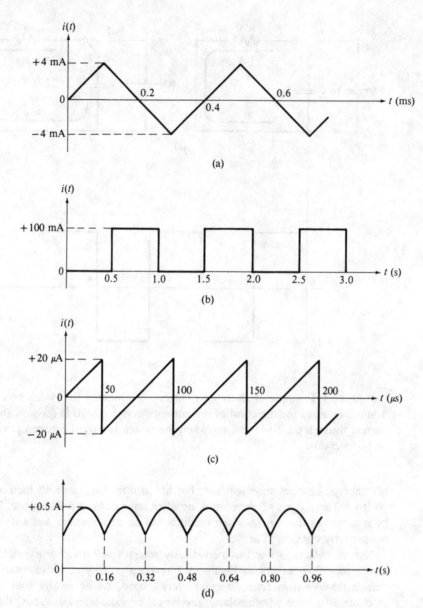

FIGURE 12.2 Examples of ac waveforms and pulsating dc waveforms. (a) An alternating current that increases and decreases linearly with time, called a *triangular* waveform. (b) Pulsating dc. Since the current does not reverse direction (go negative), it is not considered ac. (c) An ac waveform called a *sawtooth*. (d) Pulsating dc. The current is not ac, but may be regarded as ac superimposed on a dc level.

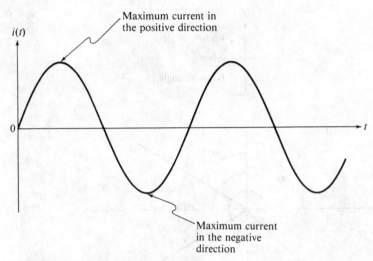

FIGURE 12.3 Sinusoidal ac current.

gradually increases in one direction, eventually reaches a maximum, then decreases until it reaches zero, then reverses direction. After passing through zero, it increases in the reverse direction until it reaches a reverse (negative) maximum, after which it returns to zero and repeats the cycle.

12.2 Review of Trigonometric Functions

We will use sine waves to illustrate many fundamental concepts in the theory of alternating current, as well as in most of the ac circuit analysis that follows. Because the sine wave is so important in ac circuit theory, we present here a brief review of the properties of the sine function and some other trigonometric relations that are essential for ac circuit analysis. Readers who are thoroughly acquainted with trigonometry may skip the material that follows, but those who are just beginning their study of that subject should commit the material to memory as soon as possible. In either case, the summary will serve as a useful reference in future discussions and examples.

The sine is a function of *angle*. Figure 12.4 illustrates how angle is measured and shows that the *x-y* plane is divided into four *quadrants*. Note that positive angle is measured in a *counterclockwise* direction from the positive *x*-axis, and negative angle is measured in a clockwise direction. In ac circuit theory, angles greater than 180° are expressed as equivalent negative angles. For example, 225° should be expressed as −135°, as illustrated in the figure. Similarly, angles that are more negative than −180° should be expressed as equivalent positive angles. Thus, angles in quadrants I and II are expressed as positive angles, and those in III and IV as negative angles. The angle of the negative *x*-axis may be expressed as either +180° or −180°. Any angle is equivalent

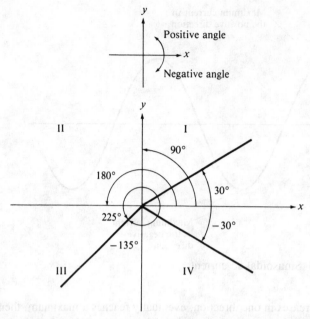

FIGURE 12.4 Positive angles are measured in a counterclockwise direction from the positive x-axis. In ac circuit theory, angles in the third and fourth quadrants should be expressed as equivalent negative angles (measured clockwise from the positive x-axis).

to the angle obtained by adding $\pm 360°$ (or a multiple of $\pm 360°$) to it. Following are some examples:

$$200° = 200° - 360° = -160°$$
$$-250° = -250° + 360° = 110°$$
$$400° = 400° - 360° = 40°$$
$$800° = 800° - 2(360°) = 80°$$

Figure 12.5 shows several examples of how the sine of an angle, written $\sin \theta$, is defined in terms of a right triangle: It is the ratio of the length of the side *opposite* the angle to the length of the hypotenuse of the triangle. Note that the length of the hypotenuse is always considered positive. The sine of any angle can be found using a "scientific"-type calculator by entering the value of the angle and then pressing the "sin" key. An important property of the sine function is

$$\sin (-\theta) = -\sin \theta \qquad (12.1)$$

For example, $\sin (-30°) = -0.5 = -\sin 30°$. As an exercise, verify this fact using a calculator.

Imagine a line rotating in a counterclockwise direction, so that its tip traces out a circle, as illustrated in Figure 12.6. The line is the radius of the circle, and as it rotates completely around the circle, the angle it generates between itself and the positive x-axis

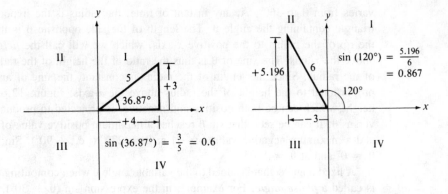

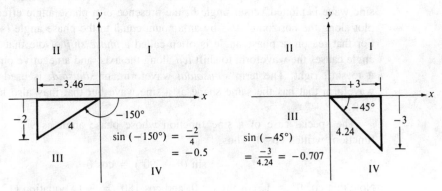

FIGURE 12.5 Examples of sine computations in each quadrant.

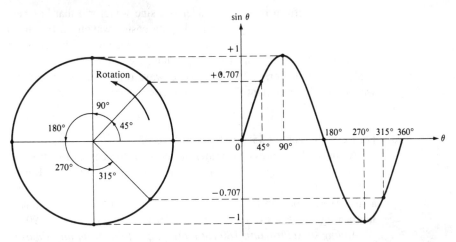

FIGURE 12.6 The projection of a rotating radius produces a plot of sin θ versus θ.

varies from 0 to 360°. At any instant of time, the radius is the hypotenuse of a right triangle containing the angle θ. The length of the side opposite θ is the distance from the tip of the radius to the positive x-axis, which we will call the *height* of the radius above the x-axis. The sine of θ is thus the ratio of the height of the radius to the length of the radius. Since the length of the radius is constant, the sine of angle θ is directly proportional to the height of the radius above the x-axis. Figure 12.6 shows how the height, which can be both positive and negative, is projected to produce a plot of $\sin \theta$ versus θ. It can be seen that $\sin \theta$ reaches a maximum positive value of $+1$ at $\theta = 90°$ and a maximum negative value of -1 at $\theta = 270°$ (i.e., $-90°$). Sin θ equals zero at $\theta = 0°$ and at $\theta = 180°$.

A fixed angle ϕ that is added to the variable angle θ when computing the sine function is called a *phase angle*. For example, in the expression $\sin (\theta + 30°)$, the phase angle is 30°, and in the expression $\sin (\theta - 60°)$, the phase angle is $-60°$. Note that the function $\sin (\theta + 30°)$ equals 0.5 when $\theta = 0°$ and equals 1.0 when $\theta = 60°$. When a sine wave is plotted versus angle θ, the presence of a phase angle effectively *shifts* the plot along the horizontal axis by an amount equal to the phase angle (see Figure 12.7). For that reason, a phase angle is often called a *phase shift*. Note that a positive phase angle causes the waveform to shift *left* along the axis, and a negative phase angle causes it to shift right. The term *sinusoidal* waveform, or *sinusoid*, is used to describe any waveform that has the same shape as a sine wave but that may also have some phase shift.

The special case of a sine function whose phase angle is 90° is called a *cosine* function, written $\cos \theta$. Thus,

$$\sin (\theta + 90°) = \cos \theta \tag{12.2}$$

Note that $\cos 0° = 1$, $\cos 90° = 0$, and $\cos 180° = -1$. Equation (12.2) is equivalent to

$$\cos (\theta - 90°) = \sin \theta \tag{12.3}$$

A cosine waveform may be considered a sine waveform that has been shifted left 90°, and a sine waveform may be considered a cosine waveform that has been shifted right 90°. In a right triangle, the cosine of an angle is the ratio of the length of the side *adjacent* to the angle to the length of the hypotenuse [see Figure 12.8(a)]. Figure 12.8(b) shows a plot of $\cos \theta$ versus θ. The cosine waveform is another example of a sinusoidal waveform. An important fact about the cosine function is that

$$\cos (-\theta) = \cos \theta \tag{12.4}$$

For example, $\cos (-45°) = 0.707 = \cos 45°$.

A sinusoidal expression written in terms of the cosine function can be converted to an equivalent expression written in terms of the sine function: Replace *cos* by *sin* and add 90°. For example, $\cos (\theta - 60°) = \sin (\theta - 60° + 90°) = \sin (\theta + 30°)$. Similarly, a sine expression can be converted to a cosine expression by replacing *sin* by *cos* and subtracting 90°. For example, $\sin (\theta + 50°) = \cos (\theta + 50° - 90°) = \cos (\theta - 40°)$.

Adding or subtracting 180° of phase angle to or from any sinusoidal function is the same as multiplying that function by -1:

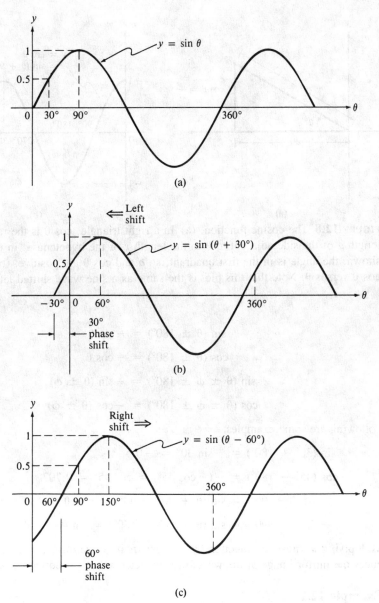

FIGURE 12.7 Effect of phase angle on the sine waveform. (a) Sine wave with zero phase shift. (b) A positive phase angle causes the sine wave to shift left along the horizontal axis. Note that $y = 0$ when $\theta = -30°$. (c) A negative phase angle causes the sine wave to shift right along the horizontal axis. Note that $y = 0$ when $\theta = 60°$.

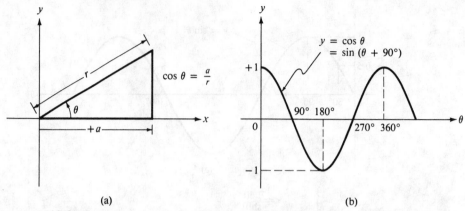

(a)   (b)

FIGURE 12.8 The cosine function. (a) In a right triangle, $\cos\theta$ is the ratio of the length a of the side adjacent to θ to the length r of the hypotenuse. In the example shown, the angle is in the first quadrant, so a and $\cos\theta$ are positive. (b) A plot of $\cos\theta$ versus θ. Note that this plot is the same as a sine wave shifted left by 90°.

$$\sin(\theta \pm 180°) = -\sin\theta \qquad (12.5)$$

$$\cos(\theta \pm 180°) = -\cos\theta \qquad (12.6)$$

$$\sin(\theta \pm \phi \pm 180°) = -\sin(\theta \pm \phi) \qquad (12.7)$$

$$\cos(\theta \pm \phi \pm 180°) = -\cos(\theta \pm \phi) \qquad (12.8)$$

Following are some examples:

$$\sin(30° + 180°) = -\sin 30° = -0.5$$

$$-\cos(45° - 180°) = -(-\cos 45°) = \cos 45° = 0.707$$

$$\cos(\theta + 90°) = \sin(\theta + 90° + 90°) = \sin(\theta + 180°) = -\sin\theta$$

$$-\sin(\theta - 90°) = \sin(\theta - 90° + 180°) = \sin(\theta + 90°) = \cos\theta$$

Multiplying a sinusoidal function by -1 *inverts* its waveform, in the sense that it produces the mirror image of the waveform, reflected about the horizontal axis.

Example 12.1

Sketch each of the following functions versus θ. Label the values of θ where each function equals zero, $+1$, and -1.

(a) $-\sin(\theta + 60°)$
(b) $-\cos(\theta - 30°)$

SOLUTION

(a) Figure 12.9(a) shows a plot of the function $y = \sin(\theta + 60°)$. Its mirror image, $y = -\sin(\theta + 60°)$, is sketched in Figure 12.9(b).

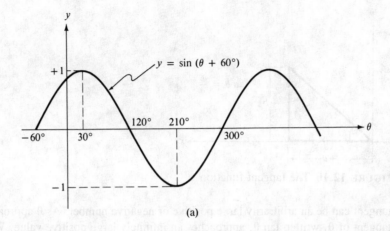

(a)

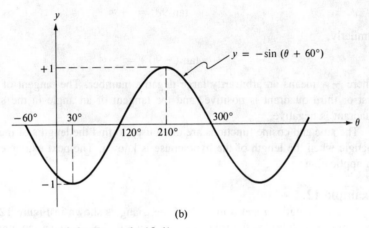

(b)

FIGURE 12.9 (Example 12.1)

(b) Using the fact that $\cos \theta = \sin (\theta + 90°)$, we can write

$$-\cos (\theta - 30°) = -\sin (\theta + 90° - 30°) = -\sin (\theta + 60°)$$

Thus, (b) is the same function as (a), and has the same plot.

Drill Exercise 12.1

If $y = -\sin (\theta - 20°)$, find a positive value of θ at which (a) $y = 0$; (b) $y = 1$; (c) $y = -1$.

ANSWER: (a) 20°; (b) 290°; (c) 110°.   ☐

The *tangent* of an angle θ in a right triangle is the ratio of the length of the side opposite the angle to the length of the side adjacent to the angle (see Figure 12.10). Unlike the sine and cosine functions, whose magnitudes never exceed 1, the value of a

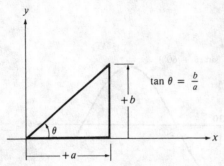

FIGURE 12.10 The tangent function.

tangent can be an arbitrarily large positive or negative number. As θ approaches 90°, the tangent of θ, written tan θ, approaches an infinitely large positive value. We symbolize this fact by writing

$$\tan 90° = +\infty \tag{12.9}$$

Similarly,

$$\tan(-90°) = -\infty \tag{12.10}$$

where $-\infty$ means an arbitrarily large negative number. The tangent of an angle in the first or third quadrant is positive, and the tangent of an angle in the second or fourth quadrant is negative.

The sine and cosine functions are often used to find the lengths of the sides of a right triangle when the length of the hypotenuse is known. The next example illustrates such an application.

Example 12.2

Find the lengths a and b in each of the triangles shown in Figure 12.11. (The units are arbitrary.) Note that the triangles are shown in the x-y plane, so lengths should be expressed as positive or negative values, depending on their location in the plane (x-values positive to the right and y-values positive upward).

SOLUTION In each of the triangles shown in Figure 12.11,

$$\sin \theta = \frac{b}{r} \quad \text{and} \quad \cos \theta = \frac{a}{r}$$

Thus,

$$b = r \sin \theta \quad \text{and} \quad a = r \cos \theta \tag{12.11}$$

(a) $b = 25 \sin 75° = 24.15$
     $a = 25 \cos 75° = 6.47$
(b) $b = 10 \sin 140° = 6.43$
     $a = 10 \cos 140° = -7.66$
(c) $b = \sqrt{2} \sin(-135°) = -1$
     $a = \sqrt{2} \cos(-135°) = -1$

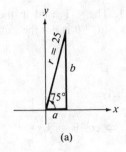

(a)

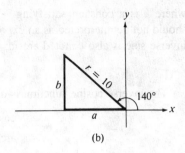

(b)

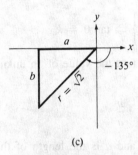

(c)

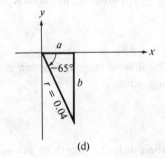

(d)

FIGURE 12.11 (Example 12.2)

(d) $b = 0.04 \sin(-65°) = -0.036$
$a = 0.04 \cos(-65°) = 0.017$

Notice that the algebraic signs of a and b in each case agree with the locations of the sides in the x-y plane, as shown in Figure 12.11.

Drill Exercise 12.2
(a) The side opposite a 20° angle has length 0.15. What is the length of the hypotenuse?
(b) The side adjacent to a 130° angle has length -4.2×10^3. What is the length of the hypotenuse?

ANSWER: (a) 0.4386; (b) 6.534×10^3. □

Inverse Trigonometric Functions

The concept of an inverse trigonometric function is best introduced by way of an example: We know that the sine of 30° is 0.5; another way of stating the same fact is to say that "the angle whose sine equals 0.5 is 30°." Still another way of expressing this fact is to say that *the inverse sine of 0.5 is 30°*. We denote the inverse sine by $\sin^{-1}$ and write

$$\sin^{-1} 0.5 = 30°$$

In general, the inverse sine function is the function that has the following property:

$$\sin^{-1} k = \theta \Rightarrow \sin \theta = k \qquad (12.12)$$

where k is a constant satisfying $-1 \le k \le 1$. The -1 used in the inverse notation should not be interpreted as an exponent; that is, $\sin^{-1} \theta$ is *not* the same as $1/\sin \theta$. The inverse sine is also denoted arcsin:

$$\arcsin k = \theta \tag{12.13}$$

The inverse cosine function is defined similarly:

$$\left. \begin{array}{l} \cos^{-1} k = \theta \\ \text{or} \quad \arccos k = \theta \end{array} \right\} \Rightarrow \cos \theta = k \tag{12.14}$$

where $-1 \le k \le 1$. Note that the inverse sine and cosine functions are not defined when $|k| > 1$. However, the inverse tangent function is defined for all values of k:

$$\left. \begin{array}{l} \tan^{-1} k = \theta \\ \text{or} \quad \arctan k = \theta \end{array} \right\} \Rightarrow \tan \theta = k \tag{12.15}$$

The inverse tangent function is often used to find the value of an unknown angle in a right triangle:

$$\theta = \tan^{-1} \frac{b}{a} \tag{12.16}$$

where b is the length of the side opposite θ and a is the length of the side adjacent to θ.

The value of an inverse trigonometric function can be computed using a scientific calculator: Enter the number, press the "inv" or "arc" key, then press the sin, cos, or tan key. Great care must be exercised when using a calculator to compute an inverse function, because there is no way to enter the quadrant in which the angle is located. Consider, for example, the problem of finding $\tan^{-1} (3/-2)$. The latter is a second-quadrant angle, because the side opposite the angle is positive ($+3$), and the side adjacent to the angle is negative (-2). However, when $\tan^{-1} (3/-2)$ is computed with a calculator, the result is the fourth-quadrant angle $-56.3°$. The reason for this erroneous result is that $(-3/2) = (3/-2) = -1.5$, and the calculator produces the angle corresponding to $(-3/2)$, rather than that corresponding to $(3/-2)$. A sketch should always be drawn to accompany the computation of an inverse trigonometric function, so that the quadrant of the angle is discernible and so that the result produced by a calculator can be corrected if necessary The next example demonstrates the procedure.

Example 12.3

Find the angle θ in each of the triangles shown in Figure 12.12.

SOLUTION
(a) $\theta = \tan^{-1} (3/4) = \tan^{-1} 0.75 = 36.87°$. This result is a first-quadrant angle, and examination of Figure 12.2(a) confirms that fact.
(b) Using a calculator to compute $\theta = \tan^{-1} (4/-1) = \tan^{-1} (-4)$ produces the result $\theta = -75.96°$, a fourth-quadrant angle. However, examination of Figure 12.12(b) clearly shows that θ is a second-quadrant angle. One way to compute the angle correctly is to find the angle ϕ shown in Figure 12.13(a), using the *absolute*

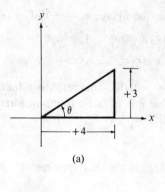

(a)

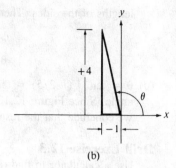

(b)

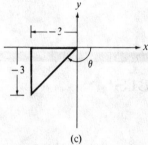

(c)

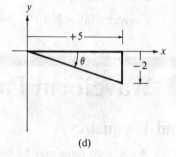

(d)

FIGURE 12.12 (Example 12.3)

values of the lengths of the sides. Then, as shown in the figure, $\theta = 180° - \phi$:

$$\phi = \tan^{-1}(4/1) = 75.96°$$

$$\theta = 180° - \phi = 180° - 75.96° = 104.04°$$

(c) Using a calculator to compute $\theta = \tan^{-1}(-3/-3) = \tan^{-1} 1$ produces the result 45°, a first-quadrant angle. However, examination of Figure 12.12(c) clearly shows that θ is a third-quadrant angle. One way to compute the angle correctly is to find the angle ϕ shown in Figure 12.13(b), using the absolute values of the

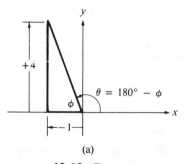

(a)

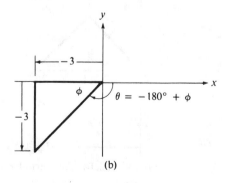

(b)

FIGURE 12.13 (Example 12.3)

lengths of the sides. Then, as shown in the figure, $\theta = -180° + \phi$:

$$\phi = \tan^{-1}(3/3) = \tan^{-1} 1 = 45°$$

$$\theta = -180° + 45° = -135°$$

(d) $\theta = \tan^{-1}(-2/5) = \tan^{-1} -0.4 = -21.8°$. This result is a fourth-quadrant angle. Since Figure 12.12(d) shows that θ is in fact a fourth-quadrant angle, we conclude that the result is correct.

Drill Exercise 12.3

Use a calculator to find (a) $\sin^{-1}(-4/6)$ in the third quadrant; (b) $\cos^{-1}(-2/7)$ in the second quadrant.

ANSWER: (a) $-138.19°$; (b) $106.6°$.   □

12.3 Waveform Parameters

Period and Frequency

An ac waveform may be considered to exist for all time. However, it is convenient to define a reference time, $t = 0$, at which the plot of the waveform begins. Among other practical uses, choosing a time reference allows us to write a mathematical expression for the waveform. (This choice does not preclude the existence of negative time: values of t lying to the left of $t = 0$ on the plot.)

A *periodic* ac waveform is one whose values are repeated at regular intervals. Virtually all ac waveforms of practical interest are periodic. The plot of a periodic waveform has the appearance of a regularly recurring pattern of values, each of which is called a *cycle*. The time required for the values to rise and fall through one complete cycle is called the *period T* of the waveform. Figure 12.14 shows an example. Note that the period can be measured between any two corresponding points on successive cycles.

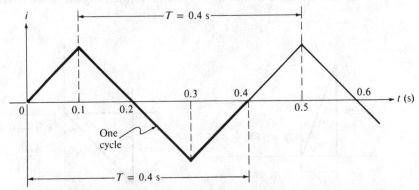

FIGURE 12.14 The period T of a periodic waveform is the time occupied by one complete cycle.

The *frequency f* of an ac waveform is the number of cycles that occur in 1 s. Clearly, the shorter the period (the smaller the time required for one cycle), the more cycles there are in 1 s. In other words, frequency is inversely proportional to period:

$$f = \frac{1}{T} \tag{12.17}$$

The units of f that best describe its nature are *cycles per second*, but the accepted SI unit is the *hertz* (Hz), named after the nineteenth-century German physicist Heinrich Hertz:

$$1 \text{ Hz} = 1 \text{ cycle/second}$$

A cycle is considered a dimensionless quantity, so the dimension of frequency is the same as $1/\text{time}$ (s^{-1}).

Example 12.4
(a) An ac waveform has a period of 2 ms. What is its frequency?
(b) What is the period of an ac waveform whose frequency is 4 MHz?
(c) 12 cycles of an ac waveform occur in 0.48 s. What is the period of the waveform?

SOLUTION

(a) $f = \dfrac{1}{T} = \dfrac{1}{2 \times 10^{-3} \text{ s}} = 500 \text{ Hz}$

(b) $T = \dfrac{1}{f} = \dfrac{1}{4 \times 10^6 \text{ Hz}} = 0.25 \text{ } \mu s$

(c) Since 12 cycles occur in 0.48 s, the frequency is

$$\frac{12 \text{ cycles}}{0.48 \text{ s}} = 25 \text{ Hz}$$

Thus,

$$T = \frac{1}{f} = \frac{1}{25} = 0.04 \text{ s}$$

Drill Exercise 12.4
(a) An ac waveform has a frequency of 16 kHz. How much time is required for 200 cycles to occur?
(b) Find the period and the frequency of the waveform in (i) Figure 12.1(b); (ii) Figure 12.2(a); (iii) Figure 12.2(c); (iv) Figure 12.14.

ANSWER: (a) 12.5 ms; (b) (i) $T = 2$ s, $f = 0.5$ Hz; (ii) $T = 0.4$ ms, $f = 2.5$ kHz; (iii) $T = 50$ μs, $f = 20$ kHz; (iv) $T = 0.4$ s, f = 2.5 Hz. $\square$

Example 12.5
Find the frequency of the ac voltage shown in Figure 12.15.

SOLUTION The period is determined by finding the total time between corresponding points on two successive cycles. For example, measuring between the points where the cycles first go positive (at $t = 0.1$ ms and at $t = 0.9$ ms), we find $T = 0.9$ ms $- 0.1$ ms $= 0.8$ ms. Alternatively, we could measure between the points where the cycles go from positive to negative (at $t = 0.4$ ms and at $t = 1.2$ ms), and again find $T = 1.2$ ms $- 0.4$ ms $= 0.8$ ms. Thus,

$$f = \frac{1}{T} = \frac{1}{0.8 \times 10^{-3} \text{ s}} = 1.25 \text{ kHz}$$

Drill Exercise 12.5

Assuming the first cycle of the waveform in Figure 12.15 goes positive at the time shown (0.1 ms), at what instant of time would it first go negative if the frequency were increased to 10 kHz?

ANSWER: 0.1375 ms. □

Radians and Angular Frequency

The *radian* (rad) is the SI unit of angle. It is related to degrees by

$$2\pi \text{ radians} = 360° \tag{12.18}$$

Angle in radians can be converted to angle in degrees, and vice versa, by using the methods of dimensional analysis described in Chapter 1.

Example 12.6

(a) Convert 45° to radians.
(b) Convert 1 rad to degrees.
(c) Convert -1.5π rad to degrees.

SOLUTION

(a) $(45°)\left(\dfrac{2\pi \text{ rad}}{360°}\right) = \dfrac{\pi}{4} \text{ rad} = 0.7854 \text{ rad}$

(b) $(1 \text{ rad})\left(\dfrac{360°}{2\pi \text{ rad}}\right) = 57.296°$

(c) $(-1.5\pi \text{ rad})\left(\dfrac{360°}{2\pi \text{ rad}}\right) = -270° = 90°$

Drill Exercise 12.6

Convert (a) 90° to radians; (b) 3π rad to degrees.

ANSWER: (a) $\pi/2$ rad $= 1.571$ rad; (b) $540° = 180°$. □

Figure 12.16 shows a plot of the sine function versus angle in radians. Note that the function reaches its maximum positive value at $\theta = \pi/2$ rad and its maximum negative value at $\theta = 3\pi/2$ rad. The function equals 0 at 0, π, and 2π rad. We showed earlier (Figure 12.6) that the sine waveform is generated by a rotating radius (sometimes called a radius vector). The number of cyles generated by that radius in 1 s depends on how

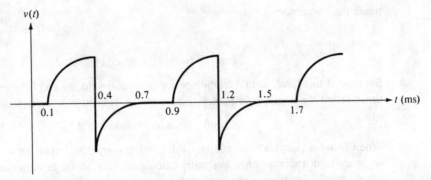

FIGURE 12.15 (Example 12.5)

fast the radius is rotating. Clearly, a rapidly rotating radius will produce more cycles per second than a slowly rotating radius. The speed of rotation is measured by the amount of angle the radius sweeps through in a given amount of time. This speed is called *angular velocity* and is designated by ω:

$$\omega = \frac{\theta}{t} \quad \text{rad/s} \tag{12.19}$$

where θ is the number of radians of angle generated in t seconds. From (12.19) we have

$$\theta = \omega t \quad \text{rad} \tag{12.20}$$

Therefore, a sine wave can be expressed as a function of time by writing

$$\sin \theta = \sin \omega t \tag{12.21}$$

In Figure 12.16 it is clear that one cycle is the same as 2π radians. Since frequency is the number of cycles produced in 1 s, the number of radians produced in 1 s is 2π

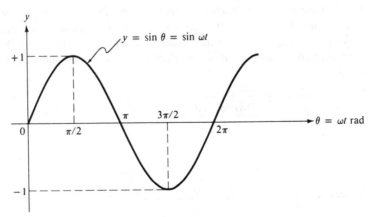

FIGURE 12.16 Plot of the sine function versus $\theta = \omega t$ radians.

times the frequency:

$$\omega = 2\pi f \quad \text{rad/s} \tag{12.22}$$

$$(\text{rad/s}) = (\text{rad/cycle})(\text{cycles/s})$$

Because of this fundamental relation between angular velocity and frequency, ω is sometimes called the *angular frequency*. Equation (12.21) can now be expressed as

$$\sin \theta = \sin \omega t = \sin (2\pi f)t \tag{12.23}$$

When using a calculator to compute values of $\sin \omega t$, it is important to remember that ωt is angle in radians. Most scientific calculators can be set to compute the values of trigonometric functions when angles are entered in radians.

Example 12.7

(a) Given that $y(t) = \sin 3141.6t$, find
    (i) the angular velocity
    (ii) the frequency
    (iii) the period of the waveform
(b) Sketch $y(t)$ versus t.

SOLUTION

(a) (i) Comparing $y(t) = \sin 3141.6t$ with equation (12.21), we see that
      $\omega = 3141.6$ rad/s.
  (ii) From equation (12.22),

$$f = \frac{\omega}{2\pi} = \frac{3141.6 \text{ rad/s}}{2\pi} = 500 \text{ Hz}$$

  (iii) $T = \dfrac{1}{f} = \dfrac{1}{500 \text{ Hz}} = 2$ ms

(b) A plot of $y(t)$ versus t can be obtained in either one of two ways. First, we note that the period (2 ms) corresponds to one full cycle, so we can start by sketching a sine-wave cycle that occupies a 2-ms time interval. One-half the cycle occupies 1 ms, and the sine wave reaches its maximum positive value at one-fourth of a cycle: 2 ms/4 = 0.5 ms. It reaches its maximum negative value in three-fourths of a cycle: $(\frac{3}{4})(2 \text{ ms}) = 1.5$ ms. Second, we can obtain points on the sine wave by computing the values of $\sin \omega t$ at selected values of t, as shown in the following table:

t (ms)	$\omega t = 3141.6t$ (rad)	$\sin \omega t$
0	0	0
0.5	1.5708	1.0
1.0	3.1416	0
1.5	4.7124	−1.0
2.0	6.2832	0

The plot is shown in Figure 12.17.

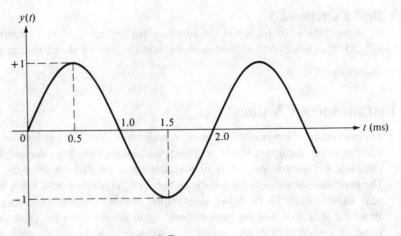

FIGURE 12.17 (Example 12.7)

Drill Exercise 12.7

What is the period of the sine wave $y(t) = \sin(2\pi \times 10^6 t)$?

ANSWER: 1 µs.   □

Example 12.8

Find the angular frequency of the ac current shown in Figure 12.18.

SOLUTION The period can be determined by finding the time between two successive peaks:

$$T = 250 \ \mu s - 50 \ \mu s = 200 \ \mu s$$

Then

$$f = \frac{1}{T} = \frac{1}{200 \times 10^{-6} \text{ s}} = 5 \text{ kHz}$$

By equation (12.22),

$$\omega = 2\pi f = 2\pi(5 \times 10^3 \text{ Hz}) = 3.1416 \times 10^4 \text{ rad/s}$$

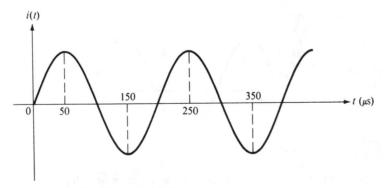

FIGURE 12.18 (Example 12.8)

Drill Exercise 12.8

If the angular frequency of the sine wave in Figure 12.18 is changed to 1.2×10^5 rad/s, at what point in time does the first cycle reach its maximum positive value?

ANSWER: 13.09 μs.   ☐

Peak and Instantaneous Values

The maximum value reached by an ac waveform is called its *peak value*. Depending on whether or not the waveform is symmetrical about the time axis, the magnitude (absolute value) of the positive peak may or may not equal the magnitude of the negative peak. The *peak-to-peak value* is the difference between the positive peak value and the negative peak value. Figure 12.19 shows an example. In this case, the waveform is symmetrical about the time axis and the peak-to-peak value equals twice the peak value. The peak value of a waveform is also called its *amplitude*, but the term "peak value" is more descriptive, and we will use it hereafter. To distinguish between a peak and a peak-to-peak value, we write *pk* or *p-p* after the value. For example, the peak value in Figure 12.19 is 3 V pk and the peak-to-peak value is 6 V p-p.

The mathematical expressions we have presented for sinusoidal waveforms up to now have all had a peak value of 1, because the maximum value of $\sin \theta = \sin \omega t$, is 1. In general, an ac voltage or current can have any peak value. If the sine function is multiplied by a fixed constant, all values of the function at all instants of time are multiplied by that constant. In particular, at the instant of time where the sine reaches its maximum value of $+1$, the product of the constant and the sine equals the constant. Thus, the multiplying constant is the peak value, and we can write

$$v(t) = V_p \sin \omega t$$

$$i(t) = I_p \sin \omega t$$

(12.24)

where V_p is the peak value of $v(t)$ and I_p is the peak value of $i(t)$. As an example, the waveform shown in Figure 12.19 has peak value $V_p = 3$ V and is therefore written

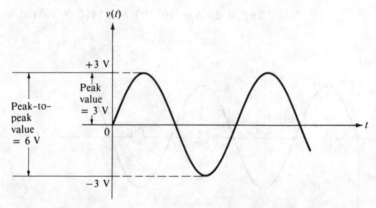

FIGURE 12.19 Peak and peak-to-peak values of an ac voltage.

$v(t) = 3 \sin \omega t$. Note that $v(t) = -3$ V when $\sin \omega t = -1$ (at $\omega t = 1.5$ rad, or $270°$). *Lowercase* letters (i and v) are used to designate ac quantities. Capital letters are used to represent dc quantities and specific constants, such as V_p and I_p.

The *instantaneous* value of an ac waveform is its value at a specific instant of time. For example, the instantaneous value of $i(t) = 3 \sin 100t$ amperes at $t = 2$ ms is $i(2 \text{ ms}) = 3 \sin (100 \times 2 \times 10^{-3}) = 3 \sin (0.2 \text{ rad}) = (3)(0.1987) = 0.596$ A.

Example 12.9

Find the instantaneous value of $v(t) = 5 \sin (2\pi \times 10^4 t)$ volts at

(a) 0 s
(b) 25 µs
(c) 65 µs

SOLUTION

(a) $v(0) = 5 \sin (2\pi \times 10^4 \times 0) = 5 \sin (0) = 5(0) = 0$ V
(b) $v(25 \times 10^{-6}) = 5 \sin (2\pi \times 10^4 \times 25 \times 10^{-6}) = 5 \sin (1.57 \text{ rad}) = 5(1) = 5$ V
(c) $v(65 \times 10^{-6}) = 5 \sin (2\pi \times 10^4 \times 65 \times 10^{-6}) = 5 \sin (4.084 \text{ rad}) = 5(-0.809) = -4.045$ V

Drill Exercise 12.9

Find the value of t at which the instantaneous value of $i(t) = 0.6 \sin (400t)$ mA equals its peak value.

ANSWER: 3.927 ms. □

Example 12.10

Write the mathematical expression for a 50-Hz sinusoidal voltage having peak value 80 V. Sketch the waveform versus t.

SOLUTION

$$\omega = 2\pi f = (2\pi)(50 \text{ Hz}) = 314.16 \text{ rad/s}$$

Since $V_p = 80$ V, we have

$$v(t) = V_p \sin \omega t = 80 \sin (314.16t) \text{ V}$$

To sketch $v(t)$, we first find the period:

$$T = \frac{1}{f} = \frac{1}{50} \text{ Hz} = 20 \text{ ms}$$

Since the period corresponds to $360°$ and the waveform reaches its peak positive value at $90°$, the instant of time at which the peak occurs is $(90/360)(20 \text{ ms}) = 5$ ms. Similarly, the waveform passes through zero at $(180/360)(20 \text{ ms}) = 10$ ms, and it reaches its peak negative value at $(270/360)(20 \text{ ms}) = 15$ ms. The waveform is shown in Figure 12.20.

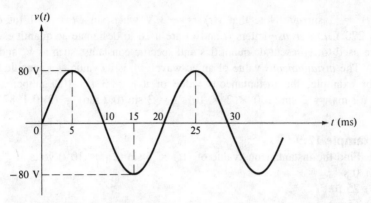

FIGURE 12.20 (Example 12.10)

Drill Exercise 12.10

What is the instantaneous value of $v(t)$ in Example 12.10 at the instant of time corresponding to $\theta = 120°$?

ANSWER: 69.28 V.   □

12.4 Phase Relations

Recall that adding angle ϕ to angle θ in the sine function: $\sin(\theta + \phi)$, causes the sine waveform to shift left or right by ϕ degrees, depending on whether ϕ is positive or negative (see Figure 12.7). Since $\theta = \omega t$, the general form of a sinusoidal voltage or current having phase angle ϕ is

$$v(t) = V_p \sin(\omega t + \phi) \text{ V}$$
$$i(t) = I_p \sin(\omega t + \phi) \text{ A}$$

(12.35)

To be rigorous, ωt and ϕ should have the same units, since ϕ is *added* to ωt. Although ωt is in radians, ϕ is frequently and customarily expressed in degrees. Thus, for example, it is acceptable to write $v(t) = 5 \sin(100 t + 30°)$ V, meaning that $v(t)$ is shifted left by 30°, or $\pi/6$ radians. To compute the instantaneous value of such an expression, it is necessary to convert ωt to degrees or to convert ϕ to radians. The next example illustrates this point.

Example 12.11

Find the instantaneous value of $i(t) = 0.5 \sin(8 \times 10^5 t + 50°)$ A at $t = 0.25 \mu s$.

SOLUTION At $t = 0.25 \mu s$,

$$i(0.25 \times 10^{-6}) = 0.5 \sin(8 \times 10^5 \times 0.25 \times 10^{-6} + 50°) \text{ A}$$
$$= 0.5 \sin(0.2 \text{ rad} + 50°) \text{ A}$$

Converting 0.2 rad to degrees, we find

$$(0.2 \text{ rad})\left(\frac{360°}{2\pi \text{ rad}}\right) = 11.46°$$

Thus,

$$i(0.25 \times 10^{-6}) = 0.5 \sin (11.46° + 50°) \text{ A}$$
$$= 0.5 \sin (61.46°) \text{ A} = 0.439 \text{ A}$$

Drill Exercise 12.11

Find the instantaneous value of $i(t)$ in Example 12.11 at $t = 2$ μs.

ANSWER: 0.31 A.
□

When two waveforms have different phase angles, the one shifted farthest to left is said to *lead* the other. For example, the voltage $v_1(t) = 6 \sin (\omega t + 50°)$ leads the voltage $v_2(t) = 0.1 \sin (\omega t + 20°)$ because v_1 is shifted left 50°, while v_2 is shifted left 20°. Since v_1 has a phase shift 30° greater than that of v_2, we say that v_1 leads v_2 by 30°. It is equivalent to say that v_2 *lags* v_1 by 30°. The lead-lag terminology is derived from observation of the relative positions of the waveforms when they are plotted versus time. The waveform having the greater positive phase reaches its peak value *first* (earliest in time) and therefore is "ahead of," or leads, the other. Figure 12.21 shows an example. In this case, one waveform is an ac voltage shifted left by 30° and the other waveform is an ac current shifted right by 45°:

$$v(t) = 10 \sin (\omega t + 30°) \text{ V}$$

$$i(t) = 10 \sin (\omega t - 45°) \text{ A}$$

The ac voltage $v(t)$ lies 75° to the left of $i(t)$, so $v(t)$ leads $i(t)$ by 75°. Equivalently, $i(t)$ lags $v(t)$ by 75°. Note that $v(t)$ reaches its peak value 75° ahead of $i(t)$. Phase comparisons

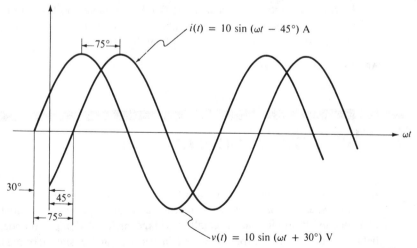

FIGURE 12.21 $v(t)$ leads $i(t)$ by 75°. Equivalently, $i(t)$ lags $v(t)$ by 75°.

of this type can be made only if the waveforms have identical frequencies, for otherwise, phase is not defined.

Example 12.12

In each of the following pairs, determine which waveform is leading and find the total angle by which it leads.

(a) $i_1 = 75 \sin(\omega t - 18°)$ A
$i_2 = 3 \sin(\omega t - 31°)$ A

(b) $v = 4 \times 10^{-3} \sin\left(10^6 t + \dfrac{\pi}{3} \text{ rad}\right)$ V

$i = 16 \sin\left(10^6 t + \dfrac{\pi}{4} \text{ rad}\right)$ A

(c) $v_1 = 27 \cos(\omega t - 40°)$ V
$v_2 = 9 \sin(\omega t + 10°)$ V

SOLUTION

(a) i_1 is shifted right 18° and i_2 is shifted right 31°. Therefore, i_1 lies $31° - 18° = 13°$ to the left of i_2. Thus i_1 leads i_2 by 13°.

(b) v is shifted left $\pi/3$ rad $= 60°$, and i is shifted left $\pi/4$ rad $= 45°$. Therefore, v lies $60° - 45° = 15°$ to the left of i. Thus, v leads i by 15°, or $\pi/12$ radians.

(c) To compare phase, both waveforms should be expressed as sine functions, or both as cosine functions. Converting v_1 to a sine function, we find

$$v_1 = 27 \cos(\omega t - 40°) = 27 \sin(\omega t + 90° - 40°)$$
$$= 27 \sin(\omega t + 50°) \text{ V}$$

Thus v_1 lies $50° - 10° = 40°$ to the left of v_2, so v_1 leads v_2 by 40°.

Drill Exercise 12.12

A certain ac voltage has peak value 0.1 V, frequency 60 Hz, and phase shift $-15°$. Write the mathematical expression for an ac current that has peak value 24 mA, frequency 60 Hz, and leads the voltage by 70°.

ANSWER: $24 \sin(377t + 55°)$ mA. ☐

12.5 The Oscilloscope

The *oscilloscope* is one of the most versatile and widely used instruments found in modern laboratories. It can be used to measure the frequency, phase, and peak value of ac waveforms, as well as dc values, transients, and many special characteristics of electronic components. It provides an especially vivid means for studying and understanding circuit behavior, and is an indispensable tool in research and design activities, as well as for troubleshooting complex systems.

The many practical uses of an oscilloscope are derived from its ability to display a time-varying waveform, that is, a plot, or graph, of voltage variations versus time. The display is created by a beam of electrons in a *cathode ray tube* (CRT). When the beam strikes a phosphorescent screen in the CRT, light is emitted, and the path of the beam as it moves across the screen, called the *trace*, is clearly visible. (This is the same principle by which images are created on a television screen and on the video display of a computer terminal.) Electronic circuitry in the oscilloscope causes the beam to *sweep* across the screen from left to right, so the horizontal position of the beam is proportional to time. The waveform to be displayed causes the beam to move up and down as it is being swept from left to right, so the vertical position of the trace at any instant of time is proportional to the instantaneous value of the waveform at that time. Causing the beam to move is called *deflecting* it. The horizontal deflection, proportional to time, and the simultaneous vertical deflection, proportional to the value of the waveform, create a waveform plot versus time. The principle is illustrated in Figure 12.22.

Synchronization and Triggering

In order for the display in Figure 12.22 to have the appearance shown, the start of the horizontal sweep (at its far left position) must coincide with the beginning of the wave-form cycle, and the sweep must reach the far right side at the instant the cycle ends. The sweep then *repeats* itself, starting again at the instant a new cycle of the waveform appears. Each new cycle of the waveform is displayed in this manner, one after the other. Since all cycles are identical, the image of each is superimposed on the image of its predecessor, and the display has the appearance of a single, stationary cycle. For this to occur, it is clear that the frequency of the sweep must be identical to the frequency

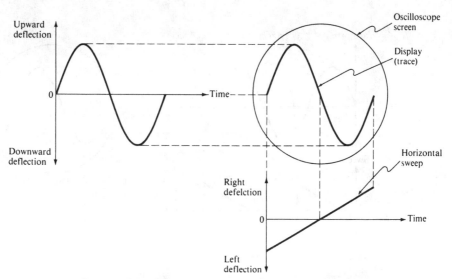

FIGURE 12.22 Principle of oscilloscope operation. As the beam is deflected from left to right across the screen by the horizontal sweep, it is simultaneously deflected up and down by the waveform.

of the waveform. Maintaining the sweep frequency equal to the waveform frequency is called *synchronization* and is accomplished by internal circuitry of the oscilloscope. The circuitry causes a new sweep cycle to be launched at the beginning of every new waveform cycle. This action is called *triggering* the sweep.

By adjusting front-panel controls on the oscilloscope, the sweep can be made to trigger at different times during a waveform cycle. Figure 12.23 shows two examples. In Figure 12.23(a), the sweep is triggered every time the waveform reaches its positive peak. In Figure 12.23(b), the sweep is triggered at a point on the negative portion of the

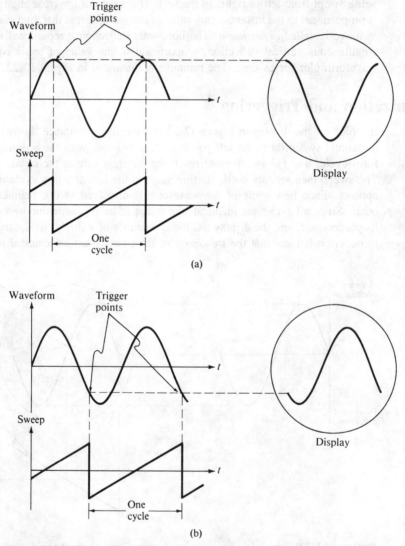

(a)

(b)

FIGURE 12.23 Operation of the triggered sweep. (a) Triggering the sweep at the positive peaks of the waveform creates a display that begins and ends at the positive peaks. (b) Triggering the sweep at a negative point on the waveform creates a display that begins and ends at those points.

cycle. In each case, a full cycle is displayed, because each begins and ends at the same point on successive cycles.

If the frequency of the waveform to be displayed is exactly twice the frequency of the sweep, one sweep cycle will correspond to two waveform cycles. Consequently, two cycles of the waveform will be displayed on the screen, as illustrated in Figure 12.24(a). Similarly, when the waveform frequency is any integer multiple of the sweep frequency, that integer number of cycles will be displayed. Figure 12.24(b) shows another example.

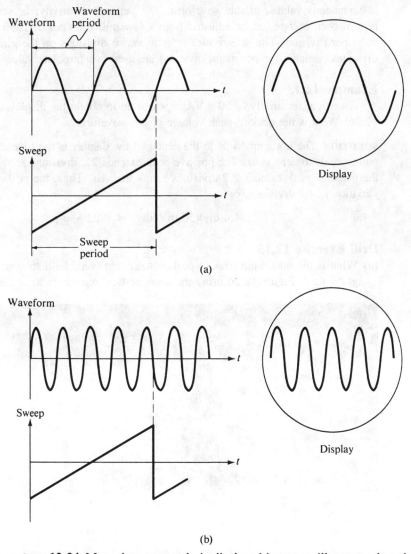

FIGURE 12.24 More than one cycle is displayed by an oscilloscope when the waveform frequency is a multiple of the sweep frequency. (a) Two cycles are displayed when the waveform frequency is exactly twice the sweep frequency. (b) Several cycles are displayed when the waveform frequency is an integer multiple of the sweep frequency.

Waveform Measurements

Figure 12.25 shows a typical laboratory-grade oscilloscope. This instrument, and others like it, have *calibrated* controls that allow the user to adjust the size of the display and to make accurate waveform measurements from it. The screen has a *graticule*, which, like graph paper, has *divisions* that form a measurement grid. Scale factors are set by adjusting front-panel controls. A calibrated control called *vertical sensitivity* adjusts the scale factor along the vertical axis and is therefore used to measure magnitudes, or instantaneous values, of the waveform. The vertical sensitivity, or scale factor, on a typical oscilloscope can be adjusted from a few millivolts per division to 100 or more volts per division. The sensitivities refer to *major* divisions on the graticule, and sub-divisions (usually five per major division) are used to interpolate values.

Example 12.13

A vertical sensitivity of 50 mV/div was used to obtain the display shown in Figure 12.26. What is the peak-to-peak voltage of the waveform?

SOLUTION The horizontal line in the center of the display is the time axis and corresponds to zero volts. The positive peak extends 2.2 divisions above the axis, and the negative peak extends 2.2 divisions below the axis. Thus, the peak-to-peak variation is 4.4 divisions, or

$$(4.4 \text{ div})(50 \text{ mV/div}) = 0.22 \text{ V p-p}$$

Drill Exercise 12.13

(a) What is the maximum peak-to-peak voltage that could be displayed on the graticule in Figure 12.26 using the same vertical sensitivity as in the example?

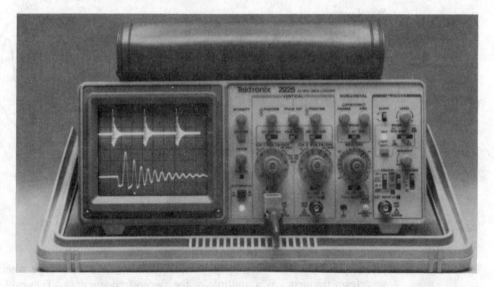

FIGURE 12.25 A typical laboratory-grade oscilloscope. (Courtesy of Tektronix, Inc.)

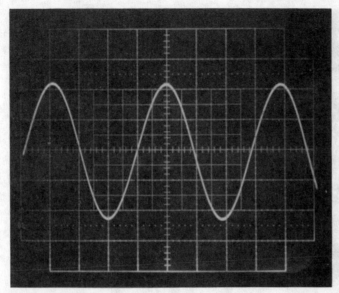

FIGURE 12.26 (Example 12.13)

(b) What would be the peak value of the waveform in Figure 12.26 if the display had been obtained with a vertical sensitivity of 0.2 V/div?

ANSWER: (a) 0.4 V p-p; (b) 0.44 V pk.   □

The horizontal axis of a waveform display represents time, so the horizontal sensitivity, called the *time base*, is the oscilloscope control used to set the scale factor for measuring increments of time. Adjusting the time base alters the frequency of the internally generated sweep. The time base of a typical oscilloscope can be adjusted from less than 1 μs per division to several seconds per division. Since time can be measured with an oscilloscope, we can find the period of a waveform and thereby determine its frequency. Most laboratory-grade oscilloscopes can display two waveforms simultaneously and are therefore called *dual-trace* oscilloscopes. By measuring the time interval between corresponding points on two waveforms, we can determine the phase shift between them. The next example demonstrates how frequency and phase measurements are made from an oscilloscope display.

Example 12.14

The waveforms shown in Figure 12.27 were obtained on a dual-trace oscilloscope using a time base of 0.1 ms/div.

(a) What is the frequency of the waveforms?
(b) What is the phase angle between them?

SOLUTION

(a) Both waveforms have the same frequency; we will use the one having the greater peak value to determine the period. Measuring between the points where the

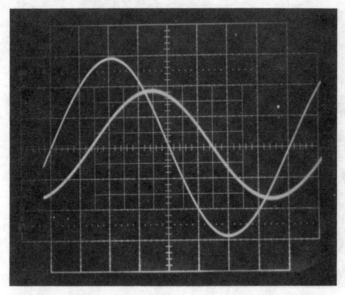

FIGURE 12.27 (Example 12.14)

sine wave crosses the time axis going in a positive direction (i.e., over one full cycle), we see that the period occupies 8 divisions along the axis. Therefore,

$$T = (8\ \text{div})(0.1\ \text{ms/div}) = 0.8\ \text{ms}$$

$$f = \frac{1}{T} = \frac{1}{0.8 \times 10^{-3}\ \text{s}} = 1.25\ \text{kHz}$$

(b) To determine the phase angle, we must measure the time interval between *corresponding* points on two cycles. A convenient pair of such points is where each waveform crosses the time axis going in a negative direction. From the display, we see that this interval is 1.5 divisions, or

$$(1.5\ \text{div})(0.1\ \text{ms/div}) = 0.15\ \text{ms}$$

Since $T = 0.8\ \text{ms} = 360°$, we find the angle corresponding to 0.15 ms by proportionality:

$$\frac{\theta}{360°} = \frac{0.15\ \text{ms}}{0.8\ \text{ms}}$$

$$\theta = \left(\frac{0.15}{0.8}\right) 360° = 67.5°$$

Thus, the waveform having the larger peak value leads the other waveform by 67.5°.

Drill Exercise 12.14

Assume that the display shown in Figure 12.27 was made with a time base of 5 μs/div and a vertical sensitivity of 10 mV/div. If the waveform having the larger peak value is assumed to have zero phase angle, write the mathematical expression for the waveform having the smaller peak value.

ANSWER: 0.019 sin (2π × 2.5 × 10⁴t − 67.5°) V.   □

12.6 Average and Effective Values

Average Value

The average value of a waveform is the average of all its values over a period of time. Computing an average over *time* means adding all the values that occur in a specific time interval and dividing the sum by that time. Given the plot of a waveform, we can sum all its values over a period of time by computing the *area* enclosed by the plot over the time period. In performing such a computation, we regard area above the time axis as positive area and area below the time axis as negative area. The algebraic signs of the areas must be taken into account when computing the total (net) area. The time interval over which the net area is computed is the period T of the waveform. The next example demonstrates these computations.

Example 12.15

Find the average value of the ac voltage whose waveform is shown in Figure 12.28.

SOLUTION One cycle of the waveform extends from $t = 0$ to $t = 0.6$ s, so the period is $T = 0.6$ s. The positive area, A_1, enclosed by the waveform during the first cycle lies between $t = 0$ and $t = 0.2$ s:

$$A_1 = (15 \text{ V})(0.2 \text{ s}) = 3 \text{ V} \cdot \text{s}$$

The negative area lies between $t = 0.2$ s and 0.6 s (i.e., over an interval of 0.4 s):

$$A_2 = (-3 \text{ V})(0.4 \text{ s}) = -1.2 \text{ V} \cdot \text{s}$$

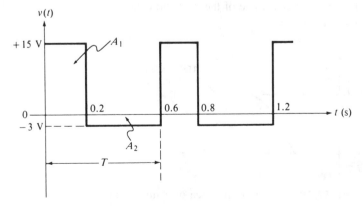

FIGURE 12.28 (Example 12.15)

Thus, the net area over the full period is

$$(3 \text{ V-s}) - (1.2 \text{ V-s}) = 1.8 \text{ V} \cdot \text{s}$$

and the average value of the waveform is

$$V_{avg} = \frac{1.8 \text{ V} \cdot \text{s}}{0.6 \text{ s}} = 3 \text{ V}$$

Drill Exercise 12.15

What would be the average value of the waveform in Example 12.15 if the frequency were doubled?

ANSWER: 3 V.   □

The waveform in Example 12.15 (Figure 12.28) is rectangular, so it is easy to compute its area. Many practical ac waveforms, such as sine waves, are not rectangular, and their areas must be computed using calculus. Note, however, that *any waveform that is symmetrical about the time axis has zero average value*, since its positive area exactly cancels its negative area. In particular, every sinusoidal voltage or current has zero average value. The concept of average value is not restricted to ac waveforms. Every periodic waveform, whether it goes negative or not, has an average value. Some practical waveforms are composed of a sequence of half-cycles of sine waves. To compute the average value of this type of waveform, it is necessary to know the area of a half-cycle of sine wave, also called a sinusoidal *pulse*, as shown in Figure 12.29. Using calculus, it can be shown that the area of such a pulse is

$$\text{area} = \frac{2V_p}{\omega} \text{ V-s} \quad \text{or} \quad \frac{2I_p}{\omega} \text{ A-s} \tag{12.26}$$

where ω is the angular frequency of the sine wave from which the half-cycle is obtained. Of course, the area of a negative half-cycle is simply the negative of the area of a positive half-cycle.

Example 12.16

Find the average value of the waveform shown in Figure 12.30 (called a *half-wave rectified* sine wave).

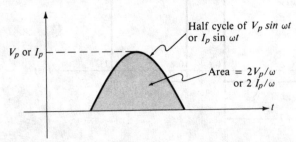

FIGURE 12.29 Computing the area of a sinusoidal pulse.

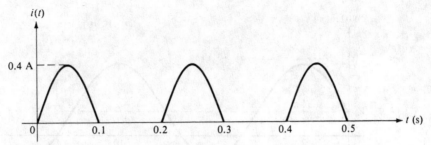

FIGURE 12.30 (Example 12.16)

SOLUTION The half-sine wave pulses shown in Figure 12.30 are derived from a sine wave whose period is $T = 0.2$ s. Therefore,

$$f = \frac{1}{T} = \frac{1}{0.2 \text{ s}} = 5 \text{ Hz}$$

and

$$\omega = 2\pi f = 2\pi(5) = 31.416 \text{ rad/s}$$

From equation (12.26), the area of a single pulse is

$$\text{area} = \frac{2I_p}{\omega} = \frac{2(0.4 \text{ A})}{31.416 \text{ rad/s}} = 25.46 \times 10^{-3} \text{ A} \cdot \text{s}$$

Note that the period of the waveform in Figure 12.30 is 0.2 s (not 0.1 s). Therefore,

$$I_{\text{avg}} = \frac{\text{area}}{T} = \frac{25.46 \times 10^{-3} \text{ A} \cdot \text{s}}{0.2 \text{ s}} = 0.127 \text{ A}$$

Drill Exercise 12.16

What peak current should the pulses in Figure 12.30 have if the waveform is to have an average value of 0.2 A?

ANSWER: 0.628 A.

The average value of a waveform is also called its *dc value*. In fact, when a waveform is measured with a dc instrument (dc ammeter or dc voltmeter), it is the average value of the waveform that is indicated by the instrument. Knowledge of the average value of a waveform is particularly important in the design of high-power devices such as power supplies and power amplifiers. The waveforms encountered in many electronic circuits are regarded as sine waves having *dc levels*, or *offsets*, which are, in fact, the average values of those waveforms. The constant dc level is simply *added* to the ac waveform, so it is easy to identify the *dc component* and the *ac component* of the combination. For example, the equation for an ac voltage having a dc component of V_{dc} volts is

$$v(t) = V_{\text{dc}} + V_p \sin (\omega t + \phi) \tag{12.27}$$

V_{dc} may be either positive or negative. Note that the values of $v(t)$ range from a minimum of $V_{\text{dc}} - V_p$ volts to a maximum of $V_{\text{dc}} + V_p$ volts.

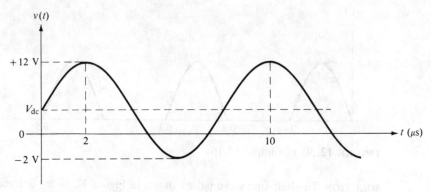

FIGURE 12.31 (Example 12.17)

Example 12.17

(a) Find the average value of the voltage shown in Figure 12.31.

(b) Write the mathematical expression for the voltage as a function of time.

SOLUTION

(a) The peak-to-peak value of the ac component is 12 V − (−2 V) = 14 V. Thus, the peak value of the ac component is $V_p = 7$ V. The maximum positive value is the sum of the dc component and the peak value of the ac component:

$$12 \text{ V} = V_{dc} + V_p = V_{dc} + 7 \text{ V}$$

Thus, the dc, or average, value is

$$V_{dc} = 12 \text{ V} - 7 \text{ V} = 5 \text{ V}$$

(b) The period can be determined from the time interval between the positive peaks:

$$T = 10 \text{ μs} - 2 \text{ μs} = 8 \text{ μs}$$

Thus,

$$f = \frac{1}{T} = \frac{1}{8 \times 10^{-6} \text{ s}} = 125 \text{ kHz}$$

$$v(t) = V_{dc} + V_p \sin \omega t = 5 + 7 \sin (2\pi \times 125 \times 10^3 t) \text{ V}$$

Drill Exercise 12.17

If the dc level in Example 12.17 were changed to −8 V, what would be the minimum and maximum values of the waveform?

ANSWER: −15 V and −1 V.   □

Effective (rms) Values

Average values are not useful for comparing sinusoidal ac waveforms, or other symmetrical waveforms, because the average value of all such waveforms is zero. Instead, a measure called the *effective* or *root-mean-square* (rms) value is used. This value is a measure of how effective the waveform is in producing *heat* in a resistance, that is, the total amount of energy it causes a resistance to dissipate. Heating, or energy dissipation, in a resistor does not depend on the direction of the current through it or on the polarity of the voltage across it. Therefore, the advantage of the rms measure is that it eliminates any consideration of waveform polarity, and the positive portion of a cycle is not canceled by the negative portion, as it is when computing an average value. Figure 12.32 illustrates the concept of effective value. Imagine that the dc voltage in Figure 12.32(b) is adjusted until the heat dissipated by resistor R is exactly the same as the heat dissipated by R in Figure 12.32(a). The two resistors are assumed to have identical resistance. The dc voltage that causes the same heating in R as the ac voltage in Figure 12.32(a) is the effective value of the ac voltage. Similarly, the dc current that causes the same heating in R as the ac current is called the effective value of the ac current.

Using calculus, it can be shown that the effective value of a sinusoidal voltage or current is found from its peak value as follows:

$$V_{\text{eff}} = \frac{\sqrt{2}}{2} V_p \approx 0.707 V_p$$

$$I_{\text{eff}} = \frac{\sqrt{2}}{2} I_p \approx 0.707 I_p$$

(12.28)

Equivalently,

$$V_p = \sqrt{2} \, V_{\text{eff}} \approx 1.414 V_{\text{eff}}$$

$$I_p = \sqrt{2} \, I_{\text{eff}} \approx 1.414 I_{\text{eff}}$$

(12.29)

To indicate that a voltage or current value is an effective value, we write *rms* after the unit. For example, 25 V rms is an effective value of 25 V.

Example 12.18
(a) Find the effective value of the ac current $i(t) = 4.2 \sin(5000t + 45°)$ A.
(b) The ac voltage supplied to convenience outlets in a home has a frequency of 60 Hz and an effective value of 120 V rms. (*Note*: Many years ago, this voltage was 110 V rms, and some people still refer to it, erroneously, as 110 V.) Write the mathematical expression for the voltage at a convenience outlet, assuming zero phase angle.

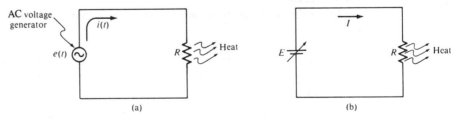

FIGURE 12.32 When E is adjusted so that the heat produced in R is the same as the heat produced by $e(t)$ in R, E equals the effective value of $e(t)$.

SOLUTION

(a) Since $I_p = 4.2$ A,

$$I_{\text{eff}} = \frac{\sqrt{2}}{2} I_p = \frac{\sqrt{2}}{2} (4.2\text{A}) = 2.97 \text{ A rms}$$

(b) $\omega = 2\pi f = 2\pi(60 \text{ Hz}) = 377 \text{ rad/s}$

$$V_p = \sqrt{2} \, V_{\text{eff}} = \sqrt{2} \, (120 \text{ V rms}) = 169.7 \text{ V pk}$$

Therefore,

$$v(t) = 169.7 \sin (377t) \text{ V}$$

Drill Exercise 12.18

Find the peak-to-peak value of an ac voltage whose effective value is 25 V rms.

ANSWER: 70.7 V p-p.   □

It must be emphasized that the relationship between peak and effective values given in equations (12.28) and (12.29) is valid *only* for sinusoidal waveforms. Conventional ac instruments (i.e., ac voltmeters and ac ammeters) are calibrated so that they can be read in either peak or rms units, because one unit is simply a multiple of the other. The calibration of these instruments is based on the multiple $\sqrt{2}/2$ that relates the peak and effective values of sinusoidal waveforms, so the instruments cannot be used to measure nonsinusoidal waveforms. Instead, specially designed instruments, such as peak-reading meters and so-called *true rms* meters must be used.

It is possible to compute the effective value of some nonsinusoidal waveforms by performing the series of calculations specified in the name "root mean square." Actually, the calculations are performed in the reverse order: We first *square* all the values of the waveform, then find the *mean* (average value) of the squared waveform, and finally, take the square *root* of that value. Symbolically, the rms value is found by performing the following computation:

$$\sqrt{\text{average } [(\text{waveform})^2]} \qquad (12.30)$$

Example 12.19

Find the effective value of the voltage whose waveform is shown in Figure 12.33(a).

SOLUTION Squaring the values of $v(t)$ produces the waveform shown in Figure 12.33(b). The period of this waveform is $T = 5$ s, and its average value is the total area of one cycle divided by T:

$$\frac{(400 \text{ V}^2)(1 \text{ s}) + (144 \text{ V}^2)(2 \text{ s})}{5 \text{ s}} = 137.6 \text{ V}^2$$

The effective value is therefore $\sqrt{137.6} = 11.73$ V rms.

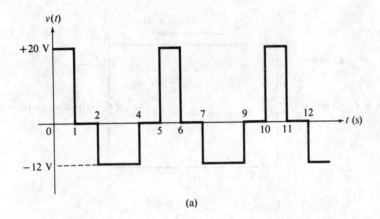

(a)

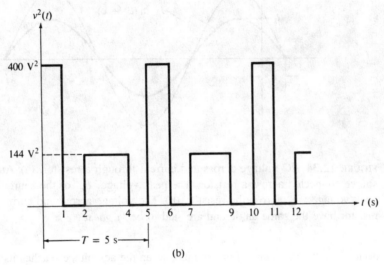

(b)

FIGURE 12.33 (Example 12.19)

Drill Exercise 12.19

What is the average value of a 2-A dc current? The effective value?

ANSWER: 2 A, in each case.   □

12.7 AC Voltage and Current in Resistance

Ohm's law can be applied to an ac circuit containing a resistance to determine the ac current in the resistance when an ac voltage is connected across it (see Figure 12.34). At every instant of time, the current in the resistor is the voltage at that instant divided by the resistance. In other words, the instantaneous current is the instantaneous voltage

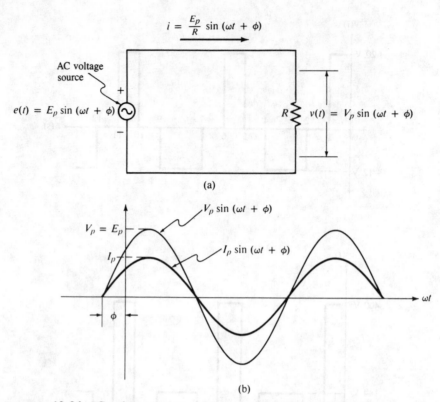

(b)

FIGURE 12.34 AC voltage across and current through a resistor. (a) An ac voltage source connected across a resistor. The peak voltage, E_p, of the source equals the peak voltage, V_p, across the resistor. (b) The voltage across and current through a resistor have the same angle and are said to be *in phase*.

divided by the resistance. For example, when the ac voltage reaches its peak value, E_p, the current in the resistance also reaches its peak value, $I_p = E_p/R$. We will continue to use the notation we used for dc circuits, whereby E designates a source voltage and V represents a voltage drop. For the circuit shown in Figure 12.34(a), the drop across the resistor clearly equals the source voltage connected across it, so $V_p = E_p$. The ac form of Ohm's law applied to a resistor is

$$i(t) = \frac{v(t)}{R} \tag{12.31}$$

For a sinusoidal voltage, (12.31) gives

$$i(t) = \frac{V_p}{R} \sin(\omega t + \phi) \tag{12.32}$$

and it is clear that $I_p = V_p/R$. Since $i(t) = I_p \sin(\omega t + \phi)$ and $v(t) = V_p \sin(\omega t + \phi)$, *the voltage across and current through a resistor have the same phase angle*. We say that the voltage and current are *in phase*.

Notice that + and − polarity symbols are attached to the ac voltage generator in Figure 12.34(a). As we know, the polarity of an ac source actually reverses every half-cycle. The polarity shown is simply the instantaneous polarity that exists when the ac current $i(t)$ has the left-to-right direction shown in the figure. In many practical applications where there is a single ac voltage source, it is not necessary to show polarity symbols, because there is no specific time point identified as $t = 0$. Note that the generator in Figure 12.34(a) would be equivalent to one in which the polarity symbols were *reversed* and for which the voltage designation was $e(t) = -E_p \sin(\omega t + \phi)$, or $E_p \sin(\omega t + \phi \pm 180°)$. Thus, polarity symbols actually provide a *phase* reference, and phase is not defined for a single source, unless the time point $t = 0$ is specified. However, as we shall see in a later discussion, polarity symbols are essential when there is more than one source in a circuit, because they specify the phase relation between the sources. Polarity symbols are also important as a phase reference when other voltages in a circuit are specified, such as ac voltage drops. As a matter of practice, we will use polarity symbols on all ac voltage sources in the examples and exercises.

Example 12.20

The current in a 2.2-kΩ resistor is $i(t) = 5 \sin(2\pi \times 100t + 45°)$ mA.

(a) Write the mathematical expression for the voltage across the resistor.
(b) What is the effective value of the resistor voltage?
(c) What is the instantaneous value of the resistor voltage at $t = 0.4$ ms?

SOLUTION
(a) From equation (12.32),

$$V_p = I_p R = (5 \text{ mA})(2.2 \text{ k}\Omega) = 11 \text{ V}$$

Since $v(t)$ across the resistor is in phase with the current,

$$v(t) = 11 \sin(2\pi \times 100t + 45°) \text{ V}$$

(b) $V_{\text{eff}} = \left(\dfrac{\sqrt{2}}{2}\right) V_p = \left(\dfrac{\sqrt{2}}{2}\right)(11 \text{ V}) = 7.78 \text{ V rms}$

(c) $v(0.4 \text{ ms}) = 11 \sin[2\pi \times 100(0.4 \times 10^{-3}) + 45°]$ V
$\qquad\qquad\quad = 11 \sin(0.2513 \text{ rad} + 45°)$ V

$$(0.2513 \text{ rad})\left(\dfrac{360°}{2\pi \text{ rad}}\right) = 14.4°$$

$v(0.4 \text{ ms}) = 11 \sin(14.4° + 45°) = 11 \sin 59.4° = 9.47$ V

Note that $i(0.4 \text{ ms}) = 5 \sin(59.4°)$ mA $= 4.304$ mA, so $v(0.4 \text{ ms})$ can also be computed by $i(t)R = (4.304 \text{ mA})(2.2 \text{ k}\Omega) = 9.47$ V, confirming that the instantaneous values obey Ohm's law at $t = 0.4$ ms.

Drill Exercise 12.20

The ac voltage across a 150-Ω resistor is $39 \sin(2\pi \times 10^3 t)$ V. At what value of t does the current through the resistor equal -0.26 A?

ANSWER: 0.75 ms.

12.8 AC Voltage and Current Relations in Capacitors and Inductors

Capacitor Voltage and Current

Unlike resistance, the ac current through a capacitor depends not only on the voltage across it but also on the *frequency* of that voltage. The property of a capacitor that causes it to resist the flow of ac current through it is called its *capacitive reactance*, denoted by X_C. Like resistance, the units of X_C are ohms, and as previously noted, its value depends on frequency:

$$X_C = \frac{1}{\omega C} = \frac{1}{2\pi f C} \quad \text{ohms} \tag{12.33}$$

Equation (12.33) shows that X_C is *inversely* proportional to frequency. Thus, the greater the frequency, the smaller the reactance, and the greater the current through the capacitor. Similarly, the lower the frequency, the greater the reactance. Figure 12.35 shows a plot of X_C versus f.

As the frequency of an ac waveform decreases, its period increases, according to $T = 1/f$. In the limiting case, as frequency approaches zero, the period approaches infinity, meaning that the waveform never goes negative. This situation corresponds to dc, so we may regard zero frequency as dc. By equation (12.33), the reactance at $f = 0$ is ∞ ohms, confirming that a capacitor is an open circuit (under steady-state conditions) when a dc voltage is connected across it. Figure 12.35 shows that the reac-

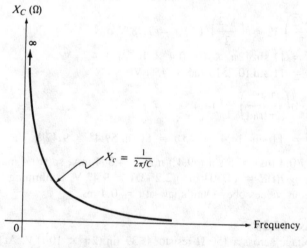

FIGURE 12.35 Plot of capacitive reactance, X_C, versus frequency, showing that X_C is inversely proportional to f.

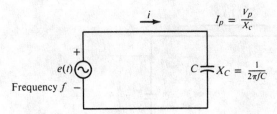

FIGURE 12.36 Ohm's law for a capacitor states that the peak current through the capacitor equals the peak voltage divided by the capacitive reactance.

tance approaches infinity as the frequency approaches zero. Note also that the capacitive reactance becomes very small (approaches zero) as the frequency becomes very large.

Figure 12.36 shows an ac voltage generator connected across a capacitor. Ohm's law for this circuit states that the peak current through the capacitor equals the peak voltage across it divided by the capacitive reactance:

$$I_p = \frac{V_p}{X_C} \frac{\text{volts}}{\text{ohms}} = \text{amperes} \qquad (12.34)$$

Example 12.21

The ac voltage across a 0.5-μF capacitor is $e(t) = 16 \sin (2 \times 10^3 t)$ V.

(a) Find the capacitive reactance of the capacitor.
(b) Find the peak value of the current through the capacitor.

SOLUTION
(a) From equation (12.33),

$$X_C = \frac{1}{\omega C} = \frac{1}{(2 \times 10^3)(0.5 \times 10^{-6})} = 1 \text{ k}\Omega$$

(b) From equation (12.34),

$$I_p = \frac{V_p}{X_C} = \frac{16 \text{ V}}{1 \text{ k}\Omega} = 16 \text{ mA}$$

Drill Exercise 12.21

The ac current through a 20-μF capacitor is $i(t) = 3 \sin (800t)$ A. What is the peak voltage across the capacitor?

ANSWER: 187.5 V. □

The peak current through a capacitor does *not* occur at the same instant of time that the voltage across the capacitor reaches its peak value. When the current and voltage are sinusoidal, *the current through a capacitor leads the voltage across it by 90°*. Equivalently, the voltage across a capacitor lags the current through it by 90°. Figure 12.37

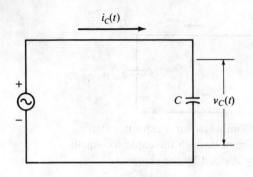

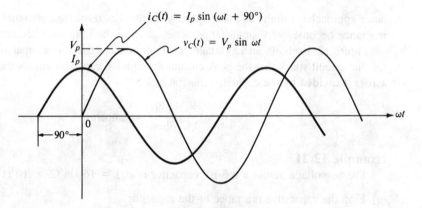

FIGURE 12.37 The current through a capacitor leads the voltage across it by 90°. Notice that $i_C(t)$ is at a positive or negative peak when $v_C(t)$ is zero, and vice versa. The current reaches its peak value 90° ahead of the voltage.

illustrates this phase relation. Note that the current reaches its peak 90° ahead of the voltage. In the figure, $v_C(t) = V_p \sin \omega t$, so $i_C(t) = I_p \sin (\omega t + 90°)$. In general, if

$$v_C(t) = V_p \sin (\omega t + \phi) \text{ V}$$

then

$$i_C(t) = I_p (\sin \omega t + \phi + 90°) \text{ A} \qquad (12.35)$$

where $I_p = V_p/X_C$.

Example 12.22

The voltage across a 0.01-μF capacitor is $v_C(t) = 240 \sin (1.25 \times 10^4 t - 30°)$ V. Write the mathematical expression for the current through it.

SOLUTION

$$X_C = \frac{1}{\omega C} = \frac{1}{(1.25 \times 10^4)(10^{-8})} = 8 \text{ k}\Omega$$

$$I_p = \frac{V_p}{X_C} = \frac{240 \text{ V}}{8 \text{ k}\Omega} = 0.03 \text{ A}$$

By equation (12.35),

$$i_C(t) = 0.03 \sin (1.25 \times 10^4 t - 30° + 90°) \text{ A}$$
$$= 0.03 \sin (1.25 \times 10^4 t + 60°) \text{ A}$$

Drill Exercise 12.22

The current through a 15-pF capacitor is $i_C(t) = 0.45 \sin (10^8 t + 45°)$ mA. Write the mathematical expression for the voltage across it.

ANSWER: $0.3 \sin (10^8 t - 45°)$ V. $\square$

Inductor Voltage and Current

Like capacitance, inductance resists, or impedes, the flow of ac current through it. This property is called *inductive reactance*, denoted X_L, and its value is directly proportional to the frequency of the current:

$$X_L = \omega L = 2\pi f L \qquad \text{ohms} \tag{12.36}$$

Figure 12.38 shows a plot of X_L versus frequency. Note that X_L decreases with frequency, approaching 0 Ω as frequency approaches zero (dc). This confirms that an inductor is a short circuit under steady-state conditions, when a dc voltage is connected across it.

Figure 12.39 shows an ac voltage generator connected across an inductor. Ohm's law for this circuit states that the peak current through the inductor equals the peak voltage across it divided by the inductive reactance:

$$I_p = \frac{V_p \text{ volts}}{X_L \text{ ohms}} = \text{amperes} \tag{12.37}$$

The voltage across an inductor leads the current through it by 90°. Thus, the voltage reaches its peak 90° ahead of the current, as illustrated in Figure 12.40. In general, if

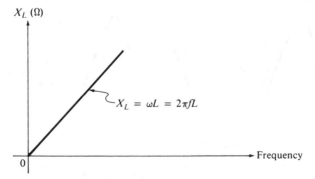

FIGURE 12.38. Plot of inductive reactance versus frequency, showing that X_L is directly proportional to f.

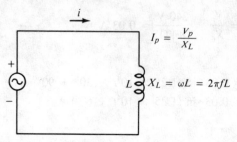

FIGURE 12.39 Ohm's law for an inductor states that the peak current through the inductor equals the peak voltage across it divided by the inductive reactance.

$$i_L(t) = I_p \sin (\omega t + \phi)$$

then

$$v_L(t) = V_p \sin (\omega t + \phi + 90°)$$

where $I_p = V_p/X_L$.

Note that the phase relation between voltage and current in an inductor is exactly the opposite of that in a capacitor. Following is a useful memory aid:

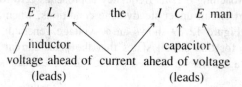

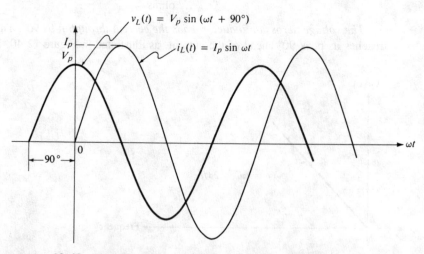

FIGURE 12.40. The voltage across an inductor leads the current through it by 90°. Compare with Figure 12.37.

Example 12.23

The current through an 80-mH inductor is 0.1 sin (400t − 25°) A. Write the mathematical expression for the voltage across it.

SOLUTION The inductive reactance is

$$X_L = \omega L = (400 \text{ rad/s})(80 \times 10^{-3} \text{ H}) = 32 \; \Omega$$

From equation (12.37),

$$V_p = I_p X_L = (0.1 \text{ A})(32 \; \Omega) = 3.2 \text{ V}$$

Since the voltage leads the current by 90°, we must add 90° to the phase angle of the current:

$$v_L(t) = V_p \sin (\omega t + \phi + 90°)$$
$$= 3.2 \sin (400t - 25° + 90°) = 3.2 \sin (400t + 65°) \text{ V}$$

Drill Exercise 12.23

The voltage across a 4-H inductor is 18 sin (2π × 10^3t − 30°) V. Write the mathematical expression for the current through it.

ANSWER: $i_L(t) = 0.716 \sin (2\pi \times 10^3 t - 120°)$ mA.   □

12.9 Average Power

Recall that the rate of energy dissipation (power) in a dc circuit can be calculated using any of the three relationships $P = VI$, $P = I^2R$, or $P = V^2/R$ watts. In ac circuits, both voltage and current are time-varying quantities, and so therefore is power. The power at any instant of time, called the *instantaneous power*, can be computed using instantaneous values of voltage and/or current:

$$p(t) = v(t)i(t)$$
$$p(t) = [i(t)]^2R \qquad (12.38)$$
$$p(t) = \frac{[v(t)]^2}{R}$$

For example, if the current through a 10-Ω resistor is $i(t) = 0.5 \sin (2\pi \times 100t)$ A, then the instantaneous power at $t = 0$ is

$$p(0) = [i(0)]^2R = (0.5 \sin 0)^2 10 = 0 \text{ W}$$

and the instantaneous power at $t = 1$ ms is

$$p(1 \times 10^{-3}) = [i(1 \times 10^{-3})]^2R = [0.5 \sin (2\pi \times 100 \times 10^{-3})]^2 10 = 0.864 \text{ W}$$

A more useful measure of power dissipation is an ac circuit is *average* power, that is, the average of the instantaneous power over a period of time. Note carefully that

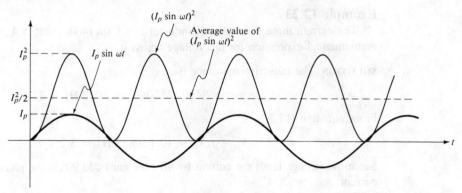

FIGURE 12.41 The square of the sine wave $I_p \sin \omega t$ is another sine wave having average value $\frac{1}{2}I_p^2$.

average power does *not* equal average voltage times average current. For example, when $v(t)$ and $i(t)$ are sinusoidal, we know that the average values of both $v(t)$ and $i(t)$ are zero. But as we shall demonstrate, the average value of $p(t) = v(t)i(t)$ is not zero.

Recall that *only* resistance is capable of dissipating power. Therefore, we can derive an expression for average power by finding the average of $p(t) = [i(t)]^2R$ over a period of time. We will restrict our investigation to the sinusoidal current $i(t) = I_p \sin(\omega t)$ A. Thus, we wish to find the average of

$$p(t) = (I_p \sin \omega t)^2 R \qquad (12.39)$$

Although the term $\sin(\omega t)$ in (12.39) goes both positive and negative, all such values are *squared*, so $p(t)$ is always positive. This fact is illustrated in Figure 12.41. Note that the squared waveform is itself sinusoidal. Using the trigonometric identity $\sin^2 \theta = \frac{1}{2} - \frac{1}{2} \cos 2\theta$, we can show that the equation of the squared sine-wave current is

$$(I_p \sin \omega t)^2 = \frac{1}{2}I_p^2 - \frac{1}{2}I_p^2 \cos 2\omega t \qquad (12.40)$$

Thus, the average value of the squared current is the "dc value" $\frac{1}{2}I_p^2$, as shown in Figure 12.41. Therefore, the average power is

$$P_{avg} = \frac{1}{2}I_p^2 R \qquad \text{watts} \qquad (12.41)$$

Since the current is sinusoidal, $I_p = \sqrt{2}I_{eff}$. Substituting in (12.41), we find an equivalent expression for average power:

$$P_{avg} = \frac{1}{2}(\sqrt{2}I_{eff})^2 R = I_{eff}^2 R \qquad \text{watts} \qquad (12.42)$$

Since $I_p = V_p/R$ and $I_{eff} = V_{eff}/R$, equations (12.41) and (12.42) give the equivalent relations:

$$P_{avg} = \frac{V_p^2}{2R} \qquad \text{watts} \qquad (12.43)$$

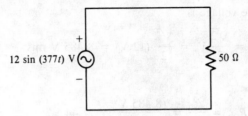

FIGURE 12.42 (Example 12.24)

and

$$P_{avg} = \frac{V_{eff}^2}{R} \quad \text{watts} \tag{12.44}$$

Also, since $V_p = I_p R$ and $V_{eff} = I_{eff} R$, we have

$$P_{avg} = \frac{V_p I_p}{2} \quad \text{watts} \tag{12.45}$$

and

$$P_{avg} = V_{eff} I_{eff} \quad \text{watts} \tag{12.46}$$

Example 12.24

Find the average power dissipated in the 50-Ω resistor shown in Figure 12.42 using each of equations (12.41) through (12.46).

SOLUTION The peak current in the resistor is

$$I_p = \frac{V_p}{R} = \frac{12 \text{ V}}{50 \text{ }\Omega} = 0.24 \text{ A pk}$$

By equation (12.41),

$$p_{avg} = \frac{I_p^2 R}{2} = \frac{(0.24 \text{ A})^2 (50 \text{ }\Omega)}{2} = 1.44 \text{ W}$$

The effective value of the current is

$$I_{eff} = \frac{\sqrt{2}}{2} I_p = \frac{\sqrt{2}}{2} (0.24 \text{ A}) = 169.7 \text{ mA rms}$$

By equation (12.42),

$$P_{avg} = I_{eff}^2 R = (169.7 \text{ mA})^2 (50 \text{ }\Omega) = 1.44 \text{ W}$$

By equation (12.43),

$$P_{avg} = \frac{V_p^2}{2R} = \frac{(12 \text{ V})^2}{2(50\Omega)} = 1.44 \text{ W}$$

The effective value of the voltage is

$$V_{eff} = \frac{\sqrt{2}}{2} V_p = \frac{\sqrt{2}}{2} (12 \text{ V}) = 8.485 \text{ V rms}$$

By equation (12.44),

$$P_{avg} = \frac{V_{eff}^2}{R} = \frac{(8.485 \text{ V})^2}{50 \text{ }\Omega} = 1.44 \text{ W}$$

By equation (12.45),

$$P_{avg} = \frac{V_p I_p}{2} = \frac{(12 \text{ V})(0.24 \text{ A})}{2} = 1.44 \text{ W}$$

By equation (12.46),

$$P_{avg} = V_{eff} I_{eff} = (8.485 \text{ V})(169.7 \text{ mA}) = 1.44 \text{ W}$$

Drill Exercise 12.24

To what value should the resistance in Example 12.24 be changed if the average power dissipated in it must be reduced to 1 W?

ANSWER: 72 Ω. $\square$

12.10 SPICE Examples

Example 12.25   (SPICE)

Use SPICE to find the current (magnitude and angle) through a 1.2-H inductor when it is connected across the voltage source $e = 60\underline{/0°}$ V. The frequency of e is 1 kHz.

SOLUTION SPICE does not permit an inductor to appear by itself in a loop with a voltage source. We must therefore insert a very small resistance in series with the source. In this example, the magnitude of the inductive reactance is $2\pi f L = 2\pi(1 \text{ kHz})$ (1.2 H) = 7.54 kΩ, so the series resistance must be much smaller than 7.54 kΩ to have negligible effect on the computations. We choose a resistance, RDUM, equal to 1 $\mu\Omega$. Figure 12.43 shows the SPICE circuit and input data file. Note that our analysis type is AC (see Section A.7 of the Appendix). VDUM is a dummy ac voltage source used to determine the current in the inductor. Note that we *must* specify its magnitude to be 0 V, since an ac source defaults to 1 V, rather than 0 V. The phase angles of both V1 and VDUM default to 0°. Execution of the program produces the result I(VDUM) = 7.958 mA and IP(VDUM) (the phase angle of the current) = -90°. These results are easily verified by hand computation.

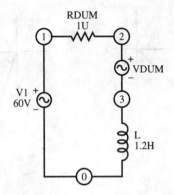

RDUM
1U

```
EXAMPLE 12.25
V1 1 0 AC 60V
RDUM 1 2 1U
VDUM 2 3 AC 0V
L 3 0 1.2H
.AC DEC 1 1KHZ 1KHZ
.PRINT AC I(VDUM) IP(VDUM)
.END
```

FIGURE 12.43 (Example 12.25)

Exercises

Section 12.2 Review of Trigonometric Functions

12.1 Express each of the following as an angle between $-180°$ and $+180°$.
(a) 200° **(b)** $-190°$ **(c)** 280°
(d) 380° **(e)** $-270°$ **(f)** 940°

12.2 Find the negative angle equivalent to each of the following.
(a) 220° **(b)** 180° **(c)** 550°

12.3 Find the sine of angle θ in each of the triangles in Figure 12.43 (include the correct algebraic sign, according to the quadrant in which the angle is located).

12.4 Using a calculator, find the sine of each of the following angles.
(a) 90° **(b)** $-250°$ **(c)** 1024°

12.5 Sketch each of the following functions versus θ. (Label the horizontal axis with the values of the angle where the function is minimum, maximum, and zero.)
(a) sin (θ + 50°) **(b)** sin (θ − 10°)
(c) sin (θ + 100°)

12.6 Sketch each of the following functions versus θ. (Label the horizontal axis with the values of the angle where the function is minimum, maximum, and zero.)
(a) sin (θ − 120°) **(b)** cos (θ − 50°)

12.7 Express each of the following as a positive sine function.
(a) cos (θ − 75°) **(b)** −sin (θ + 100°)
(c) −cos (θ + 30°) **(d)** −sin (−θ)

12.8 Sketch each of the following functions versus θ.
(a) −sin (θ − 150°) **(b)** cos (θ + 10°)
(c) −cos (θ − 45°)

12.9 Using the definitions of the sine, cosine, and tangent functions in a right triangle, show that

$$\tan \theta = \frac{\sin \theta}{\cos \theta}$$

12.10 Using the *Pythagorean* theorem ($a^2 + b^2 = c^2$ in a right triangle, where $c = $ length of the hypotenuse)

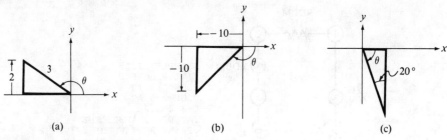

FIGURE 12.44 (Exercise 12.3)

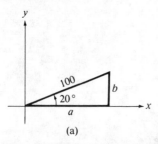

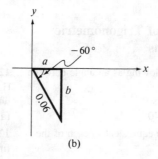

FIGURE 12.45 (Exercise 12.11)

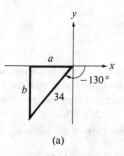

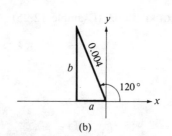

FIGURE 12.46 (Exercise 12.12)

and the definitions of the sine and cosine functions, show that

$$(\sin \theta)^2 + (\cos \theta)^2 = 1$$

12.11 Using the sine and cosine functions, find the length of sides a and b (including algebraic signs) in each of the triangles in Figure 12.45.

12.12 Repeat Exercise 12.11 for each of the triangles in Figure 12.46.

12.13 Find the length of the hypotenuse in the triangle shown in Figure 12.47(a).

12.14 Find the length of the hypotenuse in the triangle shown in Figure 12.47(b).

12.15 Find the angle θ in each of the triangles shown in Figure 12.48.

12.16 Find the angle θ in each of the triangles shown in Figure 12.49.

Section 12.3   Waveform Parameters

12.17 (a) Find the frequency of a waveform whose period is

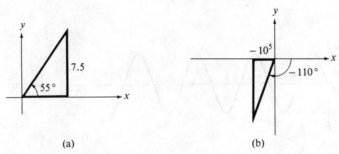

FIGURE **12.47** (Exercises 12.13 and 12.14)

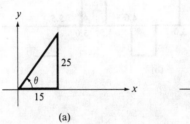
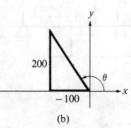

FIGURE **12.48** (Exercise 12.15)

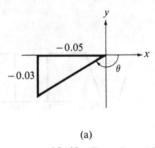
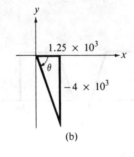

FIGURE **12.49** (Exercise 12.16)

(i) 0.02 s **(ii)** 80 μs **(iii)** 5 s **(iv)** 40 ms
(b) Find the period of a waveform whose frequency is
(i) 100 Hz **(ii)** 800 kHz **(iii)** 0.1 Hz **(iv)** 250 MHz

12.18 (a) How many cycles of a 15-kHz waveform
occur in 22 ms?
(b) What is the frequency of a waveform if 250 of its
cycles occur in 0.8 s?

12.19 Find the frequency of each of the waveforms
shown in Figure 12.50.

12.20 Find how long it would take for 10 cycles of
each of the waveforms in Figure 12.50 to occur if the
frequency of each were doubled.

12.21 Convert each of the following angles to radians.
(a) 130° **(b)** −45° **(c)** 270° **(d)** −180° **(e)** 510°

12.22 Convert each of the following angles to degrees.
(a) 0.25 rad **(b)** −π/20 rad **(c)** 3 rad **(d)** 4π rad

12.23 Find the angular frequency of
(a) $\sin 200t$
(b) a sine wave whose frequency is 25 kHz
(c) a sine wave whose period is 16 μs
(d) $\sin t$

12.24 Find the frequency (in hertz) of
(a) $\sin (3.2 \times 10^4 t)$
(b) $\sin (2\pi \times 5 \times 10^3 t)$

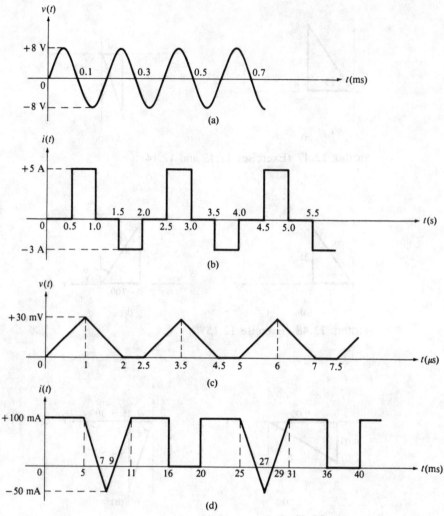

FIGURE 12.50 (Exercise 12.19)

(c) a sine wave whose angular frequency is 1687 rad/s
(d) sin t

12.25 Find the peak-to-peak value of each of the waveforms in Figure 12.50.

12.26 Write the mathematical expression for
(a) a sine-wave current whose frequency is 60 Hz and whose peak value is 0.5 A
(b) a sine-wave voltage whose period is 0.2 μs and whose peak-to-peak value is 18 mV

12.27 Find the instantaneous value of
(a) 36 sin (3000 t) V at $t = 1$ ms

(b) 55 sin ($2\pi \times 16t$) mA at $t = \frac{1}{32}$ s
(c) 0.9 sin (0.1t) V at $t = 1$ s

12.28 Find two values of t at which the instantaneous value of 100 sin ($2\pi \times 10t$) V equals 50 V.

Section 12.4   Phase Relations

12.29 Find the instantaneous value of
(a) $v(t) = 4.23$ sin (8944t + 26°) V at $t = 0.1$ ms
(b) $i(t) = 16$ sin ($2\pi \times 18 \times 10^4t$ − 15°) mA at $t = 2$ μs

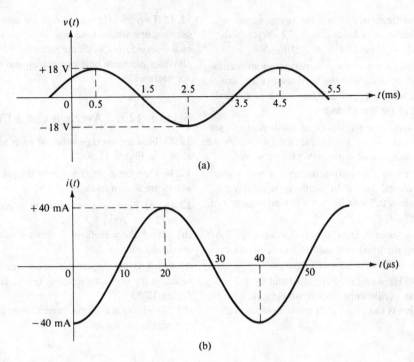

FIGURE 12.51 (Exercise 12.31)

(c) $v(t) = -100 \sin (50t + 80°)$ V at $t = 0.04$ s

(d) $i(t) = \cos (7551t - \pi/6 \text{ rad})$ mA at $t = 4$ ms

12.30 Find the instantaneous value of

(a) $v(t) = 85 \sin (10^5 t - 45°)$ V at $t = 10$ μs

(b) $i(t) = 0.003 \sin (t - 120°)$ A at $t = 5$ s

(c) $v(t) = -250 \cos (3830t + \pi/12 \text{ rad})$ mV at $t = 0.1$ ms

(d) $i(t) = -\sin (2\pi \times 60t + \pi \text{ rad})$ μA at $t = \frac{1}{240}$ s

12.31 Write the mathematical expression for each of the waveforms shown in Figure 12.51.

12.32 Sketch each of the following waveforms versus time. (Show the values of t at which the instantaneous value of each waveform is minimum, maximum, and zero.)

(a) $i(t) = 14 \sin (2\pi \times 100t + 18°)$ A

(b) $v(t) = -260 \cos (2\pi \times 25 \times 10^3 t)$ V

12.33 Determine which waveform in each of the following pairs is leading and find the angle by which it leads.

(a) $i_1(t) = 50 \sin (\omega t - 40°)$ mA

$i_2(t) = 20 \sin (\omega t + 10°)$ mA

(b) $i(t) = 3.1 \sin (10^4 t + 25°)$ A

$v(t) = 16 \sin (10^4 t + 100°)$ V

(c) $v_1(t) = 18 \cos (\omega t - \pi/4 \text{ rad})$ mV

$v_2(t) = 2 \sin (\omega t)$ mV

(d) $v(t) = 0.04 \sin (377t + 5°)$ V

$i(t) = -5.9 \sin (377t - 20°)$ A

12.34 Determine which waveform in each of the following pairs is lagging, and find the angle by which it lags.

(a) $v(t) = 169 \sin (377t - 80°)$ V

$i(t) = 5.4 \sin (377t - 60°)$ A

(b) $i_1(t) = 244 \sin (\omega t + 120°)$ mA

$i_2(t) = -38 \sin (\omega t - 50°)$ mA

(c) $v_1(t) = 0.7 \cos (42t - \pi/6 \text{ rad})$ V

$v_2(t) = -4 \cos (42t + 80°)$ V

(d) $i(t) = 75 \cos (\omega t + 10°)$ mA

$v(t) = -25 \sin (\omega t - 80°)$ μV

Section 12.5 The Oscilloscope

12.35 The sweep frequency of an oscilloscope is set to 4 kHz and the waveform displayed has a frequency of 24 kHz. How many cycles of the waveform are displayed?

12.36 To what frequency should the sweep on an oscilloscope be set in order to display 12 cycles of the waveform $v(t) = 30 \sin (2\pi \times 6 \times 10^3 t)$ V?

12.37 The peak of a waveform displayed on an oscilloscope is 2.8 divisions above the horizontal (zero) axis. If the vertical sensitivity is 0.5 V/div, what is the peak-to-peak value of the waveform?

12.38 If the vertical sensitivity of an oscilloscope is set to 20 mV/div, how many divisions are used to display the peak-to-peak variation of a 0.034-V pk sine wave?

12.39 One cycle of a waveform occupies 4.6 divisions along the horizontal axis of an oscilloscope display. If the time base is 0.5 ms/div, what is the frequency of the waveform?

12.40 Four cycles of a 320-kHz waveform occupy 2.5 divisions along the horizontal axis of an oscilloscope display. What is the time base of the display?

12.41 Two 80-Hz waveforms are separated by 1.5 divisions on an oscilloscope display whose time base is 2 ms/div. What is the phase shift between the waveforms?

12.42 Two 50-kHz waveforms are displayed on an oscilloscope whose time base is set to 10 μs/div. If one of the waveforms leads the other by 90°, how many divisions are there between corresponding points on the waveforms?

Section 12.6 Average and Effective Values

12.43 Find the average value of each of the waveforms shown in Figure 12.50.

12.44 Find the average value of the *full-wave rectified* waveform shown in Figure 12.52.

12.45 **(a)** Find the average value of $v(t) = 6 + 2 \sin (2\pi \times 100t)$ V.
(b) Sketch the waveform. Label the minimum and maximum values.

12.46 **(a)** Using the superposition principle, find the equation for the voltage across the 30-Ω resistor in Figure 12.53.
(b) Sketch the waveform versus time. Label minimum and maximum values.

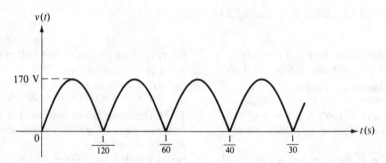

FIGURE 12.52 (Exercise 12.44)

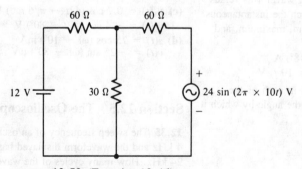

FIGURE 12.53 (Exercise 12.46)

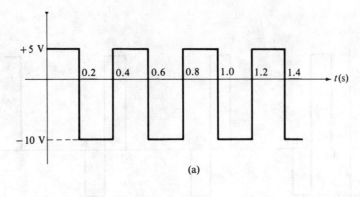

(a)

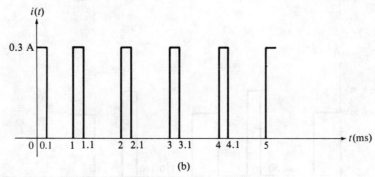

(b)

FIGURE 12.54 (Exercise 12.51)

12.47 Find the effective value of each of the following.
(a) $i(t) = 100 \sin (3840t)$ A
(b) $v(t) = 260 \sin (\omega t - 30°)$ mV
(c) $i(t) = \sqrt{2} \cos (10^4 t)$ A
(d) $v(t) = 0.13 \sin (\omega t + 120°)$ V

12.48 Find the rms value of a sinusoidal voltage whose peak-to-peak value is
(a) 338 V
(b) 19 μV

12.49 Find the peak value of a sinusoidal waveform whose rms value is
(a) 1.2 V rms
(b) 39 μA rms
(c) 480 nV rms
(d) 7.5 A rms

12.50 A 0.16-A dc current flowing in a 120-Ω resistor raises the temperature of the resistor by 15°C. A sinusoidal voltage connected across the same resistor also raises the temperature of the resistor by 15°C. What is the rms value of the sinusoidal voltage?

12.51 Find the effective value of each of the waveforms shown in Figure 12.54.

12.52 Find the rms value of each of the waveforms shown in Figure 12.55.

Section 12.7 AC Voltage and Current in Resistance

12.53 Write the mathematical expression for the current through a 2-kΩ resistor when the voltage across it is
(a) $v(t) = 18 \sin (400t)$ V
(b) $v(t) = 840 \sin (\omega t + 30°)$ mV
(c) $v(t) = 0.5 \cos (2\pi \times 10^5 t - 80°)$ V
(d) $v(t) = -\sin (1832t + 100°)$ V

12.54 Write the mathematical expression for the voltage across a 120-Ω resistor when the current through it is
(a) $i(t) = 0.1 \sin (377t)$ A
(b) $i(t) = 42 \sin (2\pi \times 60t + 45°)$ mA
(c) $i(t) = \cos (\omega t)$ A
(d) $i(t) = -185 \sin (9841t - 70°)$ μA

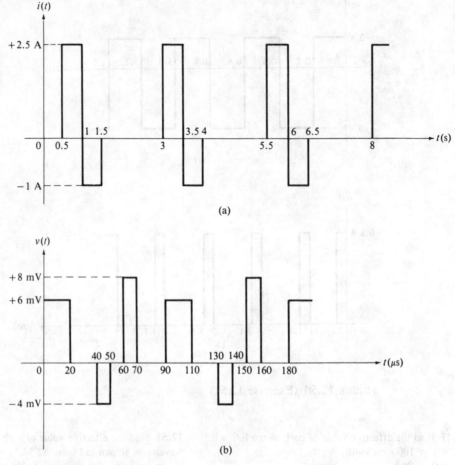

FIGURE 12.55 (Exercise 12.52)

12.55 Find the effective value of the voltage across a 22-kΩ resistor when the peak sinusoidal current through it is 1.5 mA.

12.56 Find the peak value of the current through a 47-kΩ resistor when the rms value of the sinusoidal voltage across it is 70.5 V rms.

Section 12.8   AC Voltage and Current Relations in Capacitors and Inductors

12.57 Find the capactive reactance of
(a) a 0.1-μF capacitor when the frequency of the current through it is 4.6 kHz

(b) a 150-μF capacitor when the voltage across it is 16 sin (5000t) V
(c) a 20-pF capacitor when the current through it is 95 sin ($2\pi \times 4 \times 10^6 t$) A
(d) a 0.022-μF capacitor at dc

12.58 (a) At what frequency is the capacitive reactance of a 20-μF capacitor equal to 100 Ω?
(b) What is the value of X_C when the frequency found in part (a) is doubled?
(c) What is the capacitance of a capacitor whose reactance is 100 Ω when the current through it has a frequency of 8 kHz?

12.59 Find the peak current through a 0.01-μF capacitor when the voltage across it is
(a) $v(t) = 4 \sin (10^6 t - 45°)$ V

(b) $v(t) = 169 \sin (377t)$ V

(c) $v(t) = 100 \cos (2\pi \times 10^4 t - 80°)$ V

12.60 Find the peak voltage across a 22-μF capacitor when the current through it is

(a) $i(t) = 3 \sin (10^3 t + 60°)$ A

(b) $i(t) = 15 \sin (2\pi \times 80t)$ mA

(c) $i(t) = 0.5 \cos (5000t - 10°)$ A

12.61 (a) Write the equation for the voltage across the capacitor in each part of Exercise 12.60.

(b) Sketch the current through and voltage across the capacitor (versus angle) in each part of Exercise 12.60.

12.62 (a) Write the equation for the current through the capacitor in each part of Exercise 12.59.

(b) Sketch the current through and voltage across the capacitor (versus angle) in each part of Exercise 12.59.

12.63 Find the inductive reactance of:

(a) a 20-mH inductor when the current through it has a frequency of 40 kHz

(b) a 1.5-H inductor when the voltage across it is 65 sin (320t) mV

(c) a 400-μH inductor when the current through it is 128 sin (2π × 10⁷t − 50°) mA

(d) a 0.06-H inductor at dc

12.64 (a) At what frequency is the inductive reactance of a 15-mH inductor equal to 2.5 kΩ?

(b) What is the value of X_L when the frequency found in part (a) is doubled?

(c) What is the inductance of an inductor that has a reactance of 2.5 kΩ when the current through it has a frequency of 500 Hz?

12.65 Find the peak current through a 0.04-H inductor when the voltage across it is

(a) $v(t) = 5 \sin (1800t)$ V

(b) $v(t) = 40 \sin (2\pi \times 100t + 40°)$ V

(c) $v(t) = 0.1 \sin (10^5 t + 70°)$ V

12.66 Find the peak voltage across a 0.5-H inductor when the current through it is

(a) $i(t) = 2 \sin (377t)$ A

(b) $i(t) = -10 \sin (10^3 t + 100°)$ mA

(c) $i(t) = 5.6 \cos (2\pi \times 10^6 t - 110°)$ μA

12.67 (a) Write the equation for the voltage across the inductor in each part of Exercise 12.66.

(b) Sketch the current through and voltage across the inductor (versus angle) in each part of Exercise 12.66.

12.68 (a) Write the equation for the current through the inductor in each part of Exercise 12.65.

(b) Sketch the current through and voltage across the inductor (versus angle) in each part of Exercise 12.65.

Section 12.9   Average Power

12.69 The current through a 150-Ω resistor is $i(t) = 24 \sin (377t)$ mA. Find the average power dissipated in the resistor.

12.70 The sinusoidal voltage across a 2.2-kΩ resistor has a peak-to-peak value of 180 V p-p. What is the average power dissipated in the resistor?

12.71 The average power dissipated by a 100-Ω resistor is 1.5 W. Assuming that the ac voltage across the resistor is sinusoidal:

(a) Find the effective value of the voltage across the resistor.

(b) Find the peak value of the current through the resistor.

12.72 The average power dissipated by a 1-kΩ resistor is 0.2 W. Assuming that the ac voltage across the resistor is sinusoidal:

(a) Find the effective value of the current through the resistor.

(b) Find the peak value of the voltage across the resistor.

SPICE Exercises

12.1S Use SPICE to find the current (magnitude and angle) through a 1.5-μF capacitor when it is connected across $e = 40\underline{/0°}$ V. The frequency is 2.5 kHz.

12.2S Use SPICE to find the current (magnitude and angle) through 1 kΩ of inductive reactance connected across $e = 120\underline{/-60°}$ V. (Hint: you must choose a frequency and calculate a value for L.)

12.3S Use SPICE to find the voltage (magnitude and angle) across a 25-mH inductor when it is connected across the constant-current source $i = 40\underline{/-75°}$ mA. The frequency is 20 kHz.

12.4S Use SPICE as an aid in determining the magnitude of the capacitive reactance of a 200-pF capacitor at 1.5 MHz. (Hint: Connect a voltage source across the capacitor, find the current, and use equation 12.34).

13 Complex Algebra and Phasors

13.1 Complex Numbers

Complex numbers are widely used to facilitate computations involving ac voltages and currents. Recall that the complex number system is derived from the so-called imaginary number $\sqrt{-1}$. There are no "imaginary" properties of ac voltages and currents, nor any inherent characteristics of ac quantities that *require* the use of complex numbers to analyze them. The algebra of complex numbers is simply a mathematical tool that lends itself in a very convenient way to the study and analysis of ac circuits. As we shall see, it is the basis for a particularly useful method for adding, subtracting, multiplying, and dividing time-varying quantities. Because complex numbers are so important and so widely used in ac circuit theory, we will devote a considerable portion of the present chapter to reviewing their properties and algebraic manipulations.

In most mathematics texts, i is used to denote the imaginary quantity $\sqrt{-1}$. To avoid confusion with the symbol for current, it is conventional in circuit theory to use j instead of i:

$$j = \sqrt{-1} \tag{13.1}$$

From this definition, it follows that

$$j^2 = -1 \tag{13.2}$$

Furthermore, it is easy to find a simplified form of j raised to any power. For example, $j^3 = j^2 j = -j$, and $j^4 = j^2 j^2 = (-1)(-1) = 1$.

A complex number **C** has a *real part* and an *imaginary part:*

$$\mathbf{C} = a + jb \tag{13.3}$$

where a is the real part and b is the imaginary part. Note carefully that *both a and b are real numbers.* Thus, *the imaginary part of* **C** *is the real number b.* This real number is called the imaginary part of **C** simply because it is the part that is multiplied by j. Also note that the $+$ sign in (13.3) does *not* mean addition. It is merely a convenient symbol that associates the real and imaginary parts of a particular complex number. We could just as well write $\mathbf{C} = (a, b)$.

Example 13.1

Find the real and imaginary parts of each of the following complex numbers.

(a) $6 + j2$ (d) 25
(b) $-3 + j10^{-3}$ (e) $-j400$
(c) $-0.5 - j$ (f) $j\pi$

SOLUTION
(a) real part = 6; imaginary part = 2
(b) real part = -3; imaginary part = 1×10^{-3}
(c) real part = -0.5; imaginary part = -1
(d) $25 = 25 + j0$
    real part = 25; imaginary part = 0
(e) $-j400 = 0 - j400$
    real part = 0; imaginary part = -400
(f) $j\pi = 0 + j\pi$
    real part = 0; imaginary part = π

Drill Exercise 13.1

Find the real and imaginary parts of $j^6 - 10j^3$.

ANSWER: real part = -1; imaginary part = 10.   □

13.2 The Complex Plane

The complex plane is a rectangular coordinate system in which real numbers are plotted along the horizontal (real) axis and imaginary numbers along the vertical (imaginary) axis. Thus, every point in the complex plane represents a complex number whose real part is its horizontal coordinate and whose imaginary part is its vertical coordinate. Conversely, every complex number can be represented as a point in the complex plane. Figure 13.1 shows some examples. Note the usual convention whereby positive coor-

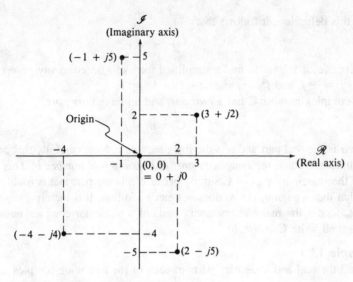

FIGURE 13.1 Examples of complex numbers plotted in the complex plane.

dinates lie to the right and above the origin, while negative coordinates lie to the left and below the origin.

The representation $a + jb$ is called the *rectangular* form of a complex number. Every complex number can also be represented in *polar* form:

$$\mathbf{C} = M \underline{/\theta} \tag{13.4}$$

where M is the *magnitude* of $\mathbf{C}$ and θ is its angle. As illustrated in Figure 13.2(a), M is the length of a line drawn from the origin to the point in the complex plane, and θ is the angle between that line and the positive real axis. The magnitude of a complex number, $\mathbf{C}$, is also called its *absolute value* and is designated by $|\mathbf{C}|$. Given the rectangular

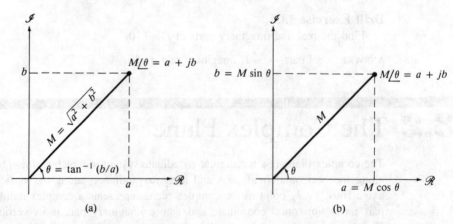

FIGURE 13.2 Converting between the polar and rectangular forms of a complex number.

form of a complex number, $a + jb$, we can convert to the equivalent polar form using the Pythagorean theorem and the definition of the tangent function [see Figure 13.2(a)]:

$$M = \sqrt{a^2 + b^2} \tag{13.5}$$

$$\theta = \tan^{-1} \frac{b}{a} \tag{13.6}$$

Similarly, given the polar form, $M \underline{/\theta}$, we can convert to the equivalent rectangular form using the sine and cosine functions:

$$a = M \cos \theta \quad \text{and} \quad b = M \sin \theta \tag{13.7}$$

Thus,

$$M \underline{/\theta} = M \cos \theta + jM \sin \theta \tag{13.8}$$

These relationships are illustrated in Figure 13.2(b).

Example 13.2
(a) Convert the following complex numbers to polar form.
   (i) $4 + j4$ (ii) $-30 + j10$
(b) Convert the following complex numbers to rectangular form.
   (i) $5 \underline{/60°}$ (ii) $2 \times 10^{-3} \underline{/-30°}$
(c) Plot each number in parts (a) and (b) in the complex plane.

SOLUTION
(a) (i) $M = \sqrt{a^2 + b^2} = \sqrt{4^2 + 4^2} = 5.66$

$$\theta = \tan^{-1} \frac{4}{4} = 45°$$

Therefore, $M \underline{/\theta} = 5.66 \underline{/45°}$.
   (ii) $M = \sqrt{a^2 + b^2} = \sqrt{(-30)^2 + (10)^2} = 31.62$

$$\theta = \tan^{-1} \frac{10}{-30} = 161.56°$$

Note that θ is a second-quadrant angle, so the precautions discussed in Section 12.2 must be observed when using a calculator to find its value.
$M \underline{/\theta} = 31.62 \underline{/161.56°}$.
(b) (i) $a = 5 \cos 60° = 2.5$
       $b = 5 \sin 60° = 4.33$
       $a + jb = 2.5 + j4.33$
   (ii) $a = 2 \times 10^{-3} \cos (-30°) = 1.73 \times 10^{-3}$
       $b = 2 \times 10^{-3} \sin (-30°) = -1 \times 10^{-3}$
       $a + jb = 1.73 \times 10^{-3} - j10^{-3}$
(c) The complex numbers are plotted in Figure 13.3.

Drill Exercise 13.2
   Convert $-3 - j4$ to polar form.
ANSWER: $5 \underline{/-126.87°}$.

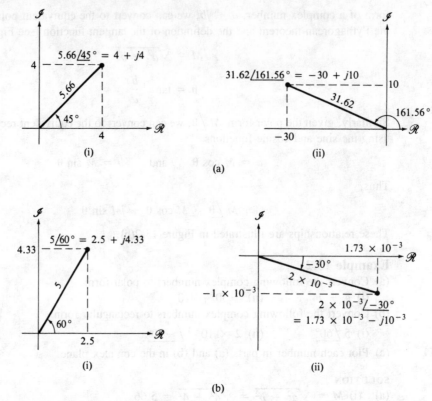

(a)

(b)

FIGURE 13.3 (Example 13.2)

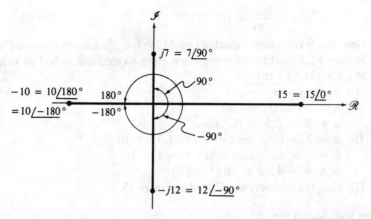

FIGURE 13.4 Examples of purely real and purely imaginary complex numbers.

Note these important special cases of complex numbers:

1. The positive real number $a = a + j0 = a \underline{/0°}$.
2. The positive imaginary number $jb = 0 + jb = b \underline{/90°}$.
3. The negative real number $-a = -a + j0 = a \underline{/180°}$.
4. The negative imaginary number $-jb = 0 - jb = b \underline{/-90°}$.

Figure 13.4 shows some examples.

Changing the sign in front of a complex number in polar form is the same as adding or subtracting 180° from its angle:

$$-M \underline{/\theta} = M \underline{/\theta \pm 180°} \tag{13.9}$$

For example, $-5 \underline{/60°} = 5 \underline{/-120°}$ and $24 \underline{/-50°} = -24 \underline{/130°}$.

13.3 Arithmetic Operations

To add two complex numbers, we add their real parts and their imaginary parts:

$$(a + jb) + (c + jd) = (a + c) + j(b + d) \tag{13.10}$$

Similarly, to subtract two complex numbers, we subtract their real and imaginary parts:

$$(a + jb) - (c + jd) = (a - c) + j(b - d) \tag{13.11}$$

Clearly, complex numbers in polar form must be converted to rectangular form before they can be added or subtracted. Algebraic operations that are valid for ordinary binomials are equally valid for complex numbers in rectangular form. For example, $-(a + jb) = -a - jb$ and $c(a + jb) = ca + jcb$.

Example 13.3

Perform the following computations.

(a) $(4 + j7) + (3 - j2)$
(b) $(25 - j5) - (-4 + j10)$
(c) $100 \underline{/30°} + 100 \underline{/-70°}$

SOLUTION

(a) $(4 + j7) + (3 - j2) = (4 + 3) + j(7 - 2) = 7 + j5$
(b) $(25 - j5) - (-4 + j10) = [25 - (-4)] + j(-5 - 10) = 29 - j15$

Note that this computation could also be performed as follows:

$$(25 - j5) + [-1(-4 + j10)] = (25 - j5) + (4 - j10) = 29 - j15$$

(c) We must convert both numbers to rectangular form before adding:

$$100 \underline{/30°} = 100 \cos 30° + j100 \sin 30° = 86.6 + j50$$

$$100 \underline{/-70°} = 100 \cos(-70°) + j100 \sin(-70°) = 34.2 - j94$$
$$(86.6 + j50) + (34.2 - j94) = (86.6 + 34.2) + j(50 - 94) = 120.8 - j44$$

Drill Exercise 13.3

Find $1.2 \times 10^6 \underline{/-40°} - 10^6 \underline{/120°}$. Express the answer in polar form.

ANSWER: $429.8 \times 10^3 \underline{/12.73°}$. ☐

To multiply two complex numbers in polar form, we multiply their magnitudes and *add* their angles:

$$(M_1 \underline{/\theta_1})(M_2 \underline{/\theta_2}) = M_1 M_2 \underline{/\theta_1 + \theta_2} \tag{13.12}$$

Two complex numbers can also be multiplied in rectangular form. The procedure is the same as that used to multiply two binomials:

$$
\begin{array}{l}
(a + jb) \\
\times (c + jd) \\
\hline
ac + jbc \\
\quad jad + j^2 bd \\
\hline
ac + j(bc + ad) - bd = (ac - bd) + j(bc + ad)
\end{array} \qquad = -bd
\tag{13.13}
$$

Example 13.4

(a) Find the product of $\mathbf{C}_1 = 4 \underline{/35°}$ and $\mathbf{C}_2 = 10 \underline{/20°}$.
(b) Find the product of $\mathbf{C}_1 = 3 + j4$ and $\mathbf{C}_2 = 8 - j4$ by
   (i) multiplying in rectangular form and converting the product to polar form
   (ii) converting $\mathbf{C}_1$ and $\mathbf{C}_2$ to polar form and then multiplying

SOLUTION

(a) $(4 \underline{/35°})(10 \underline{/20°}) = (4)(10) \underline{/35° + 20°} = 40 \underline{/55°}$

(b) (i) $(3 + j4)(8 - j4) = (3)(8) + 3(-j4) + (j4)(8) + (j4)(-j4)$
$$= 24 - j12 + j32 - j^2 16 = (24 + 16) + j(-12 + 32) = 40 + j20$$
$$= +16$$

Converting to polar form, we find

$$M = \sqrt{40^2 + 20^2} = 44.72$$

$$\theta = \tan^{-1} \frac{20}{40} = 26.56°$$

Thus,

$$\mathbf{C}_1 \mathbf{C}_2 = 44.72 \underline{/26.56°}$$

(ii) Converting C_1 and C_2 to polar form, we find

$$C_1 = \sqrt{3^2 + 4^2} \; \underline{/\tan^{-1}\frac{4}{3}} = 5 \underline{/53.13°}$$

$$C_2 = \sqrt{8^2 + (-4)^2} \; \underline{/\tan^{-1}\left(\frac{-4}{8}\right)} = 8.944 \underline{/-26.57°}$$

Then,

$$C_1 C_2 = (5 \underline{/53.13°})(8.944 \underline{/-26.57°})$$

$$= (5)(8.944) \underline{/53.13° + (-26.57°)} = 44.72 \underline{/26.56°}$$

As expected, the product is the same using both methods of computation.

Drill Exercise 13.4

If $C_1 = 0.18 \underline{/-45°}$ and $C_2 = -15 + j40$, find $C_1 C_2$ in polar form.

ANSWER: $7.69 \underline{/65.56°}$.   □

The quotient of two complex numbers in polar form, $M_1 \underline{/\theta_1}/M_2 \underline{/\theta_2}$, is found by dividing their magnitudes and *subtracting* the denominator angle from the numerator angle:

$$\frac{M_1 \underline{/\theta_1}}{M_2 \underline{/\theta_2}} = \frac{M_1}{M_2} \underline{/\theta_1 - \theta_2} \tag{13.14}$$

Complex numbers in rectangular form can be divided by converting them to polar form and using (13.14). They can also be divided using the procedure called *rationalization*. This procedure requires that we find the *complex conjugate* of the denominator. The complex conjugate, designated $\overline{C}$, of the complex number C is formed by changing the sign between the real and imaginary parts of C. Thus, if $C = a + jb$, then

$$\overline{C} = a - jb \tag{13.15}$$

For example, if $C = 8 + j5$, then $\overline{C} = 8 - j5$, and if $C = -2 - j4$, then $\overline{C} = -2 + j4$. An important and useful property of complex conjugates is that the product of a complex number and its conjugate, $C\overline{C}$, is a positive *real* number:

$$C\overline{C} = (a + jb)(a - jb) = a^2 + jab - jab - j^2 b^2 = a^2 + b^2 \tag{13.16}$$

Note that $C\overline{C} = M^2$, where M is the magnitude of C. The product of a complex number and its conjugate is always the sum of the squares of its real and imaginary parts, regardless of the algebraic signs of the parts. Following are some examples:

$$C\overline{C} = (3 + j2)(3 - j2) = 3^2 + 2^2 = 9 + 4 = 13$$

$$C\overline{C} = (6 - j6)(6 + j6) = 6^2 + (-6)^2 = 36 + 36 = 72$$

$$C\overline{C} = (-10 + j5)(-10 - j5) = (-10)^2 + 5^2 = 100 + 25 = 125$$

The complex conjugate of a number expressed in polar form is found by changing the sign of the angle. Thus, if $\mathbf{C} = M \underline{/\theta}$, then

$$\overline{\mathbf{C}} = M \underline{/-\theta} \qquad (13.17)$$

Rationalizing a fraction whose numerator and denominator are both complex numbers in rectangular form means converting the fraction to a single complex number. This conversion is accomplished by multiplying the numerator and the denominator of the fraction by the complex conjugate of the denominator:

$$\frac{a + jb}{c + jd}\left(\frac{c - jd}{c - jd}\right) = \frac{(ac + bd) + j(bc - ad)}{c^2 + d^2} = \frac{ac + bd}{c^2 + d^2} + j\left(\frac{bc - ad}{c^2 + d^2}\right) \qquad (13.18)$$

The procedure is valid because multiplying the numerator and denominator of a fraction by the same quantity does not alter the value of the fraction. As previously indicated, rationalizing amounts to find the rectangular form of the quotient of two complex numbers.

Example 13.5
Perform the following computations.

(a) $\dfrac{75 \underline{/100°}}{15 \underline{/40°}}$

(b) $\dfrac{2.8 \times 10^3 \underline{/75°}}{10^6 \underline{/-30°}}$

(c) $\dfrac{1.2 + j3}{4 - j6}$

SOLUTION

(a) $\dfrac{75 \underline{/100°}}{15 \underline{/40°}} = \dfrac{75}{15} \underline{/100° - 40°} = 5 \underline{/60°}$

(b) $\dfrac{2.8 \times 10^3 \underline{/75°}}{10^6 \underline{/-30°}} = \dfrac{2.8 \times 10^3}{10^6} \underline{/75° - (-30°)} = 2.8 \times 10^{-3} \underline{/105°}$

(c) Multiplying numerator and denominator by the complex conjugate of the denominator, we obtain

$$\left(\frac{1.2 + j3}{4 - j6}\right)\left(\frac{4 + j6}{4 + j6}\right) = \frac{1.2(4) + 1.2(j6) + j3(4) + j3(j6)}{4^2 + 6^2}$$

$$= \frac{4.8 - 18}{52} + j\left(\frac{7.2 + 12}{52}\right) = -0.254 + j0.369$$

The division can also be performed by first converting numerator and denominator to their polar forms:

$$1.2 + j3 = \sqrt{1.2^2 + 3^2} \bigg/ \tan^{-1}\frac{3}{1.2} = 3.23 \underline{/68.2°}$$

$$4 - j6 = \sqrt{4^2 + 6^2} \bigg/ \tan^{-1}\frac{-6}{4} = 7.21 \underline{/-56.3°}$$

$$\frac{1.2 + j3}{4 - j6} = \frac{3.23\underline{/68.2°}}{7.21\underline{/-56.3°}} = \frac{3.23}{7.21}\underline{/68.2° - (-56.3°)} = 0.448\underline{/124.5°}$$

Converting to rectangular form, we obtain the same result as before:

$$0.448 \cos(124.5°) + j0.448 \sin(124.5°) = -0.254 + j0.369$$

Drill Exercise 13.5

If $C_1 = 30\underline{/50°}$ and $C_2 = -5 + j2$, find C_1/C_2 in polar and rectangular form.

ANSWER: $5.57\underline{/-108.2°} = -1.74 - j5.29$. $\square$

The reciprocal of a complex number is a special case of complex division. In this case, the numerator is the real number $1 = 1 + j0 = 1\underline{/0°}$. As before, we can either convert the denominator to polar form and divide, or rationalize the fraction.

Example 13.6

Find the reciprocal of $12 - j16$

(a) in polar form, by division
(b) in rectangular form, by rationalizing

SOLUTION

(a) $M = \sqrt{12^2 + 16^2} = 20$

$$\theta = \tan^{-1}\left(\frac{-16}{12}\right) = -53.13°$$

Thus,

$$\frac{1}{12 - j16} = \frac{1}{20\underline{/-53.13°}} = \frac{1\underline{/0°}}{20\underline{/-53.13°}}$$

$$= \frac{1}{20}\underline{/0° - (-53.13°)} = 0.05\underline{/53.13°}$$

(b) Multiplying numerator and denominator by the complex conjugate of $12 - j16$, we find

$$\frac{1}{12 - j16}\left(\frac{12 + j16}{12 + j16}\right) = \frac{12 + j16}{12^2 + 16^2} = \frac{12}{400} + j\frac{16}{400}$$

$$= 0.03 + j0.04$$

Drill Exercise 13.6

Find the reciprocal of $0.02 \, \underline{/38°}$ in both polar and rectangular form.

ANSWER: $50 \, \underline{/-38°} = 39.4 - j30.8$. $\square$

A useful reciprocal is the special case $1/j$:

$$\frac{1}{j} = \frac{1 \, \underline{/0°}}{1 \, \underline{/90°}} = 1 \, \underline{/-90°} = -j$$

Thus,

$$\frac{1}{j} = -j \quad \text{and, equivalently,} \quad j = \frac{1}{-j} \qquad (13.19)$$

Equation 13.19 shows that j can be moved from the denominator to the numerator, or from the numerator to the denominator, in any expression, simply by changing its sign. Following are some examples:

$$\frac{50}{-j5} = j\left(\frac{50}{5}\right) = j10$$

$$\frac{4 + j2}{j2} = \frac{-j(4 + j2)}{2} = -j\left(\frac{4}{2}\right) - j^2\left(\frac{2}{2}\right) = 1 - j2$$

$$-j10 \, \underline{/30°} = \frac{10 \, \underline{/30°}}{j} = \frac{10 \, \underline{/30°}}{1 \, \underline{/90°}} = 10 \, \underline{/-60°}$$

When simplifying expressions containing several complex numbers in both rectangular and polar form, it is usually necessary to perform numerous conversions from one form to the other. Often the number of conversions and the computational labor can be minimized by judicious choice of the sequence and types of conversions made. For example, when evaluating the expression

$$\frac{42 \, \underline{/30°}}{3 + j4} (2 + j2)$$

we notice that multiplication and division operations are involved (by $2 + j2$ and $3 + j4$, respectively), so the computations can be performed more readily by converting $2 + j2$ and $3 + j4$ to polar form than by converting $42 \, \underline{/30°}$ to rectangular form.

13.4 Phasors

A *phasor* is a mathematical representation of an ac quantity in polar form. As we shall see, we can treat a phasor as the polar form of a complex number and can therefore convert it to an equivalent rectangular form. To represent an ac voltage or current in polar form, we let the magnitude M be the peak value of the voltage or current. (Some authors use the rms value.) The angle θ is the phase angle of the voltage or current. Following are some examples:

Sinusoidal Form	Phasor (Polar) Form
$v(t) = 170 \sin (377t + 40°)$ V	$170 \underline{/40°}$ V
$i(t) = 0.05 \sin (\omega t)$ A	$0.05 \underline{/0°}$ A
$v(t) = 10^{-3} \sin (10^6 t - 120°)$ V	$10^{-3} \underline{/-120°}$ V

Notice that the *frequency* of the sinusoidal waveform does not appear in its phasor representation. When using phasors to solve a specific problem in ac circuit analysis, we assume that all voltages and currents in that problem have the *same* frequency. Phasors and phasor analysis are used only in circuit problems where the ac waveforms are sinusoidal.

Phasors are sometimes called *vectors*, because they have both magnitude and direction (angle). Indeed, phasor analysis of ac circuits is very similar to traditional vector analysis. In keeping with this tradition, the line representing a phasor is drawn in the complex plane with an arrowhead. Drawing phasors in the complex plane is a convenient way to display phase relations among ac waveforms. Figure 13.5 shows an example. The greater the counterclockwise rotation of a phasor, the more leading is the waveform it represents. In the figure it is readily apparent that v_1 leads v_2 by 30° and that v_2 leads i by 75°.

Phasors also provide a convenient way to convert waveforms expressed as sines or cosines to equivalent waveforms expressed as cosines or sines (see Figure 13.6). Since the positive real axis has angle 0°, we can regard it as a sine reference axis; positive angle added to a sine waveform produces a counterclockwise rotation of a phasor from the positive real axis, and negative angle produces a clockwise rotation. The positive imaginary axis has angle +90°, so we can regard it as a cosine reference axis. Positive

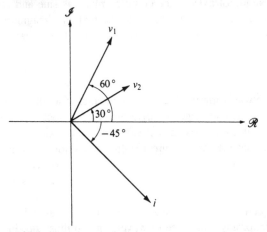

FIGURE 13.5 Examples of phasors plotted in the complex plane.

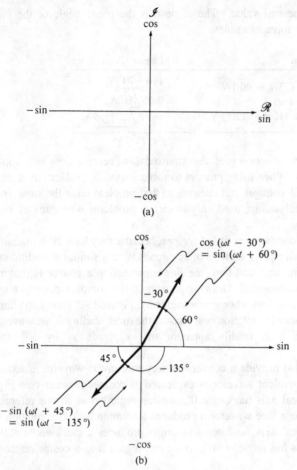

FIGURE 13.6 Interpretation of axes in the complex plane as sine and cosine reference axes. (a) Sine and cosine axes corresponding to the real and imaginary axes. (b) Using the interpretation in part (a) to convert phasors to equivalent forms.

angle added to a cosine waveform produces a counterclockwise rotation of a phasor from the positive imaginary axis, and negative angle produces a clockwise rotation from that axis. Similarly, the negative real axis can be regarded as the $-$sine axis and the negative imaginary axis as the $-$cosine axis. Figure 13.6(b) shows how these interpretations are used to convert the cosine expression $\cos(\omega t - 30°)$ to the equivalent sine expression $\sin(\omega t + 60°)$, and $-\sin(\omega t + 45°)$ to the equivalent expression $\sin(\omega t - 135°)$.

It is an interesting and useful fact that the sum (or difference) of any two sine waves having the same frequency is another sine wave with that same frequency. No such statement can be made for any other type of waveform. To illustrate, Figure 13.7 shows the sum of the two sine waves $v_1 = 10 \sin(\omega t)$ V and $v_2 = 20 \sin(\omega t + 60°)$ V. In this case, the sum is the sine wave $26.46 \sin(\omega t + 40.9°)$ V.

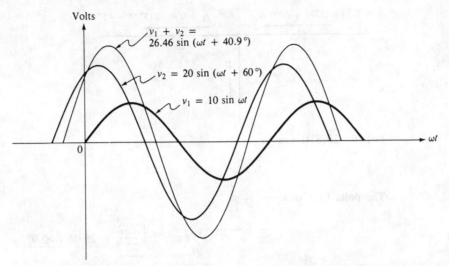

FIGURE 13.7 The sum of the two sine waves, $v_1 + v_2$, is a third sine wave having the same frequency as v_1 and v_2.

One method we could use to find the sum of two sine waves would be to add them point by point. Thus, at every instant of time we could add the instantaneous value of one waveform to the instantaneous value of the other and in that way compute all the instantaneous values of the sum waveform. That would be a very laborious procedure, and, in any case, would only give us a list of values of the sum, rather than a precise mathematical expression for it. On the other hand, adding the *phasor* forms of the sine waves is a quick way to obtain a precise expression for the sum waveform. To perform the addition, we must convert the polar forms of the phasors to their equivalent rectangular forms. The rectangular form of the sum can then be converted to polar form and, finally, back to sinusoidal form. To illustrate the procedure, we will add the waveforms v_1 and v_2 shown in Figure 13.7. The phasor forms of v_1 and v_2 are*

$$v_1 = 10 \sin \omega t = 10 \underline{/0°} \text{ V}$$
$$v_2 = 20 \sin (\omega t + 60°) = 20 \underline{/60°} \text{ V}$$

Converting each to rectangular form and adding, we find

$$v_1 = 10(\cos 0° + j \sin 0°) = 10 + j0$$
$$v_2 = 20(\cos 60° + j \sin 60°) = \underline{10 + j17.32}$$
$$v_1 + v_2 = 20 + j17.32$$

---

*Some authors prefer to use a different symbol for the phasor form of an ac voltage or current, to distinguish it from the sinusoidal form. For example, we could write $\bar{V} = 170 \underline{/40°}$ V, or $\mathbf{V} = 170 \underline{/40°}$V, when $v(t) = 170 \sin (377t + 40°)$ V. In this book we will use lower-case letters to represent both phasor and sinusoidal forms. Using different symbols for the same *physical* quantity is a needless complication, even though the mathematical representations of the quantity are, in fact, different.

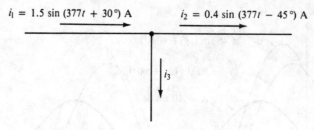

$i_1 = 1.5 \sin (377t + 30°)$ A $i_2 = 0.4 \sin (377t - 45°)$ A

i_3

FIGURE 13.8 (Example 13.8)

The polar form of $v_1 + v_2$ is

$$\sqrt{20^2 + 17.32^2} \Big/ \tan^{-1} \frac{17.32}{20} = 26.46 \underline{/40.9°}$$

Finally, converting the polar form of the sum to sinusoidal form, we have

$$v_1 + v_2 = 26.46 \sin (\omega t + 40.9°) \text{ V}$$

Example 13.7

Find the ac current i_3 at the node shown in Figure 13.8.

SOLUTION By Kirchhoff's current law,

$$i_2 + i_3 = i_1$$

Therefore,

$$i_3 = i_1 - i_2$$

Converting i_1 and i_2 to polar form, we have

$$i_1 = 1.5 \sin (377t + 30°) \text{ A} = 1.5 \underline{/30°} \text{ A}$$

$$i_2 = 0.4 \sin (377t - 45°) \text{ A} = 0.4 \underline{/-45°} \text{ A}$$

Converting the polar forms to rectangular forms, we find

$$i_1 = 1.5 (\cos 30° + j \sin 30°) = 1.3 + j0.75$$

$$i_2 = 0.4(\cos -45° + j \sin -45°) = 0.283 - j0.283$$

Then

$$i_3 = i_1 - i_2 = (1.3 + j0.75) - (0.283 - j0.283) = 1.017 + j1.033 \text{ A}$$

$$= \sqrt{(1.017)^2 + (1.033)^2} \Big/ \tan^{-1} \frac{1.033}{1.017} = 1.45 \underline{/45.45°} \text{ A}$$

Converting back to sinusoidal form, we find

$$i_3 = 1.45 \sin (377t + 45.45°) \text{ A}$$

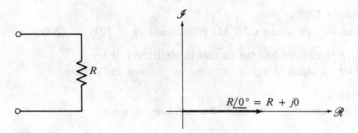

FIGURE 13.9 The phasor form of resistance.

Drill Exercise 13.7

Find the sum of $v_1 = 0.8 \sin (10^6 t + 137°)$ V and $v_2 = 0.3 \sin (10^6 t - 43°)$ V.

ANSWER: $0.5 \sin (10^6 t + 137°)$ V.   □

13.5 The Phasor Form of Impedance

Recall that reactance, like resistance, has units of ohms and is a measure of the extent to which a component hinders, or *impedes* the flow of current through it. Capacitive reactance is inversely proportional to frequency, $X_C = 1/\omega C$, inductive reactance is directly proportional to frequency, $X_L = \omega L$, and resistance does not depend on frequency. *Impedance* is the general term that applies to all three of these quantities, or to any combination of them. The symbol Z is used to represent impedance. We wish now to develop a phasor representation for each of the impedance types, resistance, capacitive reactance, and inductive reactance. In Chapter 14 we will study the phasor representation of combinations of these types.

Resistance

As discussed in Chapter 12, the voltage across a resistor, $v(t) = V_p \sin (\omega t + \theta)$ V, is in phase with the current through it: $i(t) = I_p \sin (\omega t + \theta)$ A. By Ohm's law,

$$R = \frac{v(t)}{i(t)} = \frac{V_p \sin (\omega t + \theta)}{I_p \sin (\omega t + \theta)} \tag{13.20}$$

Converting the sinusoidal expressions in (13.20) to phasors, we have

$$R = \frac{V_p \underline{/\theta}}{I_p \underline{/\theta}} = \frac{V_p}{I_p} \underline{/0°} \quad \text{ohms} \tag{13.21}$$

Equation (13.21) shows that we can regard resistance as a phasor whose magnitude is the resistance in ohms and whose angle is 0°. In the complex plane, resistance is therefore a phasor that lies along the real axis (see Figure 13.9). The rectangular form of resistance is $R + j0$.

Example 13.8

The voltage across a 2.2-kΩ resistor is $v(t) = 3.96 \sin (2000t + 50°)$ V.

(a) Use phasors to find the current through the resistor.
(b) Make a phasor diagram showing the voltage and current.

SOLUTION

(a) The phasor forms of the voltage and resistance are

$$v = 3.96 \underline{/50°} \text{ V} \quad \text{and} \quad R = 2.2 \times 10^3 \underline{/0°} \ \Omega$$

Therefore,

$$i = \frac{v}{R} = \frac{3.96 \underline{/50°} \text{ V}}{2.2 \times 10^3 \underline{/0°} \ \Omega} = 1.8 \times 10^{-3} \underline{/50°} \text{ A}$$

Converting to sinusoidal form, we have

$$i(t) = 1.8 \sin (2000t + 50°) \text{ mA}$$

(b) The phasor diagram is shown in Figure 13.10.

Drill Exercise 13.8

Use phasors to find the voltage across a 400-Ω resistor when the current through it is $i(t) = 0.06 \sin (\omega t - 30°)$ A.

ANSWER: $24 \underline{/-30°}$ V $= 24 \sin (\omega t - 30°)$ V.   □

Capacitive Reactance

Recall that the current through a capacitor leads the voltage across it by 90°. Thus, when $v(t) = V_p \sin (\omega t + \theta)$ V, $i(t) = I_p \sin (\omega t + \theta + 90°)$ A. Applying Ohm's law for capacitive reactance,

$$v(t) = X_C i(t)$$

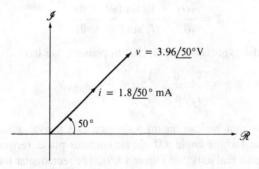

FIGURE 13.10 (Example 13.8)

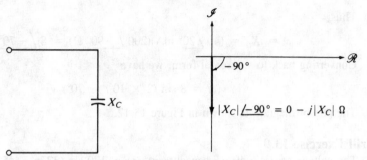

FIGURE 13.11 The phasor form of capacitive reactance.

or

$$X_C = \frac{v(t)}{i(t)} = \frac{V_p \sin(\omega t + \theta)}{I_p \sin(\omega t + \theta + 90°)} \tag{13.22}$$

Converting the sinusoidal expressions in (13.22) to phasor form, we have

$$X_C = \frac{V_p \big/\theta}{I_p \big/\theta + 90°} = \frac{V_p}{I_p} \big/-90° \; \Omega = \frac{1}{\omega C} \big/-90° \; \Omega \tag{13.23}$$

Equation 13.23 shows that capacitive reactance can be regarded as a phasor whose magnitude is $|X_C| = 1/\omega C$ ohms and whose angle is $-90°$. Thus, capacitive reactance is plotted down the negative imaginary axis in the complex plane (see Figure 13.11). Since capacitive reactance is a phasor, having both magnitude and angle, we will be careful hereafter to distinguish between its *magnitude*, denoted $|X_C|$, and its phasor form, $X_C = |X_C| \big/-90°$. The rectangular form is $X_C = 0 - j|X_C|$.

Example 13.9

The current through a 0.25-μF capacitor is $i(t) = 40 \sin(2 \times 10^4 t + 20°)$ mA.

(a) Use phasors to find the voltage across the capacitor.
(b) Make a phasor diagram showing the voltage and current.

SOLUTION
(a) We must first find the magnitude of the capacitive reactance:

$$|X_C| = \frac{1}{\omega C} = \frac{1}{(2 \times 10^4)(0.25 \times 10^{-6})} = 200 \; \Omega$$

In phasor form,

$$X_C = 200 \big/-90° \; \Omega$$

Also,

$$i = 40 \big/20° \; \text{mA}$$

Thus,

$$v = iX_C = (40 \underline{/20°} \text{ mA})(200 \underline{/-90°} \ \Omega) = 8 \underline{/-70°} \text{ V}$$

Converting back to sinusoidal form, we have

$$v(t) = 8 \sin (2 \times 10^4 t - 70°) \text{ V}$$

(b) The phasor diagram is shown in Figure 13.12.

Drill Exercise 13.9

The voltage across a 40-μF capacitor is $v(t) = 170 \sin (377t - 45°)$ V. Use phasors to find the current through the capacitor.

ANSWER: $2.56 \underline{/45°}$ A $= 2.56 \sin (377t + 45°)$ A.   ☐

Inductive Reactance

Recall that the voltage across an inductor leads the current through it by 90°. Thus, when $i(t) = I_p \sin (\omega t + \theta)$ A, $v(t) = V_p \sin (\omega t + \theta + 90°)$ V. Applying Ohm's law for inductive reactance,

$$v(t) = X_L i(t)$$

or

$$X_L = \frac{v(t)}{i(t)} = \frac{V_p \sin (\omega t + \theta + 90°)}{I_p \sin (\omega t + \theta)} \tag{13.24}$$

Converting the sinusoidal expressions in (13.24) to phasor form, we have

$$X_L = \frac{V_p \underline{/\theta + 90°}}{I_p \underline{/\theta}} = \frac{V_p}{I_p} \underline{/90°} \ \Omega = \omega L \underline{/90°} \ \Omega. \tag{13.25}$$

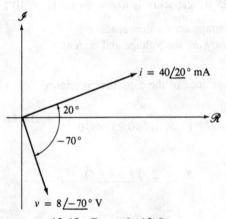

$i = 40\underline{/20°}$ mA

$20°$

$-70°$

$v = 8\underline{/-70°}$ V

FIGURE 13.12 (Example 13.9)

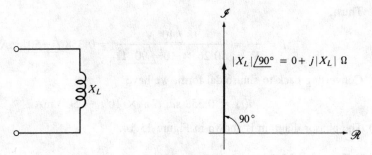

FIGURE 13.13 The phasor form of inductive reactance.

Equation (13.25) shows the inductive reactance can be regarded as a phasor whose magnitude is $|X_L| = \omega L$ ohms and whose angle is 90°. Thus, inductive reactance is plotted up the positive imaginary axis, as shown in Figure 13.13. Be careful to distinguish between the magnitude of inductive reactance, $|X_L|$, and the phasor form: $X_L = |X_L| \underline{/90°} = 0 + j|X_L|$.

Example 13.10

The voltage across an 8-mH inductor is $v(t) = 18 \sin(2\pi \times 10^6 t + 40°)$ V.

(a) Use phasors to find the current through the inductor.
(b) Make a phasor diagram showing the voltage and current.

SOLUTION

(a) The magnitude of the inductive reactance is

$$|X_L| = \omega L = (2\pi \times 10^6)(8 \times 10^{-3}) = 50.26 \text{ k}\Omega$$

In phasor form,

$$X_L = 50.26 \times 10^3 \underline{/90°} \ \Omega$$

Also,

$$v = 18 \underline{/40°} \text{ V}$$

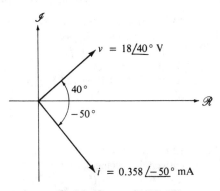

FIGURE 13.14 (Example 13.10)

Thus,

$$i = \frac{v}{X_L} = \frac{18 \;/\!40°\; V}{50.26 \times 10^3 \;/\!90°\; \Omega} = 0.358 \;/\!-50°\; mA$$

Converting back to sinusoidal form, we have

$$i(t) = 0.358 \sin (2\pi \times 10^6 t - 50°) \; mA$$

(b) The phasor diagram is shown in Figure 13.14.

Drill Exercise 13.10

The current through a 4-H inductor is $i(t) = 0.6 \sin (100t - 10°)$ A. Use phasors to find voltage across it.

ANSWER: $240 \;/\!80°\; V = 240 \sin (100t + 80°) \; V.$ □

Exercises

Section 13.1 Complex Numbers

13.1 Find the real and imaginary parts of each of the following complex numbers.
(a) $50 + j60$ (b) $3 \times 10^{-3} - j1 \times 10^{-3}$
(c) $j17$ (d) j^2
(e) 420×10^5 (f) $-2 - j$

13.2 Find the real and imaginary parts of each of the following complex numbers.
(a) $-14 - j2$ (b) $1 - j$
(c) j^3 (d) $j5 + 10$
(e) j (f) $\sqrt{2} - \sqrt{3}\, j^5$

Section 13.2 The Complex Plane

13.3 Convert each of the following complex numbers to polar form.
(a) $3 + j4$ (b) $10 - j10$
(c) $-1 + j$ (d) 56
(e) $-0.2 - 0.1j$ (f) j

13.4 Convert each of the following complex numbers to polar form.
(a) $1.2 \times 10^6 + j0.8 \times 10^6$ (b) $1 - j$
(c) $-0.003 + j0.003$ (d) $-j$
(e) $\sqrt{2} - j2\sqrt{2}$ (f) $-18 \times 10^{-3} - j12 \times 10^{-3}$

13.5 Convert each of the following to rectangular form.
(a) $200 \;/\!30°$ (b) $65 \;/\!-45°$
(c) $0.04 \;/\!130°$ (d) $8.24 \times 10^4 \;/\!-160°$
(e) $12 \;/\!0°$ (f) $3.8 \;/\!-180°$

13.6 Convert each of the following to rectangular form.
(a) $58 \;/\!75°$ (b) $1.27 \times 10^3 \;/\!175°$
(c) $330 \;/\!-125°$ (d) $1 \;/\!90°$
(e) $-5 \;/\!-80°$ (f) $\sqrt{2} \;/\!45°$

Section 13.3 Arithmetic Operations

13.7 Perform the following computations. Express all answers in rectangular form.
(a) $(3 + j5) + (3 + j7)$
(b) $(6 - j2) + (-5 + j1)$
(c) $(0.05 + j0.1) - (0.05 - j0.1)$
(d) $(-329 + j136) - (-152 - j297)$
(e) $3.4 \times 10^3 \;/\!30° + 1.7 \times 10^3 \;/\!60°$
(f) $14 \;/\!65° - 12 \;/\!140°$

13.8 Perform the following computations. Express all answers in rectangular form.
(a) $(16 - j2) + (5 + j8)$
(b) $(-395 + j160) - (200 - j321)$

(c) $(0.04 - j0.02) - (0.02 - j0.02)$

(d) $(4 - j3) + (-6 + j2) - (7 - j3)$

(e) $-16 \angle{-20°} + 20 \angle{15°}$

(f) $125 \times 10^6 \angle{180°} + 40 \times 10^6 \angle{0°} - 70 \times 10^6 \angle{-90°}$

13.9 Perform the following computations. Express all answers in polar form.

(a) $(0.3 \angle{10°})(14 \angle{96°})$

(b) $(125 \angle{-65°})(0.8 \angle{32°})$

(c) $(1.55 \times 10^{-6} \angle{0°})(0.2 \angle{-30°})(42 \times 10^3 \angle{128°})$

(d) $\dfrac{180 \angle{27°}}{1.5 \angle{85°}}$

(e) $\dfrac{3.6 \angle{-15°}}{0.18 \angle{-90°}}$

(f) $\dfrac{8.22 \angle{-170°}}{-4110 \angle{20°}}$

(g) $(1 + j2)(2 + j6)$

(h) $\dfrac{3 - j4}{0.6 - j0.8}$

13.10 Perform the following computations. Express all answers in polar form.

(a) $(12.5 \angle{-18°})(0.7 \angle{95°})$

(b) $\dfrac{1382 \angle{-40°}}{0.2 \angle{62°}}$

(c) $(-0.04 \angle{12.5°})(1.5 \angle{86.3°})$

(d) $\dfrac{6.93 \times 10^{-4}}{1.13 \angle{27.6°}}$

(e) $(5 - j7)(4 + j2)$

(f) $\dfrac{38 \angle{93°}}{8 - j8}$

(g) $\dfrac{14 \times 10^5 - j20 \times 10^5}{3 + j4}$

(h) $\dfrac{210 + j142}{3.6 \angle{-42°}}$

13.11 Perform the following computations. Express answers in both polar and rectangular forms.

(a) $\dfrac{(24 + j16)(0.5 \angle{-20°})}{3.2 \angle{40°}}$

(b) $\dfrac{j(0.2 + j0.2)}{3 \angle{90°} (0.1 - j0.05)}$

(c) $\dfrac{125 \angle{35°}}{-j(1 + j2)^2}$

13.12 Perform the following computations. Express answers in both polar and rectangular forms.

(a) $\dfrac{(1.2 \angle{-10°})^2}{2.8 + j6}$

(b) $\dfrac{140 - j60}{3 + j3} + \dfrac{60 - j140}{3 - j3}$

(c) $j(1 + j)^3 \, 6 \angle{-100°} - \dfrac{j4.2}{0.5 - j0.3}$

Section 13.4 Phasors

13.13 Write the phasor (polar) form of each of the following pairs of waveforms and sketch each pair of phasors in the complex plane.

(a) $v_1 = 16 \sin(\omega t + 35°)$ V
$v_2 = 32 \sin(\omega t - 55°)$ V

(b) $i_1 = 10 \sin(400t - 20°)$ mA
$i_2 = 15 \sin(400t - 130°)$ mA

(c) $v_1 = -0.5 \sin(10^3 t + 100°)$ V
$v_2 = 0.2 \cos(10^3 t)$

(d) $i_1 = 4.8 \sin(\omega t + 160°)$ A
$i_2 = 1.2 \sin(\omega t - 20°)$ A

13.14 Write the phasor (polar) form of each of the following pairs of waveforms and sketch each pair of phasors in the complex plane.

(a) $i_1 = 0.4 \sin(2\pi \times 10^4 t)$ A
$i_2 = 0.2 \sin(2\pi \times 10^4 t + 100°)$ A

(b) $v_1 = \sin(\omega t + 30°)$ V
$v_2 = \sin(\omega t - 150°)$ V

(c) $i_1 = 0.3 \cos(900t - 25°)$ mA
$i_2 = -0.9 \cos(900t)$ mA

(d) $v_1 = 150 \sin(2\pi \times 100t + 90°)$ V
$v_2 = 100 \sin(2\pi \times 100t - 90°)$ V

13.15 Find $e(t)$ in Figure 13.15, given the voltage drops $v_1(t)$ and $v_2(t)$ shown. Express $e(t)$ in sinusoidal form.

13.16 Find $i(t)$ in Figure 13.16. Express the answer in sinusoidal form.

13.17 Find $v_3(t)$ in Figure 13.17. Express the answer in sinusoidal form.

13.18 Find $i(t)$ in Figure 13.18. Express the answer in sinusoidal form.

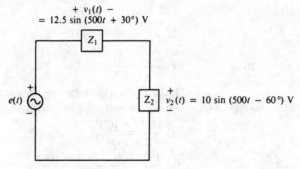

FIGURE 13.15 (Exercise 13.15)

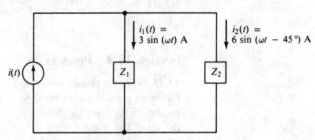

FIGURE 13.16 (Exercise 13.16)

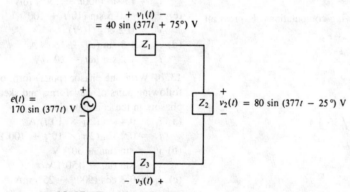

FIGURE 13.17 (Exercise 13.17)

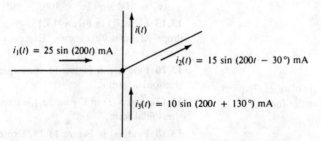

FIGURE 13.18 (Exercise 13.18)

Section 13.5 The Phasor Form of Impedance

13.19 The current through a 47-Ω resistor is $i(t) =$ 1.5 sin $(\omega t + 70°)$ A.
(a) Use phasors to find the voltage across the resistor. Express the result in sinusoidal form.
(b) Make a phasor diagram showing the voltage and current.

13.20 The voltage across a 330-kΩ resistor is $v(t) = 41.25$ sin $(4 \times 10^3 t - 150°)$ V.
(a) Use phasors to find the current through the resistor. Express the result in sinusoidal form.
(b) Make a phasor diagram showing the voltage and current.

13.21 The voltage across a 0.015-μF capacitor is $v(t) = 16.4$ sin $(3400t - 63°)$ V.
(a) Use phasors to find the current through the capacitor. Express the result in sinusoidal form.
(b) Make a phasor diagram showing the voltage and current.

13.22 The current through a 100-μF capacitor is $i(t) = 0.4$ cos $(377t)$ A.
(a) Use phasors to find the voltage across the capacitor. Express the result in sinusoidal form.
(b) Make a phasor diagram showing the voltage and current.

13.23 The current through a 0.2-H inductor is $i(t) = 0.06$ sin $(2\pi \times 10^3 t - 80°)$ A.
(a) Use phasors to find the voltage across the inductor. Express the result in sinusoidal form.
(b) Make a phasor diagram showing the voltage and current.

13.24 The voltage across a 250-μH inductor is $v(t) = 75$ sin $(3 \times 10^5 t + 120°)$ mV.
(a) Use phasors to find the current through the inductor. Express the result in sinusoidal form.
(b) Make a phasor diagram showing the voltage and current.

14 Series and Parallel AC Circuits

14.1 Series Equivalent Impedance

Recall that impedance (Z) is a general term that can refer to resistance, reactance, or combinations of resistance and reactance. When two or more impedances are connected in series, the total equivalent impedance is the sum of the series-connected impedances:

$$Z_T = Z_1 + Z_2 + \cdots + Z_n \qquad \text{ohms} \qquad (14.1)$$

It is important to remember that every impedance has both magnitude and angle. Addition of impedances must therefore be performed using the rectangular equivalents of their phasor forms. For example, it would be *incorrect* to state that the sum of 10 ohms of resistance and 10 ohms of inductive reactance is 20 Ω. Adding impedances is like adding *vectors*, in that we add the horizontal (real) components and the vertical (imaginary) components of each in order to obtain a resultant.

14.2 Series RL Circuits

Figure 14.1 shows a series RL network connected to an ac voltage source. We will find the phasor form of the total impedance of this combination. Recall that the inductive reactance of the inductor depends on the frequency of the voltage source and has the phasor form

$$X_L = \omega L \underline{/90°} = 0 + j\omega L = 0 + j|X_L| \qquad \text{ohms} \qquad (14.2)$$

The phasor form of the resistance is

$$R \underline{/0°} = R + j0 \qquad \text{ohms} \qquad (14.3)$$

Therefore, the total impedance of the series combination is

$$Z_T = (R + j0) + (0 + j\omega L) = R + j\omega L = R + j|X_L| \qquad \text{ohms} \qquad (14.4)$$

The magnitude and angle of Z_T can be found by converting (14.4) to polar form:

$$|Z_T| = \sqrt{R^2 + (\omega L)^2} \qquad (14.5)$$

$$\theta = \tan^{-1} \frac{\omega L}{R} \qquad (14.6)$$

Figure 14.2 shows a plot of Z_T in the complex plane.

Example 14.1 (Analysis)
In Figure 14.1, $R = 300 \ \Omega$, $L = 0.2$ H, and $e(t) = 17 \sin (2000t)$ V.

(a) Find the total equivalent impedance in polar and rectangular form.
(b) Sketch the impedance in the complex plane.

SOLUTION
(a) The magnitude of the inductive reactance is

$$|X_L| = \omega L = (2000)(0.2) = 400 \ \Omega$$

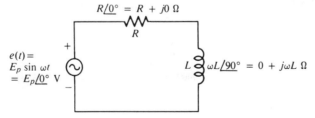

FIGURE 14.1 Series RL network connected to an ac voltage source.

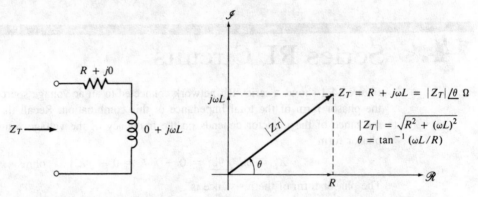

FIGURE 14.2 Total impedance of a series RL network plotted as a phasor in the complex plane.

Thus,

$$X_L = 0 + j400 \ \Omega$$

and

$$R = 300 + j0 \ \Omega$$

so

$$Z_T = (0 + j400) + (300 + j0) = 300 + j400 \ \Omega$$

$$|Z_T| = \sqrt{300^2 + 400^2} = 500 \ \Omega$$

$$\theta = \tan^{-1} \frac{\omega L}{R} = \tan^{-1} \frac{400}{300} = 53.13°$$

$$Z_T = 500 \underline{/53.13°} \ \Omega$$

(b) Z_T is plotted in Figure 14.3.

Drill Exercise 14.1

What value of resistance in Example 14.1 would make the angle of the total impedance equal 30°?

ANSWER: 692.8 Ω.   □

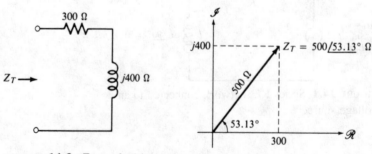

FIGURE 14.3 (Example 14.1)

Having found the phasor form of the total impedance of a circuit, we can use Ohm's law to find the total current supplied to the circuit by a voltage source:

$$i_T = \frac{v}{Z_T} \qquad (14.7)$$

Once again, it is important to remember that all mathematical operations, including the division in (14.7), must be performed using phasors, because all quantities have both magnitude and angle. The current i_T in a series circuit is the same through every series-connected component, so the ac voltage drop across each component can be found by multiplying each impedance by the current:

$$v_1 = i_T Z_1, \quad v_2 = i_T Z_2, \quad \dots \qquad (14.8)$$

Multiplication and division of phasors is most easily accomplished using polar forms, as discussed in Chapter 13.

Example 14.2 (Analysis)
In the series RL circuit shown in Figure 14.4:

(a) Find the total current in the circuit.
(b) Find the voltage drops v_R and v_L across R and L.
(c) Verify Kirchhoff's voltage law around the circuit.
(d) Make a phasor diagram showing e, v_R, v_L, and i_T.
(e) Sketch the waveforms of e, v_R, and v_L versus angle.

SOLUTION
(a) The total impedance is

$$Z_T = R + j|X_L| = 200 + j100 \ \Omega$$

$$= \sqrt{200^2 + 100^2} \ \Big/ \tan^{-1} \frac{100}{200} = 223.61 \ \underline{/26.56°} \ \Omega$$

Therefore, the total current is

$$i_T = \frac{e}{Z_T} = \frac{30\underline{/0°} \ \text{V}}{223.61 \cdot \underline{/26.56°} \ \Omega} = 0.134 \ \underline{/-26.56°} \ \text{A}$$

(b) Applying Ohm's law to each impedance, we find

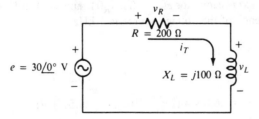

FIGURE 14.4 (Example 14.2)

$$v_R = i_T R = (0.134 \underline{/-26.56°} \text{ A})(200 \underline{/0°} \, \Omega) = 26.8 \underline{/-26.56°} \text{ V}$$

$$v_L = i_T X_L = (0.134 \underline{/-26.56°} \text{ A})(100 \underline{/90°} \, \Omega) = 13.4 \underline{/63.44°} \text{ V}$$

(c) To verify Kirchhoff's voltage law, we must show that $e = v_R + v_L$. To perform this computation, we must convert each voltage to rectangular form:

$$e = 30 \underline{/0°} \text{ V} = 30 + j0 \text{ V}$$

$$v_R = 26.8 \underline{/-26.56°} \text{ V} = 26.8 \cos(-26.56°) + j26.8 \sin(-26.56°)$$
$$= 23.97 - j11.98 \text{ V}$$

$$v_L = 13.4 \underline{/63.44°} \text{ V} = 13.4 \cos(63.44°) + j13.4 \sin(63.44°)$$
$$= 5.99 + j11.98 \text{ V}$$

$$v_R + v_L = (23.97 - j11.98) + (5.99 + j11.98)$$
$$= (29.96 + j0) \text{ V} \approx 30 + j0 \text{ V} = e$$

Except for a small round-off error, we see that the sum of the voltage drops, $v_R + v_L$, equals the source voltage e.

(d) The phasor diagram is shown in Figure 14.5(a). Notice these important points in connection with the diagram:

1. The voltage across and current through the resistor have the same angle, as expected, since these quantities are always in phase.
2. v_L leads i_T by 90°, confirming that the voltage across the inductor leads the current through it by 90°. Note carefully that it is the voltage *across* the inductor that leads the current by 90°, not the source voltage e.
3. The phasor sum $v_R + v_L$ can be found graphically by completing the parallelogram that has v_R and v_L as two of its sides. The other two sides are shown by dashed lines. This is the same graphical method that is used to find the resultant of two vectors.
4. The phasor sum of v_R and v_L equals the phasor e, graphically demonstrating the validity of Kirchhoff's voltage law. Since $|v_R|$ and $|v_L|$ are the lengths of the sides of a right triangle whose hypotenuse has length $|e|$, it follows that $|e| = \sqrt{|v_R|^2 + |v_L|^2}$.

(e) The waveforms are shown in Figure 14.5(b).

Drill Exercise 14.2

Write the sinusoidal expressions for $e(t)$, $i_T(t)$, $v_R(t)$, and $v_L(t)$ in Example 14.2, assuming that the frequency of the voltage source is 1 kHz.

ANSWER:

$$e(t) = 30 \sin(2\pi \times 10^3 t) \text{ V};$$

$$i_T(t) = 0.134 \sin(2\pi \times 10^3 t - 26.56°) \text{ A};$$

$$v_R(t) = 26.8 \sin(2\pi \times 103 \, t - 26.56°) \text{ V};$$

$$v_L(t) = 13.4 \sin(2\pi \times 10^3 t + 63.44°) \text{ V}.$$

$\square$

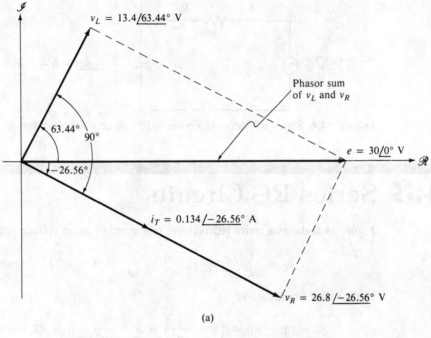

(a)

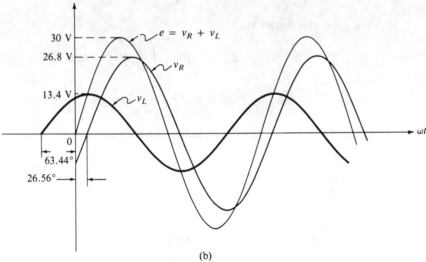

(b)

FIGURE 14.5 (Example 14.2)

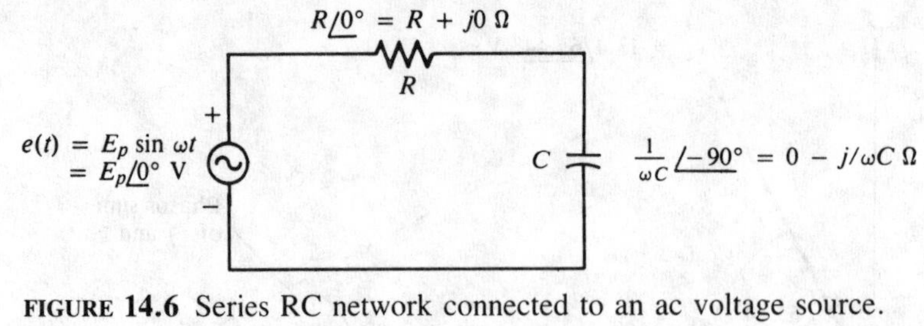

$R\underline{/0^\circ} = R + j0\ \Omega$

R

$e(t) = E_p \sin \omega t$
$= E_p\underline{/0^\circ}$ V

C

$\frac{1}{\omega C}\underline{/-90^\circ} = 0 - j/\omega C\ \Omega$

FIGURE 14.6 Series RC network connected to an ac voltage source.

14.3 Series RC Circuits

Figure 14.6 shows a series RC network connected to an ac voltage source. In this case

$$X_C = \frac{1}{\omega C}\underline{/-90^\circ}\ \Omega = 0 - \frac{j}{\omega C} = 0 - j|X_C| \qquad \text{ohms} \qquad (14.9)$$

and the total impedance is

$$Z_T = (R + j0) + \left(0 - \frac{j}{\omega C}\right) = R - \frac{j}{\omega C} = R - j|X_C| \qquad \text{ohms} \quad (14.10)$$

The polar coordinates are

$$|Z_T| = \sqrt{R^2 + \left(\frac{1}{\omega C}\right)^2} \qquad \text{ohms} \qquad (14.11)$$

$$\theta = \tan^{-1}\frac{-1/\omega C}{R} = \tan^{-1}\frac{-1}{\omega RC} \qquad (14.12)$$

Figure 14.7 shows a phasor diagram of the total impedance.

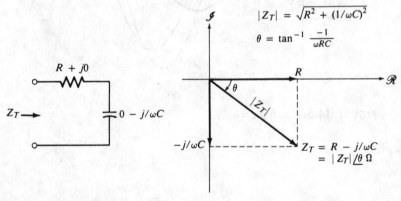

FIGURE 14.7 Total impedance of a series RC network plotted as a phasor in the complex plane.

As in the series RL circuit, the total current supplied to the RC network and the voltage drops across each component can be found using Ohm's law:

$$i_T = \frac{e}{Z_T} \qquad (14.13)$$

$$v_R = i_T R \qquad v_C = i_T X_C \qquad (14.14)$$

Again, all computations must be performed using phasors.

Example 14.3 (Analysis)
In the series RC circuit shown in Figure 14.8:

(a) Find the total current in the circuit in phasor and sinusoidal form.
(b) Find the voltage drops across the resistor and capacitor in phasor and sinusoidal form.
(c) Verify Kirchhoff's voltage law around the circuit.
(d) Make a phasor diagram showing e, i_T, v_R, and v_C.
(e) Sketch the waveforms of e, v_R, and v_C versus angle.

SOLUTION
(a) The magnitude of the capacitive reactance is

$$|X_C| = \frac{1}{\omega C} = \frac{1}{240(2.2 \times 10^{-6})} = 1894 \ \Omega$$

Thus,

$$X_C = 0 - j1894 \ \Omega = 1894 \underline{/-90°} \ \Omega$$

The total impedance of the circuit is

$$Z_T = R - \frac{j}{\omega C} = 3300 - j1894 \ \Omega$$

Converting to polar form, we have

$$Z_T = \sqrt{(3300)^2 + (1894)^2} \underline{/\tan^{-1}\left(\frac{-1894}{3300}\right)} \ \Omega$$

$$= 3805 \underline{/-29.85°} \ \Omega$$

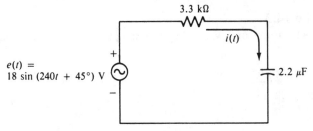

FIGURE 14.8 (Example 14.3)

Also, $e = 18 \underline{/45°}$ V. Thus,

$$i = \frac{e}{Z_T} = \frac{18 \underline{/45°} \text{ V}}{3805 \underline{/-29.85°} \text{ }\Omega} = 4.73 \times 10^{-3} \underline{/74.85°} \text{ A}$$

and

$$i(t) = 4.73 \sin(240t + 74.85°) \text{ mA}$$

(b) By Ohm's law,

$$v_R = iR = (4.73 \times 10^{-3} \underline{/74.85°} \text{ A})(3300 \underline{/0°} \text{ }\Omega) = 15.6 \underline{/74.85°} \text{ V}$$

$$v_R(t) = 15.6 \sin(240t + 74.85°) \text{ V}$$

$$v_C = iX_C = (4.73 \times 10^{-3} \underline{/74.85°} \text{ A})(1894 \underline{/-90°} \text{ }\Omega) = 8.96 \underline{/-15.15°} \text{ V}$$

$$v_C(t) = 8.96 \sin(240t - 15.15°) \text{ V}$$

(c) To verify Kirchhoff's voltage law, we convert each voltage in the circuit to rectangular form:

$$e = 18\underline{/45°} \text{ V} = 18 \cos 45° + j18 \sin 45° = 12.73 + j12.73 \text{ V}$$

$$v_R = 15.6 \underline{/74.85°} = 15.6 \cos 74.85° + j15.6 \sin 74.85°$$
$$= 4.08 + j15.06 \text{ V}$$

$$v_C = 8.96 \underline{/-15.15°} = 8.96 \cos(-15.15°) + j8.96 \sin(-15.15°)$$
$$= 8.65 - j2.34 \text{ V}$$

$$v_R + v_C = (4.08 + j15.06) + (8.65 - j2.34)$$
$$= 12.73 + j12.72 \text{ V} \approx e$$

Except for a small round-off error, $v_R + v_C = e$, verifying Kirchhoff's voltage law.

(d) The phasor diagram is shown in Figure 14.9(a). As in the case of the series RL circuit, the parallelogram graphically demonstrates that $v_R + v_C = e$. Note that the voltage, v_C, *across* the capacitor lags the current i through it by 90°, as expected. Since $|v_R|$ and $|v_C|$ are the lengths of the sides of a right triangle whose hypotenuse has length $|e|$, it follows that $|e| = \sqrt{|v_R|^2 + |v_C|^2}$.

(e) The waveforms of $e(t)$, $v_R(t)$, and $v_C(t)$ are shown in Figure 14.9(b).

Drill Exercise 14.3

Find the polar forms of i, v_R, and v_C in Example 14.3 when the frequency of e is halved.

ANSWER: $i = 3.58 \underline{/93.94°}$ mA; $v_R = 11.82 \underline{/93.94°}$ V; $v_C = 13.57 \underline{/3.94°}$ V.   □

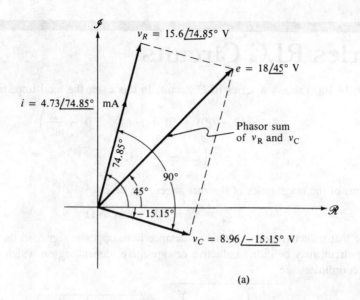

(a)

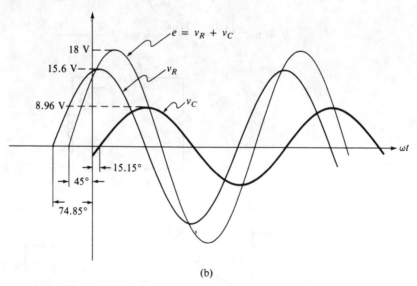

(b)

FIGURE 14.9 (Example 14.3)

14.4 Series RLC Circuits

Figure 14.10(a) shows a series RLC circuit. In this case, the total impedance is

$$Z_T = (R + j0) + (0 + j\omega L) + \left(0 - \frac{j}{\omega C}\right) \tag{14.15}$$

$$= R + j\left(\omega L - \frac{1}{\omega C}\right)\ \Omega$$

In terms of the magnitudes of the reactances, we have

$$Z_T = R + j(|X_L| - |X_C|)\ \Omega \tag{14.16}$$

Notice that inductive and capacitive reactance have opposite signs, so the *net* reactance in the circuit may be either inductive or capacitive, depending on which is larger. The polar coordinates are

$$|Z_T| = \sqrt{R^2 + \left(\omega L - \frac{1}{\omega C}\right)^2} = \sqrt{R^2 + (|X_L| - |X_C|)^2}\quad \text{ohms} \tag{14.17}$$

$$\theta = \tan^{-1}\frac{\omega L - 1/\omega C}{R} = \tan^{-1}\frac{|X_L| - |X_C|}{R} \tag{14.18}$$

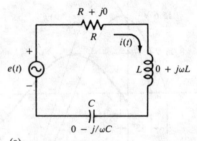

(a)

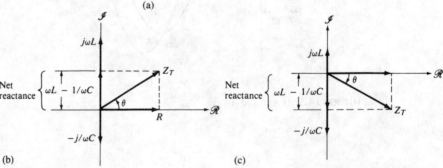

(b)   (c)

FIGURE 14.10 Impedance relations in a series RLC circuit. (a) Series RLC circuit. (b) Phasor diagram of the impedance when the inductive reactance is greater than the capacitive reactance ($|X_L| > |X_C|$). (c) Phasor diagram of the impedance when the capacitive reactance is greater than the inductive reactance ($|X_C| > |X_L|$).

The phasor diagram for the impedance of a circuit in which the inductive reactance predominates is shown in Figure 14.10(b). Figure 14.10(c) is a phasor diagram for the case where the capacitive reactance predominates.

When there is more than one resistor, capacitor, and/or inductor in a series circuit, the total impedance has a resistance component equal to the sum of the resistance values and a reactive component equal to the sum of the capacitive reactances subtracted from the sum of the inductive reactances. Note that series-connected *capacitive reactances are additive*, even though the total equivalent *capacitance* is not the sum of the capacitor values.

Example 14.4 (Analysis)
In the series RLC circuit shown in Figure 14.11:

(a) Find the current in polar form.
(b) Find the voltage drops v_R, v_L, v_C across the resistor, inductor, and capacitor, in polar form.
(c) Verify Kirchhoff's voltage law around the circuit.
(d) Make a phasor diagram showing e, i, v_R, v_L, and v_C.

SOLUTION
(a) The total impedance is

$$Z_T = 800 + j(1250 - 450) = 800 + j800 \ \Omega$$

$$= \sqrt{800^2 + 800^2} \ \bigg/ \tan^{-1} \frac{800}{800} = 1131.4 \ \underline{/45°} \ \Omega$$

Therefore, the current in the circuit is

$$i = \frac{e}{Z_T} = \frac{100 \ \underline{/0°} \ \text{V}}{1131.4 \ \underline{/45°} \ \Omega} = 88.39 \ \underline{/-45°} \ \text{mA}$$

(b) $v_R = iR = (88.39 \ \underline{/-45°} \ \text{mA})(800 \ \underline{/0°} \ \Omega) = 70.7 \ \underline{/-45°} \ \text{V}$

$v_L = iX_L = (88.39 \ \underline{/-45°} \ \text{mA})(12.50 \ \underline{/90°} \ \Omega) = 110.49 \ \underline{/45°} \ \text{V}$

$v_C = iX_C = (88.39 \ \underline{/-45°} \ \text{mA})(450 \ \underline{/-90°} \ \Omega) = 39.77 \ \underline{/-135°} \ \text{V}$

Note that the *magnitude* of the voltage across the inductor (110.49 V) is greater than the magnitude of e (100 V). As we shall see, this situation does not contradict

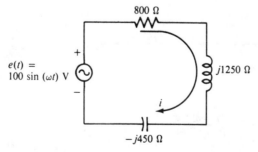

FIGURE 14.11 (Example 14.4)

Kirchhoff's voltage law, because v_L and v_C are 180° out of phase. As a consequence, the sum of the voltage drops around the circuit includes some cancellation of v_L by v_C. The situation is illustrated in Figure 14.12(a). It is often true in RLC circuits that the magnitude of the voltage across the inductor or capacitor is larger than the source voltage.

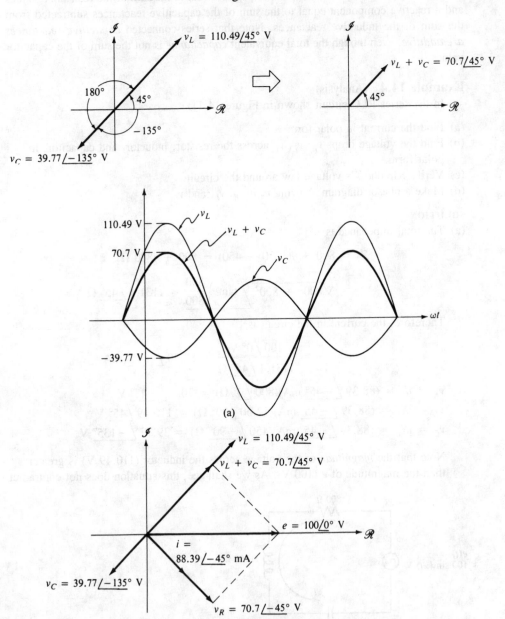

FIGURE 14.12 (Example 14.4)

(c) To verify Kirchhoff's voltage law, we convert the voltages to rectangular form:

$$e = 100 \underline{/0°} = 100 + j0 \text{ V}$$

$$v_R = 70.7 \cos(-45°) + j70.7 \sin(-45°) = 50 - j50 \text{ V}$$

$$v_L = 110.49 \cos 45° + j110.490 \sin 45° = 78.12 + j78.12 \text{ V}$$

$$v_C = 39.77 \cos(-135°) + j39.77 \sin(-135°)$$

$$= -28.12 - j28.12 \text{ V}$$

$$v_R + v_L + v_C = (50 - j50) + (78.12 + j78.12) + (-28.12 - j28.12)$$

$$= 100 + j0 = e$$

(d) The phasor diagram is shown in Figure 14.12(b). Note that v_L leads i by 90° and that v_C lags i by 90°, as prescribed by ELI the ICE man.

Drill Exercise 14.4

What would be the angle of v_R in Example 14.4 if the magnitude of X_C was changed to 1250 Ω?

ANSWER: 0°.

Example 14.5 (Design)

A series RLC circuit is connected to the ac voltage source $e(t) = 24 \sin(1000t + 20°)$ V. If $R = 15$ Ω and $L = 0.1$ H, what value of capacitance C is required to make the phase angle between $e(t)$ and the current $i(t)$ in the circuit equal to zero degrees?

SOLUTION The total impedance of the circuit is

$$Z_T = R + j\left(\omega L - \frac{1}{\omega C}\right) \tag{14.19}$$

If the total impedance were *purely resistive* (i.e., if $Z_T = R + j0$), the current in the circuit would be in phase with the voltage. For example, in the present case we would have

$$i = \frac{e}{R} = \frac{24 \underline{/20°} \text{ V}}{15 \underline{/0°} \text{ Ω}} = 1.6 \underline{/20°} \text{ A}$$

Thus, under those circumstances, i has the same angle as e, and the angle between the two is 0°.

It is clear from equation (14.19) that Z_T will equal $R + j0$ if

$$\omega L = \frac{1}{\omega C}$$

Thus, we require that

$$(1000)(0.1 \text{ H}) = \frac{1}{1000C}$$

or

$$C = \frac{1}{(1000)^2(0.1)} = 10^{-5} = 10 \ \mu F$$

14.5 The Voltage-Divider Rule

The voltage-divider rule for ac circuits is identical in concept to that for dc circuits (see Figure 14.13). The voltage v_x across any impedance, Z_x, in a series circuit can be found by multiplying the voltage across the circuit by the ratio Z_x/Z_T:

$$v_x = \frac{Z_x}{Z_T} e \qquad (14.20)$$

As in the case of dc circuits, Z_x can represent the impedance of a single component or the total impedance of several successive components. Remember that the mathematical operations implied by equation (14.20) must be performed using phasors.

Example 14.6   (Analysis)
(a) Use the voltage-divider rule to find the voltage across the 0.05-μF capacitor in Figure 14.14.
(b) Use the voltage-divider rule to find v_{ab} in Figure 14.14.

SOLUTION
(a) The magnitudes of the capacitive and inductive reactances are

$$|X_C| = \frac{1}{\omega C} = \frac{1}{(2\pi \times 10^3)(5 \times 10^{-8})} = 3183 \ \Omega$$

$$|X_L| = \omega L = (2\pi \times 10^3)(0.4) = 2513 \ \Omega$$

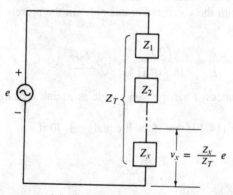

FIGURE 14.13 Voltage-divider rule for ac circuits.

Therefore, the total impedance of the circuit is

$$Z_T = R + j\left(\omega L - \frac{1}{\omega C}\right)$$

$$= 4000 + j(2513 - 3183) \ \Omega = 4000 - j670 \ \Omega$$

$$= \sqrt{(4000)^2 + (670)^2} \ \underline{/\tan^{-1}\left(\frac{-670}{4000}\right)} = 4055 \ \underline{/-9.51°} \ \Omega$$

By the voltage-divider rule,

$$v_C = \frac{X_C}{Z_T} e = \left(\frac{3183 \ \underline{/-90°}}{4055 \ \underline{/-9.51°}}\right) 30 \ \underline{/0°} \ V$$

$$= 23.55 \ \underline{/-80.49°} \ V$$

$$v_C(t) = 23.55 \sin (2\pi \times 10^3 t - 80.49°) \ V$$

(b) The total impedance between points a and b in Figure 14.14 is

$$Z_{ab} = 0 + j\left(\omega L - \frac{1}{\omega C}\right) = 0 + j(2513 - 3183) \ \Omega$$

$$= 0 - j670 \ \Omega = 670 \ \underline{/-90°} \ \Omega$$

By the voltage-divider rule,

$$v_{ab} = \frac{Z_{ab}}{Z_T} e = \left(\frac{670 \ \underline{/-90°}}{4055 \ \underline{/-9.51°}}\right) 30 \ \underline{/0°} \ V$$

$$= 4.96 \ \underline{/-80.49°} \ V$$

$$= 4.96 \sin (2\pi \times 10^3 t - 80.49°) \ V$$

Drill Exercise 14.6

Use the voltage-divider rule to find the voltage across the 4-kΩ resistor in Figure 14.14.

ANSWER: $29.59 \ \underline{/9.51°} \ V = 29.59 \sin (2\pi \times 10^3 t + 9.51°) \ V.$ ☐

The *magnitude* of the voltage across an impedance in a series circuit can be found by using the magnitudes of the quantities in the voltage-divider rule:

$$|v_x| = \frac{|Z_x|}{|Z_T|} |e| \tag{14.21}$$

Note carefully that $|Z_T| = |Z_1 + Z_2 + \cdots + Z_n|$, which is *not*, in general, the same as $|Z_1| + |Z_2| + \cdots + |Z_n|$. Magnitudes of phasor quantities can be multiplied or divided under any circumstances, but phasor magnitudes can be added *only* if the phasors have the same angle.

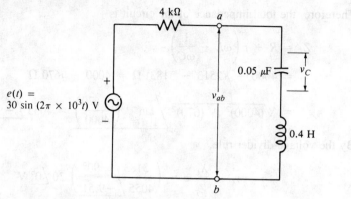

FIGURE 14.14 (Example 14.6)

Example 14.7   (Design)

Find the value of R that should be used in Figure 14.15 in order that the magnitude of the voltage across R be 40 V.

The total impedance is

$$Z_T = R + j(80 - 20 - 10) \; \Omega = R + j50 \; \Omega$$

Therefore, the magnitude of Z_T is

$$|Z_T| = \sqrt{R^2 + 50^2}$$

Using equation (14.21), we require

$$\frac{R}{|Z_T|} |e| = 40 \text{ V}$$

or

$$\left(\frac{R}{\sqrt{R^2 + 50^2}} \right) 60 = 40$$

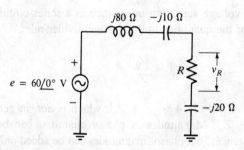

FIGURE 14.15 (Example 14.7)

Squaring both sides gives us

$$\left(\frac{R^2}{R^2 + 2500}\right) 3600 = 1600$$

$$3600R^2 = 1600R^2 + 1600(2500)$$

$$2000R^2 = (1600)(2500)$$

$$R = \sqrt{\frac{(1600)(2500)}{2000}} = 44.7 \ \Omega$$

Drill Exercise 14.7

How much additional capacitive reactance should be inserted in the circuit shown in Figure 14.15 to make the magnitude of the voltage across R equal 60 V?

ANSWER: $-j50 \ \Omega$. □

14.6 Admittance and Susceptance

Recall that the reciprocal of resistance is *conductance*, G, which has the units of siemens (S). As demonstrated in Chapter 4, the analysis of parallel dc circuits is simplified through the use of conductance. We shall see in this chapter that the analysis of parallel ac circuits is similarly simplified by using the reciprocal of impedance.

The reciprocal of impedance is called *admittance*, denoted Y:

$$Y = \frac{1}{Z} \quad \text{siemens} \tag{14.22}$$

The terms "impedance" and "admittance" are quite descriptive: Impedance is a measure of the extent to which a component *impedes* the flow of ac current through it, while admittance is a measure of how well it *admits* the flow of current. The greater the impedance, the smaller the admittance, and vice versa.

Just as "impedance" is a general term that can refer to resistance, reactance, or combinations thereof, so too can "admittance" refer to the reciprocals of resistance, reactance, or combinations of these. Resistance is one form of impedance, $Z = R$, so conductance is one form of admittance: $Y = 1/R = G$. The phasor form of conductance is

$$G = \frac{1}{R \underline{/0°}} = \frac{1}{R} \underline{/0°} \quad \text{siemens}$$

$$= \frac{1}{R} + j0 \ \text{S} \tag{14.23}$$

Reactance is another special case of impedance, so the reciprocal of reactance is a special case of admittance. The reciprocal of reactance is called *susceptance*, denoted by B:

$$B = \frac{1}{X} \quad \text{siemens} \tag{14.24}$$

Since there are two types of reactance, inductive and capacitive, there are also two types of susceptance. Inductive susceptance, B_L, is the reciprocal of inductive reactance. In phasor form,

$$B_L = \frac{1}{X_L} = \frac{1}{\omega L \underline{/90°}} = \frac{1}{\omega L} \underline{/-90°} \quad \text{siemens}$$

$$= 0 - j\left(\frac{1}{\omega L}\right) \quad \text{siemens} \tag{14.25}$$

Capacitive susceptance, B_C, is the reciprocal of capacitive reactance. In phasor form,

$$B_C = \frac{1}{X_C} = \frac{1}{(1/\omega C) \underline{/-90°}} = \omega C \underline{/90°} \quad \text{siemens}$$

$$= 0 + j\omega C \quad \text{siemens} \tag{14.26}$$

Figure 14.16 summarizes the terminology and definitions of the various special cases of impedance and admittance. Note the symmetry of the diagram and the way each special case of impedance has its counterpart in a special case of admittance. Of course, the fact that each admittance term is the reciprocal of a corresponding impedance term implies the reverse: Each impedance is the reciprocal of an admittance. In other words, $Z = 1/Y$, $R = 1/G$, $X_L = 1/B_L$, and $X_C = 1/B_C$.

Example 14.8   (Analysis)
(a) Find the admittance of
    (i) a 5-Ω resistor

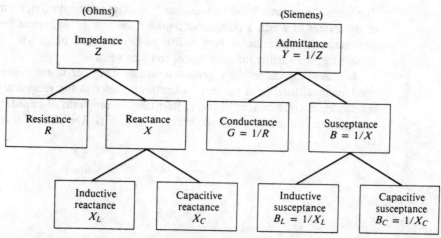

FIGURE 14.16 Impedance and admittance, and the relationships between special cases of each.

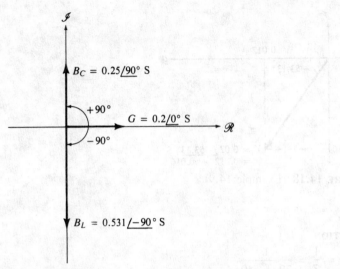

FIGURE 14.17 (Example 14.8)

(ii) a 5-mH inductor at $f = 60$ Hz
(iii) a 0.2-μF capacitor at $\omega = 1.25 \times 10^6$ rad/s
(b) Make a phasor diagram showing each admittance in the complex plane.

SOLUTION

(a) (i) $Y = G = \dfrac{1}{R} = \dfrac{1}{5\,\underline{/0^\circ}\ \Omega} = 0.2\,\underline{/0^\circ}$ S

$\qquad\qquad = 0.2 + j0$ S

(ii) $|B_L| = \dfrac{1}{2\pi f L} = \dfrac{1}{(2\pi)(60)(5 \times 10^{-3})} = 0.531$ S

$\qquad Y = B_L = 0 - j|B_L| = 0 - j0.531$ S $= 0.531\,\underline{/-90^\circ}$ S

(iii) $|B_C| = \omega C = (1.25 \times 10^6)(0.2 \times 10^{-6}) = 0.25$ S

$\qquad Y = B_C = 0 + j|B_C| = 0 + j0.25$ S $= 0.25\,\underline{/90^\circ}$ S

(b) The phasor diagram is shown in Figure 14.17.

Drill Exercise 14.8

(a) At what frequency is the capacitive susceptance of a 2.5-μF capacitor equal to 0.3 mS?

(b) At what frequency is the inductive susceptance of a 4-H inductor equal to 1 μS?

ANSWER: (a) 19.1 Hz; (b) 39.79 kHz. $\qquad\qquad\qquad\qquad\qquad\qquad$ □

Example 14.9 (Analysis)

(a) Find the admittance Y corresponding to $Z = 30 + j40$ Ω.

(b) Make a phasor diagram showing Y in the complex plane.

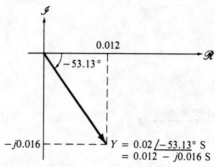

FIGURE 14.18 (Example 14.9)

SOLUTION

(a) $Y = \dfrac{1}{Z} = \dfrac{1}{30 + j40}$

Converting Z to polar form gives us

$$Z = \sqrt{30^2 + 40^2} \; \Big/ \tan^{-1} \dfrac{40}{30} = 50 \; \underline{/53.13°} \; \Omega$$

Then

$$Y = \dfrac{1}{50 \; \underline{/53.13°}} = 0.02 \; \underline{/-53.13°} \; S$$
$$= 0.02 \cos (-53.13°) + j0.02 \sin (-53.13°)$$
$$= 0.012 - j0.016 \; S$$

(b) The phasor diagram is shown in Figure 14.18.

Drill Exercise 14.9

Find the admittance Y corresponding to $Z = 4 \times 10^3 - j4 \times 10^3 \; \Omega$.

ANSWER: $1.768 \times 10^{-4} \; \underline{/45°} \; S = 1.25 \times 10^{-4} + j1.25 \times 10^{-4} \; S.$ □

14.7 Parallel AC Circuits

Figure 14.19 shows an arbitrary number of impedances connected in parallel with an ac voltage source. As in the case of parallel dc circuits, the same voltage appears across every parallel-connected component. *The total admittance of the circuit is the sum of the admittances of the parallel-connected components:*

$$Y_T = Y_1 + Y_2 + \cdots + Y_n \tag{14.27}$$

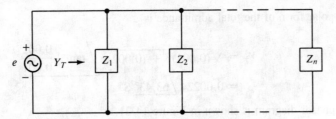

FIGURE 14.19 Parallel-connected impedances.

where

$$Y_1 = \frac{1}{Z_1}, \quad Y_2 = \frac{1}{Z_2}, \quad \ldots, \quad Y_n = \frac{1}{Z_n}$$

Example 14.10
(a) Find the total admittance of the circuit shown in Figure 14.20.
(b) Make a phasor diagram showing the total admittance in the complex plane.

SOLUTION
(a) The admittance, Y_1, of the inductor is its inductive susceptance [equation (14.25)]:

$$Y_1 = B_L = 0 - j\frac{1}{\omega L} = 0 - \frac{j}{(25 \times 10^3)(20 \times 10^{-3})} = 0 - j0.002 \text{ S}$$

The admittance, Y_2, of the resistor is its conductance [equation (14.23)]:

$$Y_2 = G = \frac{1}{R} + j0 = \frac{1}{10^3} + j0 = 0.001 + j0 \text{ S}$$

The admittance, Y_3, of the capacitor is its capacitive susceptance [equation (14.26)]:

$$Y_3 = B_C = 0 + j\omega C = 0 + j(25 \times 10^3)(0.16 \times 10^{-6}) = 0 + j0.004 \text{ S}$$

The total admittance is

$$\begin{aligned} Y_T &= Y_1 + Y_2 + Y_3 \\ &= (0 - j0.002) + (0.001 + j0) + (0 + j0.004) \\ &= 0.001 + j0.002 \text{ S} \end{aligned}$$

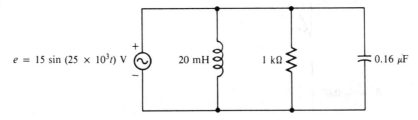

$e = 15 \sin (25 \times 10^3 t) \text{ V}$ 20 mH   1 kΩ   0.16 μF

FIGURE 14.20 (Example 14.10)

The polar form of the total admittance is

$$Y_T = \sqrt{(0.001)^2 + (0.002)^2} \ \Big/ \tan^{-1}\frac{0.002}{0.001}$$

$$= 0.00224 \ \underline{/63.43°} \ \text{S}$$

(b) The phasor diagram is shown in Figure 14.21.

Drill Exercise 14.10

Find the total admittance of the circuit in Figure 14.20 if the frequency of the voltage source is doubled.

ANSWER: $Y_T = 0.00707 \ \underline{/81.87°} \ \text{S} = 0.001 + j0.007 \ \text{S}.$ □

The total impedance of a network is the reciprocal of its total admittance, $Z_T = 1/Y_T$. For the parallel network in Figure 14.19, we have, from equation (14.27),

$$Z_T = \frac{1}{Y_T} = \frac{1}{Y_1 + Y_2 + \cdots + Y_n} \tag{14.28}$$

$$= \frac{1}{1/Z_1 + 1/Z_2 + \cdots + 1/Z_n}$$

Notice the similarity of equation (14.28) to the equation for the total equivalent resistance of parallel-connected resistors [equation (4.11)]. For the special case of a network consisting of two impedances in parallel, we find

$$Z_T = \frac{1}{1/Z_1 + 1/Z_2} = \frac{1}{(Z_1 + Z_2)/Z_1 Z_2} = \frac{Z_1 Z_2}{Z_1 + Z_2} \tag{14.29}$$

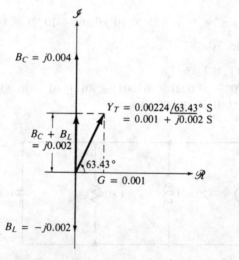

FIGURE 14.21 (Example 14.10)

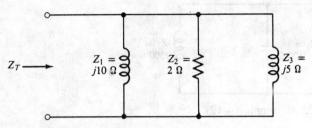

FIGURE 14.22 (Example 14.11)

Example 14.11 (Analysis)

Find the total equivalent impedance of the network shown in Figure 14.22.

SOLUTION

$$Y_1 = \frac{1}{Z_1} = \frac{1}{j10} = -j0.1 = 0 - j0.1$$

$$Y_2 = \frac{1}{Z_2} = \frac{1}{2} = 0.5 = 0.5 + j0$$

$$Y_3 = \frac{1}{Z_3} = \frac{1}{j5} = -j0.2 = 0 - j0.2$$

$$Y_T = Y_1 + Y_2 + Y_3 = (0 - j0.1) + (0.5 + j0) + (0 - j0.2)$$
$$= 0.5 - j0.3 \text{ S}$$

$$= \sqrt{0.5^2 + 0.3^3} \left/ \tan^{-1}\left(\frac{-0.3}{0.5}\right)\right. = 0.583 \left/ -30.96° \text{ S}\right.$$

$$Z_T = \frac{1}{Y_T} = \frac{1}{0.583 \left/ -30.96°\right.} = 1.715 \left/ 30.96° \ \Omega\right.$$

$$= 1.715 \cos (30.96°) + j1.715 \sin (30.96°)$$

$$= 1.47 + j0.882 \ \Omega$$

Drill Exercise 14.11

Find the total impedance of the network in Figure 14.22 if the resistance (Z_2) is changed to 1 Ω.

ANSWER: $0.958 \left/ 16.7° \ \Omega = 0.918 + j0.275 \ \Omega.\right.$ □

Example 14.12 (Analysis)

Find the total equivalent impedance of the network shown in Figure 14.23.

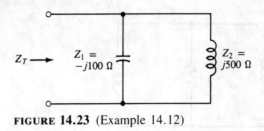

FIGURE 14.23 (Example 14.12)

SOLUTION By equation (14.29),

$$Z_T = \frac{Z_1 Z_2}{Z_1 + Z_2} = \frac{(-j100)(j500)}{-j100 + j500}$$

$$= \frac{-j^2(5 \times 10^4)}{j400} = \frac{5 \times 10^4}{j400} \ \Omega$$

Moving j from the denominator to the numerator, we have

$$Z_T = \frac{-j5 \times 10^4}{400} = -j125 \ \Omega = 125 \underline{/-90°} \ \Omega$$

The computation of Z_T could also have been performed by converting the numerator terms to polar form:

$$\frac{Z_1 Z_2}{Z_1 + Z_2} = \frac{(100 \underline{/-90°})(500 \underline{/90°})}{-j100 + j500}$$

$$= \frac{5 \times 10^4 \underline{/0°}}{400 \underline{/90°}} = 125 \underline{/-90°} \ \Omega$$

Notice that the total impedance of the network is equivalent to 125 Ω of *capacitive* reactance, despite the fact that the magnitude of the inductive reactance in the circuit is (five times) greater than the magnitude of the capacitive reactance. This situation is analogous to parallel-connected resistor networks, in which the total equivalent resistance is closest in value to the smallest resistance.

Drill Exercise 14.12

Find the total impedance of the network in Figure 14.23 when the capacitive reactance (Z_1) is made equal to $-j700$ Ω.

ANSWER: $1750 \underline{/90°}$ Ω. □

Since the same voltage, e, appears across every parallel-connected impedance, we can find the current in each impedance using Ohm's law:

$$i_1 = \frac{e}{Z_1} = eY_1$$

$$i_2 = \frac{e}{Z_2} = eY_2$$

(14.30)

$$\vdots \qquad \vdots \qquad \vdots$$

$$i_n = \frac{e}{Z_n} = eY_n$$

See Figure 14.24. By Kirchhoff's current law, the total current delivered by the voltage source equals the sum of the currents in the parallel-connected impedances:

$$i_T = i_1 + i_2 + \cdots + i_n$$

(14.31)

Furthermore,

$$i_T = \frac{e}{Z_T} = eY_T$$

Example 14.13 (Analysis)

In the parallel circuit shown in Figure 14.25:

(a) Find the current in each impedance.

(b) Show that $i_T = eY_T$.

(c) Make a phasor diagram showing e, i_T, i_1, i_2, and i_3.

SOLUTION

(a) $i_1 = \dfrac{e}{Z_1} = \dfrac{60 \underline{/0°} \text{ V}}{50 \underline{/0°} \text{ } \Omega} = 1.2 \underline{/0°} \text{ A} = 1.2 + j0 \text{ A}$

$i_2 = \dfrac{e}{Z_2} = \dfrac{60 \underline{/0°} \text{ V}}{40 \underline{/90°} \text{ } \Omega} = 1.5 \underline{/-90°} \text{ A} = 0 - j1.5 \text{ A}$

$i_3 = \dfrac{e}{Z_3} = \dfrac{60 \underline{/0°} \text{ V}}{80 \underline{/-90°} \text{ } \Omega} = 0.75 \underline{/90°} \text{ A} = 0 + j0.75 \text{ A}$

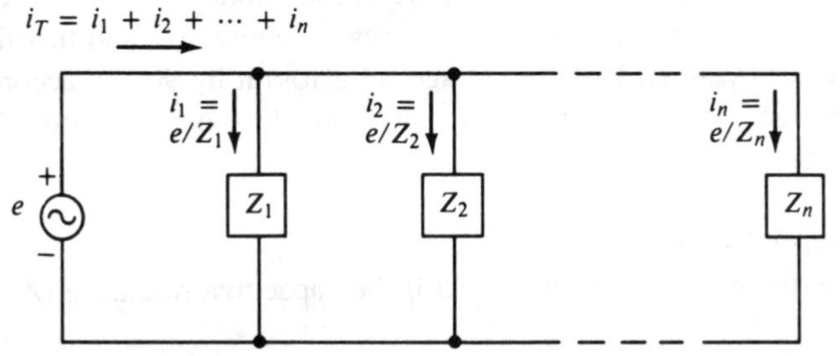

FIGURE 14.24 Current relations in a parallel ac circuit.

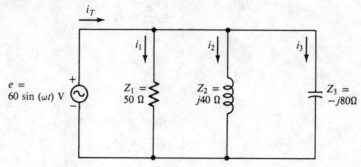

FIGURE 14.25 (Example 14.13)

(b) The total admittance of the circuit is

$$Y_T = \frac{1}{Z_1} + \frac{1}{Z_2} + \frac{1}{Z_3} = \frac{1}{50 \underline{/0°}} + \frac{1}{40 \underline{/90°}} + \frac{1}{80 \underline{/-90°}}$$

$$= 0.02 \underline{/0°} + 0.025 \underline{/-90°} + 0.0125 \underline{/90°}$$

$$= (0.02 + j0) + (0 - j0.025) + (0 + j0.0125)$$

$$= 0.02 - j0.0125 \text{ S}$$

$$= \sqrt{(0.02)^2 + (0.0125)^2} \underline{/\tan^{-1}\left(\frac{-0.0125}{0.02}\right)}$$

$$= 0.0236 \underline{/-32°} \text{ S}$$

By Kirchhoff's current law [equation (14.31)],

$$i_T = i_1 + i_2 + i_3 = (1.2 + j0) + (0 - j1.5) + (0 + j0.75)$$

$$= 1.2 - j0.75 \text{ A}$$

$$= \sqrt{1.2^2 + 0.75^2} \underline{/\tan^{-1}\left(\frac{-0.75}{1.2}\right)} = 1.415 \underline{/-32°} \text{ A}$$

To show that $i_T = eY_T$, we multiply in polar form:

$$eY_T = (60 \underline{/0°} \text{ V})(0.0236 \underline{/-32°} \text{ S}) = 1.416 \underline{/-32°} \text{ A} \approx i_T$$

(c) The phasor diagram is shown in Figure 14.26. Notice that the current in the capacitor ($i_C = i_3$) leads the voltage across it (e) by 90°, and that the current in the inductor ($i_L = i_2$) lags the voltage (e) across it by 90°, in accordance with ELI the ICE man. As a consequence, i_L and i_C are 180° out of phase. The diagram also shows that i_T is the phasor sum of i_1, i_2, and i_3.

Drill Exercise 14.13

Find i_T in the circuit in Figure 14.25 if the capacitive reactance (Z_3) is changed to $-j40$ Ω.

ANSWER: $1.2 \underline{/0°}$ A.

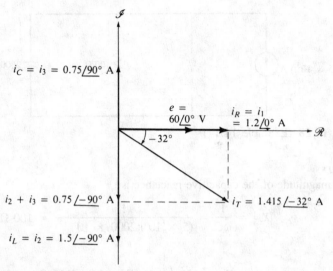

$i_C = i_3 = 0.75 \underline{/90°}$ A

$e = 60 \underline{/0°}$ V

$i_R = i_1 = 1.2 \underline{/0°}$ A

$-32°$

$i_2 + i_3 = 0.75 \underline{/-90°}$ A

$i_T = 1.415 \underline{/-32°}$ A

$i_L = i_2 = 1.5 \underline{/-90°}$ A

FIGURE 14.26 (Example 14.13)

AC Current Sources

Like a dc current source, an (ideal) ac current source supplies the same, constant current to whatever network is connected across its terminals. In the ac case, the current is constant in the sense that its peak value does not change. We will use the same symbol for an ac current source that we used for a dc current source (see Figure 14.27). Since the current reverses direction every half-cycle, the arrow in the symbol is simply a phase reference that proves useful when there is more than one ac current source in a circuit. As shown in the figure, reversing the arrow is the same as multiplying the current by -1, which is the same as adding or subtracting 180° from its phase angle.

Example 14.14   (Analysis)
    In the circuit shown in Figure 14.28:

(a) Find the voltage v_T across the ac current source.
(b) Find the currents i_1 and i_2.
(c) Show that $i = i_1 + i_2$.
(d) Make a phasor diagram showing i, v_T, i_1, and i_2.
(e) Sketch the waveforms of i, i_1, and i_2 versus angle.

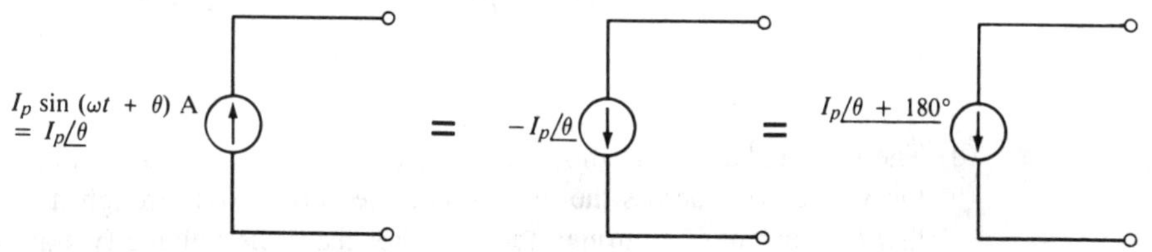

$I_p \sin (\omega t + \theta)$ A
$= I_p \underline{/\theta}$

$= -I_p \underline{/\theta}$

$= I_p \underline{/\theta + 180°}$

FIGURE 14.27 AC constant-current source.

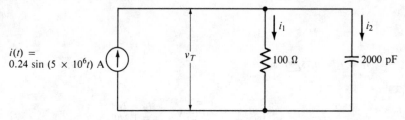

FIGURE 14.28 (Example 14.14)

SOLUTION

(a) The magnitude of the capacitive reactance is

$$|X_C| = \frac{1}{\omega C} = \frac{1}{(5 \times 10^6)(2000 \times 10^{-12})} = 100 \ \Omega$$

Thus,

$$X_C = 100 \underline{/-90°} \ \Omega = 0 - j100 \ \Omega$$

The total impedance of the circuit is therefore

$$Z_T = \frac{Z_1 Z_2}{Z_1 + Z_2} = \frac{(100 \underline{/0°})(100 \underline{/-90°})}{(100 + j0) + (0 - j100)} = \frac{10^4 \underline{/-90°}}{100 - j100}$$

$$= \frac{10^4 \underline{/-90°}}{141.4 \underline{/-45°}} = 70.7 \underline{/-45°} \ \Omega$$

By Ohm's law,

$$v_T = iZ_T = (0.24 \underline{/0°} \ \text{A})(70.7 \underline{/-45°} \ \Omega) = 16.97 \underline{/-45°} \ \text{V}$$

(b) Since v_T is the voltage across each component, we have, by Ohm's law,

$$i_1 = \frac{v_T}{R} = \frac{16.97 \underline{/-45°} \ \text{V}}{100 \underline{/0°} \ \Omega} = 0.1697 \underline{/-45°} \ \text{A}$$

$$= 0.12 - j0.12 \ \text{A}$$

and

$$i_2 = \frac{v_T}{X_C} = \frac{16.97 \underline{/-45°} \ \text{V}}{100 \underline{/-90°} \ \Omega} = 0.1697 \underline{/45°} \ \text{A}$$

$$= 0.12 + j0.12 \ \text{A}$$

(c) $i_1 + i_2 = (0.12 - j0.12) + (0.12 + j0.12)$

$$= 0.24 + j0 = 0.24 \underline{/0°} \ \text{A} = i$$

(d) The phasor diagram is shown in Figure 14.29(a). The diagram clearly shows that the voltage (v_T) across the resistor and the current (i_1) through it are in phase, and that the current (i_C) through the capacitor leads the voltage (v_T) across it by 90°. The diagram also shows that i is the phasor sum of i_1 and i_2.

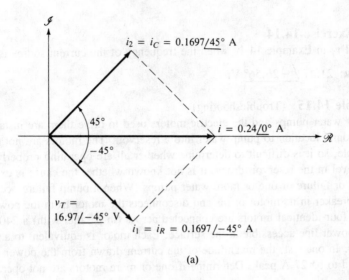

(a)

(b)

FIGURE 14.29 (Example 14.14)

(e) To sketch the waveforms of i, i_1, and i_2, we convert to sinusoidal form:

$$i_1 = 0.1697 \; \underline{/-45°} \; A = 0.1697 \sin (5 \times 10^6 t - 45°) \; A$$

$$i_2 = 0.1697 \; \underline{/45°} \; A = 0.1697 \sin (5 \times 10^6 t + 45°) \; A$$

The waveforms are shown in Figure 14.29(b).

Drill Exercise 14.14

Find v_T in Example 14.14 when the frequency of the current source is halved.

ANSWER: $21.47 \underline{/-26.56°}$ V. □

Example 14.15   (Troubleshooting)

Four water pumps and the electric motors used to drive them are installed in an underground location to pump water into a reservoir. The motors are not easily accessible, so it is difficult to determine whether all are operating properly. When the water level in the reservoir drops, it is not known whether the cause is excessive water demand or failure of one or more water pumps. When a pump failure occurs, a circuit breaker in its motor opens and disconnects the motor from the power line.

The four identical motors are connected across (in parallel with) a 240-V rms, 60-Hz power line accessible at the surface. Each motor is equivalent to a 0.1-H inductor. In one test, the magnitude of the current drawn from the power line was measured to be 27 A peak. Determine if one or more motors are not operating.

SOLUTION The equivalent circuit is shown in Figure 14.30. The magnitude of the reactance of each inductor at 60 Hz is

$$|X_L| = \omega L = 2\pi(60)(0.1) = 37.7 \ \Omega$$

The peak value of the voltage across the motors is

$$V_p = \sqrt{2} \ V_{\text{eff}} = \sqrt{2} \ (240 \ \text{V rms}) = 339.4 \ \text{V pk}$$

Therefore, the magnitude of the peak current drawn by each motor is

$$I_p = \frac{V_p}{|X_L|} = \frac{339.4 \ \text{V}}{37.7 \ \Omega} = 9 \ \text{A pk}$$

Since each motor is identical, *the currents through the motors are identical in magnitude and angle*. Therefore, the magnitude of the total current can be found by *adding the magnitudes of the individual motor currents directly*:

$$|i_T| = |i_1| + |i_2| + |i_3| + |i_4|$$
$$= 4(9 \ \text{A pk}) = 36 \ \text{A pk}$$

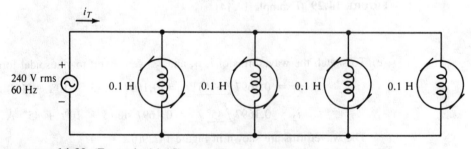

FIGURE 14.30 (Example 14.15)

Recall that magnitudes of phasor quantities can be added only if all have the same angle, as in this case.

Since the actual current drawn from the power line is 27 A pk, we conclude that one 9-A motor is not operating.

Drill Exercise 14.15

Four 20-μF capacitors are connected in parallel. Find the total equivalent capacitive reactance, X_{C_T}, at $\omega = 500$ rad/s. Since the total equivalent capacitance is 4(20 μF) = 80 μF, is the total reactance equal to four times the reactance of each capacitor?

ANSWER: $25 \underline{/-90°}$ Ω; no. □

14.8 The Current-Divider Rule

The current-divider rule for ac circuits is identical in concept to that for dc circuits (see Figure 14.31). The current i_1 in impedance Z_1 is

$$i_1 = \frac{Z_2}{Z_1 + Z_2} i \tag{14.32}$$

Similarly,

$$i_2 = \frac{Z_1}{Z_1 + Z_2} i \tag{14.33}$$

The general form of the current-divider rule is

$$i_x = \frac{Z_T}{Z_x} i \tag{14.34}$$

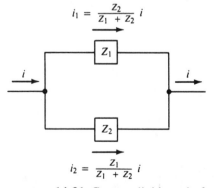

FIGURE 14.31 Current-divider rule for ac circuits.

where i_x is the current in Z_x and Z_T is the total equivalent impedance of an arbitrary number of parallel-connected impedances (*not* their sum). Substituting

$$Z_T = \frac{1}{Y_T} = \frac{1}{Y_1 + Y_2 + \cdots + Y_n}$$

and $Z_x = 1/Y_x$ into (14.34), we find

$$i_x = \frac{Y_x}{Y_1 + Y_2 + \cdots + Y_n} i = \frac{Y_x}{Y_T} i \qquad (14.35)$$

Example 14.16

Use the current-divider rule to find the currents i_R and i_L in Figure 14.32.

SOLUTION The magnitude of the inductive reactance is

$$|X_L| = 2\pi f L = 2\pi (1 \times 10^6)(30 \times 10^{-3}) = 188.5 \text{ k}\Omega$$

Thus,

$$X_L = 188.5 \underline{/90°} \text{ k}\Omega = 0 + j188.5 \times 10^3 \ \Omega$$

Letting $Z_1 = 120 \underline{/0°}$ kΩ, $Z_2 = 188.5 \underline{/90°}$ kΩ, $i_1 = i_R$, $i_2 = i_L$, and assuming that the current source has zero phase angle, we find, by the current-divider rule,

$$i_R = i_1 = \frac{Z_2}{Z_1 + Z_2} i$$

$$= \left[\frac{188.5 \times 10^3 \underline{/90°}}{(120 \times 10^3 + j0) + (0 + j188.5 \times 10^3)} \right] (40 \times 10^{-6} \underline{/0°} \text{ A})$$

$$= \frac{7.54 \underline{/90°}}{120 \times 10^3 + j188.5 \times 10^3} \text{ A} = \frac{7.54 \underline{/90°}}{223.45 \times 10^3 \underline{/57.52°}} \text{ A}$$

$$= 33.74 \underline{/32.48°} \ \mu\text{A}$$

$$i_L = i_2 = \frac{Z_1}{Z_1 + Z_2} i = \left(\frac{120 \times 10^3 \underline{/0°}}{223.45 \times 10^3 \underline{/57.52°}} \right) (40 \times 10^{-6} \underline{/0°} \text{ A})$$

$$= 21.48 \underline{/-57.52°} \ \mu\text{A}$$

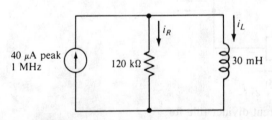

FIGURE 14.32 (Example 14.16)

Drill Exercise 14.16

Suppose that a 1.5-pF capacitor is connected in parallel with the components in Figure 14.32. Use the current-divider rule [equation (14.35)] to find the current in the capacitor.

ANSWER: $40.55 \underline{/63.74°}$ μA.   □

Example 14.17 (Design)

A 0.2-H inductor carries an ac current of 0.5 A pk at 60 Hz. It is desired to connect a capacitor in parallel with the inductor to reduce the magnitude of the current in the inductor to 0.3 A. Assuming that the installation of the capacitor does not change the total current supplied to the network (0.5 A), what value of capacitance should be used?

SOLUTION Figure 14.33 shows the situation. Since installation of the capacitor does not affect the total current, we may regard the 0.5 A as being supplied from a constant-current source. The magnitude of the inductive reactance is

$$|X_L| = 2\pi fL = 2\pi(60)(0.2) = 75.4 \ \Omega$$

Letting the inductive reactance be Z_1 and the capacitive reactance be Z_2, we have, by the current-divider rule,

$$i_L = \frac{Z_2}{Z_1 + Z_2} i$$

Now

$$Z_1 + Z_2 = j\omega L - \frac{j}{\omega C} = j\left(\omega L - \frac{1}{\omega C}\right) = j(|X_L| - |X_C|)$$

Therefore,

$$|Z_1 + Z_2| = |X_L| - |X_C|$$

Using the magnitudes of the quantities in the current-divider rule, we have

$$|i_L| = \frac{|Z_2|}{|Z_1 + Z_2|} |i| = \frac{|X_C|}{|X_L| - |X_C|} |i|$$

or

$$0.3 \text{ A} = \left(\frac{|X_C|}{75.4 - |X_C|}\right) 0.5 \text{ A}$$

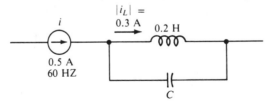

FIGURE 14.33 (Example 14.17)

Solving for $|X_C|$ gives $|X_C| = 28.28\ \Omega$. Therefore,

$$28.28\ \Omega = \frac{1}{2\pi f C} = \frac{1}{2\pi(60)C}$$

or

$$C = \frac{1}{(2\pi)(60)(28.28)} = 93.8\ \mu\text{F}$$

Drill Exercise 14.17

What value of capacitance should be used in Figure 14.33 if it is desired to make the magnitude of the voltage across the network equal to 89.48 V pk? (The current in the inductor will no longer be 0.3 A.)

ANSWER: 50 μF.

□

14.9 Equivalent Series and Parallel Networks

When analyzing or designing some ac circuits, it is convenient to replace a parallel network with an equivalent series network, or vice versa. Two networks are equivalent if the phasor forms of their impedances are identical. To illustrate this equivalency, consider the parallel network shown in Figure 14.34(a). The total impedance of this network is

$$Z_T = \frac{Z_1 Z_2}{Z_1 + Z_2} = \frac{(3\underline{/0^\circ})(4\underline{/90^\circ})}{3 + j4} = \frac{12\underline{/90^\circ}}{5\underline{/53.13^\circ}} = 2.4\underline{/36.87^\circ}\ \Omega$$

$$= 2.4\cos(36.87^\circ) + j2.4\sin(36.87^\circ) = 1.92 + j1.44\ \Omega$$

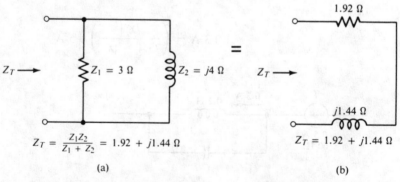

$$Z_T = \frac{Z_1 Z_2}{Z_1 + Z_2} = 1.92 + j1.44\ \Omega$$

(a)

$$Z_T = 1.92 + j1.44\ \Omega$$

(b)

FIGURE 14.34 Example of equivalent series and parallel networks.

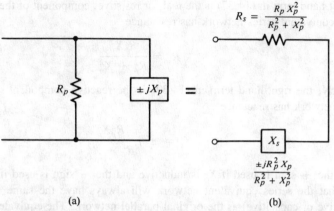

FIGURE 14.35 Converting a parallel network to an equivalent series network. (a) Parallel network. (b) Equivalent series network.

The impedance $Z_T = 1.92 + j1.44$ Ω is identical to the impedance of the series circuit shown in Figure 14.34(b), consisting of 1.92 ohms of resistance in series with 1.44 ohms of inductive reactance. Therefore, the two networks are equivalent. The inductance required in the series network can be found from knowledge of the frequency: $L = 1.44$ Ω$/2\pi f$. Of course, the equivalency is valid only for the single frequency at which the inductive reactance in the parallel network equals 4 Ω, which is the same as the frequency at which the series inductor has reactance 1.44 Ω. If the frequency is changed, a new equivalent network must be computed.

We wish now to derive general equations that can be used to convert a parallel network to an equivalent series network, and vice versa. Figure 14.35(a) shows a general parallel network consisting of resistance R_p in parallel with reactance X_p, where X_p can be either inductive or capacitive. R_p can represent a single resistor or the total equivalent resistance of several resistors. Similarly, X_p can represent a single reactance or the total equivalent reactance of several reactances. The total equivalent impedance of the parallel network is

$$Z_p = \frac{(R_p)(\pm jX_p)}{R_p \pm jX_p} \tag{14.36}$$

where the $+$ sign is used if X_p is inductive and the $-$ sign is used if X_p is capacitive. To rationalize, we multiply numerator and denominator by $R_p \mp jX_p$ (notice the opposite sign of the reactive component):

$$\frac{(R_p)(\pm jX_p)}{R_p \pm jX_p}\left(\frac{R_p \mp jX_p}{R_p \mp jX_p}\right) = \frac{\pm jR_p^2 X_p + R_p X_p^2}{R_p^2 + X_p^2}$$

$$= \frac{R_p X_p^2}{R_p^2 + X_p^2} \pm j\frac{R_p^2 X_p}{R_p^2 + X_p^2} \tag{14.37}$$

The left-hand term in (14.37) is the real, or resistive, component of the total impedance, so the equivalent series network has resistance

$$R_s = \frac{R_p X_p^2}{R_p^2 + X_p^2} \tag{14.38}$$

Similarly, the right-hand term in (14.37) is the reactive component, so the equivalent series network has reactance

$$X_s = \pm j \frac{R_p^2 X_p}{R_p^2 + X_p^2} \tag{14.39}$$

where the + sign is used if X_p is inductive and the − sign is used if X_p is capacitive. Note that the series equivalent network will always have the same *type* of reactance (inductive or capacitive) as the original parallel network. The equivalent series network is shown in Figure 14.35(b). When using equations (14.38) and (14.39), the term X_p should be treated as the *magnitude* of the impedance in the parallel network.

To demonstrate the validity of equations (14.38) and (14.39), we apply them to the example network of Figure 14.34(a), in which $R_p = 3\ \Omega$ and $X_p = 4\ \Omega$:

$$R_s = \frac{R_p X_p^2}{R_p^2 + X_p^2} = \frac{(3)(4^2)}{3^2 + 4^2} = \frac{48}{25} = 1.92\ \Omega$$

$$X_s = \frac{+ jR_p^2 X_p}{R_p^2 + X_p^2} = j\frac{(3^2)(4)}{3^2 + 4^2} = j\frac{36}{25} = j1.44\ \Omega$$

These are the same values we obtained before.

Figure 14.36(a) shows a series network consisting of resistance R_s in series with reactance $\pm jX_s$. Again, R_s and X_s may be the resistance and reactance of single components or the total equivalent resistance and reactance in a network. The total impedance of the series network is

$$Z_s = R_s \pm jX_s \tag{14.40}$$

and the total admittance is

$$Y_s = \frac{1}{Z_s} = \frac{1}{R_s \pm jX_s} \tag{14.41}$$

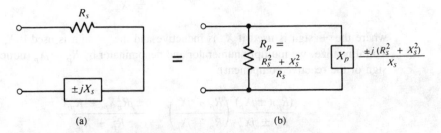

(a)   (b)

FIGURE 14.36 Converting a series network to an equivalent parallel network. (a) Series network. (b) Equivalent parallel network.

Rationalizing (14.41), we obtain

$$Y_s = \frac{1}{R_s \pm jX_s} \left(\frac{R_s \mp jX_s}{R_s \mp jX_s} \right) = \frac{R_s \mp jX_s}{R_s^2 + X_s^2}$$

$$= \frac{R_s}{R_s^2 + X_s^2} \mp j \frac{X_s}{R_s^2 + X_s^2} \tag{14.42}$$

Now the admittance of a parallel network is the sum of the admittances of its components, so the left-hand term in (14.42) is the real part of the admittance (i.e., the conductance) of the total admittance of the parallel network:

$$G_p = \frac{R_s}{R_s^2 + X_s^2} \tag{14.43}$$

Therefore,

$$R_p = \frac{1}{G_p} = \frac{R_s^2 + X_s^2}{R_s} \tag{14.44}$$

Similarly, the right-hand term in (14.42) is the imaginary, or susceptance, component of the total admittance of the parallel network:

$$\mp jB_p = \frac{\mp jX_s}{R_s^2 + X_s^2} \tag{14.45}$$

Therefore,

$$\pm jX_p = \frac{1}{\mp jX_s/(R_s^2 + X_s^2)} = \frac{\pm j(R_s^2 + X_s^2)}{X_s} \tag{14.46}$$

Once again, the plus sign is used if X_s is inductive and the minus sign is used if X_s is capacitive. The term X_S in equations (14.40) through (14.46) should be interpreted as the *magnitude* of the reactance in the original series circuit. Figure 14.36(b) shows the equivalent parallel network.

To demonstrate the validity of equations (14.44) and (14.46), we will use them to convert the series network in Figure 14.34(b) back to the parallel network in Figure 14.34(a). In this example, $R_s = 1.92 \ \Omega$ and $X_s = 1.44 \ \Omega$:

$$R_p = \frac{R_s^2 + X_s^2}{R_s} = \frac{(1.92)^2 + (1.44)^2}{1.92} = \frac{5.76}{1.92} = 3 \ \Omega$$

$$X_p = \frac{+j(R_s^2 + X_s^2)}{X_s} = \frac{j[(1.92)^2 + (1.44)^2]}{1.44} = \frac{j5.76}{1.44} = j4 \ \Omega$$

Example 14.18

Find a two-component parallel network that is equivalent to the series network shown in Figure 14.37(a). Find the value of the reactive component (inductance or capacitance) in the parallel equivalent circuit, given that the frequency is 400 Hz.

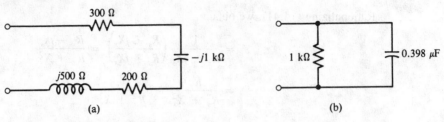

(a)   (b)

FIGURE 14.37 (Example 14.18)

SOLUTION The total impedance of the series network is

$$Z_s = 300 + 200 - j1000 + j500 = 500 - j500 \ \Omega$$

Thus, $R_s = 500 \ \Omega$ and $X_s = 500 \ \Omega$. Using equations (14.44) and (14.46),

$$R_p = \frac{R_s^2 + X_s^2}{R_s} = \frac{(500)^2 + (500)^2}{500} = 1 \ \text{k}\Omega$$

and

$$-jX_p = \frac{-j(R_s^2 + X_s^2)}{X_s} = \frac{-j[(500)^2 + (500)^2]}{500} = -j1 \ \text{k}\Omega$$

The parallel reactance is capacitive, and the capacitance, C_p, is found from

$$\frac{1}{2\pi f C_p} = 1000 \ \Omega$$

or

$$C_p = \frac{1}{2\pi f(1000)} = \frac{1}{2\pi (400)(1000)} = 0.398 \ \mu\text{F}$$

The equivalent parallel network is shown in Figure 14.37(b).

Drill Exercise 14.18

Convert the parallel network in Figure 14.37(b) to an equivalent series network when the frequency is 250 Hz. Find the value of the capacitance in the series equivalent network.

ANSWER: $R_s = 719 \ \Omega$; $C_s = 1.416 \ \mu\text{F}$. □

14.10 Power in Circuits Containing Reactance

Recall from Chapter 12 that the average power dissipated by a resistance carrying sinusoidal ac current can be found using any of the relationships:

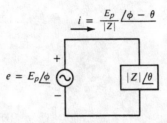

FIGURE 14.38 AC circuit containing a general impedance, $|Z| \underline{/\theta}$. The angle between e and i is θ degrees.

$$P_{avg} = \frac{I_p^2 R}{2} = \frac{V_p^2}{2R} = \frac{V_p I_p}{2}$$

$$= I_{eff}^2 R = \frac{V_{eff}^2}{R} = V_{eff} I_{eff}$$

(14.47)

where I_p and V_p are the peak values of the current through and voltage across the re-sistance, and I_{eff} and V_{eff} are their effective (rms) values. These relationships can be used to find the average power dissipated by a resistor in a circuit containing reactance, *provided* that the values used in the computations are indeed those of the voltage across and/or current through the resistor itself. In many practical circuits, it is necessary to compute the average power when the resistance in the circuit is not known, or when the resistor voltage and current are not known. Figure 14.38 shows an ac circuit containing a general impedance, $Z \underline{/\theta} \ \Omega$. If we assume that the voltage supplied by the source has angle $\phi°$, then the current in the circuit is

$$i = \frac{E_p \underline{/\phi}}{|Z| \underline{/\theta}} = \frac{E_p}{|Z|} \underline{/\phi - \theta}$$

(14.48)

The angle between the voltage and the current is therefore

$$\phi - (\phi - \theta) = \theta \text{ degrees}$$

We conclude that *the angle between the voltage applied to a network and the total current supplied to it equals the angle of the impedance of the network.*

Figure 14.39 shows that the resistive component of the impedance $|Z| \underline{/\theta}$ is

$$R = |Z| \cos \theta \qquad \text{ohms}$$

(14.49)

Since only the resistance in the network dissipates power, the average power is, from equation (14.47),

$$P_{avg} = \frac{I_p^2 R}{2} = \frac{I_p^2 |Z| \cos \theta}{2}$$

(14.50)

But $I_p |Z| = E_p$, so $I_p^2 |Z| = I_p E_p$, and

$$P_{avg} = \frac{I_p E_p \cos \theta}{2}$$

(14.51)

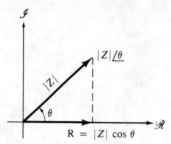

FIGURE 14.39 Resistance component of an arbitrary impedance.

In words, the average power dissipated in any network is one-half the product of the peak values of the voltage across and current supplied to the network, multiplied by the cosine of the angle between the voltage and current. (Remember that the equation is valid only for sinusoidal voltages and currents.)

Power Factor

The utility of equation (14.51) is that it gives us a means for computing average power in terms of the voltage e across an entire network and the total current i supplied to it. The term $\cos \theta$ in (14.51) is called the *power factor*. Since $\cos \theta = \cos(-\theta)$, the power factor is the same whether e leads i by θ degrees or i leads e by θ degrees. Note the following special cases:

1. If the network is purely resistive, then the voltage and current are in phase, so $\theta = 0°$, and the power factor, $\cos 0°$, equals 1. In that case, equation (14.51) reduces to $P_{avg} = E_p I_p / 2$, the same as that for the power dissipated by a resistor.
2. If the network is purely reactive, then the voltage and current are separated by 90°, and the power factor, $\cos 90°$, equals 0. In that case, $P_{avg} = (V_p I_p / 2)0 = 0$, confirming that no power is dissipated by purely reactive components (inductors and capacitors).

Since $V_p = \sqrt{2}\, V_{eff}$ and $I_p = \sqrt{2}\, I_{eff}$, equation (14.51) is equivalent to

$$P_{avg} = \frac{(\sqrt{2}\, V_{eff})(\sqrt{2}\, I_{eff})}{2} \cos \theta$$
$$= V_{eff} I_{eff} \cos \theta$$

(14.52)

Example 14.19   (Analysis)
In the circuit shown in Figure 14.40:

(a) Find the power factor.
(b) Use the power factor to find the average power dissipated in the network.
(c) Verify that the power computed in part (b) is the same as the power computed by using the voltage across and current through the resistor.

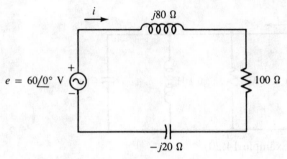

FIGURE 14.40 (Example 14.19)

SOLUTION

(a) The total impedance of the circuit is

$$Z_T = 100 + j80 - j20 = 100 + j60 \ \Omega$$
$$= \sqrt{100^2 + 60^2} \ \underline{/\tan^{-1}(60/100)} = 116.62 \ \underline{/30.96°} \ \Omega$$

The angle of the impedance is the same as the angle between e and i, so the power factor is

$$\cos \theta = \cos (30.96°) = 0.858$$

(b) The current in the circuit is

$$i = \frac{e}{Z_T} = \frac{60 \ \underline{/0°} \ \text{V}}{116.62 \ \underline{/30.96°} \ \Omega} = 0.514 \ \underline{/-30.96°} \ \text{A}$$

Note that the magnitudes of the phasor forms of e and i are, by definition, their peak values, E_p and I_p. Therefore,

$$P_{\text{avg}} = \frac{E_p I_p}{2} \cos \theta = \frac{(60)(0.514)}{2} (0.858) = 13.2 \ \text{W}$$

(c) The voltage across the 100-Ω resistor is

$$v_R = iR = (0.514 \ \underline{/-30.96°} \ \text{A})(100 \ \underline{/0°} \ \Omega) = 51.4 \ \underline{/-30.96°} \ \text{V}$$

Thus, the peak voltage across the resistor is 51.4 V, and the average power dissipated by the resistor is

$$P_{\text{avg}} = \frac{V_p I_p}{2} = \frac{(51.4)(0.514)}{2} = 13.2 \ \text{W}$$

Drill Exercise 14.19

The average power dissipated by a certain network is 0.5 W. If the effective voltage across the network is 4.1 V rms and the effective current through it is 200 mA rms, what is the power factor?

ANSWER: 0.61.   □

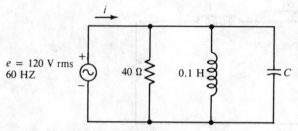

FIGURE 14.41 (Example 14.20)

Example 14.20 (Design)

(a) What value of capacitance C should be connected across the circuit in Figure 14.41 to make the power factor equal 1?

(b) What is the average power delivered to the network?

SOLUTION

(a) The magnitude of the inductive susceptance is

$$|B_L| = \frac{1}{\omega L} = \frac{1}{2\pi(60)(0.1)} = 26.53 \text{ mS}$$

The conductance is $\frac{1}{40} = 25$ mS. Therefore, the total admittance of the circuit is

$$Y_T = G + j|B_C| - j|B_L| = 25 + j(|B_C| - 26.53) \text{ mS}$$

Since the angle of Z_T is the same as the angle of Y_T, except for sign, we can make the angle of the impedance equal to zero by choosing a value for $|B_C|$ that makes the angle of Y_T equal to zero. Under those circumstances, the angle θ between e and i will be zero, and the power factor will be $\cos 0° = 1$.

To make the angle of Y_T equal zero, we require

$$\tan^{-1} \frac{|B_C| - 26.53 \times 10^{-3}}{25 \times 10^{-3}} = 0$$

or

$$\frac{|B_C| - 26.53 \times 10^{-3}}{25 \times 10^{-3}} = 0$$

$$|B_C| = 26.53 \times 10^{-3} \text{ S}$$

In other words, as we would expect, the network has angle 0 (is purely resistive) when the capacitive susceptance is exactly equal to the inductive susceptance.

$$|B_C| = 26.53 \times 10^{-3} \text{ S}$$

$$26.53 \times 10^{-3} = 2\pi f C$$

$$C = \frac{26.53 \times 10^{-3}}{2\pi(60)} = 70.37 \text{ μF}$$

(b) Whether or not the power factor is 1, the average power delivered to the network is the power dissipated by the 40-Ω resistor. Since the resistor is in parallel with the voltage source, the voltage across it is 120 V rms, regardless of the values of L and C. Thus,

$$P_{avg} = \frac{V_{eff}^2}{R} = \frac{(120)^2}{40} = 360 \text{ W}$$

Drill Exercise 14.20

Find the magnitude of i in the circuit of Figure 14.41 when the power factor is 1.

ANSWER: 4.24 A pk.

14.11 SPICE Examples

Example 14.21   (SPICE)

Use SPICE to determine the magnitude and angle of the total impedance, the total current, and the voltage across each component in Figure 14.11 (Example 14.4, p. 507). Assume the frequency is 1 kHz.

SOLUTION We first find values for L and C:

$$|X_L| = \omega L \Rightarrow L = \frac{|X_L|}{\omega} = \frac{1250 \ \Omega}{2\pi \times 1 \text{ kHz}} = 198.9 \text{ mH}$$

$$|X_C| = \frac{1}{\omega C} \Rightarrow C = \frac{1}{|X_C|\omega} = \frac{1}{(450 \ \Omega)2\pi \times 1 \text{ kHz}} = 0.3537 \ \mu\text{F}$$

As in the technique used for dc circuits, we can determine the total impedance by finding the voltage across the terminals of a 1-A ac current source connected to the circuit. The magnitude and angle of that voltage is numerically equal to the magnitude and angle of Z_T, since $v_T = i_T Z_T = (1 \ \underline{/0°})Z_T = Z_T$. However, an ac current source is an *open-circuit* to dc, and, since SPICE requires that there exist a dc path from every node to ground (node 0), the series circuit with current source connected would not meet this requirement. Therefore, we must connect a very large resistance in parallel with the current source to provide the dc paths to ground. In this example, we use RDUM = 1000 MΩ, which is much larger than the total impedance of the circuit and which therefore has negligible effect on the computations. The circuit and input data file are shown in Figure 14.42 (a). Execution of the program reveals that V(1) = 1131 V and VP(1) = 44.99°, so we conclude Z_T = 1131 $\underline{/44.99°}$ Ω, in close agreement with Example 14.4. Figure 14.42 (b) shows the circuit and input data file when the voltage source is restored. Note that RDUM is no longer required (an ac voltage source is a short circuit to dc). Also note that we are determining the total current from I(V1) and IP(V1) instead of using a dummy voltage source. The phase angle computed by SPICE, IP(V1), will therefore be 180° from the actual phase angle of i_T. Execution of the program gives:

$$I(V1) = 88.4 \text{ mA}, \quad IP(V1) = 135° \Rightarrow i_T = 88.4 \underline{/-45°} \text{ mA}$$

$$V(1,2) = 70.7 \text{ V}, \quad VP(1,2) = -44.99° \Rightarrow v_R = 70.7 \underline{/-44.99°} \text{ V}$$

$$V(2,3) = 110.5 \text{ V}, \quad VP(2,3) = 45.01° \Rightarrow v_L = 110.5 \underline{/45.01°} \text{ V}$$

$$V(3) = 39.78 \text{ V}, \quad VP(3) = -135° \Rightarrow v_C = 39.78 \underline{/-135°} \text{ V}$$

These results are in close agreement with those found in Example 14.4.

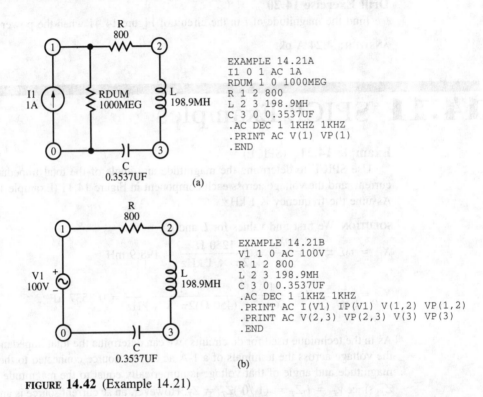

```
EXAMPLE 14.21A
I1 0 1 AC 1A
RDUM 1 0 1000MEG
R 1 2 800
L 2 3 198.9MH
C 3 0 0.3537UF
.AC DEC 1 1KHZ 1KHZ
.PRINT AC V(1) VP(1)
.END
```

```
EXAMPLE 14.21B
V1 1 0 AC 100V
R 1 2 800
L 2 3 198.9MH
C 3 0 0.3537UF
.AC DEC 1 1KHZ 1KHZ
.PRINT AC I(V1) IP(V1) V(1,2) VP(1,2)
.PRINT AC V(2,3) VP(2,3) V(3) VP(3)
.END
```

FIGURE 14.42 (Example 14.21)

Exercises

Section 14.2   Series RL Circuits

14.1 Find the total impedance of each of the networks shown in Figure 14.43. Express answers as phasors, in both rectangular and polar form. Sketch each impedance in the complex plane. Show polar and rectangular coordinates on the sketches.

14.2 Repeat Exercise 14.1 for the networks shown in Figure 14.44.

14.3 Write the sinusoidal expression for the total current in each of the circuits in Figure 14.44. Make a phasor diagram showing e and i in the complex plane.

14.4 Write the sinusoidal expression for the total current in each of the circuits in Figure 14.44. Make a phasor diagram showing e and i in the complex plane.

14.5 For the circuit shown in Figure 14.45:

(a) Find the current i in polar and sinusoidal form.

(b) Find the voltage v_R across the resistor and the voltage v_L across the inductor in polar, rectangular, and sinusoidal form.

(c) Verify Kirchhoff's voltage law around the circuit.

(d) Make a phasor diagram showing e, i, v_R, and v_L.

(e) Sketch the waveforms of e, v_R, and v_L versus angle in degrees.

14.6 Repeat Exercise 14.5 for the circuit shown in Figure 14.46.

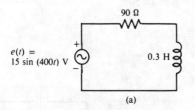

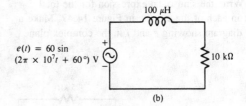

FIGURE **14.43** (Exercise 14.1)

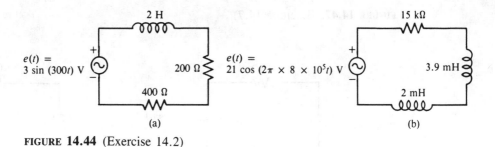

FIGURE **14.44** (Exercise 14.2)

FIGURE **14.45** (Exercise 14.5)

FIGURE **14.46** (Exercise 14.6)

Section 14.3   Series RC Circuits

14.7 Find the total impedance of each of the networks shown in Figure 14.47. Express answers as phasors, in both rectangular and polar form. Sketch each impedance in the complex plane. Show polar and rectangular coordinates on the sketches.

14.8 Repeat Exercise 14.7 for the networks shown in Figure 14.48.

14.9 Write the sinusoidal expression for the total current in each of the circuits in Figure 14.48. Make a phasor diagram showing e and i in the complex plane.

14.10 Write the sinusoidal expression for the total current in each of the circuits in Figure 14.47. Make a phasor diagram showing e and i in the complex plane.

14.11 For the circuit shown in Figure 14.49:
(a) Find the current i in polar and sinusoidal form.
(b) Find the voltage v_R across the resistor and the voltage v_C across the capacitor in polar, rectangular, and sinusoidal form.
(c) Verify Kirchhoff's voltage law around the circuit.
(d) Make a phasor diagram showing e, i, v_R, and v_C.
(e) Sketch the waveforms of e, v_R, and v_C versus angle in degrees.

14.12 Repeat Exercise 14.11 for the circuit shown in Figure 14.50.

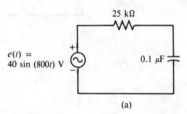

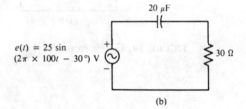

(a)

(b)

FIGURE 14.47   (Exercise 14.7)

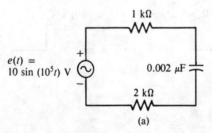

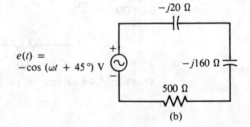

(a)

(b)

FIGURE 14.48   (Exercise 14.8)

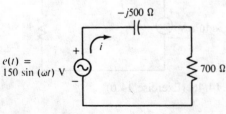

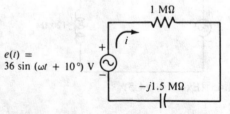

FIGURE 14.49   (Exercise 14.11)

FIGURE 14.50   (Exercise 14.12)

Section 14.4 Series RLC Circuits

14.13 For the circuit shown in Figure 14.51:
(a) Find the total impedance in polar form. Make a phasor diagram showing all components of the impedance.
(b) Find the current i in polar form.
(c) Find the voltages v_R, v_L, and v_C across the resistor, inductor, and capacitor in polar form.
(d) Verify Kirchhoff's voltage law around the circuit.
(e) Make a phasor diagram showing e, i, v_R, v_L, and v_C in the complex plane.

14.14 Repeat Exercise 14.13 for the circuit shown in Figure 14.52.

14.15 For the circuit shown in Figure 14.53:
(a) Find the total impedance Z_T in polar form.
(b) Find the current i in polar and sinusoidal form.
(c) Find the voltage across each component in polar form.
(d) Find the total equivalent inductance and the total equivalent capacitance in the circuit.

14.16 Repeat Exercise 14.15 for the circuit shown in Figure 14.54.

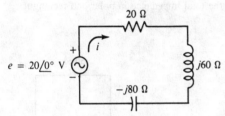

FIGURE 14.51 (Exercise 14.13)

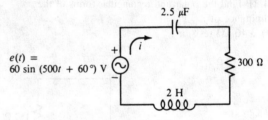

FIGURE 14.52 (Exercise 14.14)

FIGURE 14.53 (Exercise 14.15)

FIGURE 14.54 (Exercise 14.16)

Section 14.5   The Voltage-Divider Rule

14.17 Use the voltage-divider rule to find the voltage drop across each component in the circuits shown in Figure 14.55. Make a phasor diagram showing e and the voltage drops.

14.18 Use the voltage-divider rule to find v_{ab} and v_{ac} in each circuit shown in Figure 14.56. Express all voltages in polar form.

14.19 Use the voltage-divider rule to find the *magnitude only* of the voltage across each component in the circuit shown in Figure 14.56(a).

14.20 Use the voltage-divider rule to find the *magnitude only* of the voltage across each component shown in Figure 14.56(b).

Section 14.6   Admittance and Susceptance

14.21 Find the polar and rectangular forms of the admittance of
(a) a 40-kΩ resistor

(b) a 5-μF capacitor at 16 kHz
(c) a 25-mH inductor at 50×10^3 rad/s
(d) a 0.01-μF capacitor at dc

14.22 (a) At what frequency does the susceptance of a 400-pF capacitor have magnitude equal to 200 μS?
(b) What value of inductance has a susceptance whose magnitude equals 0.8 S at 400 rad/s?
(c) Find the polar and rectangular forms of the admittance of a network whose impedance is $Z_T = 12 - j8$ Ω.

Section 14.7   Parallel AC Circuits

14.23 (i) Find the total admittance of each of the circuits in Figure 14.57.
(ii) Make a phasor diagram showing each admittance in the complex plane. Identify the conductance and susceptance components of each admittance on the diagram.
(iii) Find the total impedance in polar and rectangular form.

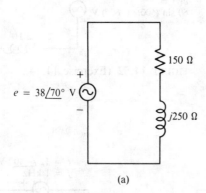

(a)

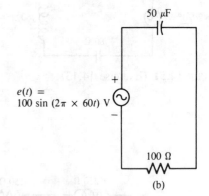

(b)

FIGURE 14.55 (Exercise 14.17)

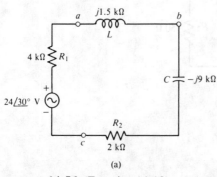

(a)

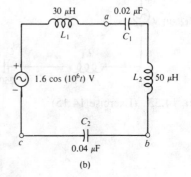

(b)

FIGURE 14.56 (Exercise 14.18)

14.24 (i) Find the total admittance of each of the circuits shown in Figure 14.58.
(ii) Make a phasor diagram showing each admittance in the complex plane. Identify the conductance and susceptance components of each admittance on the diagram.
(iii) Find the total impedance in polar and rectangular form.

14.25 Find the total current i_T in each of the circuits in Figure 14.58 in polar and sinusoidal form. Make a phasor diagram showing e and i in the complex plane.

14.26 Find the total current i_T in each of the circuits shown in Figure 14.57 in polar and sinusoidal form. Make a phasor diagram showing e and i_T in the complex plane.

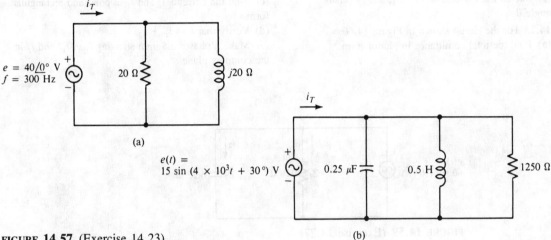

$e = 40\underline{/0°}$ V
$f = 300$ Hz

20 Ω $j20$ Ω

(a)

$e(t) =$
$15 \sin (4 \times 10^3 t + 30°)$ V

0.25 μF   0.5 H   1250 Ω

(b)

FIGURE 14.57 (Exercise 14.23)

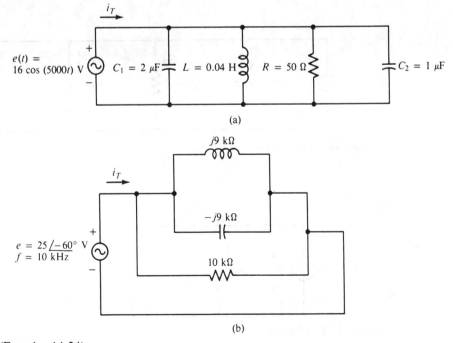

$e(t) =$
$16 \cos (5000t)$ V $C_1 = 2$ μF $L = 0.04$ H $R = 50$ Ω $C_2 = 1$ μF

(a)

$j9$ kΩ

$-j9$ kΩ

$e = 25\underline{/-60°}$ V
$f = 10$ kHz

10 kΩ

(b)

FIGURE 14.58 (Exercise 14.24)

14.27 For the circuit shown in Figure 14.59:
(a) Find the total admittance, Y_T. Verify that $Z_T = 1/Y_T = Z_1 Z_2/(Z_1 + Z_2)$.
(b) Find the total current i in polar and rectangular form.
(c) Find i_1 and i_2 and verify that $i = i_1 + i_2$.
(d) Make a phasor diagram showing i, i_1, and i_2 in the complex plane.
(e) Sketch the waveforms of i, i_1, and i_2 versus angle θ.

14.28 For the circuit shown in Figure 14.60:
(a) Find the total admittance in phasor form.

(b) Find i, i_1, i_2, i_3, and i_4 in polar and rectangular form.
(c) Verify that $i = i_1 + i_2 + i_3 + i_4$.
(d) Make a phasor diagram showing i, i_1, i_2, i_3, and i_4.

14.29 For the circuit shown in Figure 14.61:
(a) Find the total admittance in polar form.
(b) Find the polar form of the voltage v across the current source.
(c) Find the currents i_1 and i_2 in polar and rectangular form.
(d) Verify that $i = i_1 + i_2$.
(e) Make a phasor diagram showing i, v, i_1, and i_2 in the complex plane.

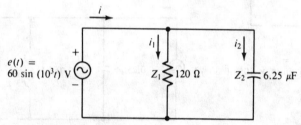

FIGURE 14.59 (Exercise 14.27)

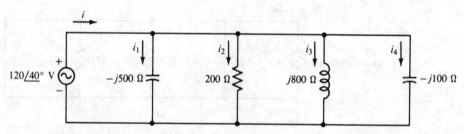

FIGURE 14.60 (Exercise 14.28)

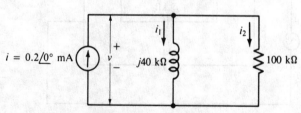

FIGURE 14.61 (Exercise 14.29)

14.30 For the circuit shown in Figure 14.62:

(a) Find the total impedance in polar form.

(b) Find the polar form of the voltage v across the current source.

(c) Find the currents i_1 and i_2 in polar and rectangular form.

(d) Use Kirchhoff's current law to find currents i_3 and i_4 in rectangular and polar forms.

Section 14.8   The Current-Divider Rule

14.31 Use the current-divider rule to find the currents i_1 and i_2 in Figure 14.63. Express each current in polar and sinusoidal form.

14.32 Use the current-divider rule (equation 14.35) to find the currents i_1, i_2, and i_3 in Figure 14.64. Express each current in polar form.

14.33 (a) Find the total current i_T in Figure 14.65. Then use the current-divider rule to find the currents i_1 and i_2. Express each current in polar form.

(b) Check your answers by finding i_1 and i_2 directly, and then adding to find i_T.

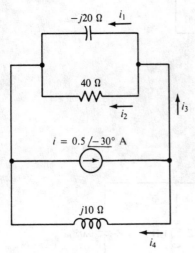

FIGURE **14.62** (Exercise 14.30)

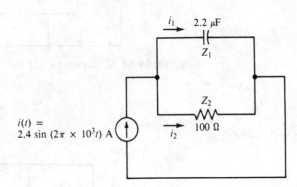

FIGURE **14.63** (Exercise 14.31)

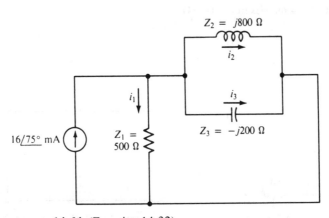

FIGURE **14.64** (Exercise 14.32)

14.34 Find the equivalent impedance of the parallel combination of Z_2 and Z_3 in Figure 14.66. Then use that impedance and the current-divider rule to find the current i_1 in polar form.

Section 14.9   Equivalent Series and Parallel Networks

14.35 Find a two-component parallel network that is equivalent to the series network shown in Figure 14.67. Draw the schematic diagram of the equivalent network.

14.36 Find a two-component series network that is equivalent to the series network shown in Figure 14.68. Find the component values in the equivalent series network, given that the frequency is 100 Hz. Draw the schematic diagram of the equivalent network. (*Hint*: Find the total equivalent impedance of the network.)

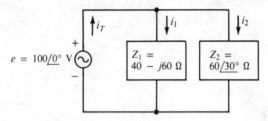

FIGURE **14.65** (Exercise 14.33)

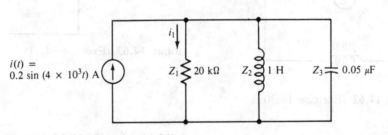

FIGURE **14.66** (Exercise 14.34)

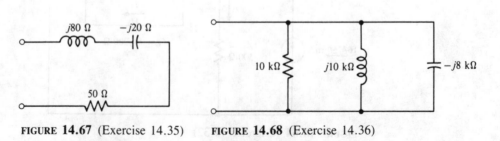

FIGURE **14.67** (Exercise 14.35)   FIGURE **14.68** (Exercise 14.36)

Section 14.10   Power in Circuits Containing Reactance

14.37 For the circuit shown in Figure 14.69:

(a) Find the power factor.

(b) Use the power factor to find the average power dissipated by the circuit.

(c) Verify that the power computed in part (b) is the same as the power dissipated by the resistor.

14.38 Repeat Exercise 14.37 for the circuit shown in Figure 14.70.

14.39 The effective value of the sinusoidal voltage across a certain network is 32 V rms and the peak value of the total current supplied to it is 24 mA pk. The average power dissipated in the network is 0.48 W. Find the power factor.

14.40 The peak value of the sinusoidal voltage across a certain network is 169 V pk. The average power dissipated in the network is 1.8 kW and the power factor is 0.92.

(a) Find the effective value of the total current supplied to the network.

(b) Find the magnitude of the phase angle between the voltage and current.

14.41 For the circuit shown in Figure 14.71:

(a) Find the power factor.

(b) Use the power factor to find the average power delivered to the network by the source.

(c) Find the power dissipated by each component in the circuit and verify that the power delivered by the source is the sum of the powers dissipated by all the components.

14.42 The block labeled × in Figure 14.72 consists of purely reactive components. The power factor for the entire circuit is 0.6. Given that the voltage across the circuit leads the current i, write the sinusoidal expression for the current.

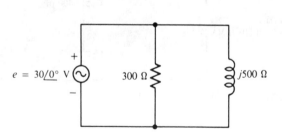

FIGURE 14.69 (Exercise 14.37)

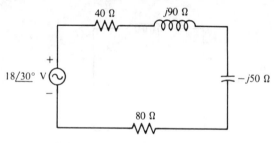

FIGURE 14.71 (Exercise 14.41)

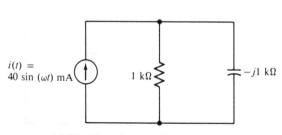

FIGURE 14.70 (Exercise 14.38)

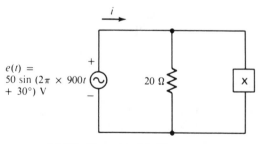

FIGURE 14.72 (Exercise 14.42)

Design Exercises

14.1D It is necessary to insert a resistance R in series with a 20-mH inductor, as shown in Figure 14.73, so that the peak value of the current through the inductor is limited to 0.5 A. What value of R should be used?

14.2D It is necessary to insert a resistance R in series with a 0.02-μF capacitor, as shown in Figure 14.74, so that the peak voltage across the capacitor is limited to 25 V. What value of resistance should be used?

14.3D What value of capacitance C should be used in the circuit shown in Figure 14.75 to make the current i have the same phase as the voltage $e(t)$?

14.4D What value of C should be used in Figure 14.75 if it is necessary to make the current i lead the voltage e by 30°? (*Hint*: To make i lead e by 30°, the angle of the total impedance must be $-30°$.)

14.5D Figure 14.76 shows a *capacitive voltage divider*. What value of C_2 should be used to make the magnitude of voltage v_2 equal 15 V?

14.6D It is necessary to connect a resistance in parallel with a 0.2-H inductor, as shown in Figure 14.77, so

that the magnitude of the current in the inductor is limited to 125 mA. What value of resistance should be used?

14.7D What should be the magnitude of the current supplied by the constant-current source in Figure 14.78 if it is necessary to produce a peak current of 40 mA in the capacitor?

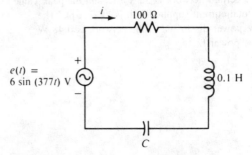

FIGURE 14.75 (Exercise 14.3D)

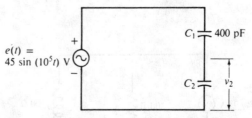

FIGURE 14.76 (Exercise 14.5D)

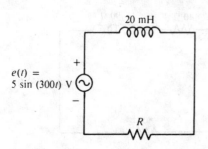

FIGURE 14.73 (Exercise 14.1D)

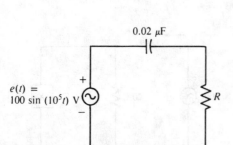

FIGURE 14.74 (Exercise 14.2D)

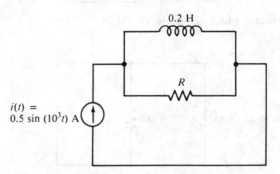

FIGURE 14.77 (Exercise 14.6D)

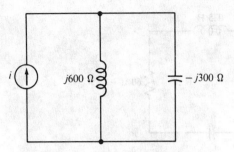

FIGURE 14.78 (Exercise 14.7D)

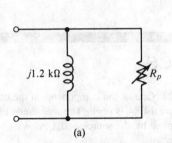

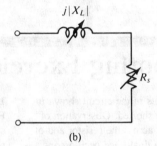

(a)   (b)

FIGURE 14.79 (Exercise 14.8D)

14.8D The equivalent circuit of a certain electronic device at a particular frequency is shown in Figure 14.79(a). Temperature changes cause the resistance, R_p, of the device to change for 500 Ω to 750 Ω, but do not affect the inductance. In an experiment designed to study the characteristics of the device, the equivalent circuit shown in Figure 14.79(b) is to be constructed. Over what range of values should it be possible to adjust $|X_L|$ and R_s in the experimental circuit?

14.9D What value of inductance L should be used in the circuit shown in Figure 14.80 in order that the average power delivered to the circuit be 25 W?

14.10D A capacitor is to be connected in parallel with a 100-Ω resistor, as shown in Figure 14.81, to reduce the power dissipated by the resistor. If the resistor has a 2-W rating, what is the minimum value of capacitance that should be used?

14.11D A capacitor is to be inserted in a series circuit, as shown in Figure 14.82, to make the power factor equal to 0.9. If the total current in the circuit is to lag the source voltage, what value of capacitance should be used? (*Hint:* The angle of the total impedance is positive when i lags e.)

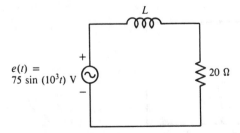

FIGURE 14.80 (Exercise 14.9D)

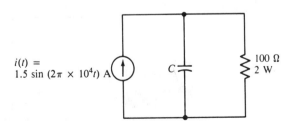

FIGURE 14.81 (Exercise 14.10D)

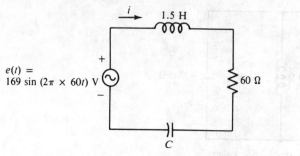

FIGURE 14.82 (Exercise 14.11D)

Troubleshooting Exercises

14.1T One of the components in the circuit shown in Figure 14.83 is known to be shorted. Observation of the voltage waveform, $v_R(t)$, across the resistor and of the source voltage, $e(t)$, on a dual-trace oscilloscope reveals that $e(t)$ leads $v_R(t)$. Which component is shorted? Explain.

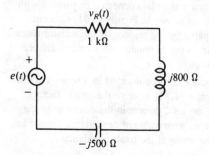

FIGURE 14.83 (Exercise 14.1T)

14.2T One of the components in the circuit shown in Figure 14.84 is open. The magnitude of the current i supplied by the source is 0.1 A. Which component is open?

14.3T As part of a manufacturer's test procedure for 20-mH inductors, a 15-V peak ac voltage with an angular frequency of 1000 rad/s is connected across each inductor. Among other measurements made in this test, the peak value of the current flowing in the inductor is determined. The resistance of the inductor is not supposed to exceed 50 Ω. Assuming that the inductance has its specified value, what minimum value of peak current should be measured in the test?

14.4T To test the capacitance of 1-μF capacitors, a manufacturer connects each unit in series with a 50-Ω resistor and a 60-V pk ac source, and measures the peak current that flows. The frequency of the voltage is 400 Hz. The manufacturer's specified tolerance on the

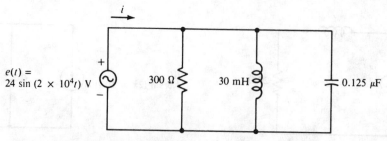

FIGURE 14.84 (Exercise 14.2T)

capacitors is $\pm 20\%$. The peak current measured in one test was 150 mA. Was the capacitor in tolerance?

14.5T The circuit in Figure 14.85 is said to be *tuned* when the capacitive reactance exactly equals the inductive reactance. Because of component tolerances and environmental factors, the inductance and/or capacitance may not be exactly those required to maintain tuning. In one test, the current through the resistor is measured to be 20 mA pk. Is the circuit tuned? Describe how observations of $e(t)$ and $V_R(t)$ on a dual-trace oscilloscope could be used to determine which reactance is greater, when the circuit is not tuned.

14.6T An electric utility company specifies that the power factor of any industrial load it serves must be at least 0.9. To test the power factor at a certain facility, a watt-hour meter was connected for a 30-minute period, during which the effective voltage and effective current supplied to the facility were a constant 240 V rms and 55 A rms, respectively. At the end of the test, the meter showed that the total energy consumption was 5.68 kWh. Does the facility meet the utility company's specification for power factor?

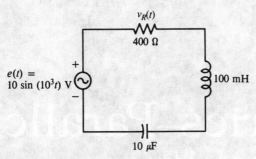

FIGURE 14.85 (Exercise 14.5T)

SPICE Exercises

14.1S Use SPICE to find the magnitude and angle of the total impedance, the total current, and the voltage across each component in the circuit shown in Figure 14.52, p. 543.

14.2S Use SPICE as an aid in verifying Kirchhoff's voltage law around the circuit in Figure 14.51, p. 543. The frequency is 500 Hz. (*Hint*: SPICE can provide the real and imaginary parts of an ac voltage; see Appendix Section A.7.)

14.3S Use SPICE to determine the magnitude and angle of the total admittance, the total current, and the current through each component of the circuit shown in Figure 14.58(a), p. 545. (*Hint*: Find the current supplied by a 1-V ac source; be careful to avoid a voltage source-inductance loop.)

14.4S Use SPICE as an aid in finding the power factor and the total power dissipated in the circuit in Figure 14.71, p. 549. The frequency is 60 Hz.

15 Series–Parallel AC Circuits

15.1 Analysis Using Simplified Equivalent Networks

AC circuits containing combinations of series- and parallel-connected components can be analyzed using the same general approach that we used for series–parallel dc circuits. Recall that our strategy is to reduce such a circuit to progressively simpler equivalent circuits by replacing series and parallel combinations with their equivalents. If necessary, the circuit can ultimately be reduced to a single equivalent impedance, and the total current can be computed. All of the analysis rules and circuit laws can then be applied to the simplified equivalent circuits until every ac voltage and current is determined. Remember that all the mathematical operations required for these computations, addition, subtraction, multiplication, and division, must be performed using the phasor forms of the voltages, currents, and impedances.

As in series–parallel dc circuits, there is usually more than one correct way to analyze an ac circuit, in the sense that different circuit laws may be applied in different sequences to produce the same solutions. The next example illustrates this point.

Example 15.1   (Analysis)

Find the polar form of the current in each component of Figure 15.1(a)

(a) using the current-divider rule

(b) using the voltage-divider rule

SOLUTION

(a) The equivalent impedance, Z_P, of the parallel combination of the inductor and capacitor is

$$Z_P = \frac{Z_L Z_C}{Z_L + Z_C} = \frac{(60 \underline{/90°})(20 \underline{/-90°})}{0 + j60 - j20} = \frac{1200 \underline{/0°}}{40 \underline{/90°}} = 30 \underline{/-90°} \; \Omega$$

$$= 0 - j30 \; \Omega$$

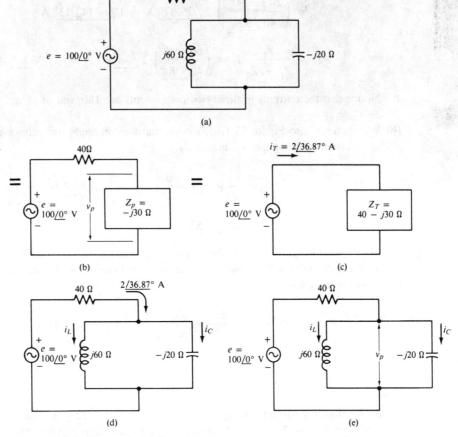

(a)

(b)

(c)

(d)

(e)

FIGURE 15.1 (Example 15.1)

Figure 15.1(b) shows the circuit with the parallel combination replaced by its equivalent impedance, Z_P. Clearly, Z_P and the 40-Ω resistor are in series, so the total impedance of the circuit is

$$Z_T = 40 - j30 \ \Omega = \sqrt{40^2 + 30^2} \ \bigg/ \tan^{-1}\left(\frac{-30}{40}\right) = 50 \ \big/ {-36.87°} \ \Omega$$

As can be seen in Figure 15.1(c), the total current in the circuit is then

$$i_T = \frac{e}{Z_T} = \frac{100 \ \big/ 0° \ \text{V}}{50 \ \big/ {-36.87°} \ \Omega} = 2 \ \big/ 36.87° \ \text{A}$$

The original circuit is redrawn in Figure 15.1(d), where we see that the current in the resistor is $i_T = 2 \ \big/ 36.87° \ \text{A}$. By the current-divider rule,

$$i_L = \frac{Z_C}{Z_L + Z_C} i_T = \left(\frac{20 \ \big/ {-90°}}{0 + j60 - j20}\right) 2 \ \big/ 36.87° \ \text{A}$$

$$= \left(\frac{20 \ \big/ {-90°}}{40 \ \big/ 90°}\right) 2 \ \big/ 36.87° \ \text{A} = 1 \ \big/ {-143.13°} \ \text{A}$$

$$i_C = \frac{Z_L}{Z_L + Z_C} i_T = \left(\frac{60 \ \big/ 90°}{40 \ \big/ 90°}\right) 2 \ \big/ 36.87° \ \text{A} = 3 \ \big/ 36.87° \ \text{A}$$

Notice that the currents in these two components are 180° out of phase, and that $i_L + i_C = i_T$.

(b) With reference to Figure 15.1(b), we see that we can apply the voltage-divider rule to determine the voltage, v_p, across Z_P:

$$v_p = \frac{Z_p}{Z_T} e = \left(\frac{30 \ \big/ {-90°}}{50 \ \big/ {-36.87°}}\right) 100 \ \big/ 0° \ \text{V}$$

$$= 60 \ \big/ {-53.13°} \ \text{V}$$

As shown in Figure 15.1(e), the voltage v_p appears across the parallel combination of the inductor and the capacitor. Therefore, by Ohm's law, the currents in those components are

$$i_L = \frac{v_p}{Z_L} = \frac{60 \ \big/ {-53.13°} \ \text{V}}{60 \ \big/ 90° \ \Omega} = 1 \ \big/ {-143.13°} \ \text{A}$$

$$i_C = \frac{v_p}{Z_C} = \frac{60 \ \big/ {-53.13°} \ \text{V}}{20 \ \big/ {-90°} \ \Omega} = 3 \ \big/ 36.87° \ \text{A}$$

By Kirchhoff's current law, $i_R = (-0.8 - j0.6) \ \text{A} + (2.4 + j1.8) \ \text{A} = 1.6 + j1.2 \ \text{A} = 2 \ \big/ 36.87° \ \text{A}$. As expected, these results are the same as those obtained using the current-divider rule.

Drill Exercise 15.1

Find the polar form of v_p in Example 15.1 when a capacitor having reactance $-j40\ \Omega$ is connected in series with the 40-Ω resistor.

ANSWER: $37.21\ \underline{/-29.74°}$ V.

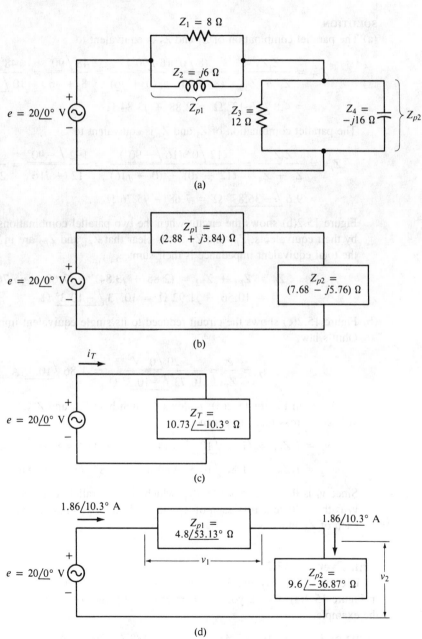

(a)

(b)

(c)

(d)

FIGURE 15.2 (Example 15.2)

Example 15.2 (Analysis)

(a) Find the total equivalent impedance of the circuit shown in Figure 15.2(a).

(b) Find the total current i_T supplied by the source to the circuit.

(c) Find the polar form of the voltage across each component.

SOLUTION

(a) The parallel combination of Z_1 and Z_2 is equivalent to

$$Z_{P1} = \frac{Z_1 Z_2}{Z_1 + Z_2} = \frac{(8 \underline{/0°})(6 \underline{/90°})}{(8 + j0) + (0 + j6)} = \frac{48 \underline{/90°}}{8 + j6} = \frac{48 \underline{/90°}}{10 \underline{/36.87°}}$$

$$= 4.8 \underline{/53.13°} \ \Omega = 2.88 + j3.84 \ \Omega$$

The parallel combination of Z_3 and Z_4 is equivalent to

$$Z_{P2} = \frac{Z_3 Z_4}{Z_3 + Z_4} = \frac{(12 \underline{/0°})(16 \underline{/-90°})}{(12 + j0) + (0 - j16)} = \frac{192 \underline{/-90°}}{12 - j16} = \frac{192 \underline{/-90°}}{20 \underline{/-53.13°}}$$

$$= 9.6 \underline{/-36.87°} \ \Omega = 7.68 - j5.76 \ \Omega$$

Figure 15.2(b) shows the circuit when the two parallel combinations are replaced by their equivalents, Z_{P1} and Z_{P2}. It is clear that Z_{P1} and Z_{P2} are in series, so the total equivalent impedance is their sum:

$$Z_T = Z_{P1} + Z_{P2} = (2.88 + j3.84) + (7.68 - j5.76)$$

$$= 10.56 - j1.92 \ \Omega = 10.73 \underline{/-10.3°} \ \Omega$$

(b) Figure 15.2(c) shows the circuit reduced to its single equivalent impedance, Z_T. By Ohm's law,

$$i_T = \frac{e}{Z_T} = \frac{20 \underline{/0°} \text{ V}}{10.73 \underline{/-10.3°} \ \Omega} = 1.86 \underline{/10.3°} \text{ A}$$

(c) As shown in Figure 15.2(d), i_T flows through both Z_{P1} and Z_{P2}. Therefore, the voltages across those impedances are

$$v_1 = i_T Z_{P1} = (1.86 \underline{/10.3°} \text{ A})(4.8 \underline{/53.13°} \ \Omega) = 8.93 \underline{/63.43°} \text{ V}$$

$$v_2 = i_T Z_{P2} = (1.86 \underline{/10.3°} \text{ A})(9.6 \underline{/-36.87°} \ \Omega) = 17.86 \underline{/-26.57°} \text{ V}$$

Since v_1 is the voltage across Z_{P1}, which is the parallel combination of Z_1 and Z_2, v_1 is the voltage across each of Z_1 and Z_2. Similarly, v_2 is the voltage across each of Z_3 and Z_4.

Drill Exercise 15.2

Use the value of i_T and the current-divider rule to find the current in each resistor in Figure 15.2(a). Check your answers using the values of v_1 and v_2 computed in the example.

ANSWER: $i_{Z1} = 1.116 \underline{/63.43°}$ A; $i_{Z3} = 1.488 \underline{/-26.57°}$ A. ☐

Example 15.3   (Analysis)

(a) Find the polar forms of the current through and voltage across each component in Figure 15.3(a).

(b) Make a phasor diagram showing all currents and another phasor diagram showing all voltages in the circuit.

SOLUTION

(a) We first compute the reactances Z_1 and Z_3:

$$|Z_1| = |X_L| = \omega L = 10^4(10 \times 10^{-3}) = 100 \ \Omega$$

$$|Z_3| = |X_C| = \frac{1}{\omega C} = \frac{1}{10^4(0.5 \times 10^{-6})} = 200 \ \Omega$$

Letting Z_S be the series combination of Z_2 and Z_3, we have

$$Z_S = Z_2 + Z_3 = 200 - j200 = 282.84 \ \underline{/-45°} \ \Omega$$

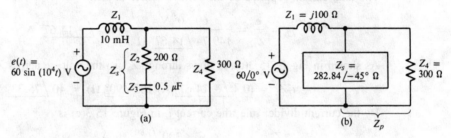

(a)   (b) Z_p

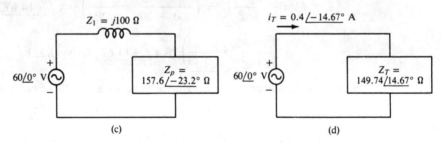

(c)   (d)

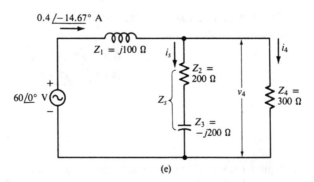

(e)

FIGURE 15.3 (Example 15.3)

Figure 15.3(b) shows the circuit when Z_2 and Z_3 are replaced by Z_s. It is clear that Z_S is in parallel with Z_4. Letting Z_P represent that parallel combination, we have

$$Z_P = \frac{Z_4 Z_s}{Z_4 + Z_s} = \frac{(300 \underline{/0°})(282.84 \underline{/-45°})}{300 + (200 - j200)}$$

$$= \frac{84{,}852 \underline{/-45°}}{500 - j200} = \frac{84{,}852 \underline{/-45°}}{538.5 \underline{/-21.8°}} = 157.6 \underline{/-23.2°}\ \Omega$$

$$= 144.86 - j62.09\ \Omega$$

Figure 15.3(c) shows that Z_P is in series with Z_1. Therefore, the total impedance is

$$Z_T = Z_1 + Z_P = (0 + j100) + (144.86 - j62.09)$$

$$= 144.86 + j37.91\ \Omega$$

$$= 149.74 \underline{/14.67°}\ \Omega$$

The total current supplied by the voltage source is then

$$i_T = \frac{e}{Z_T} = \frac{60 \underline{/0°}\text{ V}}{149.74 \underline{/14.67°}\ \Omega} = 0.4 \underline{/-14.67°}\text{ A}$$

As shown in Figure 15.3(e), i_T flows through Z_1, so the voltage across Z_1 is

$$v_1 = i_T Z_1 = (0.4 \underline{/-14.67°}\text{ A})(100 \underline{/90°}\ \Omega) = 40 \underline{/75.33°}\text{ V}$$

By the current-divider rule, the current i_s in Figure 15.3(e) is

$$i_s = \frac{Z_4}{Z_4 + Z_s}\, i_T = \left(\frac{300 \underline{/0°}}{300 + 200 - j200} \right)(0.4 \underline{/-14.67°}\text{ A})$$

$$= \left(\frac{300 \underline{/0°}}{538.5 \underline{/-21.8°}} \right)(0.4 \underline{/-14.67°}\text{ A}) = 0.223 \underline{/7.13°}\text{ A}$$

Since i_s is the current in Z_2 and Z_3, the voltages across those components are

$$v_2 = i_s Z_2 = (0.223 \underline{/7.13°}\text{ A})(200 \underline{/0°}\ \Omega) = 44.6 \underline{/7.13°}\text{ V}$$

$$v_3 = i_s Z_3 = (0.223 \underline{/7.13°}\text{ A})(200 \underline{/-90°}\ \Omega) = 44.6 \underline{/-82.87°}\text{ V}$$

The voltage v_4 across Z_4 is the same as the voltage across Z_S, since Z_4 is in parallel with Z_S. Thus,

$$v_4 = i_s Z_S = (0.223 \underline{/7.13°}\text{ A})(282.84 \underline{/-45°}\ \Omega)$$

$$= 63.07 \underline{/-37.87°}\text{ V}$$

Finally,

$$i_4 = \frac{v_4}{Z_4} = \frac{63.07 \underline{/-37.87°}\text{ V}}{300 \underline{/0°}\ \Omega} = 0.21 \underline{/-37.87°}\text{ A}$$

(b) Figure 15.4(a) shows the polar forms of all voltages and currents in the circuit. Figure 15.4(b) shows the phasor diagram of the currents. Note that $i_T = i_s + i_4$, in accordance with Kirchhoff's current law. Figure 15.4(c) shows the phasor diagram of the voltages. Note that $v_4 = v_2 + v_3$, which follows from an application of Kirchhoff's voltage law around the right-hand loop in the circuit. The diagram also shows that $e = v_1 + v_4$. Recognizing that v_4 is the same as the voltage across Z_S [see Figure 15.3(e)], this result follows from an application of Kirchhoff's voltage law around the left-hand loop.

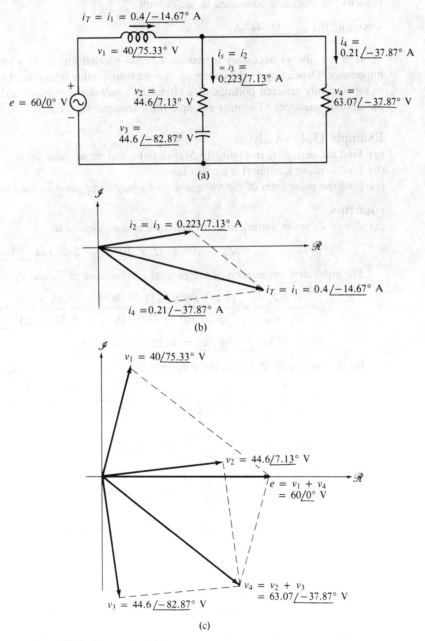

FIGURE 15.4 (Example 15.3)

As an exercise, verify that the voltage across and current through each component have the correct phase relation, as prescribed by ELI the ICE man.

Drill Exercise 15.3

Find the polar form of the current through Z_4 in Figure 15.3(a) when 300 Ω of inductive reactance is connected in series with Z_4.

ANSWER: $0.14 \angle -69.44°$ A.   □

It is not always necessary to reduce a series–parallel circuit to a single equivalent impedance. Depending on the voltage or current whose value is sought, it may be possible to replace only selected portions of a circuit by equivalent networks and then solve for the desired quantities. The next example illustrates one such case.

Example 15.4   (Analysis)

(a) Find the current i_1 in Figure 15.5(a) in polar and rectangular form.
(b) Find i_2 using Kirchhoff's current law.
(c) Find the polar form of the voltage across every component in the circuit.

SOLUTION

(a) Z_1 and Z_2 are in series, so that combination is equivalent to

$$Z_S = Z_1 + Z_2 = 10^3 + j2 \times 10^3 \ \Omega = 2.24 \text{ k}\Omega \ \angle 63.43°$$

The equivalent impedance of the parallel combination of Z_3 and Z_4 is

$$Z_P = \frac{Z_3 Z_4}{Z_3 + Z_4} = \frac{(5 \text{ k}\Omega \ \angle 0°)(3 \text{ k}\Omega \ \angle -90°)}{5 \times 10^3 - j3 \times 10^3} = \frac{15 \times 10^6 \ \angle -90°}{5.83 \times 10^3 \ \angle -30.96°}$$

$$= 2.57 \text{ k}\Omega \ \angle -59.04° = 1322 - j2204 \ \Omega$$

By the current-divider rule, the current i_1 is

$$i_1 = \frac{Z_P}{Z_S + Z_P} i = \left[\frac{2.57 \times 10^3 \ \angle -59.04°}{(1000 + j2000) + (1322 - j2204)} \right] (30 \ \angle 0° \text{ mA})$$

$$= \left(\frac{2.57 \times 10^3 \ \angle -59.04°}{2322 - j204} \right) (30 \ \angle 0° \text{ mA})$$

$$= \left(\frac{2.57 \times 10^3 \ \angle -59.04°}{2331 \ \angle -5.02°} \right) (30 \ \angle 0° \text{ mA})$$

$$= 33.1 \ \angle -54.02° \text{ mA} = (19.45 - j26.79) \text{ mA}$$

(b) By Kirchhoff's current law,

$$i_2 = i - i_1$$
$$= (30 + j0) \text{ mA} - (19.45 - j26.79) \text{ mA}$$
$$= (10.55 + j26.79) \text{ mA} = 28.79 \ \angle 68.51° \text{ mA}$$

(c) Since i_1 flows through Z_1 and Z_2, the voltages across those impedances are

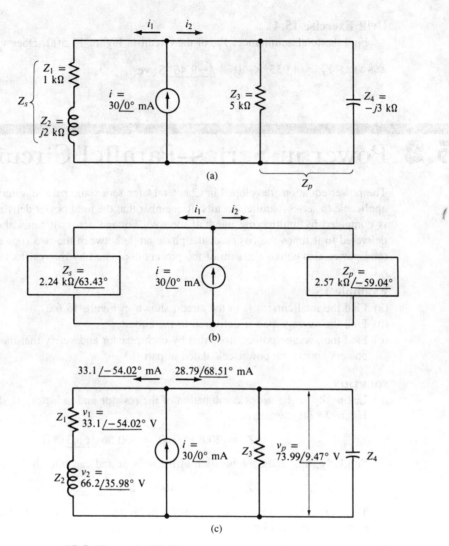

FIGURE 15.5 (Example 15.4)

$$v_1 = i_1 Z_1 = (33.1 \,\underline{/-54.02°} \text{ mA})(1 \text{ k}\Omega \,\underline{/0°}) = 33.1 \,\underline{/-54.02°} \text{ V}$$

$$v_2 = i_1 Z_2 = (33.1 \,\underline{/-54.02°} \text{ mA})(2 \text{ k}\Omega \,\underline{/90°}) = 66.2 \,\underline{/35.98°} \text{ V}$$

The voltage across the current source and across each of Z_3 and Z_4 is the same as the voltage across the parallel combination Z_P:

$$v_p = v_3 = v_4 = i_2 Z_P = (28.79 \,\underline{/68.51°} \text{ mA})(2.57 \text{ k}\Omega \,\underline{/-59.04°})$$
$$= 73.99 \,\underline{/9.47°} \text{ V}$$

The voltages and currents are shown in Figure 15.5(c). As an exercise, verify that $v_p = v_1 + v_2$.

Drill Exercise 15.4

Find the total admittance, Y_T, of the circuit in Figure 15.5(a). Does $v_p = i/Y_T$?

ANSWER: $Y_T = 4.055 \times 10^{-4} \underline{/-9.46°}$ S; yes.   □

15.2 Power in Series–Parallel Circuits

The power equations developed in Chapter 14 for series and parallel circuits are equally applicable to series–parallel circuits. Remember that the total power delivered to a circuit is computed by multiplying one-half the *total* voltage across it times the *total* current delivered to it times the cosine of the phase angle between the two (power factor). The total power also equals the sum of the powers dissipated by the resistors in the circuit.

Example 15.5

(a) Find the total current i_T in the circuit shown in Figure 15.6(a).
(b) Find the average power delivered to the circuit.
(c) Find the average power dissipated by each resistor and verify that the sum of those powers equals the power calculated in part (b).

SOLUTION

(a) Letting Z_1 be the series combination of the resistor and inductor, as shown in Figure 15.6(a), we have

$$Z_1 = 200 + j300 = 360.56 \underline{/56.31°} \ \Omega$$

Similarly, the series combination of the resistor and capacitor is

$$Z_2 = 300 - j400 = 500 \underline{/-53.13°} \ \Omega$$

The equivalent circuit is then as shown in Figure 15.6(b). It is clear that the total impedance of the circuit is the parallel combination of Z_1 and Z_2:

$$Z_T = \frac{Z_1 Z_2}{Z_1 + Z_2} = \frac{(360.56 \underline{/56.31°})(500 \underline{/-53.13°})}{(200 + j300) + (300 - j400)}$$

$$= \frac{180.28 \times 10^3 \underline{/3.18°}}{500 - j100} = \frac{180.28 \times 10^3 \underline{/3.18°}}{509.9 \underline{/-11.31°}} = 353.56 \underline{/14.49°} \ \Omega$$

As shown in Figure 15.6(c), the total current is therefore

$$i_T = \frac{e}{Z_T} = \frac{36 \underline{/0°} \text{ V}}{353.56 \underline{/14.49°} \ \Omega} = 101.82 \underline{/-14.49°} \text{ mA}$$

(b) The phase angle between the total voltage, $e = 36 \underline{/0°}$ V, and the total current, $i_T = 101.82 \underline{/-14.49°}$ mA, is 14.49°. Therefore, the average power delivered to the circuit is

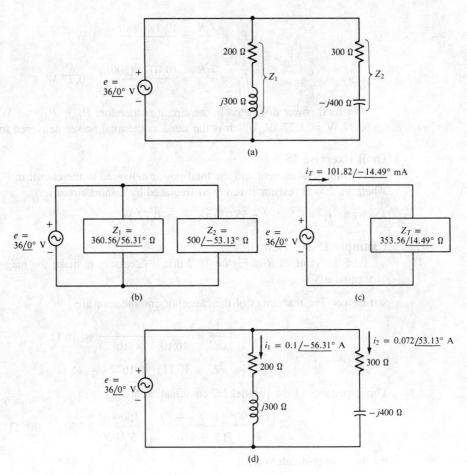

FIGURE 15.6 (Example 15.5)

$$P_{\text{avg}} = \frac{E_p I_p \cos \theta}{2} = \frac{(36)(101.82 \times 10^{-3}) \cos (14.49°)}{2}$$
$$= 1.77 \text{ W}$$

(c) Since the equivalent impedances Z_1 and Z_2 are both in parallel with the voltage source, the voltage e appears across each, and

$$i_1 = \frac{e}{Z_1} = \frac{36 \underline{/0°} \text{ V}}{360.56 \underline{/56.31°} \text{ } \Omega} = 0.1 \underline{/-56.31°} \text{ A}$$

$$i_2 = \frac{e}{Z_2} = \frac{36 \underline{/0°} \text{ V}}{500 \underline{/-53.13°} \text{ } \Omega} = 0.072 \underline{/53.13°} \text{ A}$$

These currents are shown in Figure 15.6(d). Since i_1 flows in the 200-Ω resistor and i_2 flows in the 300-Ω resistor, the average power in each resistor is

$$P_1 = \frac{I_1^2 R}{2} = \frac{(0.1)^2(200)}{2} = 1 \text{ W}$$

$$P_2 = \frac{I_2^2 R}{2} = \frac{(0.072)^2(300)}{2} = 0.77 \text{ W}$$

The total power dissipated in the circuit is therefore $P_1 + P_2 = 1$ W + 0.77 W = 1.77 W, which is the same as the total power delivered to the circuit.

Drill Exercise 15.5

Find the total current and the total power delivered to the circuit in Figure 15.6(a) when the 200-Ω resistor is removed (replaced by a short circuit).

ANSWER: $i_T = 75.89 \underline{/-55.3°}$ mA; $P = 0.77$ W. □

Example 15.6 (Design)

Find the value of R in Figure 15.7 that is necessary to make the magnitude of v equal 3 V.

SOLUTION The reactances of the capacitor and inductor are

$$|X_C| = \frac{1}{\omega C} = \frac{1}{10^6(0.1 \times 10^{-6})} = 10 \text{ } \Omega$$

$$|X_L| = \omega L = 10^6(15 \times 10^{-6}) = 15 \text{ } \Omega$$

The impedance of the parallel LC combination is

$$Z_P = \frac{(15 \underline{/90°})(10 \underline{/-90°})}{0 + j15 - j10} = \frac{150 \underline{/0°}}{5 \underline{/90°}} = 30 \underline{/-90°} \text{ } \Omega$$

By the voltage-divider rule,

$$v = \frac{Z_P}{R + Z_P} e$$

or

$$|v| = \frac{|Z_P|}{|R + Z_P|} |e|$$

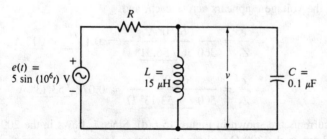

FIGURE 15.7 (Example 15.6)

where $|R + Z_P| = |R - j30| = \sqrt{R^2 + 30^2}$. Thus, to obtain $|v| = 3$ V, we require

$$3 = \left(\frac{30}{\sqrt{R^2 + 900}}\right) 5$$

Squaring both sides gives

$$9 = \left(\frac{900}{R^2 + 900}\right) 25$$

$$9R^2 + 9(900) = 900(25)$$

$$R^2 = 1600$$

$$R = 40 \ \Omega$$

Drill Exercise 15.6

Suppose that the 15-μH inductor in Figure 15.7 has a $\pm 10\%$ tolerance. Over what range of values should it be possible to adjust R to ensure that $|v| = 3$ V, taking into account the range of possible inductance values? Assume that C is exactly 0.1 μF.

ANSWER: 33.8 to 51.4 Ω. ☐

Example 15.7 (Troubleshooting)

An electric heater has three identical heating elements connected in parallel. Each element has a certain resistance and a certain inductance, and each is equivalent to a series RL network, as shown in Figure 15.8(a). Adjustable resistance R is used to control the total amount of heat generated. Because of reduced performance, it was decided to test the heater elements for proper operation. However, the terminals on the elements are inaccessible. The voltage across R was measured to be 29.2 V rms when R was set to 10 Ω. Are one or more elements shorted or open?

SOLUTION The reactance of each inductor is

$$|X_L| = \omega L = (2\pi \times 60)(0.05) = 18.85 \ \Omega$$

Therefore, the impedance of each element is

$$Z = R + j|X_L| = 60 + j18.85 \ \Omega = 62.89 \ \underline{/17.44°}$$

and the admittance of each element is

$$Y = \frac{1}{Z} = \frac{1}{62.89 \ \underline{/17.44°}}$$

The total admittance of the three parallel elements is

$$3Y = \frac{3}{62.89 \ \underline{/17.44°}}$$

Therefore, the total impedance, Z_P, of the parallel elements is

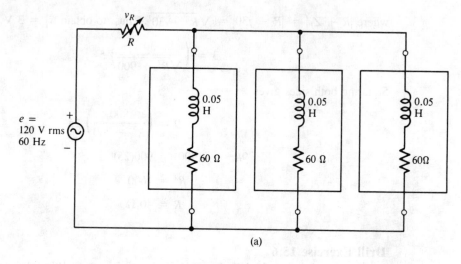

(a)

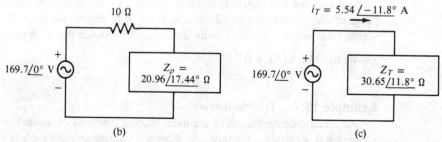

FIGURE 15.8 (Example 15.7)

$$Z_P = \frac{1}{3Y} = \frac{62.89}{3} \underline{/17.44°} = 20.96 \underline{/17.44°} \ \Omega$$

$$= 20 + j6.28 \ \Omega$$

The total impedance of the circuit is

$$Z_T = R + Z_P = 10 + 20 + j6.28 = 30 + j6.28 \ \Omega = 30.65 \underline{/11.8°} \ \Omega$$

The voltage source has peak value $\sqrt{2}$ (120) = 169.7 V and may be assumed to have zero phase. Therefore, as shown in Figure 15.8(c), the total current is

$$i_T = \frac{e}{Z_T} = \frac{169.7 \underline{/0°} \text{ V}}{30.65 \underline{/11.8°} \ \Omega} = 5.54 \underline{/-11.8°} \text{ A}$$

The voltage across the 10-Ω resistor is then

$$v_R = i_T R = (5.54 \underline{/-11.8°} \text{ A})(10 \underline{/-0°} \ \Omega) = 55.4 \underline{/-11.8°} \text{ V}$$

Thus, the effective value of the voltage across R should be

$$V_{\text{eff}} = \frac{55.4}{\sqrt{2}} = 39.17 \text{ V rms}$$

Since the effective value of v_R was measured to be 29.2 V rms, we conclude that at least one element is defective. No element is shorted, because that would make $Z_P = 0$ and $v_R = 120$ V rms. As an exercise, verify that opening any one element will make v_R equal the measured value of 29.2 V rms.

Drill Exercise 15.7

Find the effective value of v_R in Example 15.7 when two heating elements are open.

ANSWER: 16.55 V rms.   □

15.3 AC Voltage Polarities

In previous examples, we have not consistently used $+$ and $-$ polarity symbols when showing ac voltage drops across circuit components. Recall that these polarity symbols are used only for instantaneous references, because the polarities of ac voltages and currents are continually changing. In the examples, it has not been necessary to indicate polarities, because they did not affect the computations. However, in some analysis problems, particularly those involving the application of Kirchhoff's voltage law, it is necessary to show instantaneous polarities, so voltage drops can be distinguished from voltage rises. The $+$ and $-$ polarity symbols are assigned to the voltages across components in the same way they are in dc circuits: current enters the positive side and leaves the negative side. These polarities are based on the current directions that result when the active source has the instantaneous polarity shown across its terminals. The next example illustrates these ideas.

Example 15.8   (Analysis)

Find the polar form of the voltage v_{ab} between points a and b in Figure 15.9(a).

SOLUTION Let Z_1 be the impedance of the series combination of the 50-Ω resistor and the 20 Ω of inductive reactance:

$$Z_1 = (50 + j20) \ \Omega = 53.85 \ \underline{/21.8°} \ \Omega$$

Let Z_2 be the impedance of the series combination of the capacitor and inductor:

$$Z_2 = 0 - j100 + j40 = 0 - j60 \ \Omega = 60 \ \underline{/-90°} \ \Omega$$

Since Z_1 and Z_2 are both in parallel with the voltage source, the voltage across each is $e = 50 \ \underline{/0°}$ V. Therefore, the currents in Z_1 and Z_2 are

$$i_1 = \frac{e}{Z_1} = \frac{50 \ \underline{/0°} \ \text{V}}{53.85 \ \underline{/21.8°} \ \Omega}$$

$$= 0.929 \ \underline{/-21.8°} \ \text{A}$$

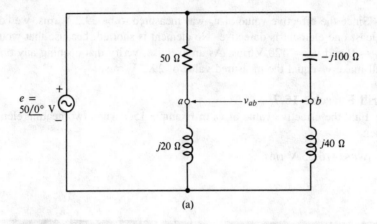

(a)

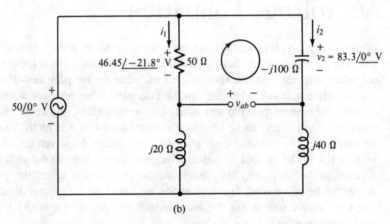

(b)

FIGURE 15.9 (Example 15.8)

$$i_2 = \frac{e}{Z_2} = \frac{50 \angle 0° \text{ V}}{60 \angle -90° \text{ } \Omega}$$

$$= 0.833 \angle 90° \text{ A}$$

The voltage across the 50-Ω resistor is

$$v_1 = i_1 R = (0.929 \angle -21.8° \text{ A})(50 \angle 0° \text{ } \Omega) = 46.45 \angle -21.8° \text{ V}$$

The voltage across the 100 Ω of capacitive reactance is

$$v_2 = i_2 X_C = (0.833 \angle 90° \text{ A})(100 \angle -90° \text{ } \Omega) = 83.3 \angle 0° \text{ V}$$

Figure 15.9(b) shows the voltages v_1 and v_2 with polarities based on the instantaneous polarity of the voltage source. In other words, the polarities are as shown at an instant when i_1 and i_2 are entering Z_1 and Z_2.

Writing Kirchhoff's voltage law in a clockwise direction around the loop shown in Figure 15.9(b), we have

$$v_{ab} + v_1 = v_2$$

or

$$v_{ab} = v_2 - v_1 = 83.3 \underline{/0°} - 46.45 \underline{/-21.8°}$$

Converting v_1 and v_2 to rectangular form so the subtraction can be performed, we find

$$v_{ab} = (83.3 + j0) - (43.13 - j17.25)$$
$$= 40.17 + j17.25 = 43.72 \underline{/23.24°} \text{ V}$$

Drill Exercise 15.8

Find the voltages across the 20 Ω of inductive reactance and the 40 Ω of inductive reactance in Figure 15.9, and write Kirchhoff's voltage law around the loop containing those reactances to determine v_{ab}.

ANSWER: $v_{j20\ \Omega} = 18.58 \underline{/68.2°}$ V $= (6.9 + j17.25)$ V; $v_{j40\ \Omega} = 33.2 \underline{/180°}$ V $= (-33.2 + j0)$ V; $v_{ab} = 43.76 \underline{/23.21°}$ V. □

15.4 Ladder Networks

Example 15.9 (Analysis)

Figure 15.10(a) shows an example of an ac *ladder network*. Recall that this type of network can be analyzed by starting at the far right end and alternately combining impedances in series and parallel.

(a) Find the total equivalent impedance of the network.
(b) Find the polar form of the current in the rightmost capacitor.

SOLUTION
(a) The reactance of each inductor is

$$X_L = j\omega L = j(10^6)(10 \times 10^{-3}) = j10^4 \ \Omega$$

The reactance of each capacitor is

$$X_C = \frac{-j}{\omega C} = \frac{-j}{10^6(50 \times 10^{-12})} = -j2 \times 10^4 \ \Omega$$

The rightmost inductive reactance is in series with the rightmost capacitive reactance, so the equivalent impedance of the combination is

$$Z_1 = (0 + j10^4) + (0 - j2 \times 10^4) = 0 - j10^4 \ \Omega$$

Figure 15.10(b) shows the circuit when Z_1 is inserted in place of the rightmost series combination. It is clear that Z_1 is in parallel with $-j2 \times 10^4 \ \Omega$, so

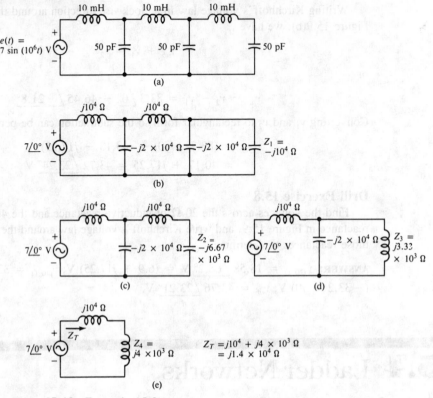

FIGURE 15.10 (Example 15.9)

that combination is equivalent to

$$Z_2 = \frac{(10^4 \, \underline{/-90°})(2 \times 10^4 \, \underline{/-90°})}{(0 - j10^4) + (0 - j2 \times 10^4)} = \frac{2 \times 10^8 \, \underline{/-180°}}{3 \times 10^4 \, \underline{/-90°}}$$

$$= 6.67 \times 10^3 \, \underline{/-90°} \ \Omega = 0 - j6.67 \times 10^3 \ \Omega$$

Figure 15.10(c) shows the circuit with Z_2 in place. It is apparent that Z_2 is in series with $j10^4 \ \Omega$, so that combination is equivalent to

$$Z_3 = (0 + j10^4) + (0 - j6.67 \times 10^3)$$
$$= 0 + j3.33 \times 10^3 \ \Omega = 3.33 \times 10^3 \, \underline{/90°} \ \Omega$$

As shown in Figure 15.10(d), Z_3 is in parallel with $-j2 \times 10^4 \ \Omega$, so the next parallel computation gives

$$Z_4 = \frac{(3.33 \times 10^3 \, \underline{/90°})(2 \times 10^4 \, \underline{/-90°})}{(0 + j3.33 \times 10^3) + (0 - j2 \times 10^4)} = \frac{6.66 \times 10^7 \, \underline{/0°}}{1.66 \times 10^4 \, \underline{/-90°}}$$

$$= 4 \times 10^3 \, \underline{/90°} = 0 + j4 \times 10^3 \ \Omega$$

Finally, as shown in Figure 15.10(e), the total impedance is the series combination of Z_4 and $j10^4$ Ω:

$$Z_T = (0 + j10^4) + (0 + j4 \times 10^3) = 0 + j1.4 \times 10^4 \ \Omega = 1.4 \times 10^4 \ \underline{/90°} \ \Omega$$

(b) The total current in the circuit is

$$i_T = \frac{e}{Z_T} = \frac{7 \ \underline{/0°} \ \text{V}}{1.4 \times 10^4 \ \underline{/90°} \ \Omega} = 0.5 \ \underline{/-90°} \ \text{mA}$$

Through successive applications of the current-divider rule, we can ultimately find the current in the rightmost capacitor. Figure 15.11(a) shows the equivalent circuit containing Z_3 [same as Figure 15.10(d)]. By the current-divider rule, we find the portion of i_T that flows in Z_3:

$$i_3 = \left[\frac{2 \times 10^4 \ \underline{/-90°}}{Z_3 + (0 - j2 \times 10^4)} \right] i_T$$

$$= \left[\frac{2 \times 10^4 \ \underline{/-90°}}{(0 + j3.33 \times 10^3) + (0 - j2 \times 10^4)} \right] 0.5 \ \underline{/-90°} \ \text{mA}$$

$$= \left(\frac{2 \times 10^4 \ \underline{/-90°}}{1.67 \times 10^4 \ \underline{/-90°}} \right) 0.5 \ \underline{/-90°} \ \text{mA} = 0.6 \ \underline{/-90°} \ \text{mA}$$

Figure 15.11(b) shows the equivalent circuit containing Z_2. Notice that i_3 computed above is the current in the $j10^4$ inductive reactance. Thus, as shown in Figure 15.11(c), we can find the current i_1 in Z_1 by another application of the current-divider rule:

$$i_1 = \left[\frac{2 \times 10^4 \ \underline{/-90°}}{Z_1 + (0 - j2 \times 10^4)} \right] i_3$$

$$= \left[\frac{2 \times 10^4 \ \underline{/-90°}}{(0 - j10^4) + (0 - j2 \times 10^4)} \right] 0.6 \ \underline{/-90°} \ \text{mA}$$

$$= \left(\frac{2 \times 10^4 \ \underline{/-90°}}{3 \times 10^4 \ \underline{/-90°}} \right) 0.6 \ \underline{/-90°} \ \text{mA} = 0.4 \ \underline{/-90°} \ \text{mA}$$

Since i_1 flows in Z_1, and Z_1 is the series equivalent impedance containing the rightmost capacitor, we conclude that the current in that capacitor is $i_1 = 0.4 \ \underline{/-90°}$ mA [see Figure 15.11(d)].

Drill Exercise 15.9

Find the polar forms of the currents in each of the other two capacitors in Example 15.9.

ANSWER: i in leftmost capacitor $= 0.1 \ \underline{/90°}$ mA; i in center capacitor $= 0.2 \ \underline{/-90°}$ mA.

□

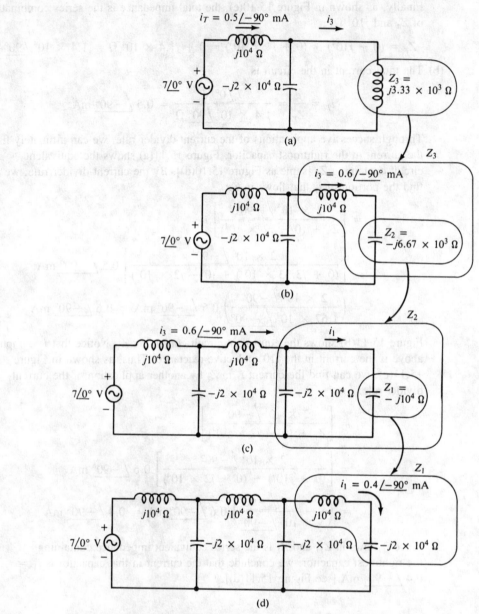

FIGURE 15.11 (Example 15.9) (a) Same as Figure 15.9(d). (b) Same as Figure 15.9(c). (c) Same as Figure 15.9(b). (d) Same as Figure 15.9(a).

15.5 Redrawing Schematic Diagrams

As in dc circuit analysis, it is often helpful to redraw ac circuit diagrams so that series and parallel combinations are readily identifiable. The common, or "ground," symbol discussed in connection with dc circuits is also widely used in schematic diagrams of ac circuits and has the same meaning as in dc circuits: It is a shorthand way of showing points that are joined electrically. Since the polarities of ac voltages periodically alternate, the common in an ac circuit will be alternately positive and negative with respect to other terminals in the circuit.

Example 15.10   (Analysis)
Find the polar form of the current in the capacitor in Figure 15.12(a).

SOLUTION The circuit is redrawn as shown in Figure 15.12(b), with solid lines joining the common points. The redrawn circuit is electrically equivalent to the original, but the rearrangement of components clearly reveals that the resistor and capacitor are in parallel, and that this combination is in series with the inductor.

The parallel combination of the resistor and capacitor is equivalent to

$$Z_P = \frac{(100 \underline{/0°})(60 \underline{/-90°})}{100 - j60} = \frac{6 \times 10^3 \underline{/90°}}{116.62 \underline{/-30.96°}}$$

$$= 51.45 \underline{/-59.04°} = 26.47 - j44.12 \ \Omega$$

As shown in Figure 15.12(c), Z_P is in series with the inductor, so we can find the voltage v_p using the voltage-divider rule:

$$v_p = \frac{Z_P}{Z_P + j20} e = \left[\frac{51.45 \underline{/-59.04°}}{(26.47 - j44.12) + (0 + j20)} \right] (12 \underline{/30°} \text{ V})$$

$$= \left(\frac{51.45 \underline{/-59.04°}}{26.47 - j24.12} \right) 12 \underline{/30°} \text{ V} = \left(\frac{51.45 \underline{/-59.04°}}{35.81 \underline{/-42.34°}} \right) 12 \underline{/30°} \text{ V}$$

$$= 17.24 \underline{/13.3°} \text{ V}$$

As shown in Figure 15.12(d), the current through the capacitor is

$$i_C = \frac{v_p}{X_C} = \frac{17.24 \underline{/13.3°} \text{ V}}{60 \underline{/-90°} \ \Omega} = 0.287 \underline{/103.3°} \text{ A}$$

Drill Exercise 15.10
Find the voltage across the inductor in Figure 15.12(a).

ANSWER: $6.7 \underline{/162.34°}$ V.   □

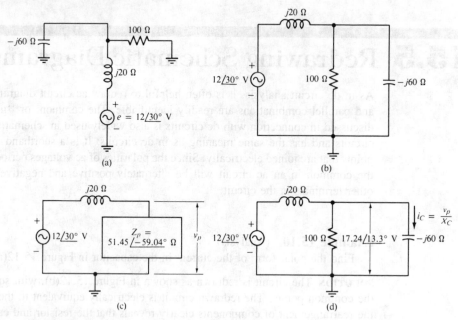

FIGURE 15.12 (Example 15.10)

15.6 Reactive and Apparent Power

Reactive Power

We have learned that resistance dissipates electrical energy and that capacitance and inductance can only store energy. Since power is defined to be the rate at which energy is dissipated, we know that there is *zero* power in both inductance and capacitance. It is nevertheless useful in many applications to define a certain quantity that is computed for inductance and capacitance in the same way that power is computed in resistance. This quantity has been given the (misleading) name *reactive power*. Recall that *true* power in a resistance, also called *real* power to distinguish it from reactive power, can be computed using any of the relations

$$P = \frac{V_P I_P}{2} = \frac{I_P^2 R}{2} = \frac{V_P^2}{2R}$$

$$= V_{\text{eff}} I_{\text{eff}} = I_{\text{eff}}^2 R = \frac{V_{\text{eff}}^2}{R} \qquad \text{watts}$$

(15.1)

where P is the average power in resistance R, V_p and I_p are the peak sinusoidal values of the voltage across and current through R, and V_{eff} and I_{eff} are the effective values. In capacitive and inductive reactance, the reactive power, designated Q, is computed similarly:

$$Q = \frac{V_P I_P}{2} = \frac{I_P^2 |X|}{2} = \frac{V_P^2}{2|X|}$$

$$= V_{eff} I_{eff} = I_{eff}^2 |X| \tag{15.2}$$

$$= \frac{V_{eff}^2}{|X|} \qquad \text{volt-amperes, reactive (vars)}$$

where $|X|$ is the magnitude of the reactance in ohms, V_p and I_p are the peak sinusoidal values of the voltage across and current through the reactance, and V_{eff} and I_{eff} are the effective values. Note that the units of reactive power are volt-amperes, reactive, or *vars*. Since pure resistance has zero reactance, there is zero reactive power in resistance.

Equations (15.1) and (15.2) are used to calculate true and reactive power in terms of the voltage across and current through a specific component (resistor, inductor, or capacitor). In terms of the *total* voltage across a circuit containing resistance and reactance, and the total current delivered to the circuit, the total real power and total reactive power can be computed from

$$P(\text{total}) = \frac{V_P I_P}{2} \cos \theta = V_{eff} I_{eff} \cos \theta \qquad \text{watts} \tag{15.3}$$

and

$$Q(\text{total}) = \frac{V_P I_P}{2} \sin \theta = V_{eff} I_{eff} \sin \theta \qquad \text{vars} \tag{15.4}$$

where θ is the magnitude of the phase angle between the total voltage and total current. Recall that $\cos \theta$ is called the power factor of the circuit. It is important to realize that the total power in (or "delivered to") a circuit is the sum of the power dissipations in all the resistors in the circuit, regardless of their series, parallel, or series–parallel configuration. Similarly, the total reactive power in a circuit is the sum of the reactive powers in all reactances, regardless of configuration. However, inductive vars (reactive power in inductive reactance) are considered positive and capacitive vars (reactive power in capacitive reactance) are considered negative. Thus, the total reactive power in a circuit is

$$Q(\text{total}) = Q_L - Q_C \qquad \text{vars} \tag{15.5}$$

where Q_L is the sum of the inductive vars and Q_C is the sum of the capacitive vars.

Example 15.11   (Analysis)
  In the circuit shown in Figure 15.13:

(a) Find the total power in the circuit by summing the power dissipations in the resistors, and the total reactive power by summing the reactive powers in the reactances.
(b) Verify that the sums computed in part (a) can be found using equations (15.3) and (15.4).

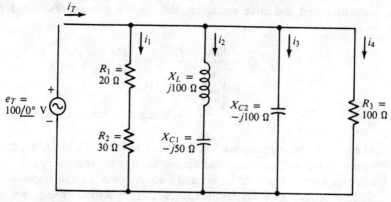

FIGURE 15.13 (Example 15.11)

SOLUTION

(a) To compute the power dissipated by R_1 and R_2, we first find the current i_1 in the series combination of those resistors:

$$i_1 = \frac{e_T}{R_1 + R_2} = \frac{100\underline{/0°}}{(20 + 30)\underline{/0°}} = 2\underline{/0°}\ \text{A}$$

Then,

$$P_{R1} = \frac{|i_1|^2 R_1}{2} = \frac{2^2(20)}{2} = 40\ \text{W}$$

$$P_{R2} = \frac{|i_1|^2 R_2}{2} = \frac{2^2(30)}{2} = 60\ \text{W}$$

The current i_2 in the series combination of X_L and X_{C1} is

$$i_2 = \frac{e_T}{X_L + X_{C1}} = \frac{100\underline{/0°}}{0 + j100 - j50} = \frac{100\underline{/0°}}{50\underline{/90°}} = 2\underline{/-90°}\ \text{A}$$

Therefore, the reactive power in each of those components is

$$Q_L = \frac{|i_2|^2|X_L|}{2} = \frac{2^2(100)}{2} = 200\ \text{vars (inductive)}$$

$$Q_{C1} = \frac{|i_2|^2|X_{C1}|}{2} = \frac{2^2(50)}{2} = 100\ \text{vars (capacitive)}$$

Since X_{C2} is in parallel with the voltage source, its reactive power can be found from

$$Q_{C2} = \frac{|e_T|^2}{2|X_{C2}|} = \frac{(100)^2}{2(100)} = 50\ \text{vars (capacitive)}$$

Similarly,

$$P_{R3} = \frac{|e_T|^2}{2R_3} = \frac{(100)^2}{2(100)} = 50 \text{ W}$$

The following table summarizes the results of these computations and shows the sums of the real and reactive powers:

Component	P (W)	Q
R_1	40	0
R_2	60	0
L	0	200 vars (ind.)
C_1	0	100 vars (cap.)
C_2	0	50 vars (cap.)
R_3	50	0
Totals	150 W	$200 - 150 = 50$ vars (ind.)

Notice that the sum of the reactive powers equals the sum of the inductive vars minus the sum of the capacitive vars, and that the sum is inductive.

(b) To use equations (15.3) and (15.4), we must compute $i_T = i_1 + i_2 + i_3 + i_4$. The values of i_1 and i_2 were found in part (a).

$$i_3 = \frac{e_T}{X_{C2}} = \frac{100 \underline{/0°} \text{ V}}{100 \underline{/-90°} \text{ } \Omega} = 1 \underline{/90°} \text{ A}$$

$$i_4 = \frac{e_T}{R_3} = \frac{100 \underline{/0°} \text{ V}}{100 \underline{/0°} \text{ } \Omega} = 1 \underline{/0°} \text{ A}$$

Thus,

$$i_T = 2 \underline{/0°} \text{ A} + 2 \underline{/-90°} \text{ A} + 1 \underline{/90°} \text{ A} + 1 \underline{/0°} \text{ A}$$
$$= (3 - j1) \text{ A} = 3.162 \underline{/-18.43°} \text{ A}$$

The magnitude of the angle between e_T and i_T is therefore 18.43°. From equation (15.3),

$$P(\text{total}) = \frac{(100)(3.162)}{2} \cos (18.43°) = 150 \text{ W}$$

From equation (15.4),

$$Q(\text{total}) = \frac{(100)(3.162)}{2} \sin (18.43°) = 50 \text{ vars}$$

As expected, these results agree with the sums calculated in part (a).

Drill Exercise 15.11

Find the total real power and the total reactive power in Figure 15.13 when X_{C1} is changed to $-j200$ Ω.

ANSWER: $P = 150$ W; $Q = 100$ vars (cap.).

Apparent Power

Another power-related concept that is useful in many applications, particularly in the electrical power industry, is called *apparent power*. Designated S, apparent power is defined to be the product of voltage and current, without regard to phase angle:

$$S = \frac{V_P I_P}{2} = V_{\text{eff}} I_{\text{eff}} \qquad \text{volt-amperes (VA)} \tag{15.6}$$

Note that the units of apparent power are volt-amperes (VA), although kilovolt-amperes (kVA) are more commonly used in the power industry. Since apparent power is calculated without using a multiplying factor of $\sin \theta$ or $\cos \theta$, it is always greater than (or at least equal to) real power and reactive power. In fact, apparent power is the hypotenuse of a right triangle whose other two sides are P and Q, as illustrated in the *power triangle* shown in Figure 15.14. From obvious trigonometric relations in the power triangle, we deduce

$$P = S \cos \theta \qquad \text{watts} \tag{15.7}$$

$$Q = S \sin \theta \qquad \text{vars} \tag{15.8}$$

$$S = \sqrt{P^2 + Q^2} \qquad \text{VA} \tag{15.9}$$

Note that equations (15.7) and (15.8) are just restatements of equations (15.3) and (15.4), respectively. Reactive power is given the symbol Q because it is in *quadrature* with (at right angles to) real power, as can be seen in the power triangle.

The total apparent power in a circuit is *not* the sum of the apparent powers in its various components. Instead, S must be computed using equation (15.9): $S = \sqrt{P^2 + Q^2}$, where P is the total real power and Q is the total reactive power. This point is demonstrated in the next example.

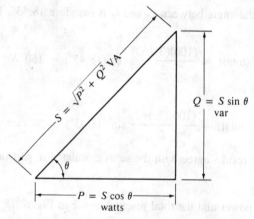

FIGURE 15.14 The power triangle. θ is the phase angle between voltage and current.

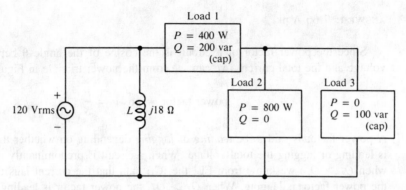

FIGURE 15.15 (Example 15.12)

Example 15.12 (Analysis)

Figure 15.15 shows a series–parallel system containing electrical loads whose real and reactive powers have the values indicated. Find the total apparent power delivered to the system.

SOLUTION To find the total apparent power, we must find the total real power and the total reactive power in the system. The reactive power in L is

$$Q_L = \frac{V_{\text{eff}}^2}{|X_L|} = \frac{(120)^2}{18} = 800 \text{ vars (ind.)}$$

The following table summarizes the real and reactive powers in the system and shows their sums. For illustration purposes, the apparent power in each load is also shown.

Load	P (W)	Q	$S = \sqrt{P^2 + Q^2}$ (VA)
L	0	800 vars (ind.)	800
1	400	200 vars (cap.)	447.2
2	800	0	800
3	0	100 vars (cap.)	100
Totals	1200	500 vars (ind.)	

The total apparent power is then

$$S_{\text{total}} = \sqrt{P_{\text{total}}^2 + Q_{\text{total}}^2} = \sqrt{(1200)^2 + (500)^2} = 1300 \text{ VA}$$

Note carefully that the total apparent power, 1300 VA, is *not* the sum of the apparent powers in the loads (2147.2 VA).

Drill Exercise 15.12

Find the magnitude of the total current delivered to the system in Figure 15.15.

ANSWER: 7.66 A pk.   ☐

Since the power factor of a circuit is the cosine of the angle θ between the total voltage and the total current, we can see from the power triangle in Figure 15.14 that

$$\text{power factor} = \cos \theta = \frac{P}{S} \tag{15.10}$$

A power factor is said to be *leading* or *lagging* depending on whether the total current is leading or lagging the total voltage. When a circuit is predominantly inductive (i.e., when $Q_L > Q_C$) we know from ELI the ICE man that the current lags the voltage, so the power factor is lagging. When $Q_C > Q_L$, the power factor is leading. The loads in most electrical power systems are predominantly inductive, so most have lagging power factors. This is an uneconomical situation for utility companies, who would prefer to have a unity power factor ($\theta = 0°$). These companies expend considerable effort selecting capacitive loads to be installed in a system as a means for achieving *power factor correction*, that is, for canceling inductive vars with capacitive vars and making the power factor as close to 1 as possible.

Example 15.13   (Design)

For the power distribution system shown in Figure 15.16:

(a) Find the total apparent power, the power factor and the magnitude of i_T without capacitance C in the system.
(b) Find the capacitive vars that must be produced by capacitance C to make the power factor of the system equal 1.
(c) Find the capacitance C necessary to achieve the power factor correction in part (b).
(d) Find the total apparent power and total current i_T after the power factor correction.

SOLUTION

(a) Q_{total} = 12 kvars (ind.) + 18 kvars (ind.) − 4 kvars (cap.)

    = 26 kvars (ind.)

   P_{total} = 25 kW + 4 kW + 32 kW = 61 kW

   $S_{total} = \sqrt{P_{total}^2 + Q_{total}^2} = \sqrt{(61 \times 10^3)^2 + (26 \times 10^3)^2}$

    = 66.31 kVA

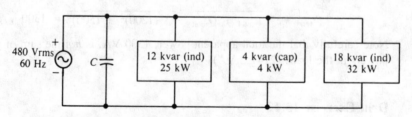

FIGURE 15.16 (Example 15.13)

From equation (15.10),

$$\cos\theta = \frac{P_{total}}{S_{total}} = \frac{61 \text{ kW}}{66.3 \text{ kVA}} = 0.92 \text{ lagging}$$

The power factor is lagging because Q_{total} is inductive. Since S is the total number of volt-amperes, $S = \dfrac{|e_T||i_T|}{2}$, the magnitude of the total current is $2S$ divided by the magnitude (peak value) of e_T:

$$|i_T| = \frac{2S}{|e_T|} = \frac{2(66.31 \times 10^3 \text{ VA})}{\sqrt{2}\,(480 \text{ V rms})} = 195.36 \text{ A pk}$$

(b) For the power factor to be 1, the total number of capacitive vars must equal the total number of inductive vars. When that is the case, $Q_{total} = 0$, $S_{total} = P_{total}$, and $\cos\theta = P_{total}/S_{total} = 1$. Since Q_{total} without power factor correction is 26 kvars (inductive), the capacitance must produce 26 kvars (capacitive) to make $Q_{total} = 0$.

(c) The reactive power in the capacitor is

$$Q_C = \frac{|e_T|^2}{2|X_C|}$$

where

$$|X_C| = \frac{1}{\omega C} = \frac{1}{(2\pi \times 60)C}$$

Since the capacitor must produce 26 kvars, we have

$$26 \times 10^3 = \frac{(\sqrt{2}\,480)^2}{2/(2\pi \times 60)C}$$

$$C = \frac{(26 \times 10^3)}{(480)^2(2\pi \times 60)} = 299.3 \text{ }\mu\text{F}$$

(d) Since $Q_{total} = 0$ after power factor correction,

$$S_{total} = \sqrt{P_{total}^2 + Q_{total}^2} = \sqrt{(61 \times 10^3)^2 + 0} = 61 \text{ kVA}$$

$$|i_T| = \frac{2S}{|e_T|} = \frac{2(61 \times 10^3 \text{ VA})}{\sqrt{2}\,(480 \text{ V rms})} = 179.72 \text{ A pk}$$

Notice that one result of power factor correction is that less current can be supplied from the source to deliver the same amount of real power to the system.

Drill Exercise 15.13

What would be the power factor of the system in Figure 15.16 if C were made equal to 800 μF?

ANSWER: 0.814 leading.   □

15.7 SPICE Examples

Example 15.14   (SPICE)

Use SPICE as an aid in determining the total average power and the total reactive power in the circuit shown in Figure 15.13 (Example 15.11). Assume the frequency is 60 Hz.

SOLUTION: We first find the values of inductance and capacitance:

$$L = \frac{|X_L|}{\omega} = \frac{100\ \Omega}{2\pi \times 60\ \text{Hz}} = 265.25\ \text{mH}$$

$$C_1 = \frac{1}{\omega |X_{C1}|} = \frac{1}{(2\pi \times 60\ \text{Hz})(50\ \Omega)} = 53.05\ \mu\text{F}$$

$$C_2 = \frac{1}{\omega |X_{C2}|} = \frac{1}{(2\pi \times 60\ \text{Hz})(100\ \Omega)} = 26.525\ \mu\text{F}$$

From equation 15.3, we have

$$P(\text{total}) = \frac{V_P I_P}{2} \cos\theta$$

where I_P is the magnitude (peak value) of the total current, i_T. Now, the rectangular form of i_T is

$$i_T = I_P\cos\theta + jI_P\sin\theta$$

That is, $I_P\cos\theta$ is the *real* part of the total current i_T: $I_P\cos\theta = \mathscr{R}_e(i_T)$. Thus,

$$P(\text{total}) = \frac{V_P \mathscr{R}_e\ (i_T)}{2} \tag{15.11}$$

Similarly,

$$Q(\text{total}) = \frac{V_P I_P \sin\theta}{2} = \frac{V_P \mathscr{I}m(i_T)}{2} \tag{15.12}$$

where $\mathscr{I}m(i_T)$ is the imaginary part of i_T.

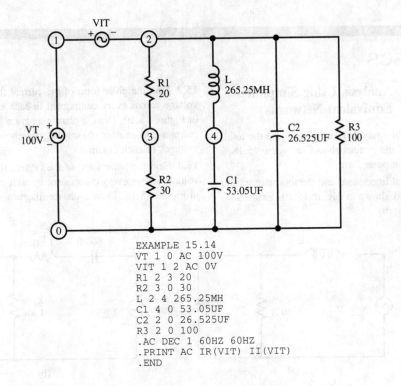

```
EXAMPLE 15.14
VT 1 0 AC 100V
VIT 1 2 AC 0V
R1 2 3 20
R2 3 0 30
L 2 4 265.25MH
C1 4 0 53.05UF
C2 2 0 26.525UF
R3 2 0 100
.AC DEC 1 60HZ 60HZ
.PRINT AC IR(VIT) II(VIT)
.END
```

FIGURE 15.17 (Example 15.14)

Since SPICE can print the real and imaginary parts of any voltage or current in a circuit simulation, we can obtain those values and use equations (15.11) and (15.12) to compute the real and reactive powers. Figure 15.17 shows the SPICE circuit and input data file. VIT is a dummy voltage source used to obtain the total current. Note that the .PRINT statement requests IR(VIT), the real part of the total current, and II(VIT), the imaginary part of the total current. Execution of the program gives IR(VIT) = 3 and II(VIT) = −1. (Thus, $i_T = 3 - j1$ A). Since i_T is in the fourth quadrant, i_T lags e_T, meaning the circuit is predominantly inductive (ELI) and the reactive power is inductive. Using equations (15.11) and (15.12), we find:

$$P(\text{total}) = \frac{100}{2}\,(3) = 150 \text{ W}$$

$$Q(\text{total}) = \frac{100}{2}\,(1) = 50 \text{ vars (ind)}$$

These results agree exactly with those found in Example 15.11.

Exercises

Section 15.1   Analysis Using Simplified Equivalent Networks

15.1 Find the total equivalent impedance and the total current in each of the circuits shown in Figure 15.18. Express answers in polar form.

15.2 Find the total impedance and the total current in each of the circuits shown in Figure 15.19. Express answers in polar form.

15.3 Find the polar form of the current through and voltage across every component in each circuit shown in Figure 15.19. Draw a phasor diagram showing all currents and another phasor diagram showing all voltages in each circuit.

15.4 Find the polar form of the current through and voltage across every component in each circuit shown in Figure 15.18. Draw a phasor diagram showing all

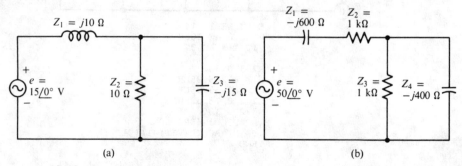

(a)   (b)

FIGURE 15.18 (Exercise 15.1)

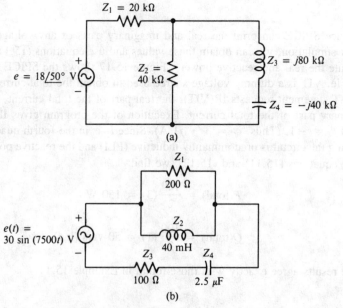

FIGURE 15.19 (Exercise 15.2)

currents and another phasor diagram showing all voltages in each circuit.

15.5 Find the polar form of voltage v_2 and current i_3 in Figure 15.20.

15.6 Find the polar form of voltage v_1 and current i_4 in Figure 15.20.

15.7 (a) Find the rectangular form of the voltage across each component in Figure 15.21.

(b) Verify Kirchhoff's voltage law around each of the loops shown.

15.8 (a) Find the rectangular form of the current in each component in Figure 15.22.

(b) Verify Kirchhoff's current law at each of the numbered nodes.

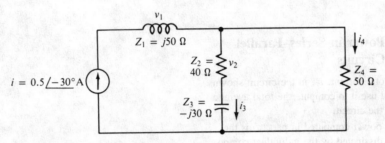

FIGURE 15.20 (Exercise 15.5 and 15.6)

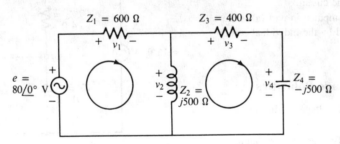

FIGURE 15.21 (Exercise 15.7)

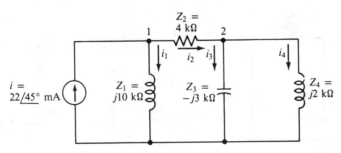

FIGURE 15.22 (Exercise 15.8)

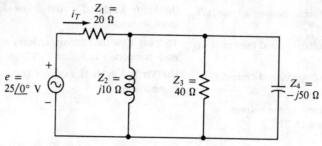

FIGURE 15.23 (Exercise 15.9)

Section 15.2   Power in Series–Parallel Circuits

15.9 (a) Find the total current, i_T, in the circuit shown in Figure 15.23 and use it to compute the total average power delivered to the circuit.

(b) Verify that the power computed in part (a) is the sum of the powers dissipated by the individual components.

15.10 (a) Find the total voltage, v_T, across the circuit shown in Figure 15.24 and use it to compute the total average power delivered to the circuit.

(b) Verify that the power computed in part (a) is the sum of the powers dissipated by the individual components.

Section 15.3   AC Voltage Polarities

15.11 Find the polar form of voltage v_{ab} in Figure 15.25.

15.12 Find the polar form of voltage v_{ab} in Figure 15.26.

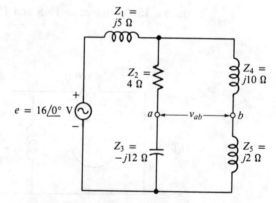

FIGURE 15.25 (Exercise 15.11)

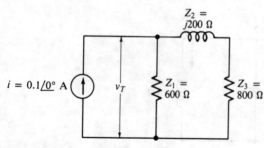

FIGURE 15.24 (Exercise 15.10)

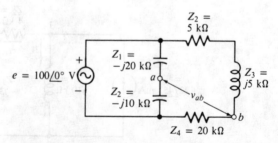

FIGURE 15.26 (Exercise 15.12)

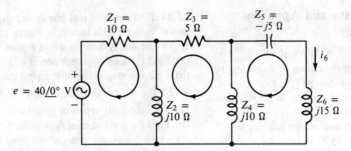

FIGURE 15.27 (Exercise 15.13 and 15.14)

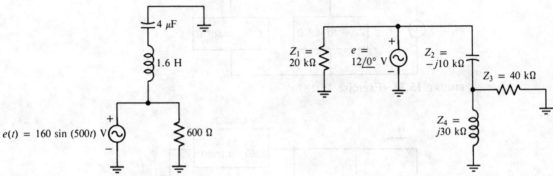

FIGURE 15.28 (Exercise 15.15)

FIGURE 15.29 (Exercise 15.16)

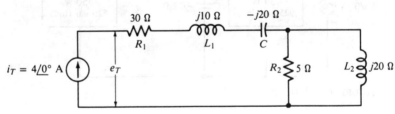

FIGURE 15.30 (Exercise 15.17)

Section 15.4 Ladder Networks

15.13 Find the polar form of current i_6 in Figure 15.27.

15.14 Verify Kirchhoff's voltage law around each of the loops shown in Figure 15.27.

Section 15.5 Redrawing Schematic Diagrams

15.15 (a) Find the total impedance of the circuit shown in Figure 15.28.

(b) Find the sinusoidal form of the total current supplied by the voltage source.

15.16 Find the total impedance and the total current in the circuit shown in Figure 15.29. Express answers in polar form.

Section 15.6   Reactive and Apparent Power

15.17 For the circuit shown in Figure 15.30:

(a) Find the total power in the circuit by summing the power dissipations in the resistors and the total reactive power by summing the reactive powers in the reactances.

(b) Find the polar form of e_T.

(c) Verify that the sums found in part (a) can be computed using equations (15.3) and (15.4).

15.18 Find the total real power and the total reactive power in the circuit shown in Figure 15.31.

15.19 For the system shown in Figure 15.32:

(a) Find the total apparent power and the power factor.

(b) Find the magnitude of the total current, i_T.

15.20 For the power system shown in Figure 15.33:

(a) Find the total apparent power and the power factor.

(b) Find the total current i_T in polar form.

(c) Draw the power triangle and label the magnitudes of each side.

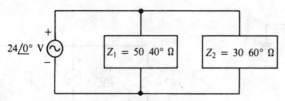

$24\underline{/0°}$ V $Z_1 = 50\ 40°\ \Omega$ $Z_2 = 30\ 60°\ \Omega$

FIGURE 15.31 (Exercise 15.18)

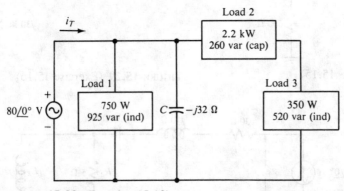

FIGURE 15.32 (Exercise 15.19)

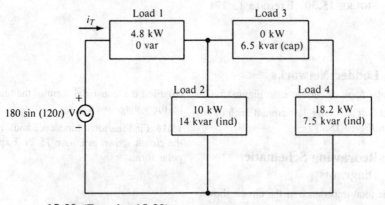

FIGURE 15.33 (Exercise 15.20)

Design Exercises

15.1D What is the maximum permissible peak voltage *e* in Figure 15.34 if the voltage *v* cannot exceed 12 V pk?

15.2D What minimum peak current *i* should the current source in Figure 15.35 maintain if the voltage *v* across the capacitor must not be permitted to fall below 12 V pk?

15.3D What value of *R* should be used in Figure 15.36 to limit the peak value of the capacitor current to 0.1 A?

15.4D What value of *R* should be used in Figure 15.36 to make the voltage across the capacitor have a phase angle of −45°?

15.5D What value of *R* should be used in Figure 15.37 to limit the peak value of the capacitor current to 0.3 A?

15.6D If resistor *R* in Figure 15.37 is made adjustable, what is the smallest possible peak current that resistor adjustment would permit to flow in the capacitor?

15.7D What minimum number of 1.25-μF capacitors must be connected in parallel in Figure 15.38 if the peak voltage across the parallel combination cannot exceed 24 V?

15.8D What value of *R* is necessary to make the power factor of the circuit shown in Figure 15.39 equal to 0.9?

15.9D What value of *R* should be used in Figure 15.40 if the average power delivered to the circuit is to be 0.5 W?

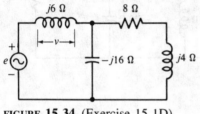

FIGURE 15.34 (Exercise 15.1D)

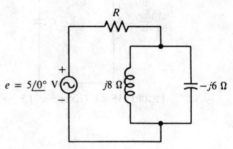

FIGURE 15.36 (Exercise 15.3D)

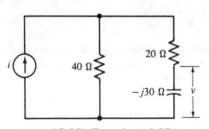

FIGURE 15.35 (Exercise 15.2D)

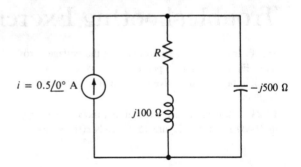

FIGURE 15.37 (Exercise 15.5D and 15.6D)

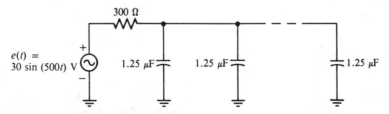

FIGURE 15.38 (Exercise 15.7D)

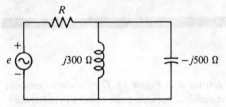

FIGURE 15.39 (Exercise 15.8D)

15.10D Find the value of capacitance C that should be used in Figure 15.41 to achieve a unity power factor for the circuit.

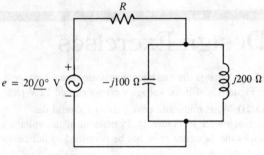

FIGURE 15.40 (Exercise 15.9D)

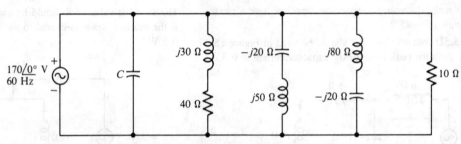

FIGURE 15.41 (Exercise 15.10D)

Troubleshooting Exercises

15.1T Test measurements reveal that the voltage across R_2 in Figure 15.42 is in phase with e. What are the possible failures (defective components) in the circuit? Explain.

15.2T Test measurements reveal that the voltage across the 1-kΩ resistor in Figure 15.43 leads $e(t)$. Is this

enough information to conclude that there is a defective component in the circuit? If so, which component is defective (shorted or open)?

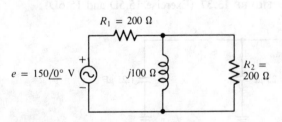

FIGURE 15.42 (Exercise 15.1T)

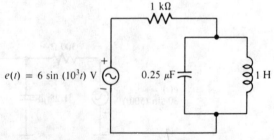

FIGURE 15.43 (Exercise 15.2T)

15.3T The average power delivered to the circuit in Figure 15.44 is 100 W. What component(s) may be defective?

15.4T When the frequency of the generator in Figure 15.45 is exactly 60 Hz, the magnitude of the inductive reactance exactly equals the magnitude of the capacitive reactance. If the peak value of the current i_T is measured to be 5.5 A, is the frequency set correctly? Explain.

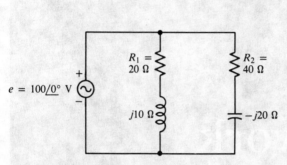

FIGURE 15.44 (Exercise 15.3T)

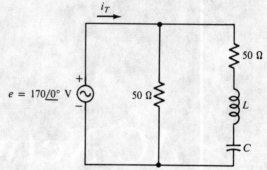

FIGURE 15.45 (Exercise 15.4T)

SPICE Exercises

15.1S Use SPICE to find the magnitude and angle of the total current supplied by the voltage source in Figure 15.28, p. 589.

15.2S Use SPICE to find the magnitude and angle of the total current supplied by the voltage source in Figure 15.29, p. 589. Assume the frequency is 1 kHz.

15.3S Use SPICE as an aid in determining the total average power and total reactive power in the circuit shown in Figure 15.23, p. 588. Assume the frequency is 500 Hz.

15.4S Repeat Exercise 15.3S for the circuit in Figure 15.24 (p. 588), except assume the frequency is 2 kHz.

$$\left(Hint:\ P(\text{total}) = \frac{I_P \mathcal{R}_e(V_T)}{2}.\right)$$

AC Network Transformations and Multisource Circuits

16.1 Wye–Delta Transformations

As in dc circuits, some ac networks contain no components in series and no components in parallel. Clearly, these networks cannot be simplified using the methods of Chapter 15, since there are no series and parallel combinations that can be reduced to equivalent impedances. However, it is often possible to transform a wye or delta configuration contained in such a network to a configuration of the opposite type. By choosing the transformation judiciously, we can solve for desired voltages and currents in untransformed components, using conventional series–parallel analysis.

Figure 16.1 illustrates wye-to-delta and delta-to-wye transformations. The equations are identical to their dc counterparts, with impedances substituted for resistances. Given the impedances Z_A, Z_B, and Z_C in a delta network, the impedances in the equivalent wye network are

$$Z_1 = \frac{Z_A Z_B}{Z_A + Z_B + Z_C} \qquad (16.1)$$

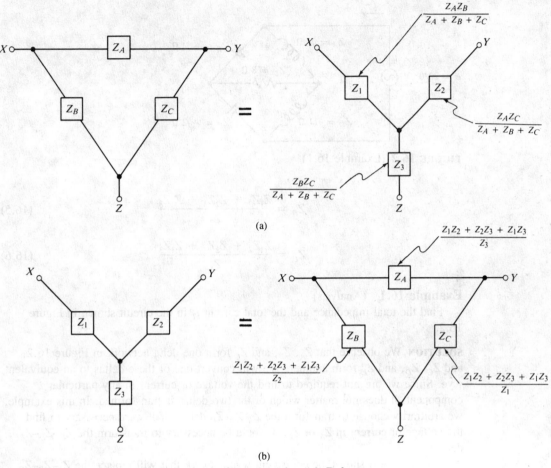

FIGURE 16.1 Wye–delta transformations of ac networks. (a) Delta-to-wye transformation. (b) Wye-to-delta transformation.

$$Z_2 = \frac{Z_A Z_C}{Z_A + Z_B + Z_C} \tag{16.2}$$

$$Z_3 = \frac{Z_B Z_C}{Z_A + Z_B + Z_C} \tag{16.3}$$

Rather than memorizing these equations, remember the pattern for associating corresponding components in wye and delta equivalents, as discussed in Chapter 6. Given the impedances Z_1, Z_2, and Z_3 in a wye network, the impedances in the equivalent delta network are

$$Z_A = \frac{Z_1 Z_2 + Z_2 Z_3 + Z_1 Z_3}{Z_3} \tag{16.4}$$

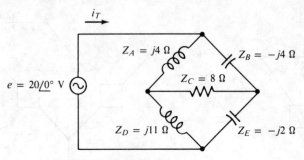

FIGURE 16.2 (Example 16.1)

$$Z_B = \frac{Z_1Z_2 + Z_2Z_3 + Z_1Z_3}{Z_2} \qquad (16.5)$$

$$Z_C = \frac{Z_1Z_2 + Z_2Z_3 + Z_1Z_3}{Z_1} \qquad (16.6)$$

Example 16.1   (Analysis)

Find the total impedance and the total current i_T in the circuit shown in Figure 16.2.

SOLUTION We observe that Z_A, Z_B, and Z_C form one delta network in Figure 16.2, and Z_C, Z_D, and Z_E form another. We will convert one of these deltas to an equivalent wye. Since we are not required to find the voltage or current in any particular component, it does not matter which of the two deltas is transformed. In this example, we arbitrarily choose to transform the Z_A–Z_B–Z_C delta. (If it were necessary to find the voltage or current in Z_A or Z_B, it would be necessary to transform the Z_C–Z_D–Z_E delta.)

Figure 16.3(a) shows the equivalent wye network that will replace the Z_A–Z_B–Z_C delta. Notice that each wye impedance is the product of the two delta impedances connected to it, divided by the sum of the delta impedances:

$$Z_1 = \frac{Z_AZ_B}{Z_A + Z_B + Z_C} = \frac{(4\,\underline{/90°})(4\,\underline{/-90°})}{(0 + j4) + (0 - j4) + (8 + j0)}$$

$$= \frac{16\,\underline{/0°}}{8 + j0} = \frac{16\,\underline{/0°}}{8\,\underline{/0°}} = 2\,\underline{/0°}\ \Omega$$

$$Z_2 = \frac{Z_AZ_C}{Z_A + Z_B + Z_C} = \frac{(4\,\underline{/90°})(8\,\underline{/0°})}{8\,\underline{/0°}} = 4\,\underline{/90°}\ \Omega$$

$$Z_3 = \frac{Z_BZ_C}{Z_A + Z_B + Z_C} = \frac{(4\,\underline{/-90°})(8\,\underline{/0°})}{8\,\underline{/0°}} = 4\,\underline{/-90°}\ \Omega$$

In this special case, Z_1 is purely resistive, Z_2 is purely inductive, and Z_3 is purely capacitive. Figure 16.3(b) shows the circuit with the wye components inserted and the

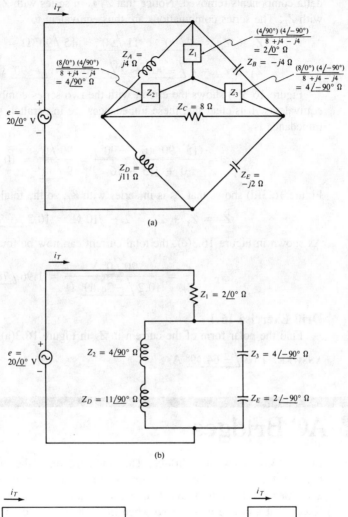

(a)

(b)

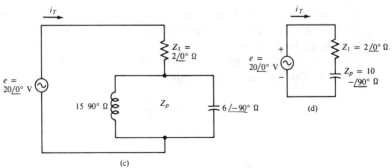

(c)

(d)

FIGURE 16.3 (Example 16.1)

delta components removed. Notice that Z_2 is in series with Z_D, and Z_3 is in series with Z_E. The series combinations are thus equivalent to

$$Z_2 + Z_D = 4 \underline{/90°} + 11 \underline{/90°} = 15 \underline{/90°} \ \Omega = 0 + j15 \ \Omega$$

$$Z_3 + Z_E = 4 \underline{/-90°} + 2 \underline{/-90°} = 6 \underline{/-90°} \ \Omega = 0 - j6 \ \Omega$$

Figure 16.3(c) shows the circuit with the two series combinations replaced by their equivalents. It is clear that these impedances are in parallel, so their equivalent impedance is

$$Z_P = \frac{(15 \underline{/90°})(6 \underline{/-90°})}{0 + j15 - j6} = \frac{90 \underline{/0°}}{9 \underline{/90°}} = 10 \underline{/-90°} \ \Omega$$

Figure 16.3(d) shows that Z_P is in series with Z_1, so the total impedance is

$$Z_T = Z_1 + Z_P = 2 - j10 \ \Omega = 10.2 \underline{/-78.69°} \ \Omega$$

As shown in Figure 16.3(e), the total current can now be found:

$$i_T = \frac{e}{Z_T} = \frac{20 \underline{/0°} \ \text{V}}{10.2 \underline{/-78.69°} \ \Omega} = 1.96 \underline{/78.69°} \ \text{A}$$

Drill Exercise 16.1

Find the polar form of the current in Z_A in Figure 16.3(a).

ANSWER: $1.63 \underline{/-64.49°}$ A. ☐

16.2 AC Bridges

Figure 16.4 shows an ac bridge. The configuration is the same as that of a dc bridge (Section 6.2), except that each component is an arbitrary impedance instead of a resistance. See Figure 6.16 to review the various equivalent ways of drawing a bridge circuit. Recall that a bridge is *balanced* when the voltage at node A is the same as the voltage at node B, both measured with respect to a common reference. Under those circumstances, $v_{AB} = 0$ and the current between nodes A and B is zero. For the bridge to be balanced, the impedances must satisfy

$$\frac{Z_1}{Z_3} = \frac{Z_2}{Z_4} \tag{16.7}$$

or, equivalently,

$$\frac{Z_1}{Z_2} = \frac{Z_3}{Z_4}$$

If the bridge is balanced, then Z_5 can be replaced by either an open or a short circuit without affecting the voltages or currents in the rest of the circuit.

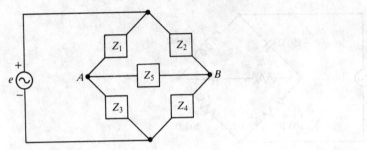

FIGURE 16.4 AC bridge circuit. Compare with Figure 6.16

Example 16.2   (Analysis)

Determine if the bridge shown in Figure 16.5 is balanced.

SOLUTION By equation 16.7, we must determine if the ratios Z_1/Z_3 and Z_2/Z_4 are equal.

$$\frac{Z_1}{Z_3} = \frac{50\ \underline{/-90°}}{100\ \underline{/0°}} = 0.5\ \underline{/-90°}$$

$$\frac{Z_2}{Z_4} = \frac{100\ \underline{/90°}}{200\ \underline{/0°}} = 0.5\ \underline{/90°}$$

Although the magnitudes of the ratios are equal, the angles are not. Therefore, the ratios are not equal, and the bridge is *not* balanced.

Drill Exercise 16.2

How could Z_1 or Z_2 in Figure 16.5 be changed to make the bridge balanced?

ANSWER: $Z_1 = j50\ \Omega$ or $Z_2 = -j100\ \Omega$.   □

Example 16.3   (Design)

What value of inductance L should be used in Figure 16.6 to balance the bridge?

SOLUTION Comparing Figure 16.6 with Figure 16.4, we see that

$$Z_1 = |X_L|\ \underline{/90°}\ \Omega$$

$$Z_2 = 10^3\ \underline{/0°}\ \Omega$$

$$Z_3 = 3 \times 10^3\ \underline{/0°}\ \Omega$$

$$Z_4 = |X_C|\ \underline{/-90°}\ \Omega$$

For balance, we require

$$\frac{Z_1}{Z_3} = \frac{Z_2}{Z_4}$$

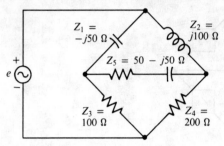

FIGURE 16.5 (Example 16.2)

or

$$\frac{|X_L| \, \underline{/90°}}{3 \times 10^3 \, \underline{/0°}} = \frac{10^3 \, \underline{/0°}}{|X_C| \, \underline{/-90°}}$$

$$\frac{|X_L|}{3 \times 10^3} \, \underline{/90°} = \frac{10^3}{|X_C|} \, \underline{/90°}$$

Since the angles of the ratios are equal under any circumstances, we need only find the value of L that makes the magnitudes equal:

$$\frac{|X_L|}{3 \times 10^3} = \frac{10^3}{|X_C|}$$

$$|X_L||X_C| = 3 \times 10^6$$

$$(\omega L) \frac{1}{\omega C} = 3 \times 10^6$$

$$L = 3 \times 10^6 C = (3 \times 10^6)(10^{-8}) = 30 \text{ mH}$$

Drill Exercise 16.3

If L in Figure 16.6 were set equal to 45 mH, what value of C would be required to balance the bridge?

ANSWER: 0.015 μF.   ☐

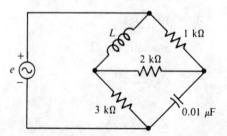

FIGURE 16.6 (Example 16.3)

The Hay Bridge

Figure 16.7 shows a special type of ac bridge, called a *Hay bridge*, used to measure the inductance L of an unknown inductor and its resistance R_L. Note that an ac instrument (ac ammeter or galvonometer) is connected in place of Z_5 in the standard bridge configuration. In applications, the values of other components in the bridge are adjusted until zero current is measured, indicating that the bridge has been balanced. When balance is achieved, the values of L and R_L can be computed in terms of the other (known) component values.

We will now use the balanced-bridge criterion [equation (16.7)] to derive equations for L and R_L in terms of the known component values. Comparing Figure 16.7 with Figure 16.4, we see that

$$Z_1 = R_1 - j|X_C|$$

$$Z_2 = R_2$$

$$Z_3 = R_3$$

$$Z_4 = R_L + j|X_L|$$

By equation (16.7), balance occurs when

$$\frac{R_1 - j|X_C|}{R_3} = \frac{R_2}{R_L + j|X_L|} \tag{16.8}$$

Cross-multiplying gives

$$(R_1 - j|X_C|)(R_L + j|X_L|) = R_2R_3 \tag{16.9}$$

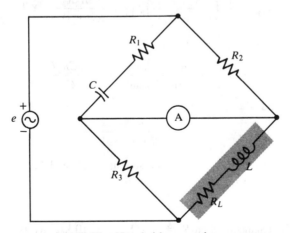

FIGURE 16.7 The Hay bridge, used to measure the inductance L of an unknown inductor and its resistance R_L.

Expanding the left side of (16.9) and collecting real and imaginary terms, we have

$$(R_1R_L + |X_L||X_C|) + j(R_1|X_L| - R_L|X_C|) = R_2R_3 + j0 \tag{16.10}$$

Note that the right side of (16.10) is the purely real number R_2R_3; that is, it has zero imaginary part. *Two complex numbers are equal if and only if their real parts are equal and their imaginary parts are equal.* (This is equivalent to stating that both their magnitudes and their angles must be equal.) Equation (16.10) states that the complex number represented by the left side equals the complex number on the right side, the latter having zero imaginary part. Therefore, we can equate the real and imaginary parts of each side:

$$R_1R_L + |X_L||X_C| = R_2R_3 \tag{16.11}$$

$$R_1|X_L| - R_L|X_C| = 0 \tag{16.12}$$

Solving (16.12) for $|X_L|$ gives

$$|X_L| = \frac{R_L}{R_1}|X_C| \tag{16.13}$$

Substituting (16.13) into (16.11) gives

$$R_1R_L + \frac{R_L}{R_1}|X_C|^2 = R_2R_3 \tag{16.14}$$

Factoring R_L out of the left side of (16.14) and solving for R_L, we find

$$R_L = \frac{R_1R_2R_3}{R_1^2 + |X_C|^2} = \frac{R_1R_2R_3}{R_1^2 + (1/\omega C)^2} \tag{16.15}$$

where ω is the angular frequency of the voltage source connected across the bridge. Substituting (16.15) into (16.13), we have

$$|X_L| = \frac{1}{R_1}\left(\frac{R_1R_2R_3}{R_1^2 + |X_C|^2}\right)|X_C| = \frac{R_2R_3}{R_1^2 + |X_C|^2}|X_C| \tag{16.16}$$

Substituting $X_L = \omega L$ and $X_C = 1/\omega C$ gives

$$\omega L = \left(\frac{R_2R_3}{R_1^2 + 1/\omega^2C^2}\right)\frac{1}{\omega C} = \frac{\omega C R_2R_3}{(\omega R_1 C)^2 + 1} \tag{16.17}$$

Solving for L, we find

$$L = \frac{CR_2R_3}{(\omega R_1 C)^2 + 1} \tag{16.18}$$

Equations (16.15) and (16.18) are those we sought for R_L and L, in terms of the other component values and the frequency of the source.

Example 16.4   (Analysis)
   A Hay bridge is balanced when $R_2 = 1$ kΩ, $R_3 = 10$ kΩ, $R_1 = 100$ kΩ, $C = 200$ pF, and the frequency of the source is 1 kHz. Find the inductance and resistance of the unknown inductor.

SOLUTION From equation (16.15),

$$R_L = \frac{R_1 R_2 R_3}{R_1^2 + (1/\omega C)^2} = \frac{(10^5)(10^3)(10^4)}{(10^5)^2 + \left[\dfrac{1}{(2\pi \times 10^3)(2 \times 10^{-10})}\right]^2} = 1.55 \ \Omega$$

From equation (16.18),

$$L = \frac{CR_2 R_3}{(\omega R_1 C)^2 + 1} = \frac{(2 \times 10^{-10})(10^3)(10^4)}{[(2\pi \times 10^3)(10^5)(2 \times 10^{-10})]^2 + 1} = 1.97 \ \text{mH}$$

In this example, the magnitude of the inductive reactance is considerably greater than the resistance of the inductor, the situation for which a Hay bridge is generally used.

Drill Exercise 16.4

If the frequency of the voltage source in Example 16.4 is changed to 2 kHz, is it possible to balance the bridge by changing the value of R_3 alone? Explain.

ANSWER: No. [Solve equations (16.15) and (16.18) for R_3.] ☐

The Maxwell Bridge

Figure 16.8 shows another bridge used to measure the inductance and resistance of an unknown inductor. This configuration is called a *Maxwell bridge* and is used when R_L is not so small in comparison to X_L that the Hay bridge must be used. The equations for R_L and L can be derived following the same general procedure that was used to derive the equations for the Hay bridge. It is an exercise at the end of this chapter to show that

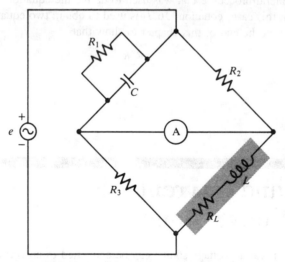

FIGURE 16.8 The Maxwell bridge, used to measure the inductance L and resistance R_L of an unknown inductor.

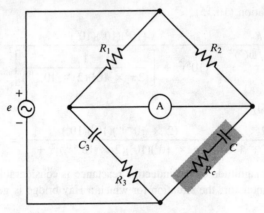

FIGURE 16.9 The capacitor comparison bridge, used to measure the capacitance C and resistance R_C of an unknown capacitor.

$$R_L = \frac{R_2 R_3}{R_1} \qquad (16.19)$$

$$L = CR_2 R_3 \qquad (16.20)$$

The Capacitor Comparison Bridge

Figure 16.9 shows a *capacitor comparison bridge*, used to measure the capacitance C and resistance R_C of an unknown capacitor. The equations for R_C and C can be derived using the same general procedure that was used to derive the equations for R_L and L in the Hay bridge. In this case, equation (16.7) is used to obtain two equations for R_C and C. It is an exercise at the end of this chapter to show that

$$R_C = \frac{R_2 R_3}{R_1} \qquad (16.21)$$

$$C = C_3 \frac{R_1}{R_2} \qquad (16.22)$$

16.3 Voltage and Current Source Conversions

As in dc circuits, a real ac voltage source can be converted to an *equivalent* ac current source, and vice versa. Recall that real dc sources have internal resistance. Similarly, real ac sources have internal *impedance*. Once a conversion has been made, it is impossible to make any measurements or to perform any computations at the source terminals that would reveal whether a given source is a voltage source or its equivalent current source. The procedure for converting from one type of source to the other is the same as that used for converting dc sources, except, as usual, the computations are performed using phasors. Figure 16.10 illustrates the process. Note that the polarities of the sources are preserved in the transformations, as shown by the polarity symbols in the figure. Remember that reversing the polarity symbols of a source is the same as adding or subtracting 180° from its phase angle.

Example 16.5

(a) Convert the voltage source shown in Figure 16.11(a) to an equivalent current source.
(b) Show that the current in a 40-Ω resistor connected across the terminals of the voltage source is the same as the current in a 40-Ω resistor connected across the terminals of the equivalent current source.

SOLUTION

(a) The impedance of the voltage source is

$$Z = 30 + j40 \ \Omega = 50 \underline{/53.13°} \ \Omega$$

Therefore, the current produced by the equivalent current source is

$$i = \frac{e}{Z} = \frac{20 \underline{/0°} \text{ V}}{50 \underline{/53.13°} \ \Omega} = 0.4 \underline{/-53.13°} \text{ A}$$

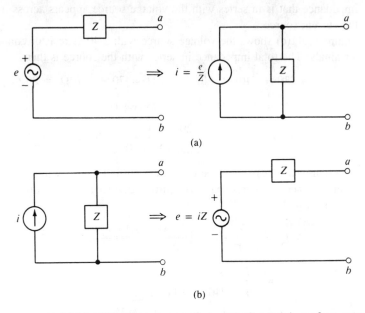

(a)

(b)

FIGURE 16.10 AC source conversions. (a) Conversion of an ac voltage source to an equivalent ac current source. (b) Conversion of an ac current source to an equivalent ac voltage source.

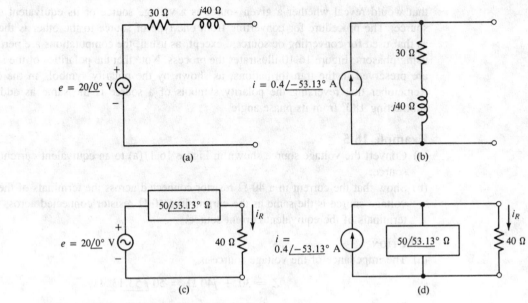

FIGURE 16.11 (Example 16.5)

The equivalent current source is shown in Figure 16.11(b). Notice that the same impedance that is in series with the voltage source appears across the terminals of the current source.

(b) Figure 16.11(c) shows the voltage source with a 40-Ω resistor connected across its terminals. The total impedance in series with the source is then

$$Z_T = (30 + j40) + (40 + j0) \ \Omega = 70 + j40 \ \Omega = 80.62 \ \underline{/29.74°} \ \Omega$$

Therefore, the current i_R in the 40-Ω resistor is

$$i_R = \frac{e}{Z_T} = \frac{20 \ \underline{/0°} \ V}{80.62 \ \underline{/29.74°} \ \Omega} = 0.248 \ \underline{/-29.74°} \ A$$

Figure 16.11(d) shows the equivalent current source with a 40-Ω resistor connected across its terminals. By current-divider rule, the current i_R in the 40-Ω resistor is

$$i_R = \frac{Z}{R + Z} i = \left(\frac{50 \ \underline{/53.13°}}{40 + 30 + j40}\right)(0.4 \ \underline{/-53.13°} \ A)$$

$$= \frac{20 \ \underline{/0°}}{70 + j40} = \frac{20 \ \underline{/0°}}{80.62 \ \underline{/29.74°}} = 0.248 \ \underline{/-29.74°} \ A$$

As expected, the current in a 40-Ω resistor connected across the current source is the same as that in a 40-Ω resistor connected across the voltage source, since the sources are equivalent.

Drill Exercise 16.5

Find the current produced by an equivalent current source when the 40 Ω of inductive reactance in the voltage source shown in Figure 16.11(a) is replaced by 20 Ω of capacitive reactance.

ANSWER: $0.55 \underline{/33.69°}$ A.   ☐

When converting a voltage source to an equivalent current source, any impedance or combination of impedances that is in series with the voltage source can be included in the conversion. However, if it is necessary to find the current through or voltage across a specific impedance in series with the voltage source, that impedance should not be included in the conversion. Similarly, when converting a current source to an equivalent voltage source, any impedances in parallel with the current source can be included in the conversion, provided that we are not seeking the current through or voltage across one of those impedances.

Example 16.6   (Analysis)

Find the current through the 1-kΩ resistor in Figure 16.12(a) by converting the current source to an equivalent voltage source.

SOLUTION Since we wish to find the current in the resistor, the resistor will not be included in the source conversion. Instead, we will use the inductor and capacitor, both of which are in parallel with the current source. Their parallel equivalent is

$$Z_P = \frac{(800 \underline{/90°})(400 \underline{/-90°})}{0 + j800 - j400} = \frac{32 \times 10^4 \underline{/0°}}{400 \underline{/90°}}$$

$$= 800 \underline{/-90°} \; \Omega = 0 - j800 \; \Omega$$

The equivalent voltage source therefore has value

$$e = iZ_P = (25 \underline{/0°} \text{ mA})(800 \underline{/-90°} \; \Omega) = 20 \underline{/-90°} \text{ V}$$

Figure 16.12(b) shows the equivalent voltage source with the 1-kΩ resistor connected across its terminals. The current in the resistor is then the total current, i_T, that flows in the series circuit. The total series impedance is

$$Z_T = 1000 - j800 \; \Omega = 1281 \underline{/-38.66°} \; \Omega$$

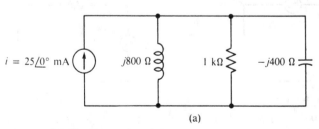
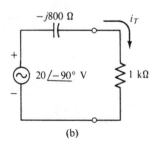

(a)   (b)

FIGURE 16.12 (Example 16.6)

Thus,

$$i_T = \frac{e}{Z_T} = \frac{20\,\underline{/-90°}\ \text{V}}{1281\,\underline{/-38.66°}\ \Omega} = 15.61\,\underline{/-51.34°}\ \text{mA}$$

Drill Exercise 16.6

Find the current in the inductor in Figure 16.12(a) by converting the current source to an equivalent voltage source.

ANSWER: $19.52\,\underline{/-141.35°}$ mA.   □

16.4 Series and Parallel AC Sources

AC voltage sources connected in a series path can be combined into a single equivalent voltage source for analysis purposes. To obtain the equivalent (resultant) voltage of such a combination, the source voltages are converted to rectangular form and added. Similarly, ac current sources in parallel can be combined by adding their rectangular forms. In both cases, care must be exercised to observe polarity symbols, remembering that opposing polarities imply subtraction rather than addition (or addition of a negative quantity). Sources can be added and subtracted in this way, to obtain a single resultant, provided that all are sinusoidal and have exactly the same frequency.

Example 16.7 (Analysis)

Find the current in the 16-Ω resistor in Figure 16.13(a).

SOLUTION The three sources are in a series path and can therefore be combined into a single equivalent source. Converting each to rectangular form, we have

$$e_1 = 4\,\underline{/0°}\ \text{V} = 4 + j0\ \text{V}$$

$$e_2 = 12\,\underline{/-60°}\ \text{V} = 6 - j10.39\ \text{V}$$

$$e_3 = 18\,\underline{/30°}\ \text{V} = 15.59 + j9\ \text{V}$$

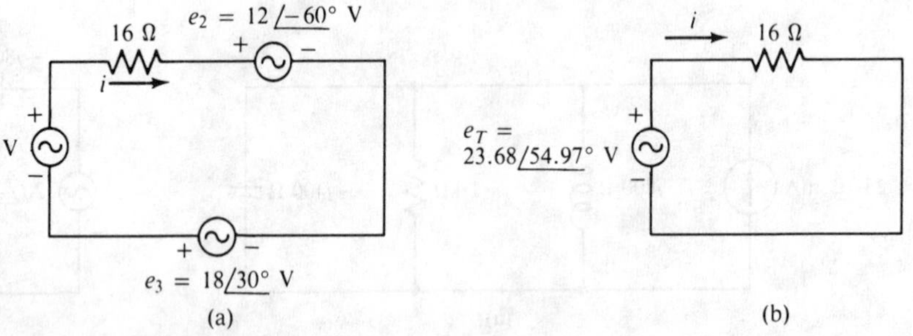

FIGURE 16.13 (Example 16.7)

Notice that the polarity symbols on e_1 and e_3 in Figure 16.13(a) indicate that those sources are series-aiding, while the polarity of e_2 opposes e_1 and e_3. Therefore, the total equivalent voltage is

$$
\begin{aligned}
e_T &= e_1 + e_3 - e_2 \\
&= (4 + j0) + (15.59 + j9) - (6 - j10.39) \\
&= (13.59 + j19.39) \text{ V} = 23.68 \,\underline{/54.97°}\ \text{V}
\end{aligned}
$$

The equivalent circuit with the three sources replaced by e_T is shown in Figure 16.13(b). The current in the 16-Ω resistor is

$$
i = \frac{e_T}{16 \,\underline{/0°}\ \Omega} = \frac{23.68 \,\underline{/54.97°}\ \text{V}}{16 \,\underline{/0°}\ \Omega} = 1.48 \,\underline{/54.97°}\ \text{A}
$$

Drill Exercise 16.7

Find the current in the 16-Ω resistor in Figure 16.13(a) when the polarity symbols on e_1 are reversed.

ANSWER: $1.26 \,\underline{/73.92°}\ \text{A}$.

16.5 AC Mesh Analysis

When a circuit contains more than one active source, and the sources cannot be combined into a single equivalent source, some other procedure must be used to analyze the circuit. Recall that one such procedure is *mesh analysis*. As in dc circuits, the approach is to write Kirchhoff's voltage law around closed loops, thereby generating equations that can be solved simultaneously for the unknown loop currents. To use this procedure, all sources in the analysis must be *voltage* sources, so any current sources present in the original circuit must be converted to equivalent voltage sources. As usual, we assume that the sources are sinusoidal and have the same frequency.

Cramer's rule, described in Chapter 6, can be used to solve simultaneous loop equations in ac as well as dc mesh analysis. However, since all voltages, currents, and impedances in an ac analysis are phasor quantities, the computations can become quite laborious. To minimize the computational effort, it is best to write the equations and set up the determinants using symbols for all terms involved (e_1, e_2, Z_1, Z_2, etc.), rather than their phasor values. Once a symbolic expression has been obtained for an unknown, the last step should be to substitute numerical values and perform the necessary phasor computations.

When writing the loop equations, it is particularly important to observe polarity symbols on the voltage sources and to distinguish between voltage rises and voltage drops. Remember that passing through a source or impedance from + to − indicates a voltage drop, while passing through it from − to + indicates a voltage rise.

Example 16.8 (Analysis)

Use mesh analysis to find the current in the capacitor in Figure 16.14(a).

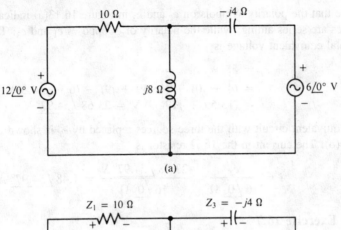

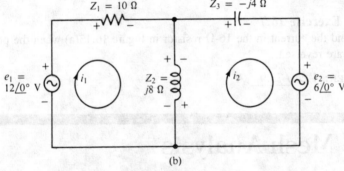

FIGURE 16.14 (Example 16.8)

SOLUTION To facilitate computations, the symbols Z_1, Z_2, and Z_3 are assigned to the impedances, as shown in Figure 16.14(b). The voltage sources are designated e_1 and e_2. We will write Kirchhoff's voltage law around the clockwise loops shown in the figure, in terms of loop currents i_1 and i_2. Notice that polarity symbols are drawn on the impedances in accordance with the assumed directions of the loop currents. When writing the equation for loop 1, the net current through Z_2 is $(i_1 - i_2)$ and the voltage drop across it is $(i_1 - i_2)Z_2$. Thus, a $+$ symbol is shown at the top of Z_2 in loop 1. When writing the equation around loop 2, the net current in Z_2 is assumed to be $(i_2 - i_1)$ and the voltage drop across it is $(i_2 - i_1)Z_2$. Since the net current in this case enters Z_2 from below, the $+$ symbol is shown at the bottom of Z_2 in loop 2. These conventions are identical to those described in Chapter 6 for dc mesh analysis.

The loop equations are obtained by setting the voltage drops equal to the voltage rises:

$$\text{loop 1: } i_1 Z_1 + (i_1 - i_2)Z_2 = e_1$$
$$\text{loop 2: } (i_2 - i_1)Z_2 + i_2 Z_3 + e_2 = 0$$

Rearranging these equations and collecting terms so they can be solved using Cramer's rule, we have

$$i_1(Z_1 + Z_2) + i_2(-Z_2) = e_1$$
$$i_1(-Z_2) + i_2(Z_2 + Z_3) = -e_2$$

The delta determinant (Δ) is the determinant of the coefficient matrix:

$$\Delta = \det \begin{bmatrix} Z_1 + Z_2 & -Z_2 \\ -Z_2 & Z_2 + Z_3 \end{bmatrix} = (Z_1 + Z_2)(Z_2 + Z_3) - Z_2^2$$
$$= Z_1Z_2 + Z_1Z_3 + Z_2^2 + Z_2Z_3 - Z_2^2$$
$$= Z_1Z_2 + Z_1Z_3 + Z_2Z_3$$

Since we are required to find the current in the capacitor, it is necessary to solve for i_2 only. By Cramer's rule,

$$i_2 = \frac{\det \begin{bmatrix} Z_1 + Z_2 & e_1 \\ -Z_2 & -e_2 \end{bmatrix}}{\Delta} = \frac{(Z_1 + Z_2)(-e_2) + e_1Z_2}{\Delta}$$

$$= \frac{-e_2Z_1 - e_2Z_2 + e_1Z_2}{Z_1Z_2 + Z_1Z_3 + Z_2Z_3}$$

We can now substitute the phasor values $e_1 = 12 \underline{/0°}$, $e_2 = 6 \underline{/0°}$, $Z_1 = 10 \underline{/0°}$, $Z_2 = 8 \underline{/90°}$, and $Z_3 = 4 \underline{/-90°}$ to obtain a phasor value for i_2:

$$i_2 = \frac{-(6 \underline{/0°})(10 \underline{/0°}) - (6 \underline{/0°})(8 \underline{/90°}) + (12 \underline{/0°})(8 \underline{/90°})}{(10 \underline{/0°})(8 \underline{/90°}) + (10 \underline{/0°})(4 \underline{/-90°}) + (8 \underline{/90°})(4 \underline{/-90°})}$$

$$= \frac{-60 \underline{/0°} - 48 \underline{/90°} + 96 \underline{/90°}}{80 \underline{/90°} + 40 \underline{/-90°} + 32 \underline{/0°}}$$

$$= \frac{-60 + j48}{32 + j40} = \frac{76.84 \underline{/141.34°}}{51.22 \underline{/51.34°}} = 1.5 \underline{/90°} \text{ A}$$

Notice that this value of i_2 has the phase reference shown by the assumed direction of i_2 in Figure 16.14(b) (i.e., from left to right). If we wished to represent i_2 as a right-to-left current, we would write $i_2 = -1.5 \underline{/90°}$ A $= 1.5 \underline{/-90°}$ A.

Drill Exercise 16.8

Find the polar form of the current in the inductor in Figure 16.14(a).

ANSWER: $1.5 \underline{/-90°}$ A (with the reference polarity of i_1). □

Example 16.9   (Analysis)

Use mesh analysis to find the voltage across the inductor in Figure 16.15(a).

SOLUTION Before writing loop equations, the current source must be converted to an equivalent voltage source. Since the current source is in parallel with a 1-kΩ resistor, it is equivalent to a voltage source having value

$$e = (10 \underline{/0°} \text{ mA})(1 \text{ k}\Omega \underline{/0°}) = 10 \underline{/0°} \text{ V}$$

Figure 16.15(b) shows the circuit with the equivalent voltage source, labeled e_1, drawn in place of the current source and in series with the 1-kΩ resistor. The figure also

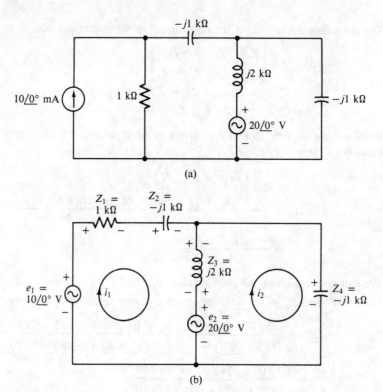

FIGURE 16.15 (Example 16.9)

shows the clockwise loop currents, the impedance designations Z_1 through Z_4, and the polarities of the voltage drops. Note that we have chosen to treat Z_1 and Z_2 as two separate impedances. We could also combine Z_1 and Z_2 into the single equivalent impedance $1 \text{ k}\Omega - j1 \text{ k}\Omega$ before writing the equations. Writing Kirchhoff's voltage law around loops 1 and 2, we have

$$i_1(Z_1 + Z_2) + (i_1 - i_2)Z_3 + e_2 = e_1$$

$$(i_2 - i_1)Z_3 + i_2Z_4 = e_2$$

Rearranging and collecting terms gives

$$i_1(Z_1 + Z_2 + Z_3) + i_2(-Z_3) = e_1 - e_2$$

$$i_1(-Z_3) + i_2(Z_3 + Z_4) = e_2$$

The delta determinant is

$$\Delta = \det \begin{bmatrix} Z_1 + Z_2 + Z_3 & -Z_3 \\ -Z_3 & Z_3 + Z_4 \end{bmatrix} = (Z_1 + Z_2 + Z_3)(Z_3 + Z_4) - Z_3^2$$

Substituting $Z_1 = 10^3$, $Z_2 = -j10^3$, $Z_3 = j2 \times 10^3$, and $Z_4 = -j10^3$, we find

$$\begin{aligned}
\Delta &= (10^3 - j10^3 + j2 \times 10^3)(j2 \times 10^3 - j10^3) - (j2 \times 10^3)^2 \\
&= (10^3 + j10^3)(10^3 \,\underline{/90°}) + 4 \times 10^6 \\
&= (1.414 \times 10^3 \,\underline{/45°})(10^3 \,\underline{/90°}) + 4 \times 10^6 \\
&= 1.414 \times 10^6 \,\underline{/135°} + 4 \times 10^6 \\
&= -10^6 + j10^6 + 4 \times 10^6 = 3 \times 10^6 + j10^6 = 3.16 \times 10^6 \,\underline{/18.43°}
\end{aligned}$$

By Cramer's rule,

$$\begin{aligned}
i_1 &= \frac{\det \begin{bmatrix} e_1 - e_2 & -Z_3 \\ e_2 & Z_3 + Z_4 \end{bmatrix}}{\Delta} \\
&= \frac{(e_1 - e_2)(Z_3 + Z_4) + e_2 Z_3}{\Delta} \\
&= \frac{(10 \,\underline{/0°} - 20 \,\underline{/0°})(0 + j2 \times 10^3 - j10^3) + (20 \,\underline{/0°})(2 \times 10^3 \,\underline{/90°})}{\Delta} \\
&= \frac{(-10 \,\underline{/0°})(10^3 \,\underline{/90°}) + 4 \times 10^4 \,\underline{/90°}}{\Delta} = \frac{3 \times 10^4 \,\underline{/90°}}{\Delta} \\
&= \frac{3 \times 10^4 \,\underline{/90°}}{3.16 \times 10^6 \,\underline{/18.43°}} = 9.49 \,\underline{/71.57°} \text{ mA} = (3 + j9) \text{ mA}
\end{aligned}$$

$$\begin{aligned}
i_2 &= \frac{\det \begin{bmatrix} Z_1 + Z_2 + Z_3 & e_1 - e_2 \\ -Z_3 & e_2 \end{bmatrix}}{\Delta} \\
&= \frac{(Z_1 + Z_2 + Z_3)e_2 + Z_3(e_1 - e_2)}{\Delta} \\
&= \frac{(10^3 - j10^3 + j2 \times 10^3)(20 \,\underline{/0°}) + (2 \times 10^3 \,\underline{/90°})(10 \,\underline{/0°} - 20 \,\underline{/0°})}{\Delta} \\
&= \frac{(10^3 + j10^3)(20 \,\underline{/0°}) + (2 \times 10^3 \,\underline{/90°})(-10 \,\underline{/0°})}{\Delta} \\
&= \frac{2 \times 10^4 + j2 \times 10^4 - 2 \times 10^4 \,\underline{/90°}}{\Delta} \\
&= \frac{2 \times 10^4 \,\underline{/0°}}{3.16 \times 10^6 \,\underline{/18.43°}} = 6.32 \,\underline{/-18.43°} \text{ mA} = (6 - j2) \text{ mA}
\end{aligned}$$

In loop 1, the net current in the inductor is

$$\begin{aligned}
i_L &= i_1 - i_2 = (3 + j9) \text{ mA} - (6 - j2) \text{ mA} = (-3 + j11) \text{ mA} \\
&= -(3 - j11) \text{ mA} = -11.4 \,\underline{/-74.74°} \text{ mA}
\end{aligned}$$

Therefore, the voltage across the inductor is

$$v_L = i_L X_L = (-11.4 \underline{/-74.74°} \text{ mA})(2 \times 10^3 \underline{/90°})$$
$$= -22.8 \underline{/15.26°} \text{ V} = 22.8 \underline{/-164.74°} \text{ V}$$

Note that v_L is expressed as the voltage that would be measured with the polarity (phase reference) shown in loop 1 of Figure 16.15(b). If we wished to represent v_L with the polarity shown in loop 2, we would write

$$v_L = -22.8 \underline{/-164.74°} \text{ V} = 22.8 \underline{/15.26°} \text{ V}$$

Drill Exercise 16.9

Find the voltage across each component in Figure 16.15(b) and verify Kirchhoff's voltage law around each loop.

ANSWER: $v_{Z1} = (3 + j9)$ V; $v_{Z2} = (9 - j3)$ V; $v_{Z4} = (-2 - j6)$ V.   □

16.6 AC Nodal Analysis

Recall that nodal analysis is the procedure used to find the voltage at each node of a circuit with respect to a reference node. All sources in the analysis must be current sources, so any voltage sources present in the original circuit must be converted to equivalent current sources. Simultaneous equations are generated by writing Kirchhoff's current law at each node. As in ac mesh analysis, the computations are facilitated by avoiding the use of numerical (phasor) values as long as possible. In nodal analysis, it is generally more convenient to write the equations in terms of admittance designations, Y_1, Y_2, . . . , than it is to use impedance designations, Z_1, Z_2, . . . After expressions have been obtained for the unknown node voltages using Cramer's rule, the last step is to substitute phasor values and evaluate the expressions.

Following is a review of the steps that are followed to perform a nodal analysis:

1. Convert any voltage sources in the circuit to equivalent current sources.
2. Identify and label all nodes, including a reference node.
3. Make an assumption about the relative magnitudes of the node voltages. Every node voltage is assumed to be greater than the voltage at the reference node (zero volts).
4. Draw arrows on the schematic diagram showing the direction of current flows, based on the assumptions made in step 3. Current always flows from a node of higher voltage to a node of lower voltage.
5. Label each arrow with an expression for the current it represents, in terms of the node voltages and the admittances between the nodes.
6. Write Kirchhoff's current law at each node, equating the currents entering the node to the currents leaving the node.
7. Solve the simultaneous equations obtained in step 6 for the unknown node voltages.

Example 16.10   (Analysis)

Use nodal analysis to find the current through the inductor in Figure 16.16(a).

SOLUTION The circuit is redrawn in preparation for nodal analysis as shown in Figure 16.16(b). Node voltages v_1, v_2 and the reference (ground) node are identified on the figure. Notice that admittance designations Y_1, Y_2, and Y_3 have been assigned to the components, and the value of each admittance has been calculated based on the impedance values shown in Figure 16.16(a). Arrows are drawn to indicate current directions based on the assumption $v_1 > v_2 > 0$. Recall that current can be expressed in terms of admittance by $i = Yv$, where v is the difference in voltage across a component and i is the current through it. Since the difference in voltage across the resistor is $(v_1 - 0) = v_1$, the current through it is $Y_1 v_1$. Similarly, the current in the capacitor is $Y_3 v_2$. The difference in voltage across the inductor is $(v_1 - v_2)$, so the current through it is $Y_2(v_1 - v_2)$. This current is the one whose value we are seeking, so we must solve for both v_1 and v_2.

Writing Kirchhoff's current law at each node, we have

$$\text{current entering} = \text{current leaving}$$

Node 1:
$$i_1 = Y_1 v_1 + Y_2(v_1 - v_2)$$

Node 2:
$$Y_2(v_1 - v_2) = Y_3 v_2 + i_2$$

Expanding these equations and rearranging them for solution by Cramer's rule, we find

$$(Y_1 + Y_2)v_1 + (-Y_2)v_2 = i_1$$

$$(-Y_2)v_1 + (Y_2 + Y_3)v_2 = -i_2$$

The delta determinant is therefore

$$\Delta = \det \begin{bmatrix} Y_1 + Y_2 & -Y_2 \\ -Y_2 & Y_2 + Y_3 \end{bmatrix} = (Y_1 + Y_2)(Y_2 + Y_3) - Y_2^2$$

$$= Y_1 Y_2 + Y_1 Y_3 + Y_2^2 + Y_2 Y_3 - Y_2^2$$

$$= Y_1 Y_2 + Y_1 Y_3 + Y_2 Y_3$$

Substituting $Y_1 = 0.25 \underline{/0°}$, $Y_2 = 0.5 \underline{/-90°}$, and $Y_3 = 0.2 \underline{/90°}$, we obtain

$$\Delta = (0.25 \underline{/0°})(0.5 \underline{/-90°}) + (0.25 \underline{/0°})(0.2 \underline{/90°}) + (0.5 \underline{/-90°})(0.2 \underline{/90°})$$

$$= 0.125 \underline{/-90°} + 0.05 \underline{/90°} + 0.1 \underline{/0°}$$

$$= 0.1 + j0.05 - j0.125 = 0.1 - j0.075$$

$$= 0.125 \underline{/-36.87°}$$

By Cramer's rule,

$$v_1 = \frac{\det \begin{bmatrix} i_1 & -Y_2 \\ -i_2 & Y_2 + Y_3 \end{bmatrix}}{\Delta} = \frac{i_1(Y_2 + Y_3) - i_2 Y_2}{\Delta}$$

Substituting values gives

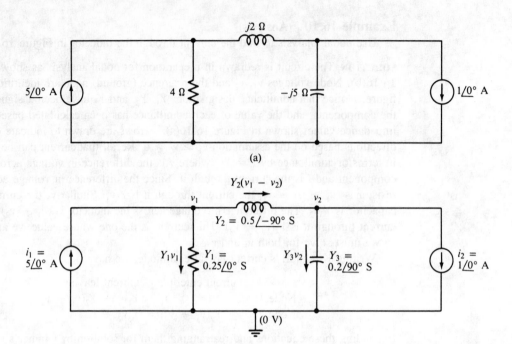

FIGURE 16.16 (Example 16.10)

$$v_1 = \frac{(5\,\underline{/0°})(0.5\,\underline{/-90°} + 0.2\,\underline{/90°}) - (1\,\underline{/0°})(0.5\,\underline{/-90°})}{\Delta}$$

$$= \frac{(5\,\underline{/0°})(0.3\,\underline{/-90°}) - 0.5\,\underline{/-90°}}{\Delta} = \frac{1\,\underline{/-90°}}{\Delta}$$

$$= \frac{1\,\underline{/-90°}}{0.125\,\underline{/-36.87°}} = 8\,\underline{/-53.13°}\ V = (4.8 - j6.4)\ V$$

Applying Cramer's rule to find v_2, we have

$$v_2 = \frac{\det\begin{bmatrix} Y_1 + Y_2 & i_1 \\ -Y_2 & -i_2 \end{bmatrix}}{\Delta} = \frac{(Y_1 + Y_2)(-i_2) + Y_2 i_1}{\Delta}$$

$$= \frac{(0.25 - j0.5)(-1\,\underline{/0°}) + (0.5\,\underline{/-90°})(5\,\underline{/0°})}{\Delta}$$

$$= \frac{-0.25 + j0.5 - j2.5}{\Delta} = \frac{-0.25 - j2}{\Delta}$$

$$= \frac{-(0.25 + j2)}{\Delta} = \frac{-2.016\,\underline{/82.87°}}{0.125\,\underline{/-36.87°}}$$

$$= -16.12\,\underline{/119.74°} = 16.12\,\underline{/-60.26°}\ V = (8 - j14)\ V$$

Finally, the current in the inductor is

$$i_L = Y_2(v_1 - v_2) = (0.5 \underline{/-90°})[(4.8 - j6.4) - (8 - j14)]$$
$$= -j0.5(-3.2 + j7.6) = (3.8 + j1.6) \text{ A}$$
$$= 4.12 \underline{/22.83°} \text{ A}$$

The computed value of i_L has the phase reference shown in Figure 16.16(b): left to right through the inductor.

Drill Exercise 16.10

Find the current in every component in Figure 16.16(a) and verify Kirchhoff's current law at each node.

ANSWER: $i_R = 2 \underline{/-53.13°}$ A; $i_C = 3.22 \underline{/60.26°}$ A.   □

In Chapter 6 we showed how simultaneous equations can be expressed in matrix form. The same methods can be used to obtain a matrix formulation of the equations resulting from an ac mesh or nodal analysis. The next example illustrates the use of matrices in ac nodal analysis.

Example 16.11 (Analysis)

(a) Write the matrix form of the equations necessary to analyze the circuit in Figure 16.17(a) using nodal analysis. Define each matrix used.
(b) Write the matrix equation for the solution to the nodal equations.

SOLUTION

(a) We must first convert the voltage source to an equivalent current source:

$$i = \frac{60 \underline{/0°} \text{ V}}{10^3 \underline{/0°} \text{ } \Omega} = 60 \underline{/0°} \text{ mA}$$

Figure 16.17(b) shows the current source, identified as i_A and drawn in place of the voltage source.

The admittance of each inductor is

$$Y = B_L = \frac{1}{\omega L} \underline{/-90°} = \frac{1}{(2 \times 10^4)(40 \times 10^{-3})} \underline{/-90°}$$

$$= 1.25 \underline{/-90°} \text{ mS} = -j1.25 \times 10^{-3} \text{ S}$$

Figure 16.17(b) shows the circuit with the admittance values identified and with node voltages v_1, v_2, and v_3 assigned. The current arrows labeled i_1 through i_6 show the currents through each admittance, based on the assumption $v_1 > v_2 > v_3 > 0$. Writing Kirchhoff's current law at each node, we obtain

$$\text{node 1:} \quad i_A = i_1 + i_2 + i_3 + i_B \tag{16.23}$$

$$\text{node 2:} \quad i_B + i_3 = i_4 + i_5 \tag{16.24}$$

$$\text{node 3:} \quad i_5 + i_C = i_6 \tag{16.25}$$

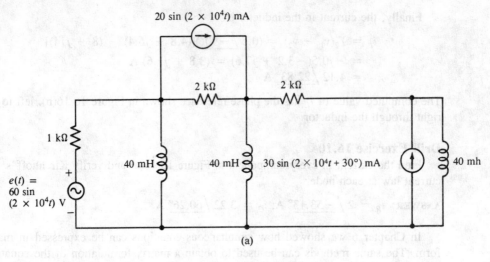

(a)

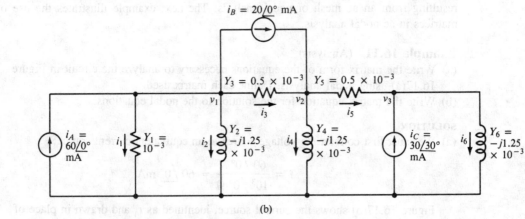

(b)

FIGURE 16.17 (Example 16.11)

In terms of admittances Y_1 through Y_6, the currents i_1 through i_6 are

$$i_1 = Y_1 v_1$$

$$i_2 = Y_2 v_1$$

$$i_3 = Y_3(v_1 - v_2)$$

$$i_4 = Y_4 v_2$$

$$i_5 = Y_5(v_2 - v_3)$$

$$i_6 = Y_6 v_3$$

Substituting these equations in (16.23) through (16.25), we obtain

$$i_A = Y_1 v_1 + Y_2 v_1 + Y_3(v_1 - v_2) + i_B$$

$$i_B + Y_3(v_1 - v_2) = Y_4 v_2 + Y_5(v_2 - v_3)$$

$$Y_5(v_2 - v_3) + i_C = Y_6 v_3$$

Rearranging the equations and collecting terms gives

$$(Y_1 + Y_2 + Y_3) v_1 + (-Y_3)v_2 = i_A - i_B$$

$$(-Y_3)v_1 + (Y_3 + Y_4 + Y_5)v_2 + (-Y_5)v_3 = i_B$$

$$(-Y_5)v_2 + (Y_5 + Y_6)v_3 = i_C$$

In matrix form, the equations can be expressed as

$$YV = I$$

where

$$Y = \begin{bmatrix} Y_1 + Y_2 + Y_3 & -Y_3 & 0 \\ -Y_3 & Y_3 + Y_4 + Y_5 & -Y_5 \\ 0 & -Y_5 & Y_5 + Y_6 \end{bmatrix}$$

$$= \begin{bmatrix} (1.5 - j1.25) \times 10^3 & -0.5 \times 10^{-3} & 0 \\ -0.5 \times 10^{-3} & (1 - j1.25 \times 10^{-3}) & -0.5 \times 10^{-3} \\ 0 & -0.5 \times 10^{-3} & (0.5 - j1.25) \times 10^{-3} \end{bmatrix}$$

$$V = \begin{bmatrix} v_1 \\ v_2 \\ v_3 \end{bmatrix}$$

$$I = \begin{bmatrix} i_A - i_B \\ i_B \\ i_C \end{bmatrix} = \begin{bmatrix} 40 \times 10^{-3} \underline{/0°} \\ 20 \times 10^{-3} \underline{/0°} \\ 30 \times 10^{-3} \underline{/30°} \end{bmatrix}$$

(b) The matrix form of the solution is

$$V = Y^{-1}I$$

Drill Exercise 16.11

A certain circuit has six nodes (including the reference node). What are the dimensions of the V, Y, and I matrices that result from a nodal analysis?

ANSWER: (5×1), (5×5), (5×1).   □

Example 16.12   (Design)

Find the magnitude and phase angle of the current i_1 that must be supplied by the current source in Figure 16.18(a) in order that the voltage drop across C_1 be $2 \underline{/0°}$ V.

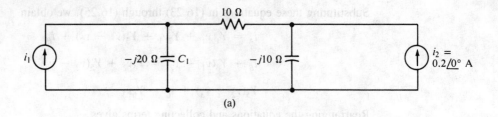

(a)

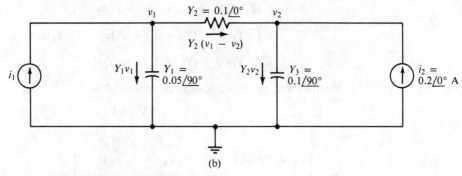

(b)

FIGURE 16.18 (Example 16.12)

SOLUTION The circuit is redrawn in Figure 16.18(b) with node voltages, admittances, and current directions indicated. Writing Kirchhoff's current law at the two nodes gives

$$i_1 = Y_1v_1 + Y_2(v_1 - v_2)$$

$$i_2 + Y_2(v_1 - v_2) = Y_3v_2$$

Rearranging the equations and collecting terms, we find

$$(Y_1 + Y_2)v_1 + (-Y_2)v_2 = i_1$$

$$(-Y_2)v_1 + (Y_2 + Y_3)v_2 = i_2$$

The delta-determinant is

$$\Delta = \det \begin{bmatrix} Y_1 + Y_2 & -Y_2 \\ -Y_2 & Y_2 + Y_3 \end{bmatrix} = (Y_1 + Y_2)(Y_2 + Y_3) - Y_2^2$$

$$= Y_1Y_2 + Y_1Y_3 + Y_2Y_3$$

$$= (0.05 \underline{/90°})(0.1 \underline{/0°}) + (0.05 \underline{/90°})(0.1 \underline{/90°}) + (0.1 \underline{/0°})(0.1 \underline{/90°})$$

$$= 0.005 \underline{/90°} + 0.005 \underline{/180°} + 0.01 \underline{/90°}$$

$$= -0.005 + j0.015 = 0.0158 \underline{/108.43°}$$

The voltage across C_1 is v_1, so, by Cramer's rule,

$$
v_1 = \frac{\det \begin{bmatrix} i_1 & -Y_2 \\ i_2 & Y_2 + Y_3 \end{bmatrix}}{\Delta}
$$

$$
= \frac{i_1(Y_2 + Y_3) + i_2 Y_2}{\Delta}
$$

$$
= \frac{i_1(0.1 + j0.1) + (0.2 \underline{/0°})(0.1 \underline{/0°})}{\Delta}
$$

$$
= \frac{i_1(0.1414) \underline{/45°} + 0.02 \underline{/0°}}{0.0158 \underline{/108.43°}}
$$

$$
= i_1 (8.95 \underline{/-63.43°}) + 1.26 \underline{/-108.43°}
$$

$$
= i_1(8.95 \underline{/-63.43°}) - 0.40 - j1.2
$$

Since v_1 must equal $2 \underline{/0°}$ V, we set the latter expression equal to $2 \underline{/0°}$ and solve for i_1:

$$
i_1(8.95 \underline{/-63.43°}) - 0.4 - j1.2 = 2 + j0
$$

$$
i_1(8.95 \underline{/-63.43°}) = 2.4 + j1.2 = 2.68 \underline{/26.58°}
$$

$$
i_1 = \frac{2.68 \underline{/26.58°}}{8.95 \underline{/-63.43°}} = 0.3 \underline{/90°} \text{ A}
$$

Drill Exercise 16.12
Find the value of i_1 that is required in Example 16.12 to obtain a voltage drop of $1 \underline{/0°}$ V across C_1.

ANSWER: $0.206 \underline{/104.03°}$ A. □

16.7 SPICE Examples

Example 16.13 (SPICE)
The Hay bridge in Figure 16.7 has $R_1 = 10$ kΩ, $C_1 = 0.01$ μF, $R_3 = 5$ kΩ, and $R_L = 10$ Ω. The voltage source is $e = 12 \underline{/0°}$ V, with frequency 159.154 Hz. Use SPICE to determine if the bridge is balanced.

SOLUTION The SPICE circuit and input data file are shown in Figure 16.19. The bridge is balanced if the voltage between nodes 1 and 4 is zero. Execution of the program gives $V(1,4) = 6.911 \times 10^{-7}$ V, which is so negligibly small in comparison to other circuit voltages that we consider it to be zero and conclude that the bridge is balanced.

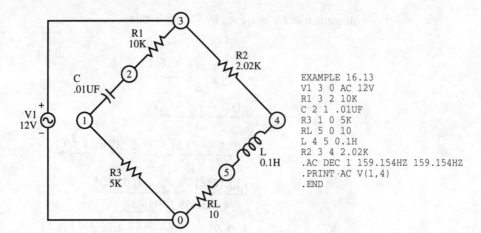

```
EXAMPLE 16.13
V1 3 0 AC 12V
R1 3 2 10K
C 2 1 .01UF
R3 1 0 5K
RL 5 0 10
L 4 5 0.1H
R2 3 4 2.02K
.AC DEC 1 159.154HZ 159.154HZ
.PRINT·AC V(1,4)
.END
```

FIGURE **16.19** (Example 16.13)

Exercises

Section 16.1 Wye–Delta Transformations

16.1 Find the wye network that is equivalent to the delta network shown in Figure 16.20. Draw the schematic diagram of the equivalent wye, label impedance values (in polar form), and show the terminals X, Y, and Z corresponding to the terminals in the delta.

16.2 Find the delta network that is equivalent to the wye network shown in Figure 16.21. Draw the schematic diagram of the equivalent delta, label

impedance values (in polar form), and show the terminals X, Y, and Z corresponding to the terminals in the wye.

16.3 Find the polar forms of the total impedance and the total current, i_T, in the circuit shown in Figure 16.22.

16.4 Find the polar form of the current i_1 in Figure 16.23.

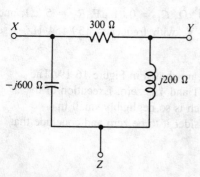

FIGURE **16.20** (Exercise 16.1)

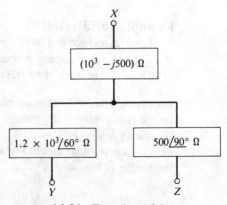

FIGURE **16.21** (Exercise 16.2)

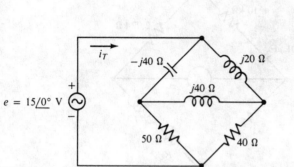

FIGURE 16.22 (Exercise 16.3)

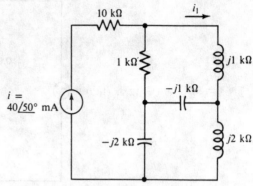

FIGURE 16.23 (Exercise 16.4)

Section 16.2 AC Bridges

16.5 Determine if the bridge shown in Figure 16.24 is balanced. If it is balanced, find the total impedance, Z_T, in polar form.

16.6 Determine if the bridge shown in Figure 16.25 is balanced. If it is balanced, find the total current, i_T, in polar forms.

16.7 The bridge in Figure 16.26 is balanced. Find L and R_L.

16.8 Derive equations (16.19) and (16.20) for the resistance and inductance of an unknown inductor in a balanced Maxwell bridge.

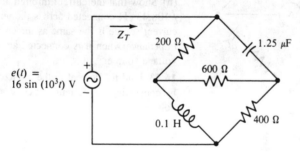

FIGURE 16.24 (Exercise 16.5)

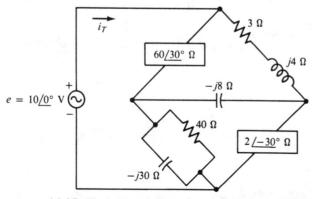

FIGURE 16.25 (Exercise 16.6)

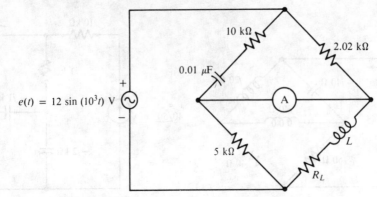

FIGURE 16.26 (Exercise 16.7)

16.9 The bridge in Figure 16.27 is balanced. Find C and R_C.

16.10 Derive equations (16.21) and (16.22) for the resistance and capacitance of an unknown capacitor in a balanced capacitor-comparison bridge.

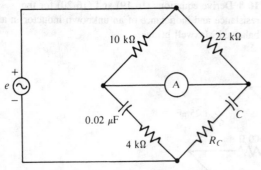

FIGURE 16.27 (Exercise 16.9)

Section 16.3 Voltage and Current Source Conversions

16.11 Convert the voltage source in Figure 16.28(a) to an equivalent current source, and the current source in Figure 16.28(b) to an equivalent voltage source. In each case, draw the schematic diagram of the equivalent source and label its impedance in polar form.

16.12 (a) Find the voltage source that is equivalent to the current source in Figure 16.29.

(b) Show that the current through an impedance of $500 \, \underline{/108.43°} \, \Omega$ connected across the terminals of the current source is the same as the current through that impedance when it is connected across the equivalent voltage source.

16.13 Find the current in the inductor in Figure 16.30 by converting the current source to an equivalent voltage source.

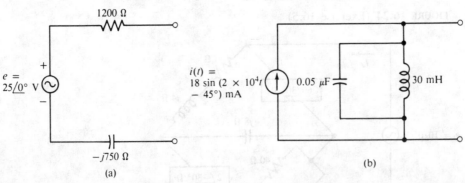

FIGURE 16.28 (Exercise 16.11)

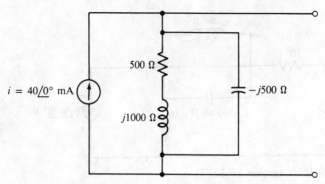

FIGURE 16.29 (Exercise 16.12)

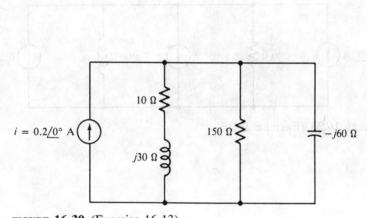

FIGURE 16.30 (Exercise 16.13)

16.14 Find the voltage across the resistor in Figure 16.31 by converting the current source to an equivalent voltage source.

Section 16.4 Series and Parallel AC Sources

16.15 Find the voltage *v* in the circuit shown in Figure 16.32. Draw + and − signs to shown the phase reference of the calculated voltage.

16.16 Find the polar form of the current through the resistor in Figure 16.33. Draw an arrow beside the resistor to show the phase reference of the calculated current.

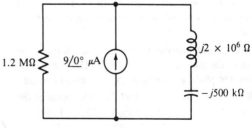

FIGURE 16.31 (Exercise 16.14)

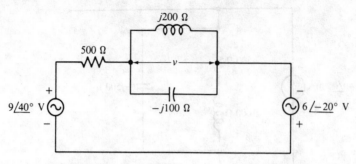

FIGURE 16.32 (Exercise 16.15)

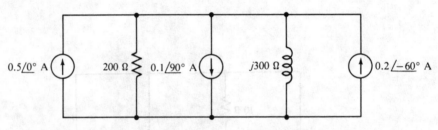

FIGURE 16.33 (Exercise 16.16)

Section 16.5   AC Mesh Analysis

16.17 Use mesh analysis to find the polar form of the current in the inductor in Figure 16.34. Draw an arrow to show the phase reference of the calculated current.

16.18 Use mesh analysis to find the polar form of the voltage across the resistor in Figure 16.34. Draw + and − signs to show the phase reference of the calculated voltage.

16.19 Use mesh analysis to find the polar form of the current in the resistor in Figure 16.35. Draw an arrow to show the phase reference of the calculated current.

16.20 Use mesh analysis to find the polar form of the voltage across the inductor in Figure 16.36. Draw + and − signs to show the phase reference of the calculated voltage.

16.21 Use the mesh analysis to find the polar form of the voltage across the 40-Ω resistor in Figure 16.36.

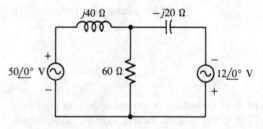

FIGURE 16.34 (Exercise 16.17)

Draw + and − signs to show the phase reference of the calculated voltage.

16.22 Write the loop equations necessary to analyze the circuit shown in Figure 16.37 using mesh analysis. Collect terms and rearrange the equations in the format

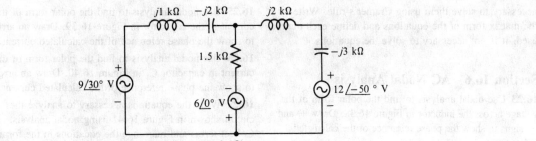

FIGURE **16.35** (Exercise 16.19)

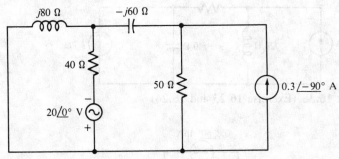

FIGURE **16.36** (Exercise 16.20 and 16.21)

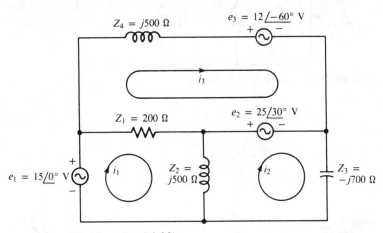

FIGURE **16.37** (Exercise 16.22)

necessary to solve them using Cramer's rule. Write the matrix form of the equations and define each matrix used. It is not necessary to solve the equations.

Section 16.6   AC Nodal Analysis

16.23 Use nodal analysis to find the polar form of the voltage across the inductor in Figure 16.38. Draw + and − signs to show the phase reference of the calculated voltage.

16.24 Use nodal analysis to find the polar form of the current in the resistor in Figure 16.38. Draw an arrow to show the phase reference of the calculated current.

16.25 Use nodal analysis to find the polar form of the current in the capacitor in Figure 16.39. Draw an arrow to show the phase reference of the calculated current.

16.26 Use nodal analysis to find the polar form of the current in capacitor C_1 in Figure 16.40. Draw an arrow to show the phase reference of the calculated current.

16.27 Write the equations necessary to analyze the circuit shown in Figure 16.41 using nodal analysis. Collect terms and rearrange the equations in the format necessary to solve them using Cramer's rule. Write the matrix form of the equations and define each matrix used. Write the matrix form of the solutions to the equations. It is not necessary to solve the equations.

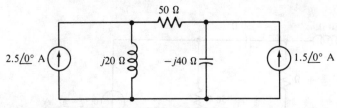

FIGURE 16.38 (Exercise 16.23 and 16.24)

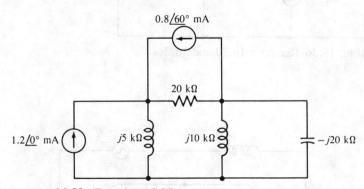

FIGURE 16.39 (Exercise 16.25)

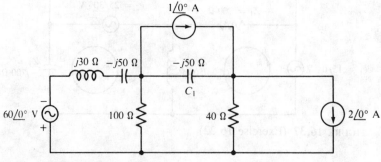

FIGURE 16.40 (Exercise 16.26)

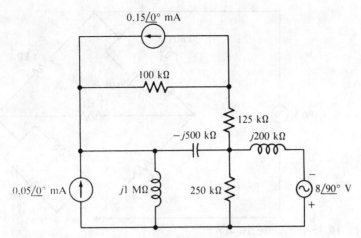

FIGURE 16.41 (Exercise 16.27)

Design Exercises

16.1D What equal-valued impedances Z will make the impedance across any pair of terminals in Figure 16.42 equal to $(12 + j8)\ \Omega$ when the third terminal is open?

16.2D What equal-valued impedances Z will make the impedance across any pair of terminals in Figure 16.43 equal to $45 \underline{/30°}\ \Omega$ when the third terminal is open?

16.3D What value of capacitance C should be used to balance the bridge shown in Figure 16.44?

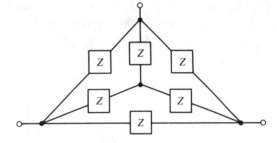

FIGURE 16.43 (Exercise 16.2D)

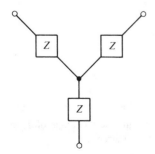

FIGURE 16.42 (Exercise 16.1D)

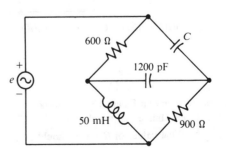

FIGURE 16.44 (Exercise 16.3D)

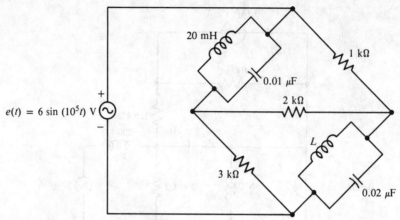

FIGURE 16.45 (Exercise 16.4D)

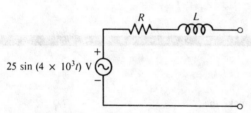

FIGURE 16.46 (Exercise 16.5D)

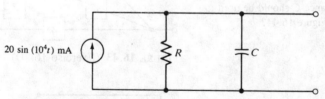

FIGURE 16.47 (Exercise 16.6D)

16.4D What value of inductance L should be used to balance the bridge in Figure 16.45?

16.5D The voltage source in Figure 16.46 is to be equivalent to a current source that supplies

$0.1 \underline{/-53.13°}$ A. What values of R and L should be used?

16.6D The current source in Figure 16.47 is to be equivalent to a voltage source that supplies

$50 \underline{/-45°}$ V. What values of R and C should be used?

16.7D What should be the magnitude and angle of e in Figure 16.48 in order that the current i

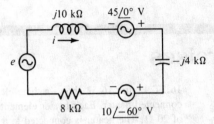

FIGURE 16.48 (Exercise 16.7D)

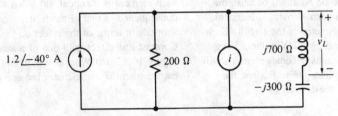

FIGURE 16.49 (Exercise 16.8D)

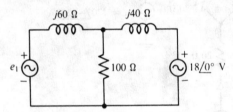

FIGURE 16.50 (Exercise 16.9D)

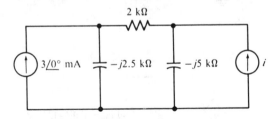

FIGURE 16.51 (Exercise 16.10D)

be $0.01 \underline{/0°}$ A? Draw $+$ and $-$ signs to show the phase reference for the calculated value of e.

16.8D What should be the magnitude and angle of i in Figure 16.49 in order that the voltage v_L across the inductor be $219.1 \underline{/-13.34°}$ V? Draw an arrow to show the phase reference for the calculated value of i.

16.9D What should be the magnitude and angle of e_1 in Figure 16.50 in order that the voltage across the 100-Ω resistor be 0 V?

16.10D What should be the magnitude and angle of i in Figure 16.51 in order that the current through the resistor be 0 A?

Troubleshooting Exercises

16.1T Figure 16.52 shows a three-element, wye-connected heater. Each heater element has a resistance of 20 Ω. The heater is connected to a *three-phase* power source, represented by e_1, e_2, and e_3. (Note that the phase angle between each voltage and any other voltage is 120°.) The load is said to be *balanced* because the magnitude of the current in each element is identical, as can be seen from the symmetry of the circuit. Each element is capable of dissipating 1 kW without being damaged. If one of the elements becomes shorted, will the other elements fail? Explain. (*Hint*: Redraw the circuit and analyze it using mesh analysis.)

16.2T Figure 16.53 shows a three-element, delta-connected heater. Each heater element has a resistance of 20 Ω. The heater is connected to a three-phase power source, represented by e_1, e_2, and e_3. As can be seen from the symmetry of the circuit, the current in each element is identical. (a) What would be the consequence of one of the heater elements becoming shorted, in terms of the effect on the power sources? Contrast this effect with that of a shorted element in Figure 16.52. (b) What power would be dissipated in each element if one element became open-circuited?

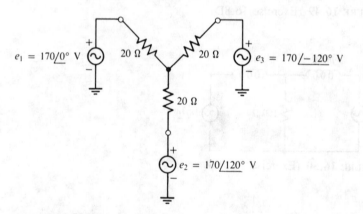

FIGURE **16.52** (Exercise 16.1T)

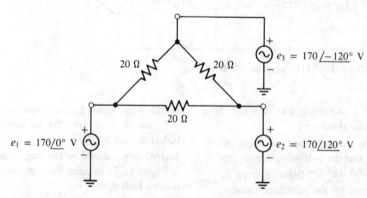

FIGURE **16.53** (Exercise 16.2T)

SPICE Exercises

16.1S In the capacitor comparison bridge in Figure 16.27 (p. 624), $C = 0.09$ µF and $R_C = 8.8$ kΩ. The voltage source is $e = 25 \underline{/0°}$ V with frequency 1 kHz. Use SPICE to determine if the bridge is balanced.

16.2S Use SPICE to determine if the bridge in Figure 16.24 (p. 623) is balanced. Also determine the total impedance Z_T of the bridge.

17 AC Network Theorems

17.1 Superposition

Recall that the superposition principle allows us to find the voltage or current in a linear circuit by summing the independent contributions of every source in the circuit. Linear circuits are those that contain linear components only. We know that a resistor is a linear component, because the current through it is directly proportional to the voltage across it. Thus, the superposition principle can be used to analyze dc circuits containing resistors, as was demonstrated in Chapter 7.

Inductors and capacitors are also linear components. When the frequency of the current through an inductor is fixed, the inductive reactance is a fixed constant, and the ac current through the inductor is directly proportional to the ac voltage across it:

$$i_L = \frac{v_L}{X_L} \tag{17.1}$$

Similarly, at a fixed frequency, capacitive reactance is constant, and the ac current through a capacitor is directly proportional to the voltage across it:

$$i_C = \frac{v_C}{X_C} \tag{17.2}$$

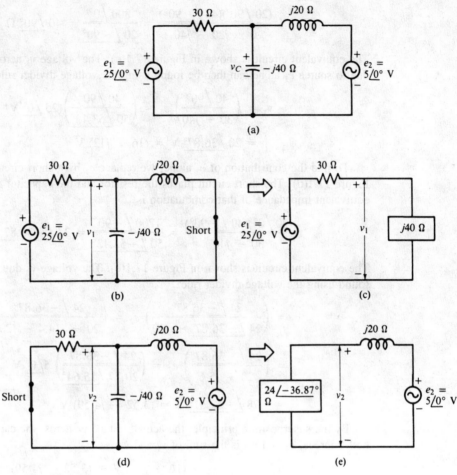

FIGURE 17.1 (Example 17.1)

AC circuits containing resistors, capacitors, and inductors can therefore be analyzed using the superposition principle.

To determine the individual contribution of each source to the voltage or current in a circuit, the contributions of all other sources must be made equal to zero. Recall that a voltage source is set equal to zero by replacing it with a short circuit, and a current source is set equal to zero by replacing it with an open circuit.

Example 17.1   (Analysis)

Use the superposition principle to find the polar form of the voltage v_C in Figure 17.1(a).

SOLUTION To find the contribution of e_1 to the voltage across the capacitor, we replace e_2 by a short circuit, as shown in Figure 17.1(b). The short circuit places the inductor in parallel with the capacitor, so the equivalent impedance of the parallel combination is

$$\frac{(20\,\underline{/90°})(40\,\underline{/-90°})}{0 + j20 - j40} = \frac{800\,\underline{/0°}}{20\,\underline{/-90°}} = 40\,\underline{/90°}\ \Omega$$

The equivalent circuit is shown in Figure 17.1(c). The voltage v_1 across the capacitor due to source e_1 alone can then be found using the voltage-divider rule:

$$v_1 = \left(\frac{40\,\underline{/90°}}{30 + j40}\right)e_1 = \left(\frac{40\,\underline{/90°}}{50\,\underline{/53.13°}}\right)(25\,\underline{/0°}\ \text{V})$$

$$= 20\,\underline{/36.87°}\ \text{V} = (16 + j12)\ \text{V}$$

To find the contribution of e_2 alone, we replace e_1 by a short circuit, as shown in Figure 17.1(d). The short circuit places the resistor and the capacitor in parallel, so the equivalent impedance of that combination is

$$\frac{(30\,\underline{/0°})(40\,\underline{/-90°})}{30 - j40} = \frac{1200\,\underline{/-90°}}{50\,\underline{/-53.13°}} = 24\,\underline{/-36.87°}\ \Omega$$

The equivalent circuit is shown in Figure 17.1(e). The voltage v_2 due to e_2 alone is found using the voltage-divider rule:

$$v_2 = \left(\frac{24\,\underline{/-36.87°}}{24\,\underline{/-36.87°} + j20}\right)e_2 = \left(\frac{24\,\underline{/-36.87°}}{20 - j14.4 + j20}\right)e_2$$

$$= \left(\frac{24\,\underline{/-36.87°}}{20 + j5.6}\right)e_2 = \left(\frac{24\,\underline{/-36.87°}}{20.77\,\underline{/15.64°}}\right)5\,\underline{/0°}\ \text{V}$$

$$= 5.78\,\underline{/-52.51°}\ \text{V} = (3.52 - j4.59)\ \text{V}$$

By the superposition principle, the actual voltage v_C across the capacitor due to both sources, e_1 and e_2, is the sum of v_1 and v_2:

$$v_C = v_1 + v_2 = (16 + j12)\ \text{V} + (3.52 - j4.59)\ \text{V}$$
$$= (19.52 + j7.41)\ \text{V} = 20.88\,\underline{/20.79°}\ \text{V}$$

Since v_1 and v_2 both have the same polarity (positive on the top side of the capacitor), their sum v_C has that polarity, as shown in Figure 17.1(a).

Drill Exercise 17.1

Use the superposition principle to find the polar form of the current in the inductor in Figure 17.1(a).

ANSWER: $i_L = 0.342\,\underline{/-78.53°}\ \overrightarrow{\text{A}}$ (directed left to right).   □

When adding the voltages or currents due to individual sources, a reference polarity must be assumed for the sum. In Example 17.1, the assumed polarity for the voltage across the capacitor was positive on the top side of the capacitor. In some cases, the polarities of the voltages or currents to be summed are the opposite of each other. In those cases, the quantities must be subtracted rather than added. One of the quantities is assumed to be positive (i.e., to have the reference polarity) and the other quantity is subtracted from it. The next example illustrates this idea.

Example 17.2 (Analysis)

Use the superposition principle to find the polar form of the current through the 120-Ω resistor in Figure 17.2(a).

SOLUTION To find the contribution of the voltage source, we open-circuit the current source, as shown in Figure 17.2(b). Since no current can flow in the capacitor, the total impedance across the voltage source is

$$Z_T = 120 + j50 \ \Omega = 130 \ \underline{/22.62°} \ \Omega$$

Therefore, the current i_1 in the 120-Ω resistor due to the voltage source alone is

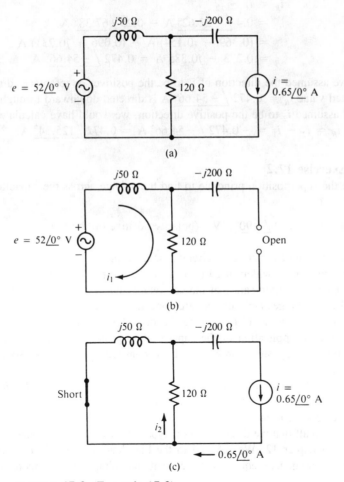

(a)

(b)

(c)

FIGURE 17.2 (Example 17.2)

$$i_1 = \frac{e}{Z_T} = \frac{52 \,\underline{/0^\circ}\; \text{V}}{130 \,\underline{/22.62^\circ}\; \Omega} = 0.4 \,\underline{/-22.62^\circ}\; \text{A}$$

Note that i_1 is directed downward through the resistor.

To find the contribution of the current source, we short-circuit the voltage source, as shown in Figure 17.2(c). Applying the current-divider rule, we find the current i_2 in the 120-Ω resistor due to the current source alone:

$$i_2 = \left(\frac{50 \,\underline{/90^\circ}}{120 + j50} \right) i = \left(\frac{50 \,\underline{/90^\circ}}{130 \,\underline{/22.62^\circ}} \right) (0.65 \,\underline{/0^\circ}\; \text{A})$$

$$= 0.25 \,\underline{/67.38^\circ}\; \text{A}$$

Note that i_2 is directed upward through the resistor. By the superposition principle, the total current in the resistor due to both sources is the sum of i_1 and i_2. However, since i_1 and i_2 have opposite directions, we must actually subtract:

$$\begin{aligned} i_R &= i_1 - i_2 \\ &= 0.4 \,\underline{/-22.62^\circ}\; \text{A} - 0.25 \,\underline{/67.38^\circ}\; \text{A} \\ &= (0.369 - j0.154)\; \text{A} - (0.096 + j0.231)\; \text{A} \\ &= 0.273 - j0.385\; \text{A} = 0.472 \,\underline{/-54.66^\circ}\; \text{A} \end{aligned}$$

Since we assumed the direction of i_1 to be the positive, or reference, direction, the calculated value $i_R = 0.472 \,\underline{/-54.66^\circ}$ A is directed downward through the resistor. If we had assumed i_2 to be the positive direction, we would have calculated the *upward* current $i_R = i_2 - i_1 = -0.472 \,\underline{/-54.66^\circ}$ A $= 0.472 \,\underline{/125.34^\circ}$ A.

Drill Exercise 17.2

Use the superposition principle to find the voltage across the capacitor in Figure 17.2(a).

ANSWER: $v_C = 130 \,\underline{/-90^\circ}\; {}^+\text{V}^-$ (polarity positive on the left).   □

In many practical electronic circuits, both dc and ac sources are present. The superposition principle is widely used to analyze these types of circuits. For example, in an ac *amplifier*, it is often said that the dc voltage sources are short circuits to ground, as far as the ac voltages and currents are concerned. The ac performance of the amplifier is then analyzed without considering the dc voltage sources. In reality, this analysis technique is an application of the superposition principle. The dc voltage sources are simply replaced by short circuits, and the contribution of the ac source(s) alone are determined.

The dc voltages in an ac amplifier circuit are designed to provide *bias* levels for the ac currents and voltages. A bias level is a dc value about which ac variations occur. The total voltage or current in such a circuit is the *sum* of the dc bias level and the ac variation. Recall that we discussed these types of waveforms in connection with average values, in Chapter 12. We referred to the bias level as the *offset*, or *dc level*, of the waveform. The general equations for current and voltage waveforms having dc levels I_{dc} and V_{dc} are

$$i(t) = I_{dc} + I_p \sin(\omega t + \phi) \tag{17.3}$$

$$v(t) = V_{dc} + V_p \sin(\omega t + \phi) \tag{17.4}$$

Figure 12.31 shows an example of a waveform whose dc level is 5 V and whose ac component has a peak value of 7 V (see Example 12.17).

The superposition principle can be used to find the total voltage or current in a circuit due to a combination of dc and ac sources. The next example illustrates such a case.

Example 17.3 (Analysis)

Figure 17.3(a) shows a simplified equivalent circuit of a certain ac amplifier having two dc sources.

(a) Use the superposition principle to find the mathematical expression for the total voltage $v_R(t)$ across the 10-kΩ resistor. Assume steady-state conditions. (The dc and ac sources have been connected to the circuit for a long period of time.)

(b) Sketch the waveform of v_R versus angle θ. Show the maximum and minimum values of the waveform on the sketch.

SOLUTION

(a) We first find the contribution v_1 of the −5-V dc source to the voltage v_R across the 10-kΩ resistor. Replacing the 15-V dc source and the ac source by short circuits, we obtain the equivalent circuit shown in Figure 17.3(b). Under steady-state conditions, the capacitor is fully charged to 5 V, and no current flows through the 10-kΩ resistor. Therefore, $v_1 = 0$ V.

To find the contribution of the 15-V dc source, we replace the −5-V dc source and the ac source by short circuits. The equivalent circuit is shown in Figure 17.3(c). Since no dc current flows through the capacitor at steady state, the dc

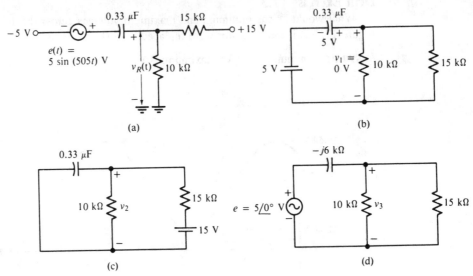

FIGURE 17.3 (Example 17.3)

voltage across the 10-kΩ resistor is found using the voltage-divider rule:

$$v_2 = \left(\frac{10 \text{ k}\Omega}{10 \text{ k}\Omega + 15 \text{ k}\Omega}\right) 15 \text{ V} = 6 \text{ V dc}$$

To find the contribution of the ac source, we replace both dc sources with short circuits. The reactance of the capacitor at the frequency of the ac source is

$$X_C = \frac{1}{\omega C} \underline{/-90°} = \frac{1}{(505)(0.33 \times 10^{-6})} \underline{/-90°}$$

$$= 6000 \underline{/-90°} \ \Omega = -j6 \text{ k}\Omega$$

The equivalent circuit is shown in Figure 17.3(d). The parallel combination of the 10-kΩ and 15-kΩ resistors is equivalent to

$$\frac{(10 \text{ k}\Omega)(15 \text{ k}\Omega)}{10 \text{ k}\Omega + 15 \text{ k}\Omega} = 6 \text{ k}\Omega$$

The contribution of the ac source is found by the voltage-divider rule:

$$v_3 = \left(\frac{6 \times 10^3 \underline{/0°}}{6 \times 10^3 - j6 \times 10^3}\right) e = \left(\frac{6 \times 10^3 \underline{/0°}}{8.485 \times 10^3 \underline{/-45°}}\right) 5 \underline{/0°} \text{ V}$$

$$= 3.54 \underline{/45°} \text{ V} = 3.54 \sin (505t + 45°) \text{ V}$$

By the superposition principle, the total voltage across the 10-kΩ resistor is

$$v_R(t) = v_1 + v_2 + v_3 = 6 + 3.54 \sin (505t + 45°) \text{ V}$$

(b) The maximum value of v is $6 + 3.54 = 9.54$ V and the minimum value is $6 - 3.54 = 2.46$ V. The waveform is sketched in Figure 17.4.

Drill Exercise 17.3

What would be the minimum and maximum voltages across the 10-kΩ resistor in Figure 17.3(a) if the capacitor were replaced by a short circuit?

ANSWER: minimum: -10 V; maximum: 0 V.

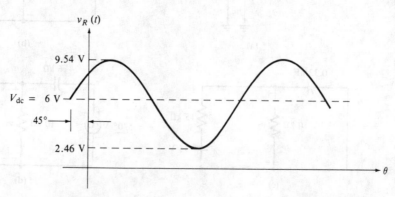

FIGURE 17.4 (Example 17.3)

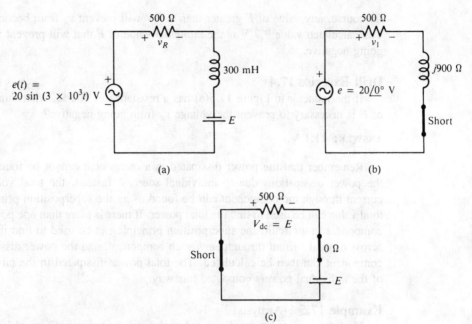

FIGURE 17.5 (Example 17.4)

Example 17.4 (Design)

What should be the value of the dc bias voltage E in Figure 17.5(a) if the voltage v_R across the resistor is not permitted to go negative?

SOLUTION The reactance of the inductor is

$$X_L = \omega L \underline{/90°} = (3 \times 10^3)(300 \times 10^{-3}) \underline{/90°} = 900 \underline{/90°} \ \Omega$$

Figure 17.5(b) shows the circuit when the dc source is replaced by a short circuit. By the voltage-divider rule, the ac voltage across the resistor is

$$v_1 = \left(\frac{500 \underline{/0°}}{500 + j900} \right) e = \left(\frac{500 \underline{/0°}}{1029.6 \underline{/60.95°}} \right) 20 \underline{/0°} \ V$$

$$= 9.7 \underline{/-60.95°} \ V = 9.7 \sin (3 \times 10^3 t - 60.95°) \ V$$

Figure 17.5(c) shows the circuit when the ac source is replaced by a short circuit. Note that the impedance of the inductor is 0 Ω at dc. Therefore, the dc voltage across the inductor is 0 V and the full value of E appears across the 500-Ω resistor.

The total voltage across the resistor is

$$v_R = V_{dc} + v_1 = E + 9.7 \sin (3 \times 10^3 t - 60.95°) \ V$$

The minimum value of v_R is $E - 9.7$ V. If v_R is not permitted to be negative, its minimum value must equal 0 V. Therefore,

$$E - 9.7 = 0 \ V$$

$$E = 9.7 \ V$$

Of course, any value of E greater than 9.7 V will prevent v_R from becoming negative. The calculated value 9.7 V is the *minimum* value of E that will prevent v_R from going negative.

Drill Exercise 17.4

If the inductor in Figure 17.5(a) has a resistance of 100 Ω, what minimum value of E is necessary to prevent the voltage v_R from going negative?

ANSWER: 11.1 V.   □

Remember that the power dissipated in a component cannot be found by summing the power dissipations due to individual sources. Instead, the total voltage across or current through the component can be found using the superposition principle, and that total value can be used to find the total power. If there is more than one power-dissipating component in a circuit, the superposition principle can be used to find the total voltage across or total current through *each* such component, and the power dissipation in *each* component can then be calculated. The total power dissipated in the circuit is the sum of the individual powers computed this way.

Example 17.5   (Analysis)

Find the average power dissipated in the circuit shown in Figure 17.6(a).

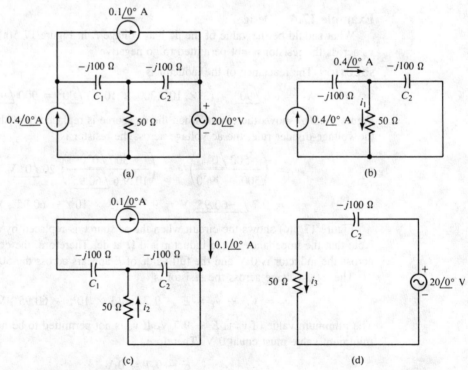

FIGURE 17.6 (Example 17.5)

SOLUTION Since the resistor is the only component in the circuit that dissipates power, we will use the superposition principle to find the total current in the resistor. The average power dissipated in the circuit can then be calculated from $P = I_p^2 R/2$.

Figure 17.6(b) shows the circuit with the 0.1-A source open-circuited and the 20-V source short-circuited. By the current-divider rule,

$$i_1 = \left(\frac{100 \,/\!\!-90°}{50 - j100} \right)(0.4 \,/\!\!\underline{0°} \text{ A}) = \left(\frac{100 \,/\!\!-90°}{111.8 \,/\!\!-63.43°} \right)(0.4 \,/\!\!\underline{0°} \text{ A})$$

$$= 0.358 \,/\!\!-26.57° \text{ A} = 0.32 - j0.16 \text{ A}$$

Figure 17.6(c) shows the circuit with the 0.4-A source open-circuited and the 20-V source short-circuited. By the current-divider rule,

$$i_2 = \left(\frac{100 \,/\!\!-90°}{50 - j100} \right)(0.1 \,/\!\!\underline{0°} \text{ A})$$

$$= \left(\frac{100 \,/\!\!-90°}{111.8 \,/\!\!-63.43°} \right)(0.1 \,/\!\!\underline{0°} \text{ A})$$

$$= 0.089 \,/\!\!-26.57° \text{ A} = 0.08 - j0.04 \text{ A}$$

Figure 17.6(d) shows the circuit with both current sources open-circuited. The total impedance across the voltage source is

$$Z_T = 50 - j100 \ \Omega = 111.8 \,/\!\!-63.43° \ \Omega$$

Thus,

$$i_3 = \frac{v}{Z_T} = \frac{20 \,/\!\!\underline{0°} \text{ V}}{111.8 \,/\!\!-63.43° \ \Omega}$$

$$= 0.179 \,/\!\!\underline{63.43°} \text{ A} = 0.08 + j0.16 \text{ A}$$

Notice that i_1 and i_3 have the same direction, downward through the resistor, and that i_2 is directed upward. Therefore, by the superposition principle, the total current in the resistor is

$$i_T = i_1 + i_3 - i_2$$
$$= (0.32 - j0.16) \text{ A} + (0.08 + j0.16) \text{ A} - (0.08 - j0.04) \text{ A}$$
$$= (0.32 + j0.04) \text{ A} = 0.322 \,/\!\!\underline{7.13°} \text{ A}$$

The average power is then

$$P_{\text{avg}} = \frac{I_p^2 R}{2} = \frac{(0.322)^2(50)}{2} = 2.59 \text{ W}$$

Drill Exercise 17.5

Use the superposition principle to find the polar form of the current in each capacitor in Figure 17.6(a).

ANSWER: $i_{C1} = 0.3 \,/\!\!\underline{0°} \ \overrightarrow{\text{A}}$ (left to right); $i_{C2} = 0.0447 \,/\!\!\underline{63.43°} \ \overleftarrow{\text{A}}$ (right to left).

□

17.2 Controlled Voltage and Current Sources

A *controlled* voltage or current source, also called a *dependent* source, is one whose value depends on the voltage or current at some other point in a circuit. All of the sources we have studied up to now have been *independent* sources. Figure 17.7 shows examples of the four types of dependent sources: the voltage-controlled voltage source, the voltage-controlled current source, the current-controlled voltage source, and the current-controlled current source. The sources in Figure 17.7(a) and (b) are controlled by voltages (designated v_1) that are *external* to the circuit containing the sources themselves. In Figure 17.7(a), for example, if the external voltage v_1 is 20 mV, then $v_o = 60(20\ \text{mV}) = 1.2\ \text{V}$. If v_1 changes to 30 mV, then v_o changes to $60(30\ \text{mV}) = 1.8\ \text{V}$. Controlled voltage sources are also found in circuits where the controlling voltage v_1 is in the same circuit as the controlled source.

Figure 17.7(c) and (d) show examples of current-controlled sources. In these examples, the controlling current, i_1, is in the same circuit as the controlled sources themselves. These examples also show that the constant that multiplies the value of voltage or current produced by a controlled source is sometimes designated by a letter, such as k or β. In Figure 17.7(d), for example, if $i_1 = 50\ \mu\text{A}$ and if the constant β equals 100, then the current produced by the controlled current source is $i = (100)(50\ \mu\text{A}) = 5\ \text{mA}$. If i_1 changes to 20 μA, then i changes to $i = (100)(20\ \mu\text{A}) = 2\ \text{mA}$.

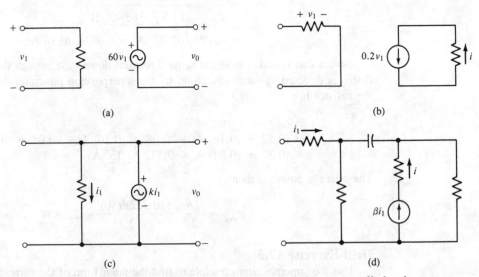

(a)

(b)

(c)

(d)

FIGURE 17.7 Examples of controlled sources. (a) A voltage-controlled voltage source. $v_0 = 60v_1$. (b) A voltage-controlled current source. $i = 0.2v_1$. (c) A current-controlled voltage source. $v_0 = ki_1$. (d) A current-controlled current source. $i = \beta i_1$.

Circuits containing controlled sources can be analyzed using the superposition principle in the same way it was used in Section 17.1, provided that the *controlling* voltages or currents are *external* to (unaffected by) the circuit containing the controlled sources. In such cases, both dependent and independent voltage sources are set equal to zero by replacing them with short circuits, and both dependent and independent current sources are set equal to zero by replacing them with open circuits. As we shall see in a later discussion, a dependent source cannot be set to zero if the value of the voltage or current that controls it is changed when a source is set equal to zero.

Example 17.6   (Analysis)

Use the superposition principle to find the current in the 500-Ω resistor in Figure 17.8(a).

SOLUTION The current produced by the current-controlled current source is $4i = 4(20\ \underline{/0°}\ \text{mA}) = 80\ \underline{/0°}\ \text{mA}$. Notice that the controlling current i is external to the circuit containing the controlled source, so the value of i will not change when any of

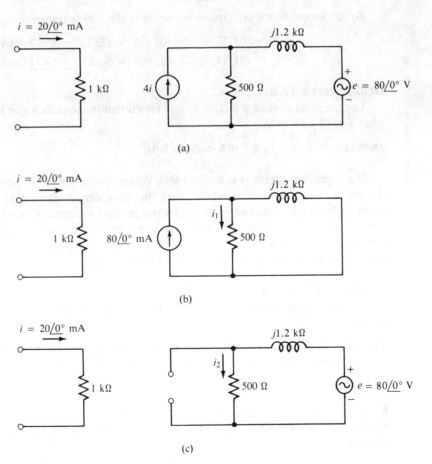

(a)

(b)

(c)

FIGURE 17.8 (Example 17.6)

the sources are set to zero. Since the circuit meets this test, it can be analyzed using the superposition principle in the same manner as before (i.e., by finding the contribution of each source when all other sources are set equal to zero).

Figure 17.8(b) shows the circuit when the voltage source is replaced by a short circuit. By the current-divider rule,

$$i_1 = \left(\frac{1200\,\underline{/90°}}{500 + j1200}\right)(80\,\underline{/0°}\text{ mA}) = \left(\frac{1200\,\underline{/90°}}{1300\,\underline{/67.38°}}\right)(80\,\underline{/0°}\text{ mA})$$

$$= 73.85\,\underline{/22.62°}\text{ mA} = (68.17 + j28.40)\text{ mA}$$

Figure 17.8(c) shows the circuit when the controlled current source is replaced by an open circuit. The contribution of the voltage source is then

$$i_2 = \frac{80\,\underline{/0°}\text{ V}}{(500 + j1200)\,\Omega} = \frac{80\,\underline{/0°}\text{ V}}{1300\,\underline{/67.38°}\,\Omega}$$

$$= 61.54\,\underline{/-67.38°}\text{ mA} = (23.67 - j56.8)\text{ mA}$$

By the superposition principle, the current in the 500-Ω resistor is

$$i_1 + i_2 = (68.17 + j28.40)\text{ mA} + (23.67 - j56.8)\text{ mA}$$
$$= (91.84 - j28.40)\text{ mA} = 96.13\,\underline{/-17.18°}\text{ mA}$$

Drill Exercise 17.6

Use the superposition principle to find the current in the inductor in Figure 17.8(a) when i is $12\,\underline{/30°}$ mA.

ANSWER: $46.47\,\underline{/-78.83°}\,\overleftarrow{\text{mA}}$ (right to left). ☐

If a dependent source is controlled by a voltage or current whose value depends on the circuit containing the dependent source, the dependent source cannot be set to zero when applying the superposition principle. Independent sources can be set equal to zero, one at a time in the usual way, but each dependent source must be included in every computation. As a consequence, it is usually necessary to analyze such a circuit by using some combination of superposition and Kirchhoff's voltage or current laws.

Example 17.7 (Analysis)

Use the superposition principle to find the polar form of current i_a in Figure 17.9(a).

SOLUTION The current-controlled current source produces a current whose value $(0.3i_a)$ depends on the current i_a in the circuit. Therefore, we cannot replace the controlled current source with an open circuit when applying the superposition principle.

Figure 17.9(b) shows the circuit when e_1 is replaced by a short circuit. We must find the current i_{a1} due to the combined contributions of e_2 and the controlled current source. Applying Kirchhoff's current law at node A, we see that the current i_L in the inductor is

$$i_L = i_{a1} - 0.3i_{a1} = 0.7i_{a1}$$

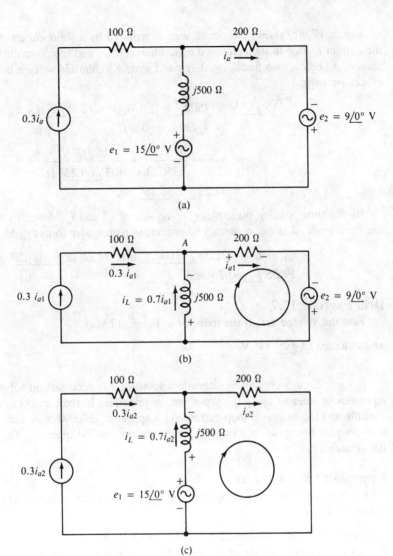

FIGURE 17.9 (Example 17.7)

Writing Kirchhoff's voltage law around the clockwise loop shown in Figure 17.9(b), we have

$$9\underline{/0°}\text{ V} = (500\underline{/90°})(0.7i_{a1}) + (200\underline{/0°})i_{a1}$$
$$= i_{a1}(200 + j350)$$

Solving for i_{a1} gives

$$i_{a1} = \frac{9\underline{/0°}\text{ V}}{(200 + j350)\text{ }\Omega} = \frac{9\underline{/0°}\text{ V}}{403\underline{/60.26°}\text{ }\Omega} = 22.33\underline{/-60.26°}\text{ mA}$$

Figure 17.9(c) shows the circuit with e_2 replaced by a short circuit. We must find the current i_{a2} due to the combined contributions of e_1 and the controlled current source. As before, we find $i_L = 0.7i_{a2}$ and write Kirchhoff's voltage law around the clockwise loop:

$$15\ \underline{/0°}\ \text{V} = (500\ \underline{/90°})(0.7i_{a2}) + (200\ \underline{/0°})i_{a2}$$
$$= i_{a2}(200 + j350)$$

$$i_{a2} = \frac{15\ \underline{/0°}\ \text{V}}{(200 + j350)\Omega} = \frac{15\ \underline{/0°}\ \text{V}}{403\ \underline{/60.26°}\ \Omega}$$
$$= 37.22\ \underline{/-60.26°}\ \text{mA}$$

By the superposition principle, i_a is the sum of i_{a1} and i_{a2}. Since i_{a1} and i_{a2} have the same angle, it is not necessary to convert to rectangular form to add them:

$$i_a = i_{a1} + i_{a2} = 22.33\ \underline{/-60.26°}\ \text{mA} + 37.22\ \underline{/-60.26°}\ \text{mA}$$
$$= 59.55\ \underline{/-60.26°}\ \text{mA}$$

Drill Exercise 17.7

Find the voltage across the inductor in Figure 17.9(a).

ANSWER: $20.84\ \underline{/29.74°}\ \text{V}$. ☐

In some circuits containing dependent sources, it is necessary to solve simultaneous equations in order to apply the superposition principle. In these circuits, the controlling variable and the values of loop current(s) all appear as unknowns in a loop equation, so it is not possible to solve a single equation for the desired quantity. The next example illustrates such a case.

Example 17.8   (Analysis)

Use the superposition principle to find the voltage v across the 20-Ω resistor in Figure 17.10(a).

SOLUTION   The voltage-controlled voltage source produces a voltage whose value $(0.5v_a)$ depends on the voltage v_a across the 50-Ω resistor. Since the resistor is in the same circuit as the controlled source, we cannot short-circuit the controlled source when applying the superposition principle. Figure 17.10(b) shows the circuit with the independent source e replaced by a short circuit. We must find the voltage v_1 across the 20-Ω resistor due to both the controlled voltage source and the independent current source. Writing Kirchhoff's voltage around loop 1 in Figure 17.10(b), we have

$$0.5v_a = 50(i_1 - i_2) + 20i_1 \tag{17.5}$$

This equation contains three unknowns. Note, however, that

$$v_a = 50(i_1 - i_2) \tag{17.6}$$

and

$$i_2 = i \tag{17.7}$$

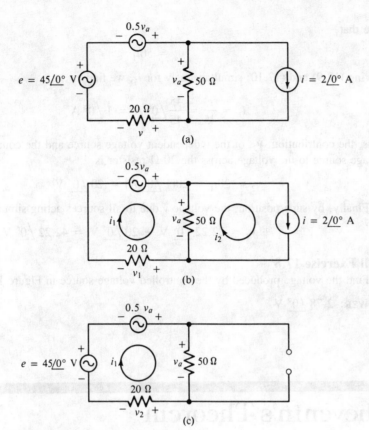

FIGURE 17.10 (Example 17.8)

Substituting (17.7) into (17.5) and (17.6) gives the simultaneous equations

$$70i_1 - 0.5v_a = 50\,i$$

$$50i_1 - v_a = 50\,i \tag{17.8}$$

Solving these equations simultaneously for i_1 gives $i_1 = 25i/45$. Thus,

$$i_1 = \frac{25i}{45} = \frac{25(2\,\underline{/0^\circ}\ A)}{45} = \frac{50}{45}\underline{/0^\circ}\ A$$

Therefore, the contribution, v_1, of the controlled voltage source and the independent current source to the voltage across the 20-Ω resistor is

$$v_1 = 20i_1 = 20\left(\frac{50}{45}\underline{/0^\circ}\ A\right) = 22.22\,\underline{/0^\circ}\ V$$

Figure 17.10(c) shows the circuit with the independent current source replaced by an open circuit. Writing Kirchhoff's voltage law around loop 1, we have

$$e + 0.5v_a = 50i_1 + 20i_1 = 70i_1 \tag{17.9}$$

Note that

$$v_a = 50i_1 \qquad\qquad (17.10)$$

Solving (17.9) and (17.10) simultaneously for i_1, we find

$$i_1 = \frac{e}{45} = \frac{45}{45} \underline{/0°}\ \text{A} = 1 \underline{/0°}\ \text{A}$$

Thus, the contribution, v_2, of the independent voltage source and the controlled voltage source to the voltage across the 20-Ω resistor is

$$v_2 = 20i_1 = 20(1 \underline{/0°}\ \text{A}) = 20 \underline{/0°}\ \text{V}$$

Finally, by superposition, the voltage v due to all sources acting simultaneously is

$$v = v_1 + v_2 = 22.22 \underline{/0°}\ \text{V} + 20 \underline{/0°}\ \text{V} = 42.22 \underline{/0°}\ \text{V}$$

Drill Exercise 17.8

Find the voltage produced by the controlled voltage source in Figure 17.10(a).

ANSWER: $2.78 \underline{/0°}\ \text{V}$.   □

17.3 Thévenin's Theorem

Recall that Thévenin's theorem allows us to replace a linear circuit by a single equivalent voltage source in series with a single equivalent resistance. The ac version of Thévenin's theorem is similar: A linear ac circuit can be replaced by a single equivalent ac voltage source in series with a single equivalent *impedance*. If all the sources in an ac circuit are independent, then the procedure used to find a Thévenin equivalent circuit is completely parallel to that described in Chapter 7 for dc circuits:

1. Identify the pair of terminals with respect to which the Thévenin equivalent circuit is to be constructed. Open-circuit the terminals (i.e., remove all components external to the circuitry that is to be replaced by a Thévenin equivalent).
2. The Thévenin equivalent impedance is the impedance looking into the open-circuited terminals when all voltage sources are replaced by short circuits and all current sources are replaced by open circuits. (Any internal impedance associated with a source remains in the circuit.)
3. The Thévenin equivalent voltage is the voltage appearing across the open-circuited terminals when all sources are present.

Example 17.9

Find the Thévenin equivalent of the circuit lying to the left of terminals *a–b* in Figure 17.11(a).

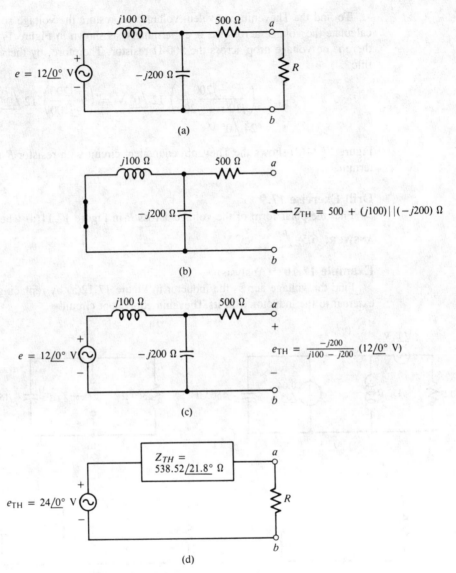

FIGURE 17.11 (Example 17.9)

SOLUTION To find the Thévenin equivalent impedance, we remove resistor R and replace the voltage source by a short circuit, as shown in Figure 17.11(b). The inductor and capacitor are then in parallel, and

$$Z_{TH} = 500 + \frac{(100 \angle 90°)(200 \angle -90°)}{0 + j100 - j200} = 500 + \frac{2 \times 10^4 \angle 0°}{100 \angle -90°}$$

$$= 500 + j200 \ \Omega = 538.52 \angle 21.8° \ \Omega$$

To find the Thévenin equivalent voltage, we restore the voltage source and calculate the voltage across the open terminals, as shown in Figure 17.11(c). Note that there is no voltage drop across the 500-Ω resistor. Therefore, by the voltage-divider rule,

$$e_{TH} = \left(\frac{-j200}{j100 - j200}\right) 12 \underline{/0°} \text{ V} = \left(\frac{-j200}{-j100}\right) 12 \underline{/0°} \text{ V}$$
$$= 24 \underline{/0°} \text{ V}$$

Figure 17.11(d) shows the Thévenin equivalent circuit with resistor R restored to the terminals.

Drill Exercise 17.9

Find the polar form of the voltage across R in Figure 17.11(a) when $R = 200 \ \Omega$.

ANSWER: $6.59 \underline{/-15.95°}$ V. ☐

Example 17.10 (Analysis)

Find the voltage across the inductor in Figure 17.12(a) by replacing the circuit external to the inductor with its Thévenin equivalent circuit.

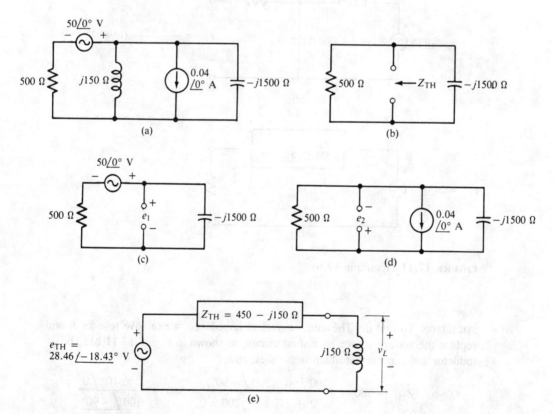

FIGURE 17.12 (Example 17.10)

SOLUTION To find the Thévenin equivalent impedance at the inductor terminals, we remove the inductor, short-circuit the voltage source, and open-circuit the current source, as shown in Figure 17.12(b). It is then clear that Z_{TH} is the parallel equivalent impedance of the resistor and capacitor:

$$Z_{TH} = \frac{(500 \ \underline{/0°})(1500 \ \underline{/-90°})}{500 - j1500} = \frac{7.5 \times 10^5 \ \underline{/-90°}}{1581 \ \underline{/-71.57°}}$$

$$= 474.38 \ \underline{/-18.43°} \ \Omega = 450 - j150 \ \Omega$$

We will use the superposition principle to find the voltage across the open terminals. Figure 17.12(c) shows the circuit with the current source replaced by an open circuit. The voltage e_1 across the terminals due to the voltage source alone is seen to be the same as the voltage across the capacitor. By the voltage-divider rule:

$$e_1 = \left(\frac{1500 \ \underline{/-90°}}{500 - j1500}\right) 50 \ \underline{/0°} \ V = \left(\frac{1500 \ \underline{/-90°}}{1581 \ \underline{/-71.57°}}\right) 50 \ \underline{/0°} \ V$$

$$= 47.44 \ \underline{/-18.43°} \ V = 45 - j15 \ V$$

Figure 17.12(d) shows the circuit with the current source restored and the voltage source replaced by a short circuit. The voltage e_3 across the open terminals due to the current source alone equals the current multiplied by the parallel equivalent impedance of the resistor and capacitor. Since the parallel impedance in this circuit is the same as Z_{TH}, already computed, we have

$$e_2 = (0.04 \ \underline{/0°} \ A)(474.38 \ \underline{/-18.43°} \ \Omega)$$

$$= 18.98 \ \underline{/-18.43°} \ V = 18 - j6 \ V$$

Note that e_2 has the opposite polarity of e_1. Therefore, by superposition,

$$e_{TH} = e_1 - e_2 = (45 - j15) \ V - (18 - j6) \ V$$

$$= 27 - j9 \ V = 28.46 \ \underline{/-18.43°} \ V$$

Figure 17.12(e) shows the Thévenin equivalent circuit with the inductor restored. By the voltage-divider rule,

$$v_L = \left(\frac{150 \ \underline{/90°}}{450 - j150 + j150}\right) 28.46 \ \underline{/-18.43°} \ V$$

$$= \left(\frac{150 \ \underline{/90°}}{450 \ \underline{/0°}}\right) 28.46 \ \underline{/-18.43°} \ V = 9.49 \ \underline{/71.57°} \ V$$

Drill Exercise 17.10

Repeat Example 17.10 when the positions of the resistor and capacitor in Figure 17.12(a) are interchanged.

ANSWER: $v_L = 8.23 \ \underline{/-148.24°} \ V.$ □

17.4 Thévenin Equivalents of Circuits Containing Controlled Sources

If the circuit to be connected to a Thévenin equivalent circuit contains controlled (dependent) sources, and if all the controlling variables are *external* to the circuit, then the method used to find e_{TH} and Z_{TH} can be the same as that used for circuits containing independent sources. However, if any controlling variable appears in the circuit for which a Thévenin equivalent is desired, a different method must be used to find Z_{TH}. One method that is applicable to all situations involves calculating the current that flows through a short circuit connected across the Thévenin terminals. Referring to Figure 17.13, it is clear that

$$Z_{TH} = \frac{e_{TH}}{i_{SC}} \tag{17.11}$$

where i_{SC} is the current that flows through a short circuit connected across the terminals. This method can be used to find Z_{TH} in any dc or ac circuit, whether it contains controlled sources or not. Of course the value of e_{TH}, which is the voltage across the open-circuited terminals, must be found using whatever methods are appropriate for the type of circuit. If the circuit contains dependent sources that are controlled by variables within the circuit, a method like that described previously for finding a voltage by superposition must be used.

Example 17.11

Find the Thévenin equivalent circuit with respect to terminals a–b in Figure 17.14(a). Assume that the ac sources have zero phase angle.

SOLUTION To find e_{TH}, we must find the voltage across the open-circuited terminals a–b. As shown in Figure 17.14(b), it is clear that this voltage is the same as the voltage v across the 5-Ω resistor. Since the circuit contains a controlled source whose value $(0.1v)$ depends on v, we cannot replace that source by a short circuit and

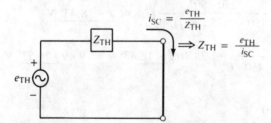

FIGURE 17.13 Z_{TH} can be computed by finding the current that flows through a short circuit connected across the terminals.

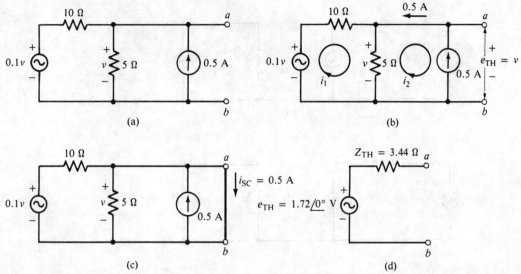

FIGURE **17.14** (Example 17.11)

apply the superposition principle. Writing Kirchhoff's voltage law around loop 1 shown in Figure 17.14(b), we have

$$0.1v = 10i_1 + 5(i_1 - i_2) \qquad (17.12)$$

It is clear that

$$v = 5(i_1 - i_2) \qquad (17.13)$$

and

$$i_2 = -0.5 \text{ A} \qquad (17.14)$$

Substituting (17.14) into (17.12) and (17.13) gives us two simultaneous equations involving v and i_1:

$$0.1v - 15i_1 = 2.5$$
$$v - 5i_1 = 2.5$$

Solving these equations for v, we find

$$v = 1.72 \text{ V} = e_{\text{TH}}$$

Since we cannot short circuit the voltage source, we must find Z_{TH} by calculating the current that flows through a short circuit connected across the a–b terminals. As shown in Figure 17.14(c), it is clear that $i_{\text{SC}} = 0.5$ A. Thus,

$$Z_{\text{TH}} = \frac{e_{\text{TH}}}{i_{\text{SC}}} = \frac{1.72 \text{ V}}{0.5 \text{ A}} = 3.44 \text{ }\Omega$$

The Thévenin equivalent circuit is shown in Figure 17.14(d).

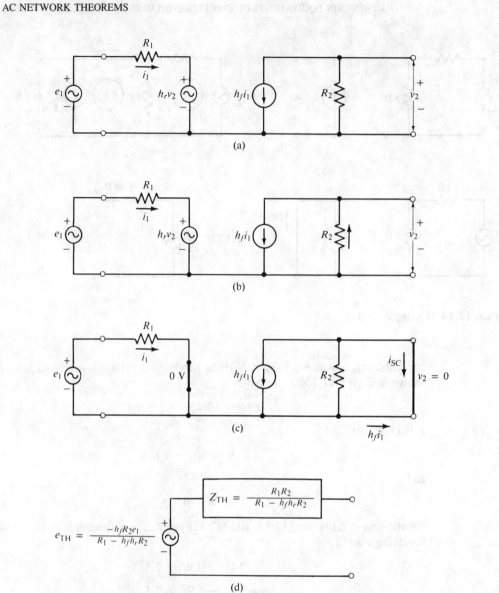

FIGURE 17.15 (Example 17.12)

Drill Exercise 17.11

Find the current through the 10-Ω resistor in Figure 17.14(a).

ANSWER: $i_1 = -0.155$ A.

Example 17.12

Figure 17.15(a) shows the equivalent circuit of a transistor. Notice that it contains a controlled voltage source, labeled $h_r v_2$, and a controlled current source, labeled

$h_f i_1$. h_r and h_f are constants called *hybrid parameters*. The independent source e_1 is the *input voltage* connected to the transistor. The terminals across R_2 are the output terminals of the transistor and v_2 is the *output voltage*. Find the Thévenin equivalent circuit with respect to the output terminals, in terms of h_f, h_r, e_1, R_1, and R_2.

SOLUTION In Figure 17.15(a), it is clear that the Thévenin equivalent voltage, e_{TH}, equals the voltage v_2 across the open-circuited output terminals. We must find v_2 in terms of h_f, h_r, e_1, R_1, and R_2. The current i_1 that controls the dependent current source is found by applying Ohm's law at the input:

$$i_1 = \frac{e_1 - h_r v_2}{R_1}$$

Therefore, the current source produces a current equal to

$$h_f i_1 = h_f \left(\frac{e_1 - h_r v_2}{R_1} \right)$$

As shown in Figure 17.15(b), all of this current flows through R_2. Notice that the polarity of the voltage drop across R_2 is the opposite of that assumed for v_2. By Ohm's law,

$$v_2 = -h_f i_1 R_2 = -h_f \left(\frac{e_1 - h_r v_2}{R_1} \right) R_2$$

Solving for v_2 yields

$$v_2 = \frac{-h_f e_1 R_2}{R_1} + \frac{h_f h_r v_2 R_2}{R_1}$$

$$v_2 \left(1 - \frac{h_f h_r R_2}{R_1} \right) = \frac{-h_f e_1 R_2}{R_1}$$

$$v_2 = \frac{-h_f e_1 R_2 / R_1}{1 - h_f h_r R_2 / R_1} = \frac{-h_f R_2 e_1}{R_1 - h_f h_r R_2} = e_{TH}$$

To find the Thévenin equivalent impedance, we short-circuit the output terminals, as shown in Figure 17.15(c). Notice that $v_2 = 0$ when the terminals are shorted, so the controlled voltage source, $h_r v_2$, equals 0 V. In that case,

$$i_1 = \frac{e_1}{R_1}$$

Therefore, the current produced by the current source is

$$h_f i_1 = \frac{h_f e_1}{R_1}$$

The assumed positive direction of i_{SC} (from the positive output terminal to the negative output terminal) is the opposite direction of the current $h_f i_1$ through the shorted terminals. Thus

$$i_{SC} = -h_f i_1 = \frac{-h_f e_1}{R_1}$$

Finally,

$$Z_{TH} = \frac{e_{TH}}{i_{SC}} = \frac{-h_f R_2 e_1 / (R_1 - h_f h_r R_2)}{h_f e_1 / R_1} = \frac{R_1 R_2}{R_1 - h_f h_r R_2}$$

Figure 17.15(d) shows the Thévenin equivalent circuit.

Drill Exercise 17.12

The *voltage gain* of the transistor in Example 17.12 is defined to be the ratio v_2/e_1. Find the voltage gain when $R_1 = 1$ kΩ, $R_2 = 5$ kΩ, $h_r = 2 \times 10^{-4}$, and $h_f = 100$.

ANSWER: -555. □

17.5 Norton's Theorem

Recall that Norton's theorem allows us to replace a linear circuit by a single equivalent current source in parllel with a single equivalent resistance. The ac version of Norton's theorem is similar: A linear ac circuit can be replaced by a single equivalent ac current source in parallel with a single equivalent *impedance*. One method that can always be used to find a Norton equivalent circuit, regardless of the types of sources it contains, is to find the Thévenin equivalent circuit and convert that to an equivalent current source.

The Norton equivalent impedance, Z_N, has exactly the same value as the Thévenin equivalent impedance and is found in the same way. Thus, if the sources in the circuit are independent, or are dependent sources controlled by external variables, Z_N is found by short-circuiting voltage sources and open-circuiting current sources. If the circuit contains a dependent source controlled by a variable within the circuit, the method described in Section 17.4 for finding Z_{TH} in such a case must be used. The Norton equivalent current, i_N, is the current that flows through a short circuit connected across the Norton terminals. As previously indicated, i_N can also be found by converting the Thévenin circuit to an equivalent current source:

$$i_N = \frac{e_{TH}}{Z_{TH}} \tag{17.15}$$

Note that (17.15) is, in fact, the short-circuit current of a Thévenin equivalent circuit.

Example 17.13

(a) Find the Norton equivalent of the circuit lying to the left of terminals *a–b* in Figure 17.16(a).
(b) Find the Thévenin equivalent of the same circuit.

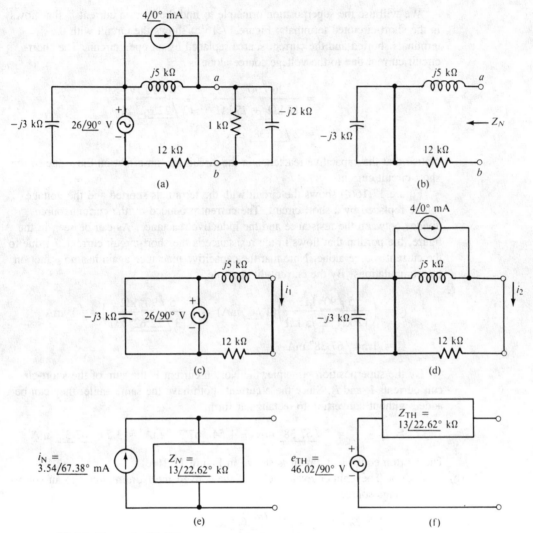

FIGURE 17.16 (Example 17.13)

SOLUTION

(a) To find the Norton equivalent impedance, we remove the resistor-capacitor combination, short-circuit the voltage source, and open-circuit the current source, as shown in Figure 17.16(b). Since the capacitive reactance in parallel with the voltage source is shorted out, the Norton impedance is simply the series combination of the resistance and inductive reactance:

$$Z_N = 12 \text{ k}\Omega + j5 \text{ k}\Omega = 13 \underline{/22.62°} \text{ k}\Omega$$

We will use the superposition principle to find the Norton current i_N that flows in the short-circuited terminals. Figure 17.16(c) shows the circuit with the terminals shorted and the current source replaced by an open circuit. The short-circuit current due to the voltage source alone is

$$i_1 = \frac{26 \underline{/90°} \text{ V}}{12 \text{ k}\Omega + j5 \text{ k}\Omega} = \frac{26 \underline{/90°} \text{ V}}{13 \underline{/22.62°} \text{ k}\Omega}$$

$$= 2 \underline{/67.38°} \text{ mA}$$

Notice that the capacitive reactance in the circuit has no effect on the value of the short-circuit current.

Figure 17.16(d) shows the circuit with the terminals shorted and the voltage source replaced by a short circuit. The current produced by the current source divides between the resistance and the inductive reactance. As can be seen in the figure, the portion that flows in the resistance is the short-circuit current, i_2, due to the current source alone. Note that the capacitive reactance again has no effect on the computations. By the current-divider rule,

$$i_2 = \left(\frac{5 \underline{/90°} \text{ k}\Omega}{12 \text{ k}\Omega + j5 \text{ k}\Omega} \right)(4 \underline{/0°} \text{ mA}) = \left(\frac{5 \underline{/90°} \text{ k}\Omega}{13 \underline{/22.62°} \text{ k}\Omega} \right) 4 \underline{/0°} \text{ mA}$$

$$= 1.54 \underline{/67.38°} \text{ mA}$$

By the superposition principle, the Norton current is the sum of the short-circuit currents i_1 and i_2. Since these currents both have the same angle, they can be added without converting to rectangular form:

$$i_N = i_1 + i_2 = 2 \underline{/67.38°} \text{ mA} + 1.54 \underline{/67.38°} \text{ mA} = 3.54 \underline{/67.38°} \text{ mA}$$

The Norton equivalent circuit is shown in Figure 17.16(e).

(b) To find the Thévenin equivalent circuit, we convert the Norton circuit to an equivalent voltage source:

$$Z_{TH} = Z_N = 13 \underline{/22.62°} \text{ k}\Omega$$

$$e_{TH} = i_N Z_N = (3.54 \underline{/67.38°} \text{ mA})(13 \underline{/22.62°} \text{ k}\Omega)$$

$$= 46.02 \underline{/90°} \text{ V}$$

The Thévenin equivalent circuit is shown in Figure 17.16(f).

Drill Exercise 17.13

Find the Norton equivalent of the circuit in Example 17.13 when the positions of the 12-kΩ resistance and the 5-kΩ inductive reactance are interchanged.

ANSWER: $i_N = 4.2 \underline{/5.87°}$ mA, $Z_N = 13 \underline{/22.62°}$ kΩ.   □

17.6 Maximum Power Transfer

In our study of dc circuits, we learned that maximum power is developed in a load when the load equals the Thévenin equivalent resistance of the circuit to which the load is connected. Recall that, speaking loosely, we say maximum power is "transferred" from the source to the load under those circumstances. In ac circuits, maximum power is transferred to a load when the load is an impedance equal to the *complex conjugate* of the Thévenin impedance of the source. For example, if the Thévenin impedance of a circuit is $(20 + j10)\ \Omega$, it will deliver maximum power to a load impedance of $(20 - j10)\ \Omega$. Notice that the dc version of the maximum power theorem is just a special case of the ac version, because the complex conjugate of a resistance (real number) is exactly equal to the resistance: $R = R + j0 = R - j0$.

Of course, it is only the resistive component of a load that dissipates power. Zero power is delivered to a load that is purely reactive. The value of the maximum power transferred to a load can be calculated using any of the methods discussed in Chapter 12 [see equations (12.41) through (12.46)]. When using the power equations, be certain that the voltage across and/or current through the *resistive* component of Z_L is used in the computations.

Example 17.14   (Design)

(a) Find the value of load impedance Z_L in Figure 17.17(a) that results in maximum power transfer to Z_L.
(b) Find the average power transferred to Z_L when
  (i) Z_L equals the Thévenin impedance of the circuit to which Z_L is connected
  (ii) Z_L equals the complex conjugate of the Thévenin impedance.

SOLUTION

(a) To find Z_{TH}, we remove Z_L and replace the voltage source by a short circuit, as shown in Figure 17.17(b). The parallel combination of Z_1 and Z_2 is equivalent to

$$\frac{(50\ \underline{/-90°})(100\ \underline{/90°})}{0 - j50 + j100} = \frac{5 \times 10^3\ \underline{/0°}}{50\ \underline{/90°}} = 100\ \underline{/-90°}\ \Omega$$

Figure 17.17(c) shows the circuit with Z_1 and Z_2 replaced by their parallel equivalent. The $-j100\ \Omega$ is in series with Z_3, so that combination is equivalent to $-j100\ \Omega - j50\ \Omega = -j150\ \Omega$. Figure 17.17(d) shows that Z_4 is in parallel with the $-j150\ \Omega$. The impedance of that parallel combination is the Thévenin impedance of the circuit:

$$Z_{TH} = \frac{(200\ \underline{/0°})(150\ \underline{/-90°})}{200 - j150} = \frac{3 \times 10^4\ \underline{/-90°}}{250\ \underline{/-36.87°}}$$

$$= 120\ \underline{/-53.13°}\ \Omega = 72 - j96\ \Omega$$

For maximum power transfer, Z_L must equal the complex conjugate of Z_{TH}.

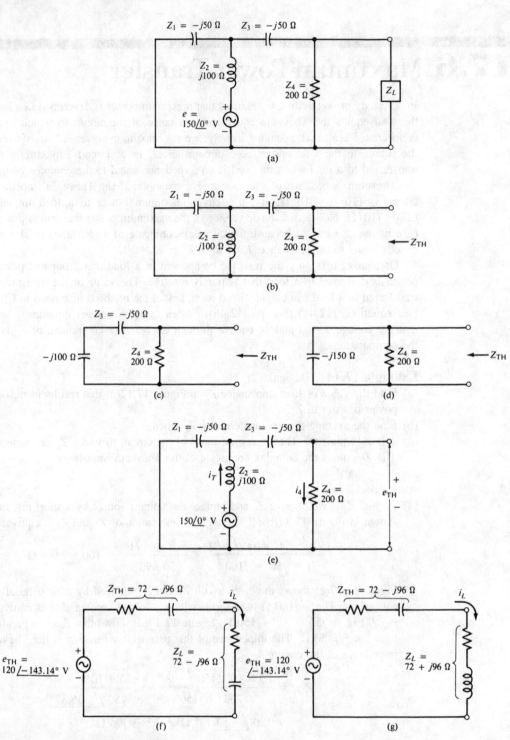

FIGURE 17.17 (Example 17.14)

Therefore, we require

$$Z_L = 120 \underline{/53.13°} \ \Omega = 72 + j96 \ \Omega$$

(b) To find the power delivered to the load, we must find the Thévenin equivalent voltage. Figure 17.17(e) shows the circuit with the load terminals open and the voltage source restored. The total impedance connected to the source is

$$Z_T = Z_2 + Z_1 \parallel (Z_3 + Z_4)$$

$$= 0 + j100 + 50 \underline{/-90°} \parallel (200 - j50)$$

$$= 0 + j100 + \frac{(50 \underline{/-90°})(206.16 \underline{/-14.04°})}{200 - j50 - j50}$$

$$= 0 + j100 + \frac{10{,}308 \underline{/-104.04°}}{223.61 \underline{/-26.56°}}$$

$$= 0 + j100 + 46.1 \underline{/-77.48°} = 0 + j100 + 10 - j45$$

$$= 10 + j55 \ \Omega = 55.9 \underline{/79.7°} \ \Omega$$

The total current is then

$$i_T = \frac{e}{Z_T} = \frac{150 \underline{/0°} \ V}{55.9 \underline{/79.7°} \ \Omega} = 2.68 \underline{/-79.7°} \ A$$

By the current-divider rule, the current in Z_4 is

$$i_4 = \frac{Z_1}{Z_1 + Z_3 + Z_4} i_T = \left(\frac{50 \underline{/-90°}}{200 - j100}\right) 2.68 \underline{/-79.7°} \ A$$

$$= \left(\frac{50 \underline{/-90°}}{223.61 \underline{/-26.56°}}\right) 2.68 \underline{/-79.7°} \ A = 0.6 \underline{/-143.14°} \ A$$

The Thévenin voltage is the voltage across Z_4:

$$e_{TH} = i_4 Z_4 = (0.6 \underline{/-143.14°} \ A)(200 \underline{/0°} \ \Omega) = 120 \underline{/-143.14°} \ V$$

(i) Figure 17.17(f) shows the Thévenin equivalent circuit with Z_{TH} connected across the load terminals. The total series impedance is then

$$Z_T = 2Z_{TH} = 2(120 \underline{/-53.13°} \ \Omega) = 240 \underline{/-53.13°} \ \Omega$$

The current in the load is therefore

$$i_L = \frac{e_{TH}}{Z_T} = \frac{120 \underline{/-143.14°} \ V}{240 \underline{/-53.13°} \ \Omega} = 0.5 \underline{/-90°} \ A$$

The average power in the 72-Ω resistive component of the load is therefore

$$P_{avg} = \frac{|i_L|^2 R_L}{2} = \frac{(0.5)^2 (72)}{2} = 9 \ W$$

(ii) Figure 17.17(g) shows the Thévenin equivalent circuit with the complex conjugate of Z_{TH} connected across the load terminals. In this case, the total impedance is

$$Z_T = Z_{TH} + Z_L = (72 - j96) \ \Omega + (72 + j96) \ \Omega = 144 + j0 \ \Omega$$

Therefore,

$$i_L = \frac{e_{TH}}{Z_T} = \frac{120 \ \underline{/-143.14°} \ V}{144 \ \underline{/0°} \ \Omega} = 0.833 \ \underline{/-143.14°} \ A$$

The average power in the 72-Ω resistive component of the load is

$$P_{avg} = \frac{|i_L|^2 R_L}{2} = \frac{(0.833)^2 (72)}{2} = 25 \ W$$

These results confirm that less power is delivered to the load when $Z_L = Z_{TH}$ than it is when Z_L is the value required for maximum power transfer: the complex conjugate of Z_{TH} (9 W versus 25 W).

Drill Exercise 17.14

Find the value of Z_L in Figure 17.17(a) that results in maximum power transfer to Z_L, and the value of the maximum power, when Z_1 is removed (replaced by an open circuit).

ANSWER: $Z_L = 11.77 - j47.06 \ \Omega$, $P_{avg} = 224.6$ W.   □

The selection of a load impedance whose value results in maximum power transfer is called *impedance matching*. When the load impedance equals the complex conjugate of the Thévenin impedance of the source, the load and the source are said to be matched to each other. Notice that it is only possible to make Z_{TH} and Z_L complex conjugates of each other at a *single* frequency. If the frequency changes, both inductive and capacitive reactance change, so the conjugate relationship is lost. In many applications (such as audio amplifiers), the voltage and current waveforms are not pure, single-frequency sine waves, but are combinations of many different frequencies. In those cases it is not possible to achieve impedance matching. As a compromise, when possible, the resistive component of a complex load is matched to the resistive component of the source.

Example 17.15   (Troubleshooting)

Figure 17.18(a) shows the equivalent circuit of an amplifier used in a certain siren to drive a loudspeaker with a 500-Hz tone. The loudspeaker has 16 Ω of resistance in series with 7 mH of inductance. When the siren was tested, it was found that the peak voltage across the loudspeaker terminals was 40 V when the controlling current i was 50 $\underline{/0°}$ mA. Is the loudspeaker properly matched to the amplifier?

SOLUTION The inductive reactance of the loudspeaker is

$$X_L = j\omega L = j(2\pi \times 500)(7 \times 10^{-3}) = j22 \ \Omega$$

Therefore, the total impedance of the loudspeaker is

$$Z_L = R + j\omega L = 16 + j22 \ \Omega$$

Since the amplifier is controlled by a current external to the circuit, we can find the Thévenin impedance by removing the loudspeaker and open-circuiting the dependent

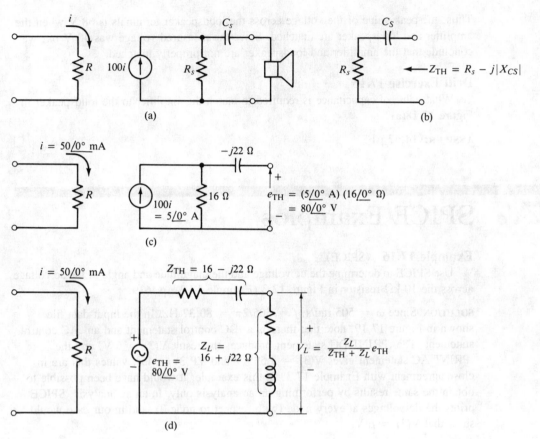

FIGURE 17.18 (Example 17.15)

current source. The resulting circuit is shown in Figure 17.18(b). It is clear that $Z_{TH} = R_s - j|X_{Cs}|$. In order for the amplifier to be matched to the load, we must have $Z_{TH} = 16 - j22\ \Omega$.

When $i = 50\ \underline{/0^\circ}$ mA, the current-controlled current source produces $100i = 100(50\ \underline{/0^\circ}$ mA$) = 5\ \underline{/0^\circ}$ A. Figure 17.18(c) shows that the Thévenin equivalent voltage of the amplifier when $Z_{TH} = 16 - j22\ \Omega$ is

$$e_{TH} = (5\ \underline{/0^\circ}\ A)(16\ \underline{/0^\circ}\ \Omega) = 80\ \underline{/0^\circ}\ V$$

Figure 17.18(d) shows the Thévenin equivalent circuit when the amplifier is matched to the load. In that case, the voltage across the speaker terminals is, by the voltage-divider rule,

$$v_L = \frac{Z_L}{Z_{TH} + Z_L}\ e_{TH} = \left[\frac{16 + j22}{(16 + j22) + (16 - j22)}\right] 80\ \underline{/0^\circ}\ V$$

$$= \left(\frac{27.2\ \underline{/53.97^\circ}}{32\ \underline{/0^\circ}}\right) 80\ \underline{/0^\circ}\ V = 68\ \underline{/53.97^\circ}\ V$$

Thus, the peak value of the votlage across the loudspeaker terminals is 68 V when the amplifier and loudspeaker are matched. Since the measured voltage was 40 V pk, we conclude that the amplifier and loudspeaker are not properly matched.

Drill Exercise 17.15

What value of capacitance is required to match the amplifier to the loudspeaker in Figure 17.18(a)?

ANSWER: 14.47 μF.   □

17.7 SPICE Examples

Example 17.16 (SPICE)

Use SPICE to determine the dc voltage and the magnitude and angle of the ac voltage across the 10-kΩ resistor in Figure 17.3 (Example 17.3), p. 639.

SOLUTION Since $\omega = 505$ rad/s, $f = 505/2\pi = 80.37$ Hz. In the input data file shown in Figure 17.19, note that there is a .DC control statement and an .AC control statement. The .PRINT DC statement produces the result V(3) = 6 V, and the .PRINT AC statement gives V(3) = 3.535 V and VP(3) = 45°, values that are in close agreement with Example 17.3. In this example, it would have been possible to obtain the same results by performing an ac analysis only. In an ac analysis, SPICE prints the dc voltages at every node (with respect to node 0), and in our case would show that V(3) = 6 V.

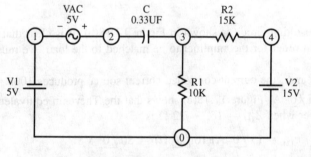

```
EXAMPLE 17.16
V1 0 1 5V
VAC 2 1 AC 5V
C 2 3 0.33UF
R1 3 0 10K
R2 3 4 15K
V2 4 0 15V
.DC V1 5V 5V 1
.AC DEC 1 80.37HZ 80.37HZ
.PRINT DC V(3)
.PRINT AC V(3) VP(3)
.END
```

FIGURE 17.19 (Example 17.16)

Example 17.17 (SPICE)

Use SPICE to find the Thévenin equivalent voltage with respect to terminals *a–b* in the circuit containing the controlled voltage source in Figure 17.14(a) (Example 17.11), p. 655.

SOLUTION The SPICE circuit and input data file are shown in Figure 17.20. Note that the voltage-controlled voltage source, E, is connected between nodes 1 and 0 and is controlled by the voltage between nodes 2 and 0. The magnitude of E is $0.1v$, so the gain is 0.1 (see Appendix, Section A.8). The Thévenin equivalent voltage we seek is the voltage appearing between nodes 2 and 0 (see Example 17.11). Since there are no reactive components in the circuit, the frequency is irrelevant, and we arbitrarily choose it to be 100 Hz. Execution of the program gives $V(2) = e_{TH} = 1.724$ V, in close agreement with Example 17.11.

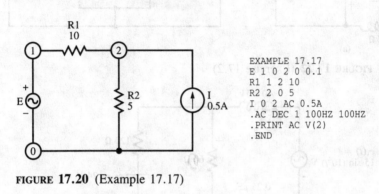

```
EXAMPLE 17.17
E 1 0 2 0 0.1
R1 1 2 10
R2 2 0 5
I 0 2 AC 0.5A
.AC DEC 1 100HZ 100HZ
.PRINT AC V(2)
.END
```

FIGURE 17.20 (Example 17.17)

Exercises

Section 17.1 Superposition

17.1 Use the superposition principle to find the polar form of the current i in each circuit in Figure 17.21.

17.2 Use the superposition principle to find the polar form of the voltage v in each circuit in Figure 17.22. Draw + and − signs to show the polarity (of v) used as the positive reference for each answer.

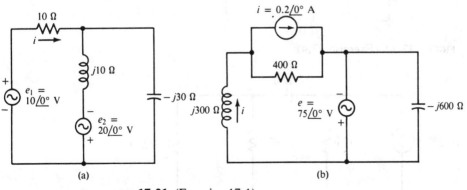

(a)

(b)

FIGURE 17.21 (Exercise 17.1)

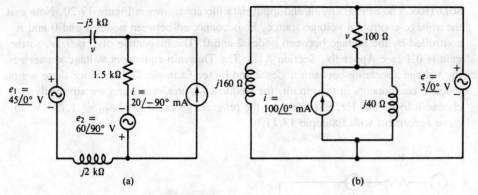

FIGURE 17.22 (Exercise 17.2)

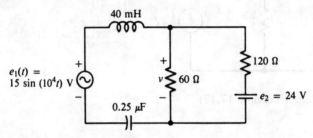

FIGURE 17.23 (Exercise 17.3)

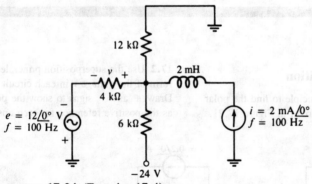

FIGURE 17.24 (Exercise 17.4)

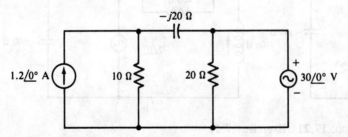

FIGURE 17.25 (Exercise 17.5)

17.3 Use the superposition principle to find the voltage v across the 60-Ω resistor in Figure 17.23. Assume steady-state conditions. Write the mathematical expression for the voltage and sketch its waveform. Label the minimum and maximum values of the waveform on the sketch.

17.4 Repeat Exercise 17.3 for the voltage across the 4-kΩ resistor in Figure 17.24.

17.5 Use the superposition principle to find the total average power dissipated in the circuit shown in Figure 17.25.

17.6 Use the superposition principle to find the total average power dissipated in the circuit shown in Figure 17.26.

Section 17.2   Controlled Voltage and Current Sources

17.7 Find the polar form of the voltage v in each circuit shown in Figure 17.27.

17.8 Find polar form of the voltage v in each circuit shown in Figure 17.28.

17.9 Use the superposition principle to find the polar form of the voltage across the 400-Ω resistor in Figure 17.29 when $e = 1.6 \sin (2 \times 10^3 t)$ V and $\mu = 20$.

17.10 Use the superposition principle to find the polar form of the current i_L in Figure 17.30 when $e = 0.05 \sin (10^4 t)$ V and $\beta = 100$.

17.11 Use the superposition principle to find the current i_A in Figure 17.31.

17.12 Use the superposition principle to find the polar form of the voltage across the 100-Ω resistor in Figure 17.32.

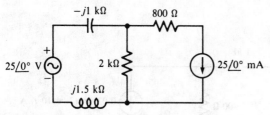

FIGURE 17.26 (Exercise 17.6)

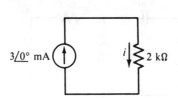

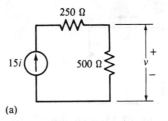

(a)

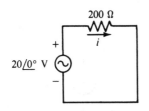

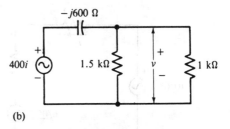

(b)

FIGURE 17.27 (Exercise 17.7)

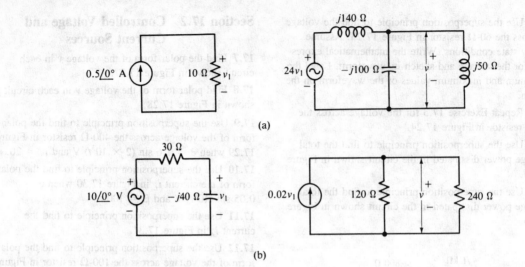

(a)

(b)

FIGURE 17.28 (Exercise 17.8)

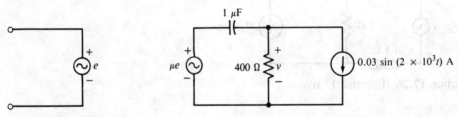

FIGURE 17.29 (Exercise 17.9)

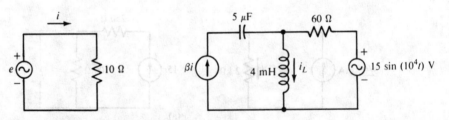

FIGURE 17.30 (Exercise 17.10)

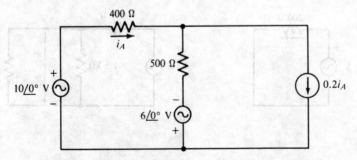

FIGURE 17.31 (Exercise 17.11)

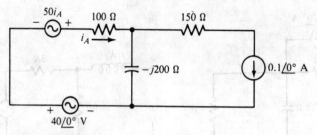

FIGURE 17.32 (Exercise 17.12)

Section 17.3   Thévenin's Theorem

17.13 Find the Thévenin equivalent circuit with respect to terminals *a–b* in each circuit shown in Figure 17.33.

17.14 Find the Thévenin equivalent circuit with respect to terminals *a–b* in each circuit shown in Figure 17.34.

17.15 Find the polar form of the current through the 12-Ω resistor in Figure 17.35 by finding the Thévenin equivalent circuit external to the resistor. Draw an arrow to show the positive reference for the calculated current.

17.16 Find the Thévenin equivalent circuit with respect to terminals *a−b* in Figure 17.36.

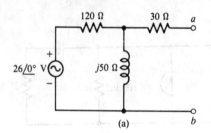

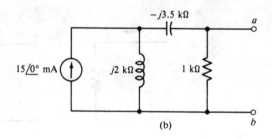

FIGURE 17.33 (Exercise 17.13)

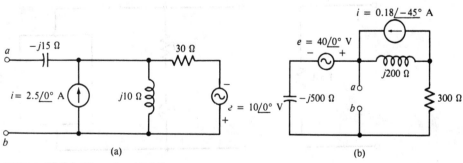

FIGURE 17.34 (Exercise 17.14)

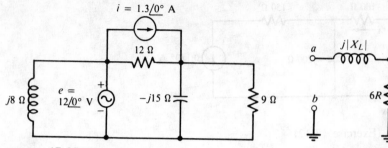

FIGURE 17.35 (Exercise 17.15)

FIGURE 17.36 (Exercise 17.16)

Section 17.4   Thévenin Equivalents of Circuits Containing Controlled Sources

17.17 Find the Thévenin equivalent of the circuit lying to the right of terminals *a–b* in Figure 17.37 when $i_1 = 0.5 \sin (40t)$ mA and $\alpha = 0.95$.

17.18 Find the Thévenin equivalent of the circuit lying

to the right of terminals *a–b* in Figure 17.38 when $i_1 = 1.5 \sin (10^6 t)$ μA and $\mu = 30$.

17.19 Find the Thévenin equivalent circuit with respect to terminals *a–b* in Figure 17.39. Assume all sources have zero phase angle.

17.20 Find the Thévenin equivalent circuit with respect to terminals *a–b* in Figure 17.40.

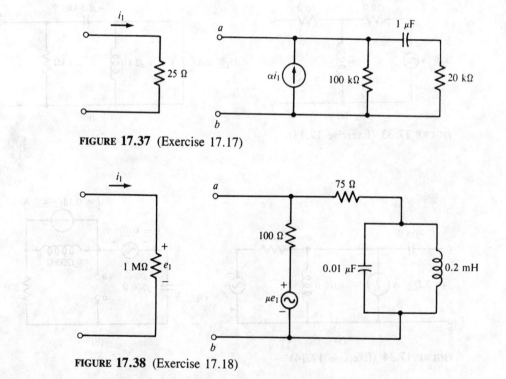

FIGURE 17.37 (Exercise 17.17)

FIGURE 17.38 (Exercise 17.18)

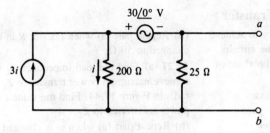

FIGURE 17.39 (Exercise 17.19)

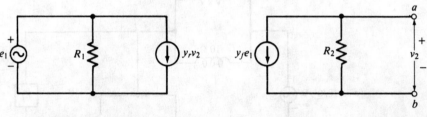

FIGURE 17.40 (Exercise 17.20)

Section 17.5 Norton's Theorem

17.21 Find the Norton equivalent circuit with respect to terminals *a–b* in each circuit in Figure 17.34 by
(i) finding the short-circuit current through the terminals
(ii) converting the Thévenin circuit to an equivalent current source.

17.22 Find the Norton equivalent circuit with respect to terminals *a–b* in each circuit in Figure 17.33 by
(i) finding the short-circuit current through the terminals
(ii) converting the Thévenin circuit to an equivalent current source.

17.23 Find the Norton equivalent circuit with respect to terminals *a–b* in Figure 17.39.

17.24 (a) Find the Norton equivalent circuit with respect to terminals *a–b* in Figure 17.41. (*Hint:* First convert the voltage sources to current sources.)
(b) Find the Thévenin equivalent circuit in Figure 17.41.
(c) The procedure used in this exercise is the ac version of the procedure used to find an equivalent dc source in connection with what previously studied theorem?

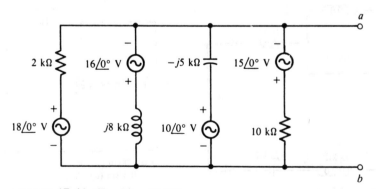

FIGURE 17.41 (Exercise 17.24)

Section 17.6   Maximum Power Transfer

17.25 Find the load impedance Z_L necessary to achieve maximum power transfer to Z_L in each of the circuits shown in Figure 17.42. In each case, find the value of the maximum power transferred.

17.26 (a) Find the load impedance Z_L necessary to achieve maximum power transfer to Z_L in Figure 17.43. Find the value of the maximum power transferred.

(b) Repeat part (a) when resistor R in the circuit is changed to 50 Ω.

17.27 (a) Find the load impedance Z_L necessary to achieve maximum power transfer to Z_L when $\omega = 500$ rad/s in Figure 17.44. Find the value of maximum power transferred to Z_L.

(b) Repeat part (a) when ω is changed to 250 rad/s.

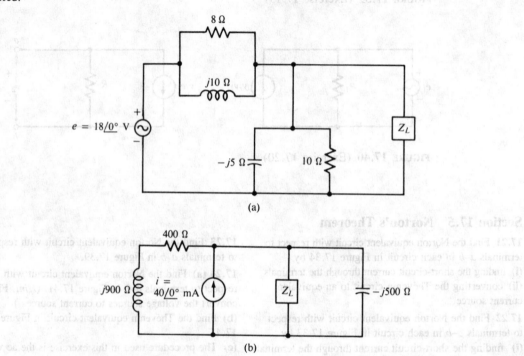

(a)

(b)

FIGURE 17.42 (Exercise 17.25)

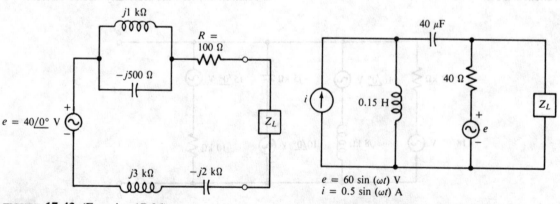

FIGURE 17.43 (Exercise 17.26)

FIGURE 17.44 (Exercise 17.27)

Design Exercises

17.1D Find the value of e_1 (magnitude and angle) in Figure 17.45 that is necessary to make the voltage across R_1 equal 0 V.

17.2D Find the value of i (magnitude and angle) in Figure 17.46 that is necessary to make the current in the inductor equal 1.5 $\underline{/0°}$ A.

17.3D What is the minimum value of E that can be used in Figure 17.47 if the steady-state voltage v across the 600-Ω resistor is not permitted to go negative? (The

capacitor has reactance $-j500$ Ω at the frequency of the ac source.)

17.4D The capacitor in Figure 17.48(a) has negligibly small reactance ($X_C \approx 0$) at the frequency of the ac voltage source. The circuit is required to produce the steady-state waveform v shown in Figure 17.48(b) across the 2-kΩ resistor. What should be the values of the peak ac source voltage e and the dc voltage E?

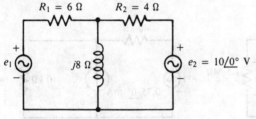

FIGURE 17.45 (Exercise 17.1D)

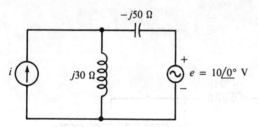

FIGURE 17.46 (Exercise 17.2D)

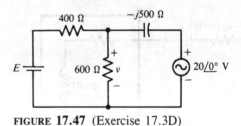

FIGURE 17.47 (Exercise 17.3D)

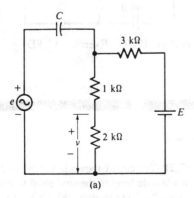

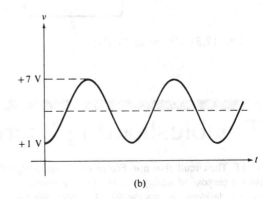

FIGURE 17.48 (Exercise 17.4D)

17.5D In Figure 17.49, find the value of β that is necessary to make the magnitude of v_o equal 26 V.

17.6D In Figure 17.50, find the value of μ that is necessary to make $|v_o|/|v_1| = 15$.

17.7D To what frequency should the voltage source in

Figure 17.51 be set in order that maximum power be delivered to load Z_L?

17.8D What should be the value of R_S in Figure 17.52 in order to achieve maximum power transfer to Z_L?

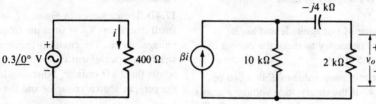

FIGURE 17.49 (Exercise 17.5D)

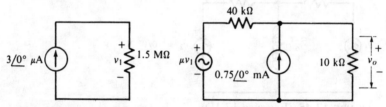

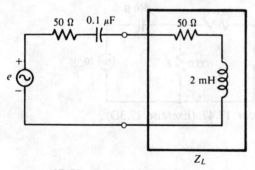

FIGURE 17.50 (Exercise 17.6D)

FIGURE 17.51 (Exercise 17.7D)

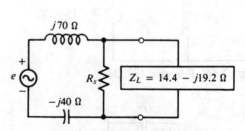

FIGURE 17.52 (Exercise 17.8D)

Troubleshooting Exercises

17.1T The circuit shown in Figure 17.53 was designed for the purpose of adding a dc level to a sinusoidal voltage developed across the 400-Ω resistor. Why is the circuit unsatisfactory?

17.2T The circuit shown in Figure 17.54 was designed to add a dc level to the sinusoidal voltage developed across the 2-kΩ resistor. Why does the circuit not accomplish this goal?

17.3T In order to match the source shown in Figure 17.55 to load Z_L, a 300-Ω resistor was connected in series with Z_L. Does this modification achieve maximum power transfer to the load? Expain.

17.4T Figure 17.56 shows the equivalent circuit of a transistor connected in an amplifier circuit. For the amplifier to operate properly, the value of the transistor parameter labeled β must be at least 50. In a test of one such circuit, the peak value of i_L was found to be 1.2 mA when the peak value of e_1 was 60 mV. Is the transistor acceptable? [*Hint:* Find the Norton equivalent circuit of Figure 17.15(d).]

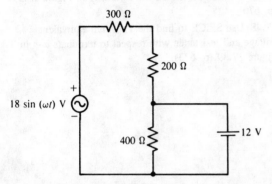

FIGURE 17.53 (Exercise 17.1T)

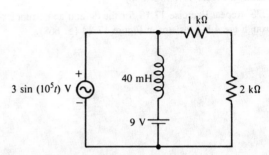

FIGURE 17.54 (Exercise 17.2T)

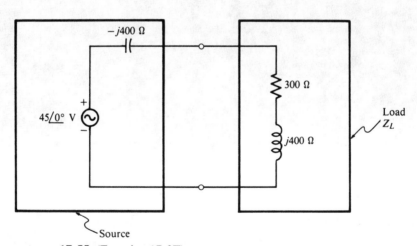

FIGURE 17.55 (Exercise 17.3T)

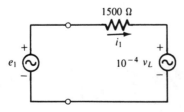

FIGURE 17.56 (Exercise 17.4T)

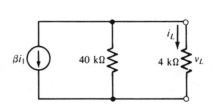

SPICE Exercises

17.1S Use SPICE to find the dc voltage and the magnitude and angle of the ac voltage across the 60-Ω resistor in Figure 17.23 (p. 668).

17.2S Repeat Exercise 17.1S for the dc and ac current through the 4-kΩ resistor in Figure 17.24 (p. 668).

17.3S Use SPICE to find the current i_A in Figure 17.31 (p. 670).

17.4S Use SPICE to find the Thévenin equivalent voltage and resistance with respect to terminals a–b in Figure 17.39 (p. 673).

18 Filters and Resonant Circuits

18.1 Filter Definitions

A *filter* is a device that modifies the amplitude, or magnitude, of an ac voltage as the frequency of the voltage changes. One pair of terminals connected to a filter is the *input*, where an ac source is connected, and another pair of terminals is the *output*, where the voltage variations occur. Figure 18.1 illustrates the action of one type of filter, called a *low-pass* filter. The figure shows that the magnitude of the output voltage decreases as frequency increases. The source is often called a *signal* source, and the low-pass filter can be regarded as one that allows low-frequency signals to pass through it, while reducing the magnitude (*attenuating*) high-frequency signals.

A good way to depict the action of a filter is to plot the magnitude of its output voltage versus frequency. This type of plot shows the *frequency response* of the filter, that is, the way it *responds* to changes in frequency. Figure 18.2 shows the frequency response of a typical low-pass filter. Note that the magnitude of v_o remains constant over a certain range of low frequencies, but then decreases steadily as frequency increases. The maximum value of the mgnitude of v_o is labeled V_m in the figure. The frequency at which the magnitude of v_o drops to $(\sqrt{2}/2)V_m \approx 0.707V_m$ is called the *cutoff frequency*

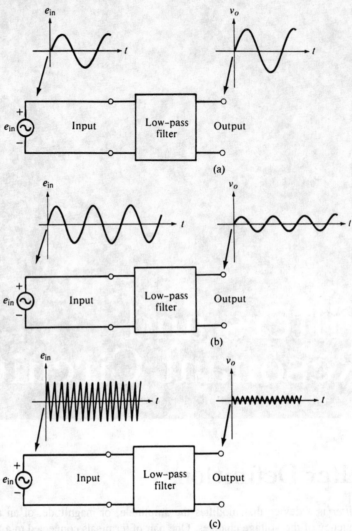

FIGURE 18.1 Low-pass filter action. Notice that the magnitude of input signal e_{in} remains constant, but the magnitude of v_0 decreases as frequency increases.
(a) Low-frequency signal. (b) Medium-frequency signal. (c) High-frequency signal.

of the filter. For reasons that will become apparent later, the cutoff frequency is labeled f_2 in Figure 18.2. The range, or band, of frequencies below the cutoff frequency is called the *passband*.

A *high-pass filter* passes signals whose frequencies are above a certain cutoff frequency and attenuates others. Figure 18.3(a) shows the frequency response of a typical high-pass filter. Note that we designate the cutoff frequency in this case by f_1. As in the low-pass filter, the cutoff frequency is the frequency where the magnitude of the output equals $\sqrt{2}/2$ times its maximum value in the passband. A *bandpass* filter passes signals

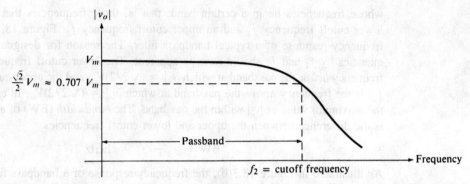

FIGURE **18.2** Plot of the frequency response of a low-pass filter.

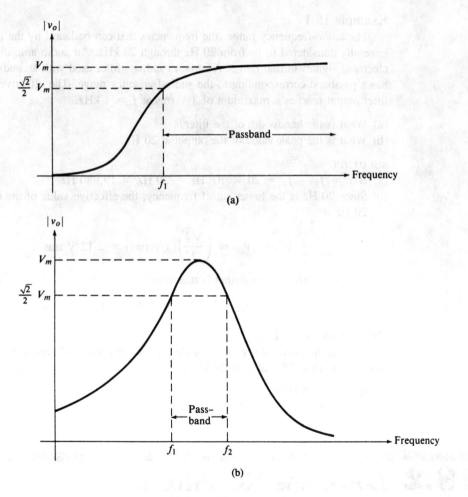

(a)

(b)

FIGURE **18.3** High-pass and bandpass filters. (a) Frequency response of a high-pass filter. (b) Frequency response of a bandpass filter.

whose frequencies lie in a certain band, that is, those frequencies that are between a lower cutoff frequency f_1 and an upper cutoff frequency f_2. Figure 18.3(b) shows the frequency response of a typical bandpass filter. The reason for designating cutoff frequencies by f_1 and f_2 should now be apparent. The lower cutoff frequency, f_1, is the frequency *below* the passband at which $|v_o| = (\sqrt{2}/2)V_m$, and the upper cutoff frequency, f_2, is the frequency *above* the passband at which $|v_o| = (\sqrt{2}/2)V_m$. In each case, V_m is the maximum value of $|v_o|$ within the passband. The *bandwidth* (BW) of a bandpass filter is the difference between the upper and lower cutoff frequencies:

$$BW = f_2 - f_1 \quad \text{hertz} \tag{18.1}$$

As illustrated in Figure 18.3(b), the frequency response of a bandpass filter is not necessarily symmetrical.

Example 18.1

The *audio*-frequency range (the frequencies that can be heard by the human ear) is generally considered to be from 20 Hz through 20 kHz. An audio amplifier amplifies electrical signals in that range. A certain bandpass filter used with an audio amplifier has a passband corresponding to the audio-frequency range. The effective value of the filter output reaches a maximum of 3 V rms at $f = 1$ kHz.

(a) What is the bandwidth of the filter?
(b) What is the peak value of the output at 20 Hz?

SOLUTION
(a) $BW = f_2 - f_1 = 20 \times 10^3 \text{ Hz} - 20 \text{ Hz} = 19{,}980 \text{ Hz}$
(b) Since 20 Hz is the lower cutoff frequency, the effective value of the output at 20 Hz is

$$V_{eff} = \left(\frac{\sqrt{2}}{2}\right)(3 \text{ V rms}) = 2.12 \text{ V rms}$$

The peak value of the output is therefore

$$V_p = \sqrt{2} \, V_{eff} = \sqrt{2} \, (2.12) = 3 \text{ V pk}$$

Drill Exercise 18.1

What is the power delivered to a 400-Ω resistor connected across the output of the filter in Example 18.1 at $f = 20$ kHz?

ANSWER: 11.25 mW. □

18.2 Low-Pass RC Filters

Figure 18.4 shows a simple low-pass filter consisting of a resistor and a capacitor. Notice that the input is connected in series with the resistor and the output is the voltage across the capacitor. The input and output have one common terminal, which is the low (ground,

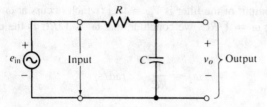

FIGURE 18.4 Low-pass RC filter.

or reference) side of each. Before analyzing the circuit using phasors, let us verify that this arrangement will behave like a low-pass filter, based on an intuitive analysis of how its impedance changes with frequency. Notice that the circuit is basically a voltage divider having one component (the resistor) whose impedance does not depend on frequency. As in any voltage divider, the magnitude of the voltage across the other component depends directly on the impedance of that component. Thus, if $|X_C|$ is large in comparison to R, the magnitude of v_o will be large. The smaller the value of $|X_C|$, the smaller the magnitude of v_o. Since $|X_C| = 1/(2\pi fC)$ *decreases* with frequency, we conclude that $|v_o|$ will decrease with frequency. Conversely, $|v_o|$ increases as frequency decreases, confirming that the circuit "passes" low-frequency signals. Consider the extreme condition where frequency approaches zero, that is, when e_{in} becomes a *dc* voltage. In that case, the capacitor is an open circuit (infinite reactance), so $v_o = e_{in}$. At the other extreme, when the frequency of e_{in} is extremely large, the capacitive reactance approaches $0\ \Omega$, that is, a short circuit. Thus v_o approaches zero. Visualizing the circuit in these two extreme cases is a helpful way to remember that it behaves like a low-pass filter.

Let us now analyze the low-pass RC circuit to obtain an expression for v_o in terms of R, C, e_{in}, and the frequency $\omega = 2\pi f$ of e_{in}. Referring to Figure 18.4, we can apply the voltage divider rule to obtain v_o:

$$v_o = \frac{-j|X_C|}{R - j|X_C|}\, e_{in} \tag{18.2}$$

Substituting $|X_C| = 1/\omega C$ into (18.2), we obtain

$$v_o = \frac{-j/\omega C}{R - j/\omega C}\, e_{in} = \frac{-j}{\omega RC - j}\, e_{in} \tag{18.3}$$

The magnitude of v_o is thus

$$|v_o| = \frac{|-j|}{|\omega RC - j|}\, |e_{in}| = \frac{1}{\sqrt{(\omega RC)^2 + 1}}\, |e_{in}| \tag{18.4}$$

Equation (18.4) confirms that $|v_o|$ approaches zero as ω becomes very large. Furthermore, when $\omega = 0$, (18.4) shows that $|v_o| = |e_{in}|$, confirming our previous discussion of the dc case.

If we evaluate (18.4) at the specific frequency $\omega = 1/RC$ rad/s, we find

$$|v_o| = \frac{1}{\sqrt{[(1/RC)(RC)]^2 + 1}}\, |e_{in}|$$

$$= \frac{1}{\sqrt{1 + 1}}\, |e_{in}| = \frac{|e_{in}|}{\sqrt{2}} = \frac{\sqrt{2}}{2}\, |e_{in}|$$

Since the maximum output of the filter is $V_m = |e_{in}|$ (which occurs at $\omega = 0$), and since $|v_o| = (\sqrt{2}/2)\,|e_{in}|$ at $\omega = 1/RC$, we conclude that $\omega = 1/RC$ is the cutoff frequency of the filter:

$$\omega_2 = \frac{1}{RC} \qquad \text{rad/s} \tag{18.5}$$

or

$$f_2 = \frac{1}{2\pi RC} \qquad \text{Hz} \tag{18.6}$$

The value of $|v_o|$ at any frequency f can be found in terms of f_2 by substituting $1/\omega_2 = RC$ in equation (18.4):

$$|v_o| = \frac{|e_{in}|}{\sqrt{(\omega/\omega_2)^2 + 1}} = \frac{|e_{in}|}{\sqrt{(f/f_2)^2 + 1}} \tag{18.7}$$

Example 18.2   (Analysis)

For the low-pass filter shown in Figure 18.5:

(a) Find the cutoff frequency, in hertz.

(b) Find the magnitude of v_o when e_{in} has frequency 500 Hz, 1 kHz, and 2 kHz.

(c) Sketch $|v_o|$ versus frequency.

SOLUTION

(a) From equation (18.6),

$$f_2 = \frac{1}{2\pi RC} = \frac{1}{2\pi(10.61 \times 10^3)(0.015 \times 10^{-6})} = 1 \text{ kHz}$$

(b) From equation (18.7), at $f = 500$ Hz,

$$|v_o| = \frac{10 \text{ V}}{\sqrt{(500/10^3)^2 + 1}} = 8.94 \text{ V}$$

at $f = 1$ kHz,

$$|v_o| = \frac{10 \text{ V}}{\sqrt{(10^3/10^3)^2 + 1}} = 7.07 \text{ V}$$

and at $f = 2$ kHz,

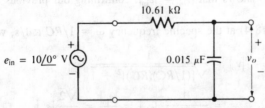

FIGURE 18.5 (Example 18.2)

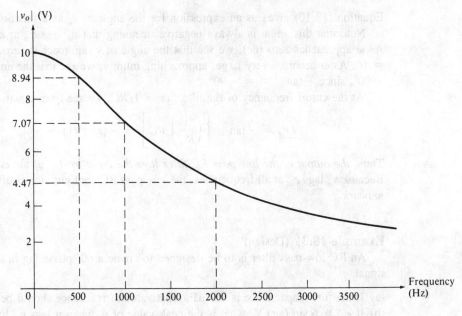

FIGURE 18.6 (Example 18.2)

$$v_o = \frac{10 \text{ V}}{\sqrt{\left(\frac{2 \times 10^3}{10^3}\right)^2 + 1}} = 4.47 \text{ V}$$

(c) Figure 18.6 shows a plot of $|v_o|$ versus frequency. The points corresponding to the three frequencies at which $|v_o|$ was calculated in part (b) are indicated on the plot.

Drill Exercise 18.2

At what frequency is $|v_o|$ in Example 18.2 equal to 1.0 V?

ANSWER: 9.95 kHz.   □

Let us now return to equation (18.3), repeated below, for v_o in terms of ω, R, C, and e_{in}:

$$v_o = \frac{-j}{\omega RC - j} e_{in} \qquad (18.8)$$

Multiplying numerator and denominator by j, we obtain

$$v_o = \frac{-j^2}{j\omega RC - j^2} e_{in} = \frac{1}{1 + j\omega RC} e_{in} \qquad (18.9)$$

If we assume that e_{in} has zero phase angle, then the angle of v_o is the same as the angle of the term $1/(1 + j\omega RC)$. Since the angle of the numerator is zero degrees, we have

$$\underline{/v_o} = 0° - \underline{/1 + j\omega RC} = -\tan^{-1} \omega RC \qquad (18.10)$$

Equation (18.10) gives us an expression for the angle of v_o as a function of frequency ω. Note that this angle is always negative, meaning that v_o lags e_{in} at all frequencies. As ω approaches zero (dc), we see that the angle of v_o approaches zero, since $\tan^{-1} 0 = 0°$. As ω becomes very large, approaching infinity, we see that the angle approaches $-90°$, since $-\tan^{-1} \infty = -90°$.

At the cutoff frequency of the filter, $\omega = 1/RC$, we see from equation (18.10) that

$$\underline{/v_o} = -\tan^{-1}\left[\left(\frac{1}{RC}\right)RC\right] = -\tan^{-1}(1) = -45°$$

Thus, *the output of the low-pass RC filter lags the input by 45° at the cutoff frequency.* Because v_o lags e_{in} at all frequencies, the low-pass RC configuration is often called a *lag network*.

Example 18.3   (Design)

An RC low-pass filter is to be designed to create a 60° phase lag in a 2.5-kHz signal.

(a) If the filter capacitance is 0.2 μF, what value of resistance should be used?
(b) If $e_{in} = 6 \sin(\omega t)$ V, what is the peak value of v_o when it lags e_{in} by 60°?

SOLUTION
(a) From equation (18.10),

$$\underline{/v_o} = -60° = -\tan^{-1}[(2\pi \times 2.5 \times 10^3)(0.2 \times 10^{-6})R]$$

or,

$$60° = \tan^{-1}(3.14 \times 10^{-3}R)$$

$$\tan 60° = 1.73 = 3.14 \times 10^{-3}R$$

$$R = \frac{1.73}{3.14 \times 10^{-3}} = 552 \ \Omega$$

(b) With $R = 552 \ \Omega$ and $C = 0.2$ μF, the cutoff frequency of the filter is

$$f_2 = \frac{1}{2\pi RC} = \frac{1}{(2\pi)(552)(0.2 \times 10^{-6})} = 1442 \text{ Hz}$$

From equation (18.7),

$$|v_o| = \frac{|e_{in}|}{\sqrt{(f/f_2)^2 + 1}} = \frac{6 \text{ V}}{\sqrt{\left(\dfrac{2.5 \times 10^3}{1442}\right)^2 + 1}} = 3 \text{ V}$$

Drill Exercise 18.3

At what frequency does the output of the filter designed in Example 18.3 lag e_{in} by 30°?

ANSWER: 832 Hz.   □

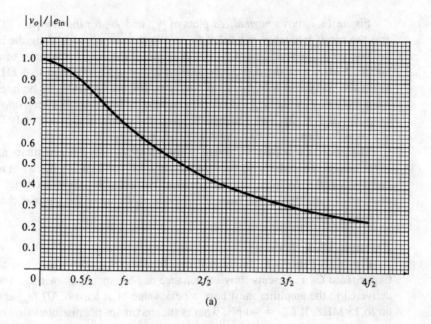

(a)

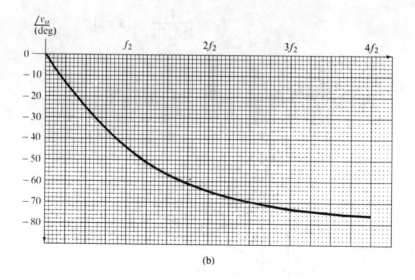

(b)

FIGURE 18.7 (a) Normalized plot of the magnitude of the output of a low-pass RC filter. (b) Normalized plot of the phase angle of the output of a low-pass RC filter. (The vertical axis can be interpreted as the angle of v_0 when $\underline{/e_{in}} = 0°$, or as $\underline{/v_0} - \underline{/e_{in}}$.)

Figure 18.7 shows *normalized* plots of $|v_o|$ and $\underline{/v_o}$ for the low-pass RC filter. Notice that the cutoff frequency is labeled f_2, so all other coordinates along the horizontal axis are interpreted as multiples of f_2. For example, if the cutoff frequency of a certain filter were 1.5 kHz, then the coordinate labeled $2f_2$ would correspond to 3 kHz. Each value along the vertical axis of the magnitude plot [Figure 18.7(a)] can be interpreted as the magnitude of v_o when $|e_{in}| = 1$ V. It can also be interpreted as the *ratio* $|v_o|/|e_{in}|$. For example, at $f = 2f_2$, we see that the magnitude ratio is approximately 0.45. Thus, if $|e_{in}| = 4$ V, then $|v_o| \approx (0.45)(4 \text{ V}) = 1.8$ V at $2f_2$.

The study of low-pass RC filters is important not only because filters are constructed in the configuration, but also because the same RC combination occurs in many practical circuits, whether by intention or otherwise. For example, the capacitance component is often *stray* capacitance (see Section 9.3) due to wiring or the internal characteristics of certain devices.

Example 18.4   (Design)

Figure 18.8 shows an amplifier driven by a signal source whose resistance is R_S. Capacitance C_W represents stray capacitance due to component wiring. The signal delivered to the amplifier must have a peak value of at least $0.707 |e_{in}|$ at frequencies up to 15 MHz. If $C_W = 40$ pF, what is the maximum permissible value of R_S?

SOLUTION The R_S–C_W combination forms a low-pass filter at the input to the amplifier. The given criteria require that the cutoff frequency be at least 15 MHz:

$$f_2 = \frac{1}{2\pi R_S C_W} \geq 15 \times 10^6$$

or

$$R_S \leq \frac{1}{2\pi C_W(15 \times 10^6)}$$

Since $C_W = 40$ pF, we have

$$R_S \leq \frac{1}{2\pi(40 \times 10^{-12})(15 \times 10^6)} = 265 \ \Omega$$

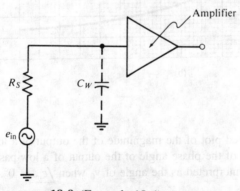

FIGURE 18.8 (Example 18.4)

Drill Exercise 18.4

If $R_S = 1\ \text{k}\Omega$ in Example 18.4, what is the maximum permissible value of C_W?

ANSWER: 10.6 pF.   □

18.3 High-Pass RC Filters

Figure 18.9 shows an RC network that behaves as a high-pass filter. Notice that the high-pass filter is the same as the low-pass filter with the positions of the resistor and capacitor interchanged. In this case, the input is in series with the capacitor and the output voltage is the voltage across the resistor. An intuitive analysis of this circuit will confirm that it performs the high-pass function: At low frequencies, the capacitive reactance is large in comparison to the resistance, so v_o is small. At high frequencies, the reverse is true. In the extreme case where the frequency is zero (dc), the capacitor is an open circuit, so $v_o = 0$. At very high frequencies, the capacitor approaches a short circuit, making $v_o = e_{\text{in}}$.

Applying the voltage-divider rule to Figure 18.9, we find

$$v_o = \frac{R}{R - j|X_C|}\,e_{\text{in}} = \frac{R}{R - j/\omega C}\,e_{\text{in}}$$

$$= \frac{\omega RC}{\omega RC - j}\,e_{\text{in}} \tag{18.11}$$

The magnitude of v_o is therefore

$$|v_o| = \frac{\omega RC}{\sqrt{(\omega RC)^2 + 1}}\,|e_{\text{in}}| \tag{18.12}$$

Note that $|v_o|$ is zero when $\omega = 0$, confirming that the high-pass filter "blocks" dc. Equation (18.12) can also be used to show that $|v_o|$ approaches $|e_{\text{in}}|$ when ω becomes very large.

The cutoff frequency of the RC high-pass filter is calculated in the same way it is for the RC low-pass filter:

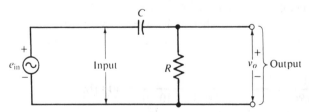

FIGURE 18.9 High-pass RC filter.

$$\omega_1 = \frac{1}{RC} \quad \text{rad/s} \tag{18.13}$$

$$f_1 = \frac{1}{2\pi RC} \quad \text{hertz} \tag{18.14}$$

It is an exercise at the end of this chapter to verify that ω_1 given by equation (18.13) is the cutoff frequency. Substituting $\omega_1 = 1/RC$ into (18.12), we obtain an expression for $|v_o|$ in terms of the cutoff frequency:

$$|v_o| = \frac{\omega/\omega_1}{\sqrt{(\omega/\omega_1)^2 + 1}} |e_{\text{in}}| \tag{18.15}$$

Algebraic manipulation leads to

$$|v_o| = \frac{|e_{\text{in}}|}{\sqrt{(\omega_1/\omega)^2 + 1}} = \frac{|e_{\text{in}}|}{\sqrt{(f_1/f)^2 + 1}} \tag{18.16}$$

To find the angle of v_o as a function of R, C, and ω, we first multiply numerator and denominator of equation (18.11) by j:

$$v_o = \frac{j}{j}\left(\frac{\omega RC}{\omega RC - j}\right) e_{\text{in}} = \frac{j\omega RC}{1 + j\omega RC} e_{\text{in}} \tag{18.17}$$

Assuming that $\underline{/e_{\text{in}}} = 0°$, the angle of v_o is then

$$\underline{/v_o} = \underline{/j\omega RC} - \underline{/1 + j\omega RC} = 90° - \tan^{-1} \omega RC \tag{18.18}$$

Since $\tan^{-1} \omega RC$ is always an angle less than 90°, equation (18.18) shows that *the angle of v_o in the high-pass RC filter is always positive*. Thus, the output of the filter leads the input, and the configuration is often called a *lead network*. Substituting the cutoff frequency, $\omega = 1/RC$, into (18.18), we find

$$v_o = 90° - \tan^{-1}\left(\frac{1}{RC} RC\right) = 90° - \tan^{-1} 1 = 45°$$

We see that the output of the filter leads the input by 45° at the cutoff frequency.

Example 18.5   (Analysis)
For the high-pass filter shown in Figure 18.10:

(a) Find the cutoff freqency in hertz.
(b) Find the magnitude of v_o when e_{in} has frequency 15 kHz, 40 kHz, and 100 kHz.
(c) Find the angle of v_o when e_{in} has frequency 20 kHz, 40 kHz, and 80 kHz.

SOLUTION

(a) $f_1 = \dfrac{1}{2\pi RC} = \dfrac{1}{2\pi(173)(0.023 \times 10^{-6})} = 40$ kHz

(b) From equation (18.16), at $f = 15$ kHz,

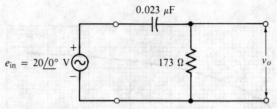

FIGURE 18.10 (Example 18.5)

$$|v_o| = \frac{20}{\sqrt{\left(\dfrac{40 \times 10^3}{15 \times 10^3}\right)^2 + 1}} = 7.02 \text{ V}$$

at $f = 40$ kHz,

$$|v_o| = \frac{20}{\sqrt{\left(\dfrac{40 \times 10^3}{40 \times 10^3}\right)^2 + 1}} = 14.14 \text{ V}$$

and at $f = 100$ kHz,

$$|v_o| = \frac{20}{\sqrt{\left(\dfrac{40 \times 10^3}{100 \times 10^3}\right)^2 + 1}} = 18.57 \text{ V}$$

(c) From equation (18.18), at $f = 20$ kHz,

$$\underline{/v_o} = 90° - \tan^{-1} (2\pi \times 20 \times 10^3 \times 173 \times 0.023 \times 10^{-6})$$
$$= 63.43°$$

at $f = 40$ kHz,

$$\underline{/v_o} = 90° - \tan^{-1} (2\pi \times 40 \times 10^3 \times 173 \times 0.023 \times 10^{-6})$$
$$= 45°$$

and at $f = 80$ kHz,

$$\underline{/v_o} = 90° - \tan^{-1} (2\pi \times 80 \times 10^3 \times 173 \times 0.023 \times 10^{-6})$$
$$= 26.57°$$

Drill Exercise 18.5

At what frequency does $|v_o|/|e_{in}| = 0.5$ in Example 18.5?

ANSWER: 23,094 Hz. □

Figure 18.11 shows normalized plots of the magnitude and angle of the output of the high-pass RC filter. Notice that the plot of the $|v_o|/|e_{in}|$ is zero at dc and approaches 1 as frequency becomes very large. The phase angle approaches 90° as frequency approaches zero, and approaches zero as frequency becomes very large. The normalized coordinates along the axes are interpreted in the same way we discussed for the plots of the low-pass filter (Figure 18.7).

$|v_o|/|e_{in}|$

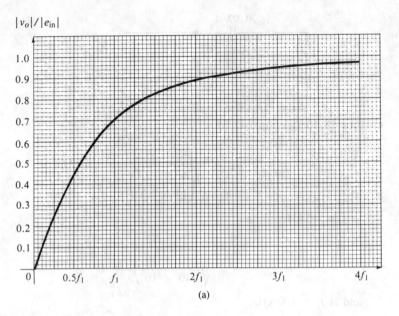

(a)

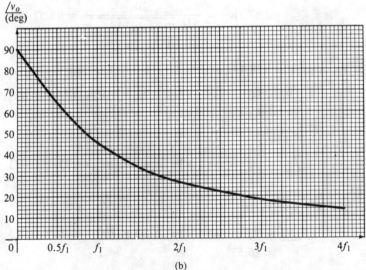

(b)

FIGURE 18.11 (a) Normalized plot of the magnitude of the output of a high-pass RC filter. (b) Normalized plot of the phase angle of the output of a high-pass RC filter. (The vertical axis can be interpreted as the angle of v_o when $\underline{/e_{in}} = 0°$, or as $\underline{/v_o} - \underline{/e_{in}}$.)

As is the case for low-pass RC networks, the high-pass configuration occurs in many practical applications where the filtering action is not necessarily desirable. One example is *RC coupling* networks, where the series capacitor is used to block the flow of dc current, but ideally, should pass any ac component. The next example illustrates such an application.

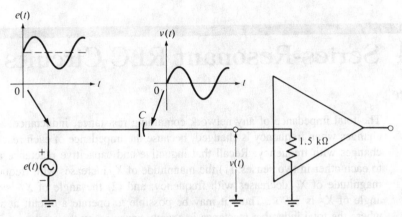

FIGURE 18.12 (Example 18.6)

Example 18.6   (Design)

Figure 18.12 shows an amplifier connected to a signal source using an RC coupling network. As illustrated in the figure, the capacitor effectively removes the dc component in the signal. The 1.5-kΩ resistance represents the resistance of the amplifier. The frequency range of signals delivered to the amplifier is from 25 Hz to 12 kHz. If the ratio $|v|/|e|$ is not permitted to fall below 0.707, what is the minimum value of capacitance C that can be used?

SOLUTION Note that $v(t)$ is the output of a high-pass RC network. As shown in Figure 18.11(a), the output of such a network decreases as frequency decreases. Therefore, in this example, the only frequency that concerns us is the lowest frequency in the range of input signals, namely, 25 Hz. Since $|v_o|/|e_{in}| = 0.707$ at the cutoff frequency, we will have satisfied the design requirement if we make the cutoff frequency *no greater than* 25 Hz:

$$f_1 = \frac{1}{2\pi RC} \le 25 \text{ Hz}$$

or

$$\frac{1}{C} \le (2\pi R)(25)$$

which is the same as

$$C \ge \frac{1}{2\pi R(25)} = \frac{1}{2\pi(1500)(25)} = 4.24 \text{ μF}$$

Drill Exercise 18.6

If a 1-μF capacitor is used in Example 18.6, what is the smallest value of amplifier resistance, R, that will satisfy the design requirement?

ANSWER: 6366 Ω.

18.4 Series-Resonant RLC Circuits

Resonance

The total impedance of any network containing resistance, inductance, and capacitance changes when frequency is changed, because the impedance of each reactive component changes with frequency. Recall that inductive and capacitive reactance are "opposite" to each other in two senses: (1) the magnitude of X_L increases with frequency, while the magnitude of X_C decreases with frequency; and (2) the angle of X_L is $+90°$, and the angle of X_C is $-90°$. Thus, it may be possible to operate a circuit at *some* frequency where the total inductive reactance is exactly canceled by the total capacitive reactance. When a circuit is operated at a frequency where the reactive (imaginary) component of the total impedance is zero, the circuit is said to be *resonant*. It follows that the total impedance of a resonant circuit is purely resistive (real).

Series RLC Circuits at Resonance

Figure 18.13 shows a series RLC circuit. The total impedance of this circuit is

$$Z_T = R + j(|X_L| - |X_C|)$$
$$= R + j\left(\omega L - \frac{1}{\omega C}\right) \tag{18.19}$$

From the definition, we know that the circuit will be resonant when the imaginary component of (18.19) equals zero. Therefore, we can find the frequency at which resonance occurs by setting the imaginary component equal to zero and solving for ω. We designate that *particular* frequency by ω_0:

$$\omega_0 L - \frac{1}{\omega_0 C} = 0$$

$$\omega_0^2 L = \frac{1}{C}$$

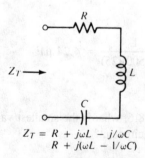

$$Z_T = R + j\omega L - j/\omega C$$
$$R + j(\omega L - 1/\omega C)$$

FIGURE 18.13 Series RLC circuit.

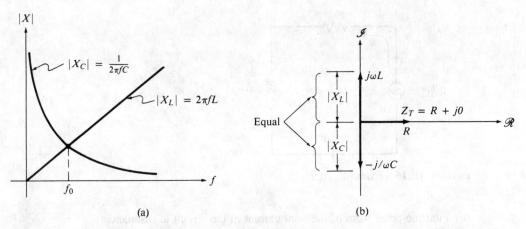

FIGURE 18.14 Impedance relations at the resonant frequency of a series RLC circuit. (a) The resonant frequency f_0 is the frequency at which the magnitude plots of X_L and X_C intersect. (b) Phasor diagram of the total impedance at resonance, showing that $Z_T = R + j0$.

$$\omega_0^2 = \frac{1}{LC}$$

$$\omega_0 = \frac{1}{\sqrt{LC}} \quad \text{rad/s} \tag{18.20}$$

Converting the angular frequency given by (18.20) to frequency in Hz, we find

$$f_0 = \frac{\omega_0}{2\pi}$$

$$= \frac{1}{2\pi\sqrt{LC}} \quad \text{hertz} \tag{18.21}$$

The frequency specified by equation 18.20 or 18.21 is called the *resonant frequency* of the series RLC circuit. Notice that the value of resistance in the circuit has no effect on the value of the resonant frequency.

Figure 18.14(a) shows the magnitudes $|X_L| = 2\pi fL$ and $|X_C| = 1/2\pi fC$ plotted versus frequency. It is clear that f_0 is the frequency where the two magnitude plots intersect, because that is the frequency where $|X_L| = |X_C|$. Figure 18.14(b) is the phasor plot of the total impedance at resonance and shows that $j|X_L|$ and $-j|X_C|$ exactly cancel each other, making $Z_T = R + j0$ Ω. At frequencies below or above resonance, the magnitude of Z_T is greater than R. Thus, $|Z_T|$ reaches a *minimum* value at the resonant frequency. A voltage source connected to the circuit would therefore supply maximum current at the resonant frequency.

Example 18.7   (Analysis)
(a) Find the resonant frequency of the circuit shown in Figure 18.15.
(b) Find the inductive reactance, capacitive reactance, and total impedance of the circuit at resonance.

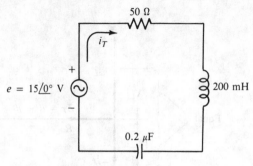

FIGURE 18.15 (Example 18.7)

(c) Find the polar form of the total current in the circuit at resonance.

SOLUTION

(a) $\omega_0 = \dfrac{1}{\sqrt{LC}} = \dfrac{1}{\sqrt{(200 \times 10^{-3})(0.2 \times 10^{-6})}} = 5000$ rad/s

$f_0 = \dfrac{\omega_0}{2\pi} = \dfrac{5000}{2\pi} = 795.8$ Hz

(b) At resonance,

$$X_L = j\omega_0 L = j5000(200 \times 10^{-3}) = j1000 \ \Omega$$

$$X_C = \frac{-j}{\omega_0 C} = \frac{-j}{5000(0.2 \times 10^{-6})} = -j1000 \ \Omega$$

$$\begin{aligned} Z_T &= R + j(|X_L| - |X_C|) \\ &= 50 + j(1000 - 1000) = 50 + j0 \ \Omega \end{aligned}$$

(c) At resonance,

$$i = \frac{e}{Z_T} = \frac{15 \underline{/0°} \text{ V}}{50 \underline{/0°} \ \Omega} = 0.3 \underline{/0°} \text{ A}$$

Drill Exercise 18.7

Find the polar form of the voltage across each component in Figure 18.15 at resonance.

ANSWER: $v_R = 15 \underline{/0°}$ V; $v_L = 300 \underline{/90°}$ V; $v_C = 300 \underline{/-90°}$ V.   ☐

The Quality Factor (Q_s)

Recall that only resistance can dissipate energy. If it were possible to construct a series LC circuit having zero resistance, then no energy would be lost in the circuit. Since every practical inductor and capacitor has some resistance, it is not possible to construct such a circuit. However, in some applications of resonant circuits, it is desirable to minimize the total resistance in the circuit (i.e., to dissipate as little energy as possible).

The *quality factor*, Q_s, of a series RLC circuit is defined to be the ratio of the reactive power in either the inductor or the capacitor to the average power at resonance:

$$Q_s = \frac{\text{reactive power}}{\text{average power}} \qquad \text{(at resonance)} \qquad (18.22)$$

The subscript s in Q_s refers to the *series* configuration. Equation (18.22) shows that the quality factor is large when the average power is small, in keeping with the notion that the quality of a circuit is great when the energy it dissipates is small. In practice, the quality factor is used as an indicator of other circuit characteristics that are more important than energy dissipation, as we shall see in later discussions.

Recall from Chapter 15 that the reactive power in inductors and capacitors can be found from

$$\text{reactive power} = \begin{cases} \dfrac{I_P^2\,|X_L|}{2} & \text{vars (inductor)} \\[2ex] \dfrac{I_P^2|X_C|}{2} & \text{vars (capacitor)} \end{cases} \qquad (18.23)$$

Since $|X_L| = |X_C|$ at resonance, the reactive power in the inductor equals the reactive power in the capacitor at resonance. Also, the current I_P at resonance is the same in every series component, including the resistor, so the average power is

$$P_{\text{avg}} = \frac{I_P^2 R}{2} \qquad \text{watts} \qquad (18.24)$$

Substituting the reactive power in the inductor from (18.23), and P_{avg} from (18.24), into the definition of Q_s [equation (18.22)], we find

$$Q_s = \frac{I_P^2|X_L|/2}{I_P^2 R/2} = \frac{|X_L|}{R} = \frac{\omega_0 L}{R} \qquad (18.25)$$

Note carefully that the resonant frequency ω_0 is used in (18.25), because Q_s is defined at resonance. Equation (18.25) shows that Q_s is the ratio of the inductive reactance at resonance to the resistance in the circuit. Substituting the reactive power in the capacitor at resonance into equation (18.22), we find an equivalent expression for Q_s:

$$Q_s = \frac{I_P^2|X_C|/2}{I_P^2 R/2} = \frac{|X_C|}{R} = \frac{1}{\omega_0 R C} \qquad (18.26)$$

Equation (18.26) shows that Q_s is also the ratio of the capacitive reactance at resonance to the resistance in the circuit.

In some RLC circuits, there is no resistor present and the total resistance is, for all practical purposes, the resistance of the inductor (coil) winding. For that reason, it is conventional to speak of the "Q of the coil," meaning the ratio of its reactance to its resistance:

$$Q \text{ (coil)} = \frac{\omega L}{R_l} \qquad (18.27)$$

where R_l is the resistance of the winding. This terminology is often used in respect to a coil alone, that is, without reference to a resonant circuit or to a resonant frequency. Thus, the frequency ω in (18.27) could be any value, and it is clear that the Q of a coil increases with frequency. To be a meaningful characteristic of a coil, the value of its Q must be given with reference to a specific frequency. In practical coils, there is an upper limit to the value that Q may reach as frequency is increased, due to a property known as *effective resistance*. Because of certain magnetic phenonema that occur in conductors at high frequencies, there is an increase in energy loss. This effect is the same as that due to an increase in resistance. At very high frequencies, the increase in effective resistance may become so great that the Q of a coil actually decreases. The maximum Q of commercially available coils is near 100.

The quality factor of a series-resonant circuit can be obtained in terms of R, L, and C by substituting for ω_0 in either equation (18.25) or (18.26). Substituting $\omega_0 = 1/\sqrt{LC}$ in (18.25), we find

$$Q_s = \frac{\omega_0 L}{R} = \frac{(1/\sqrt{LC})L}{R}$$

$$= \frac{L}{R\sqrt{LC}} = \frac{1}{R}\sqrt{\frac{L}{C}} \tag{18.28}$$

Consider the series RLC circuit shown in Figure 18.16. At resonance, $Z_T = R + j0$, so the total current in the circuit is

$$i = \frac{e}{Z_T} = \frac{e}{R} \tag{18.29}$$

The voltage across the inductor is therefore

$$v_L = iX_L = i|X_L|\underline{/90°} = \frac{e|X_L|\underline{/90°}}{R} \tag{18.30}$$

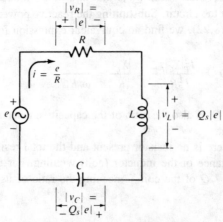

FIGURE 18.16 Voltages and current in a series RLC circuit at resonance.

Since $|X_L|/R = Q_s$, we see that the magnitude of v_L is

$$|v_L| = Q_s|e| \tag{18.31}$$

Similarly, the voltage across the capacitor is

$$v_C = iX_C = i|X_C| \underline{/-90°} = \frac{e|X_C| \underline{/-90°}}{R} \tag{18.32}$$

and

$$|v_C| = Q_s|e| \tag{18.33}$$

Since Q_s may be greater than 1, equations (18.32) and (18.33) show that the magnitudes of the voltages across L and C at resonance may be greater than the magnitude of the source voltage. In some practical circuits, the value of Q_s is much greater than 1, so the inductor and capacitor voltages can be very large. This fact is worth remembering when handling components in such a circuit, or when specifying breakdown ratings for the components. Note that

$$v_R = iR = \frac{e}{R}R = e \tag{18.34}$$

Example 18.8 (Analysis)
For the circuit shown in Figure 18.17:

(a) Find the resonant frequency in hertz.
(b) Find the value of Q_s.
(c) Find the polar forms of i, v_R, v_L, and v_C at resonance.
(d) Verify Kirchhoff's voltage law around the circuit.
(e) Make a phasor diagram showing i, v_R, v_L, and v_C.

SOLUTION

(a) $f_0 = \dfrac{1}{2\pi\sqrt{LC}} = \dfrac{1}{2\pi\sqrt{(10 \times 10^{-3})(0.1 \times 10^{-6})}} = 5033$ Hz

(b) $Q_s = \dfrac{1}{R}\sqrt{\dfrac{L}{C}} = \dfrac{1}{33}\sqrt{\dfrac{10 \times 10^{-3}}{0.1 \times 10^{-6}}} = 9.58$

(c) At resonance, $Z_T = R = 33 \underline{/0°}$ Ω. Therefore,

$$i = \frac{e}{R} = \frac{13.2 \underline{/0°} \text{ V}}{33 \underline{/0°} \text{ }\Omega} = 0.4 \underline{/0°} \text{ A}$$

$$v_R = iR = (0.4 \underline{/0°} \text{ A})(33 \underline{/0°} \cdot \Omega) = 13.2 \underline{/0°} \text{ V}$$

Note that $v_R = e$, confirming equation (18.34).
The inductive and capacitive reactances at resonance are

$$X_L = j\omega_0 L = j(2\pi)(5033)(10 \times 10^{-3}) = j316.2 \text{ } \Omega$$

$$X_C = \frac{-j}{\omega_0 C} = \frac{-j}{(2\pi)(5033)(0.1 \times 10^{-6})} = -j316.2 \text{ } \Omega$$

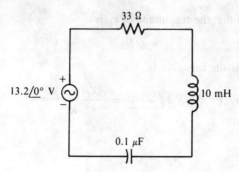

FIGURE 18.17 (Example 18.8)

Therefore,

$$v_L = iX_L = (0.4 \underline{/0°} \text{ A})(316.2 \underline{/90°} \text{ Ω}) = 126.5 \underline{/90°} \text{ V}$$

$$v_C = iX_C = (0.4 \underline{/0°} \text{ A})(316.2 \underline{/-90°} \text{ Ω}) = 126.5 \underline{/-90°} \text{ V}$$

Note that $|v_L| = |v_C| = Q_s|e| = (9.58)(13.2) = 126.5$ V, confirming equations (18.31) and (18.33).

(d) $v_R + v_L + v_C = 13.2 \underline{/0°} \text{ V} + 126.5 \underline{/90°} \text{ V} + 126.5 \underline{/-90°} \text{ V}$

$$= (13.2 + j126.5 - j126.5) \text{ V}$$

$$= (13.2 + j0) \text{ V} = e$$

Although the magnitudes of v_L and v_C are each greater than the magnitude of e, we see that those two voltages are 180° out of phase and therefore add to zero.

(e) The phasor diagram is shown in Figure 18.18. The fact that v_L and v_C are 180° out of phase is again clearly apparent. Note also that e and i have the same angle (are in phase), as will always be the case at resonance.

Drill Exercise 18.8

(a) For what new value of R in Figure 18.17 will the quality factor of the circuit equal 15?

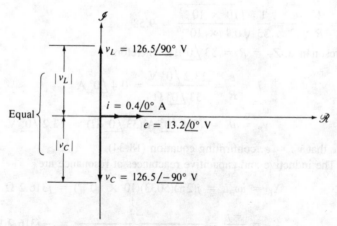

FIGURE 18.18 (Example 18.8)

(b) What will be the magnitude of the voltage across the inductor in Figure 18.17 if the value of R found in part (a) is used?

ANSWER: (a) 21.1 Ω; (b) 198 V.   □

18.5 The Series RLC Circuit as a Bandpass Filter

Figure 18.19 shows how a series RLC circuit is used as a filter, with the output voltage taken across the resistor. Before developing equations that show how v_o changes with frequency, let us perform an intuitive analysis and verify that this arrangement behaves as a bandpass filter. By voltage-divider action, we know that v_o will be maximum when the ratio $R/Z_T = R/(R + j|X_L| - j|X_C|)$ is maximum. This occurs when the denominator of the ratio is minimum, that is, when the phasor sum of X_L and X_C is as small as possible. We already know that X_L and X_C cancel each other at resonance, so we conclude that v_o is maximum at the resonant frequency. On the other hand, at very high frequencies $|X_L|$ is very large, making the ratio of R to Z_T very small. Thus, v_o is very small at high frequencies. Similarly, at very low frequencies, $|X_C|$ is large and the ratio of R to Z_T is again small. Thus, v_o is also small at low frequencies. This analysis shows that the output is small at low frequencies, small at high frequencies, and maximum at an in-between frequency (resonance), which are exactly the characteristics of a bandpass filter.

To develop a mathematical expression for the magnitude of v_o, we apply the voltage-divider rule as follows:

$$|v_o| = \frac{R|e_{in}|}{|Z_T|} = \frac{R|e_{in}|}{\sqrt{R^2 + (|X_L| - |X_C|)^2}}$$

$$= \frac{R|e_{in}|}{\sqrt{R^2 + (\omega L - 1/\omega C)^2}} = \frac{R|e_{in}|}{\sqrt{R^2 + (2\pi fL - 1/2\pi fC)^2}} \quad (18.35)$$

When $|X_L| = |X_C|$, equation (18.35) shows that

$$|v_o| = \frac{R|e_{in}|}{\sqrt{R^2 + 0}} = |e_{in}|$$

Thus, $|v_o|$ reaches its maximum value of $|e_{in}|$ when $|X_L| = |X_C|$ (i.e., at resonance).

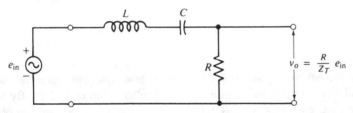

FIGURE 18.19 Series RLC circuit as a bandpass filter.

Equation (18.35) also shows that $|v_o|$ becomes very small (approaches zero) when frequency f is very small, as well as when f is very large.

We can derive an expression for $|v_o|$ in terms of Q_s and ω_0 by algebraic manipulation of equation (18.35):

$$|v_o| = \frac{R|e_{in}|}{\sqrt{R^2 + (\omega L - 1/\omega C)^2}} = \frac{R|e_{in}|}{R\sqrt{1 + (1/R^2)(\omega L - 1/\omega C)^2}}$$

$$= \frac{|e_{in}|}{\sqrt{1 + (\omega L/R - 1/\omega RC)^2}} \tag{18.36}$$

From equations (18.25) and (18.26), we have

$$\frac{L}{R} = \frac{Q_s}{\omega_0} \quad \text{and} \quad \frac{1}{RC} = \omega_0 Q_s$$

Substituting for L/R and $1/RC$ in (18.36), we obtain

$$|v_o| = \frac{|e_{in}|}{\sqrt{1 + \left(\dfrac{\omega}{\omega_0} Q_s - \dfrac{\omega_0}{\omega} Q_s\right)^2}}$$

$$= \frac{|e_{in}|}{\sqrt{1 + Q_s^2 \left(\dfrac{\omega}{\omega_0} - \dfrac{\omega_0}{\omega}\right)^2}} = \frac{|e_{in}|}{\sqrt{1 + Q_s^2 \left(\dfrac{f}{f_0} - \dfrac{f_0}{f}\right)^2}} \tag{18.37}$$

Figure 18.20 shows normalized plots of $|v_o|$ versus frequency for several different values of Q_s. Notice that the curves become narrower and more symmetrical as Q_s increases. Thus, the greater the value of Q_s, the smaller the bandwidth of the filter. Bandpass filters are *selective*, in the sense that they pass only those signals whose frequencies fall within a selected range. *Selectivity* is the term used to describe that property: The narrower the bandwidth and the more steeply the response characteristic falls on either side of the resonant frequency, the greater the selectivity of the filter. Bandpass filters are widely used in communications circuits, where they are often called *tuned* circuits. Tuning a filter means adjusting the resonant frequency (by adjusting L or C) until it corresponds to the frequency of a specified signal.

The lower and upper cutoff frequencies of the series RLC filter can be found by setting equation (18.37) equal to $(\sqrt{2}/2)|e_{in}|$ and solving for f (or ω). The two solutions to the equation are

$$f_1 = \frac{f_0\sqrt{(1/Q_s^2) + 4} - f_0/Q_s}{2} \quad \text{hertz} \tag{18.38}$$

$$f_2 = \frac{f_0\sqrt{(1/Q_s^2) + 4} + f_0/Q_s}{2} \quad \text{hertz} \tag{18.39}$$

Only the *positive* square roots are used in these equations. As usual, ω_1 and ω_2 can be found from (18.38) and (18.39) using the relationship $\omega = 2\pi f$. By substituting $f_0 = 1/2\pi\sqrt{LC}$ and $Q_s = (\sqrt{L/C})/R$ into (18.38) and (18.39), we can obtain expressions for the cutoff frequencies, in terms of the circuit components:

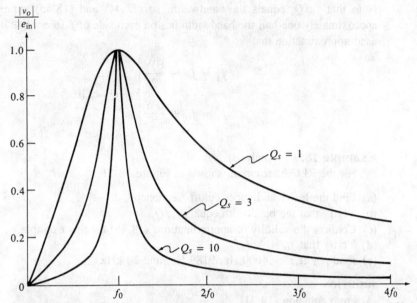

FIGURE 18.20 Normalized plots of the output of a series RLC bandpass filter for different values of the quality factor.

$$f_1 = \frac{1}{4\pi} \left(\sqrt{\left(\frac{R}{L}\right)^2 + \frac{4}{LC}} - \frac{R}{L} \right) \qquad \text{hertz} \qquad (18.40)$$

$$f_2 = \frac{1}{4\pi} \left(\sqrt{\left(\frac{R}{L}\right)^2 + \frac{4}{LC}} + \frac{R}{L} \right) \qquad \text{hertz} \qquad (18.41)$$

It is an exercise at the end of this chapter to show that the bandwidth of the filter is

$$\text{BW} = f_2 - f_1 = \frac{f_0}{Q_s} = \frac{R}{2\pi L} \qquad \text{hertz} \qquad (18.42)$$

Equation (18.42) shows that the bandwidth is inversely proportional to Q_s, confirming our previous observation that the response characteristic becomes narrower as Q_s increases. The resonant frequency f_0 is often called the *center* frequency of the filter, although it is not necessarily midway between the cutoff frequencies. In fact, f_0 is at the *geometric center* of the response, defined by

$$f_0 = \sqrt{f_1 f_2} \qquad (18.43)$$

For large values of Q_s, the geometric center very nearly corresponds to the algebraic center, that is, to the frequency midway between the cutoff frequencies. This fact is confirmed by equations (18.38) and (18.39): when Q_s is very large, $1/Q_s^2 \approx 0$ and

$$f_1 \approx f_0 - \frac{1}{2}\left(\frac{f_0}{Q_s}\right) \qquad \text{hertz} \qquad (18.44)$$

$$f_2 \approx f_0 + \frac{1}{2}\left(\frac{f_0}{Q_s}\right) \qquad \text{hertz} \qquad (18.45)$$

Note that f_0/Q_s equals the bandwidth, so (18.44) and (18.45) express the fact that approximately one-half the bandwidth lies on each side of f_0. For $Q_s \geq 10$, it is a widely used approximation that

$$
\left.
\begin{aligned}
f_1 &\approx f_0 - \frac{BW}{2} \\
f_2 &\approx f_0 + \frac{BW}{2}
\end{aligned}
\right\} \quad (Q_s \geq 10)
\tag{18.46}
$$

Example 18.9 (Analysis)

For the RLC filter circuit shown in Figure 18.21:

(a) Find the lower and upper cutoff frequencies.
(b) Verify that the bandwidth equals f_0/Q_s.
(c) Confirm the validity of approximations (18.46) for this example.
(d) Verify that $f_0 = \sqrt{f_1 f_2}$.
(e) Find $|v_o|$ at $f = 4000$ Hz, 9189 Hz, and 20 kHz.

SOLUTION

(a) From equation (18.21),

$$
f_0 = \frac{1}{2\pi\sqrt{LC}} = \frac{1}{2\pi\sqrt{(15 \times 10^{-3})(0.02 \times 10^{-6})}} = 9189 \text{ Hz}
$$

From equation (18.28),

$$
Q_s = \frac{1}{R}\sqrt{\frac{L}{C}} = \frac{1}{75}\sqrt{\frac{15 \times 10^{-3}}{0.02 \times 10^{-6}}} = 11.55
$$

From equations (18.38) and (18.39),

$$
f_1 = \frac{f_0\sqrt{\dfrac{1}{Q_s^2} + 4} - \dfrac{f_0}{Q_s}}{2} = \frac{9189\sqrt{\dfrac{1}{(11.55)^2} + 4} - \dfrac{9189}{11.55}}{2} = 8800 \text{ Hz}
$$

$$
f_2 = \frac{f_0\sqrt{\dfrac{1}{Q_s^2} + 4} + \dfrac{f_0}{Q_s}}{2} = \frac{9189\sqrt{\dfrac{1}{(11.55)^2} + 4} + \dfrac{9189}{11.55}}{2} = 9595 \text{ Hz}
$$

(b) Using the results of part (a) and the definition of bandwidth, we find

$$
BW = f_2 - f_1 = 9595 \text{ Hz} - 8800 \text{ Hz} = 795 \text{ Hz}
$$

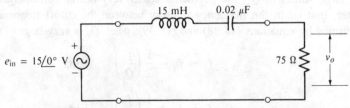

FIGURE 18.21 (Example 18.9)

Also,

$$\frac{f_0}{Q_s} = \frac{9189}{11.55} = 795 \text{ Hz} = \text{BW}$$

(c) Since $Q_s \geq 10$, f_0 should be very nearly at the algebraic center of the passband:

$$f_1 \approx f_0 - \frac{\text{BW}}{2} = 9189 - \frac{795}{2} = 8792 \text{ Hz}$$

$$f_2 \approx f_0 + \frac{\text{BW}}{2} = 9189 + \frac{795}{2} = 9587 \text{ Hz}$$

We see that the approximations yield values that are only 8 Hz less than the exact values of f_1 and f_2: 8800 Hz and 9595 Hz.

(d) Using the results of part (a), we have

$$\sqrt{f_1 f_2} = \sqrt{(8800)(9595)} = 9189 \text{ Hz}$$

(e) From equation (18.37),

$$|v_o| = \frac{|e_{\text{in}}|}{\sqrt{1 + Q_s^2 \left(f/f_0 - f_0/f\right)^2}}$$

At $f = 4000$ Hz,

$$|v_o| = \frac{15}{\sqrt{1 + (11.55)^2 \left(\dfrac{4000}{9189} - \dfrac{9189}{4000}\right)^2}} = 0.697 \text{ V}$$

At $f = 9189$ Hz,

$$|v_o| = \frac{15}{\sqrt{1 + (11.55)^2 \left(\dfrac{9189}{9189} - \dfrac{9189}{9189}\right)^2}} = 15 \text{ V}$$

At $f = 20$ kHz,

$$|v_o| = \frac{15}{\sqrt{1 + (11.55)^2 \left(\dfrac{2 \times 10^4}{9189} - \dfrac{9189}{2 \times 10^4}\right)^2}} = 0.755 \text{ V}$$

These computations show that $|v_o|$ reaches its maximum value of $|e_{\text{in}}| = 15$ V at $f = f_0 = 9189$ Hz, and that $|v_o|$ is considerably smaller at frequencies below and above f_0.

Drill Exercise 18.9

Find Q_s, f_1, f_2, and the bandwidth of the filter in Figure 18.21 when the 75-Ω resistor is replaced by a 500-Ω resistor.

ANSWER: $Q_s = 1.732$; $f_1 = 6912$ Hz; $f_2 = 12,217$ Hz; BW = 5305 Hz.   □

When designing practical RLC filters, it is necessary to consider the presence of resistance in the inductor winding and the effect of that resistance on filter performance. The equivalent series resistance, R_1, of the inductor (see Figure 11.14) must be added to any other resistance in the circuit. Consequently, R_1 contributes to a reduction in the quality factor of the filter, thus broadening the response curve and reducing the selectivity. Furthermore, the magnitude of the output voltage is reduced by the voltage divider action that takes place between R_1 and the resistance across which the output is obtained. Instead of reaching a maximum value of $|e_{in}|$ at resonance, the maximum value of the output is

$$|v_o| = \frac{R}{R_1 + R} |e_{in}| \tag{18.47}$$

where R is the resistance across which the output is obtained.

Example 18.10   (Design)

An RLC bandpass filter is to be designed using a 10-mH inductor whose resistance is 75 Ω. The center frequency of the filter is to be 25 kHz.

(a) What value of capacitance should be used?
(b) If the bandwidth of the filter must be no greater than 2500 Hz, what is the maximum value of resistance across which the output can be obtained?
(c) When the maximum resistance found in part (b) is used, what is the maximum output voltage of the filter if $|e_{in}| = 12$ V?

SOLUTION

(a) $f_0 = \dfrac{1}{2\pi\sqrt{LC}} \Rightarrow 25 \times 10^3 = \dfrac{1}{2\pi\sqrt{10 \times 10^{-3}C}}$

Squaring both sides and solving for C, we find

$$C = \frac{1}{4\pi^2 (10 \times 10^{-3})(25 \times 10^3)^2} = 4050 \text{ pF}$$

(b) If the bandwidth were equal to its maximum permissible value of 2500 Hz, the quality factor of the circuit would be

$$Q_s = \frac{f_0}{\text{BW}} = \frac{25 \text{ kHz}}{2500 \text{ Hz}} = 10$$

Thus, we require that Q_s be greater than or equal to 10. We can use equation (18.28) to find the maximum total resistance, R_T, corresponding to a minimum Q_s of 10:

$$Q_s = 10 \leq \frac{1}{R_T}\sqrt{\frac{L}{C}} = \frac{1}{R_T}\sqrt{\frac{10 \times 10^{-3}}{4050 \times 10^{-12}}} = \frac{1571}{R_T}$$

or

$$R_T \leq 157 \ \Omega$$

Since 157 Ω is the maximum permissible value of the *total* resistance in the circuit, the maximum resistance across which the output can be taken is

$$R = R_T - R_1 = 157 \ \Omega - 75 \ \Omega = 82 \ \Omega$$

(c) From equation (18.47), the maximum output is

$$|v_o| = \frac{R}{R_1 + R} |e_{in}| = \left(\frac{82}{75 + 82}\right) 12 \ V = 6.27 \ V$$

Drill Exercise 18.10

Suppose that the bandwidth of the filter in Example 18.10 cannot exceed 2 kHz. If the resistor across which the output is taken must be 100 Ω, what is the maximum permissible resistance of the 10-mH inductor?

ANSWER: 25.7 Ω.

18.6 Parallel-Resonant RLC Circuits

Figure 18.22 shows a parallel RLC circuit, often called a *tank* circuit. For the moment, we assume that the inductor is ideal, in the sense that it has zero winding resistance ($R_1 = 0$). Recall that the resonant frequency of a circuit is defined to be that frequency at which the reactive (imaginary) component of the total impedance is zero. It follows that the resonant frequency is also that frequency at which the imaginary part of the total *admittance* is zero. The total admittance of the circuit in Figure 18.22 is

$$Y_T = \frac{1}{R} + j\omega C - \frac{j}{\omega L}$$

$$= \frac{1}{R} + j\left(\omega C - \frac{1}{\omega L}\right) \tag{18.48}$$

The resonant frequency, ω_0, can therefore be found by setting the imaginary part of (18.48) equal to zero and solving for ω_0:

$$\omega_0 C - \frac{1}{\omega_0 L} = 0$$

$$\omega_0^2 = \frac{1}{LC}$$

$$\omega_0 = \frac{1}{\sqrt{LC}} \qquad \text{rad/s} \tag{18.49}$$

$$f_0 = \frac{1}{2\pi\sqrt{LC}} \qquad \text{hertz} \tag{18.50}$$

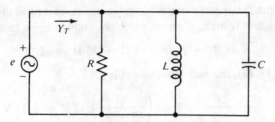

FIGURE 18.22 Parallel RLC (tank) circuit.

Equations (18.49) and (18.50) show that the resonant frequency of the parallel RLC circuit is computed in exactly the same way it is for a series RLC circuit. In both cases, f_o is the frequency at which the inductive reactance equals the capacitive reactance. However, as frequency changes, the impedance variations of the two circuits are quite different. Consider the parallel equivalent impedance of the inductor and capacitor:

$$X_L \| X_C = \frac{(j|X_L|)(-j|X_C|)}{j|X_L| - j|X_C|} = \frac{|X_L||X_C|}{j(|X_L| - |X_C|)} \qquad (18.51)$$

At resonance, $|X_L| = |X_C|$, which makes the denominator of (18.51) equal to zero. Thus, at resonance, the LC combination theoretically has infinite impedance. (This result can follow only from the unrealistic assumption of an *ideal* inductor—one having zero resistance.) At frequencies below or above resonance, the impedance of the LC combination reduces the total impedance of the circuit, since the combination is in parallel with R. Thus, the total impedance reaches a *maximum* value at resonance, and the total current reaches a minimum value. Recall that this variation is just the opposite of that in a series RLC circuit.

Of course, the voltages across all three components in a parallel RLC circuit are equal at every frequency. The currents in the components are

$$i_R = \frac{e}{R} \qquad i_L = \frac{e}{X_L} \qquad i_C = \frac{e}{X_C} \qquad (18.52)$$

At resonance, $|X_L| = |X_C|$, so $|i_L| = |i_C|$. The phasor sum of i_L and i_C is zero at resonance, since those currents are always 180° out of phase. Thus, the total current at resonance is $i_R = e/R$.

Example 18.11 (Analysis)
For the circuit shown in Figure 18.23:

(a) Find the resonant frequency in hertz and radians per second.
(b) Find the inductive and capacitive reactance and the total impedance at resonance.
    Also find the total impedance at a frequency equal to twice the resonant frequency.
(c) Find the currents i_R, i_L, i_C and i_T at resonance. Make a phasor diagram showing these quantities.

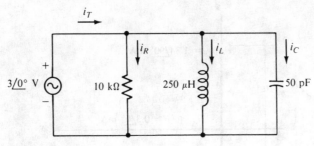

FIGURE 18.23 (Example 18.11)

SOLUTION

(a) $\omega_0 = \dfrac{1}{\sqrt{LC}} = \dfrac{1}{\sqrt{(250 \times 10^{-6})(50 \times 10^{-12})}} = 8.944 \times 10^6$ rad/s

$f_0 = \dfrac{\omega_0}{2\pi} = \dfrac{8.944 \times 10^6}{2\pi} = 1.4235$ MHz

(b) At resonance,

$$X_L = j\omega_0 L = j(8.944 \times 10^6)(250 \times 10^{-6}) = j2236 \ \Omega$$

$$X_C = -\dfrac{j}{\omega_0 C} = -\dfrac{j}{(8.944 \times 10^6)(50 \times 10^{-12})} = -j2236 \ \Omega$$

Since the total admittance at resonance is $Y_T = 1/R + j0$, the total impedance is

$$Z_T = \dfrac{1}{Y_T} = R + j0 = 10^4 + j0 \ \Omega$$

At $\omega = 2\omega_0 = 17.888 \times 10^6$ rad/s,

$$\omega C = (17.888 \times 10^6)(50 \times 10^{-12}) = 8.944 \times 10^{-4}$$

$$\dfrac{1}{\omega L} = \dfrac{1}{(17.888 \times 10^6)(250 \times 10^{-6})} = 2.236 \times 10^{-4}$$

From equation (18.48),

$$Y_T = \dfrac{1}{R} + j\left(\omega C - \dfrac{1}{\omega L}\right)$$

$$= \dfrac{1}{10^4} + j(8.944 \times 10^{-4} - 2.236 \times 10^{-4})$$

$$= 10^{-4} + j6.708 \times 10^{-4} \ \text{S} = 6.782 \times 10^{-4} \underline{/81.52°} \ \text{S}$$

Therefore,

$$Z_T = \dfrac{1}{Y_T} = \dfrac{1}{6.782 \times 10^{-4} \underline{/81.52°}} = 1474 \underline{/-81.52°} \ \Omega$$

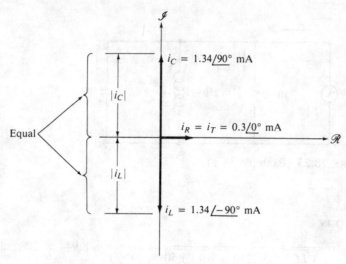

FIGURE 18.24 (Example 18.11)

This result confirms that the magnitude of the total impedance at a frequency other than resonance is smaller than it is at resonance.

(c) $i_R = \dfrac{e}{R} = \dfrac{3\,\underline{/0°}\ \text{V}}{10^4\,\underline{/0°}\ \Omega} = 0.3\,\underline{/0°}\ \text{mA}$

$i_L = \dfrac{e}{X_L} = \dfrac{3\,\underline{/0°}\ \text{V}}{2236\,\underline{/90°}\ \Omega} = 1.34\,\underline{/-90°}\ \text{mA}$

$i_C = \dfrac{e}{X_C} = \dfrac{3\,\underline{/0°}\ \text{V}}{2236\,\underline{/-90°}\ \Omega} = 1.34\,\underline{/90°}\ \text{mA}$

$i_T = i_R + i_L + i_C = (0.3 - j1.34 + j1.34)\ \text{mA} = 0.3 + j0\ \text{mA}$

The phasor diagram is shown in Figure 18.24. The cancellation of i_L and i_C at resonance is readily apparent.

Drill Exercise 18.11

Find the total current i_T in Figure 18.23 at a frequency equal to one-half the resonant frequency.

ANSWER: $2.035\,\underline{/-81.52°}\ \text{mA}$.   □

The Quality Factor (Q_p)

As in the series circuit, the quality factor (Q_p) of a parallel RLC circuit is the ratio of the reactive power in either L or C at resonance to the average power in R. Since the magnitude of the voltage is the same across every parallel component, and since $|X_L| = |X_C|$ at resonance, the reactive power in L equals the reactive power in C:

$$\frac{V_p^2}{2|X_L|} = \frac{V_p^2}{2|X_C|} \qquad \text{vars} \tag{18.53}$$

The average power in R is

$$P_{\text{avg}} = \frac{V_p^2}{2R} \qquad \text{watts} \tag{18.54}$$

Therefore,

$$Q_p = \frac{V_p^2/2|X_L|}{V_p^2/2R} = \frac{R}{|X_L|} = \frac{R}{\omega_0 L} \tag{18.55}$$

$$= \frac{V_p^2/2|X_C|}{V_p^2/2R} = \frac{R}{|X_C|} = \omega_0 CR \tag{18.56}$$

where the subscript p refers to the parallel configuration. Note carefully that Q_p is computed differently from Q_s. In terms of component values,

$$Q_p = R\sqrt{\frac{C}{L}} \tag{18.57}$$

As demonstrated in Example 18.11, i_L and i_C cancel at resonance, making the total current

$$i_T = i_R = \frac{e}{R}$$

The ratio of i_L to i_T at resonance is

$$\frac{i_L}{i_T} = \frac{e/X_L}{e/R} = \frac{R}{X_L} = \frac{R}{\omega_0 L \underline{/90°}} \tag{18.58}$$

Therefore, at resonance,

$$|i_L| = \frac{R}{\omega_0 L}|i_T| = Q_p|i_T| \tag{18.59}$$

Similarly,

$$|i_C| = \omega_0 RC|i_T| = Q_p|i_T| \tag{18.60}$$

Equations (18.59) and (18.60) show that the magnitude of the current in each reactive component at resonance is Q_p times the magnitude of the total current at resonance. Compare this result with the relationship between Q_s and the voltages in a series-resonant circuit.

The Parallel-Resonant Circuit as a Bandpass Filter

When a parallel RLC circuit is operated as a filter, the input is typically produced by a transistor, or a similar electronic device, that behaves like a constant-current source. The circuit is shown in Figure 18.25. We have already discussed the fact that the total impedance of this circuit reaches a maximum at resonance. Therefore, since the current

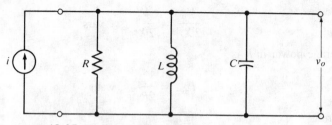

FIGURE 18.25 Bandpass RLC filter driven by a constant-current source.

is constant, the voltage across the circuit also reaches a maximum at resonance. At frequencies below and above resonance, the impedance decreases and the voltage is smaller. Thus, the configuration has the characteristics of a bandpass filter.

Since the total admittance of the parallel RLC circuit is

$$Y_T = \frac{1}{R} + j\left(\omega C - \frac{1}{\omega L}\right)$$

the output voltage is

$$v_o = \frac{i}{Y_T} = \frac{i}{(1/R) + j(\omega C - 1/\omega L)}$$

Therefore, the magnitude of the output is

$$|v_o| = \frac{|i|}{\sqrt{(1/R^2) + (\omega C - 1/\omega L)^2}} \tag{18.61}$$

Using a method similar to that used to derive equation (18.37), we can express v_o in terms of Q_p (Exercise 18.26):

$$|v_o| = \frac{|i|R}{\sqrt{1 + Q_p^2\left(\dfrac{\omega}{\omega_0} - \dfrac{\omega_0}{\omega}\right)^2}} = \frac{|i|R}{\sqrt{1 + Q_p^2\left(\dfrac{f}{f_0} - \dfrac{f_0}{f}\right)^2}} \tag{18.62}$$

Note that the numerator, $|i|R$, in equation (18.62) is the magnitude of the output voltage at resonance (i.e., the maximum value of the output). Equation (18.62) has the same form as equation (18.37) for the output of the series RLC circuit, so the frequency response curves in Figure 18.20 also apply to the parallel circuit, with Q_p substituted for Q_s. The lower and upper cutoff frequencies for the parallel circuit can be found by setting (18.62) equal to $(\sqrt{2}/2)|i|R$ and solving:

$$f_1 = \frac{f_0\sqrt{\dfrac{1}{Q_p^2} + 4} - \dfrac{f_0}{Q_p}}{2} \qquad \text{hertz} \tag{18.63}$$

$$f_2 = \frac{f_0\sqrt{\dfrac{1}{Q_p^2} + 4} + \dfrac{f_0}{Q_p}}{2} \qquad \text{hertz} \tag{18.64}$$

Note the similarity of these equations to (18.38) and (18.39). In terms of component values, the cutoff frequencies are

$$f_1 = \frac{1}{4\pi}\left(\sqrt{\frac{1}{(RC)^2} + \frac{4}{LC}} - \frac{1}{RC}\right) \qquad \text{hertz} \qquad (18.65)$$

$$f_2 = \frac{1}{4\pi}\left(\sqrt{\frac{1}{(RC)^2} + \frac{4}{LC}} + \frac{1}{RC}\right) \qquad \text{hertz} \qquad (18.66)$$

Using equations (18.63) through (18.66), we find

$$BW = f_2 - f_1 = \frac{f_0}{Q_p} = \frac{1}{2\pi RC} \qquad \text{hertz} \qquad (18.67)$$

Equation (18.67) shows that the bandwidth of the filter is inversely proportional to the quality factor, Q_p. As in the series RLC circuit, f_0 is called the *center* frequency of the circuit, and is approximately at the algebraic center of the passband for $Q_p \geq 10$. In general,

$$f_0 = \sqrt{f_1 f_2} \qquad (18.68)$$

Example 18.12   (Analysis)
For the filter circuit shown in Figure 18.26:

(a) Find the magnitude of the maximum output voltage.
(b) Find the center frequency and the quality factor.
(c) Find the cutoff frequencies and the bandwidth.

SOLUTION
(a) The output voltage is maximum at resonance and is equal to

$$|i|R = (40 \text{ mA})(500 \text{ }\Omega) = 20 \text{ V}$$

(b) $f_0 = \dfrac{1}{2\pi\sqrt{LC}} = \dfrac{1}{2\pi\sqrt{(0.1 \times 10^{-3})(0.1 \times 10^{-6})}} = 50.33 \text{ kHz}$

$Q_p = R\sqrt{\dfrac{C}{L}} = 500\sqrt{\dfrac{0.1 \times 10^{-6}}{0.1 \times 10^{-3}}} = 15.81$

(c) $f_1 = \dfrac{f_0\sqrt{\dfrac{1}{Q_p^2} + 4} - \dfrac{f_0}{Q_p}}{2}$

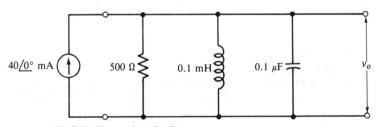

FIGURE 18.26 (Example 18.12)

$$= \frac{50.33 \times 10^3 \sqrt{\dfrac{1}{(15.81)^2} + 4} - \dfrac{50.33 \times 10^3}{15.81}}{2} = 48.76 \text{ kHz}$$

$$f_2 = \frac{f_0 \sqrt{\dfrac{1}{Q_p^2} + 4} + \dfrac{f_0}{Q_p}}{2}$$

$$= \frac{50.33 \times 10^3 \sqrt{\dfrac{1}{(15.81)^2} + 4} + \dfrac{50.33 \times 10^3}{15.81}}{2} = 51.94 \text{ kHz}$$

$$\text{BW} = f_2 - f_1 = 51.94 \text{ kHz} - 48.76 \text{ kHz} = 3.18 \text{ kHz}$$

Also, note that

$$\text{BW} = \frac{f_0}{Q_p} = \frac{50.33 \text{ kHz}}{15.81} = 3.18 \text{ kHz}$$

Since Q_p is large, the cutoff frequencies are at approximately one-half the bandwidth below and above the center frequency:

$$f_1 \approx f_0 - \frac{\text{BW}}{2} = 50.33 \text{ kHz} - \frac{3.18 \text{ kHz}}{2} = 48.74 \text{ kHz}$$

$$f_2 \approx f_0 + \frac{\text{BW}}{2} = 50.33 \text{ kHz} + \frac{3.18 \text{ kHz}}{2} = 51.92 \text{ kHz}$$

Drill Exercise 18.12

Find the magnitude of v_o in Example 18.12 at frequencies equal to one-half and twice the center frequency.

ANSWER: 0.843 V at both frequencies.   □

18.7 Series–Parallel Resonance

The results derived in the preceding section were based on the assumption that the inductor was ideal (i.e., that it had zero resistance). Of course, practical inductors do have winding resistance, and in some circuits that resistance is large enough to make the ideal equations deviate significantly from reality. In this section we will develop equations that take winding resistance into account, and investigate the conditions under which it can be neglected.

Figure 18.27 shows the inductor represented by an ideal inductor in series with winding resistance R_1. For notation purpose, any other resistance in series with L, such as a resistor, is included in the term R_1. In keeping with our discussion in the preceding section, we assume that the circuit is driven by a constant-current source. For the moment,

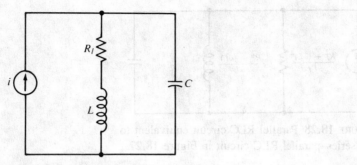

FIGURE 18.27 Series–parallel RLC circuit containing a practical inductor with winding resistance R_l.

we also assume that there is no resistor in parallel with the source (corresponding to the assumption of an ideal current source). Since the capacitor is in parallel with the series combination of the inductor and resistance, we refer to this configuration as a *series–parallel RLC* circuit.

To determine the resonant frequency of the circuit in Figure 18.27, we write an expression for its total admittance:

$$Y_T = j\omega C + \frac{1}{R_l + j\omega L} \tag{18.69}$$

Rationalizing the term $1/(R_1 + j\omega L)$ leads to

$$Y_T = j\omega C + \frac{R_l}{R_l^2 + (\omega L)^2} - \frac{j\omega L}{R_l^2 + (\omega L)^2} \tag{18.70}$$

Collecting real and imaginary parts gives

$$Y_T = \frac{R_l}{R_l^2 + (\omega L)^2} + j\left[\omega C - \frac{\omega L}{R_l^2 + (\omega L)^2}\right] \tag{18.71}$$

Setting the imaginary part of (18.71) equal to zero and solving for ω_0, we find

$$\cancel{\omega_0}C = \frac{\cancel{\omega_0}L}{R_l^2 + (\omega_0 L)^2}$$

$$R_l^2 C + \omega_0^2 L^2 C = L$$

$$\omega_0^2 = \frac{L - R_l^2 C}{L^2 C} = \frac{1}{LC} - \frac{R_l^2}{L^2}$$

$$\omega_0 = \sqrt{\frac{1}{LC} - \frac{R_l^2}{L^2}} = \frac{1}{\sqrt{LC}} \sqrt{1 - \frac{R_l^2 C}{L}} \quad \text{rad/s} \tag{18.72}$$

or

$$f_0 = \frac{1}{2\pi\sqrt{LC}} \sqrt{1 - \frac{R_l^2 C}{L}} \quad \text{hertz} \tag{18.73}$$

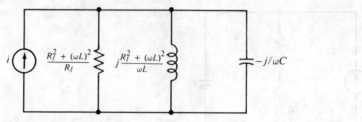

FIGURE 18.28 Parallel RLC circuit equivalent to the series–parallel RLC circuit in Figure 18.27.

Unlike series and parallel RLC circuits, we see that the resonant frequency in the series–parallel circuit depends on resistance R_l. Note that when $R_l = 0$, equation (18.73) reduces to the equation for the resonant frequency of the ideal circuit: $f_0 = 1/(2\pi\sqrt{LC})$.

Examination of equation (18.70) for the total admittance reveals three distinct components: a conductance equal to $R_l/(R_l^2 + \omega^2 L^2)$ S, a capacitive susceptance equal to $j\omega C$ S, and an inductive susceptance equal to $-j\omega L/(R_l^2 + \omega^2 L^2)$ S. The circuit is therefore equivalent to a parallel circuit containing each of these components. Figure 18.28 shows the equivalent circuit with each component labeled with its impedance (the reciprocals of the admittances). This circuit is, in fact, the equivalent circuit that results when the methods described in Section 14.9 are used to convert the series $R_l L$ combination to a parallel equivalent. At resonance, the magnitude of the equivalent inductive reactance in the circuit equals the magnitude of the capacitive reactance: $[R_l^2 + (\omega_0 L)^2]/\omega_0 L = 1/\omega_0 C$. Therefore, the magnitude of the total impedance at resonance is just the resistive component: $|Z_T| = [R_l^2 + (\omega_0 L)^2]/R_l$.

Using Figure 18.28, we can now derive an expression for the quality factor of the series–parallel circuit, which we will designate Q_{sp}. The ratio of the reactive power in the equivalent inductance to the average power in the equivalent resistance at resonance is

$$Q_{sp} = \frac{\dfrac{V_P^2}{2\left[\dfrac{R_l^2 + (\omega_0 L)^2}{\omega_0 L}\right]}}{\dfrac{V_P^2}{2\left[\dfrac{R_l^2 + (\omega_0 L)^2}{R_l}\right]}} = \frac{\omega_0 L}{R_l} \tag{18.74}$$

Equivalently, the ratio of the reactive power in the capacitor to the average power in the equivalent resistance at resonance is

$$Q_{sp} = \frac{\dfrac{V_P^2}{2(1/\omega_0 C)}}{\dfrac{V_P^2}{2\left[\dfrac{R_l^2 + (\omega_0 L)^2}{R_l}\right]}} = \frac{\omega_0 C}{R_l}\left[R_l^2 + (\omega_0 L)^2\right] \tag{18.75}$$

To develop a criterion for determining when R_l can be neglected, we wish to express equation (18.73) for the resonant frequency of the series–parallel circuit in terms of the quality factor. With some algebraic manipulation, it can be shown that

$$f_0 = \frac{1}{2\pi\sqrt{LC}} \sqrt{\frac{Q_{sp}^2}{Q_{sp}^2 + 1}} \tag{18.76}$$

When Q_{sp} is large, $Q_{sp}^2 + 1 \approx Q_{sp}^2$, so the term in the radical is approximately $\sqrt{Q_{sp}^2/Q_{sp}^2} = 1$. Under that condition, the resonant frequency becomes $f_0 \approx 1/(2\pi\sqrt{LC})$, that is, the same as in a parallel RLC circuit. The approximation is widely used when $Q_{sp} \geq 10$, and the circuit is said to be "high-Q" in that case. When $Q_{sp} = 10$, the exact and approximate values differ by less than 0.5%. However, note carefully that Q_{sp} is the quality factor of the inductor *at the resonant frequency*, whereas (18.76) is an equation for the resonant frequency in terms of Q_{sp}. Thus, (18.76) cannot be used by itself to solve for f_0.

Let us now consider the effect of resistance R_S in parallel with the current source. R_S could represent the internal resistance of the source itself. The equivalent circuit is shown in Figure 18.29. Since R_S does not affect the imaginary part of the total admittance, the value of the resonant frequency is not changed by R_S. Thus, f_0 is still given by equation (18.73). However, the quality factor of the circuit becomes

$$Q_{sp}' = \frac{\omega_0 L}{R_l + [R_l^2 + (\omega_0 L)^2]/R_S} \tag{18.77}$$

or

$$Q_{sp}' = \frac{Q_{sp}}{1 + R_p/R_S} \tag{18.78}$$

where $R_p = [R_l^2 + (\omega_0 L)^2]/R_l$.

Note the following points in connection with equations (18.77) and (18.78):

1. When $R_l = 0$, the circuit in Figure 18.29 reduces to a simple parallel R_sLC circuit and equation (18.77) reduces to

$$Q_{sp}' = \frac{R_S}{\omega_0 L} = Q_P$$

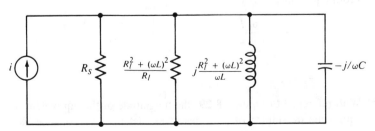

FIGURE 18.29 Series–parallel circuit with source resistance R_S.

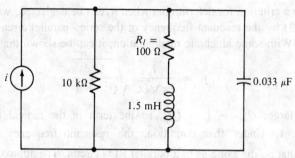

FIGURE 18.30 (Example 18.13)

2. When $R_S = \infty$, the circuit in Figure 18.29 reduces to a simple series–parallel circuit (Figure 18.27), and equation (18.78) reduces to

$$Q'_{sp} = Q_{sp} = \frac{\omega_0 L}{R_l}$$

Example 18.13 (Analysis)

For the circuit shown in Figure 18.30:

(a) Find the resonant frequency and the quality factor.
(b) Find the magnitude of the total impedance at resonance.
(c) Find the percent difference between the exact resonant frequency and the resonant frequency that is computed when R_l is neglected.

SOLUTION

(a) From equation (18.72),

$$\omega_0 = \frac{1}{\sqrt{LC}} \sqrt{1 - \frac{R_l^2 C}{L}}$$

$$= \frac{1}{\sqrt{(1.5 \times 10^{-3})(0.033 \times 10^{-6})}} \sqrt{1 - \frac{(100)^2(0.033 \times 10^{-6})}{1.5 \times 10^{-3}}}$$

$$= 125{,}529 \text{ rad/s}$$

$$f_0 = \frac{\omega_0}{2\pi} = 19.978 \text{ kHz}$$

From equation (18.77),

$$Q'_{sp} = \frac{\omega_0 L}{R_l + \dfrac{R_l^2 + (\omega_0 L)^2}{R_s}} = \frac{(125{,}529)(1.5 \times 10^{-3})}{100 + \dfrac{(100)^2 + [(125{,}529)(1.5 \times 10^{-3})]^2}{10^4}}$$

$$= 1.8$$

(b) With reference to Figure 18.29, the magnitude of the equivalent inductive reactance equals the magnitude of the capacitive reactance at resonance, so the total impedance is the parallel combination of the two resistances:

$$R_S \left\| \frac{R_l^2 + (\omega_0 L)^2}{R_l} \right\| = 10^4 \left\| \frac{(100)^2 + (125{,}529 \times 1.5 \times 10^{-3})^2}{100} \right\| \Omega$$

$$= 10^4 \| 454 \ \Omega = 434 \ \Omega$$

(c) If R_l is neglected, the resonant frequency is calculated in the same way it is in a parallel RLC circuit:

$$f_0 = \frac{1}{2\pi\sqrt{LC}} = \frac{1}{2\pi\sqrt{(1.5 \times 10^{-3})(0.033 \times 10^{-6})}} = 22.621 \ \text{kHz}$$

The percent difference between the actual and approximate resonant frequencies is therefore

$$\frac{22.621 - 19.978}{19.978} \times 100\% = 13.23\%$$

The relatively low quality factor of the circuit is responsible for this rather large difference between exact and approximate values of the resonant frequency.

Drill Exercise 18.13

Repeat Example 18.13 when $R_l = 20 \ \Omega$.

ANSWER: (a) $f_0 = 22{,}521$ Hz, $Q'_{sp} = 8.65$; (b) 1852 Ω; (c) 0.44%.   □

The Series–Parallel Bandpass Filter

Figure 18.31 shows the series–parallel circuit operated as a bandpass filter. In contrast with filters studied earlier, an important characteristic of this filter is that the frequency at which $|v_o|$ is maximum is *not* the resonant frequency f_0. This property is due to the fact that the magnitude of the total impedance is not maximum at the same frequency where the imaginary component of the impedance is zero. At the frequency where the imaginary component is zero, the impedance is purely resistive, so there is zero phase shift between i and v_o. To avoid confusion, we will refer to that frequency (which we

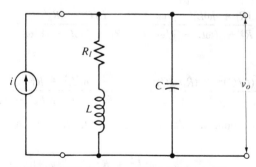

FIGURE 18.31 Series–parallel RLC circuit as a bandpass filter.

have been designating by f_0) as the *zero-phase resonant frequency*. The frequency at which the impedance, and therefore $|v_o|$, is maximum we will call the *maximum-impedance frequency*, f_m. Despite a certain reluctance to introduce yet another "Q," it is convenient to define the following quantities:

$$\omega_c = \frac{1}{\sqrt{LC}} \tag{18.79}$$

$$Q_c = \frac{\omega_c L}{R_l} \tag{18.80}$$

Note that ω_c and Q_c are computed in the same way as ω_0 and Q_s in a *series*-resonant circuit. We use the subscript c because these quantities are related to the inductor (coil) alone. Some authors simply use Q, with no subscript, to mean Q_c as defined above, in reference to the quality factor of any circuit. Do not be confused by this practice.

In terms of the quantities defined by equation (18.79) and (18.80), the zero-phase resonant frequency of the series–parallel filter is

$$\omega_0 = \omega_c \sqrt{1 - \frac{1}{Q_c^2}} \quad \text{rad/s} \tag{18.81}$$

As we have noted previously, ω_0 is approximately equal to $\omega_c = 1/\sqrt{LC}$ for large values of Q_c. The maximum-impedance frequency is found from

$$\omega_m = \omega_c \sqrt{1 - \frac{1}{4Q_c^2}} \quad \text{rad/s} \tag{18.82}$$

Note that ω_m is always greater than ω_0, but that both are approximately equal to ω_c for large values of Q_c. In general,

$$\omega_c > \omega_m > \omega_0 \tag{18.83}$$
$$\approx \omega_c \quad \text{for large } Q_c$$

The total impedance of the circuit at any frequency ω is

$$Z_T = (R_1 + j\omega L) \left\| \frac{-j}{\omega C} \right. \tag{18.84}$$
$$= \frac{(R_l + j\omega L)(-j/\omega C)}{R_l + j(\omega L - 1/\omega C)} = \frac{L/C - jR_l/\omega C}{R_l + j(\omega L - 1/\omega C)}$$

Therefore,

$$|Z_T|^2 = \frac{(L/C)^2 + (R_l/\omega C)^2}{R_l^2 + (\omega L - 1/\omega C)^2} = \frac{\omega^2 L^2 + R_l^2}{(\omega R_l C)^2 + (\omega^2 LC - 1)^2} \tag{18.85}$$

Let Z_m represent the maximum impedance of the circuit, which occurs when $\omega = \omega_m$. Then, from (18.85),

$$|Z_m|^2 = \frac{\omega_m^2 L^2 + R_l^2}{(\omega_m R_l C)^2 + (\omega_m^2 LC - 1)} \tag{18.86}$$

Since the lower and upper cutoff frequencies are defined to be those frequencies at which the magnitude of $v_o = iZ_T$ is $\sqrt{2}/2$ times its maximum value, the impedance of the circuit at each cutoff frequency is $\sqrt{2}/2$ times its maximum value:

$$|Z_T| = \frac{\sqrt{2}}{2} |Z_m| \quad \text{at cutoff} \tag{18.87}$$

or

$$|Z_T|^2 = 0.5|Z_m|^2 \tag{18.88}$$

We must therefore solve the following equation for ω:

$$\frac{\omega^2 L^2 + R_l^2}{(\omega R_l C)^2 + (\omega^2 LC - 1)^2} = 0.5|Z_m|^2 \tag{18.89}$$

Applying the quadratic formula to solve for ω^2, we finally obtain the following rather unwieldy, but exact, expressions for the cutoff frequencies:

$$\omega_1 = \sqrt{\frac{-B - \sqrt{B^2 - 4AC}}{2A}} \tag{18.90}$$

$$\omega_2 = \sqrt{\frac{-B + \sqrt{B^2 - 4AC}}{2A}} \tag{18.91}$$

where

$$A = 0.5|Z_m|^2 L^2 C^2$$

$$B = |Z_m|^2 (0.5R_l^2 C^2 - LC) - L^2$$

$$C = 0.5|Z_m|^2 - R_l^2$$

In view of the computational labor required to obtain exact values for f_1 and f_2, it is recommended that the value of Q_c be computed before using equations (18.90) and (18.91). If $Q_c \geq 10$, then f_0, f_1, and f_2 can be computed with very little error using the equations for a parallel-resonant circuit: (18.50), (18.63), and (18.64).

Remember that the true quality factor of the series–parallel circuit is Q_{sp}, given by equation (18.74) or (18.75). However, the quality factor *cannot* be used to determine the exact bandwidth; that is, the relationship BW $= f_m/Q_{sp}$ does *not* hold in general. On the other hand, if $Q_c \geq 10$, then $f_m \approx f_0$, $Q_c \approx Q_{sp}$, and

$$\text{BW} \approx \frac{f_0}{Q_{sp}} \approx \frac{f_m}{Q_c} \quad (Q_c \geq 10) \tag{18.92}$$

Example 18.14   (Analysis)

For the bandpass filter shown in Figure 18.32:

(a) Find the zero-phase resonant frequency and the maximum-impedance frequency.
(b) Find the maximum impedance and the maximum value of $|v_0|$.
(c) Find the lower and upper cutoff frequencies.

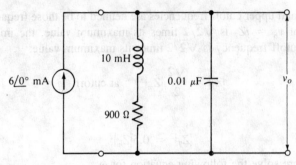

FIGURE 18.32 (Example 18.14)

SOLUTION

(a) We will use equations (18.81) and (18.82) to calculate ω_0 and ω_m, so we must first calculate the values of ω_c and Q_c. From equation (18.79),

$$\omega_c = \frac{1}{\sqrt{LC}} = \frac{1}{\sqrt{(10 \times 10^{-3})(0.01 \times 10^{-6})}} = 10^5 \text{ rad/s}$$

From equation (18.80),

$$Q_c = \frac{\omega_c L}{R_l} = \frac{10^5(10 \times 10^{-3})}{900} = 1.11$$

Therefore,

$$\omega_0 = \omega_c \sqrt{1 - \frac{1}{Q_c^2}} = 10^5 \sqrt{1 - \frac{1}{(1.11)^2}} = 43{,}589 \text{ rad/s}$$

$$f_0 = \frac{\omega_0}{2\pi} = 6937 \text{ Hz}$$

As an exercise, verify that the same value of f_0 is calculated using equation (18.73).

$$\omega_m = \omega_c \sqrt{1 - \frac{1}{4Q_c^2}} = 10^5 \sqrt{1 - \frac{1}{4(1.11)^2}} = 89{,}302 \text{ rad/s}$$

$$f_m = \frac{\omega_m}{2\pi} = 14{,}213 \text{ Hz}$$

Notice that the frequency, f_m, where $|v_o|$ is maximum is more than twice the value of the zero-phase resonant frequency.

(b) From equation (18.86),

$$|Z_m|^2 = \frac{\omega_m^2 L^2 + R_l^2}{(\omega_m R_l C)^2 + (\omega_m^2 LC - 1)^2}$$

$$= \frac{(89{,}302)^2(10 \times 10^{-3})^2 + (900)^2}{[(89{,}302)(900)(0.01 \times 10^{-6})]^2 + [(89{,}302)^2(10 \times 10^{-3})(0.01 \times 10^{-6} - 1)]^2}$$

$$= 2.3399 \times 10^6$$

Therefore, $|Z_m| = \sqrt{2.3399 \times 10^6} = 1530 \ \Omega$. The maximum value of $|v_o|$ is then

$$|v_o| \ \text{max} = |i||Z_m| = (6 \times 10^{-3})(1530) = 9.18 \ \text{V}$$

(c) Since the value of Q_c is only 1.11, we must use the exact equations for f_1 and f_2. Substituting $R_1 = 900 \ \Omega$, $L = 10 \ \text{mH}$, $C = 0.01 \ \mu\text{F}$, and $|Z_m|^2 = 2.3399 \times 10^6$ into equations (18.90) and (18.91), we find

$$A = 1.16997 \times 10^{-14}$$

$$B = -2.3922 \times 10^{-4}$$

$$C = 3.5995 \times 10^5$$

$$\sqrt{B^2 - 4AC} = \sqrt{(-2.3922 \times 10^{-4})^2 - 4(1.16997 \times 10^{-14})(3.5995 \times 10^5)}$$

$$= 2.0095 \times 10^{-4}$$

$$\omega_1 = \sqrt{\frac{2.3922 \times 10^{-4} - 2.0095 \times 10^{-4}}{2(1.16997 \times 10^{-14})}} = 40{,}441 \ \text{rad/s}$$

$$f_1 = \frac{\omega_1}{2\pi} = 6436 \ \text{Hz}$$

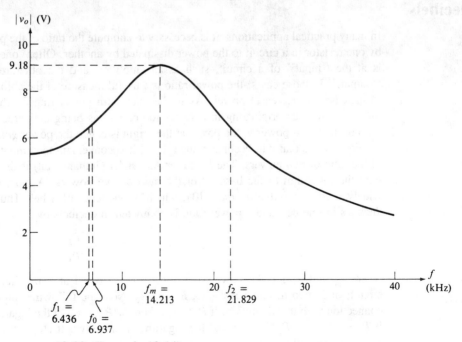

FIGURE 18.33 (Example 18.14)

$$\omega_2 = \sqrt{\frac{2.3922 \times 10^{-4} + 2.0095 \times 10^{-4}}{2(1.16997 \times 10^{-14})}} = 137{,}154 \text{ rad/s}$$

$$f_2 = \frac{\omega_2}{2\pi} = 21{,}829 \text{ Hz}$$

Note that the zero-phase resonant frequency ($f_o = 6{,}937$ Hz) is close in value to the lower cutoff frequency, so f_o is decidedly *not* the "center" frequency of the filter. Figure 18.33 shows a plot of the frequency response of this filter over the range 0 to 40 kHz.

Drill Exercise 18.14

Find exact and approximate values of f_o, f_m, f_1, and f_2 in Example 18.14 when the 900-Ω resistance is changed to 15 Ω.

ANSWER: (exact) $f_o = 15{,}913.7$ Hz, $f_m = 15{,}915$ Hz, $f_1 = 15{,}796.6$ Hz, $f_2 = 16{,}035.3$ Hz; (approximate) $f_0 \approx f_m \approx 15{,}915.5$ Hz, $f_1 \approx 15{,}796.1$ Hz, $f_2 \approx 16{,}034.9$ Hz. ☐

18.8 Decibels and Logarithmic Plots

Decibels

In many practical applications, it is necessary to compute the ratio of the power dissipated by one resistor in a circuit to the power dissipated by another. Often, one of the resistors is at the "input" of a circuit, such as an amplifier, and the other resistor is at the "output." In those cases, the power ratio is a useful measure of the ability of the circuit to increase (or decrease) power. As in the maximum power transfer theorem, we can think of power as originating in a signal source and as being delivered to a load. The ratio of delivered power to the power at the origin is called the power *gain* of the circuit.

For reasons that will become evident later, it is convenient to compute the *logarithm* of the ratio of two powers. The *bel* (after Alexander Graham Bell) is the unit associated with the logarithm to the base 10 of the ratio of two powers. A more convenient and widely used unit is the *decibel* (dB), which is one-tenth of a bel. Thus, there are 10 decibels in one bel, and a power ratio is computed in decibels by

$$\text{power ratio (dB)} = 10 \log_{10} \frac{P_2}{P_1} \tag{18.93}$$

where P_1 and P_2 are the powers developed or dissipated at two different points in a circuit, or in two different resistors, R_1 and R_2. Note the following important points in connection with this definition: If $P_2 > P_1$, then $P_2/P_1 > 1$ and the logarithm is *positive*. If $P_2 < P_1$, then $P_2/P_1 < 1$ and the logarithm is *negative*. If $P_2 = P_1$, then $P_2/P_1 = 1$ and the logarithm is *zero*. Summarizing, negative dB ⇒ power level decreases, going from point 1 to point 2

positive dB ⇒ power level increases, going from point 1 to point 2

zero dB ⇒ equal power at points 1 and 2

Example 18.15

The power P_1 dissipated in resistor R_1 in a certain circuit is 25 mW, and the power P_2 dissipated in R_2 is 0.5 W.

(a) Find the power ratio P_2/P_1 in decibels.
(b) The circuit is changed so that the power ratio becomes 22 dB. Assuming that P_1 remains the same, what is the new value of P_2?

SOLUTION

(a) Note that P_1 and P_2 must be expressed with identical units when computing the power ratio:

$$10 \log_{10} \frac{P_2}{P_1} = 10 \log_{10} \frac{0.5 \text{ W}}{25 \times 10^{-3} \text{ W}} = 10 \log_{10} 20$$

$$= 10(1.301) = 13.01 \text{ dB}$$

(b) We must solve the following equation for P_2:

$$22 = 10 \log_{10} \frac{P_2}{25 \times 10^{-3} \text{ W}}$$

$$2.2 = \log_{10} \frac{P_2}{25 \times 10^{-3} \text{ W}}$$

Taking the antilog, base 10, of both sides, we obtain

$$\text{antilog } 2.2 = \frac{P_2}{25 \times 10^{-3}}$$

$$158.49 = \frac{P_2}{25 \times 10^{-3}}$$

(Note that the antilog can be found on a calculator by entering a sequence such as 2.2, INV, LOG, or by using a y^x key on the calculator to compute $10^{2.2}$.)
Solving for P_2, we find

$$P_2 = (158.49)(25 \times 10^{-3} \text{ W}) = 3.96 \text{ W}$$

Drill Exercise 18.15

The ratio of power P_2 to power P_1 is 6 dB. If $P_2 = 0.2$ W, what is the value of P_1?

ANSWER: 0.0502 W. ☐

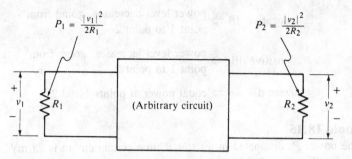

FIGURE 18.34 Computing power in terms of the voltages at two locations in a circuit.

Figure 18.34 shows an arbitrary circuit containing resistors R_1 and R_2 with voltages v_1 and v_2 across them. We can express the power ratio in terms of the voltage magnitudes as follows:

$$\frac{P_2}{P_1} = \frac{|v_2|^2/2R_2}{|v_1|^2/2R_1} = \left(\frac{|v_2|}{|v_1|}\right)^2 \frac{R_1}{R_2} \qquad (18.94)$$

If $R_1 = R_2$, then $R_1/R_2 = 1$, and

$$10 \log_{10} \frac{P_2}{P_1} = 10 \log_{10} \left(\frac{|v_2|}{|v_1|}\right)^2 = 20 \log_{10} \frac{|v_2|}{|v_1|} \qquad (18.95)$$

Equation (18.95) shows that the power ratio in decibels can be found by computing 20 times the log of the voltage ratio, *provided that the resistor values are indentical at the locations where the powers are compared.* Note that equations (18.94) and (18.95) are also applicable in dc circuits, where v_1 and v_2 are replaced by dc voltages V_1 and V_2.

It is common practice to express voltage ratios as well as power ratios in decibels. In fact, a voltage ratio is computed in decibels in the same way that a power ratio is computed using voltages:

$$\text{voltage ratio (dB)} = 20 \log_{10} \frac{|v_2|}{|v_1|} \qquad \text{decibels} \qquad (18.96)$$

Note that the computation in (18.96) is a valid way to express a *voltage* ratio in decibels *regardless of the resistors across which the voltages are developed.* The voltage ratio in decibels equals the power ratio in decibels only if the resistor values are equal.

Example 18.16

In Figure 18.34, $|v_1| = 10$ V, $R_1 = 100$ Ω, $|v_2| = 25$ V, and $R_2 = 1$ kΩ.

(a) Find P_2/P_1 in decibels.

(b) Find $|v_2|/|v_1|$ in decibels.

SOLUTION

(a) $$P_1 = \frac{|v_1|^2}{2R_1} = \frac{(10)^2}{2(100)} = 0.5 \text{ W}$$

$$P_2 = \frac{|v_2|^2}{2R_2} = \frac{(25)^2}{2(1000)} = 0.3125 \text{ W}$$

$$10 \log_{10} \frac{P_2}{P_1} = 10 \log_{10} \frac{0.3125}{0.5} = -2.04 \text{ dB}$$

The negative result confirms that there is less power in R_2 than in R_1, that is, that there is a decrease in power as one passes through the circuit from R_1 to R_2. Note that $10 \log_{10} (P_1/P_2) = +2.04$ dB.

(b) $20 \log_{10} \dfrac{|v_2|}{|v_1|} = 20 \log_{10} \dfrac{25}{10} = 7.96$ dB

This result shows that there is an increase in voltage as one passes through the circuit from R_1 to R_2. Note that the power and voltage ratios in decibels are decidedly not equal, because R_1 and R_2 are not equal.

Drill Exercise 18.16

In Example 18.16, find the ratio of $|v_1|$ to $|v_2|$ in decibels.

ANSWER: -7.96 dB.   □

The terms "decibel" and "dB" are often misused by the broadcast media and in popular publications. Remember that dB always refers to a *ratio*, so it is meaningless to state that any "volume" or level, such as a "noise level," is a certain number of dB, without specifying a *reference* level. In some technical fields, standard reference levels (values of P_1 or v_1) have been adopted, so a quantity expressed in decibels always refers to a ratio of power or voltage to another, predetermined, value. In those cases, the unit dB is modified to indicate the reference. For example, the unit dBm refers to the ratio of a certain power to the reference level 1 mW:

$$\text{dBm} = 10 \log_{10} \frac{P}{10^{-3} \text{ W}} \tag{18.97}$$

Note that $P = 1$ mW corresponds to 0 dBm. For voltage ratios, a commonly used reference is 1 V, in which case the unit becomes dBV:

$$\text{dBV} = 20 \log_{10} \frac{|v|}{1 \text{ V}} \tag{18.98}$$

Thus, a voltage of 1 V corresponds to 0 dBV.

As an aid in gaining an intuitive feel for the voltage ratios that correspond to different decibel values, it is useful to memorize that 6 dB corresponds (approximately) to a 2:1 voltage ratio, and that 20 dB corresponds (exactly) to a 10:1 ratio. Furthermore, multiplying two ratios is the same as adding their decibel values. For example, a ratio of 100:1 corresponds to $10 \times 10 = 20$ dB + 20 dB = 40 dB. A ratio of 20:1 corresponds to $2 \times 10 = 6$ dB + 20 dB = 26 dB. Remember that ratios less than 1 correspond to negative dB values. For example, a ratio of 1:40 ($\frac{1}{40}$) is $\frac{1}{2} \times \frac{1}{2} \times \frac{1}{10} = -6$ dB −

TABLE 18.1

$\dfrac{|v_2|}{|v_1|}$	$20 \log_{10} \dfrac{|v_2|}{|v_1|}$ (dB)	$\dfrac{|v_2|}{|v_1|}$	$20 \log_{10} \dfrac{|v_2|}{|v_1|}$ (dB)
0.001	−60	2	6
0.002	−54	4	12
0.005	−46	8	18
0.008	−42	10	20
0.01	−40	20	26
0.02	−34	40	32
0.05	−26	80	38
0.08	−22	100	40
0.1	−20	200	46
0.2	−14	400	52
0.5	−6	800	58
0.8	−2	1000	60
1.0	0		

6 dB − 20 dB = −32 dB. Table 18.1 lists some commonly encountered voltage ratios and their corresponding values in decibels.

Logarithmic (Log-Log) Plots

Plots of the frequency response of many practical circuits extend over a wide range of frequencies. It is impractical to construct such plots on conventional graph paper, because the resolution is very poor at low frequencies. For example, if it were necessary to plot frequency response over the range 0 to 1 MHz using typical graph paper with 100 divisions, each division would correspond to 10 kHz. If the response has a lower cutoff frequency of, say, 100 Hz, it would be impossible to locate that frequency, with any precision, on such a coarse scale. To eliminate this difficulty, frequency response is often plotted on special paper whose divisions are *logarithmically* spaced. This type of paper effectively expands the spacing between low frequencies and compresses that between high frequencies. Furthermore, every 1-to-10 range occupies the same amount of space on the paper. For example, the frequency range 1 to 10 Hz occupies the same space as the frequency range 10 kHz to 100 kHz.

Graph paper having logarithmically spaced divisions along both the horizontal and vertical axes is called *log-log* paper. Figure 18.35 shows an example. Every 1-to-10 range along both axes has the same numbers printed in the margins, and the user re-numbers the grid lines as required for a particular application. In the example shown, the horizontal axis has been relabeled to correspond to the ranges 10 to 100 and 100 to 1000, and the vertical axis has been relabeled to correspond to the ranges 0.1 to 1.0 and 1.0 to 10. Each such range is called a *cycle*, and log-log paper is identified by the number of cycles along each axis. The graph paper shown in the example is called "2-cycle by 2-cycle" paper, or simply "2 × 2." Log-log paper is available with different numbers of cycles along each axis, including 4 × 2, 5 × 3, 3 × 3, and so on. The user must

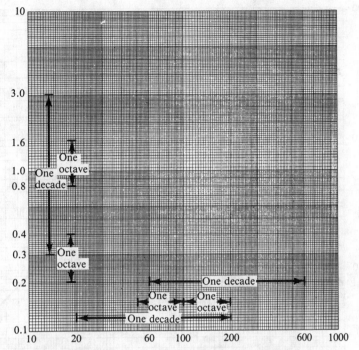

FIGURE 18.35 Example of log-log graph paper showing octaves and decades.

select graph paper that is appropriate for the range of data that he or she wishes to plot. For example, to plot an output voltage that varies from 3 to 80 V over the frequency range 140 Hz to 75 kHz would require vertical cycles covering the ranges 1–10 and 10–100 and horizontal cycles covering the ranges 100–1000, 1000–10,000, and 10,000–100,000. Thus 3 × 2 graph paper would be required. Note that the value 0 (zero) can never appear on a logarithmic scale, because log 0 = −∞.

Any 1:10 (or 10:1) ratio of values is called a *decade*. For example, each cycle on log-log graph paper represents one decade. Also, the ranges 5 to 50, 0.2 to 2, and 1500 to 15,000 are examples of decades. As previously mentioned, every decade occupies the same space on log-log paper. Several horizontal and vertical decades are shown in Figure 18.35. Any 1:2 or 2:1 ratio of values is called an *octave*, and every octave occupies the same space on log-log paper. Several horizontal and vertical octaves are shown in Figure 18.35.

It takes some practice to develop a facility for plotting data on logarithmic scales. A helpful way to acquire that skill is to practice reading values from a graph plotted on log-log paper, as in the next example.

Example 18.17

Figure 18.36 shows a plot of the frequency response of a series RLC bandpass filter having $Q_s = 1$. The plot is constructed on 3 × 3 log-log graph paper. (Compare

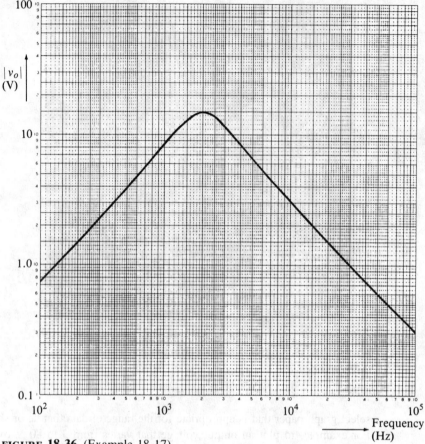

FIGURE 18.36 (Example 18.17)

this response with that plotted for $Q_s = 1$ on conventional graph paper, shown in Figure 18.20.) Use the plot to determine approximate values of

(a) the center frequency, f_0
(b) the lower and upper cutoff frequencies
(c) the output voltage one decade above the center frequency
(d) the output voltage one octave below the lower cutoff frequency

SOLUTION When reading or plotting data on a log-log plot, note carefully that the values corresponding to intervals between grid lines *changes* as we move along either axis. For example, in Figure 18.36, the interval between any two grid lines in the region from 10^3 Hz to 2×10^3 Hz represents 0.1×10^3 Hz, or 100 Hz. However, in the region from 5×10^3 Hz to 6×10^3 Hz, each interval represents 0.2×10^3 Hz, or 200 Hz, while in the interval from 10^4 Hz to 2×10^4 Hz, each interval represents 0.1×10^4 Hz or 1 kHz.

(a) The center frequency is the frequency at which $|v_o|$ is maximum; that frequency can be seen from the plot to be 2×10^3 Hz, or 2 kHz.

(b) At the center frequency, the output reaches its maximum value of 15 V. Therefore, the cutoff frequencies are those frequencies at which

$$|v_o| = \frac{\sqrt{2}}{2} (15 \text{ V}) = 10.6 \text{ V}$$

From the plot, we see that the output is 10.6 V at aproximately 1.25×10^3 Hz and 3.2×10^3 Hz. Thus, $f_1 \approx 1.25$ kHz and $f_2 \approx 3.2$ kHz. (The exact cutoff frequencies are, by calculation, $f_1 = 1.236$ kHz and $f_2 = 3.236$ kHz.)

(c) The frequency one decade above the center frequency is

$$10f_0 = 10(2 \text{ kHz}) = 20 \text{ kHz}$$

From the plot, $|v_o|$ at 20 kHz (2×10^4 Hz) is approximately 1.5 V.

(d) The frequency one octave below the lower cutoff frequency is

$$\tfrac{1}{2}f_1 = (\tfrac{1}{2})(1.25 \text{ kHz}) = 625 \text{ Hz}$$

From the plot, $|v_o|$ at 625 Hz (6.25×10^2 Hz) is approximately 4.8 V.

Drill Exercise 18.17

Using the plot in Figure 18.36, find approximate values of

(a) the frequencies at which $|v_o|$ is one-half its maximum value;
(b) the value of $|v_o|$ one octave above the upper cutoff frequency.

ANSWER: (a) 950 Hz, 4.4 kHz; (b) 8.3 V.   □

Semilog Plots

Graph paper having a logarithmic scale on one axis and a conventional linear scale on the other axis is called *semilog* paper. Semilog paper is often used to plot the phase angle of an output voltage versus frequency, with angle measured along the linear axis. (Phase angle versus frequency is sometimes called a *phase response*.) Semilog paper is also widely used to plot a voltage ratio in decibels versus frequency, with decibels along the linear axis. Since decibels are defined in terms of the logarithm of a ratio, a plot of decibels on a linear axis versus frequency on a log axis will have the same shape and appearance as a plot of the (non-log) ratios versus frequency on log-log axes. This fact is illustrated in Figure 18.37.

Since the cutoff frequency of a filter is defined to be that frequency where the magnitude of the output voltage equals $\sqrt{2}/2 \approx 0.707$ times its maximum value in the passband, the voltage ratio at cutoff is

$$\frac{(\sqrt{2}/2)|v_o(\text{max})|}{|v_o(\text{max})|} = \frac{\sqrt{2}}{2}$$

In decibels, this ratio is

$$20 \log_{10} \frac{\sqrt{2}}{2} = -3 \text{ dB} \tag{18.99}$$

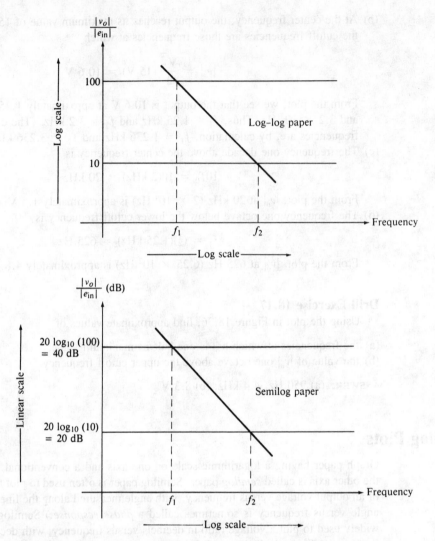

FIGURE 18.37 A frequency response has the same appearance when plotted on log-log paper as it does when decibels are plotted on semilog paper.

We say that the output of a filter is "down 3 dB" (or "3 dB down") at cutoff, and a cutoff frequency is often called a *3-dB frequency*. When the output voltage is developed across resistor R, the maximum output power is $P_o(\text{max}) = |v_o(\text{max})|^2/2R$. Since $v_o = (\sqrt{2}/2)|v_o(\text{max})|$ at cutoff, the output power at cutoff is

$$P_o(\text{cutoff}) = \frac{[(\sqrt{2}/2)|v_o(\text{max})|]^2}{2R} = \frac{0.5|v_o(\text{max})|^2}{2R} \tag{18.100}$$

Therefore, the ratio of the power at cutoff to the maximum power is

$$\frac{P_o(\text{cutoff})}{P_o(\text{max})} = \frac{0.5|v_o(\text{max})|^2/2R}{|v_o(\text{max})|^2/2R} = 0.5 \tag{18.101}$$

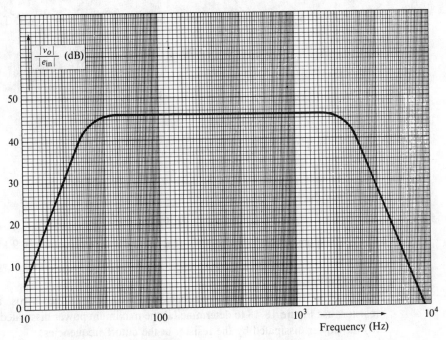

FIGURE 18.38 (Example 18.18)

We see that the output power at cutoff is one-half of its maximum value. For that reason, a cutoff frequency is also called a *half-power frequency*.

Example 18.18

Figure 18.38 shows the frequency response of a certain amplifier plotted on semilog graph paper. The vertical axis represents values of the ratio of output voltage to input voltage, $|v_o|/|e_{in}|$, in decibels. Use the plot to find approximate values of

(a) the maximum value of $|v_o|/|e_{in}|$ in decibels
(b) the lower and upper cutoff frequencies
(c) the output voltage when $|e_{in}| = 16$ mV at 15 Hz
(d) the frequency at which $|v_o| = |e_{in}|$

SOLUTION

(a) The maximum value of $|v_o|/|e_{in}|$ occurs on the response plot along the horizontal line that extends from about 50 Hz to about 1.8 kHz. From Figure 18.38 it can be seen that this line coincides with the 46-dB line, so $|v_o|/|e_{in}|$ has a maximum value of 46 dB.

(b) Since the maximum value of $|v_o|/|e_{in}|$ is 46 dB, the cutoff frequencies are the frequencies where the response falls to 46 dB − 3 dB = 43 dB. From the figure, those frequencies are found to be $f_1 \approx 30$ Hz and $f_2 \approx 2.5$ kHz.

(c) At $f = 15$ Hz, the response curve shows that $|v_o|/|e_{in}| \approx 20$ dB. Thus,

$$20 = 20 \log_{10} \frac{|v_o|}{16 \text{ mV}}$$

$$1 = \log_{10} \frac{|v_o|}{16 \text{ mV}}$$

$$10 = \frac{|v_o|}{16 \text{ mV}}$$

$$|v_o| = 160 \text{ mV}$$

(d) When $|v_o| = |e_{in}|$, $|v_o|/e_{in}| = 1$ and

$$20 \log_{10} 1 = 0 \text{ dB}$$

The response shows that $|v_o|/|e_{in}|$ reaches 0 dB at approximately 9 kHz.

Drill Exercise 18.18

If $|e_{in}| = 100$ mV pk at all frequencies and $|v_o|$ is developed across a 12-Ω resistor, use Figure 18.18 to determine (a) the maximum power dissipated by the resistor; (b) the power dissipated by the resistor at the cutoff frequencies.

ANSWER: (a) 0.332 W; (b) 0.166 W.

18.9 SPICE Examples

Appendix Examples A.8 and A.9 demonstrate the use of SPICE to obtain the frequency response of a low-pass RC filter and a bandpass RLC filter.

Exercises

Section 18.1   Filter Definitions

18.1 A low-pass filter has a cutoff frequency of 400 Hz. The input to the filter is a signal generator whose voltage has a constant amplitude as its frequency is varied. When the frequency is adjusted over a wide range, it is observed that the peak value of the filter output reaches a maximum of 12 V pk. What is the effective value of the filter output at 400 Hz?

18.2 A high-pass filter has a cutoff frequency of 2 kHz. The input to the filter is a signal generator whose voltage has a constant amplitude as its frequency is varied. When the signal generator is set to 2 kHz, the peak value of the filter output is 600 mV pk. What is the maximum effective value of the filter output at a frequency greater than 2 kHz?

18.3 The input to a certain bandpass filter is a signal generator whose voltage has a constant amplitude as its frequency is varied. When the frequency of the signal generator was set to the values listed, the magnitude of the filter output had the values shown. What is the bandwidth of the filter?

Frequency (Hz)	$|v_o|$ (V)
650	12.25
750	16.25
781	17.68
880	22.25
1000	25.00
1126	20.38
1200	20.16
1281	17.68
1540	12.25

18.4 A bandpass filter has a passband that corresponds to the audio-frequency range. At 20 kHz, the effective value of the filter output is 8.32 V rms. What maximum peak value does the output reach in the audio-frequency range?

Section 18.2   Low-Pass RC Filters

18.5 For the network shown in Figure 18.39:
(a) Identify the type of filter it represents.

(b) Find the cutoff frequency in hertz.
(c) Find the magnitude of v_o when the frequency of e_{in} is 2 kHz, 10 kHz, and 20 kHz.

18.6 For the network shown in Figure 18.40:
(a) Identify the type of filter it represents.
(b) Find the cutoff frequency in radians per second.
(c) Find the magnitude of v_o at the cutoff frequency.
(d) Find the phase angle of v_o relative to e_{in} at $\omega = 100$ rad/s, 300 rad/s, and 1000 rad/s.

18.7 For the circuit shown in Figure 18.41:
(a) Find $|v_o|/|e_{in}|$ at $\omega = 5 \times 10^3$ rad/s, 1.25×10^4 rad/s, and 3.6×10^4 rad/s.
(b) Find $\underline{/v_o}$ at $\omega = 6 \times 10^3$ rad/s, 1.25×10^4 rad/s, and 1.25×10^5 rad/s, assuming that $\underline{/e_{in}}$ remains equal to 45° at each frequency.

18.8 For the circuit shown in Figure 18.42:
(a) Find $|v_o|/|e_{in}|$ at $f = 132.6$ Hz, 1.326 kHz, and 13.26 kHz.
(b) Find $(\underline{/v_o} - \underline{/e_{in}})$ at the same frequencies.

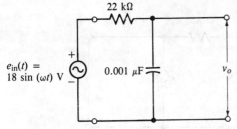

FIGURE 18.39 (Exercise 18.5)

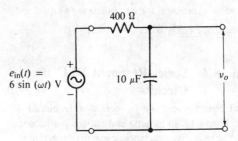

FIGURE 18.40 (Exercise 18.6)

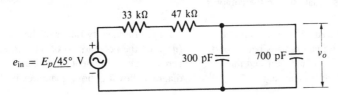

FIGURE 18.41 (Exercise 18.7)

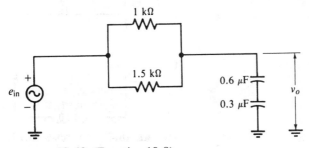

FIGURE 18.42 (Exercise 18.8)

Section 18.3   High-Pass RC Filters

18.9 For the network shown in Figure 18.43:
(a) Identify the type of filter it represents.
(b) Find the cutoff frequency in hertz.
(c) Find the magnitude of v_o at $f = 2$ kHz, 4 kHz, and 10 kHz.
(d) Find the phase angle of v_o at $\omega = 2.5 \times 10^3$ rad/s, 25×10^3 rad/s, and 10^5 rad/s.

18.10 For the network shown in Figure 18.44:
(a) Find the value of $|v_o|/|e_{in}|$ when the frequency of e_{in} is 200 Hz, 796 Hz, and 9 kHz.
(b) Find the value of $\underline{/v_o} - \underline{/e_{in}}$ at $\omega = 1000$ rad/s, 5000 rad/s, and 20,000 rad/s.

18.11 Using equation (18.12), verify that $\omega = 1/(RC)$ rad/s is the cutoff frequency of a high-pass RC filter.

18.12 For the network shown in Figure 18.45:
(a) Identify the type of filtering action it performs (low-pass or high-pass).
(b) Derive an expression for $|v_o|$ in terms of $|e_{in}|$, ω, R, and L.
(c) Derive an expression for $\underline{/v_o}$ in terms of ω, R, and L. Assume that $\underline{/e_{in}} = 0°$.

Section 18.4   Series-Resonant RLC Circuits

18.13 **(a)** Find the resonant frequency of the circuit shown in Figure 18.46 in hertz and radians per second.
(b) Find the inductive reactance, capacitive reactance, and total impedance of the circuit at resonance. Make a phasor diagram showing these quantities in the complex plane.

18.14 **(a)** Find the resonant frequency of the circuit shown in Figure 18.47 in hertz and radians per second.
(b) Find the reactive power in the inductor and the average power in the resistor at resonance.

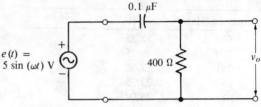

FIGURE 18.43 (Exercise 18.9)

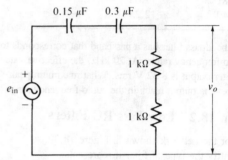

FIGURE 18.44 (Exercise 18.10)

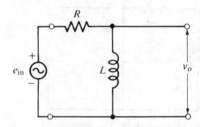

FIGURE 18.45 (Exercise 18.12)

(c) Find the quality factor of the circuit.
(d) Find the polar forms of the current and the voltage across each component at resonance. Make a phasor diagram showing those quantities in the complex plane.

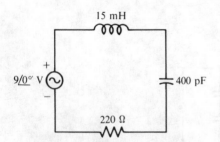

FIGURE 18.46 (Exercise 18.13)

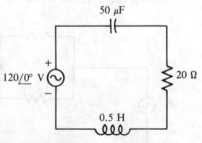

FIGURE 18.47 (Exercise 18.14)

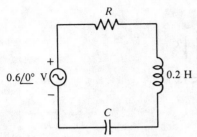

FIGURE 18.48 (Exercise 18.15)

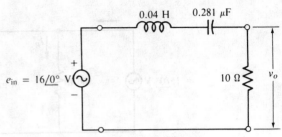

FIGURE 18.49 (Exercise 18.17)

18.15 The circuit shown in Figure 18.48 has a resonant frequency of 400 Hz and a quality factor of 8.
(a) Find the values of R and C.
(b) Find the magnitudes of the voltages across each component at resonance.

18.16 The quality factor of a 50-mH coil is 25 at a frequency of 10 kHz. What would be the quality factor of a circuit containing this coil in series with a 0.02-µF capacitor, assuming there is no separate resistor in the circuit?

Section 18.5 The Series RLC Circuit as a Bandpass Filter

18.17 For the RLC filter shown in Figure 18.49:
(a) Find the center frequency and the cutoff frequencies in hertz.
(b) Find the bandwidth and the quality factor.
(c) Find $|v_o|$ at frequencies equal to one-half and twice the center frequency.

18.18 The RLC filter shown in Figure 18.50 has a lower cutoff frequency of 1 kHz and a center frequency of 1.8 kHz.
(a) Find the upper cutoff frequency.
(b) Find the quality factor.
(c) Find $|v_o|$ at frequencies $f = 0.1f_0$ and $f = 10f_0$.

18.19 Using equations (18.38) through (18.41), show that the bandwidth of a series RLC filter is
$BW = f_0/Q_s = R/2\pi L$ hertz.

18.20 Show that the following approximation is valid for a series RLC filter having a large quality factor $(Q_s \geq 10)$:

$$f_0 \approx \frac{f_1 + f_2}{2}$$

18.21 The quality factor of a 1-H coil is 30 at $f = 1$ kHz. Find the bandwidth of a series RLC filter containing the coil in series with a 0.25-µF capacitor and a 75-Ω resistor.

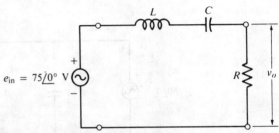

FIGURE 18.50 (Exercise 18.18)

18.22 The component values in a series RLC filter are $R = 80\ \Omega$, $L = 0.1\ H$, and $C = 0.1\ \mu F$. When the input voltage to the filter has peak value 14 V and frequency 10^4 rad/s, the output voltage across the 80-Ω resistor has peak value 8 V. What is the winding resistance of the inductor?

Section 18.6 Parallel-Resonant RLC Circuits

18.23 For the circuit shown in Figure 18.51:
(a) Find the resonant frequency in hertz and radians per second.
(b) Find the polar form of i_R, i_L, i_C, and i_T at resonance.
(c) Find the magnitude of the total impedance at the following multiples of f_0: $0.1f_0$, $0.5f_0$, f_0, $2f_0$, $10f_0$.
(d) Using the values obtained in part (c), sketch $|Z_T|$ versus frequency.

18.24 For the circuit shown in Figure 18.52:
(a) Find the resonant frequency in hertz and radians per second.
(b) Find the magnitude of i_T at the following multiples of f_0: $0.1f_0$, $0.5f_0$, f_0, $2f_0$, $10f_0$.
(c) Using the values obtained in part (b), sketch $|i_T|$ versus frequency.

18.25 For the filter circuit shown in Figure 18.53:
(a) Find the center frequency in radians per second and hertz.

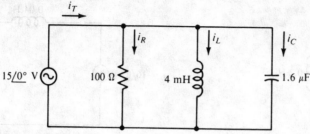

FIGURE 18.51 (Exercise 18.23)

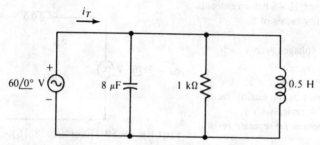

FIGURE 18.52 (Exercise 18.24)

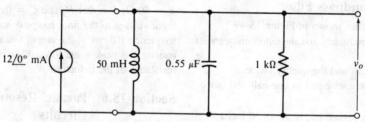

FIGURE 18.53 (Exercise 18.25)

(b) Find the quality factor and the bandwidth.
(c) Find the lower and upper cutoff frequencies.
(d) Are the approximations $f_1 \approx f_0 - BW/2$ and $f_2 \approx f_0 + BW/2$ valid for this circuit? Explain.

18.26 Derive equation (18.62) from equation (18.61).

Section 18.7   Series–Parallel Resonance

18.27 For the circuit shown in Figure 18.54:
(a) Find the resonant frequency and the quality factor.
(b) Find the magnitude of the total impedance at resonance.
(c) Find the percent difference between the exact resonant frequency and the resonant frequency that is computed when R_l is neglected.

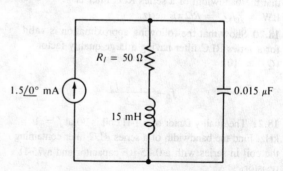

FIGURE 18.54 (Exercise 18.27)

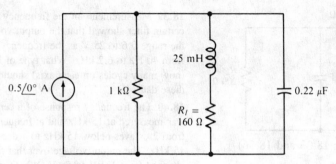

FIGURE 18.55 (Exercise 18.28)

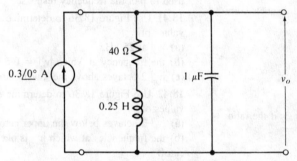

FIGURE 18.56 (Exercise 18.29)

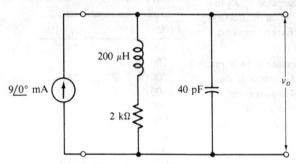

FIGURE 18.57 (Exercise 18.30)

18.28 For the circuit shown in Figure 18.55:
(a) Find the resonant frequency and the quality factor.
(b) Find the magnitude of the total impedance at resonance.
(c) Find the percent difference between the exact resonant frequency and the resonant frequency that is computed when R_1 is neglected.

18.29 For the filter circuit shown in Figure 18.56:
(a) Find the zero-phase resonant frequency.
(b) Find the maximum-impedance frequency.

(c) Find the lower and upper cutoff frequencies.
(d) Find the maximum value of the total impedance.
Approximations may be used if the approximation criterion is satisfied.

18.30 For the filter circuit shown in Figure 18.57:
(a) Find the zero-phase resonant frequency.
(b) Find the maximum-impedance frequency.
(c) Find the lower and upper cutoff frequencies.
(d) Find the maximum value of $|v_o|$.
Approximations may be used if the approximation criterion is satisfied.

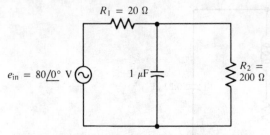

FIGURE 18.58 (Exercises 18.33 and 18.36)

Section 18.8   Decibels and Logarithmic Plots

18.31 Find the power ratio P_2/P_1 in decibels corresponding to each of the following.
(a) $P_1 = 1.5$ W, $P_2 = 14$ W
(b) $P_1 = 300$ μW, $P_2 = 55$ mW
(c) $P_1 = 0.2$ W, $P_2 = 0.05$ W
(d) $P_1 = 1$ W, $P_2 = 1000$ mW
(e) $P_2 = 100P_1$

18.32 Given that $P_1 = 0.1$ W, find P_2 if the ratio P_2/P_1 is
(a) 6 dB
(b) -10 dB
(c) 0 dB

18.33 Find the ratio of the power dissipated by R_2 in Figure 18.58 to that dissipated by R_1 when the frequency of e_{in} is **(a)** 1 kHz, and **(b)** 10 kHz. Express answers in decibels.

18.34 The voltage across a 470-Ω resistor is 18 V rms. What is the rms voltage across a 1-kΩ resistor if the ratio of the power dissipated by the 470-Ω resistor to that dissipated by the 1-kΩ resistor is 15 dB?

18.35 Find the voltage ratio $|v_2|/|v_1|$ in decibels corresponding to each of the following cases.
(a) $|v_1| = 25$ mV pk, $|v_2| = 1$ V pk
(b) $v_1 = 3.1 \sin(\omega t)$ V, $v_2 = 0.6 \sin(\omega t)$ V
(c) $|v_1| = 100$ V rms, $|v_2| = 141.4$ V pk
(d) $i_1 = 1.5 \sin(10^6 t)$ mA, $R_1 = 2.2$ kΩ; $i_2 = 3.3 \sin(10^6 t)$ A, $R_2 = 100$ Ω

18.36 Find the ratio of the voltage across R_2 in Figure 18.58 to that across R_1 when the frequency of e_{in} is **(a)** 1 kHz, and **(b)** 10 kHz. Express answers in decibels.

18.37 The output of a certain amplifier is specified to be 14.8 dBV. What is the magnitude of its output voltage?

18.38 The output power of a certain amplifier is specified to be 40 dBm. If the output power is developed across a 50-Ω resistor, what is the magnitude of the output voltage?

18.39 Measurements of the frequency response of a certain filter showed that the output voltage varied over the range 0.6 to 25 V as the frequency was changed from 40 Hz to 6.2 kHz. What type of log-log paper (how many cycles on each axis) should be used to plot these data?

18.40 The frequency response of a certain amplifier was measured at 15 kHz and at frequencies ranging from 2 octaves below 15 kHz to 1 decade above 15 kHz. The output voltage over that frequency range changed from 1.2 V to 55 V. What type of log-log graph paper (how many cycles on each axis) should be used to plot the frequency response?

18.41 Use Figure 18.36 to determine approximate values of
(a) $|v_o|$ at 25 kHz
(b) the frequency at which $|v_o| = 0.6$ V
(c) $|v_o|$ 2 octaves above the lower cutoff frequency

18.42 Use Figure 18.36 to determine approximate values of
(a) $|v_o|$ 2 octaves below the upper cutoff frequency
(b) the frequencies at which $|v_o|$ is one-half its value at cutoff

18.43 Plot the following frequency response data on appropriately selected log-log graph paper.

f (Hz)	$|v_o|$ (V)
100	19.8
200	19.4
300	18.7
500	17.0
700	15.1
1,000	12.5
2,000	7.43
3,000	5.15
5,000	3.16
10,000	1.59
20,000	0.80
50,000	0.32

Using the plot, determine the approximate cutoff frequency.

18.44 Plot the frequency response of a series RLC bandpass filter having $R = 471$ Ω, $L = 5$ mH, and $C = 5629$ pF. The input is a 10-V pk sinusoidal voltage source. Construct the plot on log-log graph paper over the frequency range 2 kHz to 0.5 MHz.

18.45 Use Figure 18.38 to determine approximate values for

(a) the value of $|v_o|/|e_{in}|$ in decibels at 20 Hz
(b) the value of $|v_o|/|e_{in}|$ in decibels 2 decades above the lower cutoff frequency
(c) the output voltage when $|e_{in}| = 44$ mV at 325 Hz

18.46 Use Figure 18.38 to determine approximate values for
(a) the frequencies at which $|v_o|/|e_{in}|$ is down 10 dB

from its value at 1 kHz
(b) the input voltage when $|v_o| = 18$ V at 5 kHz

18.47 Using the data given in Exercise 18.43, plot $|v_o|/|e_{in}|$ in decibels versus frequency on semilog graph paper. Assume that $|e_{in}| = 20$ V. (*Note:* All decibel values will be negative.)

Design Exercises

18.1D A low-pass RC filter is to be designed to have a cutoff frequency of 1.2 kHz. If a 0.05-μF capacitor is used, what value of resistance is necessary?

18.2D It is necessary to reduce the magnitude of a 15-V pk, 25-kHz signal to 6 V pk using a low-pass RC filter. If a 2.2-kΩ resistor is used, what value of capacitance is necessary?

18.3D A signal source generates a 10-kHz sine wave and it is necessary to produce a sine wave that leads the source waveform by 20°. If a 500-Ω resistor is to be used in an RC filter to create the phase shift, what value of capacitance should be used? Draw the circuit.

18.4D An amplifier is capacitor-coupled to a signal source using a 1-μF capacitor. What minimum resistance must the amplifier have if its passband must correspond to the audio-frequency range?

18.5D A series-resonant RLC circuit must have a quality factor of at least 10. If the capacitive reactance in the circuit is $-j1400$ Ω at resonance, what is the maximum resistance the circuit can have?

18.6D A series RLC circuit is resonant when a 3-V pk, 3-kHz signal source is connected to it. If the inductance in the circuit is 50 mH, what should be the circuit resistance if it is desired to develop a 15-V pk voltage across the inductor at resonance?

18.7D A series RLC bandpass filter is to be designed using an adjustable capacitor so that its center frequency can be adjusted from 570 kHz to 1.1 MHz.

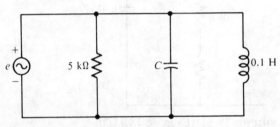

FIGURE 18.59 (Exercise 18.9D)

(a) If the inductance in the circuit is 2 mH, what range of capacitance should the adjustable capacitor provide?
(b) What is the maximum permissible resistance in the circuit if the quality factor at any center frequency must be at least 10?

18.8D A series RLC bandpass filter having a low quality factor must be designed so that its lower cutoff frequency is 12 kHz and its bandwidth is 4 kHz.
(a) If the inductance in the circuit is 30 mH, what value of capacitance should be used?
(b) What is the quality factor of the circuit?

18.9D What value of capacitance C should be used in the circuit shown in Figure 18.59 if the peak current in the inductor must be four times the peak current in the resistor at resonance?

18.10D The filter circuit shown in Figure 18.60 must produce an output whose magnitude is 18 V at resonance. What quality factor should the circuit have?

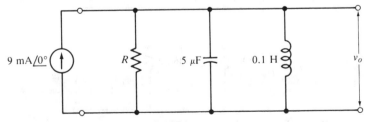

FIGURE 18.60 (Exercise 18.10D)

18.11D The winding resistance of the inductor in Figure 18.61 is $R_l = 100 \ \Omega$. What additional resistance R should be connected in series with the inductor to make the circuit resonant at $f_0 = 4$ kHz?

18.12D The filter shown in Figure 18.62 is required to produce a maximum output voltage at 36.17 kHz. If the winding resistance of the inductor is $R_l = 3.5$ kΩ, how much resistance R should be connected in series with the inductor?

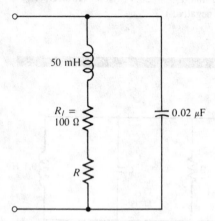

FIGURE 18.61 (Exercise 18.11D)

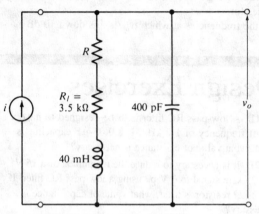

FIGURE 18.62 (Exercise 18.12D)

Troubleshooting Exercises

18.1T When the frequency response of the circuit shown in Figure 18.63 was measured, it was found that the cutoff frequency was 1989 Hz. Which components, if any, may be shorted or open?

18.2T The output of the circuit shown in Figure 18.64 was found to lead the input by 45° at 1768 Hz. Which components, if any, may be shorted or open?

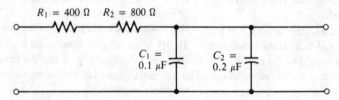

FIGURE 18.63 (Exercise 18.1T)

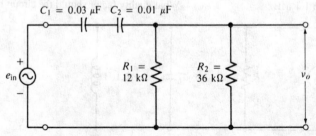

FIGURE 18.64 (Exercise 18.2T)

18.3T The specifications for the coil used in the filter shown in Figure 18.65 state that its winding resistance is 100 Ω ±20%. When one of the coils was connected in the circuit, it was found that the magnitude of the output voltage was 20 V at the center frequency of the filter. Is the coil within its tolerance specification?

18.4T When the filter shown in Figure 18.66 was tested, it was found that the output voltage did not rise to a maximum value at the resonant frequency. Instead, it was observed that the voltage increased when frequency was decreased, and continuously decreased when the frequency was increased. What are the possible faults in the circuit? Explain.

18.5T When the filter shown in Figure 18.67 was tested, it was found that the output voltage did not rise to a maximum value at the resonant frequency. Instead, it was observed that the voltage decreased when frequency decreased and increased when frequency increased. What are the possible faults in the circuit? Explain.

18.6T The manufacturer of the coil used in the filter circuit shown in Figure 18.68 specifies that the quality factor of the coil is at least 2 when the frequency is 150 × 10³ rad/s. When the coil was tested in the filter circuit, it was found that the quality factor of the circuit was 1.3. Does the coil meet specifications?

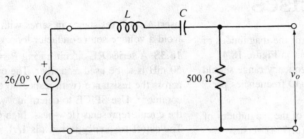

FIGURE 18.65 (Exercise 18.3T)

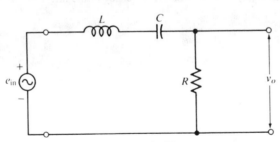

FIGURE 18.66 (Exercise 18.4T)

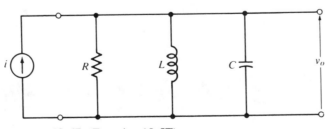

FIGURE 18.67 (Exercise 18.5T)

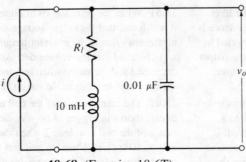

FIGURE 18.68 (Exercise 18.6T)

SPICE Exercises

18.1S Use SPICE to obtain plots of the magnitudes of the voltages across each component in Figure 18.47 (p. 736) versus frequency. The frequency range should extend from 3 Hz to 300 Hz with 10 frequencies per decade.

18.2S Use SPICE to obtain plots of the magnitudes of i_R, i_L, i_C, and i_T in Figure 18.51 (p. 738) versus frequency. The frequency range should extend from approximately one decade below resonance to approximately one decade above resonance, in one-tenth decade intervals. (It will be necessary to

insert a small resistance in series with the inductor to avoid a voltage source-inductor loop.)

18.3S A series RL circuit having $R = 1$ kΩ and $L = 50$ mH is to be used as a filter. The output is taken across the resistance (with respect to the circuit common). Use SPICE to determine what type of filter the circuit represents (low-pass, high-pass, etc.). (*Hint:* The cutoff frequency in rad/s is $1/\tau$.)

18.4S Repeat Exercise 18.3S when the output is taken across the inductance, with respect to the circuit common. Also use SPICE to determine the approximate cutoff frequency of the filter.

19 Transformers

19.1 Definitions and Basic Principles

A transformer is a magnetic circuit whose name is derived from its ability to transform (i.e., change) the magnitude of an ac voltage applied to it. It contains two windings, one of which is connected to an ac voltage source, the *input* to the transformer, and the other of which is connected to a load, or *output*. The input winding is called the *primary* winding of the transformer, and the output winding is called the *secondary* winding. The output voltage appearing across the secondary winding may be greater or less than the input voltage connected across the primary winding, depending, as we shall learn, on how the transformer is constructed. The two windings are essentially two inductors wound on a common core, as illustrated in Figure 19.1. In Figure 19.1(a), both windings are wrapped around the same cylindrical core, while in Figure 19.1(b) the windings are physically separated, but joined magnetically (coupled) through the window-shaped core. Of course in both cases, the wire used to form the winding must be *insulated*, to prevent adjoining turns from shorting each other and/or to prevent a short circuit between the primary and secondary windings.

Figure 19.2 shows the standard symbols used to represent transformers on schematic diagrams. The symbols differ only in respect to the material used in the construction of the core. The primary and secondary windings are not identified in the symbols because

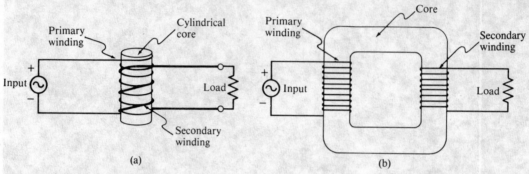

FIGURE 19.1 Basic transformer construction.

the input and output sides of most transformers can be interchanged. In other words, either winding can be used as the primary and the other winding as the secondary, although as we shall see, interchanging the windings changes the voltage transformation between primary and secondary. Figure 19.2(d) shows a schematic diagram of an iron-core transformer with the primary winding connected to a voltage source and the secondary winding connected to a load resistance, R_L. The diagram shows that the voltages across the primary and secondary windings are designated e_p and e_s, respectively, while the primary and secondary currents flowing in the windings are designated i_p and i_s. The number of turns in the primary winding is N_p and the number of turns in the secondary winding is N_s.

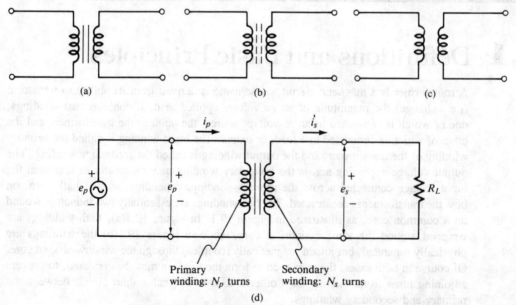

FIGURE 19.2 Transformer symbols and circuit diagram. (a) Iron core. (b) Ferrite core. (c) Air core. (d) Primary and secondary voltages and currents.

19.2 Ideal Transformers

To derive fundamental relationships between the primary and secondary voltages of a transformer, and between the primary and secondary currents, we assume two ideal conditions: (1) the primary and secondary windings have zero resistance, and (2) all of the magnetic flux produced by one winding is coupled (through the transformer's core) to the other winding. Although these assumptions are never completely valid for practical transformers, they are, in fact, good approximations of reality, and yield acceptable accuracy in most practical applications.

The Unloaded Transformer

Let us first consider the situation where the secondary winding of the transformer is open, that is, the *unloaded* transformer, illustrated in Figure 19.3. The ac source connected to the primary winding produces current in that winding and creates magnetic flux ϕ in the core. (Recall from Chapter 10 that current flowing in the winding generates magnetomotive force $\mathcal{F} = NI$.) Since the current is ac, the flux *changes* continually with time. Under the ideal assumption of zero resistance, the primary winding is pure inductance, and by Faraday's law [equation (11.1)], the voltage induced across the primary winding at any instant of time is

$$e_p = N_p \frac{\Delta\phi}{\Delta t} \tag{19.1}$$

where $\Delta\phi/\Delta t$ is the instantaneous rate of change of the flux. The voltage e_p given by (19.1) is exactly equal to the voltage produced by the source.

Since we assume that all of the flux produced by the primary winding is coupled to the secondary winding, it follows that the rate of change of flux, $\Delta\phi/\Delta t$, in the secondary winding is the same as that in the primary winding. The secondary winding is also pure inductance, so the voltage e_s induced across it by the changing flux is

$$e_s = N_s \frac{\Delta\phi}{\Delta t} \tag{19.2}$$

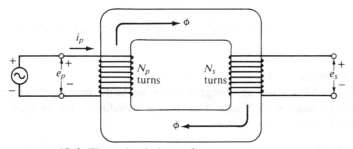

FIGURE 19.3 The unloaded transformer.

Dividing equation (19.1) by equation (19.2), we obtain the ratio of the primary voltage to the secondary voltage:

$$\frac{e_p}{e_s} = \frac{N_p(\Delta\phi/\Delta t)}{N_s(\Delta\phi/\Delta t)} = \frac{N_p}{N_s} \qquad (19.3)$$

The term N_p/N_s (often written $N_p:N_s$) is called the *turns ratio* of the transformer. Equation (19.3) shows that the ratio of primary to secondary voltage is the same as the turns ratio of the transformer. Thus, if the secondary winding has more turns than the primary winding, the secondary voltage is greater than the primary voltage. In that case, the transformer is said to be a *step-up* transformer. If there are fewer turns on the secondary than on the primary, the secondary voltage is less than the primary voltage, and the transformer is called *step-down*. It is important to remember that transformer action occurs only when the flux in the core is *changing*, for only under those circumstances can a voltage be induced in the secondary winding. The primary and secondary windings are said to be coupled by *mutual* inductance, whereby flux changes in one induce voltages in the other. As is the case in any type of inductance, flux that does not change with time does not induce a voltage, so a transformer cannot be used to step a dc voltage up or down.

Example 19.1   (Analysis)

A certain transformer has 50 turns on the secondary winding and 20 turns on the primary winding. An ac voltage having peak value 12 V is connected across the primary winding.

(a) What is the turns ratio of the transformer?
(b) What is the effective value of the secondary voltage?

SOLUTION

(a) $\dfrac{N_p}{N_s} = \dfrac{20}{50} = 0.4:1$

(b) Using voltage magnitudes (peak values) in equation (19.3), we find

$$|e_s| = \frac{|e_p|}{N_p/N_s} = \frac{12 \text{ V}}{0.4} = 30 \text{ V pk}$$

Letting E_s represent the effective value of e_s, we have

$$E_s = \frac{\sqrt{2}}{2}(30) = 21.21 \text{ V rms}$$

As this example illustrates, the turns ratio of a step-up transformer is a number less than 1.

Drill Exercise 19.1

When a 57.6 V rms ac voltage is connected to the primary of a transformer, the secondary voltage is 18 V rms. If the primary winding has 64 turns, how many turns does the secondary winding have?

ANSWER: 20.

The Loaded Transformer

In the unloaded transformer shown in Figure 19.3, there is zero current in the secondary winding, because that winding is open-circuited. The current that flows in the primary winding lags the primary voltage by 90°, since the primary is a pure inductance. The primary current is a very small value, just sufficient to establish flux ϕ in the core, and is called the *magnetizing current*. Zero power is delivered to the primary, because, again, it is purely inductive. These circumstances are very much different in the *loaded* transformer, shown in Figure 19.4. In this case, the voltage induced in the secondary winding creates current i_s that flows through load resistance R_L. Since the current through R_L is in phase with the voltage across it, we know that real power is delivered to the load. Calling this power the output power of the transformer, its instantaneous value is

$$p(t) = e_s i_s \quad \text{watts} \tag{19.4}$$

and its average value is

$$P_0(\text{average}) = E_s I_s = \frac{|e_s||i_s|}{2} \quad \text{watts} \tag{19.5}$$

where E_s and I_s are the effective values of the secondary voltage and current and $|e_s|$ and $|i_s|$ are the magnitudes (peak values) of the sinusoidal quantities.

From our discussion of energy, power and efficiency in Chapter 3, we know that output power cannot exceed input power in any real device. Therefore, the power delivered to the input (primary) side of the transformer must be at least as great as the output power. Since we have assumed that the transformer windings have zero resistance, there is zero loss in the windings, and the input power must equal the output power (100% efficient):

$$P_o = P_{\text{in}} \quad \text{(ideal transformer)} \tag{19.6}$$

The input power to the transformer is obtained from the voltage source connected across the primary winding. The primary current, i_p, is therefore very much greater than the small magnetizing current that flowed in the unloaded transformer. Furthermore, the

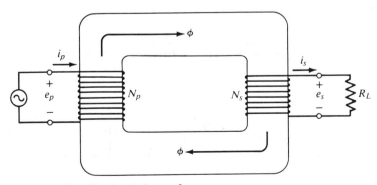

FIGURE 19.4 The loaded transformer.

primary current must now be in phase with the primary voltage, in order for real power to be delivered to the input. Thus, the input power is

$$p_{in}(t) = e_p i_p \quad \text{watts} \tag{19.7}$$

and

$$P_{in}(\text{average}) = E_p I_p = \frac{|e_p||i_p|}{2} \quad \text{watts} \tag{19.8}$$

Although it seems contradictory to state that input power is delivered to the primary winding, which is purely inductive, we must realize that no power is dissipated in that winding; power is merely transferred *through* the transformer to the load resistance.

Since output power equals input power in the ideal transformer, we have

$$e_p i_p = e_s i_s \tag{19.9}$$

or

$$\frac{e_p}{e_s} = \frac{i_s}{i_p} \tag{19.10}$$

Since $e_p/e_s = N_p/N_s$ in both the loaded and unloaded transformer, equation (19.10) can be written

$$\frac{i_s}{i_p} = \frac{N_p}{N_s} \tag{19.11}$$

Note carefully that the ratio of *secondary to primary* current is the same as the ratio of *primary to secondary* turns, which is just the inverse of the way the primary and secondary *voltages* are related. In other words, i_s is greater than i_p when e_p is greater than e_s, and vice versa. We see that a step-up transformer steps current down, and a step-down transformer steps current up. This is the perfectly logical consequence of the fact that the product of voltage and current must be the same on both sides.

Example 19.2   (Analysis)

Assuming the transformer shown in Figure 19.5 is ideal, find the magnitudes of

(a) e_s; (b) e_p; (c) i_p; (d) $P_{in}(\text{average})$; (e) $P_o(\text{average})$.

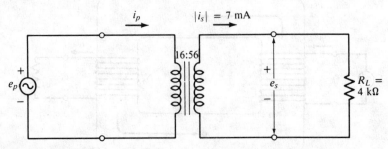

FIGURE 19.5 (Example 19.2)

SOLUTION

(a) $e_s = i_s R_L$

$|e_s| = |i_s|(4 \text{ k}\Omega) = (7 \text{ mA})(4 \text{ k}\Omega) = 28 \text{ V}$

(b) $e_p = e_s \dfrac{N_p}{N_s} = e_s \left(\dfrac{16}{56}\right)$

$|e_p| = (28 \text{ V})\left(\dfrac{16}{56}\right) = 8 \text{ V}$

(c) From equation (19.11),

$$i_p = \frac{i_s}{N_p/N_s} = \frac{i_s}{16/56}$$

$$|i_p| = \frac{7 \text{ mA}}{16/56} = 24.5 \text{ mA}$$

Notice that the transformer steps the voltage up by a factor of 3.5 (from 8 to 28 V) and steps the current down by the same factor (from 24.5 to 7 mA).

(d) $P_{in}(\text{average}) = \dfrac{|e_p||i_p|}{2} = \dfrac{(8 \text{ V})(24.5 \text{ mA})}{2} = 98 \text{ mW}$

(e) $P_o(\text{average}) = \dfrac{|e_s||i_s|}{2} = \dfrac{(28 \text{ V})(7 \text{ mA})}{2} = 98 \text{ mW}$

As expected, the output power equals the input power in this ideal transformer.

Drill Exercise 19.2

Repeat Example 19.2 if the turns ratio of the transformer is 30:20.

ANSWER: (a) 28 V; (b) 42 V; (c) 4.67 mA; (d) 98 mW; (e) 98 mW.   □

19.3 Impedance Transformation

Figure 19.6 shows an arbitrary circuit having a pair of input terminals to which a voltage source is connected. The input voltage produced by the source is designated e_{in} and the current that flows from the source into the circuit is designated i_{in}. The *input resistance*, R_{in}, of the circuit is defined to be

$$R_{in} = \frac{e_{in}}{i_{in}} \tag{19.12}$$

(In this definition, and in our subsequent discussion, we assume that the circuit is purely resistive. In the more general case, e_{in}/i_{in} is the input *impedance* of the circuit.) The input resistance of a circuit is often called the resistance that a voltage source "sees" when the source is connected to the circuit. Notice that the input resistance seen by a source does not depend on the voltage of the source, since the ratio e_{in}/i_{in} is constant.

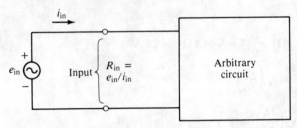

FIGURE 19.6 The input resistance of a circuit is defined to be the ratio of input voltage to input current.

Consider the loaded transformer circuit shown in Figure 19.7. The input resistance seen by the voltage source is

$$R_{in} = \frac{e_{in}}{i_{in}} = \frac{e_p}{i_p} \tag{19.13}$$

Substituting $e_p = e_s(N_p/N_s)$ and $i_p = i_s(N_s/N_p)$ into (19.13), we obtain

$$R_{in} = \frac{e_s(N_p/N_s)}{i_s(N_s/N_p)} = \left(\frac{N_p}{N_s}\right)^2 \frac{e_s}{i_s} \tag{19.14}$$

In Figure 19.7, it is clear that

$$e_s = i_s R_L \tag{19.15}$$

or

$$\frac{e_s}{i_s} = R_L \tag{19.16}$$

Substituting (19.16) into (19.14), we find

$$R_{in} = \left(\frac{N_p}{N_s}\right)^2 R_L \tag{19.17}$$

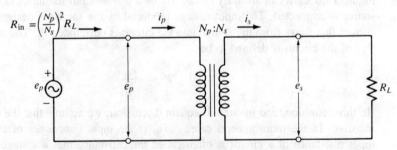

FIGURE 19.7 The input resistance looking into the primary of a transformer equals the load resistance times the *square* of the turns ratio.

Equation (19.17) expresses the very important fact that the input resistance of a transformer is the load resistance times the *square* of the turns ratio. In other words, the resistance looking into the primary side is $(N_p/N_s)^2$ times the resistance connected across the secondary. We say that the load resistance has been *transformed*, or *reflected* back to the input, by the square of the turns ratio. In general, the transformer is said to perform an *impedance* transformation. Note that a step-up transformer *reduces* the resistance looking into the primary, since $(N_p/N_s) < 1$, while a step-down transformer increases the resistance.

Example 19.3   (Analysis)

The transformer in Figure 19.7 has 48 turns on its primary winding and 12 turns on its secondary winding. A 50-Ω load is connected across the secondary. The magnitude of e_p is 16 V.

(a) What is the input resistance of the transformer?

(b) Use the input resistance calculated in part (a) to determine the magnitude of the primary current.

(c) Find the magnitude of the primary current by finding i_s and using equation (19.11).

SOLUTION

(a) From equation 19.17,

$$R_{in} = \left(\frac{N_p}{N_s}\right)^2 R_L = \left(\frac{48}{12}\right)^2 (50 \ \Omega) = 800 \ \Omega$$

(b) $i_{in} = \dfrac{e_{in}}{R_{in}} \Rightarrow i_p = \dfrac{e_p}{R_{in}}$

$|i_p| = \dfrac{16 \ V}{800 \ \Omega} = 0.02 \ A$

(c) $|e_s| = \dfrac{|e_p|}{N_p/N_s} = \dfrac{16 \ V}{48/12} = 4 \ V$

Therefore, the secondary current is

$$|i_s| = \frac{|e_s|}{R_L} = \frac{4 \ V}{50 \ \Omega} = 0.08 \ A$$

From equation (19.11),

$$|i_p| = \frac{|i_s|}{N_p/N_s} = \frac{0.08 \ A}{48/12} = 0.02 \ A$$

As expected, the primary current we calculated using the methods of Section 19.2 is the same as that calculated using the input resistance of the transformer.

Drill Exercise 19.3

Find the input resistance of the transformer in Example 19.3 and the magnitude of its primary current when the primary and secondary sides of the transformer are interchanged.

ANSWER: $R_{in} = 3.125 \ \Omega$; $i_p = 5.12$ A.   □

Impedance Matching

Because of their ability to transform impedances, transformers are widely used to achieve impedance matching between a source and load, as discussed in Chapter 17. Recall that maximum power is transferred from a source to a resistive load when the Thévenin equivalent resistance of the source equals the load resistance. By selecting a transformer with the correct turns ratio, and connecting the load across its secondary, the resistance looking from the source into the primary can be made equal to the Thévenin resistance of the source. In other words, maximum power transfer can be achieved by reflecting a load resistance through a transformer to the value required to match the source. Of course, impedance matching using a transformer can only be accomplished in ac circuits, since a transformer does not respond to dc. The next example illustrates how a transformer is selected to achieve maximum power transfer.

Example 19.4   (Design)

Figure 19.8(a) shows the equivalent circuit of a certain source connected to a load. It is clear that the Thévenin equivalent resistance of the source is 180 Ω and that the 20-Ω load is not matched to the source.

(a) Find the power transferred to the load in Figure 19.8(a).
(b) Find the turns ratio of a transformer that could be inserted between source and load, as shown in Figure 19.8(b), to achieve maximum power transfer to the load.
(c) Find the power transferred to the load when the matching transformer is used.

SOLUTION

(a) By the voltage-divider rule, the magnitude of the voltage v_L across the load resistance in Figure 19.8(a) is

$$|v_L| = \left(\frac{20}{180 + 20} \right) 90 \text{ V} = 9 \text{ V}$$

Therefore, the average power delivered to the load is

$$P_L(\text{avg}) = \frac{|v_L|^2}{2R_L} = \frac{(9)^2}{2(20)} = 2.025 \text{ W}$$

(b) To match the load to the source, the resistance R_{in} looking into the primary winding in Figure 19.8(b) must be 180 Ω. By equation (19.17), we must therefore have

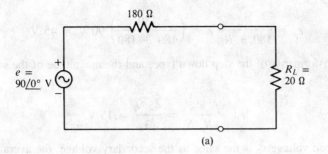

(a)

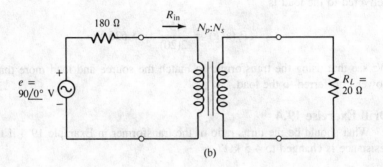

(b)

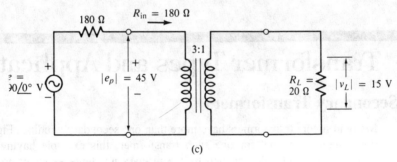

(c)

FIGURE 19.8 (Example 19.4)

$$R_{in} = 180\ \Omega = \left(\frac{N_p}{N_s}\right)^2 20\ \Omega$$

or

$$\frac{N_p}{N_s} = \sqrt{\frac{180}{20}} = 3$$

Thus, the turns ratio must be 3:1.

(c) Figure 19.8(c) shows the circuit with the 3:1 transformer inserted between source and load. *Note carefully that the primary voltage on the transformer is not the same as source voltage e.* A voltage division takes place between the 180-Ω source resistance and the value of R_{in}, so the magnitude of e_p is

$$e_p = \frac{R_{in}}{180 + R_{in}} e = \left(\frac{180}{180 + 180}\right) 90 \text{ V} = 45 \text{ V}$$

The transformer is of the step-down type, and the magnitude of the secondary voltage is

$$|e_s| = \frac{|e_p|}{N_p/N_s} = \frac{45 \text{ V}}{3} = 15 \text{ V}$$

Since the load voltage v_L is the same as the secondary voltage, the average power delivered to the load is

$$P_L(\text{avg}) = \frac{(15)^2}{2(20)} = 5.625 \text{ W}$$

We see that using the transformer to match the source and load more than doubles the power transferred to the load.

Drill Exercise 19.4

What should be the turns ratio of the transformer in Example 19.4 if the load resistance is changed to 4.5 kΩ?

ANSWER: 0.2:1, or 1:5.   □

19.4 Transformer Types and Applications

Multiple-Secondary Transformers

Many iron-core transformers have more than one secondary winding. Figure 19.9 shows the schematic symbol for one such transformer, this example having two secondary windings. Magnetic flux created by the primary winding is coupled to both secondary windings, so changes in flux ($\Delta\phi/\Delta t$) induce voltages in each secondary winding, just as they do in a transformer having one secondary winding. The ratio of the primary voltage to each secondary voltage equals the ratio of primary turns to secondary turns in each winding. Thus,

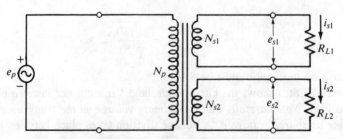

FIGURE 19.9 Transformer having two secondary windings.

$$\frac{e_p}{e_{s1}} = \frac{N_p}{N_{s1}} \quad \text{and} \quad \frac{e_p}{e_{s2}} = \frac{N_p}{N_{s2}} \tag{19.18}$$

where e_{s1}, e_{s2}, N_{s1}, and N_{s2} are the secondary voltages and secondary turns, as illustrated in Figure 19.9. In general, either or both secondaries can step up or step down the primary voltage. The secondary windings can also be connected together to form a single secondary having a total number of turns equal to the sum of the turns of the individual windings (assuming that they are connected to produce in-phase voltages, about which we will have more to say later).

The current in each secondary winding in Figure 19.9 is found, as usual, from Ohm's law:

$$i_{s1} = \frac{e_{s1}}{R_{L1}} \quad \text{and} \quad i_{s2} = \frac{e_{s2}}{R_{L2}} \tag{19.19}$$

The average power delivered to each load is

$$P_1 = \frac{|e_{s1}||i_{s1}|}{2} \quad \text{and} \quad P_2 = \frac{|e_{s2}||i_{s2}|}{2} \tag{19.20}$$

The total power delivered to the loads is the sum of P_1 and P_2, and the current in the primary winding can be found by equating the input power to the total output power (assuming an ideal transformer):

$$P_{\text{in}} = P_1 + P_2$$

$$\frac{|e_p||i_p|}{2} = \frac{|e_{s1}||i_{s1}|}{2} + \frac{|e_{s2}||i_{s2}|}{2}$$

$$|i_p| = \frac{|e_{s1}||i_{s1}| + |e_{s2}||i_{s2}|}{|e_p|} \tag{19.21}$$

Example 19.5 (Analysis)
In the circuit containing the ideal transformer shown in Figure 19.10:

(a) Find the magnitudes of e_{s1} and e_{s2}.
(b) Find the magnitude of i_p.

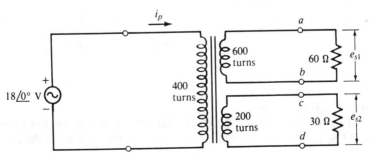

FIGURE 19.10 (Example 19.5)

(c) Assuming e_{ab} and e_{cd} are in phase, find e_{ad} when the loads are removed and terminal b is connected to terminal c.

SOLUTION

(a) From equation (19.18),

$$e_{s1} = \frac{e_p}{N_p/N_{s1}} \quad \text{and} \quad e_{s2} = \frac{e_p}{N_p/N_{s2}}$$

$$|e_{s1}| = \frac{18 \text{ V}}{400/600} = 27 \text{ V}$$

$$|e_{s2}| = \frac{18 \text{ V}}{400/200} = 9 \text{ V}$$

(b) From equation (19.21),

$$|i_p| = \frac{|e_{s1}||i_{s1}| + |e_{s2}||i_{s2}|}{|e_p|}$$

where

$$|i_{s1}| = \frac{27 \text{ V}}{60 \text{ }\Omega} = 0.45 \text{ A}$$

$$|i_{s2}| = \frac{9 \text{ V}}{30 \text{ }\Omega} = 0.3 \text{ A}$$

$$|i_p| = \frac{(27 \text{ V})(0.45 \text{ A}) + (9 \text{ V})(0.3 \text{ A})}{18 \text{ V}} = 0.825 \text{ A}$$

(c) When terminals b and c are connected, the secondary windings are equivalent to a single winding having $N_s = 600 + 200 = 800$ turns. Thus,

$$|e_{ad}| = \frac{|e_p|}{N_p/N_s} = \frac{18 \text{ V}}{400/800} = 36 \text{ V}$$

Notice that the same result can be obtained by simply adding the two secondary voltages, since they are in phase:

$$e_s = e_{s1} + e_{s2} = e_{ab} + e_{dc}$$

$$|e_s| = 27 \text{ V} + 9 \text{ V} = 36 \text{ V}$$

Drill Exercise 19.5

If the 18-V source and the 60-Ω resistor were interchanged in Figure 19.10, what would be the magnitudes of the voltages across the resistors in the circuit?

ANSWER: $e_{60\Omega} = 12 \text{ V}$; $e_{30\Omega} = 6 \text{ V}$.   □

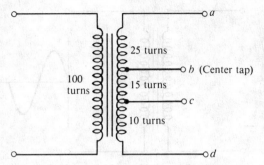

FIGURE 19.11 Example of a transformer having secondary taps, including a center tap.

Tapped Secondaries

In some transformers, a lead is connected to the secondary winding at a location somewhere between the endpoints of the winding. The lead is connected to an external terminal, so secondary voltages can be obtained between that terminal and either or both end terminals of the winding. Such a connection is called a *tap*. The number of turns between the tap and either end of the secondary can be used to determine the secondary voltage at the tap in the usual way. The only difference between a tapped secondary and a multiple secondary is that the windings in the multiple secondary are electrically isolated from each other. When the tap is made in the middle of the secondary winding, so there are an equal number of turns between it and either end of the winding, it is called a *center tap*. Figure 19.11 shows the schematic symbol of a transformer having a center tap and one other tap. In this example, voltages v_{ab} and v_{bd} would each be one-half of the total secondary voltage v_{ad}, and voltage v_{cd} would be $\frac{10}{50}$ of v_{ad}.

Isolation Transformers

One very useful characteristic of a transformer is the *isolation*, or electrical separation, it maintains between its primary and secondary circuits. Since there is no electrical connection between the primary and secondary windings, the circuits connected on each side of a transformer can have separate voltage references (commons, or grounds). One valuable consequence is that a high-voltage circuit can be isolated from the low-voltage side of a step-down transformer, for safety purposes. In many applications, a transformer having a 1:1 ratio is used specifically for the isolation it provides. A transformer of that design is called an isolation transformer.

Another type of isolation provided by a transformer is dc isolation. Since voltages are induced in the secondary only in response to current *variations* occurring in the primary, any dc component in the ac voltage connected to the primary is effectively blocked from the secondary. This blocking action is useful in many electronic circuits, such as that discussed in Example 18.6, where we showed how a capacitor is also used to block dc (bias) voltages. Figure 19.12 shows an example of a transformer that steps up an ac voltage and blocks its dc component. One limitation that must be considered

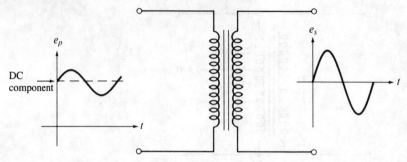

FIGURE 19.12 The dc component (bias) in the primary voltage does not appear in the secondary voltage. In the example shown, the transformer steps up the ac component.

in such applications is that a large dc current in the primary winding may *saturate*, or nearly saturate, an iron core (see Figure 10.12), to the extent that the transformer cannot respond to further increases in current caused by the ac component.

Autotransformers

An autotransformer is a transformer having a single winding, as illustrated in Figure 19.13. Notice that the winding is tapped, so the secondary winding is actually a portion of the primary winding. The transformer operates because of mutual induction between the two sections of the winding. The primary and secondary voltages are related by the turns ratio in the same way they are in a conventional transformer. The number of turns in the primary is the *total* number of turns in the winding, and the number of turns in the secondary is the number of turns between the tap and one end of the winding. With these definitions, the autotransformer is a step-down transformer. By connecting a voltage source across the tapped winding and taking the output across the full winding, the autotransformer can also be operated as a step-up transformer. Clearly, the autotransformer does not enjoy the isolation that a conventional transformer does between primary and secondary. However, the autotransformer has certain economic advantages that we will discuss shortly.

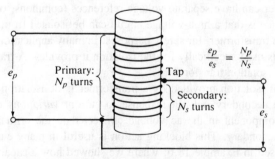

$$\frac{e_p}{e_s} = \frac{N_p}{N_s}$$

FIGURE 19.13 Construction of an autotransformer.

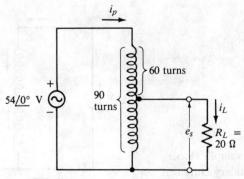

FIGURE 19.14 (Example 19.6)

Example 19.6 (Analysis)

For the autotransformer circuit shown in Figure 19.14:

(a) Find the magnitudes of the secondary voltage e_s and load current i_L.
(b) Find the average power delivered to the load.
(c) Assuming the transformer is ideal, find the average power delivered to the primary.
(d) Find the primary current.

SOLUTION

(a) The number of turns in the secondary winding is $90 - 60 = 30$ turns. Thus, $N_p = 90$ turns and $N_s = 30$ turns.

$$e_s = \frac{e_p}{N_p/N_s}$$

$$|e_s| = \frac{54\ \text{V}}{90/30} = 18\ \text{V}$$

$$|i_L| = \frac{|e_s|}{R_L} = \frac{18\ \text{V}}{20\ \Omega} = 0.9\ \text{A}$$

(b) $P_L(\text{avg}) = \dfrac{|e_s||i_L|}{2} = \dfrac{(18\ \text{V})(0.9\ \text{A})}{2} = 8.1\ \text{W}$

(c) Since the transformer is ideal, there are no losses in the winding, and the power delivered to the primary equals the power delivered to the load: 8.1 W.

(d) The primary current can be found from the primary voltage and input power:

$$8.1\ \text{W} = \frac{|e_p||i_p|}{2}$$

$$|i_p| = \frac{2(8.1)}{54} = 0.3\ \text{A}$$

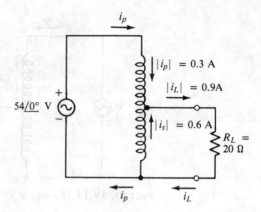

$|i_p| = 0.3$ A

$|i_L| = 0.9$A

$54\underline{/0°}$ V

$|i_s| = 0.6$ A

$R_L = 20\ \Omega$

i_p i_L

FIGURE 19.15 Circuit of Example 19.6, demonstrating that the secondary current is less than the load current.

Drill Exercise 19.6

What would be the magnitude of the primary current in Figure 19.14 if the load resistance were changed to 45 Ω?

ANSWER: 0.133 A.   ☐

In Example 19.6, we did not calculate the secondary current i_s in the autotransformer. That computation warrants special consideration, because it demonstrates a unique advantage of autotransformers. Figure 19.15 is the same circuit studied in the example, showing the calculated magnitudes of i_p and i_s at the tap, with their positive (reference) directions indicated. Since the load current leaving the tap, $|i_L| = 0.9$ A, is greater than the primary current entering the tap, $|i_p| = 0.3$ A, we know that the secondary current must be entering the tap. By Kirchhoff's current law,

$$|i_s| = 0.9\ \text{A} - 0.3\ \text{A} = 0.6\ \text{A}$$

In contrast with a conventional, two-winding transformer, we see that *the secondary current is less than the load current in an autotransformer*. As a consequence, the secondary winding of an autotransformer can be constructed with wire having smaller current-carrying capacity than a conventional transformer. The attendant savings in bulk and cost can be a significant economic advantage in autotransformers designed to furnish substantial load currents.

Transformer Phasing: The Dot Notation

In the examples discussed thus far, we have assumed that the secondary voltage of a transformer is in phase with the primary voltage. Such is not always the case. The phase relation between primary and secondary voltages (and currents) depends on how each winding is wrapped around the core, as illustrated in Figure 19.16. Notice that the primary windings in both parts of the figure are identical, but the secondary winding in (b) is

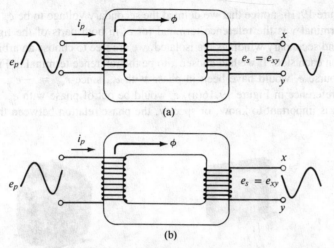

FIGURE 19.16 The phase relation between primary and secondary voltages in a transformer depends on how the windings are wrapped around the core. (a) Primary and secondary voltages in phase. (b) Primary and secondary voltages out of phase.

wound around the core in the opposite direction from the secondary winding in (a). As a consequence, the voltage induced in the secondary winding in (b) is 180° out of phase with that induced in (a). We see that the primary and secondary voltages are in phase in (a) and 180° out of phase in (b).

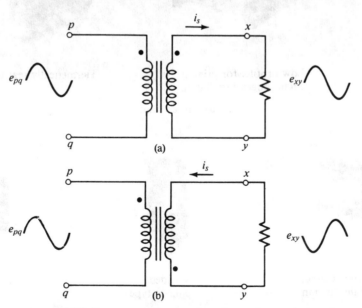

FIGURE 19.17 Dot convention used to show phase relations in a transformer.

In Figure 19.16, notice that we defined the secondary voltage to be $e_s = e_{xy}$. In other words, terminal y is the reference terminal for e_s in both parts of the figure. Since the primary and secondary windings are isolated, we are free to choose an arbitrary reference terminal on each side. If we had chosen x to be the reference terminal in Figure 19.16(b), then of course e_s would have been in phase with e_p, since $e_{yx} = -e_{xy}$. Similarly, if x were the reference in Figure 19.16(a), e_s would be out of phase with e_p. In some applications, it is important to know, or specify, the phase relation between the primary and

Power supply transformer   Autotransformer

Low profile, for printed   Hermetically sealed
circuit board installation

(a) Power transformers

For printed circuit   Shielded,   Contour molded
board installation   plug–in type

(b) Audio transformers

FIGURE 19.18 Examples of commercially available transformers. (Courtesy of Microtran Company, Inc.)

secondary voltages, and in those cases we cannot have arbitrary references. To eliminate any ambiguity in the phase relation between primary and secondary, a *dot convention* has been adopted for schematic diagrams. In this convention, dots are placed on the primary and secondary terminals that go positive at the same time (each with respect to the opposite terminal on its respective side). This convention is illustrated in Figure 19.17. In Figure 19.17(a), we see that terminals p and x are both positive at the same time, signifying that e_{pq} and e_{xy} are in phase. In Figure 19.17(b), terminals p and y are positive at the same time, meaning that e_{pq} and e_{xy} are 180° out of phase. [Note in Figure 19.17(b) that e_{xy} is negative when e_{pq} is positive, which is the same as saying that e_{yx} is negative at that time, or that y is positive with respect to x at that time.] Figure 19.18 shows examples of some commercially available transformers manufactured for a variety of applications.

19.5 Losses in Practical Transformers

Copper Losses

As we know, all materials have electrical resistance, including the wire wrapped around the core of a transformer to form its primary and secondary windings. The resistance in these windings is responsible for an average power loss, P_l, that can be calculated in the usual way:

$$P_l(\text{primary}) = \frac{|i_p|^2 R_p}{2} \quad \text{watts} \qquad (19.22)$$

$$P_l(\text{secondary}) = \frac{|i_s|^2 R_s}{2} \quad \text{watts} \qquad (19.23)$$

where R_p and R_s are the resistances of the primary and secondary windings, respectively. Since the core is usually wound with copper wire, these losses are often called *copper losses*. In most transformers, copper losses are relatively small, typically on the order of 1% of the total power transferred.

Many commercially available transformers are specified to have certain resistance ratios, such as 10 kΩ:100 Ω. These specifications are *not* winding resistances, which are typically a few ohms. The specifications refer instead to the resistance *transformation* that a transformer provides. Often the resistance transformation ratio is given instead of the turns ratio, but the turns ratio can be determined from that specification. Recalling that $R_{\text{in}} = (N_p/N_s)^2 R_L$, we have

$$\frac{N_p}{N_s} = \sqrt{\frac{R_{\text{in}}}{R_L}} \qquad (19.24)$$

For example, if the resistance transformation is 10 kΩ to 100 Ω, the turns ratio is $\sqrt{10^4/10^2} = 10{:}1$.

Eddy Currents

A transformer core itself has a certain inductance, so when flux in the core changes with time, as it does in normal transformer operation, electrical currents are induced in the core. If the core material is a conductor, such as the iron in an iron-core transformer, these currents may be large enough to cause noticeable power losses. The currents induced in a core this way are called *eddy currents*, and eddy current losses are losses caused by current flowing through the resistance of the core material. Eddy current losses increase as the frequency of the voltage applied to the transformer increases, because higher frequencies mean greater rates of change of flux, which in turn mean larger induced currents.

One popular construction method that is used to reduce eddy current losses is to assemble the core from *laminated* sheets. These laminations, or layers, are insulated from each other, so current flow is interrupted. Ferrite cores, which are made from a special type of ceramic that has a small electrical conductivity but large permeability, are also used for that purpose.

Hysteresis Losses

Recall from Chapter 10 that magnetism in ferromagnetic materials is attributable to the orientation of tiny magnetic domains. The magnetic fields produced by these domains align themselves with an externally applied magnetic field. In an inductor or transformer core, the external field is created by current flowing through the windings. When the current is ac, as it is in a transformer, the external field is continually reversing direction, so the magnetic domains must also continually reverse their orientation. There is a type of inertia, or resistance to change, which is an inherent property of magnetic domains, and which requires energy to overcome. This energy must be supplied each time the orientation of a domain is changed, so energy is consumed in the core of a transformer as it responds to continually changing alternating current. The energy consumed in that process is called *hysteresis loss* and is responsible for still another power loss in practical transformers. Hysteresis losses increase with the frequency of the applied voltage, because higher frequencies force the domains to reverse direction more often during a given interval of time.

Transformer Efficiency

The useful output power of a transformer is that which is delivered to its load, and the input power is that which is delivered from a source to its primary side. The efficiency of a transformer is defined the same way it is for other electrical devices (Chapter 3):

$$\eta = \frac{P_o}{P_{\text{in}}} = \frac{P_{\text{in}} - P_l}{P_{\text{in}}} = 1 - \frac{P_l}{P_{\text{in}}} \tag{19.25}$$

where P_l is the sum of all the power losses in the transformer. Eddy current and hysteresis losses are usually grouped into a category called *core losses*, because they are both associated with the core. Thus, $P_l(\text{total}) = P_{\text{core}} + P_{\text{copper}}$. As usual, the efficiency given by (19.25) is less than 1 and is often expressed as a percent. Of course, the *ideal* transformer we discussed earlier is 100% efficient. Practical transformer efficiencies are generally quite high in comparison to other electrical and electronic devices, on the order of 90 to 98%.

Example 19.7   (Analysis)

The transformer shown in Figure 19.19 has a primary winding resistance of 0.5 Ω and a secondary winding resistance of 0.1 Ω. The power delivered to the load resistance is 24 W and the magnitude of the primary current is 0.4 A. If the core losses are 0.9 W, find the efficiency of the transformer.

SOLUTION The magnitude of the secondary current can be found from the load power:

$$P_L = 24 \text{ W} = \frac{|i_s|^2 R_L}{2}$$

$$|i_s| = \sqrt{\frac{2(24 \text{ W})}{12 \text{ Ω}}} = 2 \text{ A}$$

Therefore, the copper loss in the secondary winding is

$$P_l(\text{secondary}) = \frac{|i_s|^2 R_s}{2} = \frac{(2)^2(0.1)}{2} = 0.2 \text{ W}$$

The copper loss in the primary winding is

$$P_l(\text{primary}) = \frac{|i_p|^2 R_p}{2} = \frac{(0.4)^2(0.5)}{2} = 0.04 \text{ W}$$

The sum of the losses, including the 0.9 W core loss, is

$$P_l(\text{total}) = 0.9 \text{ W} + 0.2 \text{ W} + 0.04 \text{ W} = 1.14 \text{ W}$$

The input power is the sum of the output power and the losses:

$$P_{\text{in}} = P_o + P_l(\text{total}) = 24 \text{ W} + 1.14 \text{ W} = 25.14 \text{ W}$$

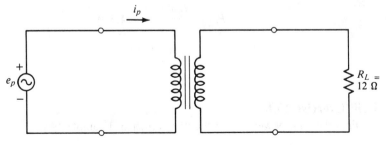

FIGURE 19.19 (Example 19.7)

Therefore, the efficiency is

$$\eta = \frac{P_o}{P_{in}} = \frac{24\ W}{25.14\ W} = 0.9546 \quad or \quad 95.46\%$$

Drill Exercise 19.7

Assuming that the transformer in Example 19.7 has the same copper losses, what are the core losses if the efficiency is 92%?

ANSWER: 1.187 W.   □

Recall that we derived the expression for the current transformation ratio, i_p/i_s, in terms of the turns ratio of a transformer, based on the assumption that the transformer was ideal. This expression, (19.12), is a good approximation for most practical transformers, but is not exact. If the load power and efficiency of a transformer are known, the primary and secondary currents can still be determined, as demonstrated in the next example.

Example 19.8   (Analysis)

A transformer is 96.5% efficient and delivers 400 W to a 12-Ω load. Find the magnitudes of the primary and secondary currents if the magnitude of the primary voltage is 210 V.

SOLUTION

$$P_L = 400\ W = \frac{|i_s|^2 R_L}{2}$$

$$|i_s| = \sqrt{\frac{2(400\ W)}{12\ \Omega}} = 8.16\ A$$

Since $\eta = P_o/P_{in}$, the input power is

$$P_{in} = \frac{P_o}{\eta} = \frac{400\ W}{0.965} = 414.5\ W$$

Then

$$P_{in} = \frac{|i_p||e_p|}{2} = 414.5\ W$$

$$|i_p| = \frac{2(414.5\ W)}{210\ V} = 3.95\ A$$

Drill Exercise 19.8

Find the total power loss in the transformer of Example 19.8.

ANSWER: 14.5 W.   □

19.6 Limitations of Practical Transformers

Leakage Flux and Coupling Coefficients

One of the assumptions we made for an ideal transformer is that all of the magnetic flux generated by one winding is coupled to the other winding. In fact, some of the flux generated by the primary winding of a practical transformer "escapes," in the sense that it is not confined entirely to the core, and therefore does not reach the secondary winding. As discussed in Chapter 10 (see Figure 10.8), magnetic flux is concentrated in a high-μ material such as iron, but disperses in low-μ air. Thus, two windings on one iron core have a good chance of sharing essentially the same flux, but are less likely to share flux in an air core. As illustrated in Figure 19.20, flux that is shared by the two windings of a transformer is called *mutual flux* ϕ_m, and that which escapes is called *leakage flux*, ϕ_l.

The *coefficient of coupling* k between the two windings of a transformer is defined to be

$$k = \frac{\phi_m}{\phi_m + \phi_l} \tag{19.26}$$

In an ideal transformer, $\phi_l = 0$ and $k = 1$. In a practical iron-core transformer, the leakage flux is very small, typically about 1% of the mutual flux. Consequently, the coupling coefficient for iron-core transformers is very nearly equal to the ideal value of 1. Such transformers are said to be *tightly coupled*. On the other hand, air-core transformers have a large leakage flux, and the coefficient of coupling may be as small as 0.01. Transformers with small values of k are said to be *loosely coupled*. The coupling coefficient can be increased by wrapping the secondary turns physically very close to the primary turns (i.e., overlapping them), as shown in Figure 19.1(a). However, in some high-frequency applications, loosely coupled, air-core transformers are desirable.

As an aid in analyzing the effect of leakage flux on the performance of a transformer, it is convenient to think of the total inductance of the primary winding, L_p, as being composed of two components: one, the *primary magnetizing inductance*, L_{pm}, that is subjected to the mutual flux, and, two, the *primary leakage inductance*, L_{pl}, that is subjected to leakage flux only. Thus,

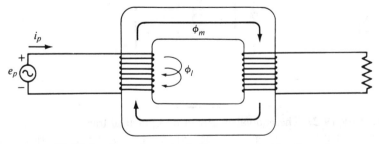

FIGURE 19.20 Leakage flux, ϕ_l, is flux generated by the primary winding that does not reach the secondary winding.

$$L_p = L_{pm} + L_{pl} \qquad (19.27)$$

If the coefficient of coupling were equal to 1, then L_{pl} would be zero, since ϕ_l would be zero. In that case, we would have $L_p = L_{pm}$. In general,

$$L_{pm} = kL_p \qquad (19.28)$$

Substituting (19.28) into (19.27) and solving for L_{pl}, we obtain

$$L_{pl} = (1 - k)L_p \qquad (19.29)$$

The secondary winding is also responsible for a certain amount of leakage. For example, if it were very loosely wound on the core and had large air gaps between its turns, we would attribute the reduced coupling to it. We can therefore regard the total inductance of the secondary winding, L_s, to be similarly composed of a secondary magnetizing inductance, L_{sm}, and a secondary leakage inductance, L_{sl}. Thus, $L_s = L_{sm} + L_{sl}$, and

$$L_{sm} = kL_s \qquad (19.30)$$

$$L_{sl} = (1 - k)L_s \qquad (19.31)$$

Figure 19.21 shows the equivalent circuit of a practical transformer in which the winding resistances and leakage inductances are identified as separate components. In reality, these quantities are intimate, inseparable parts of the entire windings and are said to be *distributed* throughout the windings. However, we can analyze their effects on the transformer by considering them to be separate entities, called *lumped parameters*, just as we treated winding resistance, R_l, of inductors in previous chapters. Note that the equivalent circuit contains an *ideal* transformer that has inductances L_{pm} and L_{sm} and that has zero winding resistance. To use the transformer theory studied earlier, we now regard

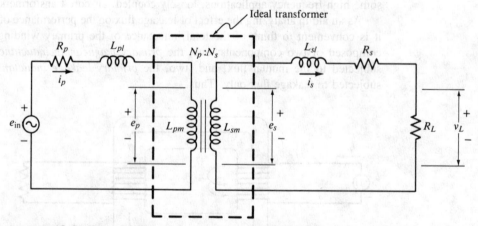

FIGURE 19.21 The equivalent circuit of a practical transformer is an ideal transformer that has winding resistance and leakage inductance in series with each winding.

the primary voltage e_p as that which appears across the terminals of the ideal primary in the equivalent circuit, and the secondary voltage as that which appears across the ideal secondary.

Loading Effects

It is apparent from Figure 19.21 that the full value of e_{in} will not appear across the primary winding. The primary voltage, e_p, will be less than e_{in} because of the voltage drop across the impedance in series with the winding. Consequently, the full value of e_{in} will not be multiplied by the turns ratio, and the secondary voltage will be less than would be predicted for an ideal transformer. Similarly, the full value of e_s does not appear at the load, because of the voltage drop appearing across the impedance in series with it. We see that winding resistance and leakage inductance make the ratio e_{in}/v_L smaller than the ratio N_p/N_s. We say that R_L *loads* the circuit, in much the same way that a load on the output of a real voltage source reduces the terminal voltage.

To calculate the actual load voltage in a practical transformer, we must first consider the voltage divider formed by the impedance in series with the primary and the impedance *reflected* from the secondary. Recall that we showed in Section 19.3 how a load resistance is reflected through a transformer by the square of the turns ratio. Similarly, any *complex* impedance in series with the secondary is reflected to the primary side by the square of the turns ratio: Z (reflected to primary) $= (N_p/N_s)^2 Z_s$, where Z_s is the total impedance across the secondary winding. It is this reflected impedance that we must use in our voltage-divider computation on the primary side. The next example illustrates the computation.

Example 19.9   (Analysis)

The transformer shown in Figure 19.22(a) has a total primary inductance of 0.3 H and a total secondary inductance of 50 mH. The winding resistances are $R_p = 2\ \Omega$ and $R_s = 1\ \Omega$. The transformer has a coupling coefficient of 0.99.

(a) Draw the equivalent circuit of the transformer.

(b) Find the magnitude of the load voltage.

SOLUTION

(a) We first find the mutual and leakage inductance in each winding. From equations (19.28) through (19.31),

$$L_{pm} = kL_p = 0.99(0.3\ \text{H}) = 0.297\ \text{H}$$

$$L_{pl} = (1 - k)L_p = (0.01)(0.3\ \text{H}) = 3\ \text{mH}$$

$$L_{sm} = kL_s = (0.99)(50\ \text{mH}) = 49.5\ \text{mH}$$

$$L_{sl} = (1 - k)L_s = (0.01)(50\ \text{mH}) = 0.5\ \text{mH}$$

Figure 19.22(b) shows the equivalent circuit.

(b) Since the frequency of e_{in} is 1.5 kHz, the reactance corresponding to each inductive component in the circuit is $X_L = j\omega L = j(2\pi f)L = j(2\pi \times 1.5 \times 10^3)L = j9425L\ \Omega$. Thus,

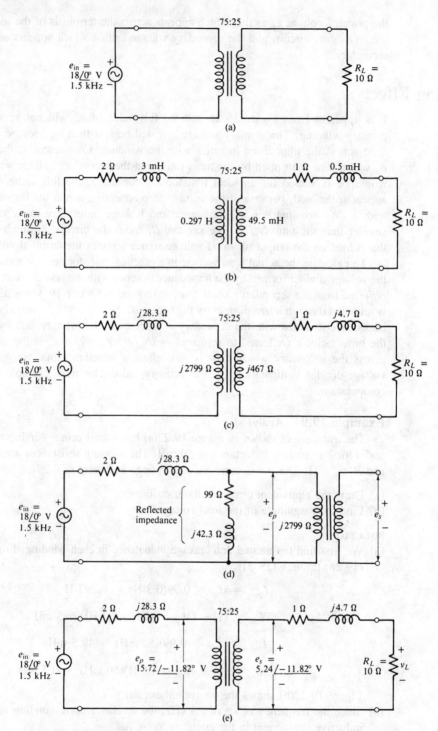

FIGURE 19.22 (Example 19.9)

$$X_{Lpm} = j9425(0.297 \text{ H}) = j2799 \ \Omega$$

$$X_{Lpl} = j9425(3 \times 10^{-3} \text{ H}) = j28.3 \ \Omega$$

$$X_{Lsm} = j9425(49.5 \times 10^{-3} \text{ H}) = j467 \ \Omega$$

$$X_{Lsl} = j9425(0.5 \times 10^{-3} \text{ H}) = j4.7 \ \Omega$$

Figure 19.22(c) shows the equivalent circuit with the reactance values labelled. The total impedance in the secondary winding is

$$Z_s = (10 + 1 + j4.7) \ \Omega = 11 + j4.7 \ \Omega$$

Therefore, the impedance reflected to the primary side is

$$Z(\text{reflected}) = \left(\frac{N_p}{N_s}\right)^2 Z_s = \left(\frac{75}{25}\right)^2 (11 + j4.7) \ \Omega$$

$$= (99 + j42.3) \ \Omega = 107.6 \ \underline{/23.13°} \ \Omega$$

Figure 19.22(d) shows the equivalent circuit with the reflected impedance drawn on the primary side. We can now use the voltage-divider rule to find the value of e_p. We will neglect the impedance of L_{pm} ($j2799 \ \Omega$) in parallel with the reflected impedance, because it is so much greater than the reflected impedance. Thus,

$$e_p \approx \frac{Z(\text{reflected})}{(2 + j28.3) + Z(\text{reflected})} e_{\text{in}}$$

$$= \left[\frac{107.6 \ \underline{/23.13°}}{(2 + j28.3) + (99 + j42.3)}\right] 18 \ \underline{/0°} \ \text{V} = 15.72 \ \underline{/-11.82°} \ \text{V}$$

The secondary voltage e_s is therefore

$$e_s = \frac{e_p}{N_p/N_s} = \frac{15.72 \ \underline{/-11.82°} \ \text{V}}{75/25} = 5.24 \ \underline{/-11.82°} \ \text{V}$$

Finally, v_L can be found from the voltage divider appearing on the secondary side, as shown in Figure 19.22(e):

$$v_L = \left(\frac{10 \ \underline{/0°}}{10 + 1 + j4.7}\right) e_s = \left(\frac{10 \ \underline{/0°}}{11.96 \ \underline{/23.14°}}\right) 5.24 \ \underline{/-11.82°} \ \text{V}$$

$$= 4.38 \ \underline{/-34.96°} \ \text{V}$$

If we had assumed that the entire transformer were ideal, we would have calculated a load voltage of

$$e_s = v_L = \frac{e_{\text{in}}}{N_p/N_s} = \frac{18 \ \underline{/0°} \ \text{V}}{75/25} = 6 \ \underline{/0°} \ \text{V}$$

We see that there is a discrepancy of 1.62 V, or about 37%, between the actual and ideal magnitudes of the load voltage. This rather large discrepancy is due to the small value of load resistance (10 Ω) in the example. Note that the smaller the load resistance, the greater its loading effect, due to voltage division on both sides of the transformer.

Drill Exercise 19.9

What would be the discrepancy between the actual and ideal magnitudes of the load voltage in Example 19.9 if the turns ratio had been 4:1? (Neglect X_{Lpm} and assume that all other component values are the same.)

ANSWER: 1 V, or 28.7%.   □

Frequency Response

The output voltage of an ideal transformer is independent of frequency. However, in a real transformer, we must consider how the inductive components affect the output when frequency changes, because the reactances of those components increase with frequency. Furthermore, every transformer has *capacitance* between the turns of its windings, and capacitive reactance is also affected by frequency. Recall that capacitance exists whenever two conductors are separated by an insulator, which is precisely the situation when the insulated conductors of a transformer winding are wrapped around the core. The capacitances of both windings are distributed throughout the windings, but it is convenient to lump their effect into equivalent capacitors in parallel with each winding. The primary

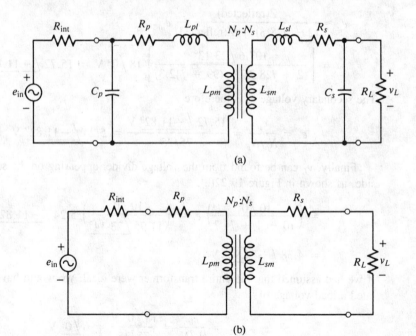

(a)

(b)

FIGURE 19.23 Equivalent circuits of a practical transformer, used to study frequency response. (a) Complete equivalent circuit, including inductance and shunt capacitance. (b) Low-frequency equivalent circuit.

and secondary capacitance, C_p and C_s, are said to be *shunt* capacitances, because they divert current that would otherwise flow in the windings. Figure 19.23(a) shows the equivalent circuit of a transformer, including the lumped capacitance and the lumped inductance we discussed earlier. Also shown is the internal resistance, R_{int}, of the voltage source connected to the primary winding, because this resistance also affects frequency response.

To understand how frequency affects the response of the circuit in Figure 19.23(a), we can study the effects of low frequencies and high frequencies separately. Let us first suppose that the frequency of e_{in} is quite low. In that case, the capacitive reactances of C_s and C_p are quite large ($|X_C| = 1/\omega C$). Since the capacitances are in parallel with the input and output, their large reactances can be neglected at a sufficiently low frequency. Furthermore, the reactances of the leakage inductances, X_{Lpl} and X_{Lsl}, become very small in comparison to R_p and R_s, since inductive reactance decreases with frequency. Figure 19.23(b) shows the low-frequency equivalent circuit when the shunt capacitances and the leakage inductances are neglected (capacitances replaced by open circuits and inductances replaced by short circuits). It is clear that the inductive reactance of L_{pm} and the total series resistance ($R_{int} + R_p$) form a voltage divider across the primary winding. This voltage divider is, in fact, a high-pass filter. The lower the frequency, the smaller the voltage drop across X_{Lpm}. Consequently, as frequency decreases, the voltage appearing at the primary decreases, and the output of the transformer decreases. In the limiting case, where $f = 0$ Hz (dc), $X_{Lpm} = 0 \ \Omega$, and there is zero voltage across the primary. This confirms our previous discussion of the fact that a transformer does not respond to dc.

Let us now consider the other frequency extreme, that is, the situation when the frequency of e_{in} becomes very large. In that case, the reactances of the shunt capacitances become very small and can no longer be neglected. Referring again to Figure 19.23(a), notice that R_{int} and C_p form a low-pass RC filter. As we know from Chapter 18, the voltage across the output capacitor in a low-pass filter decreases as frequency increases. Furthermore, the reactance of L_{pl} increases with frequency, so a greater portion of the input voltage is dropped across it. The combined effects of the shunt capacitance and the series inductance cause the voltage at the primary to decrease rapidly as frequency increases. Similarly, R_s and C_s form a low-pass RC filter on the secondary side of the

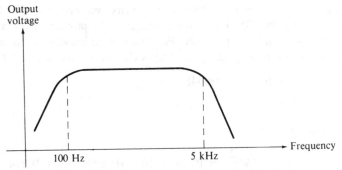

FIGURE 19.24 Frequency response of a typical audio transformer.

transformer. The effect of this filter, and the series reactance of L_{sl}, further reduce the voltage v_L appearing at the load. The reduction in output voltage due to all these effects is responsible for a frequency response that falls rapidly at high frequencies.

Figure 19.24 shows a typical frequency response for a transformer used in audio circuits (an audio transformer). Note that the cutoff frequencies are about 100 Hz and 5 kHz, which means that the passband is a relatively small portion of the audio range (20 Hz to 20 kHz). In modern high-fidelity systems, audio transformers are avoided whenever possible, because of their bulk and their limited frequency response.

Mutual Inductance

We have mentioned that mutual inductance is a measure of the ability of one winding to induce a voltage in a second winding. Contrast this concept with that of self-inductance, which is a measure of the ability of a winding to induce a voltage in itself. Of course, in both cases, a voltage is induced only when current is changing with time. In a transformer, a change of current in the primary winding induces a voltage in the secondary winding, and vice versa. Thus, mutual inductance M relates rate of change of current in each winding to voltage induced in the other winding:

$$e_s = M \frac{\Delta i_p}{\Delta t} \tag{19.32}$$

$$e_p = M \frac{\Delta i_s}{\Delta t} \tag{19.33}$$

Like self-inductance, the units of mutual inductance are henries (H).

Mutual inductance is a parameter used most frequently in connection with loosely coupled air-core transformers, because the turns ratio in those types of transformers is of little value in predicting output voltages. As might be expected, the value of mutual inductance depends on the coefficient of coupling k between the windings, as well as upon the self-inductance of each winding:

$$M = k\sqrt{L_s L_p} \tag{19.34}$$

Example 19.10   (Analysis)

The coupling coefficient between the primary and secondary windings of an air-core transformer is 0.05. The self-inductances of the primary and secondary windings are 8 mH and 12 mH, respectively. Find the voltage induced in the secondary winding when the current in the primary winding is changing at the rate of 600 A/s.

SOLUTION From equation (19.34),

$$M = k\sqrt{L_s L_p} = 0.05 \sqrt{(12 \times 10^{-3})(8 \times 10^{-3})} = 0.49 \text{ mH}$$

From equation (19.32),

$$e_s = M \frac{\Delta i}{\Delta t} = (0.49 \times 10^{-3} \text{ H})(600 \text{ A/s}) = 0.294 \text{ V}$$

Drill Exercise 19.10

The transformer in Example 19.10 is redesigned to increase the coupling coefficient. The inductance of each winding is unchanged, but it is found that a voltage of 0.5 V is now induced in the secondary winding when the current in the primary changes at the same rate as in the example. What is the new value of the coupling coefficient?

ANSWER: 0.085.   □

19.7 SPICE Examples

Example 19.11   (SPICE)

Figure 19.25(a) shows a transformer used to match a 100-kΩ source to a 250-Ω load. The secondary winding has inductance 50 mH, and the input frequency is 50 kHz. Assuming the transformer is ideal, use SPICE to find the primary and secondary voltages and currents.

SOLUTION From Appendix Section A.10, the turns ratio is related to the primary and secondary inductance, L_p and L_s, by

$$\frac{N_p}{N_s} = \sqrt{\frac{L_p}{L_s}} \quad \text{or,} \quad L_p = \left(\frac{N_p}{N_s}\right)^2 L_s$$

Therefore, since $\dfrac{N_p}{N_s}$ = 20 and L_s = 50 mH,

$$L_p = (20)^2 \, (50 \times 10^{-3} \text{ H}) = 20 \text{ H}$$

Figure 19.25(b) shows the SPICE circuit and input data file. Note that the secondary circuit cannot be isolated from the primary because of the SPICE requirement that there be a dc path to ground from every node. As shown, the secondary winding must have one side connected to node 0. Since the transformer is ideal, the coupling coefficient k is set equal to 1 in the statement K1 LP LS 1. Note that the primary reactance, $2\pi f L_p$ = 6.28 MΩ, is much greater than the 100-kΩ source impedance, and the secondary reactance, $2\pi f L_s$ = 1.57 kΩ, is much greater than the 250-Ω load impedance, as required to simulate an iron-core transformer in SPICE. Dummy voltage sources VIP and VIS are used to determine primary and secondary currents. Execution of the program gives V(3) = v_p = 4 V, V(5) = v_s = 0.2 V, I(VIP) = i_p = 40 μA, and I(VIS) = i_s = 0.8 mA. These results can be readily verified by hand computation.

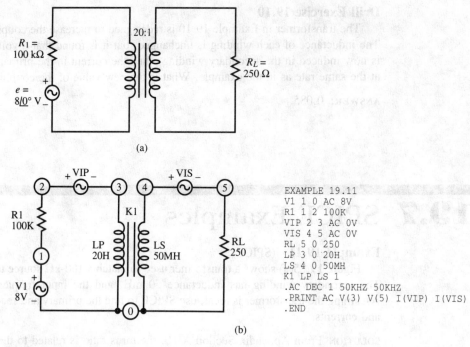

FIGURE 19.25 (Example 19.11)

```
EXAMPLE 19.11
V1 1 0 AC 8V
R1 1 2 100K
VIP 2 3 AC 0V
VIS 4 5 AC 0V
RL 5 0 250
LP 3 0 20H
LS 4 0 50MH
K1 LP LS 1
.AC DEC 1 50KHZ 50KHZ
.PRINT AC V(3) V(5) I(VIP) I(VIS)
.END
```

Exercises

Section 19.2   Ideal Transformers

19.1 The sinusoidal voltage across the primary winding of a transformer has peak value 39 V. The voltage across the secondary winding is 9.192 V rms. What is the turns ratio of the transformer?

19.2 When the voltage across the primary winding of a transformer is 6 V rms, the secondary voltage is 45 V rms. If the secondary winding has 900 turns, how many turns does the primary winding have?

19.3 If the current in the primary winding of the transformer in Exercise 19.2 is 200 mA pk when the transformer is loaded, what is the effective value of the current in the secondary winding?

19.4 The output voltage of a loaded, step-down transformer is one-sixth of its input voltage. If the secondary current is 0.42 A rms, what is the effective value of the primary current?

19.5 A loaded transformer has a turns ratio of 180:756. If the primary current is 1 A pk and the secondary voltage is 63 V pk:

(a) Find the peak value of the primary voltage.

(b) Find the peak value of secondary current.

(c) Find the power delivered to the load.

(d) Find the value of the load resistance.

19.6 The average power delivered to the primary side of the ideal transformer shown in Figure 19.26 is 80 W. If the magnitude (peak value) of the sinusoidal source e is 170 V:

(a) Find the average power delivered in the 12-Ω resistor.

(b) Find the magnitude of i_s.

(c) Find the magnitude of e_s.

(d) Find the magnitude of i_p.

(e) Find the turns ratio $N_p:N_s$.

Section 19.3 Impedance Transformation

19.7 In the circuit containing the ideal transformer shown in Figure 19.27:
(a) Find the input resistance R_{in} looking into the primary side.
(b) Find the peak value of i_p.
(c) Find the peak value of i_L.

19.8 An ideal transformer has a turns ratio of 250:40. The resistance seen by a voltage source connected across the primary winding is 46.875 kΩ.
(a) What is the value of the load resistance connected across the secondary winding?
(b) How much power is delivered to the primary side of the transformer when the primary voltage is 93.75 V pk?

19.9(a) Is the load in Figure 19.28 matched to the source?
(b) Find the magnitude of the load current i_L.

19.10 The load R_L in Figure 19.29 is matched to the source by the ideal transformer. Find the power delivered to the load.

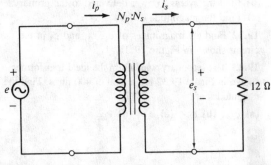

FIGURE 19.26 (Exercise 19.6)

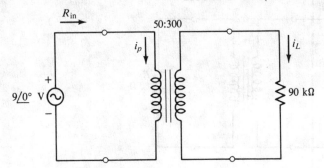

FIGURE 19.27 (Exercise 19.7)

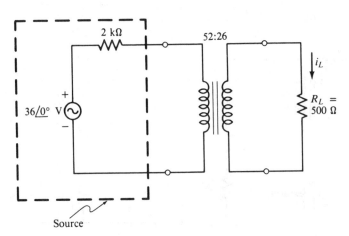

FIGURE 19.28 (Exercise 19.9)

Section 19.4   Transformer Types and Applications

19.11 In the circuit containing the ideal transformer shown in Figure 19.30.

(a) Find the magnitudes of e_{s1} and e_{s2}.
(b) Find the average power delivered to the primary.
(c) Find the magnitude of i_p.

19.12 Find the magnitudes of e_1, e_2, and e_3 in the circuit shown in Figure 19.31.

19.13 The secondary winding in the ideal transformer shown in Figure 19.32 has a total of 400 turns. Find the magnitudes of

(a) e_{ad}; **(b)** e_{bd}; **(c)** e_{bc}; **(d)** e_{ab}.

19.14 The ideal transformer shown in Figure 19.33 has the center tap (CT) shown. The magnitude of the voltage across the load resistance is 15 V.

(a) How many turns are between the center tap and each end of the secondary winding?
(b) What would be the magnitude of the primary current if the load resistance shown in the circuit were reconnected between the center tap and one end of the secondary winding?

19.15 In the circuit containing the ideal autotransformer shown in Figure 19.34:

(a) Find the magnitude of the load current i_L.
(b) Find the magnitude of the primary current i_p.

19.16 Find the magnitude of secondary current i_s in Figure 19.34 when the load resistance is changed to 50 Ω.

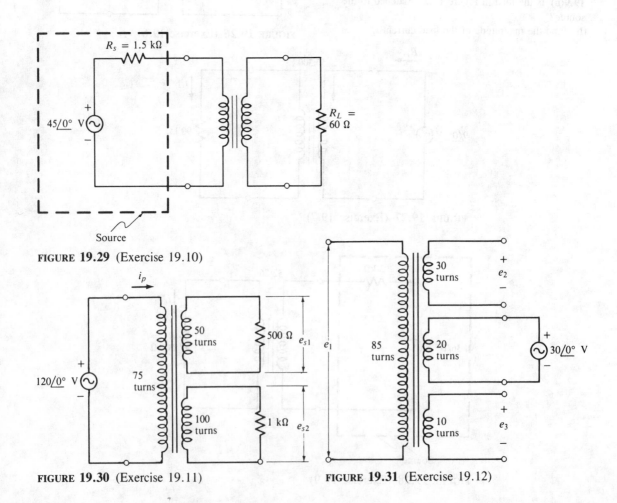

FIGURE 19.29 (Exercise 19.10)

FIGURE 19.30 (Exercise 19.11)

FIGURE 19.31 (Exercise 19.12)

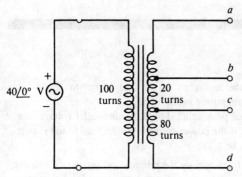

FIGURE 19.32 (Exercise 19.13)

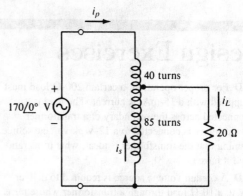

FIGURE 19.34 (Exercises 19.15 and 19.16)

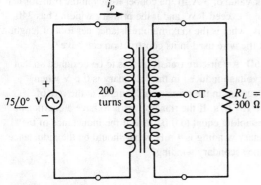

FIGURE 19.33 (Exercise 19.14)

Section 19.5 Losses in Practical Transformers

19.17 The primary and secondary winding resistances of a transformer are 0.9 Ω and 0.5 Ω, respectively. The transformer delivers 32 W to a 64-Ω load connected across its secondary. The primary current is 0.6 A. If the core losses are 1.5% of the load power, find the efficiency of the transformer.

19.18 A transformer having an efficiency of 94.3% delivers 200 W to a load.
(a) If the copper losses are 1% of the load power, what is the core loss?
(b) If the magnitude of the primary current is 6 A pk, what is the magnitude of the voltage supplied by the source connected to the primary winding?

Section 19.6 Limitations of Practical Transformers

19.19 When the mutual flux in an iron-core transformer is 6 × 10⁻⁴ Wb, the leakage flux is 3.16 × 10⁻⁵

FIGURE 19.35 (Exercise 19.20)

Wb. The total inductance of the primary winding is 0.22 H and the total inductance of the secondary winding is 140 mH. The resistance of the primary winding is 1.8 Ω and the resistance of the secondary winding is 0.5 Ω. Draw the equivalent circuit of the transformer. Be certain to label the values of all components.

19.20 The transformer shown in Figure 19.35 has a coupling coefficient of 0.95. The total inductance of the primary winding is 1 H and the total inductance of the secondary winding is 80 mH. The primary and secondary winding resistances are 1.5 Ω and 0.4 Ω, respectively. Find the magnitude of the load voltage, v_L.

19.21 The coefficient of coupling in an air-core transformer is 0.08. The primary inductance is 8 mH and the secondary inductance is 18 mH. What voltage is induced in the secondary when the current in the primary is changing at the rate of 500 A/s?

19.22 The primary and secondary inductances in an air-core transformer are each equal to 14 mH. When the current in one winding is changing at the rate of 200 A/s, the voltage induced in the other winding is 0.56 V. What is the coefficient of coupling between the windings?

Design Exercises

19.1D For proper operation, a certain 200-Ω load must be supplied with a 15-mA pk current. The load is to be connected across the secondary of a transformer whose primary is connected to a 12-V pk voltage source. Assuming that the transformer is ideal, what turns ratio should it have?

19.2D A certain voltage source is required to deliver 80 W to a 10-Ω load through a transformer whose turns ratio is 2.5:1. Assuming that the transformer is ideal, what peak sinusoidal voltage should the voltage source produce?

19.3D A transformer is to be used to transfer the maximum possible power from a voltage source having an internal resistance of 3698 Ω to a 200-Ω load. What should be the turns ratio of the transformer?

19.4D A voltage source is connected to the primary of a transformer whose turns ratio is 0.4:1. The voltage source has a Thévenin equivalent resistance of 100 Ω.
(a) What value of load resistance should be connected

across the secondary of the transformer to obtain maximum power in the load?
(b) What peak value should the sinusoidal voltage source provide if the power delivered to the load found in part (a) is to be 8 W?

19.5D Each turn on the primary and secondary winding of a 1:1 isolation transformer requires 2.5 in. of copper wire. The sinusoidal current in each winding will have a peak value of 3 A. If the copper losses in the transformer cannot exceed 1 W, and if the primary winding has 240 turns, what is the maximum resistance per foot of length that the wire used in its construction can have?

19.6D An air-core transformer is to be designed so that the voltage induced in the secondary is 0.6 V when the current in the primary is changing at the rate of 5×10^3 A/s. If the transformer is to have a coefficient of coupling equal to 0.02, and if the inductance of the primary winding is 4 mH, what should be the inductance of the secondary winding?

Troubleshooting Exercises

19.1T An ohmmeter is often used to test the "continuity" of a transformer. Since winding resistance is usually very small, the resistance through any continuous winding should measure near zero ohms. If there is an open in a winding, its resistance should measure "∞" on the ohmmeter. Of course, there should be infinite resistance between any two terminals that belong to electrically isolated windings. For the transformer shown in Figure 19.36, determine the fault condition, if any, corresponding to each of the following sets of measurements (R_{ab} means the resistance measured between terminals a and b, etc.)
(a) $R_{ab} = 0$; $R_{cd} = 0$; $R_{ef} = 0$; $R_{ed} = \infty$; $R_{eg} = \infty$
(b) $R_{ac} = \infty$; $R_{ce} = 0$; $R_{de} = 0$; $R_{ef} = 0$; $R_{fg} = 0$
(c) $R_{bc} = \infty$; $R_{de} = \infty$; $R_{cd} = 0$; $R_{ae} = 0$; $R_{bg} = 0$
(d) $R_{bg} = 0$; $R_{ce} = \infty$; $R_{fg} = 0$; $R_{ad} = \infty$

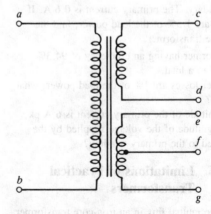

FIGURE 19.36 (Exercise 19.1T)

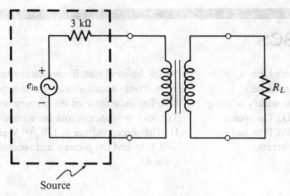

Source

FIGURE 19.37 (Exercise 19.3T)

19.2T When a 24-V rms source was connected to a transformer having a 66:22 turns ratio, it was found that the current in a 500-Ω load connected to the transformer was 144 mA rms. What is the most likely problem, if any?

19.3T The transformer shown in Figure 19.37 was installed to match load R_L to the 3-kΩ source resistance. When e_{in} was adjusted to 18 V pk, the current produced by the source was measured to be 2 mA pk. Assuming that the transformer is ideal, has it been properly selected for this application? Explain.

19.4T To verify that the coupling coefficient of a transformer is at least 0.05, the current shown in Figure 19.38 is passed through the primary winding and the

voltage across the primary and secondary windings is measured. The same current is then passed through the secondary winding and the voltage across the primary and secondary windings are measured again. In one test, the following sets of measurements were obtained:

Current Passed Through:	Primary Voltage	Secondary Voltage
Primary	6.4 V	128 mV
Secondary	128 mV	1.6 V

Did the transformer meet specifications? Explain.

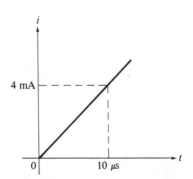

FIGURE 19.38 (Exercise 19.4T)

SPICE Exercises

19.1S An ideal transformer having turns ratio $N_p/N_s = 0.5$ is driven from a source whose impedance is $R_s = 600 \ \Omega$. The inductance of the secondary winding is 10 H and the load resistance is 2 kΩ. The source voltage is 30 $\underline{/0°}$ V at 25 kHz. Use SPICE to find the primary and secondary voltages and currents.

19.2S An ideal transformer having turns ratio $N_p/N_s = 10$ is driven from a source whose impedance is $R_s = 1$ kΩ. The inductance of the primary winding is 1000 H. The load resistance across the secondary winding is 50 Ω. The source voltage is 170 $\underline{/0°}$ V at 60 Hz. Use SPICE to find the primary and secondary voltages and currents.

20 Polyphase Circuits and Electric Power Distribution

20.1 Introduction

An ac circuit having a single ac voltage source is said to be a *single-phase* circuit. By and large, the ac circuits we have studied up to now have been examples. In these examples, electrical power is delivered from a source, such as an ac voltage generator, to a load via *two* conductors, or wires. This arrangement is called a single-phase, two-wire system. (In a later discussion, we will study the single-phase, *three*-wire system, the system most commonly used in homes.) By contrast, a *polyphase* circuit is one containing more than one ac source and three or more conductors, upon which appear ac voltages having different phase angles. (*Poly* means more than one.) The most common polyphase circuits are those containing three ac sources and three or four wires. These *three-phase* circuits are widely used in the electrical power industry to transmit power from generating stations, such as hydroelectric and nuclear power plants, to metropolitan areas and to distribute that power to individual consumers. Three-phase power is also found in manufacturing plants, large office buildings, hospitals, shopping centers, and similar complexes having high demands for power. Most large electrical motors are designed to be operated from a three-phase source.

There are numerous advantages to using the three-phase method of power distribution in situations where large amounts of power are transmitted and consumed. Several of these advantages will be demonstrated in forthcoming examples. In particular, we will learn that the total volume of the conductors required to transfer a fixed amount of power using a three-phase system is about 40–50% of that required for a single-phase system, a significant difference when considering the cost of long-distance transmission lines. Furthermore, the phase relations between the voltages in a three-phase system make it possible to operate certain types of electrical equipment, including motors, more efficiently than in a single-phase system.

Although we defined a three-phase circuit as one having three ac voltage sources, we should note that the three voltages are usually created by a single piece of equipment: a three-phase ac generator, also called an *alternator*. A three-phase generator is equivalent to three ac sources that produce three voltages having the same frequency (usually 60 Hz in the United States, 50 Hz in Europe), each of which is separated in phase by 120° from the next. Depending on how the generator is wired internally, the three voltages may appear on three or four output lines. These ideas will become clear when we examine the structure of the generator in a later discussion. As an aid in understanding how three-phase voltages are generated, we will study first the structure and operating principles of a single-phase generator.

20.2 AC Generators (Alternators)

The Elementary Single-Phase Generator

The basic principle underlying the operation of both dc and ac generators is electromagnetic induction. Recall from Chapter 11 that a voltage is induced across the ends of a conductor when the conductor moves through (cuts) the flux of a magnetic field (see Figure 11.1). To construct a practical generator, we must devise a means for maintaining *continuous* motion of a conductor through a magnetic field. As illustrated in Figure 20.1, continuous motion can be achieved by *rotating* a loop of wire through a magnetic field. In this way, each side of the loop becomes a voltage-generating conductor as it continuously cuts through the magnetic field. Using Lenz's law (Figures 11.2 and 11.3), we can determine the polarity of the voltages induced in each side of the loop. When the loop is rotated in a clockwise direction, as shown in Figure 20.1, we see that the polarities are such that the two voltages are additive and that the right terminal of the loop (*x*) is positive with respect to the left terminal (*y*). (Note that the back side of the loop does not cut flux in the plane through which it moves, so no voltage is induced in it.) In practical generators, the external magnetic field shown in the figure is created by passing current through coils called *field coils,* which are wrapped around a stationary core.

Recall that the magnitude of the voltage induced in a moving conductor depends, among other things, on the *direction* of that motion relative to the direction of the magnetic field (see Figure 11.6). In Figure 20.1, the loop is shown in a horizontal

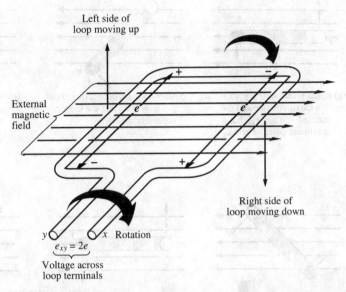

Left side of
loop moving up

External
magnetic
field

Right side of
loop moving down

y x Rotation

$e_{xy} = 2e$

Voltage across
loop terminals

FIGURE 20.1 Rotation of a loop of wire through a magnetic field creates continuous motion of the conductors on each side of the loop. Each side cuts flux and voltage e is induced in each. The polarities are additive, so the terminal voltage is $2e$.

position, so each side of the loop is moving in a direction perpendicular to the flux. As a result, flux lines are being cut at a maximum rate and the induced voltages are their maximum possible values. However, when the loop rotates beyond the horizontal position, the sides of the loop are no longer moving perpendicularly to the flux. Consequently, the induced voltage has a smaller value. Figure 20.2 shows end views of the loop and the directions that each side is moving relative to the flux at several positions throughout a 360-degree rotation. Note that the induced voltage decreases until the loop has rotated 90°. At that position, each side of the loop is moving *parallel* to the flux, so no lines are cut and the induced voltage is zero. Once the loop rotates through 90°, we see that the side that was originally on the left is now on the right, and vice versa, so the polarity of the voltage at the terminals of the loop reverses, i.e., e_{xy} is negative. (End x was originally positive with respect to end y, and is now negative with respect to end y.) When the loop rotates 180° from its original position, flux is once again cut at a maximum rate, and the induced voltage has its maximum possible value, but with a polarity opposite to the original. As the loop continues to rotate, the induced voltage rises and falls and reverses polarity every 180°. Figure 20.2(j) shows the voltage waveform (e_{xy}) that is created by one complete revolution of the loop (one cycle). The fact that the rate at which flux lines are cut is proportional to the *sine* of the angle between the loop and the magnetic field accounts for the fact that the induced voltage is sinusoidal. (In vector terms, the *perpendicular component* of the motion is proportional to the sine of the angle). Notice that reversing the direction of rotation or the direction of the external field will reverse the polarity of the induced voltage.

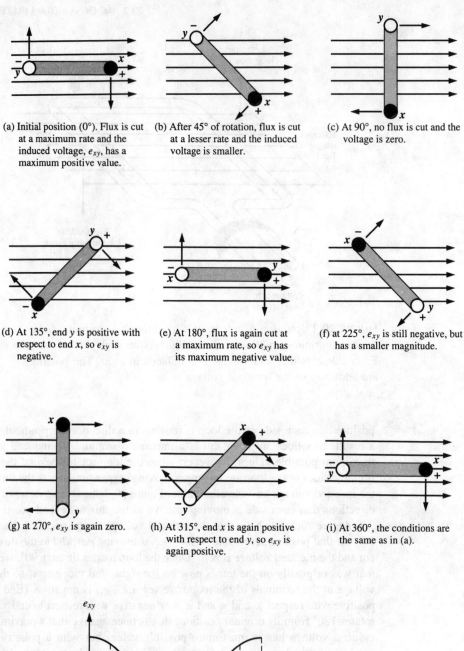

(a) Initial position (0°). Flux is cut at a maximum rate and the induced voltage, e_{xy}, has a maximum positive value.

(b) After 45° of rotation, flux is cut at a lesser rate and the induced voltage is smaller.

(c) At 90°, no flux is cut and the voltage is zero.

(d) At 135°, end y is positive with respect to end x, so e_{xy} is negative.

(e) At 180°, flux is again cut at a maximum rate, so e_{xy} has its maximum negative value.

(f) at 225°, e_{xy} is still negative, but has a smaller magnitude.

(g) at 270°, e_{xy} is again zero.

(h) At 315°, end x is again positive with respect to end y, so e_{xy} is again positive.

(i) At 360°, the conditions are the same as in (a).

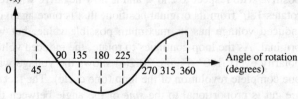

(j) Plot of the induced voltage versus degrees of rotation from the horizontal.

FIGURE 20.2 The voltage e_{xy} induced in the rotating loop as a function of the angle of rotation. Parts (a) through (i) are end views of the loop in Figure 20.1.

Example 20.1

Assume the loop in Figure 20.1 is rotating clockwise at a rate of 3600 revolutions per minute. When the loop passes through the horizontal position, each end is cutting flux at the rate of 0.4 Wb/s. Sketch the voltage waveform, e_{xy}, that is generated versus time, beginning at a point in time where the loop is 270° clockwise from the position shown in Figure 20.1.

SOLUTION The maximum (peak) value of the voltage induced in each side of the loop is found from Faraday's law (equation 11.1):

$$E = \frac{\Delta\phi}{\Delta t} = 0.4 \text{ Wb/s} = 0.4 \text{ V}$$

Since the induced voltages in each side are additive, the peak voltage at the terminals is $2 \times 0.4 \text{ V} = 0.8 \text{ V}$. The frequency of the voltage equals the number of revolutions (cycles) through which the loop rotates in one second:

$$f = (3600 \text{ rev/min})(1 \text{ min/60 s}) = 60 \text{ rev/s} = 60 \text{ Hz}$$

The period is therefore $T = 1/(60 \text{ Hz}) = 1/60$ s. The waveform is shown in Figure 20.3. Note that the initial value of the voltage is 0 because the plot begins at a point where the loop is vertical, and no flux is cut in that position [see Figure 20.2(g)].

Drill Exercise 20.1

If the speed of the angular rotation in Example 20.1 were doubled, what would be the peak value and frequency of the induced voltage?

ANSWER: 1.6 V, 120 Hz.   □

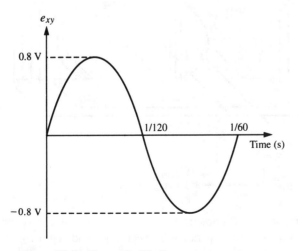

FIGURE 20.3 (Example 20.1)

Practical Single-Phase Generators

In practical generators, the loop of wire that rotates through the magnetic field has numerous turns instead of just one. Recall from Chapter 11 (Figure 11.7) that increasing the number of turns by a certain factor increases the induced voltage by the same factor. The turns are wound on a cylindrical structure called the *rotor*. See Figure 20.4. Since the ends of the winding are continuously rotating, some means must be used to connect them electrically to a pair of stationary output terminals. As shown in Figure 20.4, this connection is accomplished using *slip rings* that are mechanically attached to the ends of the winding and rotate with them. A set of stationary *brushes* makes continuous contact to the slip rings and thus provides the required electrical connection to the winding. As mentioned earlier, the external magnetic field is created by passing current through field windings wrapped around a stationary core. The field current is called the *excitation* for the generator. Since the field structure is stationary, it is called the *stator*.

Three-Phase Generators

A three-phase voltage is essentially three single-phase voltages, each separated from the next by 120° of phase angle. Therefore, the same basic structure found in the single-phase generator can be used to generate the three voltages simply by equipping the rotor with

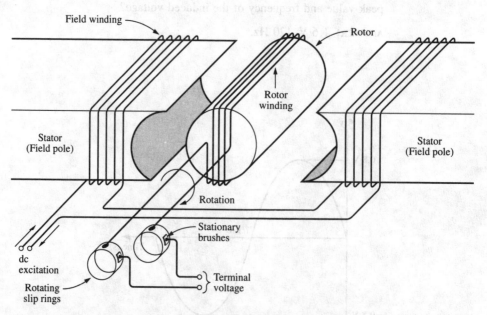

FIGURE 20.4 Components of a practical single-phase generator. (Expanded view; in practice, there is a very small separation between rotor and stator.)

three separate windings. If the windings are spaced 120° apart, then the voltages induced in them will be shifted from each other by 120° of phase, as required.

The concept just described is implemented in practical three-phase generators, but the physical structure is somewhat different. Recall that electromagnetic induction occurs when there is *relative* motion between a conductor and a magnetic field; that is, either the conductor or the field may be moving while the other is stationary. In practical three-phase generators, the three conductors (windings) are stationary and the magnetic field is rotated. See Figure 20.5(a). The windings are embedded in the stator and dc current (the excitation) is passed through brushes and slip rings (not shown) to the field winding on

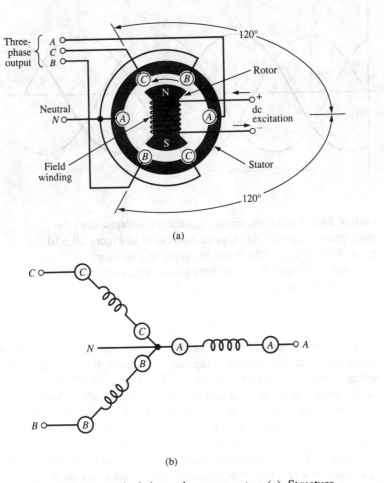

(a)

(b)

FIGURE 20.5 A practical three-phase generator. (a) Structure and wiring of a three-phase generator. Voltages are induced in the windings embedded in the stator. (b) Equivalent circuit of the stator windings.

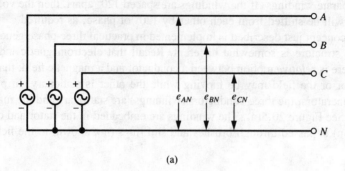

(a)

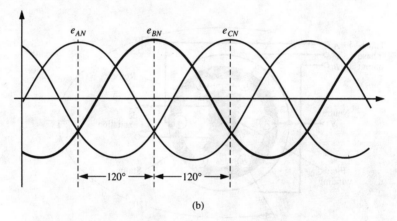

(b)

FIGURE 20.6 Equivalent circuit and output voltages from the three-phase generator. (a) Circuit equivalent to Figure 20.5(b). Each winding in 20.5(b) is shown as an ac generator. (b) Output voltages from the three-phase generator, each with respect to neutral N.

the rotor. As the rotor turns, the field it produces cuts the conductors of the three stator windings. Since the stator windings are 120° apart, the rotating magnetic field induces voltages that are separated in phase by 120°. Figure 20.5(b) shows the equivalent circuit of the stator windings. Note that in this case the windings have a common connection labelled N, called *neutral*, and the windings form an electrical wye network. Since line N is an output, the output is said to be three-phase, four-wire. The windings may also be connected in a delta configuration. Figure 20.6 shows a circuit equivalent to 20.5(b) and consisting of three ac generators. Also shown is a plot of the three voltages, e_{AN}, e_{BN}, and e_{CN}, each voltage taken with respect to the common neutral N.

20.3 Single-Phase, Three-Wire Power Distribution

Electrical power is transmitted from generating stations to consumers in the three-phase format at very high voltages (in the kV or MV range). Transformers are used to step down the voltages to levels that can be used by consumers. As mentioned earlier, the power delivered to most homes is single-phase, three-wire. The nominal voltage is 120 V rms (169.7 V pk), but the three-wire system also makes it possible to obtain 240 V rms (339.4 V pk). As shown in Figure 20.7(a), the transformer used to step down the transmission voltage has a center-tapped secondary. The center tap is the neutral for the

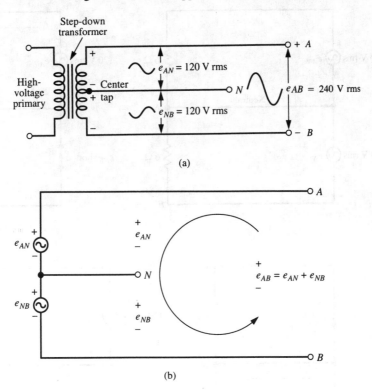

(a)

(b)

FIGURE 20.7 Single-phase, three-wire power as distributed to residential consumers. (a) A center-tapped, step-down transformer is used to produce 120 V rms and 240 V rms. (b) Equivalent circuit of the single-phase, three-wire source.

system. Since 240 V rms is developed across the entire secondary winding, 120 V rms is available between either end of the winding and the neutral. The 120-V circuits are used for lights and convenience outlets, whereas the 240-V circuit is used for heavy appliances such as clothes dryers and heaters. Figure 20.7(b) is an equivalent circuit showing how Kirchhoff's voltage law can be applied to the system to verify that $e_{AB} = e_{AN} + e_{NB}$.

When the loads connected to the circuit shown in Figure 20.7 are equal, the load on the system is said to be *balanced*. The next example demonstrates that *under a balanced-load condition, the current in the neutral conductor is zero*.

Example 20.2

Find the current in each load and the current in the neutral conductor of the circuit shown in Figure 20.8(a).

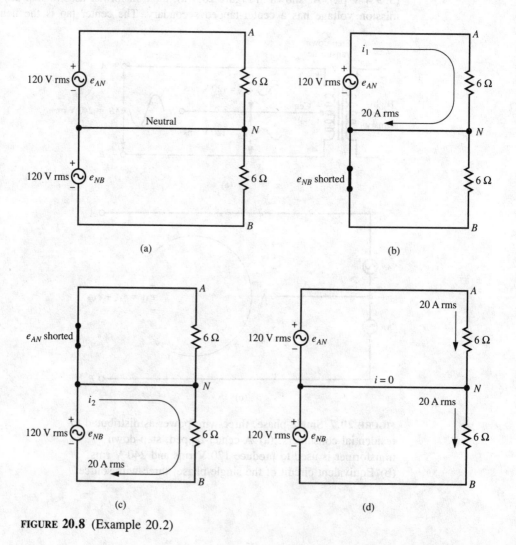

FIGURE 20.8 (Example 20.2)

SOLUTION It is convenient to use the superposition principle to analyze the circuit. Figure 20.8(b) shows the circuit when source e_{NB} is replaced by a short. Since the 6-Ω load between B and N is then shorted out, the only current that flows is

$$i_1 = \frac{e_{AN}}{6\ \Omega} = \frac{120\ \text{V rms}}{6\ \Omega} = 20\ \text{A rms}$$

Notice that i_1 flows through the neutral from right to left. Similarly, when e_{AN} is replaced by a short,

$$i_2 = \frac{e_{NB}}{6\ \Omega} = \frac{120\ \text{V rms}}{6\ \Omega} = 20\ \text{A rms}$$

and i_2 flows through the neutral from left to right (Figure 20.8(c)). Figure 20.8(d) shows the circuit after application of the superposition principle. Since i_1 and i_2 are equal in amplitude but opposite in direction through the neutral, the net current in the neutral is zero.

Drill Exercise 20.2

If the 6-Ω load between A and N in Figure 20.8(b) were changed to 4 Ω, what then would be the current in the neutral?

ANSWER: 10 A rms, from right to left.   □

Drill Exercise 20.2 demonstrates that the current in the neutral is nonzero when the load is not balanced. The next example demonstrates that the current in the neutral is zero when the loads across the 120-V circuits are equal and a load (any load) is connected across the 240-V circuit.

Example 20.3

Find the current in each load and the current in the neutral in Figure 20.9(a).

SOLUTION We again apply superposition. Figure 20.9(a) shows the equivalent circuit when e_{NB} is replaced by a short. Since 120 V rms appears across both loads, the currents by Ohm's law are 40 A rms in the 3-Ω load and 30 A rms in the 4-Ω load. By Kirchhoff's current law, the current returning to the source in the neutral conductor is 30 A rms + 40 A rms = 70 A rms, from right to left. As shown in Figure 20.9(c), the results are exactly the same when e_{AN} is replaced by a short, except that the 70-A rms current in the neutral flows from left to right. By the superposition principle, the total current in the 4-Ω load is 30 A rms + 30 A rms = 60 A rms, and the current in the neutral is zero, as shown in Figure 20.9(d).

Drill Exercise 20.3

Find the current in the neutral of Figure 20.9(a) if the 3-Ω load between A and N is replaced by a 6-Ω load.

ANSWER: 20 A rms, from left to right.   □

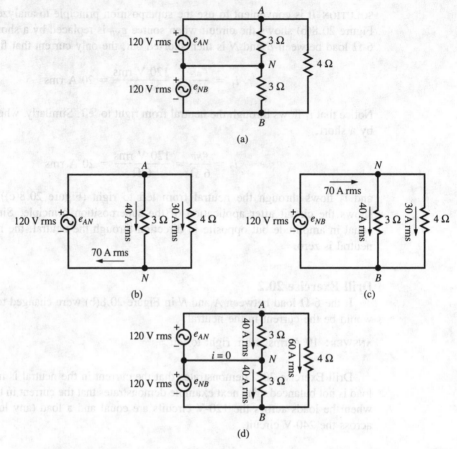

FIGURE 20.9 (Example 20.3)

20.4 Wye- and Delta-Connected Three-Phase Sources

Voltages and Currents in a Wye-Connected Source

Figure 20.10 shows a wye-connected, three-phase, four-wire source supplying current to an arbitrary load. The source is represented by the three windings of the generator, each of which is itself considered to be an ac voltage source, as previously discussed. Following is the conventional terminology used in the power industry to identify the various currents and voltages in the system:

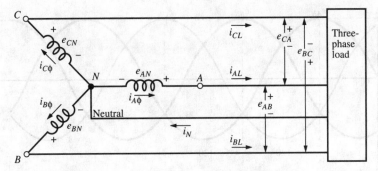

FIGURE 20.10 Phase (ϕ) and line (L) quantities in a wye-connected source driving a three-phase load.

1. The voltages e_{AN}, e_{BN}, and e_{CN} are called *phase voltages*. Notice that these voltages, taken with respect to neutral (N), are the voltages generated by each of the windings. We will use the notation e_ϕ to represent an arbitrary phase voltage.
2. The voltages e_{AB}, e_{BC}, and e_{CA} across the outputs of the generator are called *line voltages*. We will use the notation e_L to represent an arbitrary line voltage.
3. The currents $i_{A\phi}$, $i_{B\phi}$, and $i_{C\phi}$ in the generator windings are called *phase currents*. We will use the notation i_ϕ to represent an arbitrary phase current.
4. The currents i_{AL}, i_{BL}, and i_{CL} flowing through the output lines to the load are called *line currents*. We will use the notation i_L to represent an arbitrary line current.

Figure 20.11 shows the three phase voltages in sinusoidal form and the corresponding phasor diagram. The peak value of each phase voltage is E, so we see that

$$e_{AN} = E \sin \omega t = E \underline{/0°}$$

$$e_{BN} = E \sin (\omega t - 120°) = E \underline{/-120°}$$

$$e_{CN} = E \sin (\omega t + 120°) = E \underline{/120°}$$

We wish to determine the line voltages e_{AB}, e_{BC}, and e_{CA}, but let us first make a useful observation about the sum of the phase voltages. Converting each to rectangular form, we find

$$e_{AN} = E \underline{/0°} = E + j0 \tag{20.1}$$

$$e_{BN} = E \underline{/-120°} = -0.5E - j0.866E \tag{20.2}$$

$$e_{CN} = E \underline{/120°} = -0.5E + j0.866E \tag{20.3}$$

Thus,

$$e_{AN} + e_{BN} + e_{CN} = E + j0 - 0.5E - j0.866E - 0.5E + j0.866E$$

$$= 0 + j0$$

We see that the sum of the phase voltages is zero, and this is true regardless of the nature of the load.

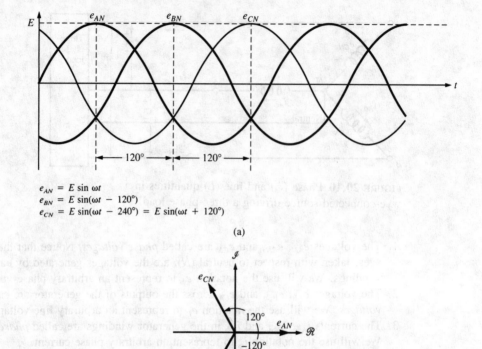

$$e_{AN} = E \sin \omega t$$
$$e_{BN} = E \sin(\omega t - 120°)$$
$$e_{CN} = E \sin(\omega t - 240°) = E \sin(\omega t + 120°)$$

(a)

$$e_{AN} = E\underline{/0°}$$
$$e_{BN} = E\underline{/-120°}$$
$$e_{CN} = E\underline{/-240°} = E\underline{/120°}$$

(b)

FIGURE 20.11 Sinusoidal and phasor forms of the phase voltages. (a) Sinusoidal form. (b) Phasor form.

Figure 20.12 shows that we can determine the line voltage e_{AB} by writing Kirchhoff's voltage law around the loop in Figure 20.10 that contains e_{AN} and e_{BN}:

$$e_{AB} + e_{BN} = e_{AN}$$
$$e_{AB} = e_{AN} - e_{BN}$$
$$= (E + j0) - (-0.5E - j0.866E)$$
$$= 1.5E + j0.866E$$
$$e_{AB} = \sqrt{3}E\ \underline{/30°} \tag{20.4}$$

In a similar fashion, we can write Kirchhoff's voltage law around appropriate loops in Figure 20.10 to find

$$e_{BC} = e_{BN} - e_{CN}$$
$$= \sqrt{3}E\ \underline{/-90°} \tag{20.5}$$

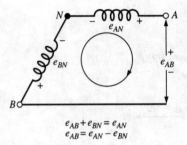

$$e_{AB} + e_{BN} = e_{AN}$$
$$e_{AB} = e_{AN} - e_{BN}$$

FIGURE 20.12 Finding line voltage e_{AB} using Kirchhoff's voltage law around a loop from Figure 20.10.

$$e_{CA} = e_{CN} - e_{AN}$$
$$= \sqrt{3}E \underline{/150°} \tag{20.6}$$

By adding the three Kirchhoff equations, we find that the sum of the line voltages is also zero. Equations 20.4 through 20.6 show that the magnitude of the line voltages is $\sqrt{3}$ times greater than the magnitude of the phase voltages:

$$E_L = \sqrt{3}E_\phi \tag{20.7}$$

where E_L is the magnitude of a line voltage and E_ϕ is the magnitude of a phase voltage. Figure 20.13 shows a phasor diagram of line and phase voltages. Notice that the line voltages, like the phase voltages, are separated from one another by 120°, and that each line voltage leads the nearest phase voltage by 30°.

Referring to Figure 20.10, it is apparent that each line current equals the phase current in the winding to which the line is connected: $i_{AL} = i_{A\phi}$, $i_{BL} = i_{B\phi}$, and $i_{CL} = i_{C\phi}$. In general,

$$i_L = i_\phi \tag{20.8}$$

Writing Kirchhoff's current law at node N in Figure 20.10, we find the current in the neutral conductor to be

$$i_N = i_{A\phi} + i_{B\phi} + i_{C\phi}$$
$$= i_{AL} + i_{BL} + i_{CL} \tag{20.9}$$

When the phase (and line) currents have equal magnitude and are spaced by 120° of phase from one another, the system is said to be *balanced*. In this special case, the current in the neutral, i_N, will be zero. The neutral conductor can then be omitted without affecting system performance. However, in most practical systems, the neutral is retained because perfect balance is difficult to achieve and maintain. The next example illustrates the special case of a balanced system.

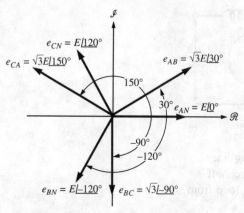

FIGURE 20.13 Line and phase voltages.

Example 20.4

The phase voltages and currents in a wye-connected, three-phase, four-wire generator are as follows: $e_{AN} = 170 \underline{/0°}$ V, $e_{BN} = 170 \underline{/-120°}$ V, $e_{CN} = 170 \underline{/120°}$ V, $i_{A\phi} = 12 \underline{/90°}$ A, $i_{B\phi} = 12 \underline{/-30°}$ A, $i_{C\phi} = 12 \underline{/-150°}$ A.

(a) Find the line voltages and currents.
(b) Find the sum of the line voltages.
(c) Find the neutral current.
(d) Draw a phasor diagram showing line voltages and currents.

SOLUTION

(a) By equations (20.4) through (20.6),

$$e_{AB} = \sqrt{3}(170) \underline{/30°} \text{ V} = 294.45 \underline{/30°} \text{ V}$$
$$e_{BC} = \sqrt{3}(170) \underline{/-90°} \text{ V} = 294.45 \underline{/-90°} \text{ V}$$
$$e_{CA} = \sqrt{3}(170) \underline{/150°} \text{ V} = 294.45 \underline{/150°} \text{ V}$$

Since the line currents equal the phase currents,

$$i_{AL} = i_{A\phi} = 12 \underline{/90°} \text{ A}$$
$$i_{BL} = i_{B\phi} = 12 \underline{/-30°} \text{ A}$$
$$i_{CL} = i_{C\phi} = 12 \underline{/-150°} \text{ A}$$

(b) Converting the line voltages to rectangular form and adding, we find:

$$e_{AB} = 294.45 \underline{/30°} \text{ V} = 255 + j147.225 \text{ V}$$
$$e_{BC} = 294.45 \underline{/-90°} \text{ V} = 0 - j294.450 \text{ V}$$
$$e_{CA} = 294.45 \underline{/150°} \text{ V} = -255 + j147.225 \text{ V}$$
$$\text{Sum} = 0 + j0 \text{ V}$$

This result verifies that the sum of the line voltages, like the sum of the phase voltages, is zero. This is true irrespective of balance.

(c) By equation 20.9, $i_N = i_{AL} + i_{BL} + i_{CL}$. Converting the line currents to rectangular form and adding, we find

$$i_{AL} = 12 \underline{/90°} \text{ A} \qquad = 0 + j12 \text{ A}$$
$$i_{BL} = 12 \underline{/-30°} \text{ A} = 10.39 - j6 \text{ A}$$
$$i_{CL} = 12 \underline{/-150°} \text{ A} = -10.39 - j6 \text{ A}$$
$$i_N = \text{Sum} = 0 + j0 \text{ A}$$

We see that the neutral current is zero. Remember that this is only true for the special case of a balanced system.

(d) The phasor diagram is shown in Figure 20.14.

Drill Exercise 20.4

Find the current in the neutral in Example 20.4 if phase current $i_{A\phi}$ is changed to 12 $\underline{/0°}$ A.

ANSWER: 16.97 $\underline{/-45°}$ A.   ☐

Voltages and Currents in a Delta-Connected Source

Figure 20.15 shows how the windings of a three-phase generator can be connected in a delta configuration instead of a wye. Also shown are two ways to draw the equivalent circuit. Note that we once again treat each winding of the generator as an individual ac voltage source. Also note that there are just three output lines, so every delta-connected source is a three-phase, three-wire source. Part (d) of the figure shows the phasor diagram for the output voltages. We see that

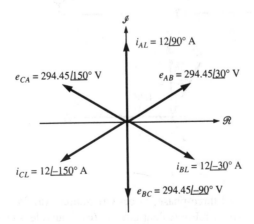

FIGURE 20.14 (Example 20.4)

$$e_{AB} = E \,\underline{/0^\circ} = E \sin \omega t$$

$$e_{BC} = E \,\underline{/-120^\circ} = E \sin (\omega t - 120^\circ)$$

$$e_{CA} = E \,\underline{/120^\circ} = E \sin (\omega t + 120^\circ)$$

where E is the peak value of the voltage induced in each winding.

Figure 20.16 shows a delta-connected source driving a three-phase load. The phase and line quantities are defined in the same way as they are for the wye-connected source: phase quantities are those within the source and line quantities are those in the output lines connected to the load. From examination of the figure, it is readily apparent that the line voltages equal the phase voltages:

$$e_L = e_\phi \tag{20.10}$$

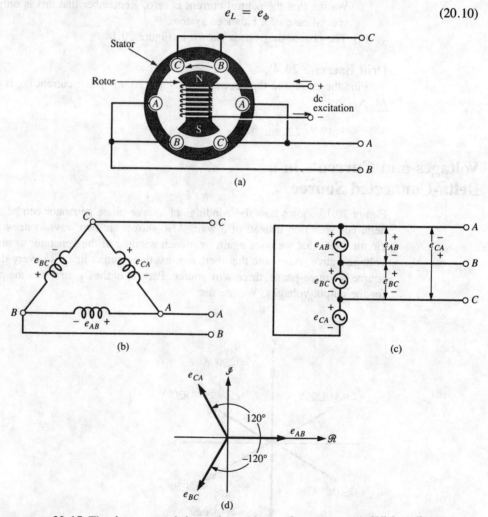

(a)

(b)   (c)

(d)

FIGURE 20.15 The Δ-connected three-phase, three-wire source. (a) Wiring diagram [compare with Figure 20.5(a)]. (b) Equivalent circuit. (c) Equivalent circuit with each winding shown as an ac voltage source. (d) Phasor diagram of output voltages.

It is also apparent that the line currents are not, in general, equal to the phase currents. Writing Kirchhoff's current law at each node, we find

$$\text{Node } A: i_{AL} = i_{A\phi} - i_{C\phi} \tag{20.11}$$

$$\text{Node } B: i_{BL} = i_{B\phi} - i_{A\phi} \tag{20.12}$$

$$\text{Node } C: i_{CL} = i_{C\phi} - i_{B\phi} \tag{20.13}$$

By adding equations (20.11), (20.12), and (20.13), we find that

$$i_{AL} + i_{BL} + i_{CL} = 0 \tag{20.14}$$

In short, the sum of the line currents in the system with the delta-connected generator is *always* zero. The sum of the phase currents is zero only when the system is balanced, that is, when the phase currents have equal magnitudes and are separated in phase by 120° from one another.

In the special case when the system in Figure 20.16 is balanced, the magnitude of each line current is $\sqrt{3}$ times the magnitude of each phase current:

$$I_L = \sqrt{3}I_\phi \tag{20.15}$$

This fact is demonstrated in the next example.

Example 20.5

In the system shown in Figure 20.16, the phase currents are $i_{A\phi} = 40\ \underline{/60°}$ A, $i_{B\phi} = 40\ \underline{/-60°}$ A, and $i_{C\phi} = 40\ \underline{/180°}$ A. Find the line currents and draw a phasor diagram showing phase and line currents.

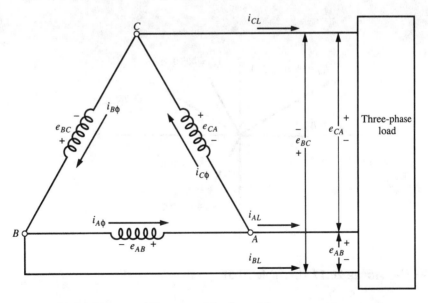

FIGURE 20.16 Phase and line quantities in a delta-connected generator driving a three-phase load.

SOLUTION From equations (20.11), (20.12), and (20.13),

$$i_{AL} = i_{A\phi} - i_{C\phi} = 40 \underline{/60°} \text{ A} - 40 \underline{/180°} \text{ A}$$

$$= (20 + j34.64) - (-40 + j0) \text{ A}$$

$$= 60 + j34.64 \text{ A} = 69.28 \underline{/30°} \text{ A}$$

$$i_{BL} = i_{B\phi} - i_{A\phi} = 40 \underline{/-60°} \text{ A} - 40 \underline{/60°} \text{ A}$$

$$= (20 - j34.64) - (20 + j34.64) \text{ A}$$

$$= 0 - j69.28 \text{ A} = 69.28 \underline{/-90°} \text{ A}$$

$$i_{CL} = i_{C\phi} - i_{B\phi} = 40 \underline{/180°} \text{ A} - 40 \underline{/-60°} \text{ A}$$

$$= (-40 + j0) - (20 - j34.64) \text{ A}$$

$$= -60 + j34.64 \text{ A} = 69.28 \underline{/150°} \text{ A}$$

Notice that the magnitude of every line current is 69.28 A, which equals $(\sqrt{3})40$ A, confirming equation (20.15). The phasor diagram is shown in Figure 20.17. Observe that each line current is separated from the nearest phase current by 30°.

Drill Exercise 20.5

In a balanced system with a delta-connected generator, each line current has magnitude 30 A. What is the magnitude of each phase current?

ANSWER: 17.32 A.   □

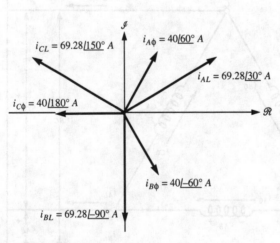

FIGURE 20.17 (Example 20.5)

Phase Sequences

The three-phase voltages we have studied up to now are said to have the *phase sequence ABC*. If we can imagine standing at a point and observing the rise and fall of the three voltages as time goes by, we would see *A* reach its peak first, followed by *B*, then *C*, then *A* again, and so forth. Depending on what point in time we began the observation, the same three voltages might peak in the sequence *BCABC. . . .* In other words, *BCA* is the same sequence as *ABC*. Also, *CAB* is the same as *ABC*. There is just one phase sequence different from *ABC*, namely *CBA*.* This sequence is equivalent to *BAC* and *ACB*.

It is important to know (or prescribe) the phase sequence in a three-phase power distribution system. In unbalanced systems, reversal of the phase sequence can greatly alter currents flowing in different loads. Also, reversal of the phase sequence will reverse the direction of rotation of any three-phase motors in the system.

Figures 20.18(a) and (b) show the sinusoidal waveforms in phase sequences *ABC* and *CBA*. The phase sequence can be determined from the phasor diagram of either the phase voltages or the line voltages. Figure 20.18(c) shows a phasor diagram of phase voltages in sequence *ABC*. Pick any point in the plane as an observation point and visualize the phasors rotating in a counterclockwise direction past the point. The sequence of first subscripts on the phasors as they rotate past will correspond to the phase sequence [*ABC* = *BCA* = *CAB* in Figure 20.18(c)]. A phasor diagram of the line voltages can be used in the same way to determine the phase sequence. The sequence of either the first or the second subscripts of the rotating phasors corresponds to the phase sequence. Figure 20.18(d) shows the line-voltage phasors in the sequence *CBA*. From an observation point in the fourth quadrant, we would observe the sequence of first subscripts to be *ACB* and the sequence of second subscripts to be *BAC*, both of which are equivalent to *CBA*. Figure 20.18(e) shows the phasors in a delta-connected generator. In this example, phase voltages and line voltages are equal, so in either case the sequence of first subscripts or of second subscripts corresponds to the phase sequence.

20.5 Balanced and Unbalanced Wye and Delta Loads

The output of a wye- or delta-connected generator may be connected to a three-phase load that has either a wye or a delta configuration. Thus, there are four different system configurations: wye generator driving wye load (Y-Y), wye generator driving delta load (Y-Δ), delta generator driving wye load (Δ-Y), and delta generator driving delta load (Δ-Δ). In each of these, the load may be balanced or unbalanced.

*Phase sequence *ABC* is often called a positive sequence and *CBA* a negative sequence.

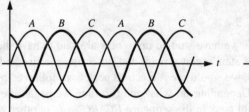

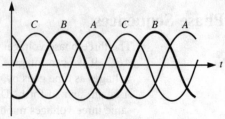

(a) Sinusoidal waveforms in the phase sequence *ABC*.

(b) Sinusoidal waveforms in the phase sequence *CBA*.

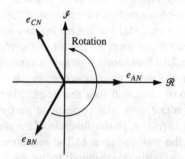

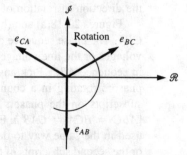

(c) Phasor diagram of the phase voltages in a Y-connected generator with the phase sequence *ABC*.

(d) Phasor diagram of the line voltages in a Y-connected generator with the phase sequence *CBA*.

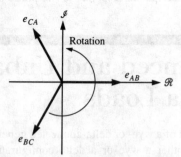

(e) Phasor diagram of the phase and line voltages in a △-connected generator with the phase sequence *ABC*.

FIGURE 20.18 Phase sequences *ABC* and *CBA*.

Y-Y Configuration

Figure 20.19 shows a wye-connected generator driving a wye load. Since the neutral is connected between source and load, this is a three-phase, four-wire system. All other system configurations (Y-Δ, Δ-Y, Δ-Δ) are of necessity three-wire systems because there are just three connection points on a delta. Four-wire systems are useful when two different load voltages are desired. In a typical application, low-voltage loads such as lights are connected between one line and neutral for operation at a phase voltage of 120 V rms, and high-voltage loads such as heaters are connected between two lines for operation at the line voltage $\sqrt{3}(120) = 208$ V rms. It is apparent in Figure 20.19 that the voltage across each impedance in the wye load equals the phase voltage of the generator. Therefore, each phase current, i_ϕ, which equals a corresponding line current, i_L, is

$$i_\phi = i_L = \frac{e_\phi}{Z} \tag{20.16}$$

where Z is the impedance in a leg of the wye load. By equation (20.9), the neutral current is the sum of the phase (or line) currents. If every impedance in the wye load is equal ($Z_A = Z_B = Z_C$), then the load is balanced and the neutral current is zero. Since the load voltages equal the phase voltages of the generator and the sum of the phase voltages is zero, the sum of the load voltages is zero whether the load is balanced or not. The next example illustrates the balanced case.

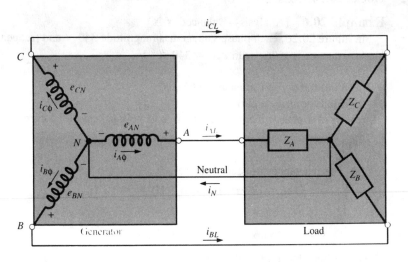

FIGURE 20.19 The Y-Y configuration.

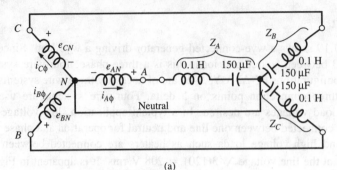

(a)

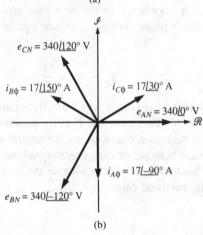

(b)

FIGURE 20.20 (Example 20.6)

Example 20.6 (Analysis—Balanced Y-Y)

In Figure 20.20, each phase voltage has magnitude 340 V and frequency 60 Hz. The phase sequence is ABC, with $e_{AN} = 340 \underline{/0°}$ V.

(a) Find the current in each load.
(b) Find the neutral current.
(c) Draw a phasor diagram showing load voltages and load currents.

SOLUTION

(a) The reactances in each load are

$$j\omega L = j(2\pi)(60 \text{ Hz})(0.1 \text{ H}) = j37.7 \ \Omega$$

$$\frac{-j}{\omega C} = \frac{-j}{(2\pi)(60 \text{ Hz})(150 \times 10^{-6} \text{ F})} = -j17.7 \ \Omega$$

Thus, the impedance of each load is

$$Z_A = Z_B = Z_C = Z = j37.7 - j17.7 \ \Omega$$

$$= j20 \ \Omega = 20 \underline{/90°} \ \Omega$$

The load, line, and phase currents are then

$$i_{A\phi} = i_{AL} = \frac{e_{AN}}{Z} = \frac{340 \,\underline{/0°}\ V}{20 \,\underline{/90°}\ \Omega} = 17 \,\underline{/-90°}\ A$$

$$i_{B\phi} = i_{BL} = \frac{e_{BN}}{Z} = \frac{340 \,\underline{/-120°}\ V}{20 \,\underline{/90°}\ \Omega} = 17 \,\underline{/-210°}\ A = 17 \,\underline{/150°}\ A$$

$$i_{C\phi} = i_{CL} = \frac{e_{CN}}{Z} = \frac{340 \,\underline{/120°}\ V}{20 \,\underline{/90°}\ \Omega} = 17 \,\underline{/30°}\ A$$

(b) Converting the currents to rectangular form and adding, we find

$$i_{A\phi} = 17 \,\underline{/-90°}\ A = 0 - j17\ A$$

$$i_{B\phi} = 17 \,\underline{/150°}\ A = -14.72 + j8.5\ A$$

$$i_{C\phi} = 17 \,\underline{/30°}\ A = 14.72 + j8.5\ A$$

$$i_N = i_{A\phi} + i_{B\phi} + i_{C\phi} = 0 + j0\ A$$

As expected, the neutral current is zero since the load is balanced.

(c) Figure 20.20(b) shows the phasor diagram of load voltages and currents. These are identified as phase voltages and currents since they are equal to the respective load voltages and currents.

Drill Exercise 20.6

Find the neutral current in Figure 20.20(a) if Z_A is changed to $20 \,\underline{/-90°}\ \Omega$.

ANSWER: $17 \,\underline{/90°}\ A$.   □

The next example illustrates an unbalanced Y-Y system and shows that balance does not exist when the loads simply have equal magnitudes. (For balance, the loads must have equal magnitudes and equal angles.)

Example 20.7   (Analysis—Unbalanced Y-Y)

In the Y-Y system shown in Figure 20.21(a), $e_{AN} = 500 \,\underline{/0°}\ V$, $e_{BN} = 500 \,\underline{/-120°}\ V$, and $e_{CN} = 500 \,\underline{/120°}\ V$.

(a) Find the current in each load and the current in the neutral.

(b) Construct a phasor diagram showing load and neutral currents.

SOLUTION

(a)

$$i_{AL} = \frac{500 \,\underline{/0°}\ V}{10 \,\underline{/0°}\ \Omega} = 50 \,\underline{/0°}\ A$$

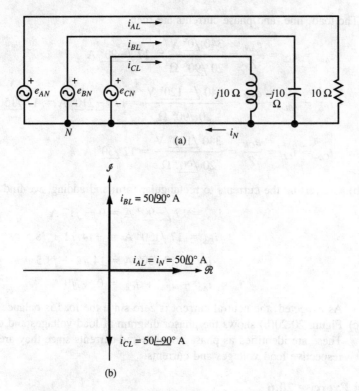

FIGURE 20.21 (Example 20.7)

$$i_{BL} = \frac{500 \;\underline{/0°}\; \text{V}}{10 \;\underline{/-90°}\; \Omega} = 50 \;\underline{/90°}\; \text{A}$$

$$i_{CL} = \frac{500 \;\underline{/0°}\; \text{V}}{10 \;\underline{/90°}\; \Omega} = 50 \;\underline{/-90°}\; \text{A}$$

$$i_N = i_{AL} + i_{BL} + i_{CL}$$

$$= 50 + j50 - j50 \;\text{A} = 50 + j0 \;\text{A} = 50 \;\underline{/0°}\; \text{A}$$

(b) The phasor diagram is shown in Figure 20.21(b). We see that the neutral
   current is nonzero despite the fact that the magnitudes of the load impedances
   are equal.

Drill Exercise 20.7
   Find the neutral current in Figure 20.21(a) if the 10-Ω resistance is changed to $j10 \;\Omega$.

ANSWER: 50 $\underline{/-90°}$ A.   □

Y-Δ Configuration

Figure 20.22 shows a wye-connected generator driving a delta load. It is apparent that the magnitude of the voltage across each of the loads Z_A, Z_B, and Z_C is equal to the magnitude of the line voltage. Recall that $E_L = \sqrt{3}E_\phi$. In a typical low-voltage system, each load voltage is $\sqrt{3}$ (120 V rms) = 208 V rms. In a system with phase sequence ABC and $e_{AN} = E\ \underline{/0°}$ V, we know that the line voltages are $e_{AB} = \sqrt{3}E\ \underline{/30°}$, $e_{BC} = \sqrt{3}E\ \underline{/-90°}$, and $e_{CA} = \sqrt{3}\ \underline{/150°}$ (Figure 20.13). Therefore, the currents in each of the loads are

$$i_{Z_A} = \frac{e_{AB}}{Z_A} = \frac{\sqrt{3}E\ \underline{/30°}}{Z_A} \tag{20.17}$$

$$i_{Z_B} = \frac{e_{BC}}{Z_B} = \frac{\sqrt{3}E\ \underline{/-90°}}{Z_B} \tag{20.18}$$

$$i_{Z_C} = \frac{e_{CA}}{Z_C} = \frac{\sqrt{3}E\ \underline{/150°}}{Z_C} \tag{20.19}$$

If the load is balanced ($Z_A = Z_B = Z_C$), then we see that these currents have equal magnitudes and are separated from one another by 120°. Writing Kirchhoff's current law at the node where Z_A is connected to Z_C, we have

$$i_{AL} = i_{Z_A} - i_{Z_C} \tag{20.20}$$

At the other two nodes,

$$i_{BL} = i_{Z_B} - i_{Z_A} \tag{20.21}$$

$$i_{CL} = i_{Z_C} - i_{Z_B} \tag{20.22}$$

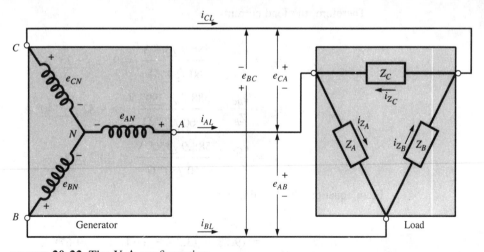

FIGURE 20.22 The Y-Δ configuration.

Adding equations (20.20) through (20.22) shows that the sum of the line currents is zero, regardless of whether the load is balanced or unbalanced. As already noted, when the load is balanced, i_{Z_A}, i_{Z_B}, and i_{Z_C} in these equations have equal magnitudes and are separated in phase by 120°. The equations therefore express exactly the same relationships as those between line and phase voltages in the wye-connected generator [equations (20.4), (20.5), and (20.6)]. Just as we did in those cases, we conclude that the magnitude of each line current, I_L, is greater by a factor of $\sqrt{3}$ than the magnitude of each load current I_Z:

$$I_L = \sqrt{3}I_Z \text{ (balanced load)} \qquad (20.23)$$

Furthermore, when the load is balanced, each line current lags the respective load current by 30° (i_{AL} lags i_{Z_A}, i_{BL} lags i_{Z_B}, and i_{CL} lags i_{Z_C}), and the sum of the load currents is zero. Remember that equations (20.17) through (20.22) are valid regardless of the load, and (20.23) is true for balanced loads only.

Example 20.8 (Analysis—Balanced Y-Δ)

In the Y-Δ system shown in Figure 20.22, $Z_A = Z_B = Z_C = 60 \underline{/50°}\ \Omega$. The phase sequence is ABC and $e_{AN} = 340 \underline{/0°}$ V.

(a) Find the current in each load.
(b) Draw a phasor diagram showing load and line currents.

SOLUTION From equations (20.4) through (20.6), the line voltages are

$$e_{AB} = \sqrt{3}(340) \underline{/30°} = 588.9 \underline{/30°} \text{ V}$$

$$e_{BC} = \sqrt{3}(340) \underline{/-90°} = 588.9 \underline{/-90°} \text{ V}$$

$$e_{CA} = \sqrt{3}(340) \underline{/150°} = 588.9 \underline{/150°} \text{ V}$$

Therefore, the load currents are

$$i_{Z_A} = \frac{e_{AB}}{Z_A} = \frac{588.9 \underline{/30°} \text{ V}}{60 \underline{/50°}\ \Omega} = 9.82 \underline{/-20°} \text{ A}$$

$$i_{Z_B} = \frac{e_{BC}}{Z_B} = \frac{588.9 \underline{/-90°} \text{ V}}{60 \underline{/50°}\ \Omega} = 9.82 \underline{/-140°} \text{ A}$$

$$i_{Z_C} = \frac{e_{CA}}{Z_C} = \frac{588.9 \underline{/150°} \text{ V}}{60 \underline{/50°}\ \Omega} = 9.82 \underline{/100°} \text{ A}$$

From equation (20.20), line current i_{AL} is

$$i_{AL} = i_{Z_A} - i_{Z_C} = 9.82 \underline{/-20°} - 9.82 \underline{/100°} \text{ A}$$

$$= (9.23 - j3.36) - (-1.71 + j9.67) \text{ A}$$

$$= 10.94 - j13.03 \text{ A}$$

$$i_{AL} = 17 \underline{/-50°} \text{ A}$$

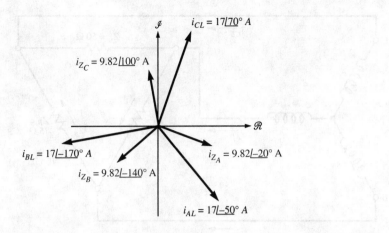

FIGURE 20.23 (Example 20.8)

We note that the magnitude of i_{AL} is $17 = \sqrt{3}(9.82)$ A, confirming equation (20.23). Since the load is balanced, we can write immediately

$$i_{BL} = \sqrt{3}(9.82) \ \underline{/-140° - 30°} \ A = 17 \ \underline{/-170°} \ A$$
$$i_{CL} = \sqrt{3}(9.82) \ \underline{/100° - 30°} \ A = 17 \ \underline{/70°} \ A$$

(b) The phasor diagram of line and load currents is shown in Figure 20.23.

Drill Exercise 20.8

If $i_{Z_A} = 40 \ \underline{/30°}$ A in a balanced Y-Δ system, what is i_{AL}?

ANSWER: $69.28 \ \underline{/0°}$ A.   □

Example 20.9 (Analysis—Unbalanced Y-Δ)

The Y-Δ system in Figure 20.24(a) has phase sequence ABC with $e_{AB} = 600 \ \underline{/30°}$ V.

(a) Find the load currents.
(b) Find the line currents.
(c) Find the sum of the line currents.
(d) Find the sum of the load currents.
(e) Draw a phasor diagram showing load and line currents.

SOLUTION

(a)

$$i_{Z_A} = \frac{e_{AB}}{Z_A} = \frac{600 \ \underline{/30°} \ V}{100 \ \underline{/0°} \ \Omega} = 6 \ \underline{/30°} \ A$$

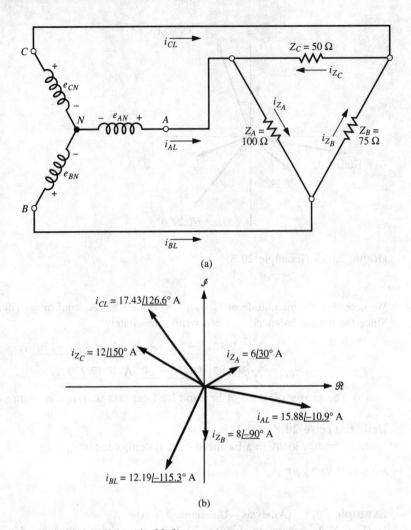

(a)

(b)

FIGURE 20.24 (Example 20.9)

$$i_{Z_B} = \frac{e_{BC}}{Z_B} = \frac{600\ \underline{/-90°}\ \text{V}}{75\ \underline{/0°}\ \Omega} = 8\ \underline{/-90°}\ \text{A}$$

$$i_{Z_C} = \frac{e_{CA}}{Z_C} = \frac{600\ \underline{/150°}\ \text{V}}{50\ \underline{/0°}\ \Omega} = 12\ \underline{/150°}\ \text{A}$$

(b) Since the load is not balanced, we must use equations (20.20) through (20.22):

$$i_{AL} = i_{Z_A} - i_{Z_C} = 6\ \underline{/30°} - 12\ \underline{/150°}\ \text{A}$$

$$= (5.20 + j3) - (-10.39 + j6)\ \text{A}$$

$$= 15.59 - j3\ \text{A} = 15.88\ \underline{/-10.9°}\ \text{A}$$

$$i_{BL} = i_{Z_B} - i_{Z_A} = 8\ \underline{/-90°} - 6\ \underline{/30°}\ \text{A}$$

$$= (0 - j8) - (5.2 + j3)\ \text{A}$$

$$= -5.2 - j11\ \text{A} = 12.17\ \underline{/-115.3°}\ \text{A}$$

$$i_{CL} = i_{Z_C} - i_{Z_B} = 12\ \underline{/150°} - 8\ \underline{/-90°}\ \text{A}$$

$$= (-10.39 + j6) - (0 - j8)\ \text{A}$$

$$= -10.39 + j14\ \text{A} = 17.43\ \underline{/126.6°}\ \text{A}$$

We see that neither the load currents nor the line currents have equal magnitudes. The line currents are not separated by 120° from one another. The load currents happen to be separated by 120° because in this example each load impedance has the same angle (0°), but that is not generally the case for an unbalanced load.

(c) The sum of the line currents is zero, despite the fact that the load is unbalanced:

$$i_{AL} = 15.59 - j3\ \text{A}$$

$$i_{BL} = -5.2 - j11\ \text{A}$$

$$i_{CL} = \underline{-10.39 + j14}\ \text{A}$$

$$\text{Sum} = 0 + j0\ \text{A}$$

(d) However, the sum of the load currents is not zero:

$$i_{Z_A} = 6\ \underline{/30°}\ \text{A} = 5.2 + j3\ \text{A}$$

$$i_{Z_B} = 8\ \underline{/-90°}\ \text{A} = 0 - j8\ \text{A}$$

$$i_{Z_C} = 12\ \underline{/150°}\ \text{A} = \underline{-10.39 + j6}\ \text{A}$$

$$\text{Sum} = -5.19 + j1\ \text{A}$$

$$= 5.29\ \underline{/169.1°}\ \text{A}$$

(e) The phasor diagram is shown in Figure 20.24(b).

Drill Exercise 20.9

Find the sum of the load currents in Example 20.9 if Z_B is changed to $50\ \underline{/0°}\ \Omega$.

ANSWER: $6\ \underline{/-150°}\ \text{A}$. □

The Δ-Δ Configuration

Figure 20.25 shows a delta-connected generator driving a delta load. It is readily apparent that the line, load, and phase voltages are all equal. The load currents are

$$i_{Z_A} = \frac{e_{AB}}{Z_A}, \ i_{Z_B} = \frac{e_{BC}}{Z_B}, \ i_{Z_C} = \frac{e_{CA}}{Z_C} \tag{20.24}$$

Writing Kirchhoff's current law at each node of the delta load yields the same relationships between line and load currents that we found in the Y-Δ configuration:

$$i_{AL} = i_{Z_A} - i_{Z_C} \tag{20.25}$$

$$i_{BL} = i_{Z_B} - i_{Z_A} \tag{20.26}$$

$$i_{CL} = i_{Z_C} - i_{Z_B} \tag{20.27}$$

Adding these equations shows that the sum of the line currents is zero regardless of the nature of the load. As discussed previously, when the load is balanced these equations yield the following relationship between the magnitudes of the load and line currents:

$$I_L = \sqrt{3} I_Z \text{ (balanced load)} \tag{20.28}$$

We also know in this situation that each line current lags its respective load current by 30° and that the sum of the load currents is zero.

Example 20.10   (Analysis—Unbalanced Δ-Δ)

The loads in Figure 20.25 are $Z_A = 10\ \underline{/0°}\ \Omega$, $Z_B = 10\ \underline{/90°}\ \Omega$, and $Z_C = 10\ \underline{/-90°}\ \Omega$. The phase sequence is ABC with $e_{AB} = 933\ \underline{/0°}$ V. Find the line currents.

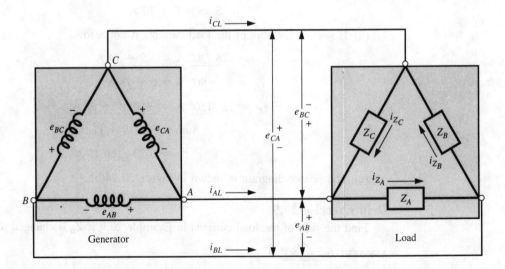

FIGURE 20.25 The Δ-Δ configuration.

SOLUTION From equation (20.24), the load currents are

$$i_{Z_A} = \frac{e_{AB}}{Z_A} = \frac{933 \underline{/0°} \text{ V}}{10 \underline{/0°} \ \Omega} = 93.3 \underline{/0°} \text{ A}$$

$$i_{Z_B} = \frac{e_{BC}}{Z_B} = \frac{933 \underline{/-120°} \text{ V}}{10 \underline{/90°} \ \Omega} = 93.3 \underline{/-210°} \text{ A} = 93.3 \underline{/150°} \text{ A}$$

$$i_{Z_C} = \frac{e_{CA}}{Z_C} = \frac{933 \underline{/120°} \text{ V}}{10 \underline{/-90°} \ \Omega} = 93.3 \underline{/210°} \text{ A} = 93.3 \underline{/-150°} \text{ A}$$

From equations (20.25) through (20.27), the line currents are

$$i_{AL} = i_{Z_A} - i_{Z_C} = (93.3 + j0) - (-80.8 - j46.65) \text{ A}$$

$$= 174.1 + j46.65 \text{ A} = 180.2 \underline{/15°} \text{ A}$$

$$i_{BL} = i_{Z_B} - i_{Z_A} = (-80.8 + j46.65) - (93.3 + j0) \text{ A}$$

$$= -174.1 + j46.65 \text{ A} = 180.2 \underline{/165°} \text{ A}$$

$$i_{CL} = i_{Z_C} - i_{Z_B} = (-80.8 + j46.65) - (-80.8 + j46.65) \text{ A}$$

$$= 0 - j93.3 \text{ A} = 93.3 \underline{/-90°} \text{ A}$$

Drill Exercise 20.10

What would be the magnitude of each line current in Example 20.10 if every load were equal to 10 $\underline{/0°}$ Ω?

ANSWER: 161.6 A.   □

The Δ-Y Configuration

Figure 20.26 shows a delta-connected source driving a wye load. It is readily apparent that the load currents equal the line currents:

$$i_{AL} = i_{Z_A}, \ i_{BL} = i_{Z_B}, \ i_{CL} = i_{Z_C} \tag{20.29}$$

If the load is balanced, the load (and line) currents have equal magnitudes, are separated by 120°, and their sum is zero. By writing Kirchhoff's voltage law around each pair of terminals in the wye load, we find the following relationships between line and load voltages:

$$e_{AB} = v_{Z_A} - v_{Z_B} \tag{20.30}$$

$$e_{BC} = v_{Z_B} - v_{Z_C} \tag{20.31}$$

$$e_{CA} = v_{Z_C} - v_{Z_A} \tag{20.32}$$

If the system is balanced, the same analysis used previously on sets of equations having the same form as these leads to

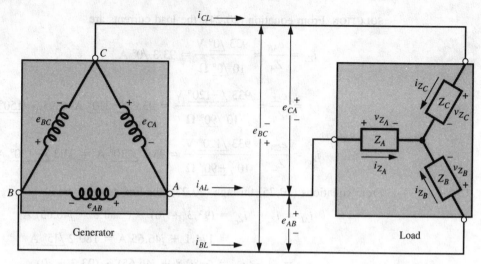

FIGURE 20.26 The Δ-Y configuration.

$$E_L = \sqrt{3}\, V_Z \text{ (balanced load)} \tag{20.33}$$

where E_L is the magnitude of a line voltage and V_Z is the magnitude of a load voltage. Each load voltage in that case lags a corresponding line voltage by 30°. Using these facts, the analysis of a balanced system is straightforward. If the load is unbalanced, the analysis can be simplified by converting the wye load to an equivalent delta. The next example illustrates the procedure.

Example 20.11   (Analysis—Unbalanced Δ-Y)

Find the current in and voltage across each load of the unbalanced Δ-Y system shown in Figure 20.27(a), given that $e_{AB} = 360\ \underline{/0°}$ V, $e_{BC} = 360\ \underline{/-120°}$ V, and $e_{CA} = 360\ \underline{/120°}$ V.

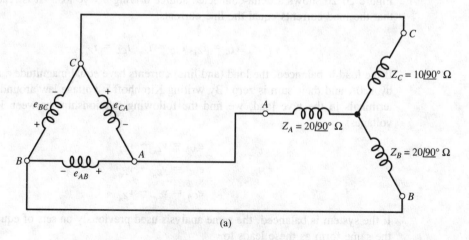

(a)

FIGURE 20.27 (Example 20.11) (Continues on next page.)

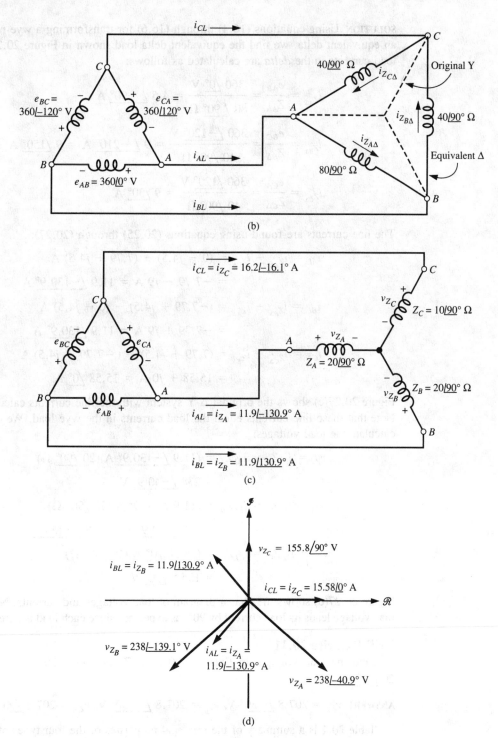

FIGURE 20.27 (Example 20.11) (continued)

SOLUTION Using equations (16.4) through (16.6) for transforming a wye network to an equivalent delta, we find the equivalent delta load shown in Figure 20.27(b). The load currents in the *delta* are calculated as follows:

$$i_{Z_{A\Delta}} = \frac{e_{AB}}{Z_{A\Delta}} = \frac{360 \; \underline{/0°} \; \text{V}}{80 \; \underline{/90°} \; \Omega} = 4.5 \; \underline{/-90°} \; \text{A}$$

$$i_{Z_{B\Delta}} = \frac{e_{BC}}{Z_{B\Delta}} = \frac{360 \; \underline{/-120°} \; \text{V}}{40 \; \underline{/90°} \; \Omega} = 9 \; \underline{/-210°} \; \text{A} = 9 \; \underline{/150°} \; \text{A}$$

$$i_{Z_{C\Delta}} = \frac{e_{CA}}{Z_{C\Delta}} = \frac{360 \; \underline{/120°} \; \text{V}}{40 \; \underline{/90°} \; \Omega} = 9 \; \underline{/30°} \; \text{A}$$

The line currents are found using equations (20.25) through (20.27):

$$i_{AL} = i_{Z_{A\Delta}} - i_{Z_{C\Delta}} = (0 - j4.5) - (7.79 + j4.5) \; \text{A}$$
$$= -7.79 - j9 \; \text{A} = 11.9 \; \underline{/-130.9°} \; \text{A}$$

$$i_{BL} = i_{Z_{B\Delta}} - i_{Z_{A\Delta}} = (-7.79 + j4.5) - (0 - j4.5) \; \text{A}$$
$$= -7.79 + j9 \; \text{A} = 11.9 \; \underline{/130.9°} \; \text{A}$$

$$i_{CL} = i_{Z_{C\Delta}} - i_{Z_{B\Delta}} = (7.79 + j4.5) - (-7.79 + j4.5) \; \text{A}$$
$$= 15.58 + j0 \; \text{A} = 15.58 \; \underline{/0°} \; \text{A}$$

Figure 20.27(c) shows the original Δ-Y system with the line currents calculated above. Note that these line currents equal the load currents in the wye load. We can therefore calculate the load voltages:

$$v_{Z_A} = i_{Z_A} Z_A = i_{AL} Z_A = (11.9 \; \underline{/-130.9°} \; \text{A})(20 \; \underline{/90°} \; \Omega)$$
$$= 238 \; \underline{/-40.9°} \; \text{V}$$

$$v_{Z_B} = i_{Z_B} Z_B = i_{BL} Z_B = (11.9 \; \underline{/130.9°} \; \text{A})(20 \; \underline{/90°} \; \Omega)$$
$$= 238 \; \underline{/220.9°} \; \text{V} = 238 \; \underline{/-139.1°} \; \text{V}$$

$$v_{Z_C} = i_{Z_C} Z_C = i_{CL} Z_C = (15.58 \; \underline{/0°} \; \text{A})(10 \; \underline{/90°} \; \Omega)$$
$$= 155.8 \; \underline{/90°} \; \text{V}$$

Figure 20.27(d) shows the phasor diagram of load voltages and currents. Note that each load voltage leads its load current by 90°, as expected, since each load is purely inductive.

Drill Exercise 20.11

Find the load voltages in Example 20.11 if Z_C in Figure 20.27(a) is changed to $20 \; \underline{/90°} \; \Omega$.

ANSWER: $v_{Z_A} = 207.8 \; \underline{/-30°} \; \text{V}$, $v_{Z_B} = 207.8 \; \underline{/-150°} \; \text{V}$, $v_{Z_C} = 207.8 \; \underline{/90°} \; \text{V}$.   □

Table 20.1 is a summary of the principal properties of the four types of three-phase configurations we have studied.

TABLE **20.1** Properties of three-phase configurations

System configuration	Either balanced or unbalanced	Balanced only
Y-Y (four-wire)	$\Sigma e_\phi = 0$ $\Sigma e_L = 0$ $\Sigma v_Z = 0$ $E_L = \sqrt{3}\, E_\phi = \sqrt{3}\, V_Z$ $i_L = i_\phi = i_Z$ $i_N = \Sigma\, i_L$	$i_N = 0$
Y-Δ	$\Sigma e_\phi = 0$ $\Sigma e_L = 0$ $E_L = \sqrt{3}\, E_\phi$ $\Sigma i_L = 0$	$\Sigma i_Z = 0$ $I_L = \sqrt{3}\, I_Z$
Δ-Δ	$\Sigma e_\phi = 0$ $\Sigma e_L = 0$ $\Sigma v_Z = 0$ $E_\phi = E_L = V_Z$	$\Sigma i_Z = 0$ $I_L = \sqrt{3}I_Z$
Δ-Y	$\Sigma e_\phi = 0$ $\Sigma e_L = 0$ $E_L = E_\phi$ $i_L = i_Z$	$\Sigma i_L = 0$ $E_L = \sqrt{3}\, V_Z$

Note: It is assumed that each generator produces a balanced set of three-phase voltages (equal magnitudes, separated in phase by 120°).

Key:

Σ = "the sum of"

e_ϕ, i_ϕ = generator phase voltage, phase current

e_L, i_L = line voltage, line current

v_Z, i_Z = load voltage, load current

E_L, I_L = magnitude of line voltage, line current

V_Z, I_Z = magnitude of load voltage, load current

20.6 Power in Three-Phase Circuits

Advantages of Three-Phase Power Distribution

To compare single-phase power distribution with a typical three-phase system, consider Figure 20.28. Part (a) of the figure shows a single-phase, 120-V rms source driving a 20-Ω load. The total power delivered to the load is

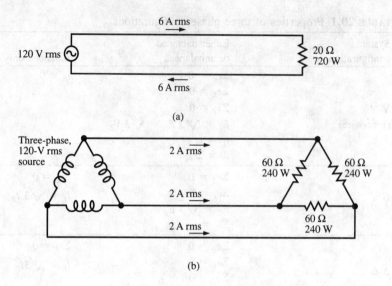

FIGURE 20.28 A comparison of single-phase and three-phase systems delivering equal power. (a) In a single-phase system delivering 720 W of power, each of two conductors must carry 6 A rms. (b) In a three-phase system delivering 720 W of power, each of three conductors must carry 2 A rms.

$$P = \frac{V_{rms}^2}{R} = \frac{(120 \text{ V})^2}{20 \ \Omega} = 720 \text{ W}$$

and the current delivered to the load is

$$I = \frac{V_{rms}}{R} = \frac{120 \text{ V}}{20 \ \Omega} = 6 \text{ A rms}$$

Figure 20.28(b) shows a balanced Δ-Δ system with a 120-V rms, three-phase generator driving three 60-Ω loads. Since the power delivered by each phase to the load is

$$P = \frac{(120 \text{ V})^2}{60 \ \Omega} = 240 \text{ W}$$

the total power delivered by the three-phase system is 3 × 240 W = 720 W, the same power delivered by the single-phase system. The current in each line of the three-phase circuit is

$$I = \frac{120 \text{ V}}{60 \ \Omega} = 2 \text{ A rms}$$

We see that the wires in the three-phase system can have one-third the current-carrying capacity of those in the single-phase system (2 A vs. 6 A). A wire with one-third the capacity of another has about one-fourth the cross-sectional area. Since there are three wires in the three-phase system versus two in the single-phase system, the total length of wire required in the three-phase system is 3/2 that of the single-phase system. The total *volume* (area × length) of conductor required, and therefore the material cost, of the three-phase installation is therefore (1/4)(3/2), or about 40%, that of the single-phase system. Even in a three-phase, four-wire installation, the conductor cost is about 50% that of the single-phase system. This reduction in material cost is a significant economic benefit in the long-distance delivery of large amounts of power.

Recall from Chapter 12 (Figure 12.41) that the *instantaneous* power delivered by a single ac circuit is continually fluctuating—it is, in fact, a sinusoidal function of time. The total instantaneous power of a balanced three-phase circuit is the sum of the powers delivered by the phases:

$$p(t) = V \sin(\omega t) I \sin(\omega t + \theta) + V \sin(\omega t - 120°) I \sin(\omega t - 120° + \theta)$$

$$+ V \sin(\omega t + 120°) I \sin(\omega t + 120° + \theta) \qquad (20.34)$$

where V and I are the peak values of voltage and current and θ is the angle of each impedance in the balanced load (and therefore the phase angle between each voltage and current). Using trigonometric identities, it can be shown that equation (20.34) can be expressed as

$$p(t) = 1.5 \, VI \cos \theta \qquad (20.35)$$

Note that unlike instantaneous single-phase power, this power is *constant;* it does not vary with time. This property is advantageous in the conversion of electrical power to mechanical power, as, for example, by a three-phase motor. Since the power is constant, the conversion is uniform and more efficient.

Another advantage of three-phase power is that it can be used to create a *rotating magnetic field*. A rotating magnetic field (inside a cylindrical housing) is one that continually fluctuates in direction and magnitude around the interior of the housing. It is required in the operation of ac motors, and is achieved by connecting a three-phase source to windings embedded in the motor housing. In a single-phase ac motor, a rotating magnetic field can only be approximated, using phase-shifting circuitry, a fact that makes it less efficient than the three-phase motor. For this reason, most large motors are designed for three-phase operation.

Power Computations

Since the total instantaneous power in a balanced three-phase circuit is constant [equation (20.35)], the total average power is equal to that constant:

$$P_{avg} = 1.5 \, V_Z I_Z \cos \theta \qquad (20.36)$$

where V_Z and I_Z are the peak values of each load voltage and load current, and $\cos \theta$ is the power factor, as discussed in Chapter 14. In terms of rms values,

$$P_{avg} = 3V_{Z(rms)} I_{Z(rms)} \cos \theta \tag{20.37}$$

In a balanced, three-phase system, one of the following sets of relationships is true: either $E_L = \sqrt{3} V_Z$ and $I_L = I_Z$ (Y-Y and Δ-Y), or $E_L = E_Z$ and $I_L = \sqrt{3} I_Z$ (Δ-Δ and Y-Δ). In either case, substituting these into equations (20.36) and (20.37) gives the same result:

$$P_{avg} = \frac{\sqrt{3} E_L I_L \cos \theta}{2} = \sqrt{3} E_{L(rms)} I_{L(rms)} \cos \theta \tag{20.38}$$

Example 20.12

Find the total average power delivered to the load in Figure 20.29.

SOLUTION Each load impedance is

$$Z = 4 + j1 \ \Omega = 4.12 \ \underline{/14°} \ \Omega$$

Since the system is balanced, we know from equation (20.33) that the magnitude of each load voltage is

$$V_Z = \frac{V_L}{\sqrt{3}} = \frac{340 \ \text{V}}{\sqrt{3}} = 196.3 \ \text{V}$$

Therefore, each load current has magnitude

$$I_Z = \frac{V_Z}{|Z|} = \frac{196.3 \ \text{V}}{4.12 \ \Omega} = 47.6 \ \text{A}$$

Since the angle between v_Z and i_Z is the angle of the impedance, $\theta = 14°$, we use equation (20.36) to compute the total power:

$$P_{avg} = 1.5(196.3 \ \text{V})(47.6 \ \text{A}) \cos 14° = 13.6 \ \text{kW}$$

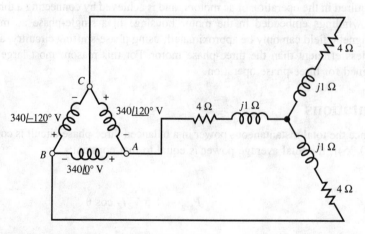

FIGURE 20.29 (Example 20.12)

Drill Exercise 20.12

What would be the total power delivered to the load in Example 20.12 if each load impedance were changed to $4 + j4 \ \Omega$?

ANSWER: 7.22 kW.   □

The average power in each impedance of a three-phase load can also be calculated using any of the ac power relationships discussed in Chapter 12. The total average power in a balanced load is then three times that in a single load impedance:

$$P_{avg} = 3 \left(\frac{I_R^2 R}{2} \right) = 3 I_{R(rms)}^2 R \tag{20.39}$$

where R is the resistive component of the load ($Z = R \pm jX$) and I_R, $I_{R(rms)}$ are the peak and rms values, respectively, of the current in R. Furthermore,

$$P_{avg} = 3 \left(\frac{V_R^2}{2R} \right) = \frac{3 V_{R(rms)}^2}{R} \tag{20.40}$$

where $V_R, V_{R(rms)}$ are the peak and rms values of the voltage across the resistive component of the load.

Similarly, the total reactive and apparent powers in a balanced load are three times the reactive and apparent powers in each impedance:

$$Q = \frac{3 V_Z I_Z \sin \theta}{2} = \frac{3 I_X^2 |X|}{2} = \frac{3 V_X^2}{2|X|}$$

$$= 3 V_{Z(rms)} I_{Z(rms)} \sin \theta = 3 I_{X(rms)}^2 |X| = \frac{3 V_{X(rms)}^2}{|X|} \tag{20.41}$$

$$S = \frac{3 V_Z I_Z}{2} = 3 V_{Z(rms)} I_{Z(rms)} \tag{20.42}$$

where V_X, $V_{X(rms)}$ are the peak and rms voltages across the reactive component of each load impedance and I_X, $I_{X(rms)}$ are the peak and rms currents through each reactive component. In terms of the total average and apparent powers, the power factor can be determined from

$$\text{power factor} = \cos \theta = \frac{P_{avg}}{S} \tag{20.43}$$

The equations we have developed in this section are for balanced systems. However, we can always find the total average or reactive power in any three-phase system by simply summing the average or reactive powers in each load. These individual powers are calculated as discussed for single-phase circuits in Chapters 12 and 15. Also, it is always true that

$$S = \sqrt{P^2 + Q^2} \tag{20.44}$$

Example 20.13

For the balanced, three-phase Δ-Δ system shown in Figure 20.30(a), find

(a) the total average power
(b) the total reactive power
(c) the total apparent power
(d) the power factor.

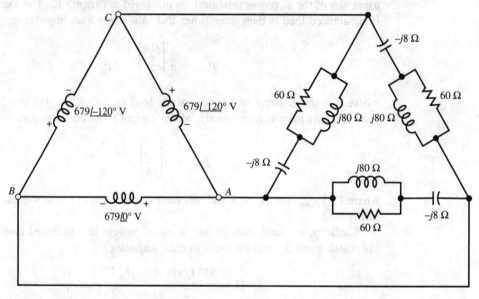

(a)

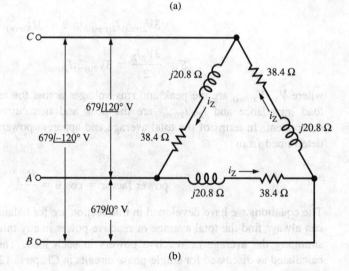

(b)

FIGURE 20.30 (Example 20.13)

SOLUTION

(a) The parallel network in each load impedance is equivalent to

$$\frac{(60 \,\underline{/0°})(80 \,\underline{/90°})}{60 + j80} = \frac{4800 \,\underline{/90°}}{100 \,\underline{/53.13°}} = 48 \,\underline{/36.87°} \;\Omega$$

$$= 38.4 + j28.8 \;\Omega$$

Thus, each load impedance is equal to

$$Z = 38.4 + j28.8 + 0 - j8 \;\Omega$$

$$= 38.4 + j20.8 \;\Omega = 43.7 \,\underline{/28.4°} \;\Omega$$

Figure 20.30(b) shows the load when redrawn with each impedance equal to the equivalent series impedance we have calculated. The magnitude of each load current is

$$I_Z = \frac{V_Z}{|Z|} = \frac{679 \text{ V}}{43.7 \;\Omega} = 15.54 \text{ A}$$

Note that 15.54 A is the magnitude of the current in each resistive and reactive component of each load impedance:

$$I_Z = I_R = I_X = 15.54 \text{ A}$$

From equation (20.39), the total average power is

$$P_{avg} = 3\left(\frac{I_R^2 R}{2}\right) = \frac{3(15.54 \text{ A})^2(38.4 \;\Omega)}{2} = 13.91 \text{ kW}$$

(b) From equation (20.41), the total reactive power is

$$Q = \frac{3I_X^2|X|}{2} = \frac{3(15.54 \text{ A})^2(20.8 \;\Omega)}{2} = 7.53 \text{ kvars (inductive)}$$

(c) From equation (20.42), the total apparent power is

$$S = \frac{3V_Z I_Z}{2} = \frac{3 \,(679 \text{ V})(15.54 \text{ A})}{2} = 15.83 \text{ kVA}$$

Note that $S = \sqrt{P_{avg}^2 + Q^2}$.

(d) From equation (20.43), the power factor is

$$\text{power factor} = \frac{P_{avg}}{S} = \frac{13.91 \text{ kW}}{15.83 \text{ kVA}} = 0.88 \text{ (lagging)}$$

Note that θ in this example is the angle of Z, namely 28.4°, and that the power factor can also be found from $\cos(28.4°) = 0.88$ (lagging).

Drill Exercise 20.13

If the frequency of the voltage in Example 20.13 is 60 Hz, what value of capacitance should be inserted in place of the capacitance shown in Figure 20.30(a) to achieve unity power factor? What is the total average power delivered to the load in that case?

ANSWER: 92.1 µF, 18 kW. □

Example 20.14   (Design)

Figure 20.31(a) shows the windings of a three-phase motor connected in a four-wire Y-Y system. Each phase voltage has magnitude 679 V and frequency 60 Hz. Capacitance C is to be installed between each line and neutral in order to make the power factor unity.

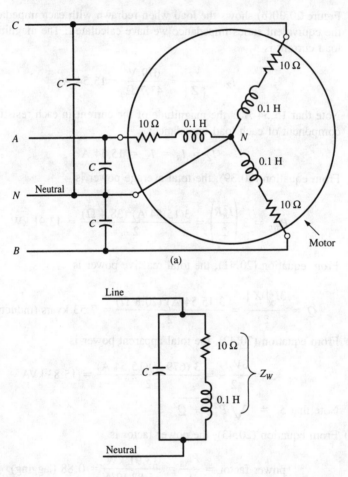

(a)

(b)

FIGURE 20.31 (Example 20.14)

(a) Find the value of capacitance required.

(b) Find the magnitude of each line current before and after installation of the capacitance.

SOLUTION

(a) Note that capacitance C is in parallel with each motor winding, Z_W. Therefore, each impedance in the circuit is equivalent to the circuit shown in Figure 20.31(b). Recall that this configuration is the series–parallel circuit studied in Chapter 18 (Figure 18.27). In the derivation for the resonant frequency of that circuit, we found that the imaginary part of the admittance could be made zero by setting the value of C according to

$$C = \frac{L}{R^2 + (\omega L)^2}$$

If the admittance has zero imaginary part, then the circuit has unity power factor. Therefore, we find the value of C as specified above:

$$C = \frac{0.1 \text{ H}}{(10 \text{ } \Omega)^2 + (2\pi \times 60 \text{ Hz} \times 0.1 \text{ H})^2} = 65.7 \text{ } \mu\text{F}$$

(b) The impedance of each motor winding before power-factor correction is

$$Z_W = 10 + j(2\pi)(60 \text{ Hz})(0.1 \text{ H}) = 10 + j37.7 \text{ } \Omega$$
$$= 39 \underline{/75.14°} \text{ } \Omega$$

Since the load is balanced, each line current has magnitude

$$I = \frac{E}{|Z|} = \frac{679 \text{ V}}{39 \text{ } \Omega} = 17.4 \text{ A}$$

After power-factor correction, the equivalent impedance of each load network is

$$Z = Z_W \,||\, (-j|X_C|)$$

where $|X_C| = \dfrac{1}{2\pi (60 \text{ Hz})(65.7 \times 10^{-6}\text{F})} = 40.37 \text{ } \Omega$

Thus,

$$Z = \frac{(39 \underline{/75.1°} \text{ } \Omega)(40.37 \underline{/-90°} \text{ } \Omega)}{10 + j37.7 - j40.37}$$

$$= \frac{1574 \underline{/-14.9°}}{10.35 \underline{/-14.9°}} = 152 \underline{/0°} \text{ } \Omega$$

Note that the impedance after power-factor correction is purely resistive, as expected. The magnitude of each line current is then

$$I = \frac{679 \text{ V}}{152 \text{ } \Omega} = 4.47 \text{ A}$$

We see that power-factor correction significantly reduces the peak current that must be supplied to the motor.

Drill Exercise 20.14

Find the total average power delivered to the motor before and after power-factor correction.

ANSWER: 4541.4 W.   □

20.7 Power Measurements

The Wattmeter

The basic instrument used to measure ac power is the *electrodynamometer* wattmeter. An electrodynamometer contains a movable coil upon which is mounted a pointer. See Figure 20.32. In a wattmeter, the movable coil is electrically connected across the voltage terminals of the circuit whose power is to be measured. Thus, the current in the coil and the magnetic field that results are directly proportional to the voltage applied to the circuit under test. Another set of (stationary) coils is connected in series with the circuit under measurement, so the current in these coils and the magnetic field they create is proportional to the current flowing into the circuit under test. The two fields interact to create a rotational force on the movable coil in much the same way that force is created on the rotor of an electric motor to make it rotate. The magnetic fields created by the coils are directly proportional to the currents flowing in them, so the rotational force on the movable coil is proportional to the product of those two currents. Hence, the movable coil rotates and deflects the pointer in direct proportion to the product of the current in the movable coil and the current in the fixed coils. (A coil spring attached to the movable coil restrains its rotation.) Since the current in the movable coil is proportional to the voltage V applied to the circuit under test, and the current in the fixed coils is proportional to the current I in the circuit, the deflection of the pointer is proportional to the product VI, or power. Figure 20.32(b) shows the symbol used to represent a wattmeter. For improved clarity, we will use the symbol shown in Figure 20.32(c), where V represents the voltage (movable) coil and I represents the current (fixed) coil. The $\pm$ symbols on the wattmeter show the polarity that must be observed for external connections. Figure 20.32(c) shows how the wattmeter is connected to measure power in a single-phase circuit. The power indicated by the meter is $V_{rms}I_{rms} \cos \theta$, where θ is the angle between the voltage and the current.

The Three-Wattmeter Method

The total power in a three-phase circuit can be found by connecting three wattmeters to measure the power in the three phases and adding the three readings. When measuring the power delivered to a three-phase load, each voltage coil may be connected across each

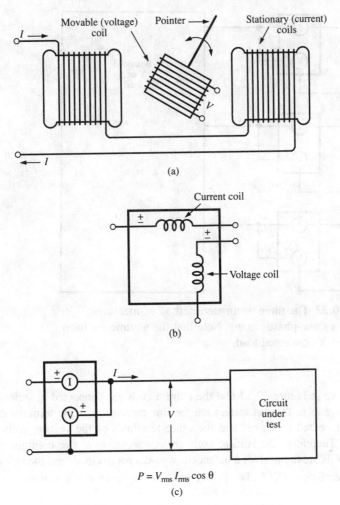

FIGURE **20.32** The electrodynamometer wattmeter.
(a) Simplified structure of the electrodynamometer.
(b) Wattmeter symbol. (c) Connections for measuring
power in a single-phase circuit.

load voltage, V_Z, and each current coil may be connected in series with each load current, I_Z. However, in practice, these connections are not always possible. For example, in a delta-connected three-phase motor winding, it would not ordinarily be possible to open the windings internally, as would be necessary to insert the current coils in series with each. Similarly, a three-wire wye load would require access to the center of the wye to make connections to the voltage coils. For this reason, a three-wattmeter method was developed so that power measurements could be made using line voltages and currents instead of load voltages and currents. Figure 20.33 shows the connections for a balanced wye or delta load.

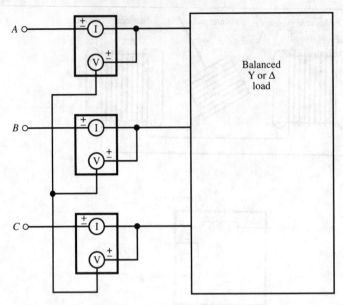

FIGURE 20.33 The three-wattmeter method of measuring power in a three-phase circuit. Note that the wattmeters form a balanced Y-connected load.

Observe in Figure 20.33 that the current coils are connected in series with the lines, so the currents in the wattmeters are the line currents, i_L. One terminal of each voltage coil is connected to a line, and the other terminals of the voltage coils are connected together. Therefore, the voltage coils are connected in a wye configuration. Thus, the voltage coils are themselves a balanced, wye-connected load, and the voltage across each coil is therefore $E_L/\sqrt{3}$. The power indicated by each meter is then

$$P = \frac{1}{2}\left(\frac{E_L}{\sqrt{3}}\right) I_L \cos\theta$$

(20.45)

where E_L and I_L are the magnitudes (peak values) of the line voltages and currents. Since the load is balanced, each wattmeter indicates the same power, and the total power is three times each:

$$P = 3\left(\frac{1}{2}\right)\left(\frac{E_L}{\sqrt{3}}\right) I_L \cos\theta$$

$$= \left(\frac{3}{2\sqrt{3}}\right)\frac{\sqrt{3}}{\sqrt{3}} E_L I_L \cos\theta$$

$$= \frac{\sqrt{3}}{2} E_L I_L \cos\theta$$

(20.46)

Equation (20.46) is precisely the same as equation (20.38), which gives the total power in a balanced three-phase load. Although we have showed only that the three-wattmeter method is valid for balanced loads, it is also valid for unbalanced loads: the total power is the sum of the wattmeter readings. This fact is demonstrated in the next example.

Example 20.15 (Analysis)

(a) Find the total power delivered to the unbalanced load in Figure 20.34 using load voltages and currents.

(b) Find the reading of each wattmeter and show that the sum of the readings is the total power found in (a).

SOLUTION

(a) The load currents are

$$i_{R_A} = \frac{e_{AB}}{R_A} = \frac{170 \; \underline{/0°} \; \text{V}}{20 \; \underline{/0°} \; \Omega} = 8.5 \; \underline{/0°} \; \text{A} = 8.5 + j0 \; \text{A}$$

$$i_{R_B} = \frac{e_{BC}}{R_B} = \frac{170 \; \underline{/-120°} \; \text{V}}{5 \; \underline{/0°} \; \Omega} = 34 \; \underline{/-120°} \; \text{A} = -17 - j29.44 \; \text{A}$$

$$i_{R_C} = \frac{e_{CA}}{R_C} = \frac{170 \; \underline{/120°} \; \text{V}}{10 \; \underline{/0°} \; \Omega} = 17 \; \underline{/120°} \; \text{A} = -8.5 + j14.72 \; \text{A}$$

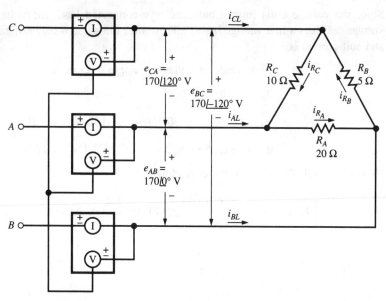

FIGURE 20.34 (Example 20.15)

The powers in the loads are then

$$P_{R_A} = \frac{I_{R_A}^2 R_A}{2} = \frac{(8.5 \text{ A})^2 (20 \text{ }\Omega)}{2} = 772.5 \text{ W}$$

$$P_{R_B} = \frac{I_{R_B}^2 R_B}{2} = \frac{(34 \text{ A})^2 (5 \text{ }\Omega)}{2} = 2890 \text{ W}$$

$$P_{R_C} = \frac{I_{R_C}^2 R_C}{2} = \frac{(17 \text{ A})^2 (10 \text{ }\Omega)}{2} = 1445 \text{ W}$$

The total average power is therefore

$$P_{R_A} + P_{R_B} + P_{R_C} = 772.5 \text{ W} + 2890 \text{ W} + 1445 \text{ W}$$

$$= 5057.5 \text{ W}$$

(b) Since the currents in the current coils are line currents, we must use the load currents to find the line currents:

$$i_{AL} = i_{R_A} - i_{R_C} = (8.5 + j0) - (-8.5 + j14.72) \text{ A}$$

$$= 17 - j14.72 \text{ A} = 22.49 \text{ } \underline{/-40.89°} \text{ A}$$

$$i_{BL} = i_{R_B} - i_{R_A} = (-17 - j29.44) - (8.5 + j0) \text{ A}$$

$$= -25.5 - j29.44 \text{ A} = 38.95 \text{ } \underline{/-130.9°} \text{ A}$$

$$i_{CL} = i_{R_C} - i_{R_B} = (-8.5 + j14.72) - (-17 - j29.44) \text{ A}$$

$$= 8.5 + j44.16 \text{ A} = 44.98 \text{ } \underline{/79.1°} \text{ A}$$

Since the voltage coils form a balanced wye-connected load, the voltage across each voltage coil lags a line voltage by 30°. Therefore, the angle θ between each coil voltage and coil current is

$$\theta = \text{angle of voltage} - \text{angle of current}$$

$$\theta_A = \quad (0° - 30°) \quad - \quad (-40.89°) = 10.89°$$

$$\theta_B = (-120° - 30°) - \quad (-130.9°) = -19.1°$$

$$\theta_C = \quad (120° - 30°) \quad - \quad (79.1°) \quad = 10.9°$$

By equation (20.45), the wattmeter readings are

$$P_A = \frac{1}{2}\left(\frac{E_L}{\sqrt{3}}\right) I_{AL} \cos \theta_A = \frac{1}{2}\left(\frac{170 \text{ V}}{\sqrt{3}}\right) (22.49 \text{ A}) \cos (10.89°) = 1083.8 \text{ W}$$

$$P_B = \frac{1}{2}\left(\frac{E_L}{\sqrt{3}}\right) I_{BL} \cos \theta_B = \frac{1}{2}\left(\frac{170 \text{ V}}{\sqrt{3}}\right) (38.95 \text{ A}) \cos (19.1°) = 1806.2 \text{ W}$$

$$P_C = \frac{1}{2}\left(\frac{E_L}{\sqrt{3}}\right) I_{CL} \cos \theta_C = \frac{1}{2}\left(\frac{170 \text{ V}}{\sqrt{3}}\right) (44.98 \text{ A}) \cos (10.9°) = 2167.5 \text{ W}$$

The sum of the wattmeter readings is then $P_A + P_B + P_C = 1083.8$ W $+ 1806.2$ W $+ 2167.5$ W $= 5057.5$ W, the same as the total average power found in (a). Note that the individual wattmeter readings do not equal the powers in the individual loads, but their sum is equal to the total power delivered to the load.

Drill Exercise 20.15

What would be the reading of each wattmeter in Example 20.15 if each load resistance were changed to 10 Ω?

ANSWER: 1445 W.   □

The Two-Wattmeter Method

Figure 20.35(a) shows how two wattmeters can be used to determine the total power delivered to a three-phase load, in this case, a balanced, wye-connected load. In this *two-wattmeter method*, each voltage coil is connected across a line voltage and each current coil is connected in series with a line current, irrespective of whether the load is a wye or delta, balanced or unbalanced. The key point to notice in Figure 20.35 is that the voltage coil of wattmeter 2 is connected across e_{CB} (not e_{BC}), and its current coil is connected in series with i_{CL} (not i_{BL}). Since $e_{BC} = E\underline{/-120°}$, then $e_{CB} = E\underline{/-120° + 180°} = E\underline{/60°}$. Recall that each load voltage v_Z in a balanced system lags a line voltage by 30°. Figure 20.35(b) shows a phasor diagram. In this example, we are assuming that the load impedances are such that each load current (and therefore each line current) lags its load voltage by angle θ. As seen in the phasor diagram, the angle between the voltage e_{AB} applied to wattmeter 1 and the current i_{AL} through it is $\theta + 30°$. The angle between the voltage e_{CB} applied to wattmeter 2 and the current i_{CL} through it is $\theta - 30°$. Therefore, the powers P_1 and P_2 indicated by wattmeters 1 and 2 are

$$P_1 = \frac{E_{AB}I_{AL}}{2} \cos (\theta + 30°) = \frac{EI}{2} \cos (\theta + 30°) \tag{20.47}$$

$$P_2 = \frac{E_{CB}I_{CL}}{2} \cos (\theta - 30°) = \frac{EI}{2} \cos (\theta - 30°) \tag{20.48}$$

where $E_{AB} = E_{CB} = E$ is the magnitude of the line voltages and $I_{AL} = I_{CL} = I$ is the magnitude of the line currents. The total power is the sum of P_1 and P_2:

$$P_1 + P_2 = \frac{EI}{2} \cos (\theta + 30°) + \frac{EI}{2} \cos (\theta - 30°)$$

$$= \frac{EI}{2} [\cos (\theta + 30°) + \cos (\theta - 30°)] \tag{20.49}$$

Using the trigonometric identities $\cos (x + y) = \cos x \cos y - \sin x \sin y$ and $\cos (x - y) = \cos x \cos y + \sin x \sin y$, we can show (Exercise 20.47) that equation (20.49) reduces to

$$P_1 + P_2 = EI \cos \theta \cos(30°) \tag{20.50}$$

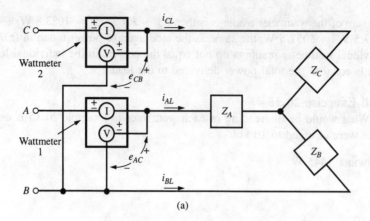

(a)

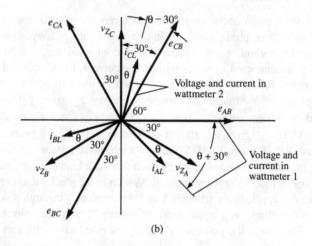

(b)

FIGURE 20.35 The two-wattmeter method. (a) Wattmeter connections to a balanced wye load. (b) Phasor diagram.

Since $\cos (30°) = \dfrac{\sqrt{3}}{2}$, equation (20.50) becomes

$$P_1 + P_2 = \frac{\sqrt{3}EI}{2} \cos \theta \qquad (20.51)$$

Equation (20.51) shows that the sum of the wattmeter readings equals the total power delivered to the load. Although we have demonstrated this fact for a balanced wye-connected load only, it is true for both balanced and unbalanced loads, wye or delta. The next example illustrates this fact for an unbalanced delta-connected load.

Example 20.16   (Analysis)

(a) Show how two wattmeters can be connected to the unbalanced delta load of Example 20.15 (Figure 20.34) to determine the total power delivered to the load.

(b) Find the power indicated by each wattmeter in (a) and show that their sum equals the total power delivered to the load.

SOLUTION

(a) Figure 20.36 shows the wattmeter connections.

(b) Wattmeter 1 is connected across e_{AB} = 170 $\underline{/0°}$ V and in series with i_{AL} = 22.49 $\underline{/-40.89°}$ A. (See Example 20.15). Therefore, the angle θ between voltage and current is 40.89°, and wattmeter 1 indicates

$$P_1 = \frac{E_{AB}I_{AL}}{2} \cos \theta = \frac{(170 \text{ V}) (22.49 \text{ A})}{2} \cos (40.89°)$$

$$= 1445.1 \text{ W}$$

Wattmeter 2 is connected across e_{CB} = 170 $\underline{/60°}$ V and in series with i_{CL} = 44.98 $\underline{/79.1°}$ A. Therefore, the angle θ between voltage and current is 79.1° − 60° = 19.1°, and wattmeter 2 indicates

$$P_2 = \frac{E_{CB}I_{AL}}{2} \cos \theta = \frac{(170 \text{ V}) (44.98 \text{ A})}{2} \cos (19.1°)$$

$$= 3612.8 \text{ W}$$

The sum of the two indications is

$$P_1 + P_2 = 1445.1 \text{ W} + 3612.8 \text{ W} = 5057.9 \text{ W}$$

Neglecting a small round-off error, this sum equals the total power found in Example 20.15 (5057.5 W).

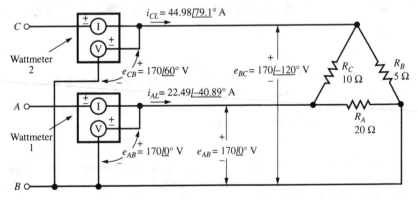

FIGURE 20.36 (Example 20.16)

Drill Exercise 20.16
What would each wattmeter in Example 20.16 indicate if each resistance in the delta were changed to 10 Ω?

ANSWER: 2167.5 W.   □

20.8 SPICE Examples

Example 20.17 (SPICE)
Use SPICE to solve Example 20.11.

SOLUTION Since no hand computations are required, it is not necessary to transform the unbalanced wye load to a delta configuration, as was done in Example 20.11. Assuming the frequency is 60 Hz, the inductances of the loads are

$$L_A = L_B = \frac{20 \ \Omega}{2\pi \times 60 \ \text{Hz}} = 53.05 \ \text{mH}$$

$$L_C = \frac{10 \ \Omega}{2\pi \times 60 \ \text{Hz}} = 26.53 \ \text{mH}$$

The SPICE input file is shown in Figure 20.37. VLA, VLB, and VLC are zero-valued voltage sources used to determine the line currents, which are equal to the load currents. Since SPICE does not permit an inductor to appear by itself in a loop with a voltage source, it is necessary to insert a small resistance in series with each ac source in the delta. As can be seen in Figure 20.37, these resistances are each given the negligibly small value 1 μΩ, which does not affect the computations. Execution of the program produces the same results found by hand computation in Example 20.11.

Exercises

Section 20.2   AC Generators (Alternators)

20.1 What would be the polarity of voltage e_{xy} at the terminals of the loop in Figure 20.1 (p. 787) in the position shown if each of the changes listed below were made?
(a) The direction of the magnetic field is reversed.
(b) Both the direction of the magnetic field and the direction of rotation are reversed.
(c) Only the direction of rotation is reversed.

20.2 List three ways of increasing the voltage induced in a coil rotating in a magnetic field.

20.3 Assume the loop in Figure 20.1 (p. 787) is rotating counterclockwise at a speed of 6000 revolutions per minute. When the loop is in a horizontal position, each side cuts 0.5×10^{-3} Wb of flux in one millisecond. Sketch the voltage e_{xy} as a function of time beginning at a point in time where the loop is in the position shown in Figure 20.1.

20.4 The loop shown in Figure 20.1 (p. 787) has a 5-Ω resistor connected across its terminals and is rotating clockwise at a speed of 377 radians/s. When the loop is in the position shown in Figure 20.1, each

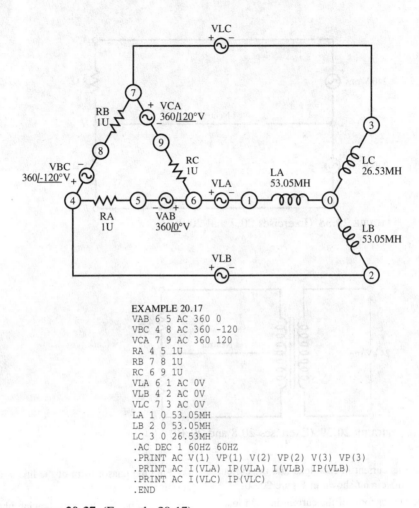

EXAMPLE 20.17
```
VAB 6 5 AC 360 0
VBC 4 8 AC 360 -120
VCA 7 9 AC 360 120
RA 4 5 1U
RB 7 8 1U
RC 6 9 1U
VLA 6 1 AC 0V
VLB 4 2 AC 0V
VLC 7 3 AC 0V
LA 1 0 53.05MH
LB 2 0 53.05MH
LC 3 0 26.53MH
.AC DEC 1 60HZ 60HZ
.PRINT AC V(1) VP(1) V(2) VP(2) V(3) VP(3)
.PRINT AC I(VLA) IP(VLA) I(VLB) IP(VLB)
.PRINT AC I(VLC) IP(VLC)
.END
```

FIGURE 20.37 (Example 20.17)

end cuts 4 μWb of flux in 2 μs. Assuming that current flowing from x to y is positive, sketch the current in the resistor versus time, beginning at a point in time where the loop is 180° clockwise from the position shown in Figure 20.1.

20.5 Briefly describe the function of each of the following components of an ac generator:
(a) slip rings
(b) brushes
(c) dc excitation

20.6 In a three-phase generator, the windings in which the output voltages are generated are stationary. Describe how these windings are made to cut magnetic flux.

Section 20.3 Single-Phase, Three-Wire Power Distribution

20.7 Find the rms current in each load and the current in the neutral conductor of the circuit shown in Figure 20.38.

20.8 Find the rms current in each load and the current in the neutral conductor of the circuit shown in Figure 20.39. The primary winding of the transformer has 1000 turns, and there are five turns between the center tap and each end of the secondary winding.

20.9 Repeat Exercise 20.7 with the load between B and N in Figure 20.38 changed to 6 Ω.

20.10 Repeat Exercise 20.8 with the load between B and N in Figure 20.39 changed to 10 Ω.

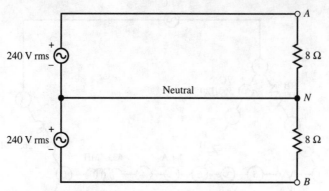

FIGURE 20.38 (Exercises 20.7 and 20.9)

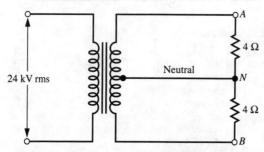

FIGURE 20.39 (Exercises 20.8 and 20.10)

20.11 Find the rms current in each load and the current in the neutral of the circuit shown in Figure 20.40.

20.12 Find the phasor form of the current in each load and the current in the neutral of the circuit shown in Figure 20.41.

Section 20.4 Wye- and Delta-Connected Three-Phase Sources

20.13 In the wye-connected generator shown in Figure 20.10 (p. 797), the phase voltages and currents are as follows:

$$e_{AN} = 679 \ \underline{/0°} \text{ V}, \ e_{BN} = 679 \ \underline{/-120°} \text{ V},$$

$$e_{CN} = 679 \ \underline{/120°} \text{ V},$$

$$i_{A\phi} = 50 \ \underline{/60°} \text{ A}, \ i_{B\phi} = 50 \ \underline{/-60°} \text{ A},$$

$$i_{C\phi} = 50 \ \underline{/180°} \text{ A}.$$

(a) Find the phasor form of the line voltages and line currents.

(b) Draw a phasor diagram showing phase and line voltages.

(c) Show that the neutral current is zero.

20.14 In the wye-connected generator shown in Figure 20.10 (p. 797), the line voltages and currents are as follows:

$$e_{AB} = 588 \ \underline{/30°} \text{ V}, \ e_{BC} = 588 \ \underline{/-90°} \text{ V},$$

$$e_{CA} = 588 \ \underline{/150°} \text{ V}, \ i_{AL} = 100 \ \underline{/90°} \text{ A},$$

$$i_{BL} = 100 \ \underline{/-30°} \text{ A}, \ i_{CL} = 100 \ \underline{/-150°} \text{ A}.$$

(a) Find the phasor form of the phase voltages and currents.

(b) Find the rms values of the phase and line voltages.

(c) Show that the neutral current is zero.

(d) Draw a phasor diagram showing phase and line voltages.

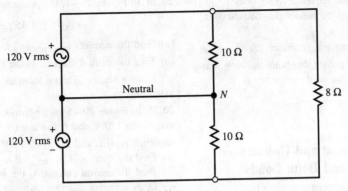

FIGURE 20.40 (Exercise 20.11)

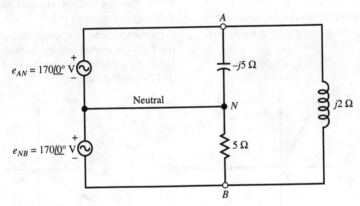

FIGURE 20.41 (Exercise 20.12)

20.15 The phase voltages and currents in the delta-connected generator in Figure 20.16 (p. 803) are as follows:

$$e_{AB} = 170 \, \underline{/0°} \text{ V}, \; e_{BC} = 170 \, \underline{/-120°} \text{ V},$$
$$e_{CA} = 170 \, \underline{/120°} \text{ V}, \; i_{A\phi} = 20 \, \underline{/30°} \text{ A},$$
$$i_{B\phi} = 20 \, \underline{/-90°} \text{ A}, \; i_{C\phi} = 20 \, \underline{/150°} \text{ A}.$$

(a) Find the phasor form of the line voltages and currents.

(b) Find the sum of the phase currents and the sum of the line currents.

(c) Draw a phasor diagram showing phase and line currents.

20.16 The line voltages and currents in the delta-connected generator in Figure 20.16 (p. 803) are as follows:

$$e_{AB} = 69 \, \underline{/0°} \text{ kV}, \; e_{BC} = 69 \, \underline{/-120°} \text{ kV},$$
$$e_{CA} = 69 \, \underline{/120°} \text{ kV}, \; i_{AL} = 250 \, \underline{/45°} \text{ A},$$
$$i_{BL} = 250 \, \underline{/-75°} \text{ A}, \; i_{CL} = 250 \, \underline{/165°} \text{ A}.$$

The system is balanced and $i_{A\phi}$ has angle 75°.

(a) Find the phase voltages and currents. [*Hint*: Use equations (20.15), (20.11), and (20.12).]

(b) Find the sum of the line currents and the sum of the phase currents.

(c) Draw a phasor diagram showing phase and line currents.

20.17 Determine the phase sequence (*ABC* or *BAC*) in each of the systems having the phasor diagrams shown in Figure 20.42.

20.18 In a certain Y-connected generator, $e_{AB} = 340 \,\underline{/60°}$ V. Draw phasor diagrams showing line and phase voltages when the phase sequence is
(a) *ABC*
(b) *BAC*

Section 20.5 Balanced and Unbalanced Wye and Delta Loads

20.19 In Figure 20.43, each phase voltage has magnitude 678.82 V. The phase sequence is *ABC* with $e_{AN} = 678.82 \,\underline{/0°}$ V.
(a) Find the current in each load.
(b) Find the neutral current. Is the system balanced?
(c) Draw a phasor diagram showing load voltages and load currents.

20.20 In Figure 20.44, the phase sequence is *BAC* and
$$e_{AB} = 294.45 \,\underline{/30°} \text{ V.}$$
(a) Find the currents i_{AL}, i_{BL}, and i_{CL}.
(b) Find the neutral current. Is the system balanced?
(c) Draw a phasor diagram showing load voltages and load currents.

20.21 In Figure 20.45 each phase voltage has magnitude 170 V and frequency 60 Hz. The phase sequence is *ABC* and $e_{AN} = 170 \,\underline{/0°}$ V.
(a) Find the currents i_{AL}, i_{BL}, and i_{CL}.
(b) Find the neutral current. Is the system balanced?
(c) Draw a phasor diagram showing the load currents and the neutral current.

20.22 In Figure 20.46, each line voltage has magnitude 69 kV and frequency 60 Hz. The phase sequence is *BAC* and $e_{AB} = 69 \,\underline{/15°}$ kV.
(a) Find the currents i_{AL}, i_{BL}, and i_{CL}.
(b) Find the neutral current. Is the system balanced?
(c) Draw a phasor diagram showing the load currents and the neutral current.

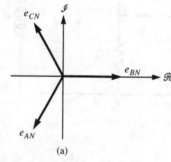
(a)

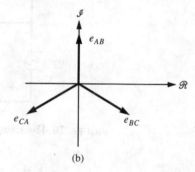
(b)

FIGURE 20.42 (Exercise 20.17)

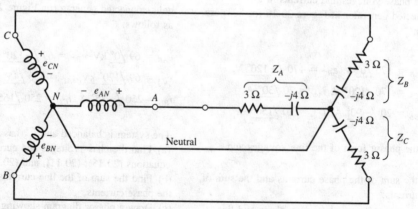

FIGURE 20.43 (Exercise 20.19)

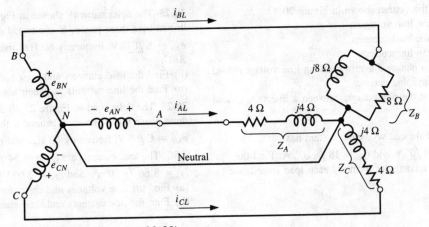

FIGURE 20.44 (Exercise 20.20)

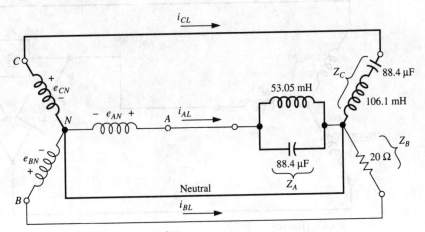

FIGURE 20.45 (Exercise 20.21)

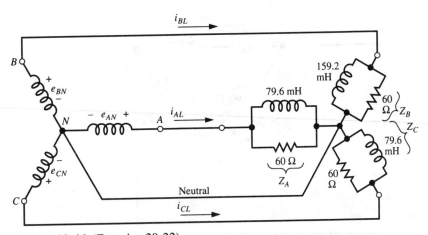

FIGURE 20.46 (Exercise 20.22)

20.23 In the system shown in Figure 20.47:
(a) Find the line voltages.
(b) Find the load currents.
(c) Find the line currents.
(d) Draw a phasor diagram showing line voltages, load currents, and line currents.
(e) What is the phase angle between a line voltage and a line current?

20.24 A balanced wye-delta system has $e_{AN} = 400 \underline{/0°}$ V and $i_{AL} = 18 \underline{/-60°}$ A. Find the polar and rectangular forms of each load impedance.

20.25 The delta network shown in Figure 20.48 is driven by a three-phase, Y-connected generator that has $e_{AN} = 8 \underline{/0°}$ kV, frequency 60 Hz, and phase sequence ABC.
(a) Find the load currents and their sum.
(b) Find the line currents and their sum.

20.26 The delta load in Figure 20.49 is driven by the three-phase wye-connected generator that has $e_{AN} = E \underline{/0°}$ V, frequency 60 Hz, and phase sequence ABC. The load currents are $i_{Z_A} = 9.54 \underline{/120°}$ A, $i_{Z_B} = 8.66 \underline{/-90°}$ A, and $i_{Z_C} = 2.953 \underline{/-145.23°}$ A.
(a) Find the line voltages and their sum.
(b) Find the line currents and their sum.

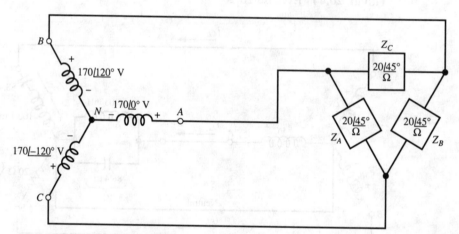

FIGURE 20.47 (Exercise 20.23)

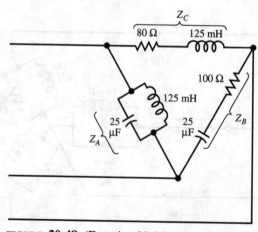

FIGURE 20.48 (Exercise 20.25)

(c) Find the phase voltages in the generator.

(d) Draw a phasor diagram showing line voltages and line currents.

20.27 Each load impedance in a three-phase Δ-Δ system is $Z = 6.5\ \underline{/20°}\ \Omega$. The effective value of each generator phase voltage is 120 V rms, and e_{AB} has phase angle 0°. The phase sequence is ABC.

(a) Find the line voltages and their sum.

(b) Find the load currents and their sum.

(c) Find the line currents and their sum.

(d) Draw a phasor diagram showing load and line currents.

20.28 In the Δ-Δ system shown in Figure 20.50:

(a) Find the load currents and their sum.

(b) Find the line currents and their sum.

(c) Is the system balanced?

(d) Draw a phasor diagram showing load and line currents.

20.29 In the Δ-Y system shown in Figure 20.51, the frequency of the three-phase voltage is 60 Hz.

(a) Find the load voltages and their sum.

(b) Find the load currents and their sum.

(c) Find the line currents and their sum.

(d) Draw a phasor diagram showing load voltages and load currents.

20.30 The generator in Figure 20.52 produces three-phase, 60-Hz voltages.

(a) Find the current in every impedance.

(b) Find the line currents.

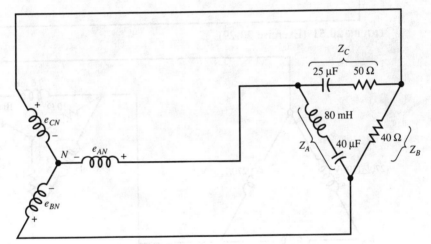

FIGURE 20.49 (Exercise 20.26)

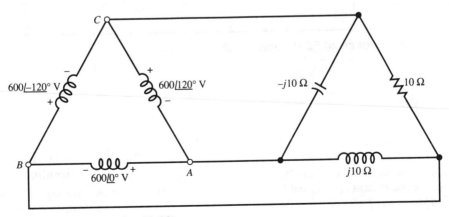

FIGURE 20.50 (Exercise 20.28)

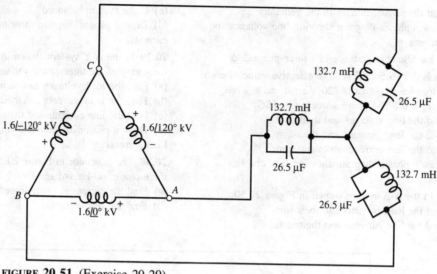

FIGURE 20.51 (Exercise 20.29)

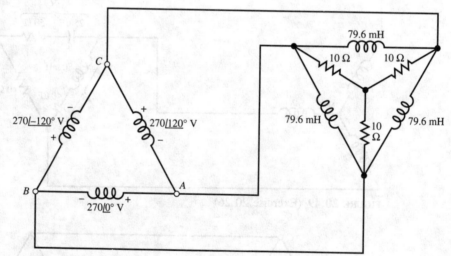

FIGURE 20.52 (Exercise 20.30)

20.31 Find the load currents and load voltages in the circuit shown in Figure 20.53.

20.32 In the circuit shown in Figure 20.54:
(a) Find the line currents i_{AL}, i_{BL}, and i_{CL}.
(b) Find the voltage across each impedance.

Section 20.6   Power in Three-Phase Circuits

20.33 List three advantages of three-phase power as compared to single-phase power.

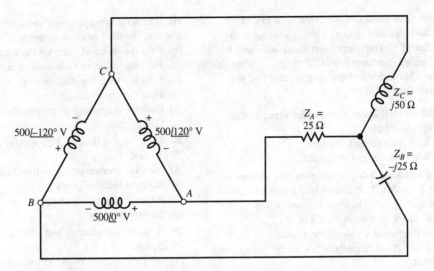

FIGURE 20.53 (Exercise 20.31)

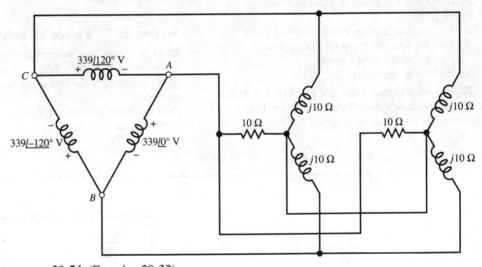

FIGURE 20.54 (Exercise 20.32)

20.34 The single-phase circuit shown in Figure 20.28(a) (p. 822) has a 1-Ω load.
(a) What current flows in each conductor?
(b) What is the total power delivered to the load?
(c) What value of each resistance in the delta-connected load in Figure 20.28(a) (p. 822) would result in the total power delivered by the three-phase system being equal to that delivered by the single-phase system?
(d) With the value of load resistances found in (c), what current would flow in each of the conductors of the three-phase system? (Continues on p. 848.)

(e) The cost C in dollars of the cable used to deliver power in these systems is $C = kll^{1.25}$, where l is the length in feet of a cable carrying rms-current I and k is a proportionality constant: $k = 1.25 \times 10^{-3}$. Find the cost of the cable in each of the systems and find the ratio of the costs.

20.35 Find the total average power delivered to the load in Exercise 20.20
(a) using load currents and voltages;
(b) using line currents and voltages.

20.36 In a balanced Δ-Δ system, each line voltage has magnitude $E_L = 500$ V and each load current has magnitude $I_Z = 42$ A. The total average power delivered to the delta load is 25.2 kW.
(a) What is the power factor of the system?
(b) What is the phase angle between a load voltage and a load current?
(c) What is the magnitude of the angle of each impedance in the load?

20.37 For the Y-Δ power distribution system shown in Figure 20.55:
(a) Find the total average power delivered to the load.
(b) Find the total reactive power in the load.
(c) Find the total apparent power.
(d) Find the power factor.

20.38 Each load impedance in a balanced Δ-Y system is a resistance in series with an inductance. Each load current has magnitude 40 A (peak) and frequency 60

Hz. The total average power delivered to the load is 6 kW and the total reactive power is 2 kvars.
(a) Find the value of each load resistance.
(b) Find the value of each load inductance.
(c) Find the total apparent power.
(d) Find the power factor.
(e) Find the phase angle between the load voltage and the load current.

20.39 In the unbalanced Y-Δ system in Exercise 20.26, find
(a) the total average power delivered to the load;
(b) the total reactive power;
(c) the total apparent power;
(d) the power factor.

20.40 In the unbalanced Y-Δ system in Exercise 20.25, find
(a) the total average power delivered to the load;
(b) the total reactive power;
(c) the total apparent power;
(d) the power factor.

Section 20.7   Power Measurements

20.41 (a) Show how wattmeters should be connected to the system in Figure 20.56 to measure the average power delivered to the load using the three-wattmeter method.
(b) Find the magnitude of the voltage and current in each wattmeter.

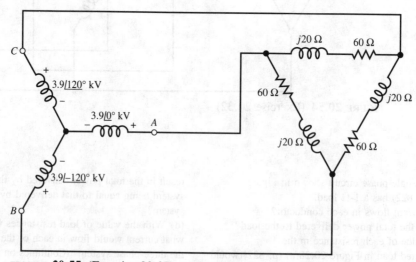

FIGURE 20.55 (Exercise 20.37)

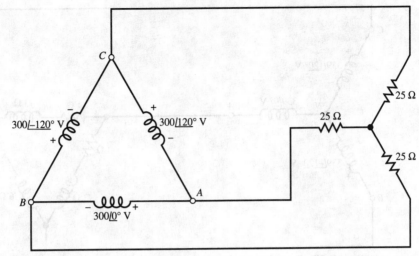

FIGURE 20.56 (Exercise 20.41)

(c) What does each wattmeter read?
(d) What is the total average power delivered to the load?

20.42 Repeat Exercise 20.41 when each impedance in the wye load is replaced by $25 - j5 \ \Omega$.

20.43 (a) Show how wattmeters should be connected to the system in Figure 20.57 to measure the total average power delivered to the load using the three-wattmeter method.
(b) Find the total average power delivered to the load using the methods of Section 20.6.
(c) Find the reading of each wattmeter and show that the sum of the readings is the total power found in (a).

20.44 Repeat Exercise 20.43 for the circuit shown in Figure 20.58.

20.45 (a) Show how wattmeters should be connected to the system in Figure 20.59 to determine the total power delivered to the load using the two-wattmeter method.
(b) Find the total power delivered to the load using the methods of Section 20.6.
(c) Find the reading of each wattmeter and show that their sum equals the total power found in (b).

20.46 Repeat Exercise 20.45 when each 4-Ω resistance in the load is replaced by $Z = 4 - j3 \ \Omega$.

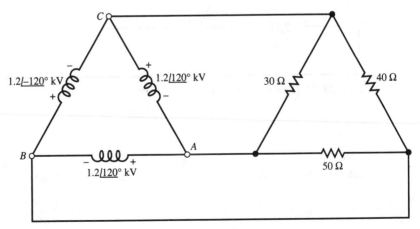

FIGURE 20.57 (Exercises 20.43 and 20.48)

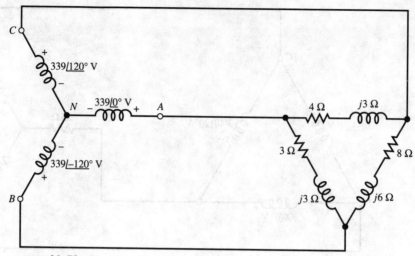

FIGURE 20.58 (Exercise 20.44)

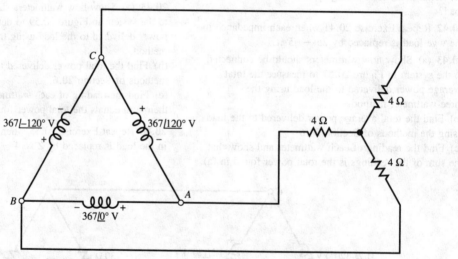

FIGURE 20.59 (Exercise 20.45)

20.47 Derive equation (20.50) for the sum of the readings in the two-wattmeter meter from equation (20.49).

20.48 Repeat Exercise 20.45 for the unbalanced system in Figure 20.57 (Exercise 20.43).

Design Exercises

20.1D The load in an unbalanced three-phase, four-wire Y-Y system is equivalent to that shown in Figure 20.60. In order to eliminate the neutral wire, the load is to be balanced by connecting resistance R in parallel with the 150-Ω resistance. What value of R should be used?

20.2D When connecting loads in a three-phase system, it is not always possible to achieve or maintain perfect balance. However, loads should be assigned in a way that results in the least imbalance possible. In a three-phase, four-wire system, the least imbalance means that the magnitude of the neutral current is as small as possible. The equivalent impedances of the loads in one such system are

$Z_A = R_A = 4 \underline{/0°} \; \Omega$, $Z_B = R_B = 6 \underline{/0°} \; \Omega$, and

$Z_C = R_C = 8 \underline{/0°} \; \Omega$. A new load, $Z = R = 12 \underline{/0°} \; \Omega$,

is to be connected to the system. It should be connected in parallel with which existing load?

20.3D As illustrated in Example 20.14, capacitive loads are connected to three-phase systems to improve power factors. These are normally connected in parallel with existing loads instead of in series, because a series connection affects the voltage and current delivered to the existing loads and a parallel connection does not. To confirm this fact, find the value of series-connected capacitors that would be required to achieve unity power factor in Example 20.14 (Figure 20.31, p. 828). Then find the voltage and current magnitudes in the loads before and after power-factor correction.

20.4D A three-phase wye-connected generator is to be selected for driving a three-phase motor with delta-connected windings. The motor is rated 12.5 hp at 208 V rms and it is 85% efficient. The power factor is 0.9.

(a) What should be the effective value of the phase voltage in the generator?

(b) How much line current should the generator be capable of providing?

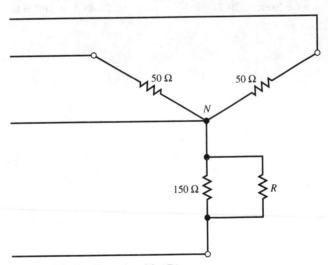

FIGURE 20.60 (Exercise 20.1D)

Troubleshooting Exercises

20.1T The electrical power service to a home is single-phase, three-wire. Approximately one-half of the lighting circuits are inoperable and none of the heavy appliance circuits is operable. No protective devices (circuit breakers, fuses) in the home or at the pole-mounted transformer serving the home have opened. It has been determined that the fault is in the transformer. What is the most likely nature of the fault?

20.2T The windings in a certain three-phase motor are connected in a four-wire wye configuration. Each winding has impedance

$$Z_A = Z_B = Z_C = Z_W = 20 \ \underline{/90°} \ \Omega.$$

The three-phase source has $e_{AN} = 170 \ \underline{/0°}$ V and phase sequence ABC. One of the windings has open-circuited. Measurements reveal that the neutral current lags e_{AN} by 150°. Which winding is open?

Spice Exercises

Note: In the exercises that follow, it will be necessary to insert small resistances to prevent the existence of loops containing only a voltage source and an inductance. See Example 20.17 (p. 838).

20.1S Use SPICE to solve Exercise 20.21.

20.2S Use SPICE to solve Exercise 20.25. It is not necessary to find the sums of the currents.

20.3S Use SPICE to solve Exercise 20.28. It is not necessary to find the sums of the currents.

20.4S Use SPICE to solve Exercise 20.32.

APPENDIX
Circuit Analysis Using SPICE

A.1 Introduction

Computers are now widely used to perform circuit computations. They have become an especially valuable tool for eliminating the many hours of computational drudgery needed to analyze complex circuits by hand. They also provide reliable, highly accurate solutions that are free of the human errors afflicting most hand computations (algebraic errors, misplaced decimal poits, etc.)

Computers can be used in one of two fundamental ways to perform circuit computations. In one method, the user writes a *program* in a so-called *high-level computer language,* such as FORTRAN, BASIC, or Pascal. The program, in effect, tells the computer the precise sequence of computations it must perform to analyze a specific circuit. The program contains instructions that define every mathematical computation (multiplication, division, etc.) that must be performed, and the order in which they must be performed, to solve one circuit problem. Obviously, the programmer must be a skilled circuit analyst, as well as a good computer programmer, to use this method. One disadvantage of the method is that the program (called the *software*) must often be modified extensively, or rewritten entirely, if it is to be used to analyze a circuit different from the one for which it was designed. The principal advantage of the method is that a solution can usually be obtained much *faster* (in terms of the time required for the computer to *execute* the program) than it can using other methods. Computer *run time* can be, and often is, an important consideration when analyzing particularly complex circuits.

The second method for analyzing circuits by computer is to use a program that has been specially designed to solve circuit problems of all kinds. This type of program is an example of *applications software*, or a program *package*: one that is designed for solving specific types of problems. When using such a package to solve circuit problems, it is only necessary to specify the values of circuit components and to supply information on how the components are connected to each other. Accurate results can be obtained by a user having very little programming skill and a minimal knowledge of circuit theory.

Using a computer program to solve a circuit problem is called circuit *simulation*, or *modeling*, because we are effectively creating a mathematical model that simulates an actual circuit. One very powerful program package used for circuit simulation is called SPICE. SPICE is an acronym for Simulation Program with Integrated Circuit Emphasis, and, as the name implies, it was originally developed (at the University of California, Berkeley) for analyzing electronic integrated circuits. However, this versatile software can also be used to analyze conventional electric circuits containing no electronic devices at all. As we shall learn in subsequent discussions, SPICE is capable of analyzing both dc and ac circuits to determine the voltage across and current through every circuit component, as well as to perform many special kinds of analyses, including transient analysis and frequency response.

One of the most widely used versions of SPICE designed for operation on micro-computers is PSpice (a registered trademark of the MicroSim Corporation). It has numerous features and options that make it more versatile and somewhat easier to use than the original Berkeley version of SPICE. However, with a few minor exceptions, any circuit simulation file written to run on Berkeley SPICE will also run on PSpice. In the discussions that follow, features of PSpice are highlighted immediately after the corresponding capabilities or requirements of Berkeley SPICE are described.

A.2 Defining a Circuit for Analysis by SPICE

It is important to understand that SPICE is a *program*: one that has been written by others and that is already stored in computer memory. To use this program, we must define the circuit we wish to analyze, by specifying all component values and telling how the components are connected, and we must specify the type of *output* we desire, whether it be a list of voltages and/or currents, a graph of voltage versus time, or data in some other format. The information we supply for this purpose is called the *input* to the SPICE program, often referred to as an *input file*. The input must be entered, typically from a keyboard at a computer terminal, in a special format. The format is easy to learn, because it consists simply of a sequence of lines (called *records* by computer scientists), each of which specifies certain input data. The input file is often called a "SPICE program," although, strictly speaking, it merely supplies data describing a particular circuit problem to the built-in SPICE program.

The first line in every SPICE input file must be a *title*. The title is any descriptive word or words that the user cares to choose, such as

DEMO

or

TEST CIRCUIT

or

PROBLEM ONE

or

VOLTAGE VS TIME

In the documentation prepared by the developers of SPICE (University of California, Berkeley), each line of input file is referred to as a *card*, a carryover from the times when punched cards were a popular means for entering data into a computer. Thus, the line containing the title is called a *title card*.

Following the title, the input file contains lines (called *element cards*) that define circuit components. Before writing these lines, it is necessary to identify and *number* all the junctions in the circuit to be analyzed (more precisely, all the circuit *nodes*—see Section 6.6). It makes no difference what sequence of numbers is used to number the nodes, but all numbers must be positive integers. One node must be assigned the number zero (0), which is the *reference node* (common, or ground). For the time being, we will consider circuits that contain only dc voltage sources and resistors. Each resistor must be assigned a different R-designation, such as R1, R2, and so on. Any numbers or letters can follow the letter R; for example: RIN, RDELTA, RX2, etc. Similarly, each dc voltage source is assigned a V-prefix, such as V1, VOUT, etc. It is good practice to draw a circuit diagram showing the node numbers and the component designations before writing any lines of the SPICE input file. Figure A.1 shows several examples of circuits whose nodes and components have been assigned designations in preparation for writing such lines.

To define a circuit for SPICE, we write one line for each and every component ("element") in the circuit. Each line specifying a resistor begins with its R-designation, followed by the numbers of the nodes between which it is connected, followed by its resistance value in ohms. For example, the line

R1 3 4 100

specifies that the 100-Ω resistor R1 is connected between nodes 3 and 4. The order in which the node numbers are listed and the spaces left between each entry are arbitrary. For example,

R1   4 3   100

is equivalent to the previous specification of R1. In general, a resistor is defined by

R******* *N1* *N2* *VALUE*

where ******* is an arbitrary sequence of characters, *N1* and *N2* are the numbers of the nodes between which the resistor is connected, and *VALUE* is the resistance value in ohms. *VALUE* may not be zero.

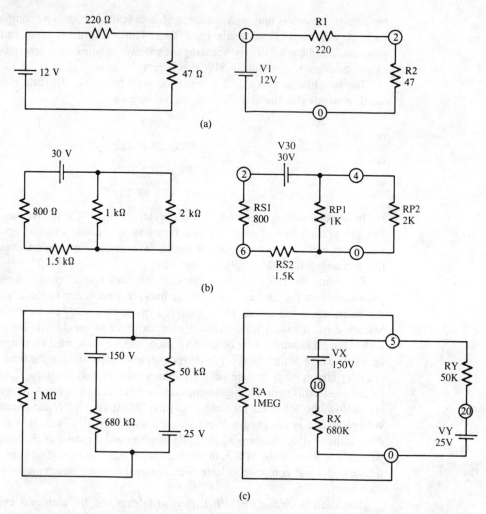

FIGURE A.1 Examples of circuits with assignments of node numbers and component designations for analysis by SPICE.

A dc voltage source is defined by the format

$$V\!*\!*\!*\!*\!*\!*\!* \quad N\!+ \ N\!- \ <\!DC\!> \ VALUE$$

where $*\!*\!*\!*\!*\!*\!*$ is an arbitrary sequence of characters, $N\!+$ is the number of the node to which the positive terminal of the source is connected, and $N\!-$ is the number of the node to which the negative terminal is connected. The bracketed entry $<\!DC\!>$ means that we may *optionally* write the characters DC to specify a dc source. If DC is omitted, SPICE assumes that the source is dc. (This is an example of what is called a *default* assumption.) *VALUE* is the value of the source in volts. It may be zero, positive, or negative. (As we shall see in a later discussion, zero-valued voltage sources are used as "ammeters" by SPICE.) Making *VALUE* a negative number has the same effect as interchanging $N\!+$ and $N\!-$. Following are two equivalent ways to define a 12-V dc source whose positive terminal is connected to node 4 and whose negative terminal is connected to node 0:

```
V1  4  0  DC  12
V1  0  4  -12
```

In the foregoing examples, all letters used in the input data files are capitalized, a requirement of Berkeley SPICE. In PSpice, either lowercase or capital letters (or both) can be used. For example, resistor R1 can be listed as r1 in one line and referred to as R1 in another line, and PSpice will recognize both as representing the same component.

Node numbers in PSpice do not have to be integers; they can be identified by any set of alphanumeric characters (letters or numbers) up to 31 characters in length. However, one node must be node 0.

Numerical Value Specifications

Numerical values, such as voltage and resistance values, can be written in conventional form, with decimal points and plus or minus signs. As in ordinary algebra, omission of a sign implies that a value is positive. No commas are permitted in a value specification. Examples of valid value specifications include 17, -25.4, $+18000$, and 0.00063. Numerical values can also be expressed using a suffix that indicates a power of 10 by which a number is multiplied. The power, which must be a positive or negative integer, is written after the letter E. Following are some examples of equivalent, valid ways of writing numerical values:

```
2640 = 2.64E3 = 2.64E+3 = 26.4E2
0.0095 = 9.5E-3 = 0.095E-1
100000 = 1E5 = 10E4 = 100E+3
```

Note that the suffix E (for exponent) is used in the same way it is for entering values in scientific notation on an electronic calculator.

SPICE also permits the use of certain letters to represent SI suffixes, such as pico, kilo, etc. Following are the letters used, the powers of 10 they represent, and the corresponding SI suffixes:

T	10^{12}	tera
G	10^{9}	giga
MEG	10^{6}	mega
K	10^{3}	kilo
M	10^{-3}	milli
U	10^{-6}	micro
N	10^{-9}	nano
P	10^{-12}	pico
F	10^{-15}	femto

SPICE permits a user to write any letters after a numerical value, or any letters after an SI suffix. Unless the letter(s) following a number are one of the SI suffixes listed above, SPICE simply ignores the letters and uses only the numerical value in performing computations. Letters are often added to numerical values to indicate units, although it is not necessary to do so. Following are some examples of equivalent, valid value specifications:

```
1200 = 1.2K = 1.2KOHM
0.003 = 0.003V = 3MV = 3E-3V
60P = 60PF = 60PFARADS = 60E-12F
5.1MEG = 5100K = 5.1MEGOHMS = 51E5
```

Note carefully that M is used to designate milli (10^{-3}), while MEG is used for mega (10^6). Also note that no spaces are permitted between a numerical value and any letters added to it.

A circuit is completely defined when a separate line has been written to specify every component in the circuit. The order in which the components are defined (the order in which the lines appear) is arbitrary.

Example A.1

Write the lines necessary to define the circuit shown in Figure A.2(a) for a SPICE input file.

SOLUTION Figure A.2(b) shows the circuit after it has been redrawn with node numbers and component designations assigned in *one* permissible way. Following is *one* permissible sequence of lines (called a *listing*) that defines the circuit. Note that the title chosen for this example is EXAMPLE A1:

```
EXAMPLE A1
V1  1  0  28V
R1  1  2  1800
R2  2  0  1MEG
R3  2  3  47KOHMS
R4  3  0  22E3
```

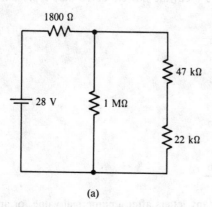

(a)

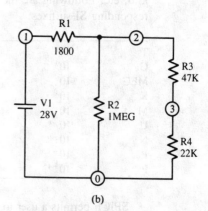

(b)

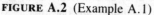

FIGURE A.2 (Example A.1)

A.3 Control Statements

Once a circuit has been completely defined, we must tell SPICE what we wish to know about the circuit, that is, what type of analysis is to be performed and what type of output data is desired (voltage values, currents, etc.). *Control statements* are written in the SPICE input file for this purpose. Every control statement is preceded by a period. Each statement occupies a separate line of the input file, and the statements are written immediately following the last line of the circuit definition. (We should note again that SPICE documentation refers to control statements as control *cards*.)

One control statement is used to specify the kind of analysis desired. Analysis types include *dc, ac,* and *transient,* among others. The dc control statement has the following format:

```
.DC SCRCNAM VSTART VSTOP VINCR
```

where *SCRCNAM* is the designation (name) of any voltage source in the circuit. *VSTART*, *VSTOP*, and *VINCR* are used to *step* the voltage source through a series of values beginning with the value *VSTART* and ending with *VSTOP*, in increments of *VINCR*. This feature is useful for studying certain electronic devices, but for conventional dc circuit analysis, we simply make *VSTART* and *VSTOP* both equal to the value of the dc source, and arbitrarily make *VINCR* equal to 1. Following is an example of a control statement specifying a dc analysis for a circuit containing the 18-V source V1:

```
.DC V1 18 18 1
```

If a circuit contains more than one source, it makes no difference which of them is referenced in the dc control statement.

To obtain a list of circuit voltages, we use the .PRINT control statement. In this statement, we specify the voltage(s) desired by using either of the formats

```
V(N1)   or   V(N1,N2)
```

where *N1* and *N2* are node numbers. V(*N1*) is voltage at node *N1 with respect to node zero,* and V(*N1,N2*) is the voltage at node *N1* with respect to node *N2*. Up to eight different voltages can be specified in one .PRINT statement. The analysis type (DC, AC, etc.) must appear in the .PRINT statement immediately after the word PRINT. Following is an example of a .PRINT statement used to obtain four different voltages in a dc analysis:

```
.PRINT DC V(1) V(2,3) V(7) V(5,6)
```

The printed output resulting from a .PRINT DC statement will be titled DC TRANSFER CURVES, a title that is related to the voltage-stepping capability mentioned earlier, but one that is irrelevant to our simple dc analysis.

The last statement in every input file must be the control statement .END. Our knowledge of how to prepare a SPICE input file is now sufficient to compute the voltages in any dc circuit containing resistors and voltage sources. The only additional computer skills that are needed are the mechanics of gaining access to the SPICE software in a given computer system. The procedures for "logging on" to a particular computer, for preparing the system to run a SPICE program, and for obtaining output produced by a program run, will vary from system to system and must be learned from independent instruction. Many versions of SPICE, some of them having abbreviated capabilities, are now also available for running on microcomputers. Example 3.14 shows an example of a complete input data file and illustrates a SPICE analysis of a simple dc circuit. Example A.2, which follows, shows a SPICE analysis of a series–parallel dc circuit and illustrates the format in which printed output is received.

Example A.2

Use SPICE to find the voltage across each resistor in Figure A.3(a).

SOLUTION Figure A.3(b) shows the circuit after node numbers and component designations have been assigned. Following is the SPICE input file used to perform a dc analysis of the circuit:

```
EXAMPLE A2
V1 1 0 36V
R1 1 2 10
R2 2 3 5
R3 3 0 15
R4 2 0 20
.DC V1 36 36 1
.PRINT DC V(1,2) V(2,3) V(2) V(3)
.END
```

Note the following points in connection with this file:

1. The value of V1 (36V) could have been omitted in the line defining V1, because that value is assigned by the .DC control statement.

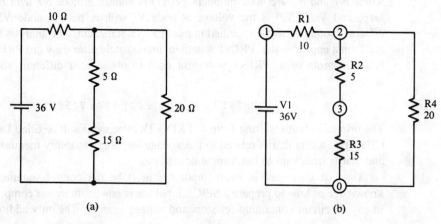

(a)   (b)

FIGURE A.3 (Example A.2)

2. The voltages we are required to find appear in the .PRINT statement:

Voltage Across Resistor	Voltage Designation in .PRINT Statement
R1	V(1,2)
R2	V(2,3)
R3	V(3)
R4	V(2)

Printout 1 is the output produced by a SPICE program run.

```
****      DC TRANSFER CURVES               TEMPERATURE =    27.000 DEG C

*************************************************************************


  V1           V(1,2)      V(2,3)      V(2)        V(3)


  3.600E+01    1.800E+01   4.500E+00   1.800E+01   1.350E+01
```
PRINTOUT 1

We see that $V_{R1} = 18$ V, $V_{R2} = 4.5$ V, $V_{R3} = 13.5$ V, and $V_{R4} = 18$ V. Note that the temperature of the circuit was assumed to be the (default) value 27°C. A special control statement can be used to specify circuit temperature, as discussed in the next paragraph.

The .TEMP Statement

All SPICE computations are performed under the assumption that the circuit temperature is 27°C, unless a different temperature is specified in the input data file. The .TEMP control statement is used to request an analysis at one or more different temperatures:

$$.\text{TEMP T1 < T2 T3 . . . >}$$

where T1 is the temperature in degrees C at which an analysis is desired and T2, T3 . . . are (optional) additional temperatures at which SPICE will automatically repeat the analysis, once for each temperature.

The values of resistors in a circuit are not changed by SPICE when analysis at a different temperature is requested unless either (or both) first- or second-order temperature coefficients of resistance are given in resistor specifications:

$$R^{*******} \text{ N1 N2 VALUE TC =   TC1, TC2}$$

where *TC1* and *TC2* are the first- and second-order temperature coefficients. When these are specified, SPICE adjusts the value of the resistance according to

$$R_T = R_{27}[1 + TC1 \, (T - 27°) + 1 + TC2 \, (T - 27°)]$$

where R_T is the resistance at new temperature T and R_{27} is the resistance at temperature 27°C. If *TC1* or *TC2* is not specified, SPICE assumes their value to be 0. Example 8.16 illustrates the use of the .TEMP control statement.

A.4 DC Current Values and Current Sources

Current Values

The .PRINT statement can be used to obtain the value of the current flowing in any voltage source in a circuit. Current in a voltage source is represented by

$$I(VNAME)$$

where *VNAME* is the name (V-designation) of any voltage source. For example, to obtain the dc currents in dc voltage sources VX and VY, we would write

.PRINT DC I(VX) I(VY)

SPICE assumes that positive current flows *into* the positive terminal of a voltage source. Note that this assumption contradicts the conventional current assumption. For example, if we had included I(V1) in the .PRINT statement of Example A.2, the corresponding output would have been printed as -1.8, since 1.8 A of conventional current flows out of the positive terminal of V1.

To find the current in a circuit path where there is no voltage source, we can simply insert a zero-valued source. Such a "dummy" voltage source has no effect on circuit behavior. Of course, it is necessary to define new nodes between which the dummy source is connected, since it is inserted in the circuit like an ammeter.

PSpice   In PSpice, the current through any component can be requested directly, without using a dummy voltage source. For example, I(R1) is the current through resistor R1. The reference polarity of the current is from the first node number of R1 to the second node number. In other words, positive current is assumed to flow from the first node given in the description of R1 to the second node. For example, if 5 A flows from node 1 to node 2 in a circuit whose input data file contains

R1 1 2 10

.PRINT DC I(R1)

then the .PRINT statement will produce 5 A. On the other hand, if the description of the same R1 were changed to R1 2 1 10, then the same .PRINT statement would produce -5 A.

Also, voltages across components can be requested directly, as, for example, V(RX), the voltage across resistor RX. The reference polarity is from the first node listed in the description of RX to the second node. For example, if the voltage across RX from node 5 to node 6 is 12 V in a circuit whose input data file contains

```
RX 5 6 1K
.PRINT DC V(RX)
```

then the .PRINT statement will produce 12 V. On the other hand, if the description of the same resistor were changed to RX 6 5 1K, the same .PRINT statement would produce −12V.

Current Sources

Constant-current sources can be modeled in a SPICE input file in much the same way that voltage sources are modeled. The format used to define a current source is

$$I******* \; N+ \; N- \; <DC>$$

where ******* is an arbitrary sequence of characters, $N+$ and $N-$ are the node numbers of the positive and negative terminals of the source, and <DC> is an optional specification of a dc source. *Conventional current flows out of the negative terminal of a SPICE current source.* This convention is the reverse of what we would normally regard as the positive terminal under a conventional current assumption. Figure A.4 is an example, showing equivalent, valid ways of defining the type of current source we have studied in earlier chapters. In this example, note carefully that SPICE regards $N+$ to be 2 and $N-$ to be 3. Example 3.15 illustrates the specification of a current source in a SPICE simulation.

Any current source can be specified instead of a voltage source in a .DC control statement:

$$.DC \; SRCNAM \; I \; I \; 1$$

where I is the constant current supplied by the current source named *SRCNAM*.

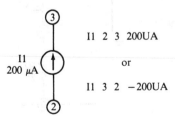

```
I1  2  3  200UA
        or
I1  3  2  −200UA
```

FIGURE A.4 An example of how a constant-current source is defined in SPICE. Note that node 2 is the "positive" node.

Example A.3

Use SPICE to find the voltage across and the current through each resistor in Figure A.5(a).

SOLUTION Figure A.5(b) shows the circuit after it has been redrawn to include node numbers and component designations. Note that a zero-valued, dummy voltage source (VA) has been inserted in series with R2 to determine the current in that resistor. The current in R1 is the same as the current in V1. Following is the SPICE input file:

```
EXAMPLE A3
V1  1  0  9V
I1  0  2  6MA
VA  3  2  0V
R1  1  2  1K
R2  3  0  2K
.DC V1 9 9 1
.PRINT DC V(1,2) V(3) I(V1) I(VA)
.END
```

Shown in Printout 2 are the results of a program run. We see that the voltage across the 1-kΩ resistor, V(1,2) is − 1 V (so node 2 is positive with respect to node 1), and that the voltage across the 2-kΩ resistor is 10 V. The current in the 1-kΩ resistor,

```
****      DC TRANSFER CURVES                 TEMPERATURE =    27.000 DEG C

***************************************************************************


V1             V(1,2)      V(3)         I(V1)       I(VA)

9.000E+00    −1.000E+00  1.000E+01   1.000E-03  −5.000E-03
```

PRINTOUT 2

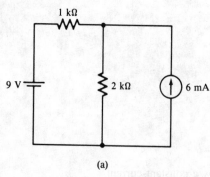
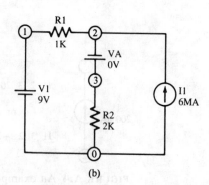

(a)   (b)

FIGURE A.5 (Example A.3)

$I(V1)$, is $+1$ mA, meaning that the positive conventional current flows through it from right to left (enters node 1 and the positive terminal of V1). The current in the 2-kΩ resistor, $I(VA)$, is -5 mA, meaning that positive conventional current enters node 3 (leaves the positive terminal of VA).

Example 4.24 shows how a current source can be used to determine the total equivalent resistance of a circuit in a SPICE simulation. Example 5.17 shows another example of a series–parallel circuit analyzed by SPICE.

A.5 Inductors and Capacitors

Inductors and capacitors are defined in a SPICE input file using the following formats:

(inductor): `L******* N+ N- VALUE`

(capacitor): `C******* N+ N- VALUE`

where `*******` is an arbitrary sequence of characters. $N+$ and $N-$ are the nodes between which the components are connected. Their positive and negative identifications have meaning only when *initial conditions* (initial voltage on a capacitor and initial current in an inductor) are specified for a *transient* analysis, about which we will have more to say later. For our purposes, we can disregard these polarities and specify the nodes in any order.

SPICE requires that there be a dc path to ground (node 0) from every node in a circuit. A dc path is a path through a resistor, an inductor, a voltage source, or a combination thereof. Thus, if all of the components connected to a particular node are capacitors, the simulation program will not run. Figure A.6 shows an example of a circuit that cannot be analyzed by SPICE because there is no dc path to ground from node 2.

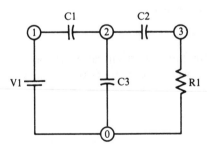

FIGURE A.6 Circuit that cannot be analyzed by SPICE, because there is no dc path from node 2 to node 0.

One way to force SPICE to perform an approximate analysis of a circuit containing such a node is to insert a very large resistance in parallel with one of the capacitors connected to the node. The resistance should be large enough to approximate an open circuit under the operating conditions of the circuit. For example, if the impedance of C3 in Figure A.6 is 10 kΩ at the frequency of an ac analysis, we could connect a resistance of, say 10^{12} Ω in parallel with it.

A.6 Transient Analysis

The .TRAN control statement is used to obtain a transient analysis of a circuit, that is, to obtain the voltage across and/or current through any component as a function of *time*, after one or more sources are switched into the circuit. Switches are not modelled in SPICE, but we can duplicate their effect by specifying *pulse*-type sources, which we will describe presently, instead of conventional dc sources. SPICE can also generate plots of transient voltages or currents versus time, so we will discuss the .PLOT control statement used for that purpose.

.TRAN Control Statement

SPICE analyzes transients by computing the value of a voltage or current *at specific instants of time*, over a total time period that must be specified by the user. Therefore, the user must have some insight into the nature and duration of a transient under investigation, so as not to specify too short or too long a time period. If too little time is specified, the complete transient waveform will not be obtained. If too long a time is specified, the transient portion will not be printed with enough resolution to reveal how it changes with time. For example, if the transient voltage across a capacitor is computed for 100 s in a circuit whose time constant is 1 μs, the voltage variation during the first five time constants (5 μs) will be lost, because the time interval between each computed value might well be 1 s. Sometimes it is necessary to perform trial-and-error runs of a SPICE program to learn the optimum time over which a transient should be computed.

The basic format of the .TRAN statement is

.TRAN *TSTEP TSTOP*

where *TSTEP* is the time increment between which values are printed or plotted, and *TSTOP* is the total time period over which the transient is computed. For example, the statement

.TRAN 5M 0.2

will cause SPICE to compute a transient over the time interval from $t = 0$ s to $t = 0.2$ s, and the output will be printed or plotted in 5 ms intervals. SPICE will not print or plot more than 201 values, so the combination of *TSTEP* and *TSTOP* must be chosen keeping that limit in mind. (The limit can be increased to any value L by using the .OPTIONS control statement and specifying LIMPTS = L.)

In some investigations, it is important to realize that the time increment between which transient values are *computed* is not necessarily the same as the time increment, *TSTEP*, between which values are printed or plotted. Letting Δt represent the time interval between which values are computed, SPICE automatically selects Δt according to the criterion:

$$\Delta t = \min\{TSTEP, (TSTOP - TSTEP)/50\} \qquad (A.1)$$

For example, in the statement .TRAN 5M 0.2, we have

$$TSTEP = 5 \text{ ms}$$

and

$$(TSTOP - TSTEP)/50 = \frac{0.2 \text{ s} - 5 \text{ ms}}{50} = 3.9 \text{ ms}$$

Thus, Δt is selected to be the minimum of 5 ms and 3.9 ms, namely, 3.9 ms. The reason this knowledge is important is that we occasionally study circuits in which a source is switched in and then out again (a pulse-type input), and Δt must always be smaller than the total time that the source remains in the circuit. The .TRAN statement can be modified to include a specification for the maximum permissible value of Δt, *TMAX*, as well as for specification of a time other than $t = 0$ at which the first output value is printed or plotted:

$$.TRAN \; TSTEP \; TSTOP \; TSTART \; TMAX$$

This statement causes SPICE to print or plot values over the time interval from $t = TSTART$ to $t = TSTOP$ in increments of *TSTEP*, and to compute values at time increments no greater than *TMAX*. The default value of *TSTART* is 0, and the default value of *TMAX* is given by equation (A.1).

Pulse and Step Inputs

As noted previously, we cannot model a switch in SPICE. To study transients resulting from a voltage source that is switched into a circuit, we must model the voltage source as a *pulse*. Figure A.7(a) shows a very general pulse, one having a time delay (*TD*) that elapses before the voltage begins to rise linearly with time, a *rise time* (*TR*) that represents the time required for the voltage to change from *V1* volts to *V2* volts, a *pulse width* (*PW*), and a *fall time* (*TF*), during which the pulse falls from *V2* volts to *V1* volts. In some applications, we study repetitive pulses, and the period (repetition interval) of the pulses is represented in the figure by *PER*.

The general form used to define a voltage pulse generator is

$$V******* \; N+ \; N- \; PULSE(\; V1 \; V2 \; TD \; TR \; TF \; PW \; PER\;)$$

where ******* is an arbitrary sequence of characters, $N+$ and $N-$ are the node numbers of the positive and negative terminals of the pulse generator, and the other parameters are as defined in Figure A.7(a). A current pulse can also be modelled by changing V******* to I******* and by specifying appropriate current levels *I1* and *I2*, instead of *V1* and *V2*.

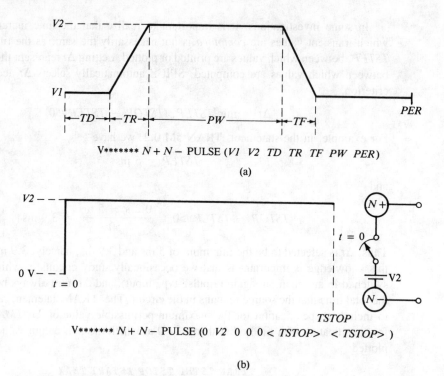

V********* N + N − PULSE (*V1* *V2* *TD* *TR* *TF* *PW* *PER*)

(a)

V********* N + N − PULSE (0 *V2* 0 0 0 < *TSTOP*> < *TSTOP*>)

(b)

FIGURE A.7 Modeling pulse and step voltages in SPICE. (a) General pulse. (b) A pulse (step) that is switched at $t = 0$ and remains in the circuit for the duration of the transient computation. *TSTOP* is the vaue specified in a .TRAN statement; *PW* and *PER* default to *TSTOP*.

When a dc source is switched into a circuit at $t = 0$ by an ideal switch, there is no time delay, the rise time is zero, and the voltage can remain connected to the circuit indefinitely. The voltage we have described is called a *step*, and is illustrated in Figure A.7(b). Note that we assume that $V1 = 0$ V. For modeling purposes, we wish to leave this modified pulse connected to the circuit for the entire duration of the transient com-

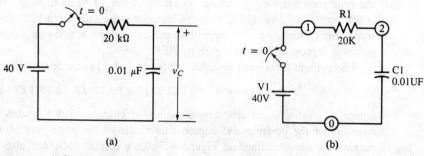

FIGURE A.8 (Example A.4)

putation, so the period is the same value of *TSTOP* that is specified in the .TRAN statement. In other words, the pulse is *not* repeated during the computation. For the same reason, the pulsewidth (*PW*) equals *TSTOP*. Thus, a step voltage is specified by

```
V******* N+ N- PULSE(0 V2 0 0 0 <TSTOP> <TSTOP>)
```

Note that the pulse width *PW* and the period *PER* can be optionally specified as *TSTOP*, since SPICE will default those parameters to *TSTOP* if they are omitted. To illustrate, suppose that we wish to switch a 15-V dc source named VS into a circuit at $t = 0$ and obtain values of a transient waveform from $t = 0$ to $t = 0.1$ s in 1-ms intervals. Assuming the dc source is connected between nodes 1 and 0, the following combination can be used:

```
VS 1 0 PULSE(0 15 0 0 0)
.TRAN 1M 0.1
```

In this example, the pulse width and period of the pulse both default to 0.1 s.

Example A.4

Use SPICE to compute and print values of the capacitor voltage v_C in Figure A.8(a) after the switch is closed at $t = 0$.

SOLUTION Figure A.8(b) shows the circuit after node numbers and component designations have been assigned. The time constant of the circuit is

$$\tau = RC = (20 \times 10^3)(0.01 \times 10^{-6}) = 0.2 \text{ ms}$$

Therefore, to obtain values of v_C over a full five time constants, we should make *TSTOP* equal to 5(0.2 ms) = 1.0 ms. We will make *TSTEP* equal to 1.0 ms/20 = 50 µs, and thus obtain 21 values of v_C (counting the value at $t = 0$). Following is the SPICE input file:

```
EXAMPLE A4
V1 1 0 PULSE(0 40 0 0 0)
R1 1 2 20K
C1 2 0 .01UF
.TRAN 50US 1MS
.PRINT TRAN V(2)
.END
```

Note that we must use TRAN in the .PRINT statement to specify the analysis type. When the program is run, we find that the results of certain preliminary computations are printed before the transient voltage values are listed. These results are not relevant to our investigation, so only the table of time points and voltage values printed by SPICE is shown in Printout 3.

```
 ****        TRANSIENT ANALYSIS                TEMPERATURE =   27.000 DEG C

****************************************************************************


      TIME           V(2)

   0.000E+00      0.000E+00
   5.000E-05      4.596E+00
   1.000E-04      1.243E+01
   1.500E-04      1.851E+01
   2.000E-04      2.328E+01
   2.500E-04      2.697E+01
   3.000E-04      2.987E+01
   3.500E-04      3.210E+01
   4.000E-04      3.386E+01
   4.500E-04      3.521E+01
   5.000E-04      3.627E+01
   5.500E-04      3.710E+01
   6.000E-04      3.774E+01
   6.500E-04      3.824E+01
   7.000E-04      3.863E+01
   7.500E-04      3.893E+01
   8.000E-04      3.917E+01
   8.500E-04      3.935E+01
   9.000E-04      3.950E+01
   9.500E-04      3.961E+01
   1.000E-03      3.969E+01
```

PRINTOUT 3

We see that V(2) (i.e., v_C) is 0 V at $t = 0$ and rises towards 40 V with the passage of time, as expected. Its value after five time constants (at $t = 1$ ms) is 39.69 V.

.PLOT Statement

The .PLOT control statement can be used to generate many different types of plots, depending on the *plot type* specified in the statement. The general format is

```
.PLOT PLTYPE OV
```

where *PLTYPE* is the plot type and *OV* is an output variable, such as a voltage or current. As in the .PRINT statement, more than one output variable can be specified (up to 8). To obtain a plot of a transient voltage or current versus time, we make the plot type TRAN. To illustrate, the statement

```
.PLOT TRAN V(1) I(VP)
```

will produce plots of the voltage at node 1 and the current in voltage source VP, each versus time, on one graph. To obtain two separate plots of these variables, two separate .PLOT statements are used. Although permissible, it is not necessary to write both a .PRINT and a .PLOT statement for the same variable, since SPICE prints a list of the values it plots.

As illustrated in the next example, SPICE automatically scales all plots, and prints scale values along the margin.

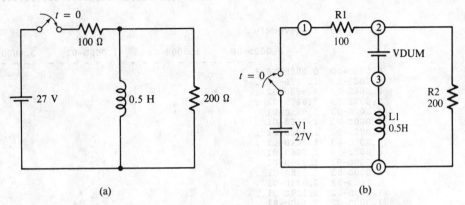

FIGURE A.9 (Example A.5)

Example A.5

Use SPICE to obtain separate plots of the transient voltage across the 200-Ω resistor and the transient current in the 0.5-H inductor in Figure A.9(a). The plots should cover the time period from $t = 0$ to $t = 25$ ms in 1-ms intervals.

SOLUTION Figure A.9(b) shows the circuit after node numbers and component designations have been assigned. Notice that a dummy voltage source, VDUM, has been inserted in series with the inductor so that the current in the inductor can be obtained. Following is the SPICE input file:

```
EXAMPLE A5
V1 1 0 PULSE(0 27 0 0 0)
R1 1 2 100
VDUM 2 3
L1 3 0 0.5H
R2 2 0 200
.TRAN 1M 25M
.PLOT TRAN I(VDUM)
.PLOT TRAN V(2)
.END
```

The SPICE-generated plots are shown in Printouts 4 and 5. It can be seen that the current in the inductor rises with time (approaching a steady-state value of 27 V/100 Ω = 0.27 A). Notice the scale factors, ranging from 0 to 0.4. The plot of the voltage across the 200-Ω resistor, V(2), which is the same as the voltage across the inductor, is misleading in one respect. Since the inductor theoretically behaves like an open circuit at the instant the switch is closed, V(2) should jump *immediately* to (200/300)27 V = 18 V. However, the SPICE results show that V(2) is 0 at $t = 0$ and is 16.85 V 1 ms later. Recall that SPICE computes transient values at increments of Δt seconds. Since Δt cannot be zero, it is impossible to compute a value that changes instantaneously. Thus, SPICE output must be examined with caution in the vicinity of a time point where a variable theoretically jumps from one value to another in zero time. If we had used a smaller value of *TSTEP* in the .TRAN statement, we would have obtained a value of V(2) closer to 18 V after the first time increment.

```
****        TRANSIENT ANALYSIS              TEMPERATURE =   27.000 DEG C

***********************************************************************************

     TIME      I(VDUM)

                   0.000D+00   1.000D-01   2.000D-01   3.000D-01  4.000D-01
              - - - - - - - - - - - - - - - - - - - - - - - - - - - - - - -
  0.000D+00   0.000D+00 *          .           .           .          .
  1.000D-03   1.719D-02 . *        .           .           .          .
  2.000D-03   4.863D-02 .     *    .           .           .          .
  3.000D-03   7.627D-02 .        * .           .           .          .
  4.000D-03   1.005D-01 .          *           .           .          .
  5.000D-03   1.216D-01 .          .   *       .           .          .
  6.000D-03   1.402D-01 .          .       *   .           .          .
  7.000D-03   1.564D-01 .          .         * .           .          .
  8.000D-03   1.706D-01 .          .           *           .          .
  9.000D-03   1.830D-01 .          .           . *         .          .
  1.000D-02   1.938D-01 .          .           .   *       .          .
  1.100D-02   2.034D-01 .          .           .     *     .          .
  1.200D-02   2.117D-01 .          .           .      *    .          .
  1.300D-02   2.190D-01 .          .           .       *   .          .
  1.400D-02   2.253D-01 .          .           .        *  .          .
  1.500D-02   2.309D-01 .          .           .         * .          .
  1.600D-02   2.358D-01 .          .           .          *.          .
  1.700D-02   2.401D-01 .          .           .           *          .
  1.800D-02   2.438D-01 .          .           .           .*         .
  1.900D-02   2.471D-01 .          .           .           . *        .
  2.000D-02   2.499D-01 .          .           .           .  *       .
  2.100D-02   2.524D-01 .          .           .           .   *      .
  2.200D-02   2.546D-01 .          .           .           .    *     .
  2.300D-02   2.566D-01 .          .           .           .    *     .
  2.400D-02   2.582D-01 .          .           .           .     *    .
  2.500D-02   2.597D-01 .          .           .           .     *    .
              - - - - - - - - - - - - - - - - - - - - - - - - - - - - - - -
```

PRINTOUT 4

```
****        TRANSIENT ANALYSIS              TEMPERATURE =   27.000 DEG C

***********************************************************************************

     TIME      V(2)

                   0.000D+00   5.000D+00   1.000D+01   1.500D+01  2.000D+01
              - - - - - - - - - - - - - - - - - - - - - - - - - - - - - - -
  0.000D+00   0.000D+00 *          .           .           .          .
  1.000D-03   1.685D+01 .          .           .           . *        .
  2.000D-03   1.476D+01 .          .           .         *. .         .
  3.000D-03   1.292D+01 .          .           .       *   .          .
  4.000D-03   1.130D+01 .          .           .     *     .          .
  5.000D-03   9.891D+00 .          .        *  .           .          .
  6.000D-03   8.656D+00 .          .      *    .           .          .
  7.000D-03   7.575D+00 .          .    *      .           .          .
  8.000D-03   6.629D+00 .          .  *        .           .          .
  9.000D-03   5.801D+00 .          . *         .           .          .
  1.000D-02   5.077D+00 .          *           .           .          .
  1.100D-02   4.443D+00 .        * .           .           .          .
  1.200D-02   3.888D+00 .       *  .           .           .          .
  1.300D-02   3.403D+00 .      *   .           .           .          .
  1.400D-02   2.978D+00 .     *    .           .           .          .
  1.500D-02   2.606D+00 .    *     .           .           .          .
  1.600D-02   2.281D+00 .   *      .           .           .          .
  1.700D-02   1.996D+00 .   *      .           .           .          .
  1.800D-02   1.747D+00 .  *       .           .           .          .
  1.900D-02   1.528D+00 . *        .           .           .          .
  2.000D-02   1.338D+00 . *        .           .           .          .
  2.100D-02   1.171D+00 .*         .           .           .          .
  2.200D-02   1.024D+00 .*         .           .           .          .
  2.300D-02   8.965D-01 .*         .           .           .          .
  2.400D-02   7.846D-01 .*         .           .           .          .
  2.500D-02   6.862D-01 .*         .           .           .          .
              - - - - - - - - - - - - - - - - - - - - - - - - - - - - - - -
```

PRINTOUT 5

Example 9.18 is another example of a transient analysis. In this case, the transient current in an RC circuit is plotted by SPICE. Example 9.19 shows how SPICE can be used to determine the steady-state voltages across the capacitors in a series–parallel RC circuit. Example 11.13 shows how a SPICE transient analysis can be used to determine steady-state voltages and currents in a series–parallel RLC circuit.

A.7 AC Analysis

The .AC control statement tells SPICE to perform an ac analysis of a circuit and specifies the frequency(ies) at which the analysis is to be performed. In order to obtain an ac analysis, it is necessary that the circuit contain at least one ac voltage source. An ac voltage source is defined by

$$V******* \ N+ \ N- \ AC \ <ACMAG> \ <ACPHASE>$$

where $*******$ is an arbitrary sequence of characters, $N+$ and $N-$ are the node numbers of the positive and negative terminals of the source (see Section 15.3), and $<ACMAG>$ and $<ACPHASE>$ are optional specifications of the magnitude and phase angle of the source. $ACPHASE$ is in degrees. If omitted, the values of $ACMAG$ and $ACPHASE$ default to 1 V and 0°, respectively. (In PSpice, $ACMAG$ does not default, so a value must be specified for it. An ac current source can be defined by substituting $I*******$ for $V*******$. All ac sources are assumed to be sinusoidal.

Using the .AC control statement, it is possible to specify a *range* of frequencies, and an increment between frequencies, at each of which an analysis is to be performed. This feature is useful for investigating the *frequency response* of a circuit, which we will describe in Section A.9. For our purposes now, we will show how the statement can be written to obtain an ac analysis at a *single* frequency. The general form of the .AC statement is

$$.AC \ VARTYPE \ N \ FSTART \ FSTOP$$

where *VARTYPE* is the type of frequency variation desired, N is related to the number of frequencies used in the computation, and *FSTART* and *FSTOP* are the smallest and largest frequencies in the range over which the analysis is to be performed. *VARTYPE* must be one of DEC, OCT, or LIN. For analysis at a single frequency, it does not matter which of these variation types is specified. Also for analysis at a single frequency, we set $N = 1$ and make *FSTART* and *FSTOP* both equal to the frequency at which we wish the analysis to be performed. For example, the control statement

$$.AC \ DEC \ 1 \ 2K \ 2K$$

will tell SPICE to perform an ac analysis at a frequency of 2 kHz.

The .PRINT control statement in an ac analysis must specify AC as the analysis type. The variables whose values are printed in an ac analysis can be specified in the .PRINT statement in the same way they are for a dc analysis. For example, the statement

$$.\texttt{PRINT AC V(2) I(VDUM)}$$

will tell SPICE to print the magnitude of the ac voltage between node 2 and node 0 and the magnitude of the ac current in dummy voltage source VDUM. In addition, each variable designation can be modified to obtain special values, according to the following scheme:

VR	real part
VI	imaginary part
VM	magnitude, $|V|$
VP	phase (degrees)
VDB	$20 \log_{10} |V|$

The same values can be obtained for ac currents by substituting I for V in the list above. Note that .PRINT defaults to the magnitude of a variable (VM or IM) if no special designation is given. To illustrate, the statement

$$.\texttt{PRINT AC VR(2) II(VDUM) VP(3)}$$

will tell SPICE to print the real part of the ac voltage at node 2, the imaginary part of the current in VDUM, and the phase angle of the voltage at node 3.

Example 12.25 illustrates a SPICE analysis of a simple ac circuit. Example 14.21 shows a SPICE analysis of a series RLC ac circuit. Example A.6, which follows, shows a SPICE analysis of a series–parallel ac circuit.

Example A.6

Use SPICE to find the magnitude and angle of the ac voltage across each component in Figure A.10(a). The frequency of both ac sources is 3 kHz.

SOLUTION Figure A.10(b) shows the circuit with node numbers and component designations assigned. Following is the SPICE input file:

```
EXAMPLE A6
V1 1 0 AC 15 45
I1 0 3 AC 20M 0
R1 1 2 470
L1 2 0 30MH
C1 2 3 0.1UF
R2 3 0 330
.AC DEC 1 3K 3K
.PRINT AC V(1,2) V(2) V(2,3) V(3)
.PRINT AC VP(1,2) VP(2) VP(2,3) VP(3)
.END
```

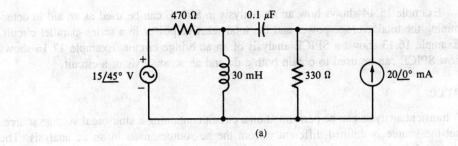

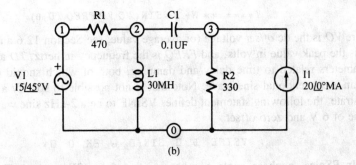

FIGURE A.10 (Example A.6)

As in a transient analysis, SPICE prints the results of certain preliminary computations when an ac analysis is performed. These results have been omitted from the printed output shown in Printout 6.

```
****     AC ANALYSIS                    TEMPERATURE =   27.000 DEG C

***********************************************************************

     FREQ      V(1,2)      V(2)      V(2,3)      V(3)

  3.000E+03   2.954E+00  1.409E+01  9.983E+00  4.189E+00
*******06/30/86 ********  SPICE 2G.6   3/15/83 *******08:44:50*****

EXAMPLE A6

****     AC ANALYSIS                    TEMPERATURE =   27.000 DEG C

***********************************************************************

     FREQ      VP(1,2)     VP(2)     VP(2,3)     VP(3)

  3.000E+03  -2.150E+01  5.609E+01  5.197E+01  6.594E+01
```
PRINTOUT 6

Example 15.14 shows how an ac analysis in SPICE can be used as an aid in determining the total average power and the total reactive power in a series–parallel circuit. Example 16.13 shows a SPICE analysis of an ac bridge circuit. Example 17.16 shows how SPICE can be used to obtain both a dc and an ac analysis of a circuit.

The SIN Source

A transient analysis can be performed on a circuit containing a sinusoidal voltage source, but the source is defined differently from the ac sources used in an ac analysis. The general form is

$$V******* \ N+ \ N- \ SIN(\ VO \ VP \ FREQ \ TD \ \theta)$$

where *VO* is the dc *offset* voltage (or average value; see Section 12.6 and Figure 12.31), *VP* is the peak value in volts, and *FREQ* is the frequency in hertz. *TD* and Θ are special parameters related to time delay and damping, both of which should be set to zero to obtain a conventional sine wave. Note that is not possible to specify a phase angle. To illustrate, the following statement defines VSINE to be a 2-kHz sine wave having a peak value of 6 V and zero offset:

$$VSINE \ 1 \ 0 \ SIN(0 \ 6 \ 2K \ 0 \ 0)$$

In PSpice, a phase angle can be specified for a SIN source. The format (for a sinusoidal voltage) is

$$V******* \ N \ + \ N- \ SIN(\ VO \ VP \ FREQ \ TD \ \theta \ PH)$$

where *PH* is phase angle in degrees and all other parameters are the same as in the SIN source description in Berkeley SPICE.

A.8 Controlled (Dependent) Sources

Any of the four types of controlled sources described in Section 17.2 can be modeled in SPICE. Figure A.11 shows the four types of sources and the general format for defining each.

A voltage-controlled current source is defined by

$$G******* \ N+ \ N- \ NC+ \ NC- \ TRANSCON$$

where $*******$ is an arbitrary sequence of characters, $N+$ and $N-$ are the node numbers of the positive and negative terminals of the controlled current source, $NC+$ and $NC-$ are the node numbers of the positive and negative terminals of the controlling voltage (which can be a voltage drop or a voltage source), and *TRANSCON* is the value of the transconductance of the source:

$$TRANSCON = \frac{\Delta I_{out}}{\Delta V_{in}}$$

The transconductance is the slope of the straight line that relates output (controlled) current to input (controlling) voltage, as illustrated in Figure A.11(a). Note carefully that

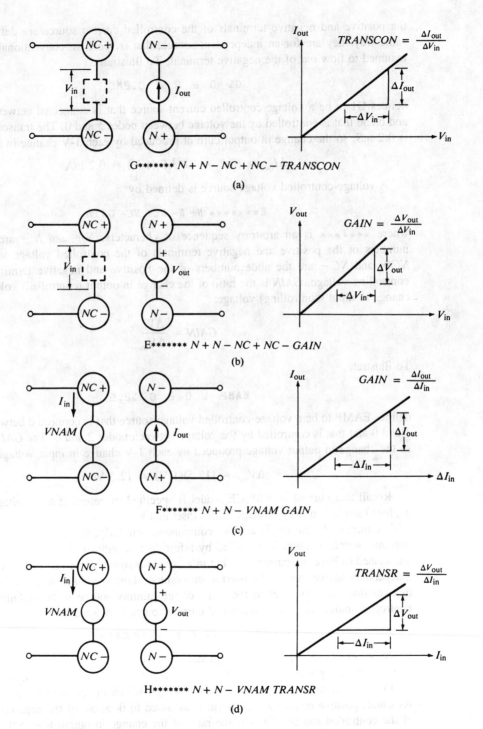

FIGURE A.11 Controlled sources and their SPICE definitions. (a) Voltage-controlled current source. (b) Voltage-controlled voltage source. (c) Current-controlled current source. (d) Current-controlled voltage source.

the positive and negative terminals of the controlled current source are defined in the same way they are for an independent source; that is, positive conventional current is assumed to flow *out* of the negative terminal. To illustrate,

$$\text{G1} \quad 0 \quad 6 \quad 5 \quad 0 \quad 0.2\text{MS}$$

defines G1 to be a voltage-controlled current source that is connected between nodes 0 and 6 and that is controlled by the voltage between nodes 5 and 0. The transconductance is 0.2 mS, so the change in output current produced by each 1-V change in input is

$$\Delta I_{\text{out}} = (0.2 \times 10^{-3} \text{ S})(1 \text{ V}) = 0.2 \text{ mA}$$

A voltage-controlled voltage source is defined by

$$\text{E}{*}{*}{*}{*}{*}{*}{*} \quad N+ \quad N- \quad NC+ \quad NC- \quad GAIN$$

where $*{*}{*}{*}{*}{*}{*}$ is an arbitrary sequence of characters, $N+$ and $N-$ are the node numbers of the positive and negative terminals of the controlled voltage source, and $NC+$ and $NC-$ are the node numbers of the positive and negative terminals of the controlling voltage. *GAIN* is the ratio of the change in output (controlled) voltage to the change in input (controlling) voltage:

$$GAIN = \frac{\Delta V_o}{\Delta V_{\text{in}}}$$

To illustrate,

$$\text{EAMP} \quad 1 \quad 0 \quad 2 \quad 0 \quad 12.5$$

defines EAMP to be a voltage-controlled voltage source that is connected between nodes 1 and 0 and that is controlled by the voltage between nodes 2 and 0. The *GAIN* is 12.5, so the change in output voltage produced by each 1-V change in input voltage is

$$\Delta V_{\text{out}} = (12.5)(1 \text{ V}) = 12.5 \text{ V}$$

Recall that current in a SPICE model is specified by referring to a voltage source, as, for example, a dummy voltage source inserted like an ammeter for the express purpose of determining the current in another component. Similarly, the controlling current in a current-controlled source is specified by referring to a voltage source. For example, if we wished to have the current flowing in a certain resistor be the controlling current for a dependent source, we would insert a zero-valued, dummy voltage source in series with that resistor and then refer to the name of the dummy source in the definition of the controlled source. A current-controlled current source is defined by

$$\text{F}{*}{*}{*}{*}{*}{*}{*} \quad N+ \quad N- \quad VNAM \quad GAIN$$

where $*{*}{*}{*}{*}{*}{*}$ is an arbitrary sequence of characters, $N+$ and $N-$ are the node numbers of the positive and negative terminals of the controlled current source, and *VNAM* is the name of the voltage source through which the controlling current flows. As usual, positive conventional current is assumed to flow out of the negative terminal of the controlled source. *GAIN* is the ratio of the change in output (controlled) current to the change in input (controlling) current:

$$GAIN = \frac{\Delta I_{out}}{\Delta I_{in}}$$

To illustrate,

```
FX 3 7 VDUM 0.4
```

defines FX to be a current-controlled current source that is connected between nodes 3 and 7 and that is controlled by the current flowing in the voltage named VDUM. The gain is 0.4, so the change in output current for each 1-A change in input current is

$$\Delta I_o = (0.4)(1 \text{ A}) = 0.4 \text{ A}$$

A current-controlled voltage source is defined by

```
H******* N+ N- VNAM TRANSR
```

where ******* is an arbitrary sequence of characters, $N+$ and $N-$ are the node numbers of the positive and negative terminals of the controlled voltage source, and *VNAM* is the name of the voltage source through which the controlling current flows. *TRANSR* is the so-called *transresistance* of the source, which relates change in output (controlled) voltage to change in input (controlling) current:

$$TRANSR = \frac{\Delta V_{out}}{\Delta I_{in}}$$

To illustrate,

```
H2 3 12 VSENS 18
```

defines H2 to be a current-controlled voltage source that is connected between terminals 3 and 12 and that is controlled by the current in voltage source VSENS. The trans-resistance is 18 Ω, so the change in output voltage for each 1-A change in input current is

$$\Delta V_{out} = (18 \text{ }\Omega)(1 \text{ A}) = 18 \text{ V}$$

Example A.7

Figure A.12(a) shows the equivalent circuit of a certain transistor amplifier, containing a voltage-controlled voltage source and a current-controlled current source. The input voltage to the amplifier is $v_{in} = 20 \underline{/0°}$ mV. Use SPICE to find the magnitude of the output voltage, v_{out}.

SOLUTION Figure A.12(b) shows the circuit after node numbers and component designations have been assigned. Notice that a dummy voltage source has been inserted in series with R1 so that we can define the current i_1 controlling dependent current source Fl. We will use SPICE to perform an ac analysis at the (arbitrary) frequency 1 kHz. Following is the SPICE input file:

```
EXAMPLE A7
VIN 1 0 AC 20MV 0
VDUM 1 2
```

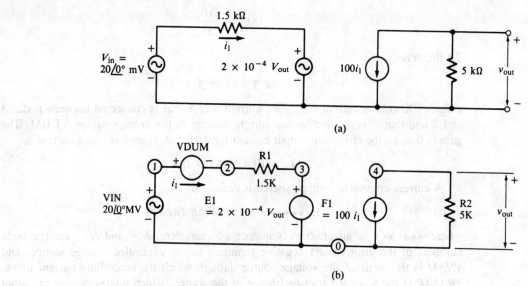

FIGURE A.12 (Example A.7)

```
R1 2 3 1.5K
R2 4 0 5K
E1 3 0 4 0 2E-4
F1 4 0 VDUM 100
.AC DEC 1 1K 1K
.PRINT AC V(4)
.END
```

The results of the analysis, listed in Printout 7, show that the magnitude of v_{out} is 7.143 V.

```
****    AC ANALYSIS                       TEMPERATURE =   27.000 DEG C
****************************************************************************


        FREQ        V(4)

        1.000E+03   7.143E+00
```
PRINTOUT 7

Example 17.17 shows how SPICE can be used to find the Thévenin equivalent circuit of a circuit containing a controlled voltage source.

A.9 Frequency Response

By specifying a frequency range in the .AC control statement, we can obtain the frequency *response* of a circuit, that is, a list (or plot) of voltages, currents and/or phase angles at any point in a circuit, as frequency changes. Recall that the general format of the .AC control statement is

.AC VARTYPE N FSTART FSTOP

where *VARTYPE* is one of DEC, OCT, or LIN and *N* is related to the number of frequencies at which the frequency response is computed over the range from *FSTART* to *FSTOP*. When the *VARTYPE* is LIN, the frequency variation is linear and N is the total number of frequencies at which the response is computed (counting *FSTART* and *FSTOP*). Thus, the interval between frequencies is $(FSTOP - FSTART)/(N - 1)$. For example, the statement

.AC LIN 21 100 1000

will tell SPICE to compute the frequency response at 21 equally spaced frequencies over the range from 100 Hz through 1 kHz. The frequency interval between each computation will be $(1000 - 100)/20 = 45$ Hz, so computations will be performed at 100 Hz, 145 Hz, 190 Hz, . . . , 1000 Hz.

When the DEC (decade) or OCT (octave) variation types are specified, the frequency response is computed at logarithmically spaced intervals. In these cases, N is the total number of frequencies *per decade* or *per octave*. For example, the statement

.AC DEC 10 100 10K

will tell SPICE to compute the frequency response at ten frequencies in each of the decades 100 Hz to 1 kHz and 1 kHz through 10 kHz. The frequencies will be in *one-tenth decade intervals,* so each interval will have a different value. The frequencies at one-tenth decade intervals are 10^x, $10^{x+0.1}$, $10^{x+0.2}$, . . . , where $x = \log_{10} (FSTART)$. In general, the frequencies at which SPICE computes frequency response for the DEC variation type are

$$10^x, \ 10^{x+1/N}, \ 10^{x+2/N}, \ . \ . \ .$$

where $x = \log_{10} (FSTART)$. The last frequency in the sequence above is not necessarily *FSTOP*, but is the first frequency greater than or equal to *FSTOP*. The frequencies at which SPICE computes frequency response for the OCT variation type are

$$2^x, \ 2^{x+1/N}, \ 2^{x+2/N}, \ . \ . \ .$$

where $x = \log_2 (FSTART)$.

Using the .PLOT statement, we can obtain a variety of linear, semilog, and log-log plots of frequency response. If the variation type specified in the .AC statement is LIN, then the frequency axis of the plot will be linear. If the variation type is DEC or OCT, the frequency axis will be logarithmic. The values of a variable specified in the .PLOT statement will be plotted on a logarithmically scaled axis, unless dB, real part, imaginary part, or phase values are requested. Note that plotting dB values on a linear axis will produce a frequency response that has the same appearance as plotting magnitude values on a logarithmic axis. Following are some examples of .AC and .PLOT statements, and the types of frequency response plots that each combination produces:

```
.AC DEC 10 50 5K        log voltage magnitude vs. log frequency
.PLOT AC V(2)
.AC LIN 20 100 20K      linear phase vs. linear frequency
.PLOT AC VP(3,4)
.AC LIN 5 20 10K        log current magnitude vs. linear frequency
.PLOT AC IM(3)
.AC OCT 6 15K 2MEG      linear dB voltage vs. log frequency
.PLOT AC VDB(1,4)
```

Example A.8

Use SPICE to obtain the following frequency response plots for the low-pass RC filter shown in Figure A.13(a).

(a) the magnitude of v_o versus frequency, on logarithmically scaled axes (log-log plot)
(b) the phase angle of v_o relative to v_{in} on a linear axis versus logarithmically scaled frequency (semilog plot)

The frequency range of both plots should be from 100 Hz through 10 kHz.

SOLUTION Figure A.13(b) shows the circuit after the node numbers and component designations have been assigned. The SPICE input file is as follows.

```
EXAMPLE A8
VIN 1 0 AC 18 0
R1 1 2 850
C1 2 0 0.2UF
.AC DEC 10 100 10K
```

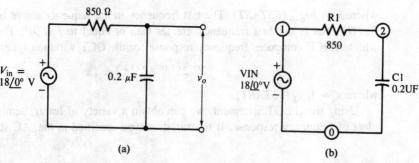

(a)   (b)

FIGURE A.13 (Example A.8)

```
.PLOT AC V(2)
.PLOT AC VP(2)
.END
```

Shown in Printouts 8 and 9 are the plots produced by the ac analysis. Note that VP(2), which is the phase angle of v_o, reaches $-45°$ at a frequency somewhere between 794.3 Hz and 1 kHz, indicating that the cutoff frequency is in that interval. If it were necessary to find the exact cutoff frequency, additional computer runs could be performed with narrower frequency ranges specified in the .AC statement. For example, a second computer run specifying 20 frequencies in the interval from 700 Hz to 1 kHz would reveal a frequency at which the phase angle is very close to $-45°$. [Of course, in the simple low-pass circuit of this example, we know that the exact cutoff frequency is $f_2 = 1/(2\pi RC) = 936.2$ Hz.]

Example A.9

For the bandpass RLC circuit shown in Figure A.14(a), use SPICE to obtain

(a) a log-log plot of the magnitude of v_o and a semilog plot of the phase angle of v_o on the same frequency axis
(b) log-log plots of the magnitudes of v_L and v_C on the same set of axes

The frequency range of the plots should extend from about one decade below the center frequency to about one decade above the center frequency of the filter.

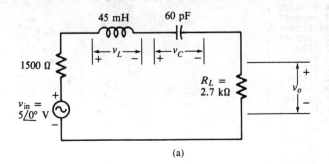

(a)

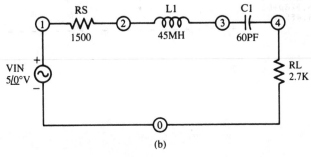

(b)

FIGURE A.14 (Example A.9)

```
    ****        AC ANALYSIS                    TEMPERATURE =    27.000 DEG C

        FREQ       V(2)

                              1.000D+00      3.162D+00      1.000D+01      3.162D+01  1.000D+02
                        - - - - - - - - - - - - - - - - - - - - - - - - - - - - - - - - - - -
    1.000D+02    1.790D+01  .                                             *            .
    1.259D+02    1.784D+01  .                                             *            .
    1.585D+02    1.775D+01  .                                             *            .
    1.995D+02    1.760D+01  .                                            *             .
    2.512D+02    1.739D+01  .                                           *              .
    3.162D+02    1.705D+01  .                                          *               .
    3.981D+02    1.656D+01  .                                        *                 .
    5.012D+02    1.587D+01  .                                      *                   .
    6.310D+02    1.493D+01  .                                   *                      .
    7.943D+02    1.373D+01  .                                *                         .
    1.000D+03    1.230D+01  .                             *                            .
    1.259D+03    1.074D+01  .                          *                               .
    1.585D+03    9.155D+00  .                      *                                    .
    1.995D+03    7.646D+00  .                  *                                        .
    2.512D+03    6.286D+00  .               *                                          .
    3.162D+03    5.110D+00  .            *                                             .
    3.981D+03    4.121D+00  .         *                                                .
    5.012D+03    3.305D+00  .      *                                                   .
    6.310D+03    2.642D+00  .    *                                                     .
    7.943D+03    2.107D+00  .  *                                                       .
    1.000D+04    1.678D+00  . *                                                        .
                        - - - - - - - - - - - - - - - - - - - - - - - - - - - - - - - - - - -
```

PRINTOUT 8

```
        FREQ       VP(2)

                          -1.500D+02     -1.000D+02     -5.000D+01     0.000D+00  5.000D+01
                        - - - - - - - - - - - - - - - - - - - - - - - - - - - - - - - - - - -
    1.000D+02   -6.097D+00  .                          .              .      *       .
    1.259D+02   -7.659D+00  .                          .              .     *        .
    1.585D+02   -9.608D+00  .                          .              .    *         .
    1.995D+02   -1.203D+01  .                          .              .   *          .
    2.512D+02   -1.502D+01  .                          .              . *            .
    3.162D+02   -1.866D+01  .                          .            *                .
    3.981D+02   -2.304D+01  .                          .         *                   .
    5.012D+02   -2.816D+01  .                          .      *                      .
    6.310D+02   -3.398D+01  .                          .   *                         .
    7.943D+02   -4.031D+01  .                          *                             .
    1.000D+03   -4.689D+01  .                       *  .                             .
    1.259D+03   -5.336D+01  .                    *     .                             .
    1.585D+03   -5.943D+01  .                 *        .                             .
    1.995D+03   -6.486D+01  .              *           .                             .
    2.512D+03   -6.956D+01  .            *             .                             .
    3.162D+03   -7.351D+01  .          *               .                             .
    3.981D+03   -7.677D+01  .         *                .                             .
    5.012D+03   -7.942D+01  .        *                 .                             .
    6.310D+03   -8.156D+01  .      *                   .                             .
    7.943D+03   -8.328D+01  .     *                    .                             .
    1.000D+04   -8.465D+01  .    *                     .                             .
                        - - - - - - - - - - - - - - - - - - - - - - - - - - - - - - - - - - -
```

PRINTOUT 9

SOLUTION From equation (18.21),

$$f_o = \frac{1}{2\pi\sqrt{LC}} = \frac{1}{2\pi\sqrt{(45 \times 10^{-3})(60 \times 10^{-12})}} = 96.858 \text{ kHz}$$

The frequencies one decade below and above f_o are therefore 9.6858 kHz and 968.58 kHz, respectively. We will obtain an ac analysis over the range from 10 kHz to 1 MHz. Figure A.14(b) shows the circuit after node numbers and component designations have been assigned. Following is the SPICE input file:

```
EXAMPLE A9
VIN 1 0 AC 5 0
RS 1 2 1500
L1 2 3 45MH
C1 3 4 60PF
RL 4 0 2.7K
.AC DEC 20 10K 1 MEG
.PLOT AC V(4) VP(4)
.PLOT AC V(2,3) V(3,4)
.END
```

Note that variables V(4) and VP(4) appear in the same .PLOT statement, as do V(2,3) and V(3,4). Thus, we will obtain two plots on each of two sets of axes.

The plots resulting from a run of the program are shown in Printouts 10 and 11. Note that SPICE uses different symbols to plot points corresponding to different variables. In the plots of V(4) and VP(4), two different axes are printed: a log axis for V(4) and a linear axis for VP(4). If two graphs intersect, as in the plots of V(2,3) and V(3,4), an *x* is printed at the point of intersection. Note also that plots showing more than one variable print the values of just one variable; the first one specified in the .PLOT statement. Of course, a .PRINT statement could be used to obtain a list of values of other variables, if desired.

```
LEGEND:

*:  V(4)
+:  VP(4)

     FREQ      V(4)

*)------------- 3.162D-02    1.000D-01    3.162D-01    1.000D+00   3.162D+00
                                       - - - - - - - - - - - - - - - - - - - - - - -
+)------------- -1.000D+02   -5.000D+01   0.000D+00    5.000D+01   1.000D+02

1.000D+04  5.144D-02  .            *                                   +    .
1.122D+04  5.787D-02  .              *                                 +    .
1.259D+04  6.516D-02  .               *                                +    .
1.413D+04  7.343D-02  .                *                               +    .
1.585D+04  8.285D-02  .                 *                              +    .
1.778D+04  9.362D-02  .                  *                             +    .
1.995D+04  1.060D-01  .                   *                            +    .
2.239D+04  1.203D-01  .                     *                          +    .
2.512D+04  1.369D-01  .                      *                         +    .
2.818D+04  1.565D-01  .                       *                        +    .
3.162D+04  1.799D-01  .                         *                       +   .
3.548D+04  2.081D-01  .                          *                      +   .
3.981D+04  2.431D-01  .                            *                    +   .
4.467D+04  2.876D-01  .                              *                  +   .
5.012D+04  3.463D-01  .                                *                +   .
5.623D+04  4.279D-01  .                                  *              +   .
6.310D+04  5.496D-01  .                                    *           +    .
7.079D+04  7.521D-01  .                                       *       +     .
7.943D+04  1.152D+00  .                                         *     +   * .
8.913D+04  2.177D+00  .                                           *  +      .
1.000D+05  2.967D+00  .                                            +     *  .
1.122D+05  1.482D+00  .                            +                    *   .
1.259D+05  8.929D-01  .                        +                     *      .
1.413D+05  6.258D-01  .                   +                        *        .
1.585D+05  4.756D-01  .          +                                *         .
1.778D+05  3.791D-01  .       +                                 *           .
1.995D+05  3.116D-01  .     +                                  *            .
2.239D+05  2.615D-01  .    +                                 *              .
2.512D+05  2.227D-01  .   +                                 *               .
2.818D+05  1.918D-01  .   +                               *                 .
3.162D+05  1.664D-01  .  +                               *                  .
3.548D+05  1.453D-01  .  +                             *                    .
3.981D+05  1.274D-01  . +                            *                      .
4.467D+05  1.121D-01  . +                           *                       .
5.012D+05  9.892D-02  . +                         *                         .
5.623D+05  8.747D-02  . +                        *                          .
6.310D+05  7.748D-02  . +                      *                            .
7.079D+05  6.871D-02  . +                     *                             .
7.943D+05  6.101D-02  . +                    *                              .
8.913D+05  5.421D-02  . +                  *                                .
1.000D+06  4.819D-02  . +               *                                   .
```

PRINTOUT 10

```
LEGEND:

*: V(2,3)
+: V(3,4)

   FREQ        V(2,3)

*+)------------  1.000D-02    1.000D-01    1.000D+00    1.000D+01  1.000D+02
- - - - - - - - - - - - - - - - - - - - - - - - - - - - - - - - - - - - - -
1.000D+04   5.386D-02   .              *    +                        .
1.122D+04   6.800D-02   .             *     +                        .
1.259D+04   8.590D-02   .            *       +                       .
1.413D+04   1.086D-01   .          *         +                       .
1.585D+04   1.375D-01   .         *          +                       .
1.778D+04   1.743D-01   .        *            +                      .
1.995D+04   2.215D-01   .       *             +                      .
2.239D+04   2.820D-01   .      *               +                     .
2.512D+04   3.602D-01   .     *                +                     .
2.818D+04   4.620D-01   .    *                  +                    .
3.162D+04   5.956D-01   .   *                   +                    .
3.548D+04   7.733D-01   .  *                    +                    .
3.981D+04   1.013D+00   .   *                   +                    .
4.467D+04   1.345D+00   .    *                   +                   .
5.012D+04   1.818D+00   .       *                +                   .
5.623D+04   2.520D+00   .          *             +.                  .
6.310D+04   3.632D+00   .              *         .+                  .
7.079D+04   5.576D+00   .                   *    . +                 .
7.943D+04   9.587D+00   .                        .*  +               .
8.913D+04   2.032D+01   .                        .      * +        X .
1.000D+05   3.107D+01   .                        .          * +      .
1.122D+05   1.741D+01   .                        .    *   +          .
1.259D+05   1.177D+01   .                        .  *   +            .
1.413D+05   9.257D+00   .                        *  +                .
1.585D+05   7.893D+00   .                    *   .+                  .
1.778D+05   7.059D+00   .                  *     + .                 .
1.995D+05   6.511D+00   .                *      +  .                 .
2.239D+05   6.131D+00   .               *      +   .                 .
2.512D+05   5.859D+00   .              *     +     .                 .
2.818D+05   5.660D+00   .             *     +      .                 .
3.162D+05   5.510D+00   .            *    +        .                 .
3.548D+05   5.397D+00   .            *   +         .                 .
3.981D+05   5.310D+00   .           *   +          .                 .
4.467D+05   5.244D+00   .           *  +           .                 .
5.012D+05   5.192D+00   .           * +            .                 .
5.623D+05   5.151D+00   .           *+             .                 .
6.310D+05   5.119D+00   .          *+              .                 .
7.079D+05   5.094D+00   .          +               .                 .
7.943D+05   5.075D+00   .         +*               .                 .
8.913D+05   5.059D+00   .        + *               .                 .
1.000D+06   5.047D+00   .       +  *               .                 .
- - - - - - - - - - - - - - - - - - - - - - - - - - - - - - - - - - - - - -
```

PRINTOUT 11

A.10 Transformers

A transformer is modeled in SPICE using three statements: one to specify the inductance and nodes of the primary winding, one to specify the inductance and nodes of the secondary winding, and one to specify the coefficient of coupling, k, between the windings. The primary and secondary windings are specified in the input data file in the same way as ordinary inductors are: using an L prefix, node numbers and the inductance in henries. The coefficient of coupling is specified in a statement that begins with K:

$$K^{*******} \quad LNAM1 \quad LNAM2 \quad k$$

where $LNAM1$ and $LNAM2$ are the names of the inductors that comprise the primary and secondary windings and k is the value of the coefficient of coupling $(0 < k \le 1)$. In an ideal transformer all magnetic flux in the primary is coupled to the secondary and $k = 1$. In order for SPICE to simulate an ideal iron-core transformer, the reactance of the primary winding must be much greater than any impedance in series with the voltage source driving it. Also, the reactance of the secondary winding must be much greater than any load impedance connected across it. To meet these requirements, it may be necessary to specify unrealistically large inductance values for the primary and secondary (depending on the frequency, since $|X_L| = 2\pi f L$). Following is an example of a SPICE specification for an ideal transformer (named KXFRMR) whose primary winding is LPRIM and whose secondary winding is LSEC:

```
LPRIM 5 0 20H

LSEC 6 0 0.2H

KXFRMR LPRIM LSEC 1
```

If operation of the transformer specified above is to be simulated at, say, 1 kHz, then any impedance in series with the primary winding must be much smaller than $2\pi f$(LPRIM) $= 2\pi(10^3 \text{ Hz})(20 \text{ H}) = 125.7 \text{ k}\Omega$, and any load impedance across the secondary must be much smaller than $2\pi f$(LSEC) $= 2\pi(10^3 \text{ Hz})(0.2 \text{ H}) = 1.257 \text{ k}\Omega$. Since there must be a dc path from every node to ground, it is *not* possible to isolate the primary and secondary windings from each other, as is done in real transformers. Both the primary and secondary windings must be connected directly or through a dc path to node 0.

The turns ratio of an ideal transformer is related to the inductance of the primary and secondary windings by

$$\frac{N_p}{N_s} = \sqrt{\frac{L_p}{L_s}}$$

where L_p and L_s are the inductances of the primary and secondary, respectively. For example, the turns ratio of the transformer in the foregoing example is found to be

$$\frac{N_p}{N_s} = \sqrt{\frac{20 \text{ H}}{0.2 \text{ H}}} = \sqrt{100} = 10$$

Example 19.25 illustrates the use of SPICE to simulate a transformer.

A.11 Subcircuits

In large, complex circuits, a network, or a portion of a network, may be duplicated several times. SPICE permits a user to define the components of a *subcircuit* and then to use that subcircuit as many times as needed, without having to redefine all of its components at every location where it appears. For example, if a circuit contained four identical RLC networks, we could define the nodes and the components of the RLC network just once, in a subcircuit, and then specify that subcircuit at each location where it appears in the original circuit. A subcircuit is similar in concept to what is called a *subroutine* in the terminology of computer science. When we wish to specify a subcircuit (or subroutine) in a program, we say that we *call* it.

To define a subcircuit, we use the statement

```
.SUBCKT NAM N1 <N2 . . . .>
```

where *NAM* is an arbitrary name, and *N1*, *N2*, . . . , are the numbers of the subcircuit nodes that are to be connected in the original circuit (which we will hereafter call the *main* circuit). None of *N1*, *N2*, . . . , can be 0. Components in the subcircuit are defined in the lines immediately following the .SUBCKT statement. The last line in the subcircuit definition must be the statement

```
.ENDS <NAM>
```

where *NAM* is the name of the subcircuit being ended.

To call a subcircuit, that is, to insert it in a main program, we write

```
XN N1 <N2. . .> NAM
```

where *N* is a different integer number for each call. Node numbers *N1*, *N2*, . . . , are the *main circuit* nodes to which the subcircuit nodes given in the .SUBCKT statement are connected, in the same order. For example, the combination

```
X1 5 9 NETWK
       .
       .
       .
.SUBCKT NETWK 3 6
```

specifies that subcircuit nodes 3 and 6 are to be connected to main circuit nodes 5 and 9, respectively.

Node numbers and component designations in a subcircuit do not have to be different from any of those used in a main circuit. SPICE treats a subcircuit as being completely independent of a main circuit. The exception is node 0, which is assumed to be the same node in both the subcircuit and the main circuit. (In the terminology of computer science, subcircuit nodes are said to be *local* to the subcircuit, and node 0 is said to be *global*.) Remember that node 0 cannot appear in the .SUBCKT statement; if node 0 appears in the subcircuit *definition*, SPICE will automatically assume that it is connected to node 0 in the main circuit.

Control statements cannot appear within a subcircuit definition (between .SUBCKT and .ENDS). However, *other* subcircuit definitions can appear within the definition of a given subcircuit, as can other subcircuit calls. Inserting a subcircuit within another subcircuit is called *nesting*. If a subcircuit name is omitted from an .ENDS statement, SPICE assumes that the subcircuit and all nested subcircuits are ended.

Example A.10

Use SPICE to obtain a plot of the transient voltage across R_L in the ladder network shown in Figure A.15(a), after the switch is closed at $t = 0$. The plot should show values of the transient from $t = 0$ through $t = 4$ μs, in 0.1-μs intervals.

SOLUTION We notice that the circuit contains three identical RLC networks, so each network can be represented by the same subcircuit, which we have named RUNG. Figure A.15(b) shows how each RUNG is connected in the main circuit. The node numbers inside each RUNG are those that are local to RUNG. The node numbers outside the subcircuits are those of the main circuit. Figure A.15(c) shows the RUNG subcircuit. Following is the SPICE input file:

```
EXAMPLE A10
V1 1 0 PULSE(0 12 0 0 0)
RL 6 7 100
X1 1 2 0 5 RUNG
X2 2 3 5 6 RUNG
X3 3 4 6 7 RUNG
.SUBCKT RUNG 1 3 4 6
R1 1 2 1
L1 2 3 2MH
R2 4 5 1
L2 5 6 2MH
C1 3 6 50PF
.ENDS
.TRAN 0.1US 4US
.PLOT TRAN V(4,7)
.END
```

Notice how the main-circuit nodes specified in X1, X2, and X3 correspond to the subcircuit nodes in the .SUBCKT statement, in conformance with Figure A.15(b). Shown in Printout 12 is the plot produced by a program run.

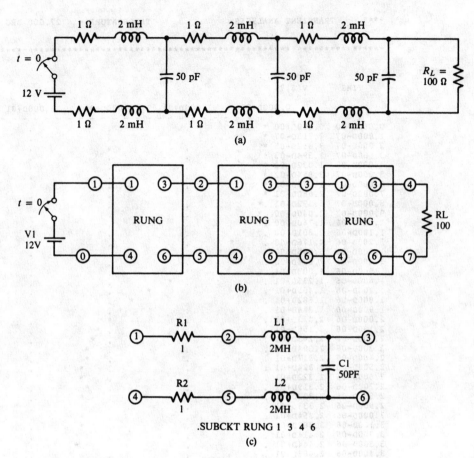

FIGURE A.15 (Example A.10)

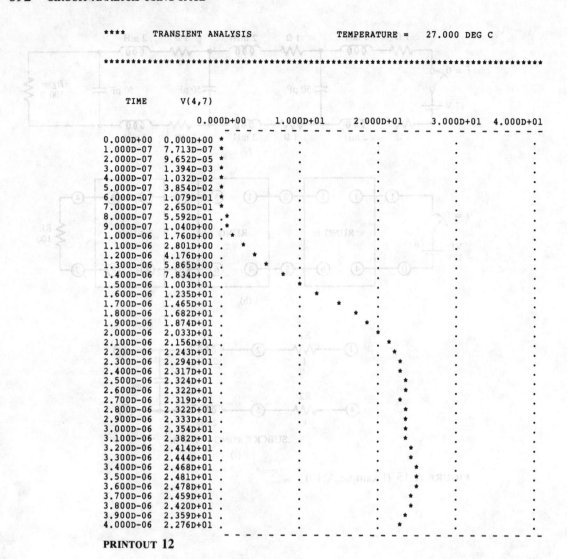

```
****       TRANSIENT ANALYSIS                    TEMPERATURE =    27.000 DEG C

*******************************************************************************

        TIME       V(4,7)

                           0.000D+00     1.000D+01     2.000D+01     3.000D+01   4.000D+01
                           - - - - - - - - - - - - - - - - - - - - - - - - - - - - - - - -
    0.000D+00   0.000D+00  *             .             .             .             .
    1.000D-07   7.713D-07  *             .             .             .             .
    2.000D-07   9.652D-05  *             .             .             .             .
    3.000D-07   1.394D-03  *             .             .             .             .
    4.000D-07   1.032D-02  *             .             .             .             .
    5.000D-07   3.854D-02  *             .             .             .             .
    6.000D-07   1.079D-01  *             .             .             .             .
    7.000D-07   2.650D-01  *             .             .             .             .
    8.000D-07   5.592D-01  .*            .             .             .             .
    9.000D-07   1.040D+00  .*            .             .             .             .
    1.000D-06   1.760D+00  . *           .             .             .             .
    1.100D-06   2.801D+00  .   *         .             .             .             .
    1.200D-06   4.176D+00  .      *      .             .             .             .
    1.300D-06   5.865D+00  .        *    .             .             .             .
    1.400D-06   7.834D+00  .           * .             .             .             .
    1.500D-06   1.003D+01  .             *             .             .             .
    1.600D-06   1.235D+01  .             .   *         .             .             .
    1.700D-06   1.465D+01  .             .      *      .             .             .
    1.800D-06   1.682D+01  .             .         *   .             .             .
    1.900D-06   1.874D+01  .             .            *.             .             .
    2.000D-06   2.033D+01  .             .             .  *          .             .
    2.100D-06   2.156D+01  .             .             .     *       .             .
    2.200D-06   2.243D+01  .             .             .       *     .             .
    2.300D-06   2.294D+01  .             .             .        *    .             .
    2.400D-06   2.317D+01  .             .             .         *   .             .
    2.500D-06   2.324D+01  .             .             .         *   .             .
    2.600D-06   2.322D+01  .             .             .         *   .             .
    2.700D-06   2.319D+01  .             .             .         *   .             .
    2.800D-06   2.322D+01  .             .             .         *   .             .
    2.900D-06   2.333D+01  .             .             .          *  .             .
    3.000D-06   2.354D+01  .             .             .          *  .             .
    3.100D-06   2.382D+01  .             .             .           * .             .
    3.200D-06   2.414D+01  .             .             .            *.             .
    3.300D-06   2.444D+01  .             .             .             *             .
    3.400D-06   2.468D+01  .             .             .             .*            .
    3.500D-06   2.481D+01  .             .             .             .*            .
    3.600D-06   2.478D+01  .             .             .             .*            .
    3.700D-06   2.459D+01  .             .             .             *             .
    3.800D-06   2.420D+01  .             .             .            *.             .
    3.900D-06   2.359D+01  .             .             .           * .             .
    4.000D-06   2.276D+01  .             .             .         *   .             .
                           - - - - - - - - - - - - - - - - - - - - - - - - - - - - - - - -
```

PRINTOUT 12

A.12 Probe and Control Shell

Probe

Probe is a PSpice option that makes it possible to obtain output plots having greater
resolution than those produced by .PLOT statements. Instead of plotting a sequence of
asterisks, Probe produces a virtually solid line on a high-resolution monitor. Also, Probe

has other features that make it very useful for analyzing output data. It behaves much like a high-quality oscilloscope that allows the user to position a cursor on a trace and to obtain a direct readout of the value of the variable displayed at the position of the cursor.

If the statement

```
.PROBE
```

appears anywhere in an input data file that requests a DC, AC, or TRAN analysis, then Probe automatically generates plotting data for the voltage (with respect to ground) at every node in the circuit and for the current entering every device in the circuit. Probe stores this data in an *output data file* named PROBE.DAT. To initiate a Probe run, the user enters the command PROBE directly from the keyboard. If more than one analysis type appeared in the input data file (called the *circuit file* in PSpice), a "start-up menu" appears, and the user selects a single analysis type for the Probe run. If only one analysis type appeared in the circuit file, then a set of axes and a menu are displayed immediately after PROBE is entered. Selecting the option ADD TRACE from this menu prompts the user to enter the variable to be plotted, using the same format used to specify outputs in PSpice (such as V(2), I(R1), etc.). The plot is scaled automatically and displayed on the monitor. Additional variables can be plotted simultaneously by repeatedly selecting the ADD TRACE option. Also, plots can be deleted by selecting the DELETE TRACE option.

Selecting the CURSOR option from the Probe menu creates two sets of crosshairs, identified as cursor 1 (C1) and cursor 2 (C2) on the display. These crosshairs can be moved along the plot using direction keys ($\leftarrow$ and $\rightarrow$) on the keyboard. Using the direction keys alone moves C1, and using the direction keys with SHIFT depressed moves C2. The numerical values of the variable at the positions of the crosshairs on the plot are displayed in a window, along with the difference in the values. Figure A.16 shows an example.

If desired, Probe plots can be limited to specific variables by giving their names in the .PROBE statement in the circuit file. For example, the statement

```
.PROBE V(2)
```

would cause Probe to create an output data file containing only plotting data for the voltage at node 2. This option is useful when memory capacity is too small to store all voltages and currents.

Another feature of Probe is that it can create plots of mathematical functions of the variables. For example, after selecting ADD TRACE, entering the expression

```
I(R1)*I(R1)*1K
```

would create a plot of the power (I^2R) dissipated by 1-kΩ resistor R1. Numerous mathematical functions, including trigonometric functions, logarithms, average values, square roots, and absolute values of the variables can also be plotted. One difference between Probe and PSpice is that suffixes in expressions written for Probe must use lowercase m to represent milli and capital M to represent meg (as opposed to M for milli and MEG for meg).

It should be noted that the high-resolution plots created by Probe are *not* the result of Probe increasing the number of output values computed in a circuit simulation. Rather,

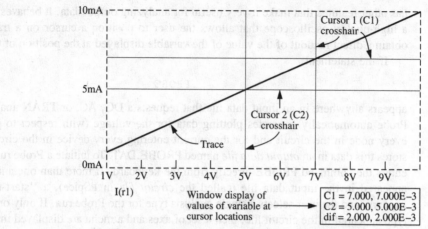

FIGURE A.16 (Example of a display produced by Probe.)

Probe *interpolates* values between the data points that a circuit file specifies are to be computed. Thus, if the output variable has a sudden change in value (such as a sharp peak) and the circuit file does not specify a fine enough increment between data points to detect that change, Probe cannot be expected to detect it either.

Control Shell

Control Shell is a PSpice option that coordinates and simplifies writing, editing, and running input circuit files and using the various options and features available in PSpice. For example, Control Shell can be used to make Probe run automatically after every circuit simulation. Options are selected by the user from various menus displayed by Control Shell.

In a system equipped with Control Shell, the first (main) menu displayed contains (among others) selections entitled Files, Analysis, and Probe. These are the principal choices that will be used in creating and running most circuit simulations. When Files is selected, another menu is displayed with options that include Edit, Browse Output, and Current File. Edit and Current File allow the user to edit an existing circuit file or to create and name a new circuit file. Selecting Browse Output allows the user to scroll through the entire output created after a circuit simulation has been run.

To perform a program run of an input circuit file that has already been created, the option Analysis is selected from the main menu. This selection creates another menu that allows the user to specify the analysis type (AC, DC, or transient). Since at least one analysis type is (presumably) already specified in the input circuit file, an arrow in this menu points to one analysis type and the user can simply press the ENTER key to initiate the run. Alternatively, another analysis type specified in the circuit file can be selected, and the user is prompted to enter (or change) analysis parameters (such as start and stop voltages and step size in a DC analysis). It should be noted that if there is an error in the input circuit file, Control Shell does not allow the user to select the Analysis option.

Selecting Probe from the main menu creates another menu that allows the user to request automatic running of Probe after a PSpice simulation run. In this option, it is not necessary to include a .PROBE statement in the input circuit file.

Answers to Odd-Numbered Exercises

Chapter 1

1.1 $\dfrac{\text{kg} \cdot \text{m}^2}{\text{s}^2}$

1.3 400 Pa

1.5 (a) 2 (b) 10^{-6} (c) a (d) 2500 (e) 10^{10}
(f) 2 (g) 25 (h) 0.909×10^{-6} (i) 1 (j) 10^{-9}

1.7 (a) 0.6 m (b) 4 cm (c) 0.15 μs (d) 0.5 MW
(e) 0.5 mV (f) 24 μV (g) 30 nA (h) 7 kW

1.9 (a) 0.2 m/s (b) 0.21 m²

1.11 (a) $x \geq 1$ (b) $x < 5.5$ (c) $x > 0.608$
(d) $x \geq 0.75R$ (e) $x \leq 125a$

1.13 (a) $R \approx R_2$ (b) $R \approx R_1$

1.15 (a) 10% (b) 1%

Chapter 2

2.1 1.09781×10^{30} electrons

2.3 k:2, l:8, m:4

2.5 $+5.12 \times 10^{-4}$ C

2.7 0.03 A

2.9 0.115 C

2.11 400 s

2.13 750 h

2.15 37.5 Ω

2.17 (a) 87.532 kΩ (b) 0.087532 MΩ (c) 40 Ω
(d) 1.25 MΩ (e) 0.84 kΩ (f) 188.5 kΩ

Chapter 3

3.1 4 A   37.6 V
     34.5 V   50 kΩ
     25 Ω   15 mA
     1.44 mA   250 Ω
     80 µA   9.62 mV

3.3

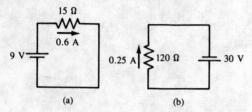

(a)   (b)

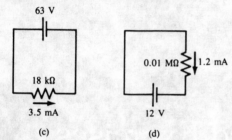

(c)   (d)

3.5 25.8 V − 45 V

3.7 180 V

3.9

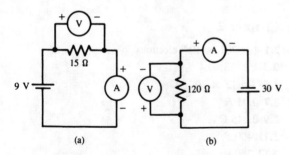

(a)   (b)

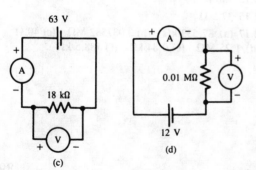

(c)   (d)

3.11 13.5 s

3.13 (a) 1.21 kW   (b) 8.51 W   (c) 1 W

3.15 25 V

3.17 5×10^6 s

3.19 1.89 kWh

3.21 55 24-h days

3.23 V ≤ 11.62 V

3.25 (a) $\frac{1}{4}$ W   (b) 2 W   (c) 1 W   (d) $\frac{1}{8}$ W

3.27 (a) 8.2 kΩ nominal; 7790–8610 Ω
(b) 1 MΩ nominal; 0.9–1.1 MΩ
(c) 39 Ω nominal; 35.1–42.9 Ω
(d) 2.2 kΩ nominal; 1760–2640 Ω

3.29 (a) Yellow, violet, brown, gold
(b) Brown, gray, red, silver
(c) Gray, red, yellow
(d) Brown, green, green, gold

3.31 3.6 kΩ, 8.12%

3.33 2.026×10^{-4} S

3.35 $P = V^2G$

3.37 (a) 0.888   (b) 2.4×10^6 J

3.39 0.559

3.41 5 hp

3.43 0.858 A

3.45 0.25 mA − 0.25 A

3.47 18.36 V

3.49 0.6 V − 60 V

3.51 45 V

3.1S 0.6 A, 3.5 mA

3.3S 60 V

Chapter 4

4.1 (a) R_1, E (b) R_1, R_2, E (c) R_1 (d) None
(e) R_2, E (f) R_2, R_3

4.3 (a) $R_T = 240$ Ω, $I_T = 0.2083$ A
(b) $R_T = 5.5$ kΩ, $I_T = 2$ mA
(c) $R_T = 1.69$ MΩ, $I_T = 7.10$ µA
(d) $R_T = 6.42$ kΩ, $I_T = 15.58$ mA

4.5 (a) $V(150 \; \Omega) = 37.5$ V, $V(220 \; \Omega) = 55$ V,
$V(330 \; \Omega) = 82.5$ V
(b) $V(75 \; k\Omega) = 37.5$ V, $V(21 \; k\Omega) = 10.5$ V
(c) $V(390 \; k\Omega) = 5.89$ V, $V(680 \; k\Omega) = 10.27$ V,
$V(0.27 \; M\Omega) = 4.08$ V
(d) $V(16 \; \Omega) = 19.2$ V, $V(24 \; \Omega) = 28.8$ V,
$V(91 \; \Omega) = 109.2$ V

4.7 (a) 37.5 V + 55 V + 82.5 V = 175 V = E
(b) 37.5 V + 10.5 V = 48 V = E
(c) 5.89 V + 10.27 V + 4.08 V = 20.24 V ≈ E
(d) 19.2 V + 28.8 V + 109.2 V = 157.2 V = E

4.9 (a) 6 V **(b)** −72 V **(c)** 20 V **(d)** −13.49 V

4.11 (a) 72 V **(b)** −60 V

4.13

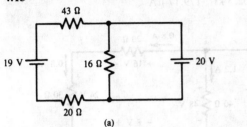

(a)

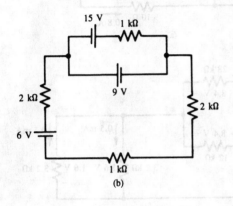

(b)

4.15 (a) 12 V **(b)** 21.83 V **(c)** 100 V **(d)** −3.5 V

4.17 $\dfrac{ER_1}{R_T} + \dfrac{ER_2}{R_T} = \dfrac{ER_1 + ER_2}{R_T} = \dfrac{E(R_1 + R_2)}{R_T} =$

$\dfrac{ER_T}{R_T} = E$.

4.19 (a) $V_{MIN} = 0$ V, $V_{MAX} = 9$ V

(b) $V_{MIN} = 0$ V, $V_{MAX} = 24.24$ V

(c) $V_{MIN} = 0$ V, $V_{MAX} = 2$ V

(d) $V_{MIN} = 3.69$ V, $V_{MAX} = 16$ V

4.21 (a) E, R_1 **(b)** None **(c)** R_2 **(d)** R_1, R_2, R_3

(e) E, R_2 **(f)** R_2, R_4

4.23 (a) $G_T = 0.095$ S, $R_T = 10.526$ Ω

(b) $G_T = 0.25$ mS, $R_T = 4$ kΩ

(c) $G_T = 3.33$ mS, $R_T = 300$ Ω

(d) $G_T = 8$ μS, $R_T = 125$ kΩ

(e) $G_T = 1.11$ mS, $R_T = 900.09$ Ω

(f) $G_T = \dfrac{1}{2R}$ S, $R_T = 2R$ Ω

4.25 (a) 1.6 A ↖

(b) $I_1 = 60$ mA →, $I_2 = 35$ mA →

(c) 1.8 A ↑

(d) $I_1 = 175$ μA →, $I_2 = 75$ μA →, $I_3 = 75$ μA ←,

$I_4 = 175$ μA ←, $I_5 = 250$ μA ←

4.27

(a) (i) $I(12\ \Omega) = 0.75$ A, $I(20\ \Omega) = 0.45$ A,

$I(10\ \Omega) = 0.9$ A

(ii) $I_T = 2.1$ A

(iii) $I_T = 9$ V/4.286 Ω = (9 V)(0.233 S) = 2.1A

(b) (i) $I(9\ k\Omega) = 6.667$ mA, $I(18\ k\Omega) = 3.333$ mA

(ii) $I_T = 10$ mA

(iii) $I_T = 60$ V/6 kΩ = (60 V)(1.667 × 10⁻⁴ S) =

10 mA

(c) (i) $I(36\ k\Omega) = 0.5$ mA, $I(18\ k\Omega) = 1$ mA,

$I(36\ k\Omega) = 0.5$ mA

(ii) $I_T = 2.0$ mA

(iii) $I_T = 18$ V/9 kΩ = (18 V)(1.111 × 10⁻⁴ S) =

2 mA

(d) (i) $I(300\ \Omega) = 0.05$ A, $I(150\ \Omega) = 0.1$ A,

$I(300\ \Omega) = 0.05$ A, $I(150\ \Omega) = 0.1$ A

(ii) $I_T = 0.3$ A

(iii) $I_T = 15$ V/50 Ω = (15 V)(0.02 S) = 0.3 A

(e) (i) $I(1\ k\Omega) = 0.001$ A, $I(100\ \Omega) = 0.01$ A,

$I(10\ \Omega) = 0.1$ A, $I(1\ \Omega) = 1.0$ A

(ii) $I_T = 1.111$ A

(iii) $I_T = 1$ V/0.9 Ω = (1 V)(1.111 S) = 1.111 A

(f) (i) $I(10R) = 0.025$ A

(ii) $I_T = 0.1$ A

(iii) $I_T = (0.25R)/(2.5R) = (0.25R)(1/2.5R) = 0.1$ A

4.29 (a) 0.9 A **(b)** 2.75 mA **(c)** 100 μA

(d) 10 mA

4.31 $I_1 + I_2 = I\dfrac{R_2}{R_1 + R_2} + I\dfrac{R_1}{R_1 + R_2}$

$= I\dfrac{R_1 + R_2}{R_1 + R_2} = I$

4.33 (a) $I_1 = 250$ mA, $V_1 = 0$ V

(b) $I_1 = 20$ mA, $V_1 = 20$ V

(c) $I_1 = 0$ A, $V_1 = 0$ V

4.35 4.5 A, E_2 charged

4.37 (a) 0.4 A, 20 V **(b)** 50 mA, −50 V

4.1D 300 Ω

4.3D 4 kΩ

4.5D (a) 1.7 kΩ **(b)** 10.113 V

4.7D ↓⊤ 5 V

4.9D 2 kΩ

4.11D 200 Ω

4.13D 100 Ω

4.15D 4 mA ↓

4.1T $I_{MIN} = 70.59$ mA, $I_{MAX} = 92.3$ mA

4.3T R_3

4.5T No

4.7T 20 Ω

4.9T 6 V

4.11T R_3

4.13T The 4-kΩ resistance of the rheostat is shorted.

4.1S R_T = 6.42 kΩ, I_T = 15.58 mA, V(920 Ω) = 14.33 V, V(1.5 kΩ) = 23.36 V, V(1.8 kΩ) = 28.04 V, V(2200 Ω) = 34.27 V

4.3S R_T = 9 kΩ, I_T = 2 mA, I(36 kΩ) = 0.5 mA, I(18 kΩ) = 1 mA

4.5S 50 mA, −50 V

Chapter 5

5.1 (a) 5 Ω, 1 A (b) 80 Ω, 0.2 A (c) 4 kΩ, 3 mA (d) 502.3 kΩ, 179.17 μA

5.3

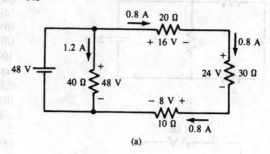

(a)

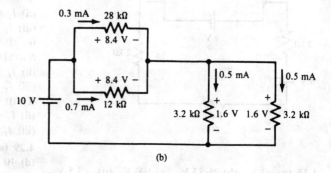

(b)

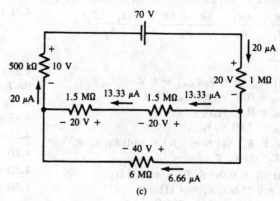

(c)

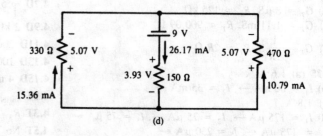

(d)

5.5 (a) 10 W, 10 V **(b)** 4.608 W, 38.4 V
(c) 0.1 W, 5 V **(d)** 5.4 mW, 18V

5.7 -3 V

5.9 (a) $E = 6$ V, $I_A = 0.2$ A, $I_B = 0.1$ A, $V_1 = 6$ V, $V_2 = 2$ V, $I_2 = 0.1$ A, $V_3 = 4$ V, $R_3 = 80\ \Omega$, $V_4 = 4$ V, $R_4 = 80\ \Omega$

(b) $R_1 = 500\ \Omega$, $V_2 = 25$ V, $I_2 = 6.25$ mA, $V_3 = 25$ V, $I_3 = 6.25$ mA, $I_4 = 8.33$ mA, $V_5 = 2.085$ V, $I_5 = 4.17$ mA, $V_6 = 10.425$ V, $I_6 = 4.17$ mA, $V_7 = 12.51$ V, $I_7 = 4.17$ mA, $I_8 = 25$ mA, $R_8 = 2.5$ kΩ

5.11 (a) 8 V, 0.2 W **(b)** 84 V, 72 mW
(c) 2 V, 0.5 W

5.13

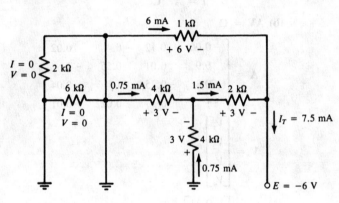

5.15 (a) 7.5 kΩ **(b)** 240 Ω **(c)** 3 kΩ **(d)** 774.2 kΩ
5.17 (a) 6 V **(b)** 5 V **(c)** 45 V **(d)** 7.157 V
5.19 (a) 250 Ω **(b)** 4.7 kΩ
5.21 (a) 0.22 W **(b)** 2.5 mW
5.1D 2 kΩ
5.3D 20 Ω
5.5D 6 kΩ
5.7D 500 Ω
5.9D 1 MΩ
5.11D 6 V
5.13D 100 kΩ, 9.6 V
5.1T (a) 5 V **(b)** 12 V
5.3T R_1 or R_4 shorted; R_2 or R_3 open
5.5T R_2 and R_3 open; R_5 shorted
5.7T R_2, R_4

5.9T

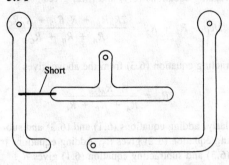

5.1S $I_T = 1$ mA

Resistance	Voltage	Current
12 kΩ	8.4 V	0.7 mA
28 kΩ	8.4 V	0.3 mA
3.2 kΩ (both)	1.6 V	0.5 mA

5.3S $V_T = 12$ V

Resistance	Voltage	Current
1 kΩ	1.33 V	1.33 mA
5 kΩ	6.67 V	1.33 mA
3 kΩ	4.00 V	1.33 mA
4.5 kΩ	12.00 V	2.67 mA

Chapter 6

6.1

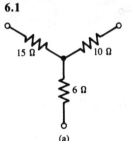

(a)

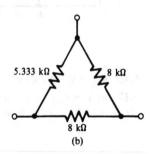

(b)

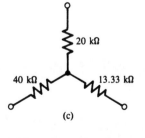

(c)

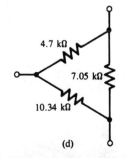

(d)

6.3 Adding equations (6.1) and (6.2) gives

$$2R_1 + R_2 + R_3 = \frac{2R_A R_B + R_A R_C + R_B R_C}{R_A + R_B + R_C}$$

Subtracting equation (6.3) from the above gives

$$R_1 = \frac{R_A R_B}{R_A + R_B + R_C}$$

Similarly, adding equations (6.1) and (6.3) and subtracting equation (6.2) gives R_2. Adding equations (6.2) and (6.3) and subtracting equation (6.1) gives R_3.

6.5 (a) $I_T = 4$ A, $I(R_1) = 2$ A
(b) $I_T = 5$ mA, $I(R_1) = 2$ mA
6.7 (a) Not balanced (b) Balanced, 1 mA
(c) Balanced, 1.67 mA (d) Not balanced
6.9 2 Ω
6.11 12.8 mA
6.13

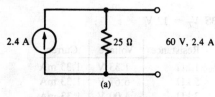

(a)

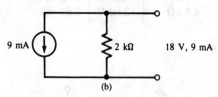

(b)

6.15 (a) 20 V (b) 30 V
6.17 (a) $x = 4$, $y = 1$ (b) $V_1 = 2$, $V_2 = -2$
(c) $I_1 = -0.12$, $I_2 = 0.46$

6.19 (a) **AI = C**

$$\mathbf{A} = \begin{bmatrix} -3 & 7 & 3 \\ 0 & 1 & 1 \\ 2 & -5 & -10 \end{bmatrix}$$

$$\mathbf{I} = \begin{bmatrix} I_1 \\ I_2 \\ I_3 \end{bmatrix}$$

$$\mathbf{C} = \begin{bmatrix} 40 \\ 6 \\ -6 \end{bmatrix}$$

$$\mathbf{I} = \mathbf{A}^{-1}\mathbf{C}$$

(b) **AV = C**

$$\mathbf{A} = \begin{bmatrix} -0.06 & 0.12 & -0.4 & 0.02 \\ 0.9 & 0.01 & 0.55 & -1 \\ 0.8 & -0.25 & 1 & 0.04 \\ -1 & 0 & 0.022 & -0.01 \end{bmatrix}$$

$$\mathbf{V} = \begin{bmatrix} V_1 \\ V_2 \\ V_3 \\ V_4 \end{bmatrix}$$

$$\mathbf{C} = \begin{bmatrix} -0.24 \\ 0 \\ 0.75 \\ -0.5 \end{bmatrix}$$

$$\mathbf{V} = \mathbf{A}^{-1}\mathbf{C}$$

6.21 (a)

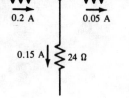

(b) $I(200\ \Omega) = 0.1$ A →, $I(150\ \Omega) = 0.2$ A ↓,
$I(100\ \Omega) = 0.1$ A ↑
(c) $I(1\ \text{k}\Omega) = 5.07$ mA ←, $I(2\ \text{k}\Omega) = 5.07$ mA ↓,
$I(4\ \text{k}\Omega) = 1.2$ mA →, $I(6\ \text{k}\Omega) = 3.87$ mA ↓,
$I(3\ \text{k}\Omega) = 3.87$ mA ↑
(d) $I(50\ \Omega) = 0.171$ A →, $I(250\ \Omega) = 0.237$ A ←,
$I(100\ \Omega) = 0.408$ A ↓

6.23 (a) AI = C

$$A = \begin{bmatrix} 1150 & 0 & -680 & 0 \\ 0 & 490 & -120 & -150 \\ -680 & -120 & 800 & 0 \\ 0 & -150 & 0 & 250 \end{bmatrix}$$

$$I = \begin{bmatrix} I_1 \\ I_2 \\ I_3 \\ I_4 \end{bmatrix}$$

$$C = \begin{bmatrix} -25 \\ 15 \\ 0 \\ 18 \end{bmatrix}$$

$$I = A^{-1}C$$

(b) AI = C

$$A = \begin{bmatrix} 10 \times 10^3 & -3 \times 10^3 & -1 \times 10^3 \\ -3 \times 10^3 & 9 \times 10^3 & -4 \times 10^3 \\ -1 \times 10^3 & -4 \times 10^3 & 10 \times 10^3 \end{bmatrix}$$

$$I = \begin{bmatrix} I_1 \\ I_2 \\ I_3 \end{bmatrix}$$

$$C = \begin{bmatrix} 9 \\ 5 \\ 35 \end{bmatrix}$$

$$I = A^{-1}C$$

6.25 In each answer, V_1 is the voltage at the leftmost node in the circuit, and V_2 is the voltage at the rightmost node, with respect to the common shown in the figure.
(a) $V_1 = 13.5$ V, $V_2 = 6$ V
(b) $V_1 = 5$ V, $V_2 = 16.67$ V
(c) $V_1 = -13.75$ V, $V_2 = -32.5$ V
(d) $V_1 = 11$ V, $V_2 = 17$ V
6.27 (b) AV = C; after dividing the equations through by 10^3,

$$A = \begin{bmatrix} 1.7 & -1 & -0.5 \\ 1 & -2 & -0.5 \\ 0.5 & 0.5 & -2 \end{bmatrix}$$

$$V = \begin{bmatrix} V_1 \\ V_2 \\ V_3 \end{bmatrix}$$

$$C = \begin{bmatrix} 3 \\ -9 \\ 9 \end{bmatrix}$$

$$V = A^{-1}C$$

6.1D $R_1 = 50\ \Omega$, $R_2 = 150\ \Omega$, $R_3 = 250\ \Omega$
6.3D 0 to 150 Ω
6.5D 25 V
6.7D 9.22 mA
6.1T R_3 or R_4 shorted
6.3T R_2 or R_3, or both, open
6.5T R_3 shorted

6.1S

Resistance	Voltage
6 Ω	8.25 V
3 Ω	5.813 V
9 Ω	9.562 V
12 Ω	3.75 V

6.3S 29.62 mW

Chapter 7

7.1 (a) 0.125 A → **(b)** 2 mA ←
(c) 4.6 A ← **(d)** 0.1 mA ↓

7.3 (a) $-(11$ V$)+$ **(b)** $-(7.5$ V$)+$

7.5

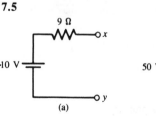

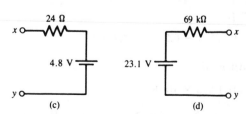

7.7 (a) $-(9.33$ V$)+$ **(b)** 0.5 V∓

7.9

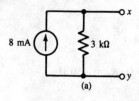

(a)

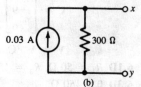

(b)

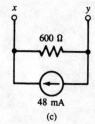

(c)

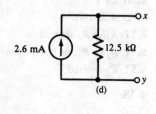

(d)

7.11 (a) 50 Ω, 0.72 W **(b)** 220 Ω, 2.2 W

7.13 33.06%

7.15

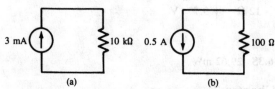

(a)   (b)

7.17 (a) 4 V± **(b)** 15.6 V± **(c)** 28 V∓
(d) 6.8 V±

71.D 4.5 V

7.3D 50 Ω

7.5D 18.75 V

7.7D 14 Ω

7.9D 336 Ω

7.11D 16 Ω

7.13D 20 Ω

7.1T 6-V source

7.3T Supply is operating properly.

7.5T No

7.7T No

7.1S 0.625 A − 0.25 A = 0.375 A

Chapter 8

8.1 (a) 1.688×10^5 N

(b) 39,050 N

8.3 94,868 m

8.5

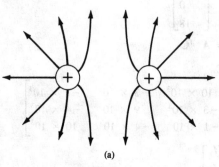

(a)

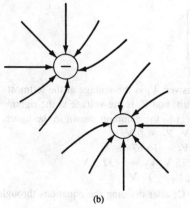

(b)

8.7 1.62×10^{-5} C

8.9 2.86 N

8.11 7.5×10^{-11}

8.13 5.625 mm

8.15 8.33×10^{-5} m

8.17 (a) 920 V **(b)** 490 V **(c)** −4 V

8.19 0.3822 μΩ

8.21 23.45 μm

8.23 0.405 Ω

8.25 AWG No. 2

8.27 50 Ω

8.29 50 × 10⁻⁶ m

8.31 (a) 13.47 Ω   (b) 9.69 Ω

8.33 327 Ω

8.35 −51.5°C

8.1S 1.9113 mW

Chapter 9

9.1 (a) 800 μC   (b) 40 V

9.3 600 pC

9.5 0.007 μF

9.7 3.75

9.9 0.002652 μF

9.11 1.105 × 10⁻⁴ m

9.13 0.02376 μF

9.15 (a) B (b) 30 V

9.17 (a) $C_T = 6$ μF, $V_1 = 3$ V, $V_2 = 9$ V, $V_3 = 18$ V

(b) $C_T = 27.5$ pF, $V_1 = 2.5$ V, $V_2 = 2.5$ V, $V_3 = 5$ V, $V_4 = 10$ V

9.19 $C_T = \dfrac{1}{1/C_1 + 1/C_2} = \dfrac{1}{(C_2 + C_2)/C_1C_2} = \dfrac{C_1C_2}{C_1 + C_2}$

9.21 (a) $Q_T = 1665$ μC, $Q_1 = 15$ μC, $Q_2 = 150$ μC, $Q_3 = 1500$ μC

(b) $Q_T = 660$ pC, $Q_1 = 360$ pC, $Q_2 = 180$ pC, $Q_3 = 120$ pC

9.23 (a) $V_1 = 3.75$ V, $Q_1 = 450$ μC
$V_2 = 11.25$ V, $Q_2 = 450$ μC
$V_3 = 10$ V, $Q_3 = 900$ μC
$V_4 = 5$ V, $Q_4 = 900$ μC

(b) $V_1 = 25$ V, $Q_1 = 62.5$ μC
$V_2 = 25$ V, $Q_2 = 187.5$ μC
$V_3 = 12.5$ V, $Q_3 = 250$ μC
$V_4 = 12.5$ V, $Q_4 = 250$ μC

9.25 (a) (i) $i(t) = 0.4e^{-t/0.4}$ mA

(ii) $i(0) = 0.4$ mA, $i(0.2) = 0.243$ mA,
$i(0.4) = 0.147$ mA, $i(0.8) = 0.054$ mA,
$i(1.2) = 0.02$ mA

(iii)

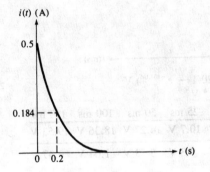

i(t) (mA)

0.4

0.147

0   0.4   → *t* (s)

(b) (i) $i(t) = 0.5e^{-t/0.2}$ A

(ii) $i(0) = 0.5$ A, $i(0.2) = 0.184$ A,
$i(0.4) = 0.0677$ A, $i(0.8) = 9.16$ mA,
$i(1.2) = 1.24$ mA

(iii)

i(t) (A)

0.5

0.184

0   0.2   → *t* (s)

9.27 (a) (i) 5 ms

(ii) $i(\tau) = 3.68$ mA, $i(2\tau) = 1.35$ mA,
$i(3\tau) = 0.498$ mA, $i(5\tau) = 0.067$ mA

(iii) 0.693τ

(b) (i) 8 ms

(ii) $i(\tau) = 0.11$ mA, $i(2\tau) = 0.041$ mA,
$i(3\tau) = 0.015$ mA, $i(5\tau) = 2.02$ μA

(iii) 0.693τ

9.29 (a) (i) $v_C(t) = 25(1 - e^{-t/0.025})$ V
$v_R(t) = 25e^{-t/0.25}$ V

(ii)

t	0	25 ms	50 ms	100 ms	150 ms
$v_C(t)$	0 V	15.8 V	21.62 V	24.54 V	24.94 V
$v_R(t)$	25 V	9.2 V	3.38 V	0.458 V	0.062 V

(iii)

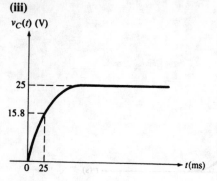

$v_C(t)$ (V)

25

15.8

0   25 t(ms)

$v_R(t)$ (V)

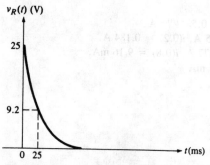

25

9.2

0   25 t(ms)

(b) (i) $v_C(t) = 20(1 - e^{-t/0.04})$ V
$v_R(t) = 20e^{-t/0.4}$ V

(ii)

t	0	25 ms	50 ms	100 ms	150 ms
$v_C(t)$	0 V	10.7 V	14.27 V	18.36 V	19.53 V
$v_R(t)$	20 V	9.3 V	5.73 V	1.64 V	0.47 V

(iii)

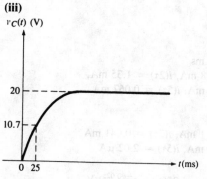

$v_C(t)$ (V)

20

10.7

0   25 t(ms)

$v_R(t)$ (V)

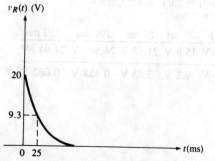

20

9.3

0   25 t(ms)

9.31 (a) (i) 25 μs

(ii)

t	τ	2τ	3τ	5τ
$v_C(t)$	18.96 V	25.94 V	28.5 V	29.8 V
$v_R(t)$	8.28 V	3.05 V	1.12 V	0.152 V

(iii) 1.204τ

(b) (i) 15 μs

(ii)

t	τ	2τ	3τ	5τ
$v_C(t)$	11.38 V	15.56 V	17.1 V	17.88 V
$v_R(t)$	1.32 V	0.487 V	0.179 V	0.024 V

(iii) 1.204τ

9.33 (a) (i) $v_C(t) = 60e^{-t/0.75}$ V
$i(t) = -6e^{-t/0.75}$ mA

(ii) $v_C(2.5\tau) = 4.93$ V, $i(2.5\tau) = -0.49$ mA

(iii)

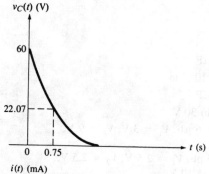

$v_C(t)$ (V)

60

22.07

0   0.75 t (s)

$i(t)$ (mA)

0   0.75 t (s)

−2.21

−6

(b) (i) $v_C(t) = 5e^{-t/3 \times 10^{-8}}$
$i(t) = -0.25e^{-t/3 \times 10^{-8}}$ A

(ii) $v_C(2.5\tau) = 0.41$ V, $i(2.5\tau) = 0.021$ A

(iii)

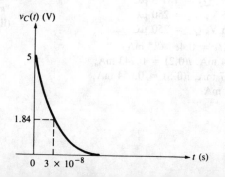

$v_C(t)$ (V)

5

1.84

0   3×10^{-8} t (s)

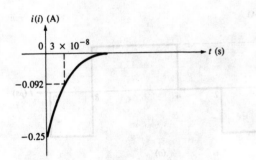

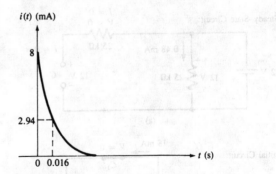

9.35 (a) $v_C(t) = 12(1 - e^{-t/0.02})$ V
$i(t) = 0.6e^{-t/0.02}$ mA
(b) $v_C(t) = 12e^{-t/0.06}$ V
$i(t) = -0.2e^{-t/0.06}$ mA
(c)

(b) (i) $v_c(t) = 7.5(1 - e^{-t/0.75 \times 10^{-6}})$ V
$i(t) = 0.1e^{-t/0.75 \times 10^{-6}}$ A
(ii)

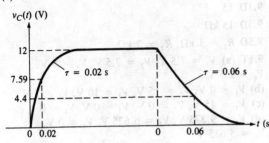

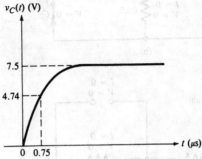

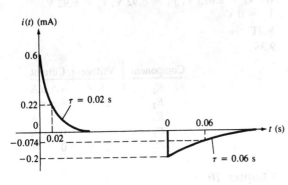

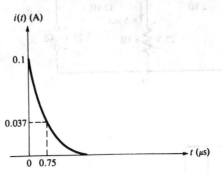

9.37 (a) (i) $v_C(t) = 6.4(1 - e^{-t/0.016})$ V
$i(t) = 8e^{-t/0.016}$ mA
(ii)

9.39

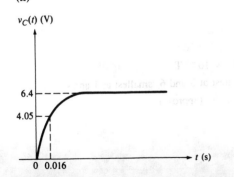

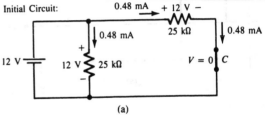

Steady–State Circuit:

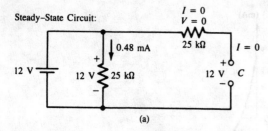

(a)

Initial Circuit:

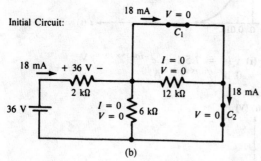

(b)

Steady–State Circuit:

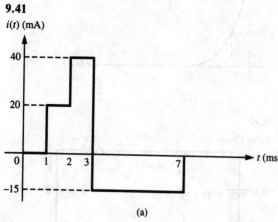

(b)

9.41

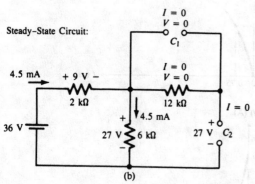

(a)

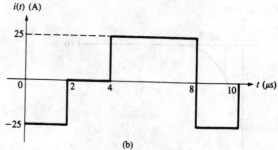

(b)

9.43 (a) 7.2×10^{-3} J
(b) 4×10^{-3} J
9.45 (a) $W_1 = 0.5$ J, $W_2 = 0.75$ J
(b) $W_1 = 0.776$ μJ, $W_2 = 1.16$ μJ
9.1D 15
9.3D 15 kΩ
9.5D $R_1 = 1$ kΩ, $R_2 = 2$ kΩ
9.1T (a) $V_1 = 7.5$ V, $V_2 = 7.5$ V, $V_3 = 5$ V, $V_4 = 2.5$ V
(b) $V_1 = 0$ V, $V_2 = 15$ V, $V_3 = 10$ V, $V_4 = 5$ V
(c) $V_1 = 15$ V, $V_2 = 0$ V, $V_3 = 0$ V, $V_4 = 0$ V
(d) $V_1 = 9.375$ V, $V_2 = 5.625$ V, $V_3 = 0$ V, $V_4 = 5.625$ V
(e) $V_1 = 8.075$ V, $V_2 = 6.92$ V, $V_3 = 6.92$ V, $V_4 = 0$ V
9.3T No
9.3S

Component	Voltage	Current
R_1	9 V	4.5 mA
R_2	27 V	4.5 mA
R_3	0 V	0 A
C_1	0 V	0 A
C_2	27 V	0 A

Chapter 10

10.1 0.0477 T
10.3 1 m
10.5 417.4 A/m
10.7 0.1 T
10.9 3979 A/m
10.11 4×10^6 A/Wb
10.13 2.51×10^{-5} T
10.15 Greatest at 3 and 6, smallest at 1 and 4
10.17 450 A/m (approx.)

10.19 8.4 A (approx.)

10.21 21 turns

10.23 43.14 A (approx.)

10.25 (a) 16.98 A   **(b)** 91.9%

Chapter 11

11.1 (a) ⊙   **(b)** ⊗   **(c)** ⊗

11.3 (a) 51.2×10^{-3} Wb/s   **(b)** 12×10^{-3} Wb/s

11.5 5.83×10^{-4} Wb/s

11.7 0.737 mH

11.9

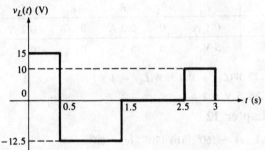

11.11 84.2 Ω

11.13 (a) 1.6×10^{-4} s

(b) $v_L(t) = 20e^{-t/1.6 \times 10^{-4}}$ V

(c) $v_L(0.1$ ms$) = 10.7$ V, $v_L(1.5\tau) = 4.46$ V,

$v_L(0.5$ ms$) = 0.879$ V, $v_L(5\tau) = 0.135$ V

(d)

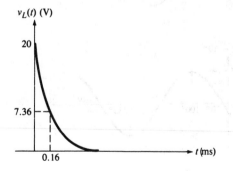

11.15 (a) $v_L(t) = 16e^{-t/5 \times 10^{-6}}$ V

$i(t) = 0.4(1 - e^{-t/5 \times 10^{-6}})$ A

$v_R(t) = 16(1 - e^{-t/5 \times 10^{-6}})$ V

(b)

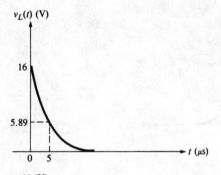

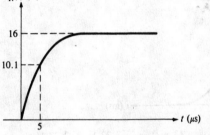

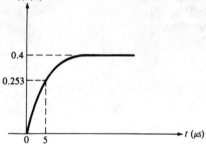

(c) $v_L(7.5 \ \mu s) + v_R(7.5 \ \mu s) = 3.57$ V $+ 12.43$ V $=$ 16 V $=$ E

11.17 (a) $v_L(t) = -10e^{-t/2 \times 10^{-3}}$ V

$i(t) = 0.2e^{-t/2 \times 10^{-3}}$ A

$v_R(t) = 10e^{-t/2 \times 10^{-3}}$ V

(b)

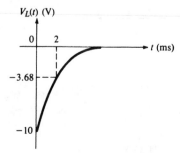

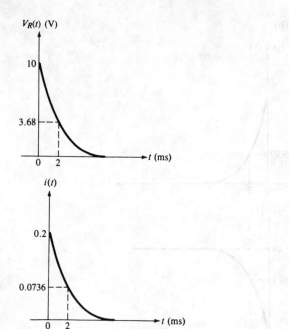

$V_R(t)$ (V)

10

3.68

0   2   → t (ms)

$i(t)$

0.2

0.0736

0   2   → t (ms)

(c) $v_L(3 \text{ ms}) + v_R(3 \text{ ms}) = -2.23 \text{ V} + 2.23 \text{ V} = 0$

11.19 (a) $i(t) = 6e^{-t/0.5}$ A
$v_R(t) = 240e^{-t/0.5}$ V
$v_L(t) = -240e^{-t/0.5}$ V

(b)

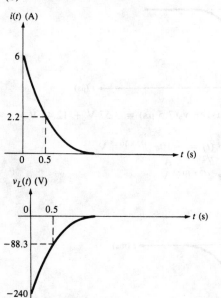

$i(t)$ (A)

6

2.2

0   0.5   → t (s)

$v_L(t)$ (V)

0   0.5   → t (s)

−88.3

−240

(c) $v_L(1 \text{ s}) + v_{10\Omega}(1 \text{ s}) + v_{30\Omega}(1 \text{ s}) =$
$-32.48 \text{ V} + 8.12 \text{ V} + 24.36 \text{ V} = 0$

11.21 (a) 26.4 mH   **(b)** 13.6 mH
11.23 $v_L(t) = 8e^{-t/15 \times 10^{-3}}$ V
$i(t) = 1 - e^{-t/15 \times 10^{-3}}$ A
11.25 Initial:

	R_1	R_2	R_3	L_1	L_2
I	0	0.04 A	0.04 A	0	0
V	0	4 V	1 V	5 V	4 V

Steady-state:

	R_1	R_2	R_3	L_1	L_2
I	0.1 A	0	0.2 A	0.1 A	0.2 A
V	5 V	0	5 V	0	0

11.27 $W(L_1) = 0.9$ J, $W(L_2) = 1.8$ J
11.3S 5 V, 0.2 A

Chapter 12

12.1 (a) $-160°$ **(b)** $170°$ **(c)** $-80°$
(d) $20°$ **(e)** $90°$ **(f)** $-140°$
12.3 (a) 0.667   **(b)** -0.707 **(c)** -0.9397
12.5

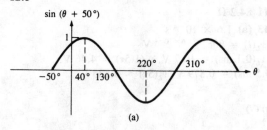

$\sin(\theta + 50°)$

1

−50°   40° 130°   220°   310°   → θ

(a)

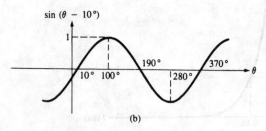

$\sin(\theta - 10°)$

1

10° 100°   190°   280°   370°   → θ

(b)

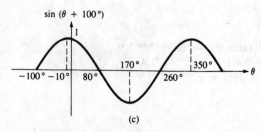

$\sin(\theta + 100°)$

1

−100° −10°   80°   170°   260°   350°   → θ

(c)

12.7 (a) sin (θ + 15°) (b) sin (θ − 80°)
(c) sin (θ − 60°) (d) sin θ

12.9 $\dfrac{\sin \theta}{\cos \theta} = \dfrac{b/c}{a/c} = b/a = \tan \theta$

12.11 (a) $a = 93.97, b = 34.2$
(b) $a = -0.002, b = 0.00346$

12.13 9.156

12.15 (a) 59.04° (b) 116.57°

12.17 (a) (i) 50 Hz (ii) 12.5 kHz (iii) 0.2 Hz
(iv) 25 Hz
(b) (i) 0.01 s (ii) 1.25 μs (iii) 10 s
(iv) 4×10^{-9} s

12.19 (a) 5 kHz (b) 0.5 Hz (c) 400 kHz
(d) 50 Hz

12.21 (a) 2.269 rad (b) −0.785 rad (c) 4.712 rad
(d) −3.1416 rad (e) 8.901 rad

12.23 (a) 200 rad/s (b) 157.08×10^3 rad/s
(c) 392.7×10^3 rad/s (d) 1 rad/s

12.25 (a) 16 V p-p (b) 8 A p-p
(c) 30 mV p-p (d) 150 mA p-p

12.27 (a) 5.080 V (b) 0 (c) 0.0898 V

12.29 (a) 4.126 V (b) 14.55 mA (c) 25.19 V
(d) −0.164 mA

12.31 (a) 18 sin (2π × 250t + 45°) V
(b) 40 sin (2π × 25 × 10³t − 90°) mA

12.33 (a) $i_2(t)$ leads by 50°
(b) $v(t)$ leads by 75°
(c) $v_1(t)$ leads by 45°
(d) $i(t)$ leads by 155°

12.35 6 cycles

12.37 2.8 V p-p

12.39 434.78 Hz

12.41 86.4°

12.43 (a) 0 (b) 0.5 A (c) 12 mV (d) 57.5 mA

12.45 (a) 6 V
(b)

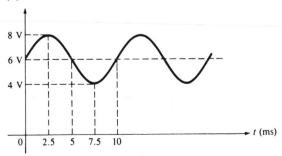

12.47 (a) 70.7 A rms (d) 183.85 mV rms
(c) 1 A rms (d) 0.0919 V rms

12.49 (a) 1.697 V pk (b) 55.15 μA pk
(b) 678.8×10^{-9} V pk (d) 10.61 A pk

12.51 (a) 7.906 V rms (b) 94.87 mA rms

12.53 (a) $i(t) = 9 \sin (400t)$ mA
(b) $i(t) = 0.42 \sin (\omega t + 30°)$ mA
(c) $i(t) = 0.25 \cos (2\pi \times 10^5 t − 80°)$ mA
(d) $i(t) = -0.5 \sin (1832t + 100°)$ mA

12.55 23.33 V rms

12.57 (a) 345.99 Ω (b) 1.33 Ω
(c) 1.989 kΩ (d) ∞

12.59 (a) 0.04 A (b) 0.637 mA (c) 62.8 mA

12.61 (a) $v(t) = 136.36 \sin (10^3 t − 30°)$ V

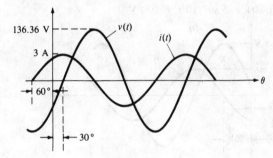

(b) $v(t) = 1.36 \sin (2\pi \times 80t − 90°)$ V

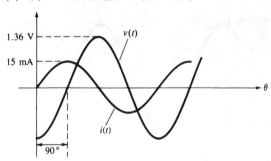

(c) $v(t) = 4.55 \sin (5000t − 10°)$ V

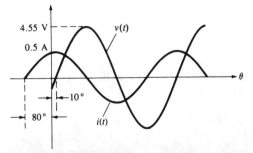

12.63 (a) 5.026 kΩ   (b) 480 Ω
(c) 25.13 kΩ   (d) 0

12.65 (a) 69.4 mA   (b) 1.59 A   (c) 25 μA

12.67 (a) $v(t) = 377 \cos (377t)$ V

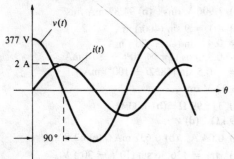

(b) $v(t) = 5 \sin (10^3 t + 10°)$ V

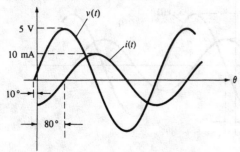

(c) $v(t) = 17.6 \sin (2\pi \times 10^6 t + 70°)$ V

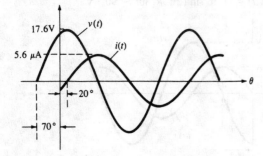

12.69 43.2 mW

12.71 (a) 12.25 V rms   (b) 0.173 A pk

12.1S 0.942 $\underline{/90°}$ A

12.3S 125.66 $\underline{/15°}$ V

Chapter 13

13.1

	Real part	Imaginary part
(a)	50	60
(b)	3×10^{-3}	-1×10^{-3}
(c)	0	17
(d)	−1	0
(e)	420×10^5	0
(f)	−2	−1

13.3 (a) 5 $\underline{/53.13°}$ (b) 14.14 $\underline{/-45°}$
(c) 1.414 $\underline{/135°}$ (d) 56 $\underline{/0°}$
(e) 0.224 $\underline{/-153.4°}$ (f) 1 $\underline{/90°}$

13.5 (a) $173.2 + j100$ (b) $45.96 - j45.96$
(c) $-0.0257 + j0.0306$
(d) $-77.43 \times 10^3 - j28.18 \times 10^4$
(e) $12 + j0$ (f) $-3.8 + j0$

13.7 (a) $6 + j12$ (b) $1 - j1$ (c) $0 + j0.2$
(d) $-177 + j433$ (e) $3.79 \times 10^3 + j3.17 \times 10^3$
(f) $15.11 + j4.98$

13.9 (a) 4.2 $\underline{/106°}$ (b) 100 $\underline{/-33°}$
(c) 1.302×10^{-2} $\underline{/98°}$ (d) 120 $\underline{/-58°}$
(e) 20 $\underline{/75°}$ (f) 2×10^{-3} $\underline{/-10°}$
(g) 14.14 $\underline{/135°}$ (h) 5 $\underline{/0°}$

13.11 (a) 4.51 $\underline{/-26.31°} = 4.04 - j2$
(b) 0.842 $\underline{/71.56°} = 0.266 + j0.799$
(c) 25 $\underline{/-1.86°} = 24.99 - j0.81$

13.13

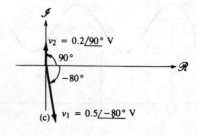

(a)

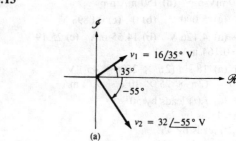

(b)

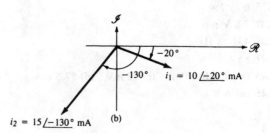

(c)

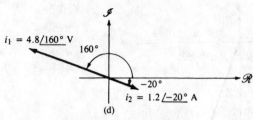

$i_1 = 4.8\underline{/160°}$ V

$160°$

$-20°$

$i_2 = 1.2\underline{/-20°}$ A

(d)

13.15 16.01 sin (500t − 8.66°) V

13.17 87.28 sin (377t − 3.17°) V

13.19 (a) 70.5 sin (ωt + 70°) V

(b)

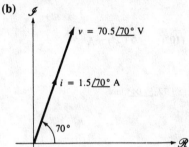

$v = 70.5\underline{/70°}$ V

$i = 1.5\underline{/70°}$ A

$70°$

13.21 (a) 0.836 sin (3400t + 27°) mA

(b)

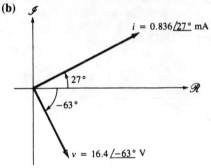

$i = 0.836\underline{/27°}$ mA

$27°$

$-63°$

$v = 16.4\underline{/-63°}$ V

13.23 (a) 75.4 sin (2π × 10³t + 10°) V

(b)

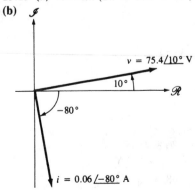

$v = 75.4\underline{/10°}$ V

$10°$

$-80°$

$i = 0.06\underline{/-80°}$ A

Chapter 14

14.1 (a) 90 + j120 Ω = 150 $\underline{/53.13°}$ Ω

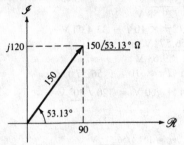

j120

$150\underline{/53.13°}$ Ω

150

$53.13°$

90

(b) $10^4 + j6.283 × 10^3$ Ω = 1.18 × 10⁴ $\underline{/32.14°}$ Ω

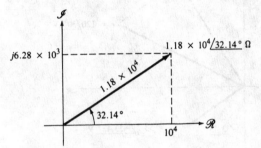

$j6.28 × 10^3$

$1.18 × 10^4\underline{/32.14°}$ Ω

1.18 × 10⁴

32.14°

10⁴

14.3 (a) 3.54 sin (300t − 45°) mA

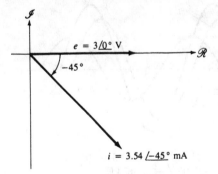

$e = 3\underline{/0°}$ V

$-45°$

$i = 3.54\underline{/-45°}$ mA

(b) 0.632 sin (2π × 8 × 10⁵t + 26.83°) mA

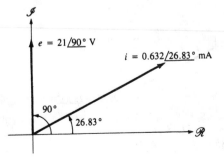

$e = 21\underline{/90°}$ V

$i = 0.632\underline{/26.83°}$ mA

$90°$

$26.83°$

14.5 (a) $0.895 \,\underline{/-33.43°}$ A =
$0.895 \sin (2\pi \times 10^3 t - 33.43°)$ A

(b) $v_R = 53.7 \,\underline{/-33.43°}$ V
$\quad = 44.82 - j29.58$ V
$\quad = 53.7 \sin (2\pi \times 10^3 t - 33.43°)$ V
$\quad v_L = 107.4 \,\underline{/56.57°}$ V
$\quad = 59.17 + j89.63$ V
$\quad = 107.4 \sin (2\pi \times 10^3 t + 56.57°)$ V

(c) $v_R + v_L = 104 + j60$ V $= 120 \,\underline{/30°}$ V $= e$

(d)

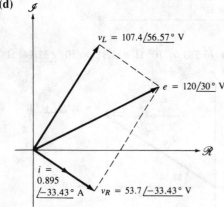

(e)

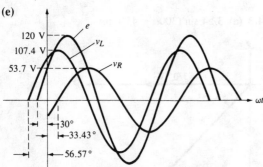

14.7 (a) $Z_T = 25 \times 10^3 - j12.5 \times 10^3$ Ω
$\quad = 27.95 \,\underline{/-26.56°}$ kΩ

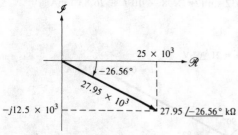

(b) $Z_T = 30 - j79.58$ Ω
$\quad = 85.05 \,\underline{/-69.34°}$ Ω

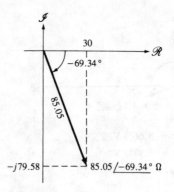

14.9 (a) $1.72 \sin (10^5 t + 59.04°)$ mA

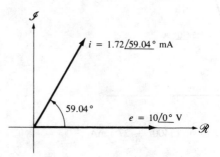

(b) $1.88 \sin (\omega t - 25.2°)$ mA

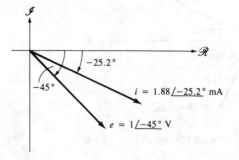

14.11 (a) $0.174 \,\underline{/35.54°}$ A =
$0.174 \sin (\omega t + 35.54°)$ A

(b) $v_R = 121.8 \,\underline{/35.54°}$ V
$\quad = 99.11 + j70.8$ V
$\quad = 121.8 \sin (\omega t + 35.54°)$ V
$\quad v_C = 87 \,\underline{/-54.46°}$ V
$\quad = 50.57 - j70.8$ V
$\quad = 87 \sin (\omega t - 54.46°)$ V

(c) $v_R + v_C = 149.68 + j0 \approx 150 \,\underline{/0°} = e$

(d)

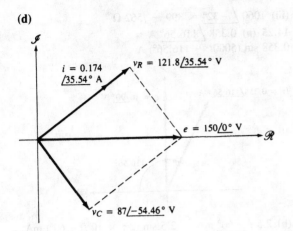

(e)

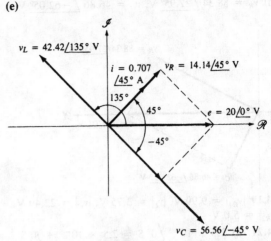

(e)

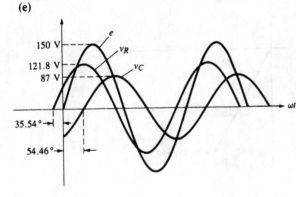

14.13 (9a) $28.28 \underline{/-45°}$.

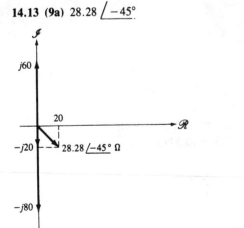

(b) $0.707 \underline{/45°}$ A

(c) $v_R = 14.14 \underline{/45°}$ V, $v_L = 42.42 \underline{/135°}$ V,
$v_C = 56.56 \underline{/-45°}$ V

(d) $v_R + v_L + v_C = 20 + j0$ V $= 20 \underline{/0°}$ V $= e$

14.15 (a) $984.89 \underline{/23.96°}$

(b) $40.61 \underline{/-23.96°}$ mA $=$
$40.61 \sin (200t - 23.96°)$ mA

(c) $v_{L1} = 50.76 \underline{/66.04°}$ V
$v_{R1} = 6.09 \underline{/-23.96°}$ V
$v_{C1} = 36.55 \underline{/-113.96°}$ V
$v_{R2} = 30.46 \underline{/-23.96°}$ V
$v_{C2} = 10.15 \underline{/-113.96°}$ V
$v_{L2} = 12.18 \underline{/66.04°}$ V

(d) 7.75 H, 4.35 μF

14.17 (a) $v_R = 19.55 \underline{/10.96°}$ V, $v_L =$
$32.58 \underline{/100.96°}$ V

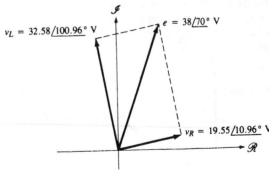

(b) $v_R = 88.34 \underline{/27.95°}$ V, $v_C = 46.86 \underline{/-62.05°}$ V

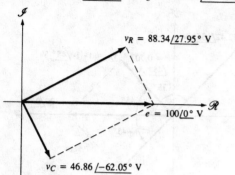

14.19 $|v_{R1}| = 9.99$ V, $|v_L| = 3.75$ V, $|v_C| = 22.49$ V, $|v_{R2}| = 5.0$ V

14.21 (a) $2.5 \times 10^{-5} \underline{/0°}$ S $= 2.5 \times 10^{-5} + j0$ S

(b) $0.503 \underline{/90°}$ S $= 0 + j0.503$ S

(c) $8 \times 10^{-4} \underline{/-90°}$ S $= 0 - j8 \times 10^{-4}$ S

(d) 0 S $= 0 + j0$ S

14.23 (a) (i) $0.05 - j0.05$ S $= 0.0707 \underline{/-45°}$ S

(ii)

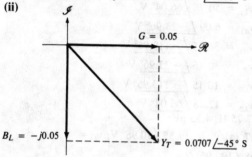

(iii) $14.14 \underline{/45°}$ Ω $= 10 + j10$ Ω

(b) (i) $0.8 \times 10^{-3} + j0.5 \times 10^{-3}$ S $= 0.943 \times 10^{-3} \underline{/32°}$ S

(ii)

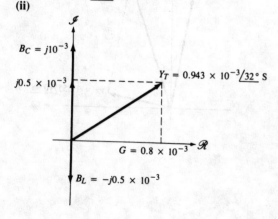

(iii) $1060 \underline{/-32°} = 899 - j562$ Ω

14.25 (a) $0.358 \underline{/116.56°}$ A $= 0.358 \sin(5000t + 116.56°)$ A

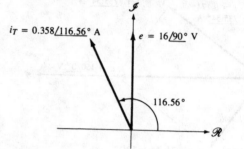

(b) $2.5 \underline{/-60°}$ mA $= 2.5 \sin(2\pi \times 10^4 t - 60°)$ mA

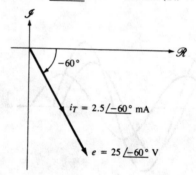

14.27 (a) $8.33 + j6.25$ mS $= 10.42 \underline{/36.86°}$ mS

(b) $0.625 \underline{/36.86°}$ A $= 0.5 + j0.375$ A

(c) $i_1 = 0.5 + j0$ A $= 0.5 \underline{/0°}$ A

$i_2 = 0 + j0.375$ A $= 0.375 \underline{/90°}$ A

$i_1 + i_2 = 0.5 + j0.375$ A $= i$

(d)

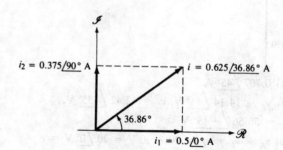

(e)

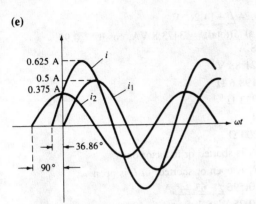

14.29 (a) $2.69 \times 10^{-5} \,\underline{/-68.2°}$ S

(b) $7.43 \,\underline{/68.2°}$ V

(c) $i_1 = 185.8 \,\underline{/-21.8°}$ µA $= 172.5 - j69$ µA

$i_2 = 74.3 \,\underline{/68.2°}$ µA $= 27.59 + j68.99$ µA

(d) $i_1 + i_2 = 200.09 - j0.01$ µA $\approx 200 \,\underline{/0°}$ µA $= i$

(e)

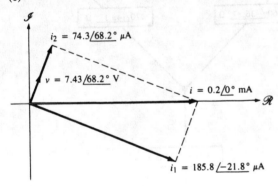

14.31 (a) $i_1 = 1.94 \,\underline{/35.88°}$ A $=$
$1.94 \sin (2\pi \times 10^3 t + 35.88°)$ A
$i_2 = 1.41 \,\underline{/-54.12°}$ A $=$
$1.41 \sin (2\pi \times 10^3 t - 54.12°)$ A

14.33 (a) $i_1 = 1.39 \,\underline{/56.31°}$ A

$i_2 = 1.67 \,\underline{/-30°}$ A

(b) $i_1 = e/Z_1$, $i_2 = e/Z_2$, $i_1 + i_2 =$
$2.24 \,\underline{/8.2°}$ A $\approx i_T$

14.35

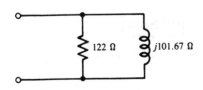

14.37 (a) 0.857 **(b)** 1.5 W **(c)** $V_p^2/2R = 1.5$ W

14.39 0.884

14.41 (a) 0.949 **(b)** 1.21 W

(c) $P(40 \,\Omega) = 0.403$ W, $P(90 \,\Omega) = 0$ W, $P(80 \,\Omega) =$
0.807 W, $P(50 \,\Omega) = 0$ W, $P(\text{total}) = 1.21$ W

14.1D 8 Ω

14.3D 70.36 µF

14.5D 800 pF

14.7D 0.02 A

14.9D 43 mH

14.11D 4.94 µF

14.1T capacitor, $e(t)$ leads i.

14.3T 0.2786 A

14.5T No, $|X_C| > |X_L|$ if $v_R(t)$ leads $e(t)$, $|X_L| > |X_C|$ if
$v_R(t)$ lags $e(t)$.

14.1S $Z_T = 360.56 \,\underline{/33.69°}$ Ω,

$i_T = 0.166 \,\underline{/26.31°}$ A, $v_C = 133.1 \,\underline{/-63.69°}$ V,

$v_R = 49.9 \,\underline{/26.31°}$ V, $v_L = 166.4 \,\underline{/116.31°}$ V

14.3S $Y_T = 22.36 \,\underline{/26.56°}$ mS,

$i_T = 0.358 \,\underline{/116.56°}$ A, $i_{C_1} = 0.16 \,\underline{/180°}$ A,

$i_{C_2} = 0.08 \,\underline{/180°}$ A, $i_L = 0.08 \,\underline{/0°}$ A,

$i_R = 0.32 \,\underline{/90°}$ A

Chapter 15

15.1 (a)
$Z_T = 8.76 \,\underline{/37.86°}$ Ω, $i_T = 1.71 \,\underline{/-37.86°}$ A

(b) $Z_T = 1479 \,\underline{/-39.7°}$ Ω,

$i_T = 33.81 \,\underline{/39.7°}$ Ω

15.3 (a)

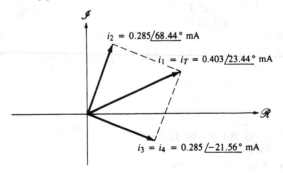

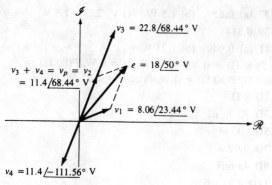

(b)

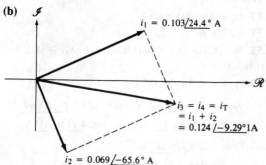

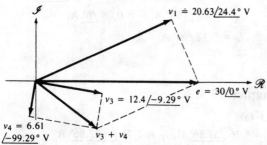

15.5 $v_2 = 10.4 \underline{/-11.57°}$ V, $i_3 = 0.26 \underline{/-11.57°}$ A

15.7 (a) $v_1 = 33.59 - j13.71$ V
$v_2 = 46.41 + j13.71$ V
$v_3 = 11.47 + j28.11$ V
$v_4 = 35 - j14.28$ V
(b) $v_1 + v_2 = e$, $v_3 + v_4 = v_2$

15.9 (a) $i_T = 0.955 \underline{/-25.8°}$ A, $P_{\text{avg}} = 10.75$ W
(b) $P(Z_1) = 9.12$ W, $P(Z_2) = P(Z_4) = 0$ W, $P(Z_3) = 1.63$ W, $P(\text{total}) = 10.75$ W

15.11 $12.07 \underline{/-29.2°}$ V

15.13 $1 \underline{/0°}$ A

15.15 (a) $268 \underline{/63.44°}$ Ω
(b) 86.8 A

15.17 (a) $P(\text{total}) = 277.64$ W
$Q(\text{total}) = 70.59$ vars (cap)

(b) $143.24 \underline{/-14.26°}$ V

15.19 (a) $S(\text{total}) = 3473.8$ VA, $\cos \theta = 0.95$
(b) 86.8 A

15.1D 21.88 V

15.3D 198.6 Ω

15.5D 273 Ω

15.7D 4

15.9D 200 Ω

15.1T R_1 is shorted or inductor open.

15.3T R_1 is open or shorted, or L is open.

15.1S $0.596 \underline{/-63.44°}$ A

15.3S 10.75 W, 5.198 vars (ind.)

Chapter 16

16.1

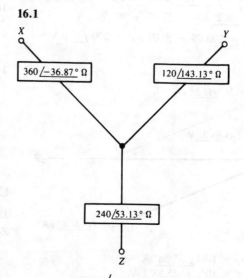

16.3 $Z_T = 60.92 \underline{/-29.16°}$ Ω
$i_T = 0.246 \underline{/29.16°}$ A

16.5 Balanced; $Z_T = 216.9 \underline{/12.53°}$ Ω

16.7 $L = 0.1$ H, $R_L = 10$ Ω

16.9 $C = 0.00909$ μF, $R_C = 8.8$ kΩ

16.11

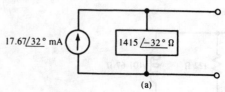

(a)

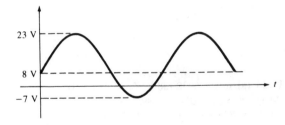

(b)

16.13 $0.296 \,\underline{/-32.9°}$ A

16.15 $+(v)- \,= 4.86 \,\underline{/-51.58°}$ V

16.17 $2.67 \,\underline{/-71.36°}$ A $\rightarrow$

16.19 $8.82 \,\underline{/8.74°}$ mA $\downarrow$

16.21 $10.72 \,\underline{/-24.89°}$ V $\pm$

16.23 $75.41 \,\underline{/59.8°}$ V $\pm$

16.25 $0.35 \,\underline{/-6.85°}$ mA $\downarrow$

16.27 $\mathbf{YV} = \mathbf{I}$

$$\mathbf{Y} = \begin{bmatrix} (10 - j0.5) \times 10^{-6} & -10 \times 10^{-6} & -j0.5 \times 10^{-6} \\ -10 \times 10^{-6} & 18 \times 10^{-6} & -8 \times 10^{-6} \\ -j0.5 \times 10^{-6} & -8 \times 10^{-6} & (12 - j5.5) \times 10^{-6} \end{bmatrix}$$

$$\mathbf{V} = \begin{bmatrix} v_1 \\ v_2 \\ v_3 \end{bmatrix} \quad \mathbf{I} = \begin{bmatrix} 0.2 \times 10^{-3} \\ -0.15 \times 10^{-3} \\ -0.04 \times 10^{-3} \end{bmatrix}$$

$$\mathbf{V} = \mathbf{Y}^{-1}\mathbf{I}$$

16.1D $30 \,\underline{/30°}$ Ω

16.3D 0.0926 μF

16.5D $R = 150$ Ω, $L = 50$ mH

16.7D $65.08 \,\underline{/52.08°}$ V $\pm$

16.9D $-27 \,\underline{/0°}$ V or $27 \,\underline{/180°}$ V

16.1T Yes, power in each unshorted element becomes 2167 W.

16.1S Not balanced.

Chapter 17

17.1 (a) $2.21 \,\underline{/-56.24°}$ A   **(b)** $0.31 \,\underline{/-36.87°}$ A

17.3 $8 + 15 \sin (10^4 t)$ V

17.5 37.29 W

17.7 (a) $22.5 \,\underline{/0°}$ V   **(b)** $28.28 \,\underline{/45°}$ V

17.9 $22.09 \,\underline{/76.44°}$ V

17.11 $0.02 \,\underline{/0°}$ A

17.13 (a) $Z_{TH} = 64 \,\underline{/41.74°}$ Ω

$e_{TH} = 10 \,\underline{/67.38°}$ V

(b) $Z_{TH} = 831.95 \,\underline{/-33.69°}$ Ω

$e_{TH} = 16.64 \,\underline{/146.31°}$ V

17.15 $0.324 \,\underline{/68.74°}$ A $\rightarrow$

17.17 $Z_{TH} = 26 \times 10^3 \,\underline{/-40°}$ Ω

$e_{TH} = 12.45 \,\underline{/-40°}$ V

17.19 $Z_{TH} = 33.33 \,\underline{/0°}$ Ω

$e_{TH} = 10 \,\underline{/0°}$ V

17.21 (a) (i) and (ii) $Z_N = 6.71 \,\underline{/-63.43°}$ Ω,

$i_N = 3.07 \,\underline{/135°}$ A

(b) (i) and (ii) $Z_N = 425 \,\underline{/-11.31°}$ Ω,

$i_N = 0.14 \,\underline{/45.49°}$ A

17.23 $Z_N = 33.33 \,\underline{/0°}$ Ω, $i_N = 0.3 \,\underline{/0°}$ A

17.25 (a) $Z_L = 4.06 \,\underline{/23.96°}$ Ω, $P(\text{max}) = 4.6$ W

(b) $Z_L = 870 \,\underline{/69.96°}$ Ω, $P(\text{max}) = 0.485$ W

17.27 (a) $Z_L = 21.2 \,\underline{/-58°}$ Ω, $P(\text{max}) = 45$ W

(b) $Z_L = 33.69 \,\underline{/32.62°}$ Ω, $P(\text{max}) = 7.19$ W

17.1D $8.94 \,\underline{/26.56°}$ V

17.3D 14.4V

17.5D 21.9

17.7D 11.254 kHz

17.1T The 400-Ω resistor is shorted for the ac component.

17.3T No, greater power delivered to load without additional resistor.

17.1S $15 \,\underline{/0°}$ V ac, 8 V dc

17.3S 0.02 A

Chapter 18

18.1 6 V rms

18.3 500 Hz

18.5 (a) Low-pass (b) 7234 Hz
(c) 2 kHz: 17.35 V, 10 kHz: 10.55 V, 20 kHz: 6.12 V
18.7 (a) 5×10^3 rad/s: 0.928,
1.25×10^4 rad/s: 0.707,
3.6×10^4 rad/s: 0.328
(b) 6×10^3 rad/s: 19.36°
1.25×10^4 rad/s: 0°
1.25×10^5 rad/s: $-39.29°$
18.9 (a) High-pass (b) 3979 Hz
(c) 2 kHz: 2.24 V, 4 kHz: 3.54 V, 10 kHz: 4.65 V
(d) 2.5×10^3 rad/s: 84.29°
25×10^3 rad/s: 45°
10^5 rad/s: 14.04°

18.11 Substituting $\omega = 1/RC$ in equation (18.12) gives

$$|v_o| = \frac{1}{\sqrt{1 + 1}} |e_{in}| = 0.707 \, |e_{in}|$$

18.13 (a) $\omega_0 = 408{,}248$ rad/s, $f_0 = 64{,}975$ Hz
(b) $X_L = j6124\ \Omega$, $X_C = -j6124\ \Omega$, $Z_T = 220 + j0\ \Omega$

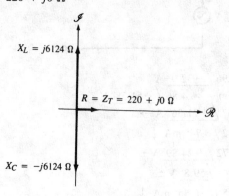

18.15 (a) $C = 0.792\ \mu F$, $R = 62.8\ \Omega$
(b) $|v_R| = 0.6$ V, $|v_L| = |v_C| = 4.8$ V
18.17 (a) $f_0 = 1.5$ kHz, $f_1 = 1481$ Hz,
$f_2 = 1521$ Hz
(b) BW = 40 Hz, $Q_s = 37.5$
(c) 0.284 V at both frequencies
18.19 $f_2 - f_1 = (f_0/2Q_s) - (-f_0/2Q_s) = f_0/Q_s$
18.21 45.2 Hz
18.23 (a) $\omega_0 = 12{,}500$ rad/s, $f_0 = 1989$ Hz
(b) $i_R = 0.15 \underline{/0°}$ A, $i_L = 0.3 \underline{/-90°}$ A, $i_C = 0.3 \underline{/90°}$ A, $i_T = 0.15 \underline{/0°}$ A
(c) $0.1f_0$: 5.04 Ω, $0.5f_0$: 31.62 Ω, f_0: 100 Ω,
$2f_0$: 31.62 Ω, $10f_0$: 5.04 Ω

18.23 (d)

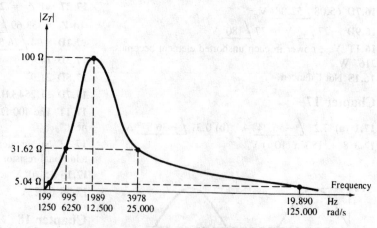

18.25 (a) $\omega_0 = 6030$ rad/s, $f_0 = 959.7$ Hz
(b) $Q_p = 3.317$, BW = 289.3 Hz
(c) $f_1 = 825.9$ Hz, $f_2 = 1115.2$ Hz
(d) No, Q_p is small.

18.27 (a) $f_0 = 10.597$ kHz, $Q_{sp} = 19.97$
(b) 20 kΩ **(c)** 0.126%
18.29 (a) 318.3 Hz **(b)** 318.3 Hz
(c) $f_1 = 305.6$ Hz, $f_2 = 331$ Hz **(d)** 6.27 kΩ
18.31 (a) 9.7 dB **(b)** 22.63 dB **(c)** -6.02 dB
(d) 0 dB **(e)** 20 dB
18.33 (a) 5.88 dB **(b)** -12 dB
18.35 (a) 32.04 dB **(b)** -14.26 dB **(c)** 0 dB
(d) 40 dB
18.37 5.5 V
18.39 3×3
18.41 (a) 1.2 V **(b)** 50 kHz **(c)** 6.5 V
18.43

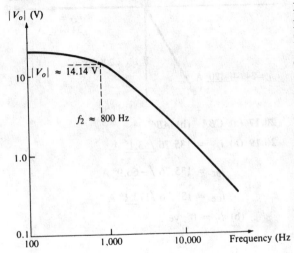

18.45 (a) 30 dB **(b)** 37 dB **(c)** 8.78 V
18.47

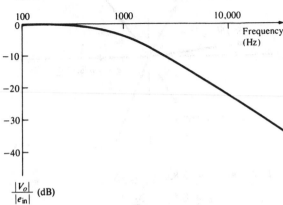

18.1D 2652 Ω
18.3D 0.0875 μF

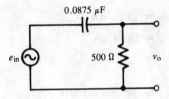

18.5D 140 Ω
18.7D (a) 10.47 to 38.98 pF
(b) 716 Ω
18.9D 0.064 μF
18.11D 859.6 Ω
18.1T R_2 shorted and C_1 open, or R_1 shorted and C_2 open
18.3T No
18.5T Capacitor is open.
18.3S Low-pass.

Chapter 19

19.1 $N_p:N_s = 3:1$
19.3 18.86 mA rms
19.5 (a) 15 V pk **(b)** 0.238 A pk **(c)** 7.5 W
(d) 264.7 Ω
19.7 (a) 2.5 kΩ **(b)** 3.6 mA pk **(c)** 0.6 mA pk
19.9 (a) yes **(b)** 18 mA
19.11 (a) $|e_{s1}| = 80$ V, $|e_{s2}| = 160$ V
(b) 19.2 W
(c) 0.32 A
19.13 (a) 160 V **(b)** 40 V **(c)** 8 V **(d)** 120 V
19.15 (a) 5.78 A **(b)** 3.93 A
19.17 97.29%
19.19

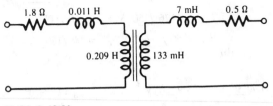

19.21 0.48 V
19.1D $N_P:N_s = 4:1$
19.3D $N_P:N_s = 4.3:1$
19.5D 2.22×10^{-3} Ω/ft
19.1T (a) Open between f and g
(b) Secondary windings shorted to each other
(c) Primary winding shorted to secondary winding e-g
(d) Same as part (c)
19.3T No, $|i_p| = 3$ mA when matched.

19.1S $v_P = 13.64$ V, $i_P = 27.27$ mA

$v_S = 27.27$ V, $i_S = 13.64$ mA

Chapter 20

20.1 (a) negative   (b) positive   (c) negative

20.3

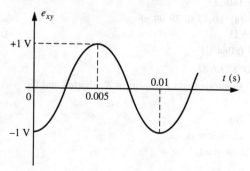

20.5 (a) Rings that rotate with the rotor and make electrical contact with the rotor winding so that electrical connection can be made to stationary terminals.

(b) Stationary set of contacts between the rings and the terminals.

(c) Current supplied to windings to create a magnetic field.

20.7 $i_{AN} = i_{NB} = 30$ A rms; $i_N = 0$

20.9 $i_N = \overline{10}$ A rms; $i_{AN} = \overline{30}$ A rms, $i_{NB} = \overline{40}$ A rms

20.11 Current in each 10-Ω load = 12 A rms; $i_{AB} = 30$ A rms; $i_N = 0$

20.13 (a) $e_{AB} = 1176 \underline{/30°}$ V, $e_{BC} = 1176 \underline{/-90°}$ V,

$e_{CA} = 1176 \underline{/150°}$ V, $i_{AL} = 50 \underline{/60°}$ A,

$i_{BL} = 50 \underline{/-60°}$ A, $i_{CL} = 50 \underline{/180°}$ A

(b)

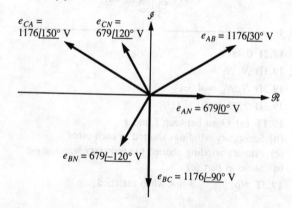

20.15 (a) $e_{AB} = 170 \underline{/0°}$ V, $e_{BC} = 170 \underline{/-120°}$ V,

$e_{CA} = 170 \underline{/120°}$ V, $i_{AL} = 34.64 \underline{/0°}$ A,

$i_{BL} = 34.64 \underline{/-120°}$ V, $i_{CL} = 34.64 \underline{/120°}$ A

(b) $i_{A\phi} + i_{B\phi} + i_{C\phi} = 0 + j0$;

$i_{AL} + i_{BL} + i_{CL} = 0 + j0$

(c)

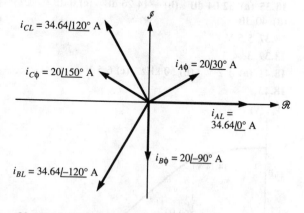

20.17 (a) *CBA*   (b) *ABC*

20.19 (a) $i_{AL} = 135.76 \underline{/53.1°}$ A,

$i_{BL} = 135.76 \underline{/-66.9°}$ A,

$i_{CL} = 135.76 \underline{/173.1°}$ A

(b) $i_N = 0$; yes.

(c)

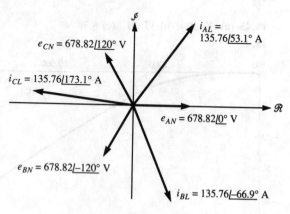

20.21 (a) $i_{AL} = 2.833 \underline{/-90°}$ A,

$i_{BL} = 8.5 \underline{/-120°}$ A,

$i_{CL} = 17 \underline{/30°}$ A

(b) $i_N = 10.61 \; \underline{/-9.17°}$ A; no.

(c)

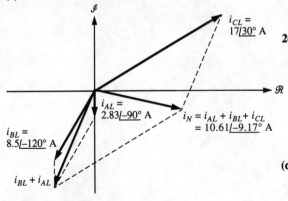

20.23 (a) $e_{AB} = 294.4 \; \underline{/30°}$ V,

$e_{BC} = 294.4 \; \underline{/-90°}$ V,

$e_{CA} = 294.4 \; \underline{/150°}$ V

(b) $i_{Z_A} = 14.72 \; \underline{/-15°}$ A,

$i_{Z_B} = 14.72 \; \underline{/-135°}$ A,

$i_{Z_C} = 14.72 \; \underline{/105°}$ A

(c) $i_{AL} = 25.5 \; \underline{/-45°}$ A,

$i_{BL} = 25.5 \; \underline{/-165°}$ A

$i_{CL} = 25.5 \; \underline{/75°}$ A

(d)

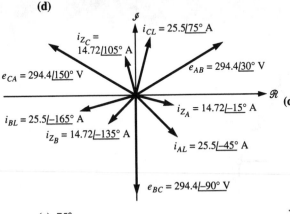

(e) 75°

20.25 (a) $i_{Z_A} = 163.6 \; \underline{/-60°}$ A,

$i_{Z_B} = 95.1 \; \underline{/-43.3°}$ A,

$i_{Z_C} = 149.3 \; \underline{/119.5°}$ A; Sum = 77.5 − j77 A

(b) $i_{AL} = 312.9 \; \underline{/-60.2°}$ A,

$i_{BL} = 77.5 \; \underline{/99.4°}$ A,

$i_{CL} = 241.7 \; \underline{/126.2°}$ A; Sum = 0 + j0.

20.27 (a) $e_{AB} = 169.7 \; \underline{/0°}$ V, $e_{BC} = 169.7 \; \underline{/-120°}$ V,

$e_{CA} = 169.7 \; \underline{/120°}$ V; Sum = 0 + j0,

(b) $i_{Z_A} = 26.1 \; \underline{/-20°}$ A, $i_{Z_B} = 26.1 \; \underline{/-140°}$ A,

$i_{Z_C} = 26.1 \; \underline{/100°}$ A; Sum = 0 + j0.

(c) $i_{AL} = 45.2 \; \underline{/-50°}$ A, $i_{BL} = 45.2 \; \underline{/-170°}$ A,

$i_{CL} = 45.2 \; \underline{/70°}$ A; Sum = 0 + j0.

(d)

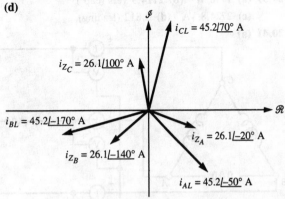

20.29 (a) $v_{Z_A} = 923.7 \; \underline{/-30°}$ V, $v_{Z_B} = 923.7 \; \underline{/-150°}$ V,

$v_{Z_C} = 923.7 \; \underline{/90°}$ V; Sum = 0 + j0.

(b) $i_{Z_A} = 9.24 \; \underline{/-120°}$ A, $i_{Z_B} = 9.24 \; \underline{/120°}$ A,

$i_{Z_C} = 9.24 \; \underline{/0°}$ A; Sum = 0 + j0.

(c) $i_{AL} = 9.24 \; \underline{/-120°}$ A, $I_{BL} = 9.24 \; \underline{/120°}$ A,

$i_{CL} = 9.24 \; \underline{/0°}$ A; Sum = 0 + j0.

(d)

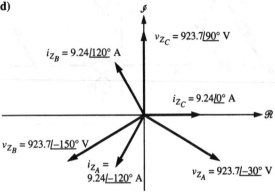

20.31 $i_{Z_A} = 15.49\ \underline{/93.4°}$ A, $i_{Z_B} = 26.02\ \underline{/-126.5°}$ A,

$i_{Z_C} = 17.27\ \underline{/18.5°}$ A, $v_{Z_A} = 387.3\ \underline{/93.4°}$ V,

$v_{Z_B} = 650.5\ \underline{/143.2°}$ V, $v_{Z_C} = 863.5\ \underline{/108.5°}$ V

20.33 1. Less material cost to deliver equal power.

2. Total power transmitted is constant.

3. Three-phase voltage can be used to create a rotating magnetic field.

20.35 **(a)** 5409.4 W   **(b)** 5409.4 W

20.37 **(a)** 1.0266 MW   **(b)** 342.1 kvars (ind.)

(c) 1082 kVA   **(d)** 0.9487 (lagging)

20.39 **(a)** 1718 W   **(b)** 2114.9 vars (cap.)

(c) 2724.8 VA   **(d)** 0.812 (leading)

20.41 **(a)**

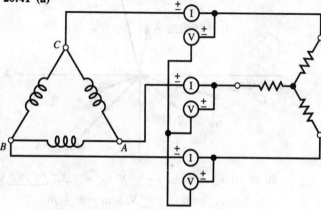

(b) 173.2 V,   6.93 A

(c) 600 W   **(d)** 1800 W

20.43 **(a)**

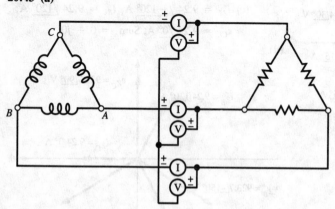

(b) 56.4 kW

(c) $P_A = 19.2$ kW, $P_B = 16.199$ kW,

$P_C = 20.997$ kW; Sum $= 56.4$ kW

20.45 (a)

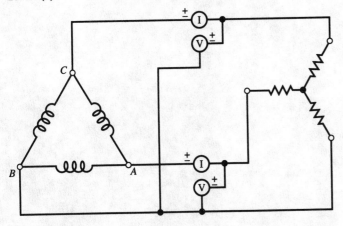

(b) 16.84 kW

(c) $P_1 = P_2 = 8.407$ kW

20.47 $P_1 + P_2 = \dfrac{EI}{2}$ (cos θ cos 30° $-$ sin θ sin 30°

$+$ cos θ cos 30° $+$ sin θ sin 30°)

$= \dfrac{EI}{2}$ (2 cos θ cos 30°) $= EI$ cos θ cos 30°

20.1D 75 Ω

20.3D 70.36 μF; Before correction: 679 V, 17.4 A;
After correction: 2648.1 V, 67.9 A.

20.1T An open in the secondary winding.

20.1S See answer to Exercise 20.21.

20.3S (a) $i_{Z_A} = 60 \,\underline{/-90°}$ A, $i_{Z_B} = 60 \,\underline{/-120°}$ A,

$i_{Z_C} = 60 \,\underline{/-150°}$ A

(b) $i_{AL} = 60 \,\underline{/-30°}$ A, $i_{BL} = 31 \,\underline{/165°}$ A,

$i_{CL} = 31 \,\underline{/135°}$ A

(a) 30.4 kW

(c) $P_a = 14.9$ kW, $P_b = 16.49$ kW

$P_c = 20.997$ kW, Sum = 50.4 kW

20.13 (a)

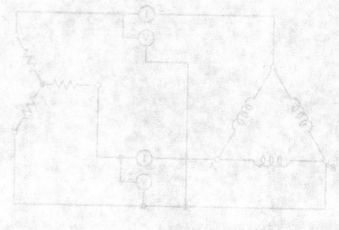

(b) 77.66 kW

(c) $P_a = P_b = 2.90$ kW

20.37 $E_a = E_b = \frac{E_L}{2}(\cos\theta + \cos 30° - \sin\theta \sin 30°$

$-\cos\theta \cos 30° - \sin\theta \sin 30°)$

$\frac{E_L}{2}(2\cos\theta \cos 30°) = E_L\cos\theta \cos 30° = ...$

20.1D 76 Ω

20.3D 20.36 μF. Series connection 679 V, 43.9 A;

Allic connection 2048.1 V, 43.9 A ...

20.17 An open in the secondary winding?

20.15 See answer to Exercise 20.4.

20.39 (a) $I_a = 60\angle-60°$ A, $I_b = 60\angle-120°$ A,

$I_c = 60\angle 180°$ A

(b) $I_a = 69\angle-30°$ A, $I_b = 34.7/65$ A,

$I_c = 31.7/65$ A

Index